BEEF PRODUCTION AND MANAGEMENT PRACTICES

SIXTH EDITION

Thomas G. Field

Upper Saddle River, New Jersey
Columbus, Ohio

330 Hudson Street, NY, NY 10013

Vice President, Portfolio Management: Andrew Gilfillan
Portfolio Manager: Pamela Chirls
Editorial Assistant: Lara Dimmick
Senior Vice President, Marketing: David Gesell
Field Marketing Manager: Thomas Hayward
Marketing Coordinator: Elizabeth MacKenzie-Lamb
Senior Marketing Assistant: Les Roberts
Director Courseware and Content Producers: Brian Hyland
Managing Producer: Cynthia Zonneveld
Managing Producer: Jennifer Sargunar
Content Producer: Rinki Kaur
Manager, Rights Management: Johanna Burke
Manufacturing Buyer: Deidra Smith, Higher Ed, LSC Communications
Cover Designer: Studio Montage
Cover Photo: EcoView/Fotolia
Full-Service Project Management and Composition: Joy Deori, iEnergizer Aptara®, Ltd.
Printer/Binder: LSC Communications
Cover Printer: LSC Communications
Text Font: Garamond 3 LT Pro (11/13)

Pearson Education Ltd.
Pearson Education Singapore Pte. Ltd.
Pearson Education Canada, Ltd.
Pearson Education—Japan
Pearson Education Australia Pty, Limited
Pearson Education North Asia Ltd.
Pearson Educación de Mexico, S.A. de C.V.
Pearson Education Malaysia, Pte. Ltd.

Library of Congress Cataloging-in-Publication Data
Names: Field, Thomas G. (Thomas Gordon)
Title: Beef production and management decisions / Thomas G. Field, University of Nebraska, Lincoln, NE
Description: Sixth Edition. | Hoboken : Pearson, [2016] | Revised edition of the author's Beef production and management decisions, c2007.
Identifiers: LCCN 2016032525 | ISBN 9780134602691 | ISBN 0134602692
Subjects: LCSH: Beef industry—United States. | Beef industry. | Beef cattle. | Beef.
Classification: LCC HD9433.U4 T39 2018 | DDC 636.2/13068—dc23 LC record available at https://lccn.loc.gov/2016032525

ISBN 10: 0-13-460269-2
ISBN 13: 978-0-13-460269-1

6 2022

DEDICATION

To be among ranchers and farmers, to be shaped by the beauty and harshness of the high country, to have been given the privilege of stewardship for cattle, horses, landscape, people, and heritage; and to be guided by a mosaic of communities—rural and urban, scholars, educators, students, cowboys, innovators, bridge builders, entrepreneurs, advocates, activists, and difference makers have been the great blessings of my life. Together, these influences coupled with the lasting influence of being in community with my family have provided the framework to write about this most marvelous, complex, imperfect, and valuable beef business.

Ultimately, the core of the beef industry is defined by those who give their energy, creativity, determination and life's purpose to providing an exceptional source of protein to the world's consumers while embracing their role as caretakers of natural resources, livestock, and people. I am hopeful that this book honors their work and the potential for future generations to find new opportunities and better paths forward.

There are so many who taught me through lesson and example to develop a systems management mindset and it would not be practical to list them all, so I offer a deep sense of gratitude to several in the hopes that they will be representative of the whole:

Bart and Mary Strang, ranchers and community builders who modeled the power of combining a deep passion for cattle breeding, the wonder that comes from being fully invested in husbandry, and the strength that comes from the enduring love of family.

Bob Taylor who was the originator of this work and was the greatest teacher I have ever known. As my mentor, he opened doors, asked tough questions, challenged and nurtured me, and always invited me into the process of discovery. I cannot imagine my life without his influence.

It was my mother, Mary, whose commitment to education, learning, and having experiences beyond the mountains that surrounded our ranch laid the foundation for all that was to follow.

My father, Fred Field, helped me develop work ethic, determination, and a deep sense of responsibility for the well-being of community—what I would not give for one more hour horseback on our ranch with him.

My father-in-law, Coleman Locke, has been a model of service to industry, of a life founded on faith, and the value of life-long learning.

The grandest moments of my life have been born from God's gifts of my talented, beautiful, and resolute wife Laura and five spectacularly unique children who have provided me joy, wonder, and discovery. For Justin, Sean, Trae, Kate, and Coleman, I hope that this book might provide a few lessons about your roots and remind you that opportunity abounds if only you have the courage and grit to run the race.

For all who labor to feed others and to practice daily the timeless principals of stewardship, I am so very grateful.

contents

Preface ix

 1

An Overview of the U.S. Beef Industry 1

General Overview 1
Contribution to the U.S. Economy 4
Beef Industry Segments 4
Beef Industry Organizations 18
Beef Industry Issues 23
Animal Well-Being 24
Selected References 25

 2

Retail Beef Products and Consumers 26

Retail Beef Products 26
Beef Consumption and Expenditures 28
Beef Palatability and Consumer Preferences 35
By-Products 56
Selected References 58

 3

Management Systems: Integrated and Holistic Resource Management 60

Resources and Principles 60
The Human Resource 61
The Financial Resource 71
Land and Feed Resources 74
The Cattle Resource 74
The Market Resource 78
Management Systems 78
Selected References 87

 4

Management Decisions for Seedstock Breeders 88

Breeding Program Goals and Objectives 89
Production Records 90
Sire Selection 93
Selecting Replacement Heifers 108
Selecting Cows 108
Marketing Decisions 112
Selected References 115

 5

Commercial Cow-Calf Management Decisions 117

Creating the Vision 119
Information Systems 120
Profit-Oriented Management Decisions 122
Factors Affecting Pounds of Calf Weaned 124
Managing Percent Calf Crop 126
Managing Weaning Weights 129
Managing Annual Cow Costs and Returns 135
Matching Cows to Their Economical Environment 145
Weaning Management 150
Marketing Decisions 151
Establishing a Commercial Cow-Calf Operation 155
Summary of Cow-Calf Operational Types and Management Practices 157
Selected References 159

 6

Stocker Management Decisions 162

Computing Breakeven Prices 165
The Budgeting Process 166

Management Considerations 168

Pasture Leases 169

Selected References 177

7 Feedlot Management Decisions 178

Types of Cattle Feeding Operations 178

Non-Finishing Feeding 180

Managing a Feedlot Operation 181

Selected References 202

8 The Beef Supply Chain 204

Trust 207

Sustainability 208

Value Creation 210

Source Verification, Traceback, and Identification Systems 214

Structural Changes in the Beef Industry 216

Supply Chain Coordination/Business Relationships 218

Selected References 221

9 The Marketing System 222

Market Classes and Grades 222

Marketing Cows and Bulls 230

Market Channels 232

Grid Pricing 238

Major Factors Affecting Cattle Prices 243

Assessing Marketing Costs 251

Forecasting Beef Prices and Managing Price Risks 251

The Futures Market 252

Advertising and Promoting Retail Beef 260

Selected References 261

10 The Global Beef Industry 263

Numbers, Production, Consumption, and Prices 264

International Trade 265

Beef Suppliers 270

NAFTA Partners 276

Buyers 281

Trade Policy and Other Factors 285

Selected References 287

11 Reproduction 288

Structure and Function of the Reproductive Organs 289

Breeding 295

Artificial Insemination 299

Synchronization of Estrous 306

Embryo Transfer 315

Cloning 317

Pregnancy 317

Calving 319

Rebreeding 326

Adaption of Reproductive Management Technologies 328

Selected References 329

12 Genetics and Breeding 331

Genetic Principles 331

Mating Strategies 337

Traits and Their Measurement 342

Improving Beef Cattle Through Breeding Methods 350

Selection Programs 355

National Sire Evaluation 357

Genetic Testing and Marker-Assisted Selection 358

Selected References 360

13 Cattle Breeds 362

Breed Variation 363

Breed Evaluation for Commercial Producers 371

Breed Evaluation for Low-Cost Production 390

Breed Evaluations to Improve Consumer Market Share 391
Integrated Production Systems 392
Selected References 393

14
Nutrition 395
Nutrients 395
Proximate Analysis of Feeds 396
The Ruminant Digestive System 398
Digestibility of Feeds 400
Energy Evaluations of Feeds 400
Feeds: Classification and Composition 402
Nutrient Requirements of Beef Cattle 411
Ration Formulation 411
Pricing Feedstuffs 412
Cow-Calf Nutrition 420
Yearling-Stocker Cattle Nutrition 429
Feeding Feedlot Cattle 429
Selected References 431

15
Managing Forage Resources 433
Grazed Forage Resources 433
Plant Types and Their Distribution 439
Grazing Management 441
Major Grazing Regions of the United States 455
Hays 465
Crop Residues 467
Drought Management 468
Market Cattle Production on Grazed Forage 468
Health Problems Associated with Grazing Plants 474
Selected References 479

16
Herd Health 481
Diseases and Health Problems 482
Immunity 490
Establishing a Herd Health Program 497
Stress and Health 498
Parasites 504
Cow-Calf Health Management Programs 509
Selected References 517

17
Growth, Development, and Beef Cattle Type 519
Growth and Development 519
Beef Type 524
Parts of the Beef Animal 528
Carcass Conformation 528
Conformation of Market Cattle 529
Conformation of Feeder Cattle 530
Conformation of Breeding Cattle 532
The Livestock Show 538
Selected References 539

18
Cattle Behavior, Facilities, and Equipment 540
Behavior 540
Gathering Cattle on Pasture and Training Cattle 552
Cattle Transport 552
Facilities and Equipment 553
Auditing Animal Well-Being 560
Selected References 562

Appendix 563
The Metric System 564
Other Weights, Measures, and Sizes 564
Volumes and Weights of Stacked and Baled Hay 564
Round Grain Bin Volumes 568
Measuring Irrigation Water Flow 568
Land Description for Legal Purposes 570
Major Organizations Within or Affecting the Beef Industry 570

Glossary 591

Index 608
Additional Resources 623

Preface

The domestication of beef cattle initiated an opportunity for human beings to apply their creativity to the formation of the modern beef cattle industry. Beef cattle provide a source of livelihood and fulfillment to producers, feeders, packers, processors, and retailers of beef products and by-products while providing a satisfying source of protein and other goods to the world's consumers. The relationship between cattle and humans provides not only a rich past but also a hopeful future to many societies and cultures throughout the world.

PURPOSE OF THIS BOOK

This book serves three primary purposes: (1) to identify the significant biological and economic principles that contribute to the profitable and sustainable production of beef cattle; (2) to systematically integrate these principles to enhance effective decision making; and (3) to serve as a model for developing a systems mindset for future leaders of the beef industry.

ORGANIZATION

The first seven chapters of the book emphasize management principles for the industry and each of the segments of the supply chain. Chapters 8, 9, and 10 are focused on the coordination of the segments, the creation of profitable marketing systems, and international trade of beef and beef products. Chapters 11 through 18 deal with the biological and economic principles required to understand beef cattle productivity and to develop management systems that are both productive and sustainable.

NEW TO THIS EDITION

- This edition has been substantially revised to include a detailed discussion of the significant shifts that have altered the structure of the beef industry, an in-depth discussion of the trends in inventories, demand, production practices, and philosophical changes that impact management decisions, as well as emerging technologies and protocols that offer opportunities to improve productivity, profitability, and sustainability. The book focuses on both the scientific basis for management decisions as well as the practicality of implementing them. While an in-depth discussion of the appropriate management specific to particular geographic regions is beyond the scope of the work, every effort has been made to more adequately describe regional differences.
- The text is focused on meeting the multiple outcomes of a sustainable enterprise—healthy profits, community, and environment. Management decisions are addressed through multiple lens—impacts at the enterprise level, the full supply chain, as well as the challenges of addressing the multiple and often conflicting messages from consumers, policy makers, and other externalities. The work focuses on sound principles without attempting to provide a one-size-fits-all approach to beef production.
- The concepts of total quality management, beef quality assurance, strategic planning, and market focus have been integrated into the text. Furthermore, the text provides a holistic perspective about management of people, natural resources, financial resources, and risk in profitable beef production systems. This edition was created with the input of professional cattle producers, educators, researchers, extension specialists, and students.

About the Author

Tom Field, Ph.D., is the holder of the Paul Engler Chair in Agribusiness at the University of Nebraska and the Director of the Engler Agribusiness Entrepreneurship Program. His program's mission is to empower enterprise builders. A noted author, commentator, and speaker at agricultural and business events in the United States and abroad, Tom has consulted with numerous enterprises and organizations and has served on advisory boards related to agriculture, education, and athletics. He is the co-owner of Field Land and Cattle Company, LLC, an enterprise with its roots in the westward migration that followed the Civil War.

He grew up on a seedstock and commercial cow-calf business in western Colorado and attended Colorado State University where he earned his B.S., M.S., and Ph.D. in Animal Sciences. Under the tutelage of Dr. Robert Taylor, he pursued a multidisciplinary approach to learning that encompassed animal sciences, economics, genetics and animal breeding, rangeland science, beef cattle management systems, human development, public policy, history, and communications.

He is a passionate advocate for free enterprise, the potential of young people, and opportunities in agriculture and rural places. Married to Laura and the father of five children—Justin, Sean, Trae, Kate, and Coleman, Tom's passion for ranching and the beef business is centered on faith, family, and country.

1

An Overview of the U.S. Beef Industry

Since the time that early humans first painted pictures of cattle on cave walls and took their first taste of beef, the bovine has played a role in the existence of humankind. Whether as a source of wealth, food, clothing, or draft power, cattle have evolved in a symbiotic relationship with people.

The Europeans who first imported cattle to the Americas could not have envisioned the size and scope of the beef industry that would eventually develop in the New World. The cowboys on the trail drives of the late 1800s would not have been able to foresee the changes in the infrastructure and marketing system that allowed the beef industry to move away from a commodity paradigm and toward that of a value-added, consumer-driven business.

The evaluation of agricultural systems has been ongoing for centuries, and the emergence of the beef cattle industry resulted from the recognition that domestication of cattle and other livestock would result in a consistent supply of food, fiber, and draft power. The organizational structure of the industry became increasingly complex when the advances of the industrial age allowed rapid increases in production efficiency, permitting people to pursue vocations other than producing their own food supply.

The scientific and information breakthroughs of the last century have heightened agricultural productivity to the point that fewer than 2% of U.S. citizens are directly employed in production agriculture.

However, as fewer people understand or participate in food production, leaders of the beef industry and the agricultural commodities find it important to increase the level of communication between producers, processors, retailers, and consumers. The beef industry has always faced challenges and the present and future are no different. Nonetheless, the relationship between the stock producer and his or her cattle is one with the potential to yield enormous benefits for humans. The thoughtful and diligent study of the beef industry, its associated infrastructure, marketplace, and management offers people the opportunity to apply creativity and energy to a fundamentally important endeavor.

GENERAL OVERVIEW

The beef industry includes breeding, feeding, and marketing cattle with the eventual processing and merchandising of retail products to consumers. The process involves many people and utilizes numerous biological and economic resources. Most important, however, is the time involved: depending on the production alternatives, approximately 2 to 3 years are required from breeding time until a beef product can be made available to consumers (see Figure 1.1).

Leaders and managers across the beef supply chain must have a comprehensive knowledge of the industry if they are to be most successful both individually and

Figure 1.1
Beef production cycle.

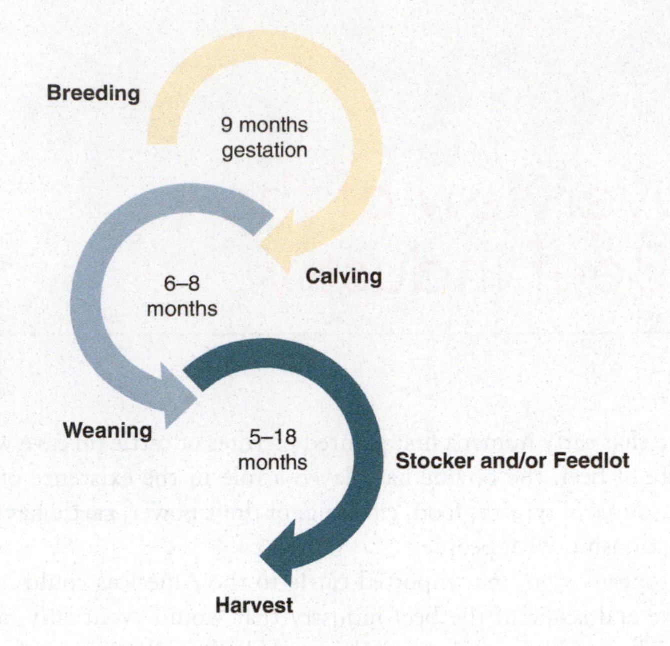

collectively. Developing a systems management mindset is critical to develop profitable and sustainable enterprises in the beef industry.

Numbers, Prices, and Consumption

The beef industry involves people (cattle producers, processors, and consumers), products (number of cattle, pounds produced and consumed), prices, and profitability. Figure 1.2 shows the cattle inventory in the United States over the past 60 years. Beef and dairy numbers are combined because the dairy industry contributes a significant amount of production to the U.S. industry. Cattle numbers increased rapidly from the early 1900s until the mid-1970s, when a dramatic decline occurred. The cattle inventory of approximately 90 million head in 2015 represents 68% of the peak numbers of 132 million head in 1975. While a modest herd rebuilding process began in 2014, the national cattle herd is likely to remain between 89 and 93 million head for the foreseeable future. There have been significant peaks and valleys in cattle numbers resulting from factors influencing the supply and demand of beef. These cycles reflect the profitable and unprofitable periods of the cattle industry. The influence of cattle cycles in marketing cattle is discussed in detail in Chapter 9.

Figure 1.2
Total cattle inventory (1955–2015).
Data source: USDA National Agricultural Statistics Service; compiled by Livestock Marketing Information Center.

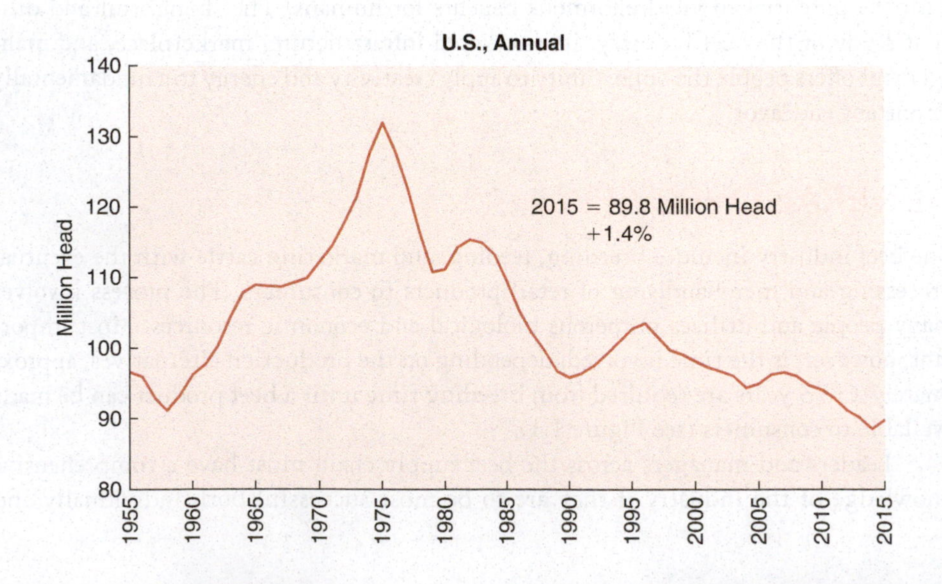

Table 1.1
Cattle Numbers and Prices, Human Population, and Beef Consumption in the United States, 1925–2015

Year	Human Population (mil.)	No. Cattle (mil.)	No. Beef Cows (mil.)	Carcass Beef Produced (bil. lb)	Per Capita Retail Beef Consumption (lb)	Choice Fed Steer Price ($/cwt)	Retail Choice Beef Price ($/lb)
1925	115.0	63.4	11.2	6.9	44	10.16	0.30
1930	122.8	61.0	9.1	5.9	36	10.95	0.35
1935	126.9	68.8	11.1	6.6	39	12.32	0.30
1940	131.8	68.3	10.7	7.2	41	11.86	0.29
1945	139.2	85.6	16.5	10.3	37	17.30	0.33
1950	151.1	78.0	16.7	9.4	47	28.88	0.75
1955	164.0	96.6	25.7	13.2	62	26.93	0.67
1960	179.3	96.2	26.3	14.4	63	25.90	0.80
1965	193.0	109.0	33.4	18.3	75	24.99	0.80
1970	201.9	112.4	36.7	21.5	85	29.45	1.00
1975	213.8	132.0	45.4	23.7	88	45.21	1.52
1980	227.2	111.2	37.1	21.6	77	65.64	2.34
1985	237.9	109.6	35.4	23.7	79	62.99	2.29
1990	249.4	95.8	32.4	22.7	68	74.71	2.81
1995	262.8	102.8	35.2	25.2	67	65.01	2.84
2000	282.2	98.2	33.6	26.9	68	69.64	3.07
2005	295.7	95.0	32.7	24.7	66	87.27	4.09
2010	310.2	94.1	31.4	26.4	60	95.03	4.38
2015	321.8	89.1	29.3	24.3	54	157.74	6.29

Source: USDA.

Table 1.1 highlights some important data about the cattle industry from 1925 to 2015. Changes in population, cattle numbers, product consumption, live cattle prices, and average retail prices impact the industry.

Figure 1.3 illustrates the relationship between cattle numbers and carcass beef production. Since 1979, cattle numbers have decreased significantly. Carcass beef production declined initially, remained relatively stable in the 1980s, and then increased in the last half of the 1990s, partly due to Canadian imports. The disruptions caused by

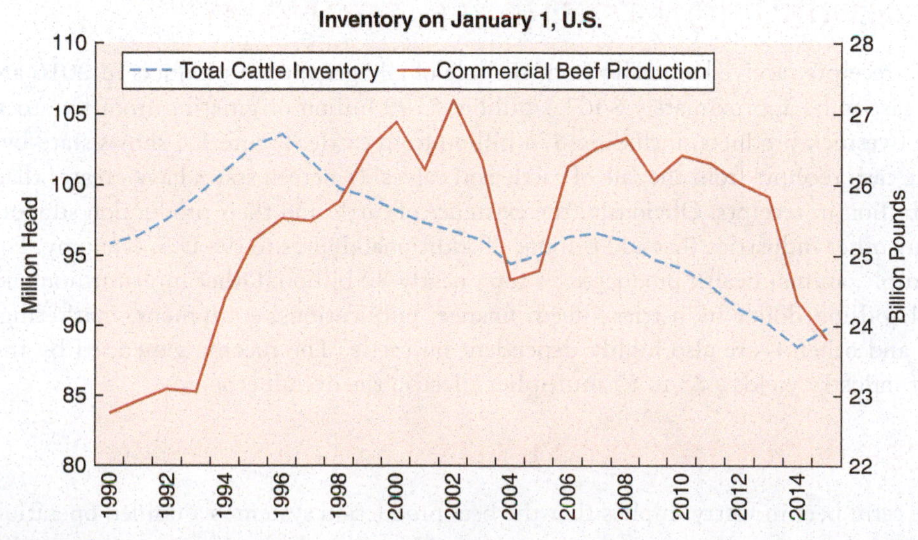

Figure 1.3
Beef production versus cattle inventory (1990–2015).
Data source: USDA National Agricultural Statistics Service; compiled by Livestock Marketing Information Center.

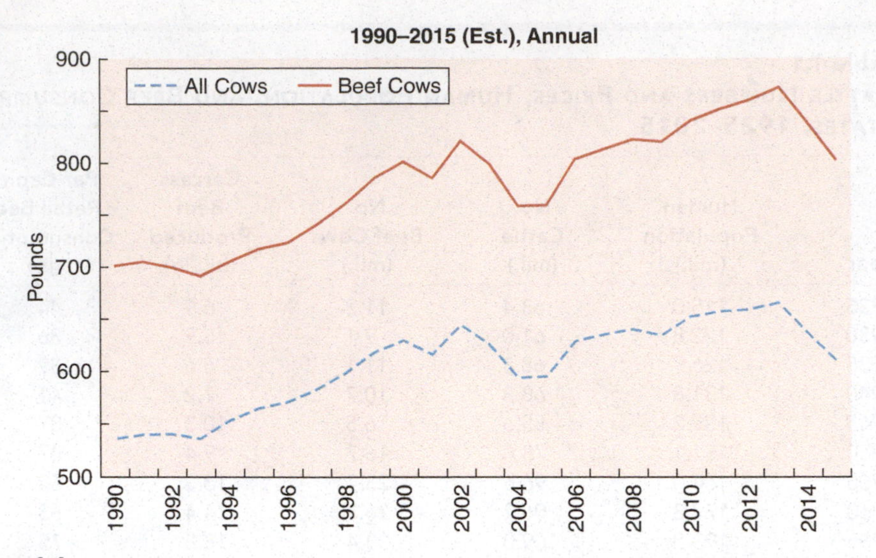

Figure 1.4

Beef production per cow (1990–2015).

Data source: USDA National Agricultural Statistics Service; analysis and compiled by Livestock Marketing Information Center.

Bovine spongiform encephalopathy-related trade restrictions shut off Canadian imports for a period of time in the mid-2000s, resulting in the decline in beef production. Supplies then normalized before falling as the result of a declining beef herd in both Canada and the United States. Total beef production has been maintained despite inventory changes for the most part over the past 40 years. There are several reasons for these trends: (1) most importantly, average carcass weights have increased from 613 lb in 1970 and 635 lb in 1980 to 750 lb in 2004; (2) the feedlot turnover rate has increased from 2 times capacity to 2.4 times capacity, resulting in more cattle available for slaughter; (3) the slaughter age of fed cattle has decreased; (4) the genetic base for heavier cattle at a given age has increased (due to more crossbreeding, increased emphasis on growth in British breeds, and utilization of more Continental breeds by commercial breeders); and (5) increased importation of cattle and beef. As a result, there has been an increase in the amount of beef produced per cow in the breeding herd (Figure 1.4). The slight decline in per animal production since 2013 is partly due to short-term supply and demand conditions.

CONTRIBUTION TO THE U.S. ECONOMY

Cash receipts received annually from the sale of all agricultural products in 2016 are forecast to be approximately $367.5 billion: $190 billion originating from livestock and livestock products, including $74 billion from cattle. Figure 1.5 shows state-by-state cash receipts from the sale of cattle and calves. Nineteen states have greater than $1 billion in receipts. Obviously, the existence of cattle and their production support many other industries that add billions of additional dollars to the U.S. economy. For example, animal health product sales total nearly $2 billion. Other multimillion and multibillion dollar industries—feed, finance, publications, equipment, marketing, AI, and others—are also highly dependent on cattle. The income generated by the beef industry yields a $3 to $5 multiplier effect in the overall economy.

BEEF INDUSTRY SEGMENTS

The term beef industry implies that the beef production system is a unified operation subject to an overall management program. However, the beef industry is actually made up of several different segments (Table 1.2) that are linked together through

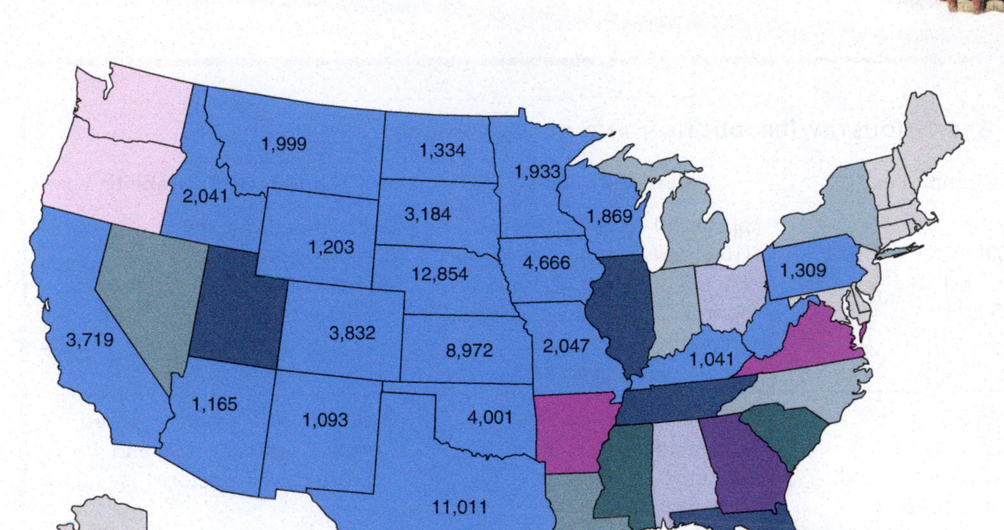

Figure 1.5
Cash receipts (million $) from sale of cattle.
Source: Adapted from USDA (*Agricultural Statistics*, 2015).

beef animals and products, yet the segments operate somewhat independently from each other. Each segment has different economic parameters and management problems and markets different products. In some cases, segments are in direct competition with one another. In some respects, the various beef industry segments can be considered separate industries because of their distinctly different characteristics.

The Seedstock Segment

Seedstock breeders, sometimes referred to as *purebred breeders* or *registered breeders*, are specialized cow-calf producers. Seedstock breeders are predominantly responsible for identification and propagation of genetics that contribute to the profitability of the industry.

Seedstock breeders sell genetic information, breeding animals, semen, and embryos to other breeders and commercial cow-calf producers. Their function is one of service—to provide the genetics that can be economically utilized by the beef industry. The breeders sell breeding animals primarily to commercial cow-calf producers within a 100- to 150-mile radius of the breeders' operations. Choice of breed—whether one or a combination of breeds—is important in developing a production and marketing program that can best serve the commercial producers in any given area.

The seedstock segment is discussed in further detail in Chapter 4. Chapters 12 and 13 cover the biological relationships for making genetic changes in the economically important traits of cattle.

The Commercial Cow-Calf Segment

Commercial cow-calf producers maintain cowherds and raise calves from birth to weaning. Under ideal conditions, each cow is expected to produce one calf annually. Calves are the primary source of revenue for the commercial producer as well as the source of heifers to replace breeding cows that are culled.

Table 1.2
Overview of the U.S. Beef Industry (production and consumption)

Segment	People/Companies	Cattle/Products	Tenderness/Palatability
Seedstock	*Marketings:* Top 25: 29,600 hd Top 10: 20,065 hd Top 5: 14,395 hd 8 AI studs	Approx. 80 cattle breeds (10 breeds are most important, while 5 breeds contribute approx. 60% of the genetics); primarily yearling bulls, semen, and AI certificates	British breeds highest Brahman breed lowest Genetic variation for tenderness exists within a breed
Cow-calf (yearling stocker)	727,906 producers cow herds <50 head have 20% of cows but 80% of all operations; 45% of cow inventory in herds >100 hd Top 25: 259,400 hd Top 10: 177,300 hd Top 5: 124,700 hd	29.7 mil. hd beef cows 9 mil. head dairy cows 88% calf crop 525 weaning wt	Highly variable based on breed, implant protocol, age, and assorted other factors
Feedlot	1,781 feedlots with >1,000 hd capacity in 12 major states Top 20 capacity: 4.7 mil. hd Top 10 capacity: 3.4 mil. hd Top 5 capacity: 2.4 mil. hd	13.1 mil. fed cattle capacity	
Packer	Top 10 daily harvest: 106,275 Top 5 daily harvest: 96,075	30.2 mil. cattle slaughtered 23.7 bil. lb carcass wt *Quality graded (2004)* Prime (4%), Choice (66%), Select (30%) *Yield graded (2004)* 1 (8%), 2 (36%), 3 (46%), 4 (9%), 5 (1%) avg. carcass wt (750 lb,all cattle)	min. fat (0.3 in.) prevents cold shortening; electrical stimulation; improves tenderness, aging (14–21 days) increases tenderness
Retailer	38,015 supermarkets with more than $2 mil. in annual sales	Annual per-capita distribution: hamburger (28 lb); steaks/roasts (30 lb); processed (9 lb)	
Purveyor	More than 300 companies	Center-of-the-plate products	Emphasize high palatability
Consumer	Population: United States (293 mil.) World (5.9 bil.)	*Per-capita consumption (2015)* *Retail* / *Boneless* Beef 56 / 52 Pork 49 / 46 Poultry 105 / 66 per-capita U.S. expenditures for beef: $324	steaks cooked higher than "medium" tend to be less tender and drier; per capita total fat consumption at all time high (fat and taste preferences are highly related)
Exports	Primarily to (1) Japan, (2) Mexico, (3) South Korea	2.5 bil. lb (carcass beef) valued at $6.5 bil.	
Imports	Primarily from (1) Australia, (2) Canada, (3) New Zealand	3.0 bil. lb valued at $2.4 bil.	Mostly in the form of ground or manufacturing beef

Table 1.2
(Continued) Current U.S. Beef Industry (financial and economic)

Segment	Costs	Prices	Profits/Returns
Seedstock		$2,000–8,000 (to commercial producers); semen $5–50/unit; AI certificates ($10–150)	
Cow-calf	*Annual cow cost* High 1/3 ($800) Avg. ($650) Low 1/3 ($400)	*450 lb* 1997 ($89/cwt) 2004 ($129/cwt) 2014 ($271/cwt)	1991 (+$55/cow) 1996 (–$80/cow) 2000 (+$80/cow) 2004 (+125/cow) 2014 (+$550/hd) 2017 proj. (+$250/hd)
Yearling/stocker		*750 lb* 1993 ($85/cwt) 1997 ($74/cwt) 2000 ($86/cwt) 2004 ($104/cwt) 2015 ($207/cwt)	*Summer Program* 1986 (+$25/hd) 1991 (–$5/hd) 1996 (+$40/hd) 2000 (+$20/hd) 2004 (+$125/hd) 2014 (+$375/hd) 2015 (–$75/hd)
Feedlot		*Fed Steer* 1993 ($76/cwt) 1997 ($66/cwt) 2004 ($84/cwt) 2015($150/cwt)	1986 (+$26/hd) 1991 (–$40/hd) 1996 (–$10/hd) 2000 (–$5/hd) 2004 (+$25/hd) (+$25) 2015 (+$300/hd)
Packer		*By-product Value* ($/cwt of live weight) 1995 ($9.60/cwt) 1997 ($10.30/cwt) 2003 ($10.12/cwt) 2004 ($9.27/cwt) 2015 ($12.85/cwt)	*Wholesale Boxed Value ($/lb)* 1986 ($0.95 lb) 1991 ($1.18 lb) 1996 ($1.03 lb) 2000 ($1.17 lb) 2003 ($1.44 lb) 2015 ($2.37 lb)
Retailer		Retail price (Choice; Avg. of all cuts) $6.28/lb	Supermarket Sales *1983* *1993* *2002* *2015* Beef (bil. $) 26.8 21.1 22.1 24.6
Consumer		Market Share % of meat expenditures *1990* *2004* *2014* Beef 42 43 46 Pork 27 27 27 Poultry 31 29 27	Beef expenditures were $324 per capita. Total consumer spending on beef exceeded $60 bil. for the first time in 2002 as well as in 2003, $70 bil. in 2004

Changes in cow numbers over the past 20 years reflect a reduction in beef cows and a relatively stable inventory of dairy cows (Figure 1.6). Distribution of beef cows by state is shown in Figure 1.7. Note the concentration of beef cows in the Great Plains. This area—which covers Texas north through North Dakota and the eastern parts of New Mexico, Colorado, Wyoming, and Montana—accounts for approximately 50% of the total U.S. beef cow population. The Corn Belt and southeastern states also have significant numbers of cows.

Changes in beef cow numbers by states over the past decade are reflected in Figure 1.8. The changes are primarily related to market conditions and forage supply.

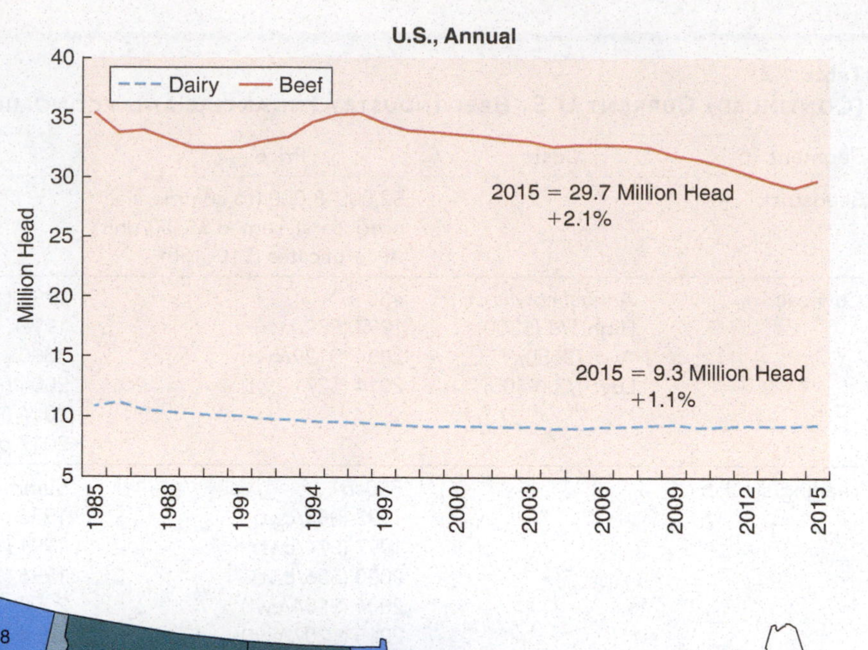

Figure 1.6
U.S. cow inventory (1985–2015).
Data source: USDA National Agricultural Statistics Service; compiled by Livestock Marketing Information Center.

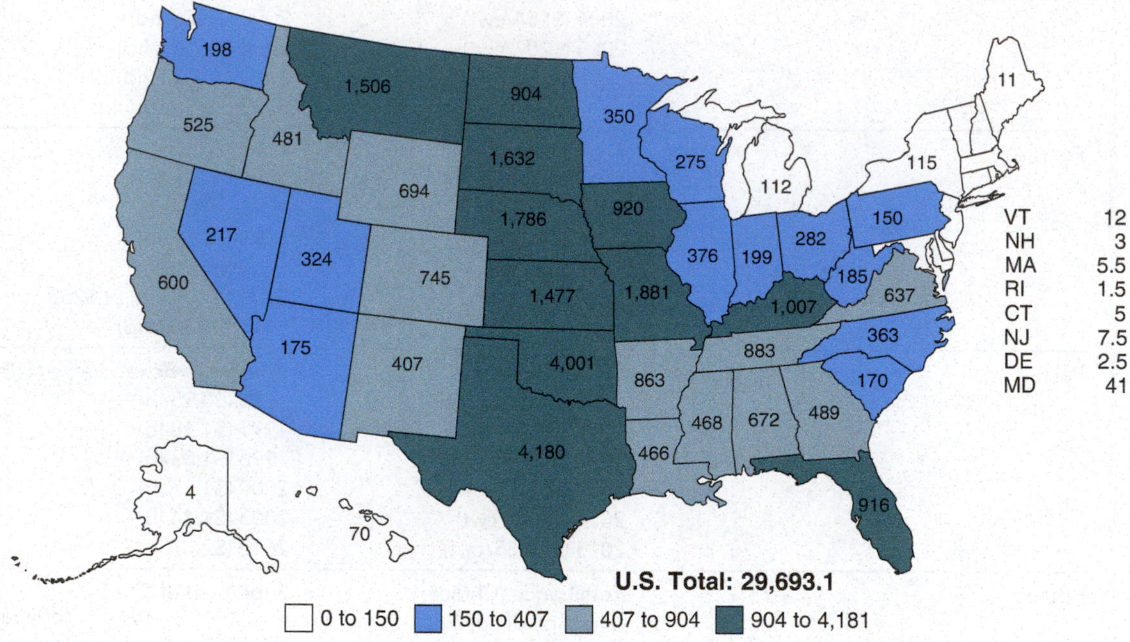

Figure 1.7
Beef breeding cows, January 2015 (1,000 head).
Data source: USDA National Agricultural Statistics Service; compiled by Livestock Marketing Information Center.

The latter can be dramatically influenced by drought, renovation of previously unproductive land, water development, and shifts in land use between crops and forages. The significant increase in crop prices that occurred from 2010 to 2013 coupled with the growth of the ethanol business increased the cost of production for the cattle industry and motivated many agricultural producers to swap pasture acres for row crops. Varying sections of the United States endured drought conditions from 2000 to 2015 that also drove down the size of the U.S. cow herd.

Figure 1.9 shows that nearly 80% of beef cow operations have fewer than 50 head of cows, while controlling less than 30% of the cow inventory. The small herd size is not surprising because part-time farmers operate about one-half of U.S. farms, and many of their farms are less than 50 acres in size. Approximately 55% of beef cows are in herds of greater than 100 head. Yet, only 10% of enterprises are in this size category.

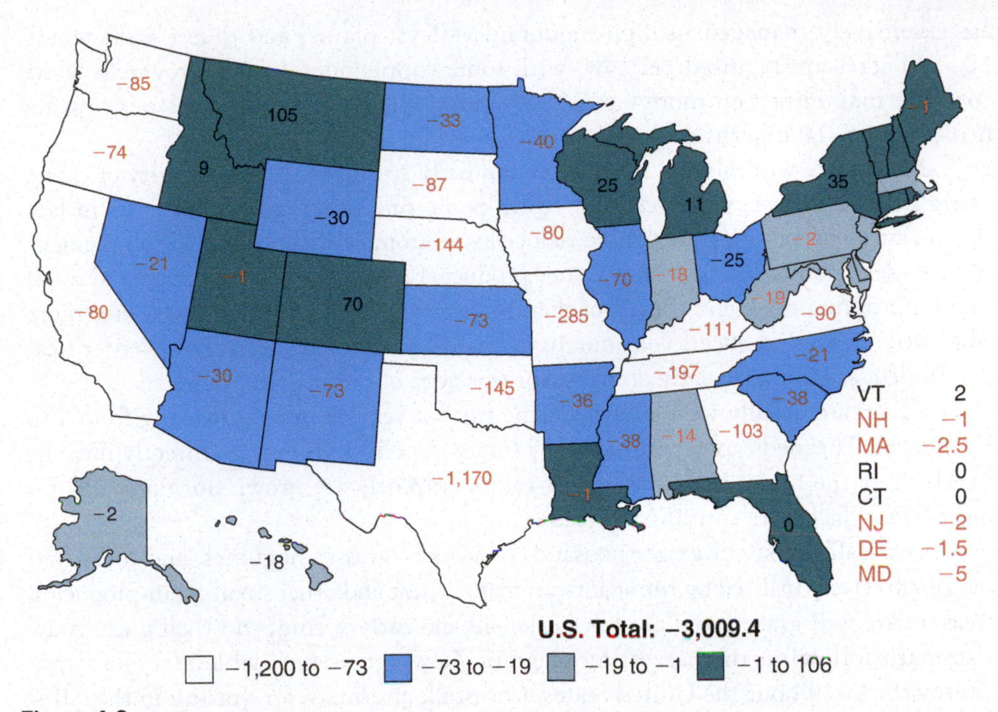

Figure 1.8
Change in beef cow numbers (2006–2015) per 1,000 head.
Data source: USDA National Agricultural Statistics Service; compiled by Livestock Marketing Information Center.

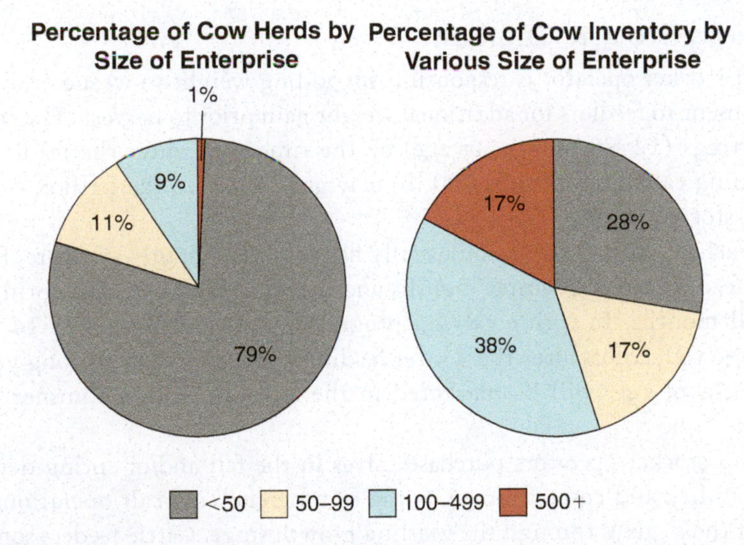

Figure 1.9
Percentage of cow herds in various size groups—2004; percent of cow inventory in various size groups—2004.
Source: Adapted from USDA-NASS.

Herd sizes of 300 cows or more are considered to be an economic unit, so there are numerous small beef cow operations that are supplemented with outside income. Increase in cow herd size does not always imply an increased efficiency of production. However, several studies demonstrate that there is a greater return per head as cow herd size increases toward 1,000 head. Although there are intensive cow-calf operations where cows are maintained under conditions comparable to large dairies year-round, the vast majority are extensively managed operations where cows are maintained on grazed and harvested forage throughout the entire year. Many cow-calf operations

are extensively managed in high mountain valleys, plains, and desert areas where 30–100 acres are required per cow, with some supplemental feeds provided. Some cows are maintained on more intensively grazed areas where 1–5 acres per cow are utilized for 5–10 months during the year.

Most cows will calve in late winter and early spring with the majority of calves being born in February, March, and April. Some producers calve their cows in late spring, summer, or fall, primarily to reduce losses from calf scours and to complement their forage production program. Other producers may have both a spring and a fall calving program to extend the use of their bulls and to use their labor and forage more efficiently. A few producers continue to calve on a year-round basis; however, critical economic assessments usually do not favor this type of calving program.

Calves are usually weaned at the same time of year, their ages ranging from 5 to 10 months. Weaned calves that are heavy (more than 500 lb) may go directly into the feedlot, but the majority of the lighter calves currently are grown out on forage for several months before entering the feedlot.

Cow-calf pairs will graze thousands of acres of grasses, legumes, and forbs that can be effectively utilized by ruminants. In many wheat and other small grain-producing areas, cattle will graze green growth in the fall and early spring, and then graze straw aftermath following the harvesting of grain. Cows graze untillable acres and crop aftermath throughout the United States. Cornstalk aftermath for grazing in the fall is very important in the major corn growing regions.

A more detailed discussion of the commercial cow-calf segment is provided in Chapter 5.

The Yearling-Stocker Segment

The yearling-stocker operator is responsible for adding weight to weaned calves prior to their shipment to feedlots for additional weight gain prior to harvest. The calves are usually yearlings (12–20 months of age) by the time they enter the feedlot. Some heavier weaning calves (more than 500 lb) may go directly to the feedlot, bypassing the yearling-stocker phase.

The yearling-stocker operation usually has available forage—pasture, hay, and silage—for feeding during winter months and grazable forage for the spring, summer, and fall months. In spring calving programs, short yearlings (10–14 months of age) may go to feedlots after the winter feeding program, whereas long yearlings (15–20 months of age) will be marketed in the fall following a summer grazing program.

Yearling-stocker operators purchase calves in the fall and/or spring depending on the availability and cost of forage. Some commercial cow-calf operations retain ownership of their calves through the yearling growth stage. Cattle feeders sometimes purchase calves and maintain ownership through both the growing and feedlot phases. These two alternatives are increasing in frequency, making the traditional yearling-stocker operation a minor beef industry segment.

The yearling-stocker segment of the beef industry is discussed in more detail in Chapter 6.

The Feedlot Segment

Feedlots are confinement feeding operations where cattle are fed primarily finishing (high-energy) rations prior to harvest. Most feedlot operations feed relatively high grain rations for 100 to 200 days for economically efficient gains and to improve the palatability of the retail product. Some operations background cattle by feeding them primarily roughage rations prior to the finishing phase.

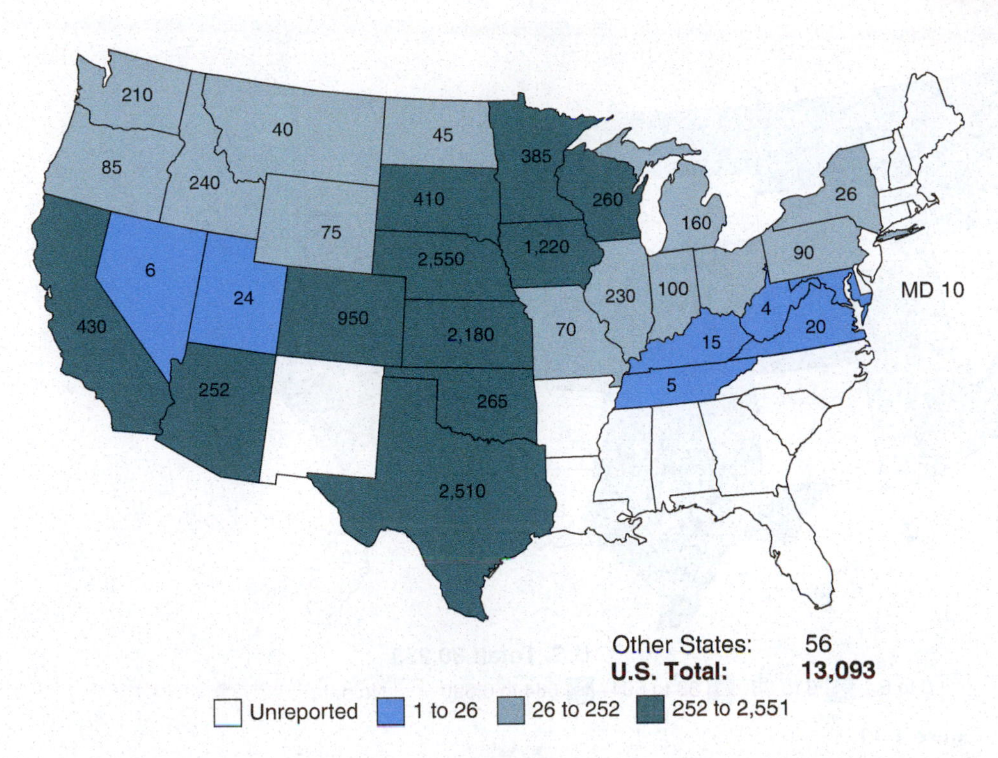

Figure 1.10
Cattle on feed, January 1, 2015 (1,000 head).
Data source: USDA National Agricultural Statistics Service; compiled by Livestock Marketing Information Center.

The number of cattle on feed is shown in Figure 1.10. Commercial cattle feeders annually feed and market approximately 2.5 times the one-time feedlot capacity. Cattle-On-Feed reports from the USDA usually give information for only the top 12 states as these 12 states feed >98% of the cattle. Note the concentration of cattle feeding in the Plains states. The primary reasons for this distribution of fed cattle are the availability of feed grains, the locations of packing plants, and the climatic and geographic conditions that favor cattle feeding.

Feedyards located in the eastern half of the upper midwestern states and east into the Corn Belt tend to have smaller capacities per feedlot as compared with the western and southern regions of the Great Plains and southwestern states. The southern and western tiers of feedlot states have larger feed yards due to more arid and consistent climatic conditions. The larger yards are also typically the primary enterprise of focus by management, whereas, in the Corn Belt region, the feeding enterprise is typically part of an integrated farming business.

The number of lots, inventory, and marketings by size of feedyard are provided in Table 1.3.

Table 1.3
Number, Cattle on Feed, and Annual Fed Cattle Marketings by Size of Feedyard, 2015

Feedyard Capacity (*N* of head)	Lots (*N*)	Inventory (1,000 hd)	Marketings (1,000 hd)	Turnover[1]
<1,000	25,000	2,602	2,895	1.11
1,000–1,999	820	355	650	1.83
2,000–3,999	580	660	1,230	1.86
4,000–7,999	340	900	1,740	1.93
8,000–15,999	180	1,210	2,250	1.86
16,000–32,999	141	1,890	3,430	1.81
32,000+	128	5,560	10,360	1.86

[1]Turnover would be higher when cattle inventories are higher.

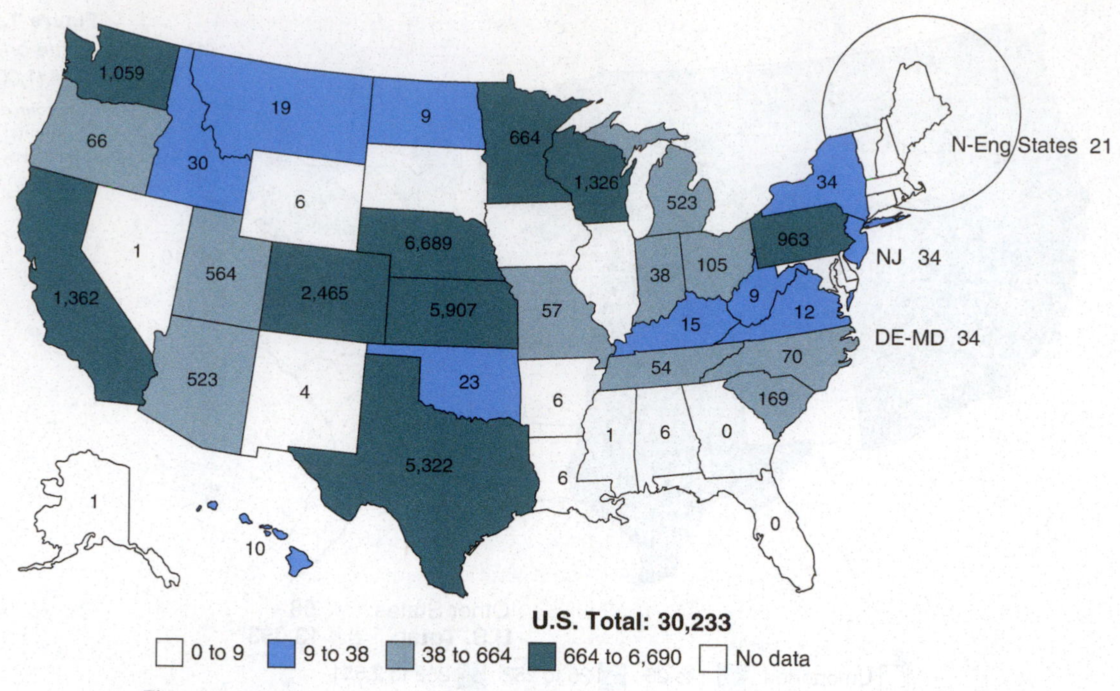

Figure 1.11

Commercial cattle harvest, 2014 (1,000 head).

Data source: USDA National Agricultural Statistics Service; compiled by Livestock Marketing Information Center.

The Packing Segment

The distribution of fed cattle harvested in the various states is shown in Figure 1.11. Cattle are harvested in the same geographical areas where feedyards are located (compare Figures 1.10 and 1.11). Table 1.4 identifies the major packing companies and their capacities. The packing industry is one of the most regulated businesses in the United States and has highly volatile margins. These factors coupled with capitalization and labor challenges led to the closing of nearly 200 beef packing plants between 1975 and 2015.

Table 1.4

LEADING 10 BEEF PACKERS IN THE UNITED STATES, 2015/2016

Name	Daily Harvest Capacity	Sales (bil. $)	All Plants (*N*)[1]	Total Employees[1]
Tyson Foods, Inc.	28,950	19 (est.)	95	63,500
JBS Beef Company	27,125	18	45	113,000
Cargill Meat Solutions	23,000	11.5	30	26,600
National Beef	12,000	8.2	6	8,100
American Food Group	NA	3.2	10	4,000
Greater Omaha Packing Company, Inc.	2,800	1.4	1	900
Nebraska Beef, Ltd.	2,800	NA	1	1,000
Caviness Beef Packers, Ltd.	1,700	0.17	2	750
Agri Beef Company	1,600	NA	1	900
Sam Kane Beef Processors, Inc.	1,600	NA	1	650

[1]All operations including beef, pork, lamb, and poultry.

Source: Adapted from multiple sources.

Packers, purveyors, and retailers harvest, process, and distribute approximately 24 billion lb of beef. The magnitude of the beef packing industry as reflected by the number of cattle harvested is shown in Figure 1.12. Of the total annual beef harvest, approximately 80% are fed steers and heifers. The smaller number of non-fed steers and heifers along with cull cows and bulls (that are harvested) receive little or no concentrate feeding prior to slaughter. Their rations have been primarily grass and other forages. The number of non-fed heifers, cows, and bulls harvested varies in response to climatic and profit conditions at the cow-calf level.

Beef sold from packing plants is primarily boxed (>80% of the beef slaughtered). The boxed beef is primal and subprimal cuts from which much of the bone and excess fat has been removed. The cuts are vacuum packaged for a longer shelf life. Boxing of beef has proven to be more cost efficient at the packing plant level because (1) labor rates are usually lower at packing plants than those at retail stores, (2) cutting is usually faster and more efficient as it is done on a moving "disassembly" line by specialized meat cutters, (3) a larger volume of retail product can be handled in less space, (4) more effective use can be made of bone and fat by-products, and (5) transportation costs are reduced, with a more valuable product that can be more easily handled than carcass beef. Retailers have an advantage in buying only those beef cuts that can be more easily merchandised without accepting the entire carcass. In addition, retailers have less spoilage because vacuum-packaged meat has a longer shelf life than carcass beef.

Packers have a preference for carcasses weighing in excess of 800 lb. Given the high fixed costs associated with a packing plant, a key strategy to assure profit is to drive high production volume. Increasing carcass weights is one means to increase plant productivity. Packers merchandise case-ready beef that has been fabricated and packaged either fresh or precooked as a means to capture value. The primary advantages of case-ready products are improved control over food safety, lowered labor costs, improved consistency and yield, enhanced inventory control, and direct delivery of products oriented to consumer preference. The move to case-ready products is a significant development in the industry.

The Purveyor Segment

A meat wholesaler, sometimes called a "jobber," is an operator who purchases beef and sells it to a retailer or to another wholesaler. **Purveyors** and distributors are two types

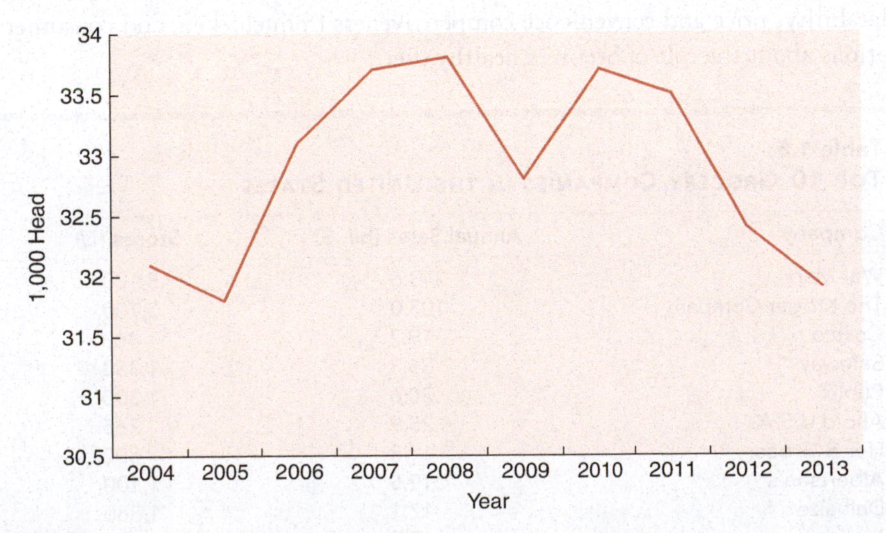

Figure 1.12
Annual cattle harvest, 2004–2013.
Source: Adapted from USDA.

of beef wholesalers. Purveyors buy beef and perform some fabrication, while distributors buy and sell beef without cutting or changing the product. Purveyors sell almost exclusively to the food service industry (which cooks and sells food for away-from-home or take-out consumption). Purveyors are specialized meat processors who provide highly palatable center-of-the-plate products to food service operators, retail stores, and mail order customers.

Purveyors handle about 5% of total beef. The number of distributors and purveyors continues to decline, however; so they are becoming less important as a separate beef industry segment. Packers are increasing their sales directly to retailers or through brokers. Fabrication of beef carcasses continues to increase at the packer level.

The Retail Segment

U.S. food retailing accounts for nearly $638 million in annual sales with approximately 38,000 stores with more than $2 million in per year sales. The grocery business employs 3.4 million people. Supermarkets stock over 42,000 items per store, ring up an average transaction of $30, and are visited by the average customer 1.5 times per week. The top 10 U.S. grocers in this highly competitive industry are listed in Table 1.5.

Historically, almost all retail cuts were prepared at the store level by in-house butchers fabricating sides or quarters of beef. Today most of the beef received by grocery stores is in the form of boxed beef primals, boneless subprimals, or beef for grinding. The movement to case-ready products was an innovative shift that improved the ability of retail stores to order specific cuts best suited to their customer demographics, improved efficiencies by reducing dead air space in refrigerated trucks that once carried hanging carcasses, and improved distribution margins. Sales by meat, poultry, and fish departments comprise approximately 14% of all grocery store sales, with fresh beef accounting for approximately one-third of meat department sales (Table 1.6).

The Consumer Segment

Figure 1.13 illustrates consumer demand by demonstrating per capita expenditures for beef and competitive proteins. Beef consumption has varied over time with significant retail beef consumption growth on a per capita basis from 1960 to 1976. This period of demand expansion was followed by a steady decline into the early 1990s. The primary reasons for the decline were excessive fat production, inconsistency in palatability, price and convenience competitiveness from chicken, and consumer perceptions about the role of beef in a healthy diet.

Table 1.5
TOP 10 GROCERY COMPANIES IN THE UNITED STATES

Company	Annual Sales (bil. $)	Stores (*N*)
Wal-Mart	343.6	5,100
The Kroger Company	103.0	3,700
Costco	79.7	460
Safeway	36.3	1,330
Publix	30.6	1,300
Ahold U.S.A.	25.9	765
H-E-B Grocery	19.8	300
Albertson's	19.5	1,100
Delhaize	17.1	1,360
Meijer	15.7	200

Source: Adapted from multiple sources.

Table 1.6
Distribution of Consumer Expenditures at Supermarkets

Item	Percentage of Sales
Perishables	54
Fresh meat and seafood	14
Produce	11
Dairy	9
Frozen foods	6
Deli (full service and self-serve)	4
Bakery and baked goods	3
Grocery (non-perishable foods)	24
Grocery (non-food items)	6
Other (general merchandise, beauty, etc.)	10

Source: Based on Progressive Grocer, Food Marketing Institute, and the Association of Retailers.

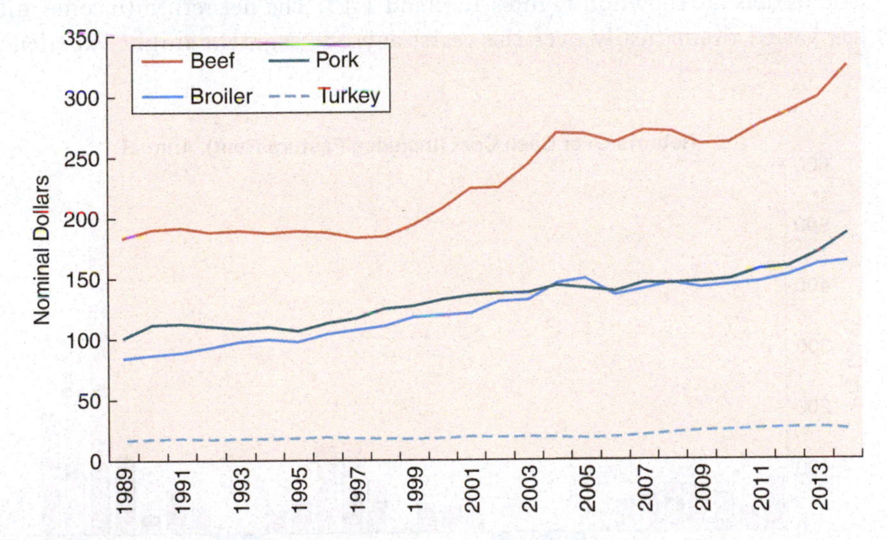

Figure 1.13
Expenditure for meat and poultry, 1989–2014.
Data source: USDA Economic Research Service; analysis and compiled by Livestock Marketing Information Center.

The industry undertook a strategic effort to reverse demand losses with investments in producer education focused on improving beef's attributes, consumer outreach through effective advertising campaigns, investments in food safety enhancements, nutritional studies to evaluate the role of beef in the diet, and a host of other initiatives designed to improve beef's competitiveness. Through these efforts, the industry stabilized demand and began to rebuild market share.

Consumers continue to demand more service and convenience in their food products as their incomes rise or as they have less time to cook, prepare, and eat meals. Time has become a precious commodity as single-parent and two-income families have increased. Time and convenience are reflected in increasing away-from-home meals. Consumers eating at home also want more convenience: they desire products that require minimal preparation time but with a significant amount of choices in regard to flavor. Consumers still want a feeling of having participated in home meal preparation and so "meal kit" and other meal packaging concepts have gained favor in the supermarket. The active lifestyles of consumers have led to supermarkets accounting for 20% of takeout food sales. A more detailed discussion of consumer trends is provided in Chapter 2.

Beef Industry Goals

Given the divergent and unique roles of each sector of the beef supply chain, it is difficult to arrive at a unifying goal for the industry. However, it is clear that the industry and its participants must accomplish several outcomes to assure its viability:

- Provide a source of protein that delivers a satisfying eating experience for its customers.
- Operate enterprises that provide meaningful careers and generate profits.
- Conduct business in such a way that landscapes, natural resources, and communities are enhanced.

The creation of effective management systems, business relationships, and enterprise models is fundamental to long-term success. A more detailed discussion of these topics is provided in Chapter 3.

Profitability

Enterprise profitability trends based on cash costs and returns for cow-calf producers and cattle feeders are shown in Figures 1.14 and 1.15. The net return (income minus cost) has varied dramatically over the years; any one segment might experience a

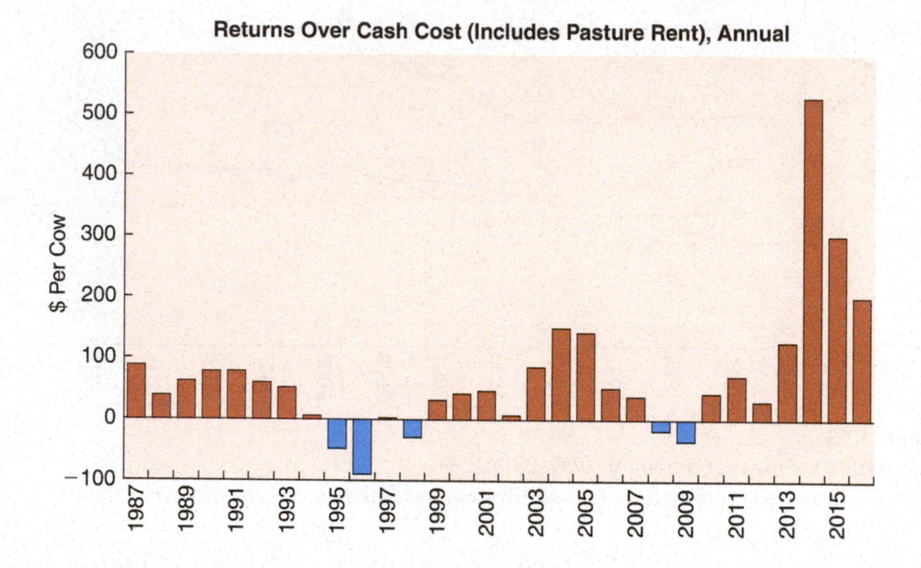

Figure 1.14
Estimated annual average cow-calf returns over cash costs.
Major data sources include: USDA Agricultural Marketing Service (Market News) and National Agricultural Statistics Service; analysis and compiled by Livestock Marketing Information Center.

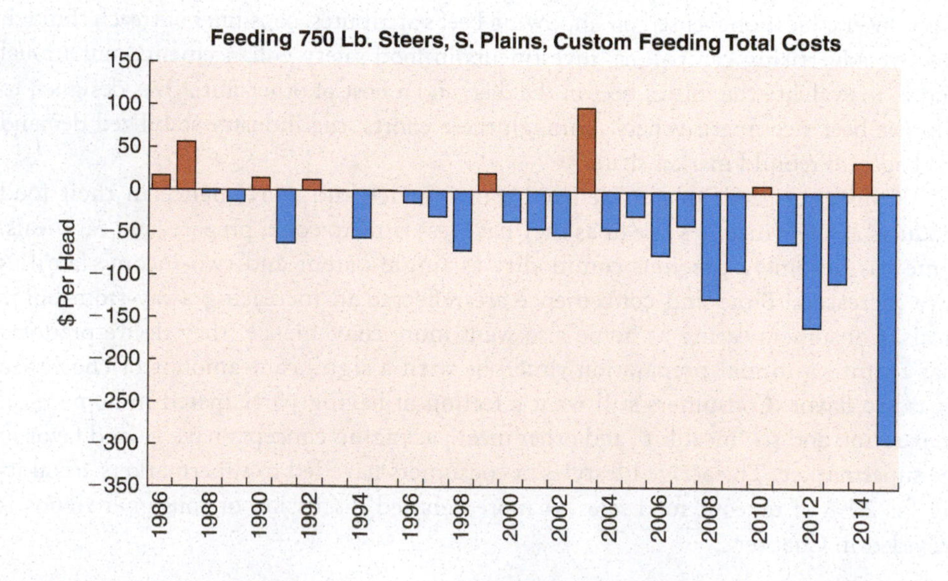

Figure 1.15
Annual average fed cattle profit and loss.
Major data sources include: USDA Agricultural Marketing Service (Market News) and National Agricultural Statistics Service; analysis and compiled by Livestock Marketing Information Center.

$100 per-head loss to more than a $100 per-head profit at a given point in time. These trends explain the need for effective risk management strategies to be employed by industry participants.

During the past several decades, there have been few years in which all segments of the beef industry have made a profit during the same year (Figure 1.16). The late 1980s and early 1990s were profitable years for most segments, but more often profit in one beef industry segment came from a loss in another segment.

There are a number of factors that affect profitability for each sector of the industry. For example, returns for cow-calf producers tend to fluctuate with changes in total cattle inventory (Figure 1.17). The U.S. cattle inventory is responsive to cost of production, profitability of alternative enterprises, and cattle prices. Increases in the cost of production impact herd retention rates and if they rise high enough to make alternative uses of land and resources attractive, then producers will cut production. Per-cow costs over time (Figure 1.18) show that costs were reasonably stable from 1988 to 2000, after which cost trends increased substantially. This increase was largely due to rising costs of feed and fuel. Rising costs coupled with historic levels of profitability in grain production led to a reduction in the U.S. beef herd. In a commodity business,

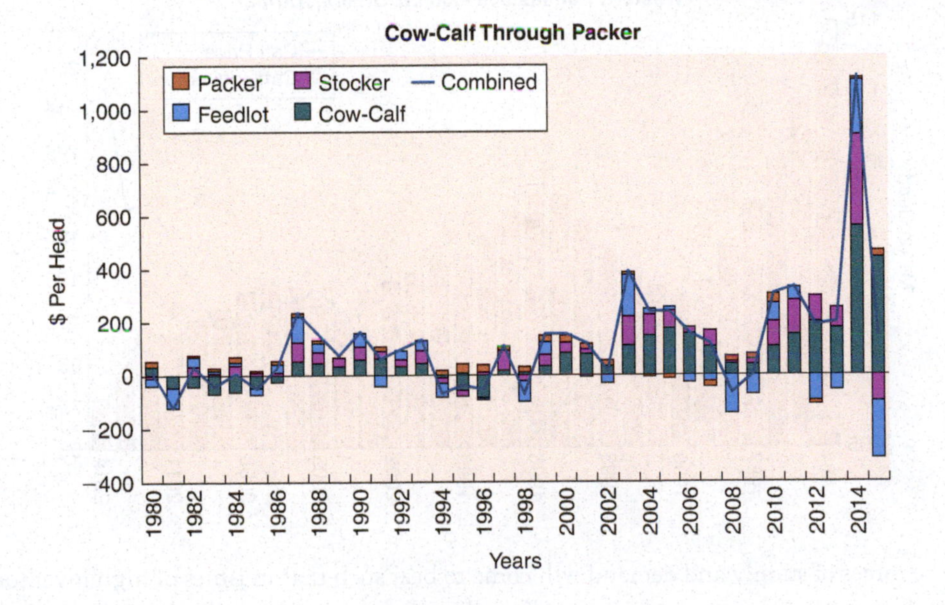

Figure 1.16
Industry profitability.
Source: Cattle-Fax.

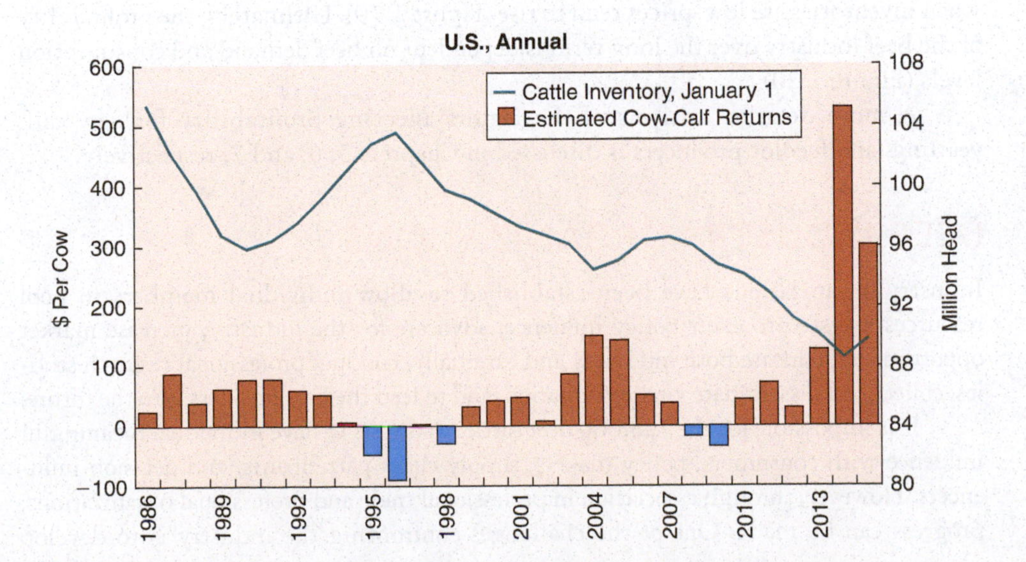

Figure 1.17
Cow-calf returns and cattle inventory.
Major data sources include: USDA Agricultural Marketing Service (Market News) and National Agricultural Statistics Service; analysis and compiled by Livestock Marketing Information Center.

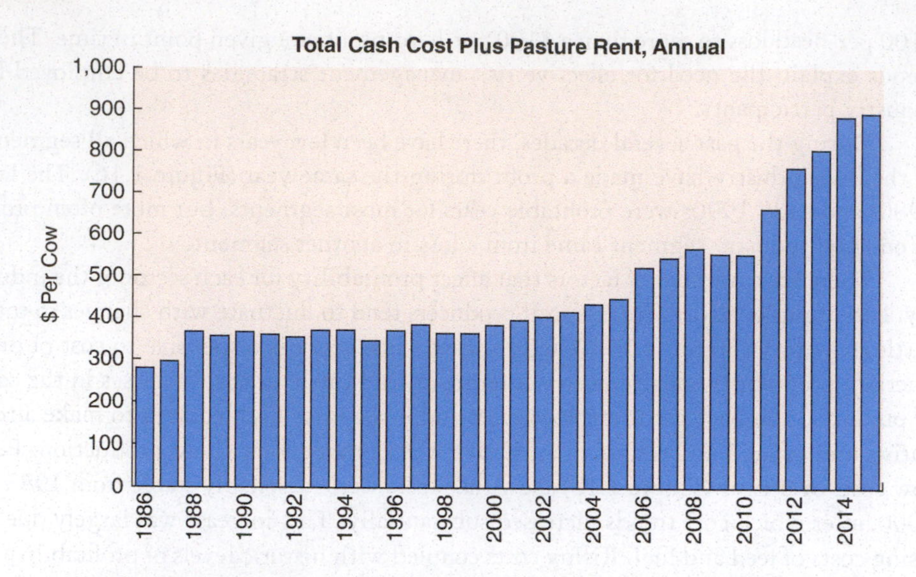

Figure 1.18
Estimated average annual cow costs.
Major data sources include: USDA Agricultural Marketing Service (Market News) and National Agricultural Statistics Service; analysis and compiled by Livestock Marketing Information Center.

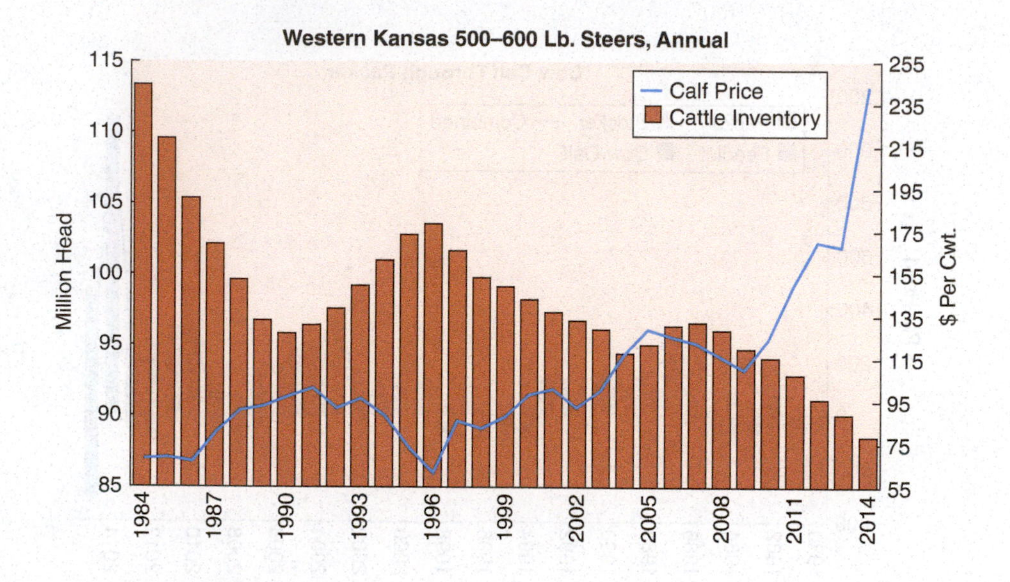

Figure 1.19
Calf prices and cattle inventory.
Data sources: USDA National Agricultural Statistics Service and Agricultural Marketing Service (Market News); compiled by Livestock Marketing Information Center.

the forces of supply and demand will come to bear such that in times of high inventory and thus relatively large beef supplies, the price for cattle declines while in times when inventories are low, prices tend to rise (Figure 1.19). Ultimately, the profitability of the beef industry over the long term is dependent on beef demand and consumption levels (Figure 1.20).

A more detailed analysis of the factors affecting profitability for cow-calf, yearling, and feedlot producers is discussed in Chapters 5, 6, and 7, respectively.

BEEF INDUSTRY ORGANIZATIONS

Industry organizations have been established to allow individual members to pool resources invested to exert policy influence, advocate for the industry, increase market opportunities, educate both internally and externally, conduct professional research studies, collect and disseminate vital information, and to lead the beef business into the future.

It is impossible for the 728,000 beef cattle producers to have individual meaningful influence with consumers, policy makers, supply chain participants, and decision influencers. However, through association in professional trade and professional organizations, progress can be made. One of the challenges confronting the industry is to develop

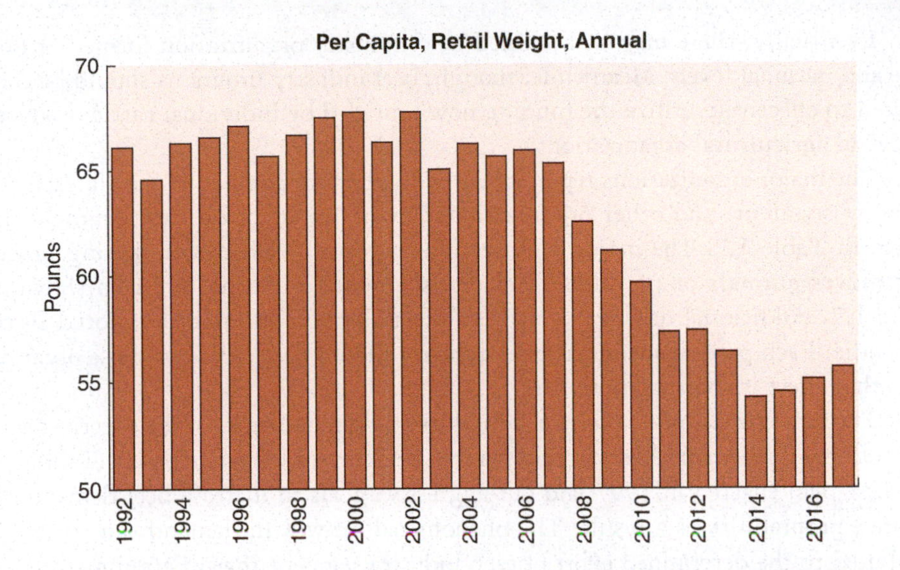

Figure 1.20 *Average annual beef consumption.*
Data sources: USDA National Agricultural Statistics Service and Economic Research Service; analysis and compiled by Livestock Marketing Information Center.

organizations that unify industry participants and that are effective at attracting membership investments of time, talent, and treasure.

There are many organizations that represent or influence cattle producers. The traditional independent philosophy of cattle producers can serve as a barrier to effective planning and adjustment to change. However, those individuals who align with others to direct change are more likely to find success. The complexity of modern agriculture, world trade, and mounting regulation create an atmosphere conducive to increasing frustration for producers. An unfortunate effect of this frustration is that some producers become angry and allow emotion to control their decision making. When organizations cater to emotion and anger, adoption of self-defeating policies that may seem appropriate in the short term but ultimately lead to a loss of competitive position for the beef industry is likely to emerge.

The future of beef as a product and of the beef industry is directed by group action. The influence of organizations on the beef industry will be determined by (1) the industry's capacity for linking and coordinating the actions of its organizations and for forming alliances with other groups having common interests, and (2) effective organizational leaders who can effectively represent their members.

Cattle producers must work together for the following reasons:

1. The impact of government on the cattle industry will remain great. This is especially true because politicians represent an urban society with less than 2% of the U.S. population directly involved in production agriculture. Government affairs will continue as a major focus of the beef industry and its representative organizations.
2. To make beef more competitive with other meats, the industry must guide, encourage, and support research and development efforts in production technology and product development and marketing.
3. Cattle producers need a source of information to help them make sound management decisions along with effective educational and training programs for themselves and their employees.
4. The industry needs an effective beef marketing program that includes consumer and market research, product development, product information, promotion, and merchandising. In addition to improving the efficiency of beef production and distribution, the industry needs to support programs that will stabilize or improve beef acceptance.
5. Public information efforts should be expanded to improve public and government acceptance and understanding of beef economics, production, and marketing methods.

Eventually, there may be a unified beef industry organization formed at both state and national levels. Meanwhile, though, beef industry programs should be coordinated to effectively utilize the funding now provided by individual cattle producers to several agricultural organizations.

The major organizations representing individuals or companies within each beef industry segment, and other organizations having an effect on each segment, are shown in Table 1.7. The major organizations involved with the marketing process that moves animals or products from one segment to another are also shown in Table 1.7. Additional organizations that affect the beef industry are noted in the Appendix. Each participant in the beef industry should participate in those organizations that affect its business activities.

The power of a unified effort in the industry is demonstrated in the effectiveness of the National Cattlemen's Beef Association (NCBA) current long-range plan (Table 1.8). The first plan enacted in 1997 laid out aggressive goals to increase beef demand and enhance profitability as a result. The phenomenal growth in demand can largely be attributed to the determined effort of each industry segment toward attainment of the goals and objectives of the long-range plan. Modifications have continued to be made to allow the industry to proactively direct resources to the areas of greatest need and impact.

Table 1.7
MAJOR ORGANIZATIONS REPRESENTING OR AFFECTING THE BEEF INDUSTRY[1]

Segments	Major Organizations Influencing Each Segment	Other Organizations Directly Affecting Each Segment[2]
Seedstock producers	American National Cattlewomen Beef Improvement Federation Breed Associations National Cattlemen's Beef Association State Beef Councils State Cattlemen's Associations U.S. Beef Breeds Council	(1) State Department of Agriculture (2) USDA Animal and Plant Health Inspection Service[3] (3) USDA Packers and Stockyards Administration (4) American Association of Bovine Practitioners (5) Occupational Safety and Health Administration
Commercial cow-calf producers	American Farm Bureau American National Cattlewomen Beef Improvement Federation Breed Associations Cattle-Fax County Livestock Growers organizations National Cattlemen's Beef Association National Grange State Beef Councils State Cattlemen's Association	(6) Environmental Protection Agency (7) American Society of Animal Science (8) USDA-Forest Service (9) USDA-Soil Conservation Service (10) USDA-Agricultural Research Service (11) USDA-Extension Service (12) USDA-Statistical Reporting Service (13) Society for Range Management (14) American Registry of Professional Animal Scientists (15) Livestock Publications Council (16) International Embryo Transfer Society (17) Beef AI organizations and National Association of Animal Breeders (18) Animal Rights Organizations and environmental groups
Feeders	American Farm Bureau American National Cattlewomen Cattle-Fax National Cattlemen's Beef Association State Beef Councils State Cattle Feeder's Association	(1), (2), (3), (4), (5), (6), (10), (11), (12), (14), (15), (18)[4] (19) American Feed Manufacturers Association (20) National Feed Ingredient Association (21) Food and Drug Administration (22) USDA Office of Transportation (23) Animal Health Institute

Table 1.7 (Continued) Major Organizations Representing or Affecting the Beef Industry[1]

Segments	Major Organizations Influencing Each Segment	Other Organizations Directly Affecting Each Segment[2]
Packers and processors	American Association of Meat Processors American Meat Institute American Meat Science Association Institute of Food Technologists National Cattlemen's Beef Association National Food Processors Association National Meat Canners Association State Beef Councils State Meat Dealers Association	(1), (2), (3), (4), (5), (6), (12), (14), (15), (18), (21), (22) (24) Federal Trade Commission (25) Labor Unions (26) National Perishable Transport Association (27) USDA Food Safety and Inspection Service (28) USDA Agricultural Marketing Service
Meat retailers/ food service organizations	American Association of Meat Processors American Institute of Food Distributors National Association of Meat Purveyors National Association of Retail Grocers of the United States National Frozen Food Association National Restaurant Association State Meat Dealers Association State Restaurant Association	(1), (2), (3), (5), (6), (18), (21), (22), (24), (25), (26), (27), (28) (29) Food and Drug Law Institute (30) USDA-Food and Nutrition Service (31) Joint Labor Management Commission of the Retail Food Industry (32) National Restaurant Association (33) Food Service and Lodging Institute
Marketing points between segments	American Stockyards Association Cattle-Fax National Auctioneers Association National Cattlemen's Beef Association National Livestock Grading and Marketing Association National Livestock Producers Association U.S. Meat Export Federation	(12), (18), (21), (22), (24) (37) Agriculture Trade Council (38) USDA-Economic Management Staff (39) USDA-Marketing and Inspection Management (40) USDA-World Agricultural Outlook Board (41) USDA-Animal Air Transport Association

[1] Addresses and descriptions of these and other organizations are provided in the Appendix.
[2] Organizations (1) through (18) are applicable to both seedstock and commercial cow-calf segments.
[3] USDA organizations can be accessed through www.usda.gov.
[4] Repeated numbers refer to the same organizations identified earlier in the table.

Table 1.8
NCBA Long-Range Plan Elements

Industry Mission:
"a beef community dedicated to growing beef demand by producing and marketing the safest, healthiest, most delicious beef that satisfies the desires of an increasing global population while responsibly managing our livestock and natural resources."

Industry Vision:
"To responsibly produce the most trusted and preferred protein in the world."

Primary Performance Metric:
"increase the wholesale beef demand index by 2 percent annually over the next five years."

Four core strategies required to attain vision:
- Drive growth in beef exports.
- Protect and enhance the business and political climate for beef.
- Grow consumer trust in beef and beef production.
- Promote and strengthen beef's value proposition.

Source: Adapted from NCBA Long-Range Plan.

National Cattlemen's Beef Association (www.beef.org; www.beefusa.org). The NCBA, with headquarters in Washington, DC, and Denver, is the national spokesperson for all segments of the nation's beef cattle industry, including cattle breeders, producers, and feeders. The nonprofit trade association was originally formed in 1898. In 1996, the National Cattlemen's Association and the National Live Stock and Meat Board merged to create the NCBA. The NCBA represents approximately 175,000 cattle professionals throughout the country via individual memberships and through membership in the state and breed affiliates aligned with NCBA. Membership includes 28,000 individual members, 46 affiliated state cattle associations, and 27 affiliated national breed organizations. NCBA programs are financed by funds contributed by individual members and affiliated associations.

The NCBA provides services cattle producers cannot perform satisfactorily as individuals. Pursuant to this goal, the NCBA performs three basic functions: (1) primarily through its Washington, DC, office, it represents the beef cattle industry in the legislative and administrative branches of the federal government; (2) it interprets beef production and beef economics for the public and economic, social, and political developments for the industry; and (3) it provides information to aid members in planning and management decisions.

American National Cattlewomen (ANCW) (www.ancw.org). The ANCW is a group established for participation in the promotion, education, and legislation of beef.

U.S. Meat Export Federation (USMEF) (www.usmef.org). The USMEF is a nonprofit trade association that works with the U.S. meat and livestock industry to identify and develop overseas markets for U.S. beef, veal, pork, lamb, and variety meats. It is based in Denver, with overseas market development offices in Beijing, Beirut, Brussels, Cairo, Hong Kong, Lima, Mexico City, Monterey, Moscow, Saint Petersburg, Seoul, Shanghai, Singapore, Taipei, and Tokyo. USMEF also has representation in the Caribbean. Through these offices, the MEF coordinates market development programs. Its programs are designed to identify new markets, create widespread product awareness, secure fair market access, provide trade servicing, and assist and educate overseas buyers and U.S. suppliers alike. Established in 1976, the MEF is a cooperator with the Foreign Agricultural Service (FAS) of the U.S. Department of Agriculture. It represents livestock producers and feeders, meat packers, purveyors and exporters, agribusiness and agriservice interests, farm organizations, and grain promotional groups. The MEF has several sources of funding: its members, overseas private sector interests, beef checkoff money, and the Foreign Agricultural Service.

State Beef Councils. Most states have a beef council that is funded with checkoff dollars. Their primary objective is to educate consumers about the nutritional aspects of beef and how to best select and prepare beef. The councils communicate with health professionals and food service industry personnel and work to promote beef to consumers.

North American Meat Institute (NAMI) (www.meatinstitute.org). The NAMI was created in 2015 through the merger of the American Meat Institute and the North American Meat Association to represent meat packers and processors who produce more than 95% of the red meat in the United States as well as suppliers of meat equipment, products, and services. Activities include marketing, research, congressional and legislative relationships, improved operating methods and products, conservation, spoilage prevention, and industrial education.

American Association of Meat Processors Association (AAMP) (www.aamp.com). AAMP is an international trade association composed of meat processing companies and associations that provide the finest center-of-the-plate products and service to food service, retail stores, and mail order customers. AAMP is also the publisher of the world-renowned publication *The Meat Buyer's Guide*.

Food Marketing Institute (FMI) (www.fmi.org). Domestic and international food retailers and wholesalers are members of the FMI, which maintains a liaison with both the government and consumers. FMI conducts programs in research, education, industry relations, and public affairs.

BEEF INDUSTRY ISSUES

The beef industry has faced numerous issues during past years. Some of these issues are addressed here; others are covered in later chapters. Issues facing the beef industry change frequently, sometimes daily.

The U.S. population continues to increase, but Americans actively involved in agricultural production represent less than 2% of the total population. This disparity in numbers reflects the communication problems between urban populations and agricultural producers. One of the significant outcomes of the shift from a rural employment pool to an urbanized economy has been the difficulty in maintaining a representative voice for agriculture in state and national policy making. Most Americans are now at least one, if not several, generations removed from the farm and ranch and as such have less connection to and understanding of the food production system. Perceptions and personal beliefs get confused with true relationships. The lack of communication between agriculture and the rest of society is rooted in a media culture that favors sensationalism, a national population that prefers to be entertained rather than enlightened, and an agricultural industry that has too frequently responded to issues in a defensive or reactive manner.

Cattle producers need an organizational structure like the National Cattlemen's Beef Association to manage issues of national and international scope. Such an organization can objectively project a positive industry image by (1) coordinating a proactive approach to issues affecting the industry; (2) developing sound technical information to be used by industry leaders; (3) encouraging producers to implement responsible production practices; (4) conducting research among consumers and industry influencers to understand their needs and opinions; and (5) providing credible information to opinion influencers. It is important that the issues facing the beef industry be addressed before they reach a crisis stage.

No better example of this can be given than the management of the single case of "mad cow disease" diagnosed in an imported dairy cow in the state of Washington and reported on December 23, 2003. In what could have been an economic disaster for the beef industry and related businesses, a rational, science-based response by NCBA, USDA, industry leaders, and health professionals filled the void with facts rather than speculation, reason instead of hysteria, and openness rather than misinformation.

The incident resulted in the loss of most exports of U.S. beef, a myriad of new regulations, and increased pressure to develop a national animal identification system to facilitate traceback. However, the NCBA management plan largely helped to maintain consumer confidence in the midst of the crisis.

Environmental Issues

The American public is generally not familiar with the economics of the food production chain. Most people, however, are concerned with the use and preservation of natural resources. There are influential consumer, nutrition/health, and environmental groups with multimillion dollar budgets that focus on these issues.

Cattle have the unique ability to graze untillable acres and convert plants that humans cannot eat into highly palatable human food. However, while some people know that cattle can effectively use the land, others feel that cattle abuse the land. Cattle producers that implement proper grazing practices prevent overgrazing along streams and rivers (riparian areas). Their grazing management also fosters compatible relationships between livestock and wildlife. The issues of grazing fees on public lands, wetlands, and inferences that overgrazing is the major cause of rangeland desertification are discussed in Chapter 15. Other specific environmental issues will be discussed in Chapters 5 and 7.

ANIMAL WELL-BEING

Beef producers have been concerned with the use and welfare of their animals for centuries. Animals were domesticated to give nomadic people a consistent supply of food and companionship. Draft animals were domesticated for transportation and power. People soon learned that the productive response of animals is greater when they are given proper care.

Today, the nutrition, health, and management needs of farm animals are well known and scientifically based. Evidence suggests that many domesticated animals in the United States receive a more nutritious diet than some humans consume. The veterinary medical profession provides on-farm services, health clinics, and hospital care that are in many ways equal to human health-care services. The members of the NCBA adopted a statement of principles that affirms that cattlemen are united in their philosophy that proper and humane care of the animals they are responsible for is a moral obligation as well as an economic necessity. The tenets of this statement of principle follow:

- I believe in the humane treatment of farm animals and in continued stewardship of all natural resources.
- I believe my cattle will be healthier and more productive when good husbandry practices are used.
- I believe that my and future generations will benefit from my ability to sustain and conserve natural resources.
- I will support research efforts directed toward more efficient production of a wholesome food supply.
- I believe it is my responsibility to produce a safe and wholesome product.
- I believe it is the purpose of food animals to serve mankind, and it is the responsibility of all human beings to care for animals in their charge.

Proactive approaches to assure that cattle are handled, managed, transported, marketed, and harvested humanely are provided throughout the following chapters.

Diet Health and Food Safety Issues

Consumers have become increasingly more aware of diet and health and the nutritional and safety aspects of food. These issues are discussed in Chapter 2.

Marketing Issues

A major challenge for the beef industry is to generate and maintain high-quality products. This is extremely difficult in a segmented industry where cattle and products are not well identified as they move from one segment to another segment.

Quality is an ambiguous word that must be more clearly defined if it is to be effectively communicated and achieved. The production and market specifications presented in Chapter 4 (Table 4.5) are a starting place in putting numerical values on traits that can be used to identify quality. Consumers' definition of quality eventually goes beyond the color of lean meat and the amount of marbling to include products that are reasonably priced, safe, nutritious, consistent, and healthful, and that are consistently high in palatability.

Quality is best defined with Total Quality Management (TQM), which is "meeting or exceeding your customers' expectations at a cost that represents value to them every time."

Following are some of the major marketing issues related to marketing slaughter cattle:

1. The 2011 Beef Quality audit estimated that $43 per slaughter steer/heifer was lost to the industry because of quality shortfalls. These per-head losses were from waste fat and cutability issues ($6); palatability, such as tenderness and marbling ($25); offal and hide defects ($6); and excessive carcass weights ($7). To remain competitive with other meats and food products, the beef industry will need to address these costs of production and quality assurance needs.
2. Surveys show that 20% of the cuts from the loin and rib and 60% of retail cuts from the round have unacceptable tenderness problems. Because marbling accounts for only 10–20% of the tenderness differences in beef, quality assurance for tenderness and overall palatability must be addressed more directly. The most frequent consumer complaints of inconsistencies of tenderness and juiciness cannot be ignored without having an effect on the entire beef industry. Investments in research that addresses these issues are critical to the future of the business.
3. A comprehensive beef quality assurance program is needed so that processors, retailers, and consumers can have confidence in the product quality of both cattle and carcasses. Cattle producers need to control their health programs in order to maintain consumer confidence that animal drug and medication procedures are well managed. The creation of supply chain trust depends on an effective program focused on the development and implementation of quality standards and best practices across the various sectors.

The beef industry needs to cooperate to consistently supply high-quality products. The ultimate goal of quality assurance for producers, processors, and retailers is to assure consumers that they are receiving beef products that are safe, healthful, and highly palatable.

SELECTED REFERENCES

Publications

Agricultural census. 2012. Washington, DC: USDA National Agricultural Statistics Service.

Annual agricultural statistics. 2014. Washington, DC: USDA National Agricultural Statistics Service.

Directions. 2015. Englewood, CO: National Cattlemen's Beef Association. 18:3, pp. 33–43.

Livestock and poultry situation and outlook report. Washington, DC: USDA.

2

Retail Beef Products and Consumers

Ultimately, the consumer determines the profitability of the beef industry. The cumulative purchase decisions of consumers at retail food stores, fast-food outlets, or in a restaurant impact the viability of each and every enterprise that comprises the beef industry. The consumer's choice of beef in amount and price gives eventual direction and momentum to the beef supply chain.

Beef producers and purveyors must be knowledgeable of the retail beef products they and others produce, how consumers accept these products, and what role they play in creating a positive consumer image for beef. Very simply, if the beef industry fails to stabilize or increase consumer market share, then the beef industry will experience significant downsizing.

RETAIL BEEF PRODUCTS

Consumer products obtained from beef cattle or made from by-products of cattle are shown in Figure 2.1. The carcass and the resulting wholesale (Figure 2.2) and **retail cuts** are products of primary value, while products of lesser value are classified as **by-products**. By-products are discussed later in the chapter. See Chapter 9 for more detail on yield grades and quality grades.

Retail Cuts

Middle meats come from the rib and loin. Although they are the higher valued cuts, they comprise only about 25% of the carcass weight. The round and chuck, known as the end meats, represent over 50% of the carcass weight. Cuts from the end meats historically posed problems to the beef industry because they are lower in palatability compared to the middle meats, and because most consumers do not consider roasts convenient products. As a result, end meats lost considerable value in the market place—as much as 20%. End meats can be ground to make excellent hamburger and processed meats. However, these are lower-valued products and more value-added products are needed that are of intermediate value and that are priced between hamburger and steaks. Thus, the industry has made a concerted effort in product development to improve end meat utilization.

New product and cut development has been a primary tactic of the industry funded by beef checkoff dollars to increase versatility and appeal to both retail and foodservice customers. The predominant focus of these efforts is described below:

1. Value-added ground beef such as seasoned, precooked beef crumbles for use as an ingredient in tacos, pizza, pasta, and casseroles or precooked hamburger patties.
2. Versatile quick beef cuts from the chuck and round utilizing marinated and precooked presentation strategies.
3. Beef finger foods targeted to capture a share of the appetizer market. Cheeseburger fries are a breaded blend of ground beef and cheese designed to compete with chicken nuggets.

4. Fresh prepared foods for use in the in-store deli. While beef leads other animal proteins in fresh meat case sales, it lags in fourth place at the deli behind chicken, pork, and turkey. A fully cooked pot roast complete with vegetables and gravy is a direct competitor to rotisserie chicken.
5. Beef whole-muscle cuts as an ingredient in salads, soups, and pasta dishes designed to capture increased consumption from diners who are typically low beef consumers.

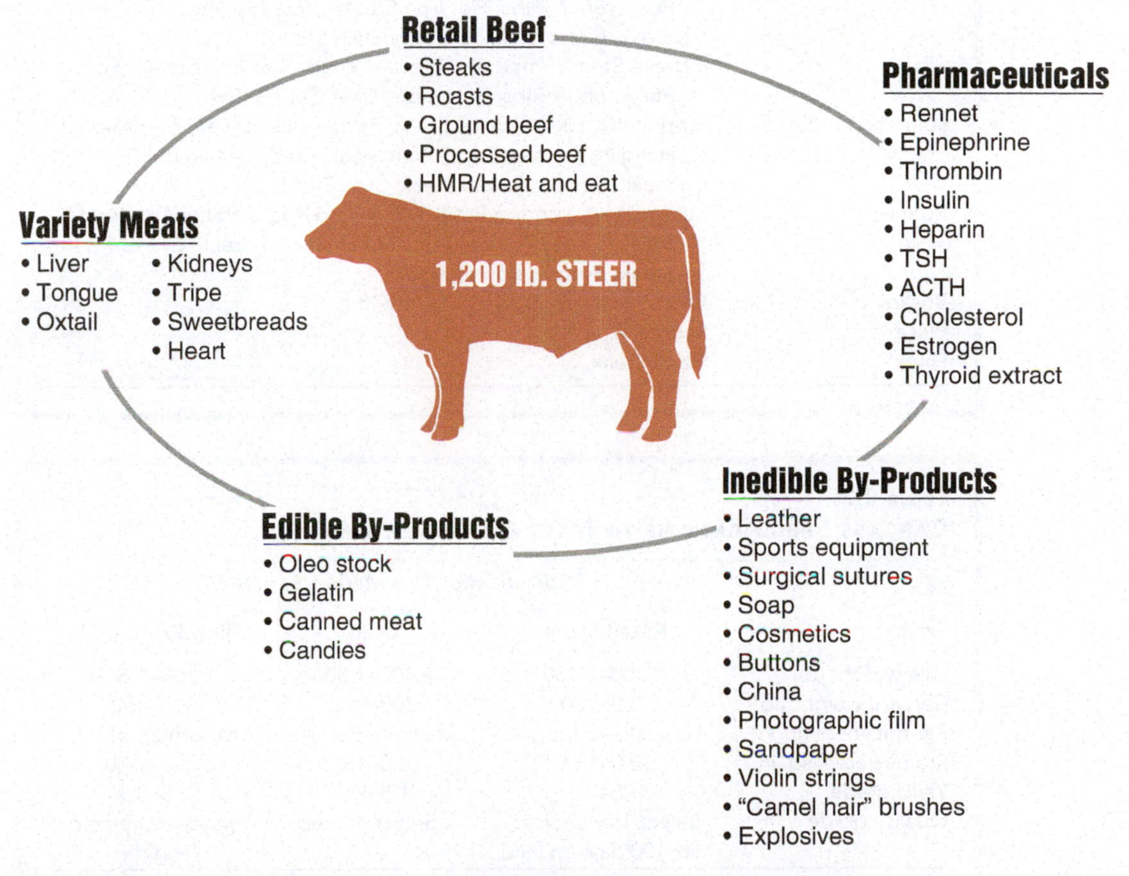

Figure 2.1
Products generated by the beef industry include retail beef, pharmaceuticals, variety meats, and by-products (edible and inedible).

Figure 2.2
The location of wholesale beef cuts from market cattle.

Table 2.1
Major Retail Cuts Originating from the Primary Wholesale Beef Cuts

Wholesale Cuts	Retail Cuts
Chuck	Chuck Eye Roast, Blade Steak, Pot Roast, Mock Tender, Blade Roast, Short Ribs, Flat Iron Steak, Cross Rib Roast, Denver Steak, Sierra Steak, Country Style Ribs
Rib	Rib-eye Steak, Prime Rib, Rib-eye Roast, Cowboy Steak, Rib Steak, Short Ribs, Back Ribs, Chef Cut Rib Eye
Loin and Sirloin	Strip Steak, Filet, T-Bone Steak, Porterhouse Steak, Filet Mignon, Hanging Tender, Top Sirloin Steak, Tri-tip Steak, Ball Tip Steak, Bottom Sirloin Steak, Strip Roast
Round	Rump Roast, London Broil, Top Round Roast, Petite Tender Steak, Sirloin Roast, Butterfly Top Round Steak, Eye of Round Steak, Bottom Round Roast
Brisket	Brisket Flat Cut, Brisket Dinkle-Off
Plate	Short Ribs
Flank	Flank Steak

Table 2.2
Carcass Specifications to Meet Consumer Demand

	Specifications by Market Segment[1]		
Trait	**Retail Store**	**Lean**	**High Palatability**
Live weight (lb)	1,200–1,450	1,300–1,500	1,100–1,350
Carcass weight (lb)	750–900	800–999	700–850
Fat thickness (in.)	0.2–0.5	Maximum of 0.3	Maximum of 0.8
Rib-eye area (sq. in.)	12.0–16.0	13.0–16.0	12.0–14.0
Yield grade	1.5–3.5	1.5–2.9	2.5–3.9
Quality grade	Select/low choice; Min. 100 days on feed	Select and up	Average choice or Higher

[1]Some overlap exists between the market segments.

Producers must understand live animal composition and carcass specifications in terms of consumer demand for size of retail cuts, palatability, and consumer lifestyle (convenience and health). Table 2.2 identifies carcasses specifications that meet different consumer demands. Shown in Table 2.2 are the three primary consumer markets: retail store, lean beef, and high palatability beef. If the carcasses identified in Table 2.2 come from cattle that have been selectively bred, fed, and processed, then their properly cooked retail cuts will meet specific consumer preferences identified later in the chapter.

BEEF CONSUMPTION AND EXPENDITURES

Consumption patterns for different foods vary depending on availability, cost, cultural preferences, disposable income for food purchases, technological advances in food processing, and new information on nutrition, health, and food safety—and, to a large extent, on consumer perception about food items and how they fit into lifestyle. When they are readily available and the standard of living allows consumers to purchase them, animal food products are preferred over plant food products by most consumers.

For many years, Americans considered beef the "king of the meats" by choosing beef in higher amounts over all other meats. However, in the mid-1970s, poultry began to capture additional market share by focusing on price competitiveness, convenience, and versatility. While beef consumption on a retail weight basis peaked in 1976 followed by a prolonged period of decline until the mid-1990s, beef demand stabilized and began to increase in the late 1990s. Beef's share on a tonnage basis then remained in the 65 to 68 pound range until 2008 at which time retail weight consumption declined below 62 pounds primarily as the result of shrinking beef supplies (Figure 2.3).

Since 1990, per-capita poultry consumption has been higher than per-capita beef consumption on a retail weight basis, but the differences narrow significantly when compared on a boneless weight basis (Table 2.3). The effect of price cannot be overlooked in consumption trends. Figures 2.3 and 2.4 illustrate the price–quantity relationship of beef (Figure 2.4) and poultry (Figure 2.5). Notice that while increases in price have resulted in reduced per-capita consumption on a tonnage basis, the outcome has been increased total dollars expended for beef. Figure 2.5 illustrates that consumers responded to incremental declines in poultry price by increasing their consumption.

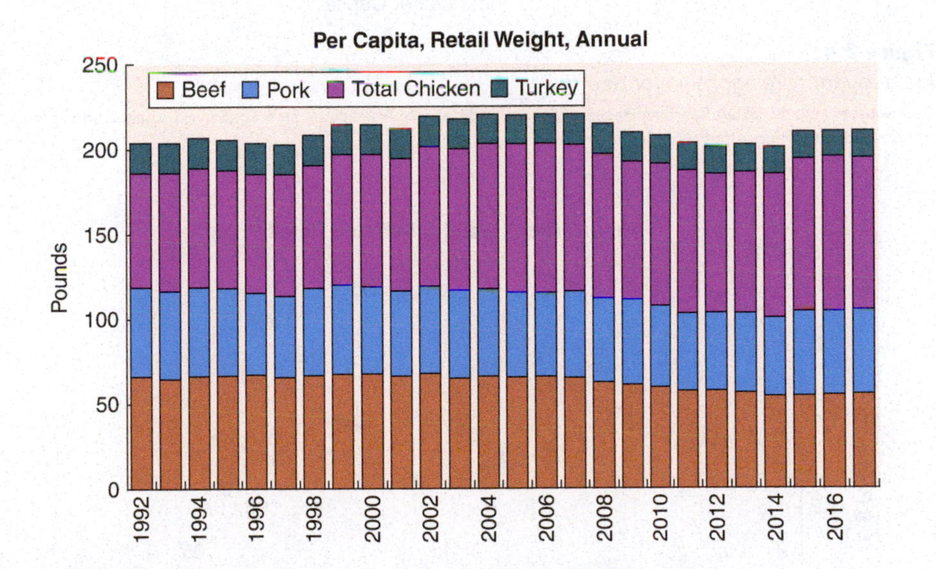

Figure 2.3
Per-capita meat and poultry consumption (retail weight).
Data sources include: USDA National Agricultural Statistics Service and Economic Research Service; analysis and compilation by Livestock Marketing Information Center.

Table 2.3
Per-capita Annual Meat, Poultry, and Fish Consumption (lb), 1966–2016

	Red Meat[1]			Poultry[1]			Total (meat, poultry, and fish)[3]
Year	Beef	Pork	Total[2]	Chicken	Turkey	Total	
1966	78 (74)	55 (44)	140 (124)	36 (25)	8 (6)	43 (31)	NA (166)
1976	94 (89)	54 (39)	153 (132)	43 (29)	9 (7)	52 (37)	219 (182)
1986	78 (74)	59 (42)	140 (118)	59 (41)	13 (10)	72 (51)	218 (184)
1996	68 (64)	49 (46)	119 (112)	72 (49)	18 (14)	90 (63)	224 (191)
2000	69 (66)	54 (50)	125 (118)	82 (56)	18 (14)	100 (70)	240 (203)
2004	66 (62)	51 (48)	119 (112)	84 (58)	17 (13)	102 (71)	237 (199)
2008	62 (60)	49 (46)	113 (107)	85 (59)	18 (14)	102 (73)	216 (184)
2012	57 (55)	46 (43)	104 (99)	82 (57)	16 (13)	98 (70)	202 (171)
2016	55 (52)	49 (46)	105 (100)	94 (66)	15 (12)	109 (78)	214 (180)

[1]Retail weight with boneless weight given in parentheses.
[2]Includes veal and lamb for which per-capita consumption has been approximately 1 lb (boneless) of each annually since 1980.
[3]Per-capita consumption of fish and shellfish was 11 lb in 1966 and 16 lb in 2004.
Source: USDA.

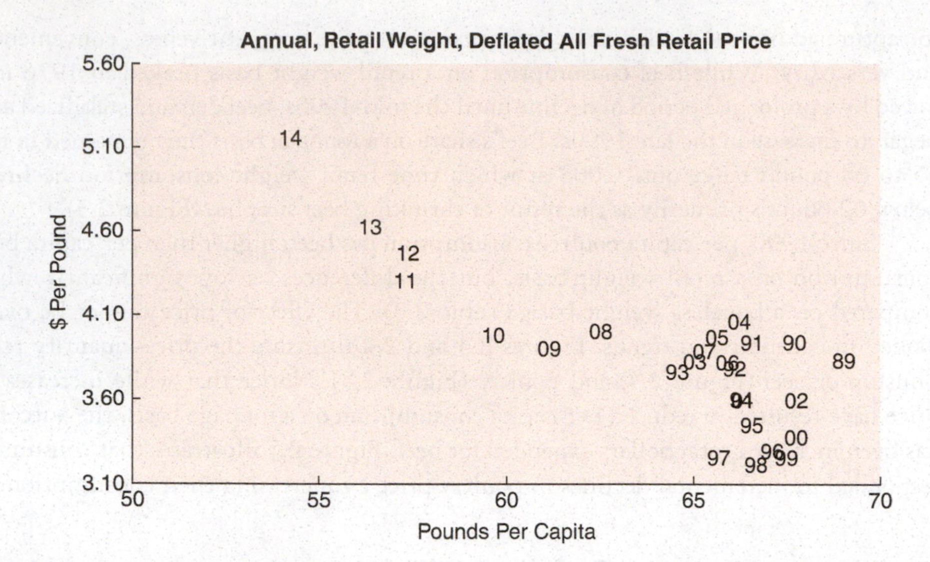

Figure 2.4

Price–quantity relationship for beef.

Data sources include: USDA Economic Research Service and National Agricultural Statistics Service; analysis and compilation by Livestock Marketing Information Center.

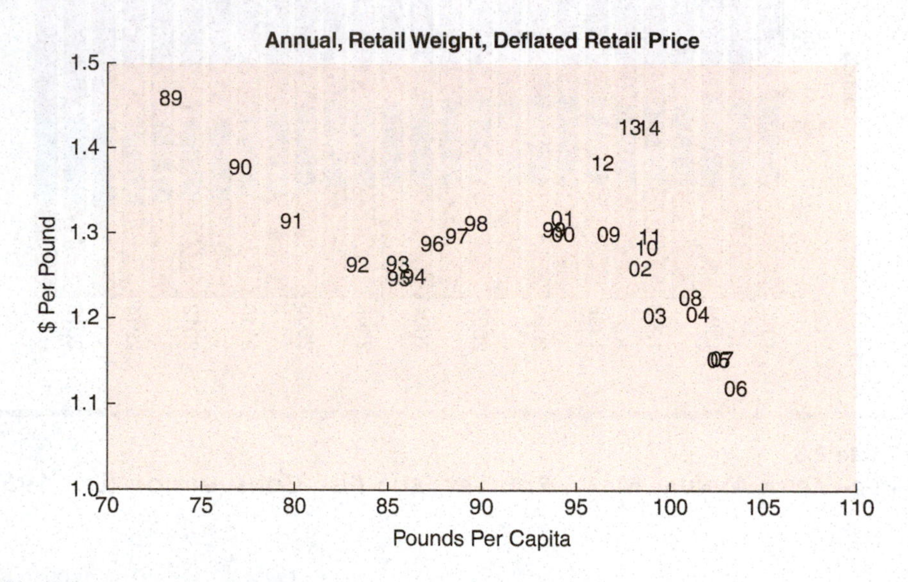

Figure 2.5

Price–quantity relationship for chicken and turkey.

Data sources include: USDA Economic Research Service and National Agricultural Statistics Service; analysis and compilation by Livestock Marketing Information Center.

Figure 2.6 shows the annual per-capita consumption of various types of beef products. Ground beef sales comprise approximately 40% of the current beef consumed. Ground beef sales have increased over time at retail due to price competitiveness as compared to whole-muscle cuts, ease of preparation, and versatility as a recipe ingredient. Foodservice sales of ground beef have expanded significantly as the hamburger has evolved from just a fast food staple to gourmet entrée.

The price of food has risen over time as has the price for beef cuts and ground beef (Figure 2.7). However, American consumers have been fortunate to experience continued increases in disposable income while spending proportionally less of that income on food. Today's U.S. consumer spends less than 11% of income for food with

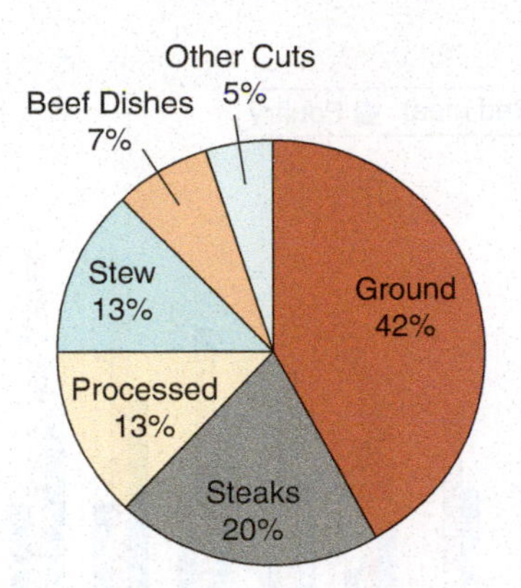

Figure 2.6
U.S. consumer utilization of various beef cuts.
Source: Adapted from USDA data.

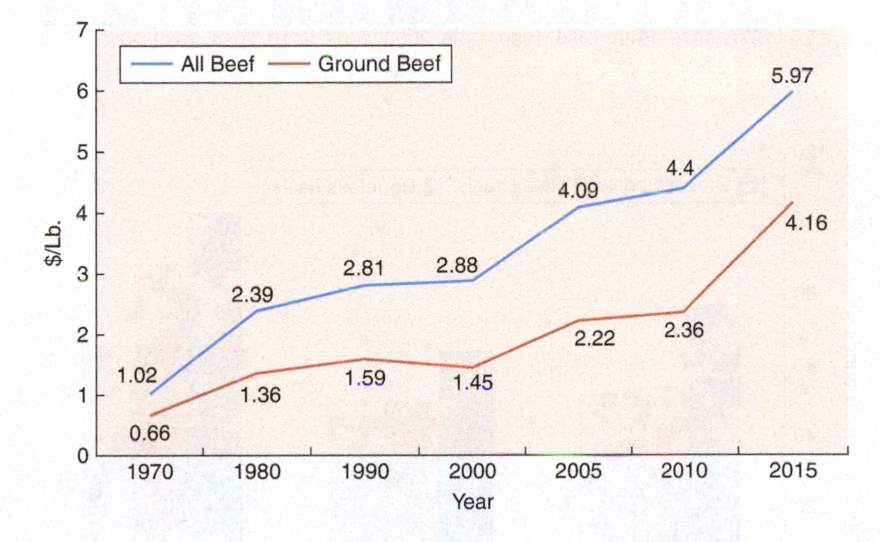

Figure 2.7
Beef price trends.
Source: Adapted from USDA.

less than 1% spent for beef and less than 2% for purchases of all meat and poultry. As Figure 2.8 illustrates, the beef industry has been able to grow per-capita expenditures capturing approximately 46% of all meat and poultry sales over the past several decades.

Beef consumption data can be misleading depending on how the information is reported. For example, the data could be based on carcass weight (used in international trade), retail weight, edible weight, or cooked weight basis. Beef "disappearance" is a more valid term than beef "consumption" because, even on a boneless weight basis, there may be fat trim, cooking, and plate waste losses that represent beef not consumed or ingested. Figure 2.9 shows the impact of various methods of describing consumption and how conversion from carcass to boneless basis changes the differences in per-capita consumption. Cooking would further reduce the boneless weight by 15–30%, depending on the type of meat and cooking method. Americans waste a large amount of food, throwing out 240 lb a year for every man, woman, and child accounting for nearly 30–40% of the total food supply.

Retail

The market for beef is generally segmented into three categories: at-home preparation, foodservice, and export (Chapter 10). Food sales in the United States account for nearly $1.2 trillion with $700 billion spend for at-home consumption and $530 billion for away-from-home dining. Less than 11% of personal disposable income is spent on food on a per-capita basis, down from nearly 25% in the 1930s (Figure 2.10).

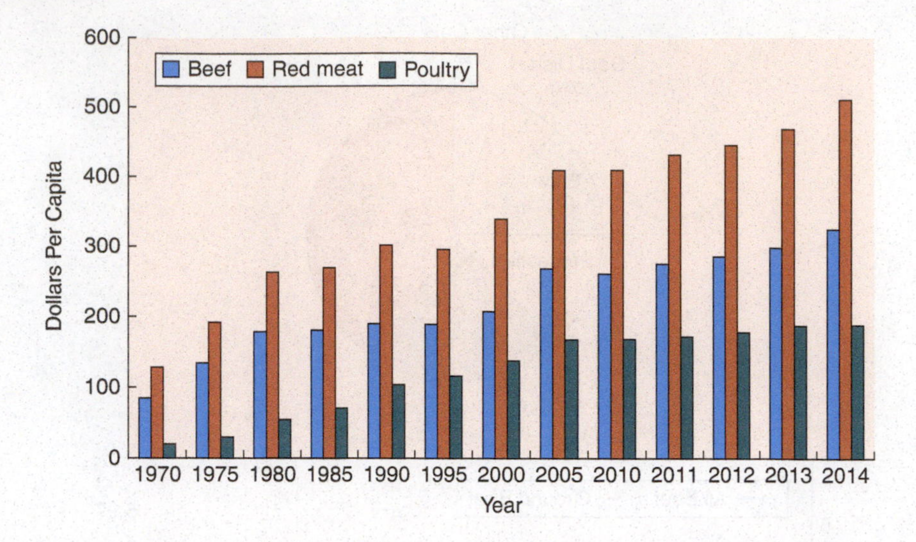

Figure 2.8
Per-capita expenditures for beef, red meat, and poultry.
Source: Adapted from USDA.

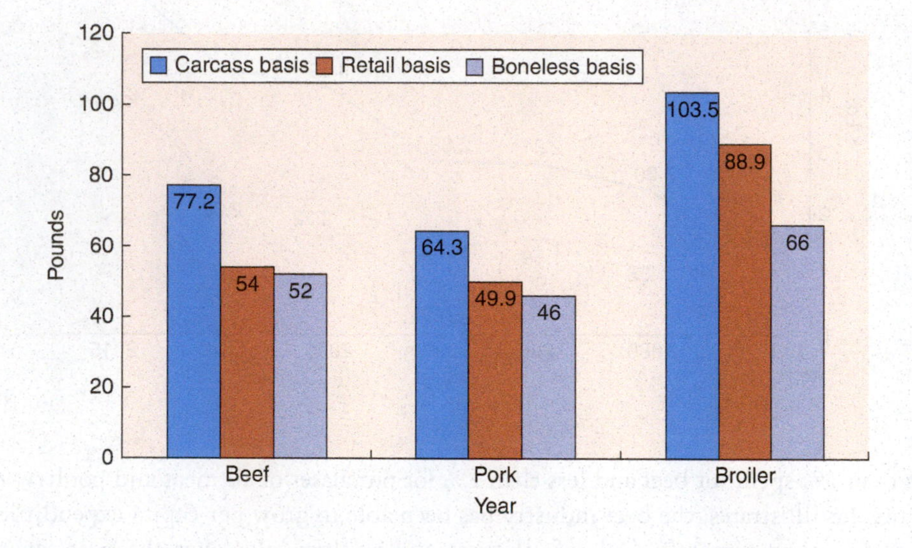

Figure 2.9
Per-capita disappearance on a carcass, retail and boneless basis.
Source: Adapted from USDA.

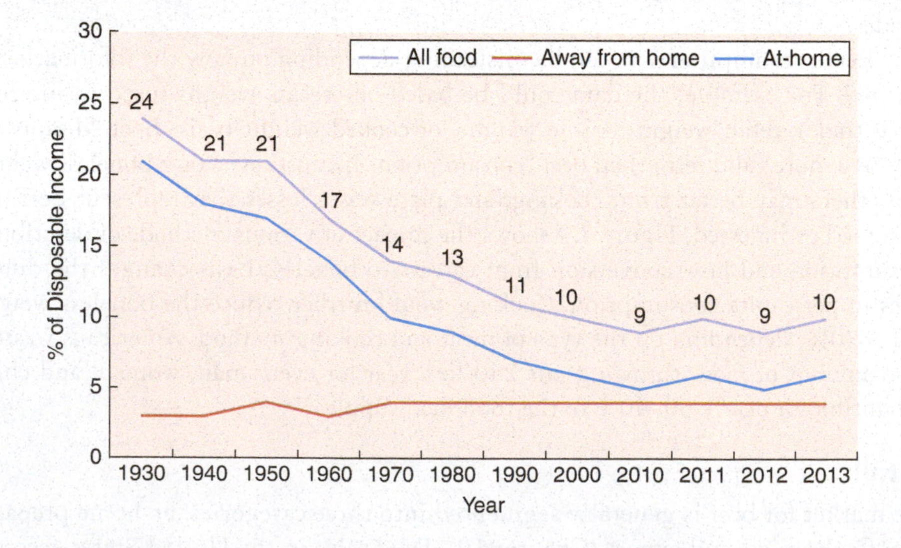

Figure 2.10
Per-capita expenditures on food.
Source: Adapted from USDA.

Ninety-three percent of beef is sold in the self-service case and 7% via full-service. Natural and organic labeled beef comprise 6% of the sales volume and 5% of the beef pounds purchased at retail, up from less than 2% a decade previous. Significant changes in beef retailing include increased use of store brands, more nutritional

Table 2.4
Top 10 Food Retailers and Category

Rank by Sales Volume	Name	Category
1	Wal-Mart	Discount Supermarket
2	Kroger	Traditional Supermarket
3	Costco	Club Supermarket
4	Albertsons	Traditional Supermarket
5	CVS Health	Drug Store
6	Target Corporation	Discount Supermarket
7	Walgreens	Drug Store
8	Publix	Traditional Supermarket
9	Ahold U.S.A.	Traditional Supermarket
10	H-E-B Grocers	Traditional Supermarket

Source: Adapted from multiple sources including company annual reports.

information and preparation instructions on packaging, and increased rate of presenting information in both English and Spanish. Food retailing has continued to evolve from a grocery or supermarket dominated interface with consumers to a much more diversified offering. The top 10 U.S. food retailers in 2015 are listed in Table 2.4. While the drug store-style outlets do not offer fresh beef for sale, the industry must respond with innovative merchandising to respond to these shifts.

Conventional presentation of beef at retail involved fresh meat items cut, weighed, wrapped, and priced by the in-house butcher shop. These packaging approaches provide an effective refrigerated shelf-life of 5–7 days. Black or white Styrofoam trays, lined with soaker pads to absorb the purge or meat juice, are overwrapped with a thin plastic film that is placed in contact with the meat surface. This packaging protocol allows oxygen to move across the film barrier to ensure that fresh beef maintains its bloom or bright, cherry red color. This packaging style also prevents escape of moisture that leads to the drying of the meat surface. Thus, conventionally plastic-wrapped beef cuts can be frozen for up to 2 months while maintaining product integrity. Excessive evaporation and drying of the meat often causes a distinct brown or grey discoloration. This discoloration as a result of dryness is called freezer burn.

The major disadvantage of this packaging system is that it fails to result in a complete seal and thus leakage can result in contamination of other foods, cutting surfaces, or storage areas. Furthermore, this approach adds a cost layer to the retail operation.

In recent years, largely driven by the desire to reduce the costs associated with in-store processing and packaging, the retail industry has shifted toward a **case-ready** package that has been prepared for sale at a central location. As such, a retailer merely places case-ready products directly into the meat case with no requirements for cutting, weighing, or pricing. Approximately 30% of all retail beef and 70% of ground beef is sold in this form.

The most frequently used tray in this system is made of Styrofoam or polypropylene and provides a refrigerated shelf-life of 14 days for ground beef and 10 days for whole-muscle beef cuts. These packages are fully sealed and contain an elevated oxygen level to promote favorable meat color. However, case-ready packaging is not appropriate for freezing beef due to a high incidence of freezer burn.

In an effort to improve the convenience of fresh beef cuts, retailers have expanded their offering to include seasoned or marinated fresh beef cuts as well as a number of precooked product lines. Marinated and seasoned products are frequently flavored

with teriyaki, peppers, Cajun spices, burgundy, mesquite, or garlic. Fully cooked products often are presented in a gravy or sauce to provide differentiation. While the precooked items can be safely consumed without cooking, most consumers reheat them prior to serving.

There are also significant regional differences in the beef cuts favored by consumers (Table 2.5). Managing the distribution chain to accommodate these regional differences is important to the overall sales volume of U.S. domestic beef.

Foodservice

Foodservice outlets in the United States number more than 1 million locations and account for 4% of gross domestic product. Employing more than 14 million people, the foodservice sector provides nearly 10% of jobs in the United States with a total economic impact of $2 trillion with sales of $780 billion in 2016, nearly double of that recorded in 2000. Foodservice purchases are responsible for nearly half of total food purchases made by consumers (Table 2.6). The foodservice sector accounts for more than $32 billion in beef sales moving nearly 8 billion pounds of beef products through fast food, family dining, fast casual, fine dining, and institutional trade. The total wholesale value of food purchased by foodservice operators is $145 billion and beef leads the list accounting for nearly 25% of those purchases. Given this impact, the quality, value, and price of beef are critical to the profitability of the foodservice sector.

Table 2.5
REGIONAL VARIATION IN PREFERRED TOP FIVE MIDDLE MEAT CUTS OF BEEF

United States	West Coast	Great Lakes	Northeast
Strip Steak (boneless)	Bone-in Rib-eye Roast	Strip Steak (boneless)	Bone-in Rib-eye Roast
Bone-in Rib-eye Roast	Tri-tip Roast	Rib-eye Steak (boneless)	Strip Steak (boneless)
Rib-eye Steak (boneless)	Strip Steak (boneless)	Rib-eye Steak (bone-in)	Rib-eye Steak (boneless)
Top Sirloin	Rib-eye Steak (bone-in)	T-Bone Steak	Tenderloin
Petite Sirloin	Strip Steak (bone-in)	Top Sirloin	Top Sirloin

Source: Adapted from National Beef Board and NCBA.

Table 2.6
PERCENTAGE OF FOOD EXPENDITURES SPENT ON FOOD PREPARED AWAY FROM HOME

Year	Percent
1960	20
1970	26
1980	32
1990	36
2000	40
2010	42
2016	45

Source: Adapted from ERS-USDA data.

Table 2.7
PRIMARY FORMS OF BEEF SOLD THROUGH FOODSERVICE OUTLETS—TONNAGE AND SALES (%)

Item	% of Weight	% of Sales
Ground	65	38
Steaks	13	32
Roasts	10	18

Source: Based on beef checkoff data available at www.beef.org.

Foodservice sales typically grow when the economy is strong and consumers continue to demand high degrees of convenience. Beef products preferred by foodservice consumers are diverse, but clearly dominated by ground beef offerings on a volume basis while whole-muscle cuts account for half the value with only one-quarter of the volume (Table 2.7). The most critical factors influencing consumer demand in the foodservice market are convenience, value, and variety. Wholesale beef prices are a critical factor in the menus offered by restaurant owners and their willingness to pay ultimately impacts the price of cattle and calves.

BEEF PALATABILITY AND CONSUMER PREFERENCES

Consumers obviously eat food for its life-sustaining nutrients, but they typically prefer foods that provide eating satisfaction (taste), convenience, consistency, and a positive contribution to good health. Consumers prefer beef that is tender, flavorful, and juicy, with a high ratio of lean to fat and a high ratio of lean to bone. Consumer demand is not constant as illustrated in Figure 2.11. Shifts in demand result from changing consumer demographics, economic conditions and buying power, and the value that consumers perceive they receive from their purchases. Producers must understand the primary factors affecting beef palatability so that they can assist in designing beef products that have high consumer acceptance.

Perhaps the initiator of the industry concentrating its efforts on improving palatability came from the 1991 Beef Quality Audit that yielded the startling observation that one out of four (25%) cooked beef steaks did not "eat right" (have

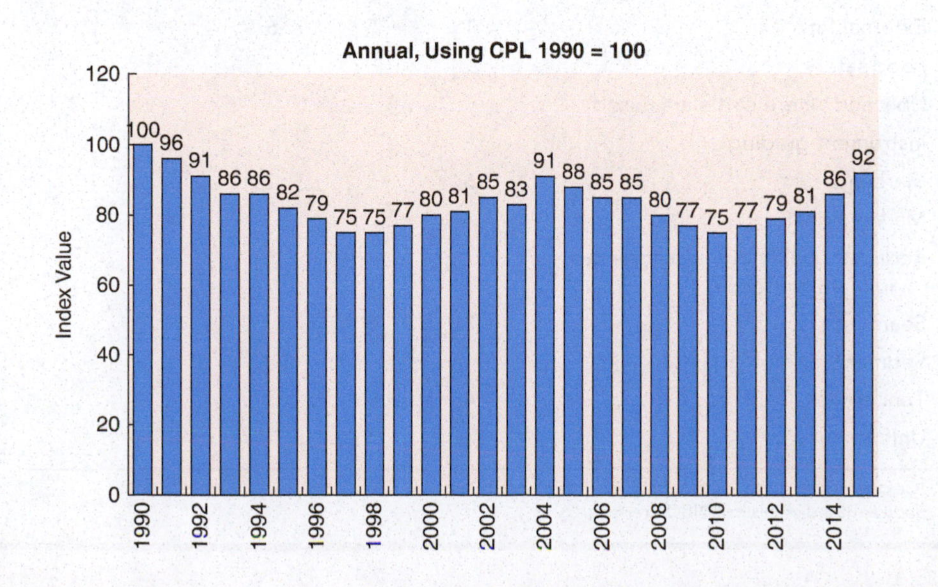

Figure 2.11
Retail all fresh beef demand index.
Data sources: USDA Economics Research Service and Livestock Marketing Information Center (LMIC); analysis and compilation by LMIC.

inadequate flavor, juiciness, tenderness, and/or overall palatability problems). A Texas A&M study showed that one carcass, poor in palatability characteristics, could negatively affect more than 540 consumers. These and several other studies vividly demonstrate that improvement in beef palatability must occur if consumer market share is to be positively influenced. The National Beef Quality Audit funded by the Beef Checkoff has been conducted multiple times with the intent of better characterizing concerns about beef quality as perceived by beef supply chain participants. Those concerns are summarized in Table 2.8. Clearly the industry has been able to use this data to make effective changes in cattle and beef, but it is also clear that the focus on quality is never going to end in light of the emergence of new and varied concerns over time.

Consumer cooking habits and meal preference are being transformed:

- From 1980 to 2016, the average time spent preparing dinner declined by half—to 34 minutes.
- Meat shifting from center of plate item to key ingredient resulting from an increased interest in convenience, affordability, and growth in ethnic recipes.
- Ground beef demand is at an all-time high with more than 60% share of beef consumed at home. Ninety percent of ground beef is being used as an ingredient in meals cooked at home. Nine billion burgers were served in U.S. restaurants and foodservice establishments in 2014.
- The prices of all meats have increased and the ratio of beef to pork and poultry prices have also risen.

Table 2.8
TOP SIX QUALITY CHALLENGES (OPPORTUNITIES) IDENTIFIED AND RANKED BY THE NATIONAL BEEF QUALITY AUDITS CONDUCTED BETWEEN 1991 AND 2011

Quality Concern	1991	1995	2000	2005	2011
Carcass composition (lean, fat, bone)	5				4
Carcass weight			2	6	5[1]
Cattle genetics					6
Cut weight		6			5[1]
Eating satisfaction/palatability	3	2	3		2
External fat	1	5	6		
Food safety					1
How and where cattle are raised					3
Instrument grading				3	
Marbling	6	3	4		
Market signals				4	
Reduced quality due to aggressive implant protocols			5		
Seam fat	2	5			
Segmentation of industry				5	
Tenderness	4	4	3		
Uniformity		1	1	2	

[1]Carcass weight and cut size combined in 2011 Audit.
Source: Based on historic NBQA reports.

- Nearly one-half of consumers who have reduced beef purchases for at-home consumption identify price as the major cause. Middle class retail consumers indicate that they wait for price specials and that they have moved to less expensive cuts as the price of beef has risen. However, more affluent customers continue to prefer and pay for premium beef cuts and experiences.
- Thirty-three percent of U.S. consumers are millennials and this demographic will have a major impact on beef marketing. The millennial generation is very interested in culinary topics and skills and views cooking as a preferred activity. Unfortunately, the millennials are confused about the beef case and the diversity of cuts available to them. However, they would purchase greater amounts of beef if they have a better understanding.
- The use of social media and hand-held electronic devices to aid in shopping, searching for discounts, comparing recipes, and sharing food choices is increasing at a rapid rate. Developing marketing strategy to engage consumers virtually will be at the heart of future beef promotions.

Tenderness

Consumers rate tenderness as the most important palatability characteristic of beef. The 1990 National Beef Tenderness Survey showed that 20% of the cuts from the loin and rib, and 40% and 50% of the retail cuts from the chuck and round were not satisfactory in tenderness. Consumers have overwhelmingly indicated a preference for tender beef with 88% indicating that tenderness was either "very important" or "important" to their purchasing decisions. One-third of consumers surveyed in one study indicated that they would pay up to 50 cents a pound more for a guaranteed tender steak. Improper cooking and serving methods can compound these tenderness problems. The beef industry must identify and utilize quality control procedures to ensure more consistency in beef tenderness.

Some knowledge of muscle structure is needed to comprehend tenderness. The structural differences among muscles that help determine the tenderness include:

a. *The amount of connective tissue.* Connective tissue surrounds the myofibrils, while another connective tissue layer covers the muscle fiber. Additional connective tissue layers cover the muscle bundles and also the entire muscle. As the amount of connective tissue increases, the less tender the beef when it is cooked. Cattle differ in the amount of connective tissue in their total musculature.

b. *Sarcomere length.* The sarcomere is the individual unit of the myofibril that allows the muscle to contract or relax. Longer sarcomeres in myofibrils result in more tender cooked beef. In living muscle, myofibrils shorten during contraction and lengthen during relaxation. Rate of cooling of the carcass during the first few hours postmortem determines sarcomere length and thus partially determines the tenderness of the cooked meat. "Cold shortening" occurs if the muscles stay contracted. The amount of outside fat and marbling help insulate muscle tissue against cold shortening. Research has demonstrated that a minimum of 0.3 in. of fat over the rib eye is sufficient to prevent cold shortening.

c. *Sarcomere degradation.* Beef that is stored (aged) for several days under refrigerated conditions will become more tender. Most of the tenderization will be achieved with in 7–14 days of aging. This increase in tenderness due to "aging" is achieved through the activity of proteolytic enzymes (calpains) by causing degradation or a weakened structure of the sarcomeres. **Bos taurus** cattle have less calpastatin (inhibitor of calpain) than **Bos indicus** cattle. Calpastatin prevents calpain from tenderizing the muscle during the aging process.

d. *Marbling.* Marbling makes the cooked beef more tender because fat is less resistant to shear force than muscle fibers or connective tissue. Also, fat lubricates the mouth during chewing, causing the consumer to perceive a higher degree of tenderness than in beef containing less marbling. Tenderizing by mechanical means (needling and blade tenderizing) can increase the perceived tenderness of beef. Both the myofibrillar and connective tissue components of tenderness are modified by mechanical tenderization. The mechanical tenderizer physically disrupts the tissue through penetration of small blades into the meat. Institutions that purchase beef that is variable in tenderness use mechanical tenderization most frequently. Blade tenderization reduces this variability. There are some food safety concerns with mechanical tenderization because the blades could transfer microorganisms on the outside of the muscle to deep within the muscle. The latter increases the risk that cooking temperature would not as easily destroy the microorganisms.

Table 2.9 summarizes the major factors affecting the tenderness of beef. Producing beef that is consistently tender is not easy; however, some in the industry are accomplishing it. Many more producers, processors, and retailers need to produce and serve more consistently tender and highly palatable beef. Better ways of measuring beef tenderness are needed from production through consumption. Examples of these improvements include the use of DNA markers to identify superior (inferior) gene types and development of objective carcass evaluation systems.

Flavor

Beef flavor is determined by specific compounds that are in the intramuscular fat (primarily in the form of marbling fat) of beef muscle.

The primary determinants of flavor desirability of marbling fats as identified by Smith (1995) are:

a. *Grain-fed versus forage-fed beef.* Most consumers greatly prefer the taste and aroma of the fat from grain-fed, as compared to forage-fed, cattle.

b. *Fed grain for approximately 100 days.* Research has demonstrated that the fat from cattle fed high-concentrate diets for 100 days was of optimal flavor desirability.

c. *Flavor increases as marbling increases.* As the amount of marbling increases from "practically devoid" (characteristic of U.S. Standard) to "slight" (characteristic of U.S. Select) to "small" (characteristic of the lower third of U.S. Choice) to "slightly abundant" (characteristic of U.S. Prime), there are progressive increases in the desirability of flavor in cooked beef.

The characteristic flavor of "aged" beef (beef stored for periods from 7 to 42 days following slaughter) is not attributed to components of fat; rather, the acid-like, tart, nut-like flavor of aged beef arises from "ripening" that generates amines from protein as well as by-products of the breakdown of nucleic acids. Beef can be "wet-aged" (held for long periods of time in vacuum packages) or "dry-aged" (held for long periods of time with no protection or packaging). There is no difference in the extent of tenderization that is achieved in "wet-aged" or "dry-aged" beef that is held, postmortem, for the same period of time. Juiciness is more desirable in "wet-aged" beef while flavor is more intense in "dry-aged" beef because of dehydration and thus concentration of flavor-eliciting compounds and/or growth of molds (Smith, 1995).

Juiciness

Smith (1995) summarized the two major components of juiciness as (1) the release of meat fluids during the first few chews of the meat, and (2) the potential stimulating effect of fat on the salivary flow. The second component appears to be more important than the first.

Table 2.9
Major Factors Affecting Beef Tenderness

Factor	Effects and Relationships
Breed and biological type within a breed	Brahman and >50% Brahman cross cattle produce beef that is less tender than other breeds. This difference is partially due to less proteolytic enzyme activity in the muscle of *Bos indicus* cattle. Brahman breeding (when 3/8 or less) combined with other appropriate breeds/biological types (e.g., British breeding) have shown satisfactory evidence of producing beef that is tender. The Senepol, Romosinuano, and Tuli breeds are being evaluated as substitutes for the Brahman influence in cross-breeding heat-tolerant cattle. British breeds rank highest in tenderness(Chapter 13). There are differences in tenderness within breeds that show apparent genetic difference between sires in producing tender beef.
Age of animal	As an animal grows older, there is an increase in the cross-linking within and between collagen molecules that makes them less soluble (less susceptible to softening during cooking) and thus less tender. Fed steers and heifers produce beef with the most desirable levels of tenderness between 12 and 24 months of age. Young bulls that are fed concentrates can produce reasonably tender beef up to 15–16 months of age, and then tenderness decreases.
Feedlot gain	Rapid live-weight gains associated with 100 days of high-concentrate feeding increases tenderness. A high daily rate of gain generates more protein turnover, which in turn is associated with higher myofibril fragmentation and collagen solubility. Cattle gaining 0.5 lb/day or less have been shown to produce beef that is less tender than that from cattle with more rapid average daily gains.
Rate of carcass cooling, carcass weight, and fat cover	Too-rapid cooling of the carcass results in greater shortening of sarcomeres. Heavier carcasses and those with more fat cover have decreased cold shortening of the sarcomeres.
Aging of carcass or retail cuts	"Aging" is the holding of beef under refrigeration for >7 days to increase tenderness. During this time, there is a proteolytic weakening of the myofibril structures. Storage temperature may be increased during aging to hasten tenderization, as proteolytic enzymes are more active at higher temperatures. Because of food safety concerns, high-temperature aging is not recommended. Whole-muscle cuts are aged almost twice as long by foodservice vendors as opposed to retail outlets.
Electrical stimulation	Administering electrical shocks (high or low voltage) to the carcass during slaughter speeds up rigor mortis, thus reducing cold shortening. There is evidence that electrical stimulation also causes weakening of the myofibril structure, thus improving tenderness. Strength of voltage and duration of treatment both influence effectiveness of electrical stimulation. Increases in both factors improve tenderness.
Marbling	Higher marbled beef is generally more tender. Marbling accounts for 10–20% of the tenderness differences in beef when research results are averaged.
Location of retail cuts	Limb muscles have more collagen to support muscles used for locomotion, so they are less tender. Loin muscles have relatively low levels of collagen, thus they are more tender. Some muscles with large amounts of collagen are ground or put through a mechanical tenderizer.
Method of cooking	Collagen softens during moist heat cooking. Under steam cooking, collagen will usually turn to gelatin. Retail cuts with high collagen content are usually cooked with moist heat. Dry heat is used on tender cuts.
Cooking temperature	When internal temperature of beef exceeds 145°F for whole-muscle cuts during the cooking process, the meat is less tender because myofibrillar proteins harden with high temperature. USDA recommends 160°F for hamburger; however, this is for food safety reasons and decreases tenderness.
Serving method	Tenderness is greater if beef is served hot (immediately after cooking) rather than allowed to cool to room temperature prior to eating it.

Juiciness is influenced largely by the amount of marbling and degree of doneness. Degree of doneness relates to the amount of intramuscular water (moisture) retained in the meat after it is cooked. More highly marbled beef and beef prepared as "rare" to "medium" would be considered more juicy by consumers; however, if beef cuts are to be cooked to 145°F for safety reasons, then "medium" would be considered the minimum degree of doneness. Cooking procedure is very important in influencing the juiciness of beef, as the cooking process that retains the most fat and fluids will yield the juiciest meat. For many consumers, beef that is cooked to the "well-done" state is considered unacceptably dry.

The ability of a muscle to hold its own or added moisture against the forces of heat and pressure is termed *water-holding capacity*. Beef that is excessively pale in color usually loses excessive amounts of its moisture during cooking and is dry and powdery when eaten; dark-cutting beef has unusually high water-holding capacity and remains juicy even when cooked to the "well-done" state (Smith, 1995).

Juiciness increases as marbling increases, almost in a linear relationship. Also, beef with higher amounts of marbling can be cooked to higher degrees of doneness (higher temperature endpoints), and yet remains acceptably juicy when compared to beef of lower degrees of marbling.

Tenderness and juiciness are related. The more tender the meat, the more quickly the juices are released by chewing and the juicier the meat seems to be.

USDA Quality Grade, Fat, and Overall Palatability

Smith et al. (1987) showed that the palatability of loin steaks was rated higher as quality grade increased—undesirable palatability ratings for each of the USDA grades were 59% for Standard, 26% for Select, 11% for Choice, and 5% for Prime. The variability of eating satisfaction among steaks within a quality grade was reduced as quality grades increased from Standard to Prime.

Numerous surveys have shown that consumers purchase beef and other foods primarily for their taste. The amount of fat in food appears to be highly related to its taste. Research shows that fat consumption may satisfy several physiological and psychological needs. The sensory pleasure response appears to be mediated by the release of endogenous opiate peptides—pleasure-enhancing molecules—manufactured by the human brain.

Because opiates mediate the body's response to stress, it may be that preference for fat is nature's way of ensuring an adequate supply of energy. On the other hand, because opiate peptides are also involved in mediating the pleasure response to foods, it may be that fat consumption affects the mental health and well-being of the individual. Foods rich in sugar and fat are sought after, not for what they do to the body, but for what they may do to the brain. Taste preferences for sugar and fat appear to be under opiate control. In the news release for the 1995 National Beef Quality Audit, Dr. Gary C. Smith stated, "America's beef producers and feedlots—through no fault of their own—have been doing exactly the opposite of what the consumer wants. The 1995 audit revealed consumers want juiciness, flavor, and tenderness in their beef cuts, qualities that only can come from having a certain amount of fat marbled into the flesh. Meanwhile, the feedlots have been turning out leaner, more muscular animals with less of the fat that provides these qualities. I expect a shift away from overly muscled cattle that are too large, back to a medium-sized animal with more marbling."

The 1995 Beef Quality Audit demonstrated that outside fat in cattle had been reduced by selecting leaner, more muscular animals, and by packers trimming fat. However, quality grade was being reduced as noted in Figure 2.12. Producers began to place more emphasis on marbling in their selection strategies resulting in an increase in the percent of beef carcasses graded USDA Prime and Choice. However, the increase in marbling came with a tradeoff of increased overall carcass fat as indicated by shifts in USDA Yield Grade distribution with the percent of Yield Grade 4

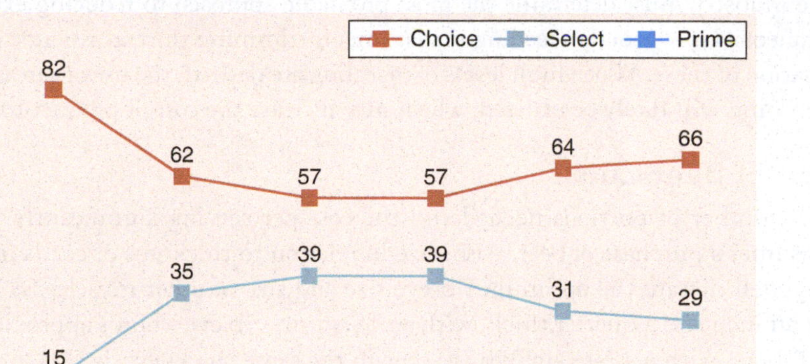

Figure 2.12
USDA Beef Quality Grade Trends.
Source: Adapted from USDA-AMS.

and 5 carcasses growing from approximately 4% in 1990 to 11% in 2013. Managing cattle to optimize USDA Quality and Yield grades will remain a critical objective of the industry for the foreseeable future. Research results support the continued use of USDA quality grades as a beef palatability critical control point, with efforts made to add critical control points that can augment, not replace, quality grades—particularly at lower marbling levels. More effective measurements of tenderness are needed to improve prediction of beef palatability. There is reason to believe that if beef is made more consistent in tenderness, juiciness, and flavor, the decline in beef consumption will be corrected. Furthermore, the industry needs to develop and adopt more objective measures of quality and tenderness using technologies, such as camera grading.

Lean to Fat

Consumer surveys give evidence that consumers use amount of fat as a selection criterion. The 1988 Market Basket Survey showed that closer trimming of external fat on retail cuts resulted in a substantial reduction in the amount of fat purchased by consumers. In that survey, more than 42% of the retail beef cuts had no external fat and the overall fat thickness for all retail cuts in the beef case was 1/8 inch. This was a marked change from the 1/4-inch trim in 1986 to the 1/2-inch trim prior to 1986. The beef industry responded to consumer preference by reducing the fat content presented to consumer beef products by 25–35% during 1985–1995. However, as previously noted, as overall fat was reduced there was a negative impact on quality grade distribution.

Proper breeding and feeding programs, with excess fat trimmed and bones removed at the packing plant or retail level, can most effectively control the lean-to-fat and lean-to-bone ratios in beef. Concentrate feeding, without overfeeding, will keep the lean-to-fat ratio at the desired level, resulting in beef having desired consumer palatability characteristics if fed beef has been processed and cooked correctly. A goal of achieving 70% Choice or better, 70% USDA Yield grade of 2 or better, and 0% carcass nonconformation is appropriate for the industry.

Beef that is too lean will not be as palatable as beef with at least a minimum level of marbling. Researchers developed the "window of acceptability" of 3–7% fat content in beef, which is equivalent to beef cuts from the lower part of the Select quality grade to the higher range of the Choice quality grade. Beef cuts with 3% fat or less content will likely not meet palatability expectations, and cuts having more than 7% fat will exceed total fat content for diet and health preferences. The statement "remove the waste fat but keep the taste fat" has considerable merit when optimizing the amount of trimmable fat with the amount of intramuscular fat needed to enhance palatability.

The industry must determine the most profitable approach to reducing external or subcutaneous fat—genetics, feeding management, trimming the carcass/cuts, or some combination of these. When high levels of marbling are desired, the use of carcass or retail cut trimming will likely be utilized, which may increase the cost of product to consumer.

Size of Beef Cuts

The number of portions needed and the cost per serving significantly influence the consumer's purchase of beef. Also, size in relation to thickness of cut is important and may well dictate the optimum rib-eye size and size of other muscles. A T-bone steak cut an inch and a quarter thick, with an 18-sq.-in. rib eye weighs approximately 32 oz. A T-bone with a 12-sq.-in. rib eye, cut to the same thickness, is a 21-oz steak. There can be palatability problems if steaks are cut too thin because the meat can be easily overcooked and then will be tough and dry. For the mainstream retail and foodservice markets featuring whole-muscle cuts, the preferred rib-eye size is 12 to 14 square inches. As the industry has continued to increase per animal productivity resulting in increased carcass weights over time, it has become difficult to maintain cut sizes in line with the desires of beef purveyors and foodservice operators in particular.

Color

Color of meat in retail cases is an important selection criterion for most consumers, as they perceive color to be a measure of freshness. Their preference is for a bright, cherry red color in contrast to a dark, less bright, red color. In past years, this was a useful criterion to evaluate beef because color can be used to separate meat from younger versus older cattle. Beef from older cattle is a darker color and is usually less tender than beef from a younger animal. Currently, however, most meat from older cattle goes into processed meats, where—by mechanical means used in producing ground beef—the tenderness problem is largely eliminated. It should be noted that there is a growing market for whole-muscle cuts from non-fed cattle. Some of the darker-colored beef in retail cases results from meat remaining in the retail case for a period of time. After such meat is cooked, the darker color disappears and eating qualities may be similar to those of brighter-colored meat.

Consumers prefer to buy fresh beef, and attempts to market frozen beef have met with little success. Consumers indicate that the inability to judge the freshness of frozen beef by color is their primary reason for not purchasing it in the frozen condition.

Addition of vitamin E to feedlot diets increases beef's color retention and extends shelf life in the retail meat case. Feedlots are implementing this recent research finding. Vitamin E costs an average of $1 per head and can return $35 through extended shelf life.

Improving the Value of the Chuck and Round

As Figure 2.13 illustrates, the value differences of primal cuts with varying marbling levels as reflected by USDA quality grades are highly irregular. Ribs and loins of increasing levels of marbling and quality indeed sell for a premium while the value differences in the other primals are nearly nil. This creates a challenge for the industry to determine a means to create better value from the round and chuck in particular.

Development of new fabrication protocols to separate individual muscles that previously were sold as a component of a roast can significantly improve the value of a subprimal. Enhanced fabrication yields higher-valued cuts from the shoulder, knuckle, and bottom round. For example, the shoulder yields 14% steaks and 58% roasts when conventionally fabricated. Under the beef value fabrication system, all muscles are separated to create a 47% steak yield. Similarly, the bottom round yields 37% and 40% for steaks and roasts, respectively, with conventional fabrication. However, utilization

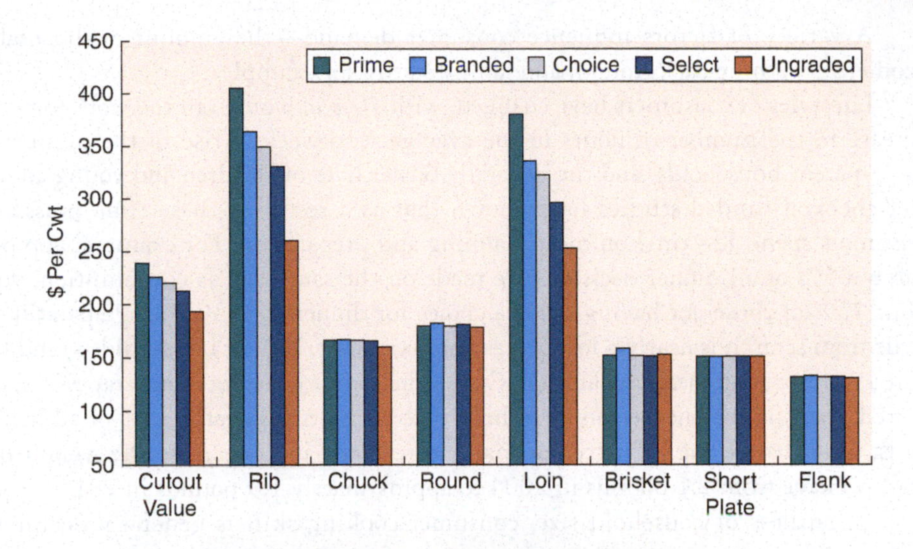

Figure 2.13
Comparison of primal beef cut values by differing levels of marbling.
Data source: USDA Agricultural Marketing Service (Market News); compiled by Livestock Marketing Information Center.

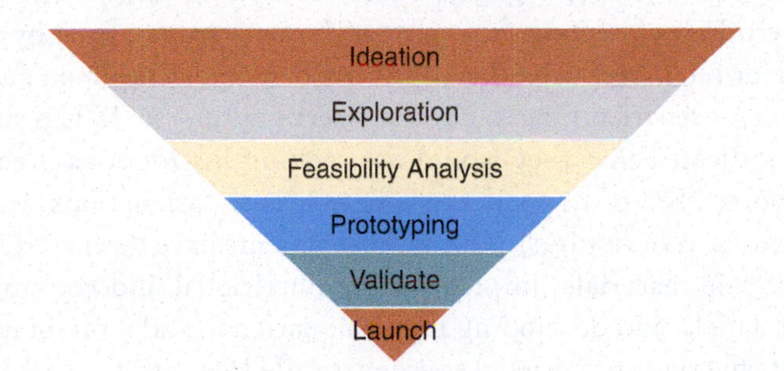

Figure 2.14
Product development process.

of the beef value fabrication protocol shifts the yield of steaks and roasts to 55% and 13%, respectively.

New product development follows a general path of creation (Figure 2.14) beginning with ideation to yield a broad set of potential solutions to the pain points in a process or product. Exploration follows to sort through the credibility and costs of the pool of possibilities identified in step one. Step three is designed to test the concept against the marketplace to determine demand and enthusiasm as well as to quantify the potential size of the market. An assessment of competitors is typical at this stage as well. In step four, prototyping is conducted to develop the concept for a larger scale market test in step five. Finally, the product is launched into the market place to compete. This process is time consuming and requires investment of talent and resources. However, product development and process refinement yield a significant value for the U.S. beef industry via improved yield, value-added enhancements, and better delivery of products suited to contemporary consumer demand and higher profits.

Identifying Consumer Attitudes and Preferences

Industry leaders identified consumer confidence and development of consumer-friendly products as the key factors influencing beef demand. Consumer confidence is the result of perceptions about food safety, nutrition, and the overall enjoyment of an eating experience. Delivery of consumer-friendly products is dependent on meeting customer expectations in regard to convenience and value.

Beginning in the 1980s, when beef demand losses were recognized as significant, the industry began to transform itself into a consumer-driven entity. The first step of this process was to quantify and understand the wants and needs of customers. Study after study has shown that the leading drivers of consumer decision making relative to food choices are taste; value defined as a composite of quality, safety, and nutrition; and price.

A variety of factors influence consumer demand—the amount of disposable income, age, ethnic background, and family status, for example.

Lifestyles of consumers have changed, with 70% of women in the work force, an increase in the number of hours in the average workweek, a rise in the number of single-parent households, and the generally busier lives of children and young adults. Beef checkoff-funded studies have shown that as a result of these time pressures, consumers spend less time on meal planning and preparation. For example, approximately 65% of all dinner decisions are made on the same day as consumption, with about 75% of those not having made a choice for dinner by 4:30 P.M. Preparation of meals from scratch is nearly a lost art, as approximately 75% of households spend less than 45 minutes on meal preparation. One of the fastest growing demographics in the United States is the one-person household that accounts for nearly 30% of all households. One change the industry has made in response is to decrease the weight of a retail package from 2.1 pounds in 2004 to approximately 1.9 pounds in 2015.

Regardless of household size, consumer cooking skill is generally declining, with 1 in 3 customers admitting that a lack of cooking knowledge prevents them from buying certain beef cuts. In fact, most consumers are confused by the use of anatomical names for retail beef cuts, the lack of cooking information on packages of fresh beef, and the poor user-friendliness of the self-serve meat case. When surveyed, nearly all consumers indicated that they would use cooking instructions if they were available. Furthermore, 78% of respondents believed these instructions would encourage increased beef sales. As a result, significant investments have been made in developing better point-of-sale materials, improving the nutritional and cooking instructions information on labels, and developing regional, national, and store brands.

The most effective promotional strategy is called "featuring" that involves reducing a product's price to attract buyer attention. Nearly 50% of shoppers prefer this approach over all other promotions. Even higher favorability is assigned by shoppers who prefer natural and organic products (which are typically expensive compared to other alternatives) as well as among single-person households.

The consumer market is highly differentiated and, thus, a multitude of opportunities are available for producers to develop specialty beef products designed to meet the unique requirements of a specific niche market. A classic example is *Certified Angus Beef*™ that focuses on the consumer segment that desires the flavor and palatability characteristics associated with relatively high levels of marbling. As Figure 2.15 points out, the demand curve for the product has been substantial.

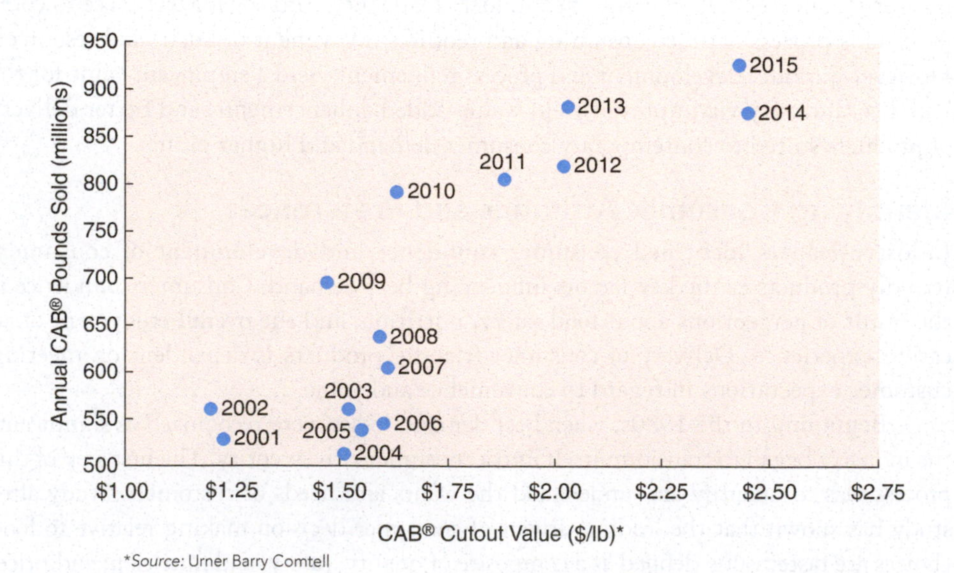

Figure 2.15
Demand curve from a well-executed beef brand strategy.
Source: Adapted from Certified Angus Beef Annual Reports.

Building Consumer Trust in Beef Products

Consumers have choices and they vote for those choices with purchasing power. The great challenge for those involved in the food industry is to understand the dramatic diversity of the consumer and to understand the various segments of buyers so that specific expectations can be met. Specific product attributes, such as tenderness, flavor, and color, have been discussed previously in this chapter and can be grouped into classification identified as "perceived value." However, there are a number of other categories that influence consumer choices (Figure 2.16) including healthfulness, food safety, husbandry, and stewardship, plus production technologies.

Healthy Beef Products

Most nutritionists agree that a healthy diet should contain all the required nutrients and enough calories to balance energy expenditure. Unfortunately, nutrient deficiencies or excesses exist in much of the world. Therefore, food choices—or lack of choices—determines the nutritional status of most people. Obesity caused by caloric excesses has become a leading nutritional problem in the United States, with approximately one-third of the population classified as obese.

The nutritional advantages of beef simply cannot be ignored (Table 2.10).

The most compelling reason to include beef as a component of the diet is its nutrient density. Nutrient-dense foods are defined as those that offer a higher proportion of several recommended nutrients than it does calories. Beef is a particularly good source of

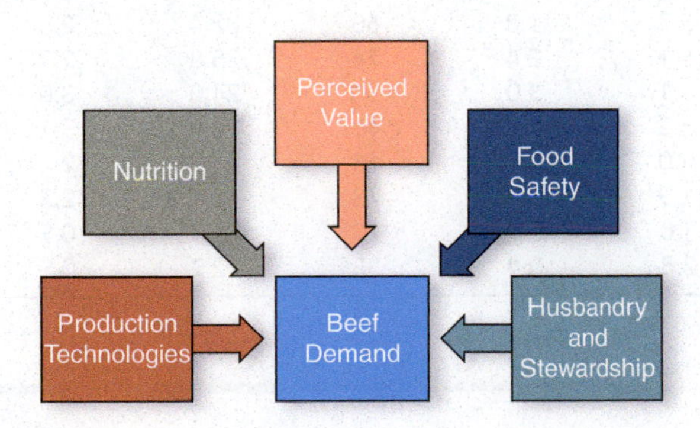

Figure 2.16
Factors affecting beef demand.

Table 2.10
KEY NUTRIENTS FOUND IN BEEF AND THEIR ASSOCIATED FUNCTIONS

Nutrient	Function
Zinc	Enhances immune function, promotes wound healing, essential for normal growth and cognitive abilities through childhood
Iron	Oxygen carrier, energy enhancer
Protein (beef is a complete protein that provides all the essential amino acids)	Building block for muscle development, required for repair of cells and regulation of metabolic processes
B Vitamins (riboflavin, niacin, and B_{12})	Assist in energy metabolism, promote skin health, aid digestion, foster normal appetite, promote normal nerve function, lower homocysteine that increases risk for dementia and heart disease

zinc, iron, protein, and the B vitamins (Table 2.11). A 3-oz serving of beef provides less than 10% of the calories in a 2,000-calorie diet while contributing 10–48% of the recommended daily allowances for protein, iron, zinc, niacin, selenium, phosphorus, choline, and vitamins B_6 and B_{12} (Figure 2.17). Beef provides a degree of nutrient density that makes it highly competitive with other protein sources as a component in a rational caloric, fat, and cholesterol content diet (Table 2.11). Table 2.12 illustrates comparative amounts of several foods to acquire the same nutrients available in a 3-oz serving of beef.

Ease of nutrient absorption is also of concern. For example, iron is present in two forms—heme and non-heme. Heme iron is the most easily absorbed and is the predominant form of iron in red meats.

Unfortunately, many American citizens are deficient in iron (40%) and zinc (73%). These deficiencies are particularly noticeable in the female population. Approximately 75% of females between the ages of 12 and 49 are iron and zinc deficient. This group of consumers, particularly teenagers, tends to be low consumers of beef despite the fact that beef provides 67% and 58% of zinc and iron, respectively, available in the food supply.

Table 2.11
Nutrient Comparison of 3 oz of Various Cooked Beef Cuts, Chicken Breast, and Halibut

Product	Calories	Total fat (g)	SFA[1] (g)	Cholesterol (mg)	Protein (g)	Iron (mg)	Zinc (mg)
Daily Value	2,000	65	20	300	50	18	15
Top round (broiled)	153	4.2	1.4	71	26.9	2.4	4.7
Boneless pot roast	147	5.7	1.8	60	22.4	2.6	5.4
Top sirloin (broiled)	166	6.1	2.4	76	25.8	2.9	5.5
Tenderloin (broiled)	175	8.1	3.0	71	24.0	3.0	4.8
T-bone steak (broiled)	172	8.2	3.0	48	23.0	3.1	4.3
Ground beef (95/5)	139	5.0	2.2	65	21.9	2.4	5.5
Ground beef (85/15)	197	11.9	4.7	73	20.9	2.3	5.3
Chicken breast (skinless)	140	3.0	0.9	72	26.4	0.9	0.9
Halibut (dry heat)	119	2.5	0.4	35	22.7	0.9	0.5

[1]Saturated fatty acids.
Source: USDA.

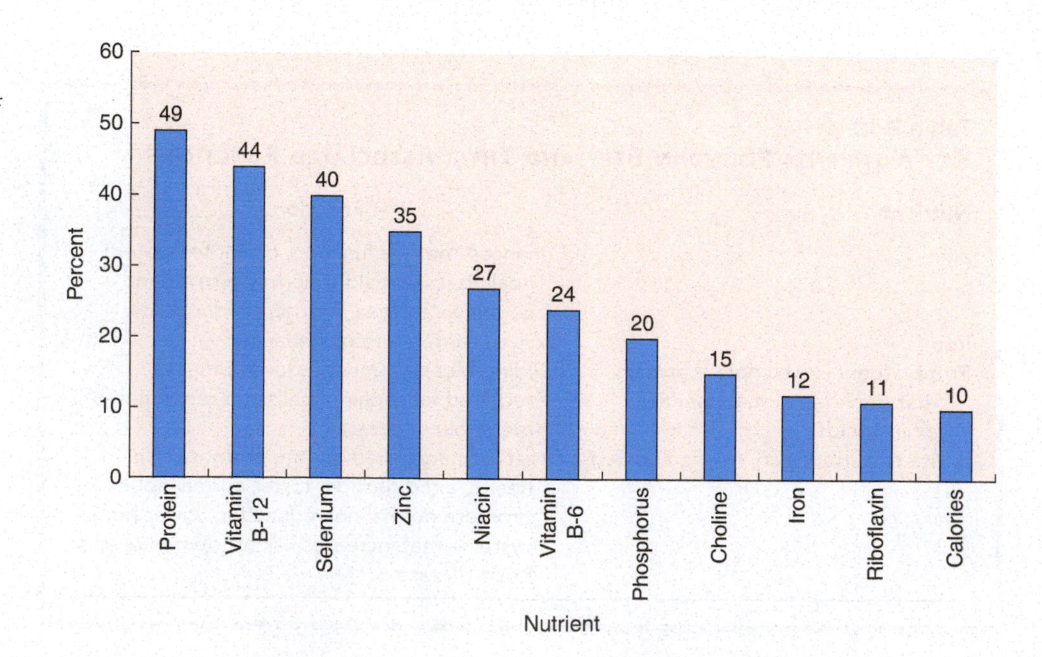

Figure 2.17
Percent nutrients provided by a 3-ounce serving of beef in a 2,000 calorie diet.
Source: USDA.

Table 2.12
Equivalent Servings Required to Equal the Amount of Nutrient Found in a 3-oz Serving of Beef

Nutrient	Equivalent Serving
Zinc	12 (3.25 oz) cans of tuna
Vitamin B_{12}	7 chicken breasts
Iron	3 cups of spinach
Riboflavin	2 1/3 chicken breasts
Thiamin	2 chicken breasts

Source: USDA.

In regard to meeting their protein requirement, only 45% of teenage males and 32% of teenage females actually consume sufficient protein. At the same time, teenagers consume 6 to 10 daily servings from the sugar, fat, and oil category, with 40% of their energy requirements being met by foodstuffs considered to be low in total nutritional value.

Despite the evident dietary advantages of beef, concerns have been raised in regard to the role of meat in the incidence of some human diseases.

The relationship between diet and human health is a controversial and complex topic. The consumption of beef has been linked to two of the most dreaded human diseases: coronary heart disease (CHD) and cancer. Consumer perceptions have been influenced to accept these alleged relationships, and consumers have reduced their consumption of some animal products accordingly.

Before examining the root causes of human disease, it is worthwhile to understand the concept of risk analysis. American citizens are enjoying an ever-increasing life span, better health, and a higher standard of living than their ancestors could have imagined. For example, the average life span of a U.S. citizen has doubled since 1900, infant mortality has declined 45% since 1980 in the United States, those living in poverty conditions have been halved since 1960, and inflation-adjusted per-capita income has doubled since 1960. Despite this evidence of well-being, widespread public worries about diet–health relationships persist. These worries are perpetrated when reports fail to account for the following:

- The wholesale extrapolation of results obtained from lab animal models to humans.
- The fact that natural compounds may contribute significantly more risk than man-made compounds.
- The effect of dose rate on disease incidence.

In a media-dominated environment where sensationalistic headlines are all too common, consumers are advised to be wary of "junk science." The following are warning signs that results of a study are being incorrectly represented.

1. The results recommend changes that offer quick-fix promises.
2. Foods are described as "good" versus "bad."
3. Simplistic conclusions are offered from a complex study.
4. The study was not peer reviewed.
5. Recommendations ignore differences between individuals or among groups.
6. Results are interpreted to offer significant negative consequences from a specific food item or diet selection.

Sound, data-based research often appears to be lost in an emotionally charged issue. Many organizations and individuals have occasionally based judgments and decisions on emotion rather than on the best accumulated research facts. Scientific principles should be identified so that decisions are based on true relationships. It is also important to recognize that epidemiological studies that attempt to characterize influences on disease occurrence often times are only able to determine associated risk factors as opposed to causative factors. Risk factors should not be assumed to cause disease but rather should be evaluated from the opportunity to alter dietary and lifestyle habits to improve overall health and well-being.

The major known risk factors associated with CHD are genetics (a family history of CHD), high blood cholesterol, smoking, hypertension, physical inactivity, and obesity. Obesity caused by excess caloric intake is a major nutritional problem in the United States. Of the leading 10 causes of death in the United States, obesity is considered a risk factor in five (CHD, stroke, hypertension, type-II diabetes, and some forms of cancer). Over 50% of U.S. adults exceed their recommended weights. Interestingly enough, survey results suggest that consumers are less concerned about caloric and fat intake than they were in 1990. In a 10-year span, the percentage of consumers who reported that they were always conscious of caloric intake fell from 40 to 25%. Consumers reporting that they were always cautious about their fat intake fell from 51 to 33% over the same time period. These trend lines motivated significant expenditures by USDA and other federal government agencies to increase consumer awareness but to date there is not clear evidence of success by these campaigns.

It is generally accepted that consumption of animal fat by humans causes an increase in the level of cholesterol in the blood, while consumption of vegetable oils (polyunsaturated fats) causes a decrease in blood cholesterol concentration. These relationships led to the theory that there is a relationship between consumption of animal fats and the incidence of atherosclerosis ("plugging" of the arteries with fatty tissue), which in turn results in an increased likelihood of CHD.

Studies in which dietary fat intake has been modified, either in kind or amount, did not show significantly reduced mortality rates. Changing diets from animal fats to vegetable fats has not improved the heart disease record. There are data that suggest that poor health can also result from eating diets high in polyunsaturated fats. Most consumers do not know the difference between saturated and polyunsaturated fats. Through many margarine and vegetable oil commercials, however, they have been informed that saturated is "bad" and unsaturated is "good." In a review of the diet and heart disease relationship, some medical doctors argue that the diet–heart hypothesis became popular because a combination of the urgent pressure of special interest groups or health agencies, oil-food companies, and ambitious scientists had transformed that fragile hypothesis into treatment dogma. In fact, recently published studies that reanalyzed the data set used by Keyes at the University of Minnesota that brought saturated fat intake under fire found that only incomplete data were used, and in fact, there were no differences in mortality that could be attributed to fat intake.

Senator George McGovern's committee on Nutrition and Human Needs in the late 1960s and early 1970s shifted its emphasis from developing policy aimed at eliminating malnutrition to dealing with the issue of consuming too many calories. Largely influenced by a self-proclaimed diet expert, McGovern's committee released a document titled "Dietary Goals for the United States" based on 2 days of testimony. Written by a journalist with absolutely no background in science, nutrition, or health, the report would become the basis for a national policy focused on dietary fat. Twenty-five years later, there is still no compelling and clear evidence as to the effect of dietary cholesterol and fat intake on human longevity.

The relationship between cholesterol intake and death from CHD is by no means absolute. All saturated fatty acids are not equal in terms of their effect on serum cholesterol. It is important to recognize that less than half of the fatty acids found in beef fat are saturated. The 18-carbon length fatty acid (stearic acid) has either a neutral effect on serum cholesterol or may actually lower serum cholesterol concentrations when substituted for other saturated fatty acids. Stearic acid comprises almost one-third of the saturated fatty acids in beef.

For example, a Select grade steak will have about 50% of its total fat in monounsaturated form of which nearly 90% is oleic acid (the beneficial fat found in olive oil). The remaining one-half of the total fat is saturated, but one-third of that is stearic acid, which is potentially beneficial, but at the very worst neutral in its effect. In total, more than one-half and as much as three-quarters of the fat in the steak will lower cholesterol levels (Taubes, 2001). Trans fatty acids are unsaturated fatty acids structurally characterized by a trans arrangement of alkyl chains. Trans fatty acids are formed during the hydrogenation of vegetable oils and have been linked to an increase in blood cholesterol. Foods that may contain trans fatty acids include margarine, vegetable cooking oils, many bakery goods and prepackaged mixes, french fries, and packaged popcorn. Food manufacturers voluntarily lowered the amounts of trans fats in their food products by more than 73% since 2005, mostly by reformulating products. The Food and Drug Administration reported that the average daily intake of trans fats by Americans fell from 4.6 grams in 2003 to 1 gram in 2012. Yet, in 2013, FDA proposed an outright ban of trans fatty acids in food processing and preparation. While some welcomed the proposed regulation, many view this as significant government overreach with limited impacts on public health in light of the aforementioned trends.

Cholesterol is a naturally occurring substance in the human body. Every cell manufactures cholesterol on a daily basis. The average human turns over (uses and replenishes) 2,000 mg of cholesterol daily. Average dietary consumption of cholesterol is approximately 600 mg daily. Therefore, the body makes 1,400 mg each day to meet its needs.

The body cannot use cholesterol unless it is joined with a water-soluble protein, creating complexes known as lipoproteins. There are several different types of lipoproteins. Research workers have identified two of these lipoproteins: HDL (high-density lipoprotein) and LDL (low-density lipoprotein). High blood levels of the LDLs have been generally associated with increased cardiovascular problems, while some research data show that higher blood levels of HDLs may reduce heart attacks by 20%. Additional research with laboratory animals has shown that those fed beef had HDL levels 33% higher than animals fed soybean diets.

There is evidence of genetic differences in the proportion of HDLs and LDLs in individuals. People who are overweight, nonexercisers, and cigarette smokers have higher proportions of LDLs than those who are lean, exercisers, and nonsmokers.

Although the exact roles of dietary cholesterol and blood levels in the development of CHD are not known, it would appear logical from the existing information to use prudence in implementing drastic changes in dietary habits. Certainly, those individuals with high health risks primarily due to genetic background should take the greatest precautions.

Some individuals argue that consumers eat too much red meat and that excessive red meat consumption causes cancer. Evidence to support this statement is questionable, and "the term excessive consumption" must be defined. Average daily per-capita beef consumption estimates range from 1.8 to 2.6 oz. The American Heart Association recommends 3.5 oz of cooked meat per person on a daily basis. Based on this recommendation, the average U.S. per-capita consumption of beef and other red meat is not excessive.

Based on a comprehensive review of the scientific literature investigating the role of meat consumption and cancer, Alexander et al. concluded that red meat consumption has not been shown to be a causative factor for cancer (Alexander, 2010; Alexander et al., 2010a, 2010b, 2010c). Their conclusion was based on the facts that red/processed meat intake and cancer incidence were weak in magnitude and not statistically significant, results of studies were highly inconsistent across studies and within different samples of the population, and results are confounded by lifestyle and other dietary choices.

Austin and McBean (1997) concluded that red meat consumption has not been clearly shown to be a risk factor for cancer. If an association is shown between red meat and cancer, one of three explanations is possible: (1) the association is due to chance or some bias in the study method; (2) the association is due to confounding (i.e., meat and cancer are associated only because they both are related to some common underlying condition such as low intake of cancer-protective fruits and vegetables); (3) the association is real. Whether or not it can be concluded that red meat is a risk factor for cancer depends on several criteria: consistency of the association, strength of the association (relative risk), specificity of the association, and congruence with existing knowledge (e.g., is there an explanation or biological mechanism).

The criterion of "consistency" requires that the association be repeated under different circumstances, such as among various population groups and among individuals within a population. In the case of red meat and cancer, the relationship is not consistently demonstrated within the population. When a positive association between red meat and a specific cancer is demonstrated, it generally is weak. Moreover, the association between consumption of red meat of different types (beef, pork, lamb, and processed meats) and specific cancers is inconsistent. Although it has been suggested that meat components, such as fat, protein, or iron, or chemicals, formed during the cooking of meat might be carcinogenic, these hypotheses remain unproven. On the contrary, meat contains some components such as conjugated linoleic acid, a fatty acid, which may protect against cancer.

Inferring a relationship from epidemiological studies of diet and chronic disease is particularly difficult due to several characteristics of both diet and chronic disease. First, there are several problems with accurately quantifying dietary intake. This was a major problem in most of the epidemiological studies reviewed by Alexander (2010), Alexander et al. (2010a, 2010b, 2010c), and Austin and McBean (1997). Second, diseases such as cancer are caused by a variety of genetic and environmental factors. Diet (including red meat intake) is only one of many lifestyle factors that may influence risk of developing a disease. Also, chronic diseases, such as cancer, tend to have a long latency period during which time changes in many factors may occur. Epidemiological investigations can identify risk factors, not a cause-and-effect relationship. Only when epidemiological findings are supported by information from other types of scientific studies, such as experimental animal studies and human clinical trials, can a decision regarding a causal relationship be made on firmer ground.

The U.S. government has released several reports that outline dietary guidelines and goals for American citizens. Some argue that it is the government's responsibility to provide people with information about diet; others support scientifically based guidelines, but believe Americans should have freedom of choice.

The *Dietary Guidelines* published by the USDA and Department of Health and Human Services (HHS) includes broad recommendations that are reasonable attempts to initiate sound nutritional practices among individuals. The revised 1990 dietary guidelines, which draw heavily on two diet and health reports by the National Academy of Sciences and the U.S. Surgeon General, recommended that not more than 30% of daily calories come from fat and that less than 10% come from saturated fat. In 1992, the USDA began use of the Food Guide Pyramid (in conjunction with the nutritional education programs). The pyramid replaced the wheel graphic used to display the four basic food groups that has been used in nutritional educational

programs since the 1950s. In 2011, USDA replaced the food pyramid with a concept known as "My Plate" to demonstrate the importance of a balanced diet (Table 2.13).

Obesity has been identified as a significant challenge to the overall health of human beings. For the past quarter of a century, per-capita red meat consumption has declined 16%, consumption of other fats and oils has grown 23%, sugar consumption grew by 3%, and soft drink consumption rose by 32%. These data suggest that the overconsumption of fats and oils from processed foods, sugary foods, and soft drinks must be addressed to address the obesity issue.

The beef industry has met consumer demands by providing beef trimmed to a fat level of 1/8 inch or less, a reduction of 35% compared to the 1960s. Furthermore, ground beef is now offered in a variety of lean-to-fat ratios and purchases of 90% and higher lean ground beef increased nearly 30% from 2005 to 2015. As a case in point, the industry offered six times more whole-muscle cuts that meet USDA standards for lean in 2013 as compared to 1989 (Figure 2.18). The fatty acid profile of beef is far more healthful than most consumers perceive (Figure 2.19). Adherence to suggested dietary guidelines can be easier by following these suggestions of the Dietary Guidelines Alliance:

Table 2.13
Age Group-Specific Dietary Recommendations

	Fruit[1]	Vegetables[1]	Grain[2]	Protein[2]	Dairy[1]	Oils[3]
Children 2–3 yrs.	1	1	3	2	2	3
Children 4–8 yrs.	1–1.5	1–1.5	5	4	2.5	4
Girls 9–13 yrs.	1.5	2	5	5	3	5
Girls 14–18 yrs.	1.5	2.5	6	5	3	5
Boys 9–13 yrs.	1.5	2.5	6	5	3	5
Boys 14–18 yrs.	2	3	8	6	3	6
Women 19–30 yrs.	2	2.5	6	5	3	6
Women 31–50 yrs.	1.5	2.5	6	5	3	5
Women 51+ yrs.	1.5	2	5	5	3	5
Men 19–30 yrs.	2	3	8	6	3	7
Men 31–50 yrs.	2	3	7	6	3	6
Men 51+ yrs.	2	2.5	6	5	3	6

[1]Number of cups.
[2]Ounce equivalent (half to be whole grain).
[3]Teaspoons.
Source: USDA.

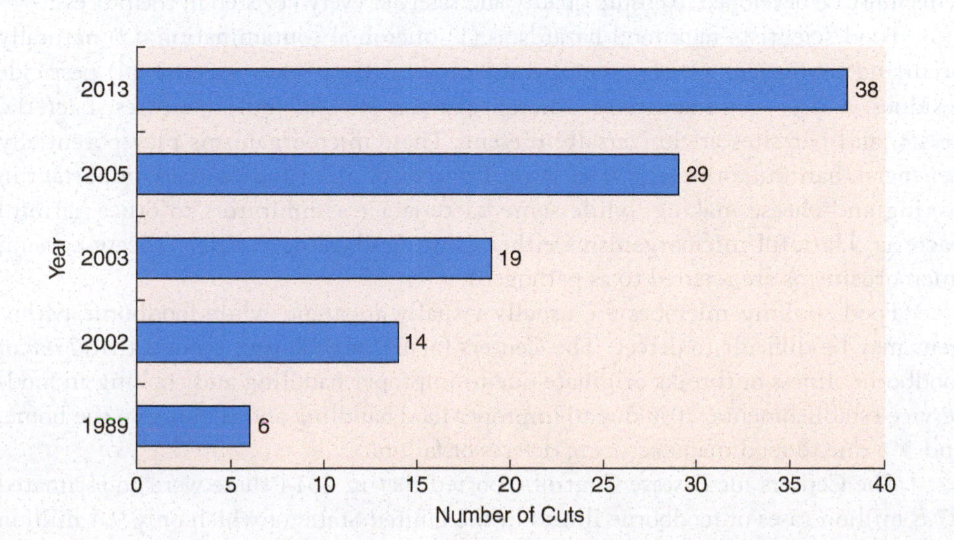

Figure 2.18
Number of beef cuts meeting USDA lean guidelines.
Source: USDA.

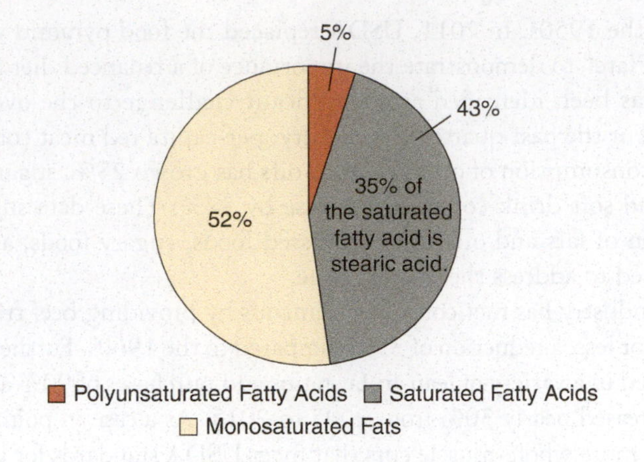

Figure 2.19
Fatty acid profile of lean beef.
Source: Adapted from USDA.

1. *Be realistic*—make small, incremental changes in eating and exercise habits.
2. *Be adventurous*—expand your tastes to enjoy a variety of foods.
3. *Be flexible*—balance food consumption and exercise regime over several days instead of focusing on one meal or one day.
4. *Be sensible*—enjoy all foods—just do not overdo it. Choose sensible portion sizes.
5. *Be active*—exercise is a key to maintaining appropriate weight.

Vegetarian and vegan diets have been advocated as healthier than diets containing animal products. However, diets devoid of meat are relatively low in protein, saturated fat, long-chain *n*-3-fatty acids, retinol, vitamin B_{12}, and zinc while vegan diets are also low in calcium. A study by Key et al. (2009) reported that there were no significant differences between meat-eaters and non-meat-eaters for incidence of hypertension, colorectal cancer, breast cancer, prostate cancer, and total mortality rate.

Beef's nutrient density, amino acid profile, flavor, and versatility provide consumers a healthy and highly palatable source of protein.

Safe Beef Products

Food safety assurance is a joint responsibility that requires the attention of consumers, producers, processors, retailers, and foodservice outlets. Historically, federal and state food inspection systems have been expected to ensure food safety. Supply chain interventions can be implemented at various stages that significantly reduce the risk of foodborne illness. These efforts are founded on total quality management systems based on the premise that quality cannot be inspected into a product. Instead, intentional, science-based systems must be developed to ensure quality and safety at every key step in the process.

Food scientists rank food hazards as (1) microbial contamination; (2) naturally occurring toxins; (3) environmental contaminants (e.g., heavy metals); (4) pesticide residues; and (5) food additives. Microorganisms such as molds, viruses, bacteria, yeasts, and parasites are universally present. These microorganisms play potentially beneficial, harmful, or neutral roles in regard to food. Mold and yeast are important in baking and cheese making, while some bacteria act as inhibitors to other harmful bacteria. Harmful microorganisms either cause spoilage or disease. Disease-causing microorganisms are referred to as pathogens.

Food spoiling microbes are usually visually apparent, while foodborne pathogens may be difficult to detect. The Centers for Disease Control reports that 77% of foodborne illness outbreaks originate due to improper handling and cooking in foodservice establishments, 20% due to improper food handling and cooking in the home, and 3% due to food manufacturing defects or failure.

The Centers for Disease Control reported that in 2014 there were an estimated 47.8 million cases of foodborne illness in the United States of which only 9.4 million were linked to one of 31 specific pathogens. Less than 1% of these cases resulted in hospitalization ($N = 127{,}800$), and mortalities resulting from foodborne illness equalled approximately 3,000 in 2014. The vast majority of these illnesses are often referred to as a "24-hour bug" but the incidence rate of one in six Americans requires attention by the full scope of the food supply chain.

Sanitation and temperature control are critical elements in controlling potentially harmful bacteria. Processors, retailers, and foodservice operators can minimize contamination via proper equipment sanitation, disciplined personnel hygiene practices, proper handling and storage, cooking to appropriate temperatures, and systematically assessing food safety efforts.

One of the most critical control points is management of food temperature. The danger zone—40 to 140°F—offers conditions under which bacterial pathogens can multiply. Therefore, foodservice establishments and consumers should strive to maintain raw and cooked foods outside the danger zone.

Additional steps that consumers should take include cooking and reheating food thoroughly, storing food at correct temperatures, avoiding cross-contamination between foods during storage and preparation, maintaining a high degree of personal sanitation, and keeping food preparation surfaces meticulously clean. These steps are most critical for establishments serving large numbers of consumers as most foodborne illness outbreaks are traced to foods served at dining establishments, catering facilities, or institutional foodservice (schools, hospitals, jails, etc.).

Examples of food safety issues for the livestock industry include infection by *Escherichia coli 0157:H7* and other foodborne pathogens (Table 2.14). Concerns about *E. coli* O157:H7 bacterial infection outbreaks peaked in 1993, when 1,000 cases were reported in the United States. Over three-quarters of the 1993 incidents were attributed to improperly prepared and cooked ground beef, most of which was traced to a single foodservice supplier. Ground beef is more susceptible to contamination because as beef is ground, the surface area is greatly increased and handling is extensive. In addition, any bacteria on the exterior of meat prior to grinding is distributed throughout the product as it is ground and blended; thus, ground beef is more likely to have pathogens in its interior than are steaks and roasts. By cooking whole-muscle cuts (e.g., steaks and roasts) to a safe internal temperature of at least 145°F, the bacteria that are present on the meat's surface will be destroyed by cooking. It is recommended to cook hamburgers to an internal temperature of 160°F to destroy bacteria embedded in the meat.

Data reported by the Centers for Disease Control comparing foodborne infections occurring in 2014 to the baseline years of 1996–1998 documented a 22% decline in foodborne illnesses with specific reductions due to *Campylobacter* (down 24%), *Listeria* (down 45%), *STEC 0157 (E. coli 0157:H7)* (down 47%), and *Salmonella* (down 5%). While the goal must be to eliminate foodborne illness outbreaks, these data show that agricultural and food industries can make significant progress in food safety.

A variety of new processing technologies designed to improve food safety have been implemented by the beef supply chain. Some of the innovations include steam vacuuming, hot water washing, weak acid rinses (sodium chlorite, lactic acid), and steam pasteurization. Most processors are working to implement a multiple set of hurdles or blockades to microbial growth. Such systems have tremendous potential to enhance food safety. Irradiation is another technology used by some food processors and food distributors to minimize the risk of foodborne illness. Relative to the beef industry, irradiation is targeted to ground beef.

The U.S. government established the Food Safety Modernization Act in 2011 as a means to "prevent food safety hazards, to detect and respond to food safety problems, and to improve the safety of imported foods." The act specifically increased the scope of FDA authority to increase the number of federal inspections, enhance pathogen detection

Table 2.14
SUMMARY OF THE MAJOR PATHOGENS ASSOCIATED WITH FOODBORNE ILLNESS

Name of Organism	Description	Infective Dose[1]	Symptoms	Annual Occurrence in United States	Associated Foods
Salmonella	Rod-shaped, motile bacterium; widespread in animals, especially poultry and swine; gram-negative	As few as 15–20 cells	Nausea, vomiting, abdominal cramps, fever	2–4 million cases	Major: raw poultry, meats, eggs, dairy foods. Minor: coconuts, sauces, salad dressings, cake mixes, peanut butter, chocolate
Campylobacter jejuni	Slender, curved motile rod; gram-negative, relatively fragile	400–500 bacteria	Diarrhea (watery or sticky), may contain blood and white cells; fever, abdominal pain, headache, muscle pain	Leading cause of bacterial diarrhea	Major: raw chicken. Minor: unpasteurized milk, nonchlorinated water, sea-food, hamburger, cheese, pork, eggs
Listeria monocytogenes	Bacterium, non-spore forming, very hardy; gram-positive	1,000 total organisms	Flu-like symptoms may be followed by sep-ticemia, meningitis, encephalitis, or cervical infections in pregnant women	350 with 48 fatalities	Major: soft cheese and ground meat. Minor: poultry, dairy products, hot dogs, seafood, vegetables
Escherichia coli O157:H7(STEC)	*E. coli* is a normal inhabitant of the gut of all animals. There are four classes of virulent *E. coli*. Majority of *E. coli* serve to suppress harmful bacteria and to synthesize vitamins. O157:H7 produces a potent toxin(s)	10 organisms	Severe cramping, diarrhea (may become bloody); typically self-limiting with 8-day duration; young children and the elderly may develop hemolytic uremic syndrome and renal failure	8,000 confirmed with 20 fatalities	Major: undercooked ground beef, human-to-human contact, environmental exposure to organisms. Minor: poultry, apple cider, raw milk, vegetables, hot dogs, mayonnaise, salad bar items

[1]Dose varies depending on immune state of the individual.
Source: USDA.

systems, and order food recalls in response to potential contamination. The new rules also required all domestic food companies to write and implement new safety protocols to control or eliminate potential foodborne hazards. While of good intent, the consequence of the rule has been to increase costs for small processors and companies to the point of reducing competition and further speeding the concentration of food processing and distribution. Furthermore, the rule exempted those who sell at farmer's markets making it the most unregulated segment of food retailing and potentially exposing consumers to increased risk, thus undermining the expressed intent of the regulation.

Production Technologies

Production technologies have been developed over time to help improve the efficiency of agricultural production and as a result people are less likely to experience hunger, fewer inputs are required to meet global food demand, and agriculturalists have been better able to manage natural resources. Technologies that are utilized in food production in the United States undergo a vigorous and expensive review process that requires years of testing and the investment of millions of dollars. To assure an abundance of caution, the federal government also implements a thorough inspection system so that food delivered to consumers is safe and wholesome. Beyond government requirements, individual companies have developed and utilized training, sampling and testing protocols, best practice protocols, and other critical control point methodologies to protect the integrity of their products. Some of the technologies most common to the beef industry are the utilization of hormone-based growth implants, antimicrobial compounds, and post-harvest technologies to improve efficiency.

Currently, many feeder calves and feedlot cattle receive implants containing growth stimulating compounds. These compounds are hormonal in nature and act to stimulate additional muscle development. Estrogenic implants are designed for slow release at low-dose levels. Implants do not create residues (Table 2.15).

An emerging concern in both human and livestock medicine is antibiotic resistance. While there is no direct data to support concerns that beef originating from cattle that have been treated with approved antibiotics under documented label requirements has led to an increase in illness in humans due to antibiotic-resistant bacteria, a plethora of new government regulations are being implemented that reduce availability of antibiotics to the livestock industry, increase veterinary oversight requirements, and implement significant penalties for failures to comply with the rules. The specifics of these regulations are discussed in Chapter 16.

Consistent with good management practices, cattle producers use FDA-approved antibiotics to prevent and treat diseases. In fact, many of the antibiotics used with cattle are not used in human medicine. The cattle industry's Beef Quality Assurance program involves veterinarians and cattle managers to ensure proper withdrawal times are followed when antibiotics are administered. In fact, the impact of industry-led efforts to

Table 2.15
ESTROGEN PRODUCTION IN HUMANS AND CONTENT IN SELECTED FOODS

Human/Food	Nanograms[1] of Estrogen
Daily human estrogen production	
Child (before puberty)	50,000
Adult male	168,000
Adult female	480,000
Pregnant female	20,000,000
Estrogen in one serving	
Steak from implanted steer	1.9
Steak from nonimplanted steer	1.2
Steak from nonimplanted intact bull	22.0
Steak from nonimplanted heifer	1.3
Coleslaw	2,724
Split pea soup	908
Chocolate ice cream	1,387
Soybean Oil	189,133

[1]One nanogram equals one-billionth of a gram.

Table 2.16
Violative Antibiotic Residue Rates in U.S. Beef Carcasses

Cattle Class	Number of Violative Samples and Total Samples	Violative Rate (%)
Scheduled sampling		
Cows	2 of 70,158	0.00002851
Steers	0 of 38,925	0.0
Heifers	0 of 39,246	0.0
Inspector-driven sampling		
Cows	81 of 20,092	0.00403146
Bulls	14 of 2,470	0.00566802
Steers	1 of 10,431	0.00009587
Heifers	21 of 3,536	0.00593891

Source: USDA-FSIS.

ensure that antibiotic residues do not undermine the integrity of beef has been tremendously positive. Residue testing of the beef supply is predominately structured around a two-prong protocol of sampling by USDA Food Safety Inspection Service—scheduled sampling based on sophisticated statistical models and inspector-driven sampling at packing plants. Table 2.16 shows the summary of 2013 results of government sampling.

Lean finely textured beef (LFTB) is a beef product developed by using high-technology food processing equipment to separate lean meat from fat beyond the capability of the human hand. LFTB products are a key sustainability effort designed to prevent the waste of valuable, lean, nutritious, safe beef. To make LFTB, beef trimmings (small portions of fat and lean resulting from fabricating beef carcasses into wholesale- and retail-sized portions) are warmed to about 100°F in equipment that looks like a large, high-speed mixing bowl that spins these trimmings to separate meat from the fat that has been liquefied. The resulting product is very low in fat content (95+% lean). This process is comparable to homogenization used to separate cream from milk.

During the process, a puff of ammonium hydroxide gas is used to slightly raise the pH of the product as an additional step to eliminate microbial contamination and reduce the risk of foodborne illness. The USDA and FDA have determined that this use of ammonium hydroxide is safe. Ammonium hydroxide has been used for decades in the production of puddings and baked goods.

Norman Borlaug, the father of the green revolution, once said that "we have the capability to defeat hunger, but only if farmers and ranchers are allowed to use available production technologies." Responsibility for thoughtful consideration and appropriate use of technologies is clearly the domain of the food supply chain. However, consumers and those who influence them have a shared responsibility to engage in an informed and rational evaluation of the facts,

BY-PRODUCTS

The hide is the best-known by-product and usually the highest valued. The hide provides three types of leather (latigo, suede, and tooling) used in sports equipment, luggage, boots, and shoes. Leather also provides felt, certain textiles, a base for many ointments and insulation, and is used as a binder for plaster and asphalt. The hair from the hide is used to produce insulation and rug pads. Fine hair from the ear is used to make artist's brushes—so-called camel hair brushes. Gelatin from hides is used in foods, film, and glues. Artificial skin for severely burned humans has been made from cowhide, shark cartilage, and plastic.

The primary edible by-products are called **variety meats** and include the liver, heart, kidney, brain, tripe (walls of the stomach), sweetbread (thymus), and tongue. These products have long been known for their high nutritive value and are considered gourmet items by some consumers. An average 1,100-lb slaughter steer produces approximately 34 lb of variety meats. Because the U.S. per-capita consumption (disappearance) of variety meats is only about 9 lb, surplus variety meats are exported to countries that have a preference for them. A more detailed discussion of the export markets for variety meats is presented in Chapter 10.

Other edible by-products come from fats (e.g., oleo stock and oil are used for making margarine and baker's shortening, while oleo stearin is used for making chewing gum and certain candies); bones, horns, and skins (e.g., gelatin for making marshmallows, yogurt, ice cream, mayonnaise, canned meats, and gelatin dessert); and intestines (e.g., natural sausage casings and surgical sutures). Other inedible by-products, besides the hide and hair, come from inedible fats and fatty acids (e.g., antifreeze, binding agent for asphalt in roads, stearic acid to produce tires, candles, cellophane, ceramics, cosmetics, crayons, deodorants, detergent, insecticides, insulation, linoleum, freon, perfumes, paints, plastics, shoe cream, shaving cream, soaps, textiles, pet foods, and floor wax); from bones, horns, and hooves (e.g., animal feeds, buttons, bone china, combs, piano keys, and bone charcoal, which is used in production of high-grade steel ball bearings); from collagen-based adhesives (e.g., glues, adhesives, bandages, wallpaper, sheet rock, and emery boards); and from nonedible gelatin for photographic film.

Table 2.17 shows the source of the primary pharmaceuticals from beef cattle and their value to humans. To obtain 1 lb of dry insulin, processors must obtain pancreas

Table 2.17
THE PHARMACEUTICALS FROM BEEF CATTLE—THEIR SOURCE AND UTILIZATION

Pharmaceutical	Source	Uses
Epinephrine	Adrenal gland	Relief of hay fever, asthma, and other allergies; heart stimulation
Thrombin	Blood	Assists in blood coagulation; treatment of wounds; skin grafting
Fibrinolysin	Blood	Dead tissue removal; wound-cleansing agent; healing of skin from ulcers or burns
Desoxycholic acid	Bile	Used in synthesis of cortisone for asthma and arthritis
Liver extract	Liver	Treatment of anemia
Ox bile extract	Liver	Treatment of indigestion, constipation, and bile tract disorders
Heparin	Lungs	Anti-coagulant
Insulin	Pancreas	Treatment of diabetes
Chymotrypsin	Pancreas	Removes dead tissue; treatment of localized inflammation and swelling
Glucagon	Pancreas	Counteracts insulin shock; treatment of some psychiatric disorders
Trypsin	Pancreas	Cleansing of wounds
Rennet	Stomach	Assists infants in digesting milk; cheesemaking
Ovarian hormone	Ovary	To treat painful menstruation and prevent abortion
Parathyroid hormone	Parathyroid gland	Treatment of human parathyroid deficiency
Corticotrophin (ACTH)	Pituitary gland	Diagnostic assessment of adrenal gland function; treatment of psoriasis, allergies, mononucleosis, and leukemia
Hyaluronidase	Testicle	Enzyme that aids drug penetration into cells
Thyrotropin (TSH)	Pituitary gland	Stimulates functions of thyroid gland
Vasopressin	Pituitary gland	Control of renal function
Cholesterol	Nervous tissue	Male sex hormone synthesis
Thyroid extract	Thyroid gland	Treatment of cretinism
Amfetin (trade name)	Amniotic fluid	Reduces postoperative pain and nausea and enhances intestinal peristalsis

glands from approximately 60,000 cattle. One cow's pancreas can provide a diabetic patient with a 2-day supply of insulin. Synthetic insulin and other synthetic pharmaceuticals may reduce the demand for certain cattle by-products.

Iron (from the blood), vitamin B_{12} and liver extract (from the liver), and calcium and phosphorus (from bone meal) are nutrients that are used in human and livestock nutrition.

This summary of beef by-products is not complete. Research has identified, and no doubt will continue to identify, useful by-products from beef cattle.

SELECTED REFERENCES

Publications

Aberle, E.D., Reeves, E.S., Judge, M.D., Hunsley, R.E., & Perry, T.W. 1981. Palatability and muscle characteristics of cattle with controlled weight gain: Time on a high energy diet. *Journal of Animal Science.* 52:757.

Alexander, DD. 2010. *Red meat and processed meat consumption and cancer: A technical survey of the epidemiologic evidence.* Health Sciences Practice: Exponent Inc. Alexandria, VA.

Alexander, D.D., Morimoto, L.M., Mink, P.J., & Cushing, C.A. 2010a. A review and meta-analysis of red and processed meat consumption and breast cancer. *Nutrition Reserve Reviews.* 23(2):349–365.

Alexander, D.D., Mink, P.J., Cushing, C.A., & Sceurman, B.A. 2010b. A review and meta-analysis of red and processed meat consumption and prostate cancer. *Nutrition Journal.* 9(50):1475–1491.

Alexander, D.D., Miller, A.J., Cushing, C.A., & Love, K.A. 2010c. Processed meat and colorectal cancer: a quantitative review of epidemiological studies. *European Journal of Cancer Prevention.* 19:328–341.

Austin, H., & McBean, L.D. 1997. *Red meat and cancer: A review of current epidemiological findings.* Chicago, IL: National Cattlemen's Beef Association.

Belk, K.E., Sofos, J.N., Scanga, J.A., & Smith, G.C. 2001. *U.S. red meat: A pledge to minimize risk to public health.* Proceedings paper for United States Meat Export Federation.

Brooks, J.C., Griffin, D.B., Hale, D.S., Henning, W.R., Morgan, J.B., Parrish, F.C., & Savell, J.W. 1999. *The 1999 National Beef Tenderness Survey.* Centennial, CO: National Cattlemen's Beef Association.

Brown, P., Will, R.G., Bradley, R., Asher, D.M., & Detwiler, L. 2001. *Bovine spongiform encephalopathy and variant Creutzfeldt-Jakob disease: Background, evolution, and current concerns.* Centers for Disease Control (www.cdc.gov).

Cannell, R.C., Tatum, J.D., Belk, K.E., Wise, J.W., Clayton, R.P., & Smith, G.C. Dual-component video image analysis system as a predictor of beef carcass red meat yield percentage and for augmenting application of USDA yield grades. *Journal of Animal Science.* 77:2942.

Cohen, J.T., Duggar, K., Gray, G.M., & Kreindel, S. 2001. Evaluation of the Potential for Bovine Spongiform Encephalopathy in the United States. Harvard Center for Risk Analysis and Harvard School of Public Health.

Contributions of Animal Products to Healthful Diets. 1997. Council for Agricultural Science and Technology. Ames, IA.

Dietary Guidelines for Americans. 2014. USDA and U.S. Department of Health and Human Services. Washington, DC.

Drewnowski, A. 1995. Impact of taste preferences on dietary choices and food consumption patterns. *Food and Nutrition News.* 67(15).

Field, T.G., Garcia, J., & Ahola, J. 1996. *Quantification of the utilization of edible and inedible beef by-products.* Final Report to NCBA. Colorado State University. Fort Collins, CO.

Food Consumption, Prices, and Expenditures, 1970–2014. 2015. Washington, DC: USDA.

Food Marketing Institute. 2015. *Trends in the United States: Consumer attitudes and the supermarket.* Washington, DC.

Gutherie, J.F., & Roper, N. 1992. Animal products: Their contribution to a balanced diet. *Food Review.* 15(29).

Hedrick, H.B., Aberle, E.D., Forrest, J.C., & Judge, M.D. 1994. *Principles of Meat Science.* San Francisco: W.H. Freeman.

Higgs, J.D. 2000. The changing nature of red meat: 20 years of improving nutritional quality. *Trends in Food Science & Technology.* 11:85–95.

Hiza, H.A.B., & Bente, L. 2007. Nutrient content of the U.S. food supply, 1909–2004: A summary report. (Home Economics research report No. 57). U.S. Department of Agriculture, Center for Nutrition Policy and Promotion.

Key, T.J., Appleby, P.N., Spencer, E.A., Travis, R.C., Roddam, A.W., Allen, N.E. 2009. Mortality in British vegetarians: Results from the European prospective investigation into cancer and nutrition. *American Journal of Clinical Nutrition.* 89(5):1613S–1619S.

McNeill, S.H., Harris, K.B., Field, T.G., & Van Elswyk, M.E. 2012. The evolution of lean beef: Identifying lean beef in today's U.S. marketplace. *Meat Science.* 90:1–8.

Morgan, J.B., et al. 1990. *National Beef Tenderness Survey: Beef Cattle Research in Texas.* College Station, TX: Texas Agric. Expt. Sta. PR 4819–4865.

National Meat Case Study. 2010. Executive Summary. Cattlemen's Beef Board. Centennial, CO.

National Research Council, Committee on Diet and Health, Food and Nutrition Board 1989. *Diet and Health: Implications for Reducing Chronic Disease Risk.* Washington, DC: National Academy Press.

NCBA. 2015. Beef Industry Long Range Plan. Centennial, CO.

NCBA. 2002. Customer Satisfaction II: The Value of Tender Beef. Centennial, CO.

NCBA. 2003. Nutrient Facts: Trans Fatty Acids. Centennial, CO.

New Variant CJD. 2003. BSEInfo.org.

Niyo, K.A. (ed.) 1997. Contribution of Animal Products to Healthful Diets. CAST Report 131, Ames, IA: Council for Agricultural Science and Technology.

Restaurant Industry Forecast. 2015. National Restaurant Association. Washington, DC.

Retail Profitability. 2003. National Cattlemen's Beef Association R & D Ranch. Centennial, CO.

Roeber, D.L., Cannell, R.C., Belk, K.E., Tatum, J.D., & Smith, G.C. 2000. Effects of a unique application of electrical stimulation on tenderness, color, and quality attributes of the beef longissimus muscle. *Journal of Animal Science.* 78:1504.

Romans, J.R., Jones, K.W., Costello, W.J., Carlson, C.W., & Ziegler, P.T. 2000. *The Meat We Eat.* Danville, IL: Interstate Printers and Publishers.

Siegel, D.G. 2000. *Beef packaging facts.* National Cattlemen's Beef Association Fact Sheet. Centennial, CO.

Smith, G.C. 1995. *Meat Science: The Palatability Piece.* Certified Angus Beef Program, South Sioux City, IA.

Smith, G.C., et al. 1987. Relationships of USDA quality grades to cooked beef palatability. *Journal of Food Quality.* 10:269.

Tatum, J.D., George, M.H., Belk, K.E., & Smith, G.C. 1997. *An overview of a TQM approach for improving beef tenderness.* Fort Collins, CO: Colorado State University.

Taubes, G. 2001. The soft science of dietary fat. 2001. *Science.* 90:2536.

The National Beef Quality Audit. 1991. Englewood, CO: National Cattlemen's Beef Association.

The National Beef Quality Audit. 1995. Englewood, CO: National Cattlemen's Beef Association.

The National Beef Quality Audit. 2001. Centennial, CO: National Cattlemen's Beef Association.

The National Beef Quality Audit. 2005. Centennial, CO: National Cattlemen's Beef Association.

The National Beef Quality Audit. 2011. Centennial, CO: National Cattlemen's Beef Association.

Tonsor, G.T., Minter, J.R., & Schroeder, T.C. 2010. U.S. meat demand: Household dynamics and media information impacts. *Journal of Agricultural and Resource Economics.* 35:1, pp. 1–17.

Wiebe, S.L., Bruce, V.M., & McDonald, B.E. 1984. A comparison of the effect of diets containing beef protein and plant proteins on blood lipids of healthy young men. *American Journal of Clinical Nutrition.* 40:982–989.

Winkler, D., & Harris, K.B. 2009. Antimicrobial interventions for beef. Beef Industry Food Safety Council. Centennial, CO.

Young, M.K., & Korpolinski, R. (eds.). 1998. *Food and Nutrition News.* 70:1, pp. 2–3.

Zanovec, M., O'Neil, C.E., Keast, D.R., Fulgoni, V.L., & Nicklas, T.A. 2010. Lean beef contributes significant amounts of key nutrients to the diets of US adults: National Health and Nutrition Examination Survey 1999–2004. *Nutrition Research.* 30: 375–381.

3

Management Systems: Integrated and Holistic Resource Management

The beef industry is one of the most complex business models at work in the global economy. Whether viewed from a supply chain perspective or under the lens of an enterprise analyses, the beef production process is an interwoven and highly interconnected system. Effective decision making requires that leaders and managers have the capacity to understand not only the parts but also the whole. Management systems, also referred to as integrated resource management (IRM) and holistic resource management (HRM), involve (1) understanding the available resources and the principles associated with managing each resource, and (2) knowing how the resources are interrelated so that integrated and holistic management decisions can be effectively implemented.

RESOURCES AND PRINCIPLES

Each enterprise of the beef production and supply chain can be divided into individual components or resource areas. Resources take many forms—human capability, finances, land and livestock, technologies, etc. Successful managers are able to develop a systematic approach to developing and leveraging these resources to move an organization toward attainment of a predetermined set of goals and objectives.

Figure 3.1 provides an overview of a manager's challenge to continuously focus on development of the capability of key resources including self, time, information, risk, and enterprise-specific categories in pursuit of a specific goal. Some resources are perpetual (e.g., sun, wind, and water); others are nonrenewable (e.g., fossil fuels and phosphates); and still others are renewable (e.g., air, water, fertile soil, plants, and animals). A manager must be knowledgeable about each type of resource if the goals are to be accomplished. Furthermore, maximum net profit may not always be the ultimate goal because the sacrifice of quality of life may be too great (e.g., working 16–20 hours a day versus 8–10 hours). However, a goal that approaches maximum net profit is often realistic for many beef operations.

Successful beef producers and beef industry organizations identify and apply important principles that assure their individual and collective success. Principles are truths or natural laws that can be applied universally. A truth is a fact, law, or verified hypothesis. Truth is knowledge of things as they are, as they were, and as they are to come. Thus, principles transcend time and humans are continually motivated to identify new principles and effectively utilize time-proven principles that will improve their well-being.

Principles are the same across cultures. However, the application of these truths, through production practices, can vary from one place to another. For example, reproduction principles involve hormones and target organs that function to create

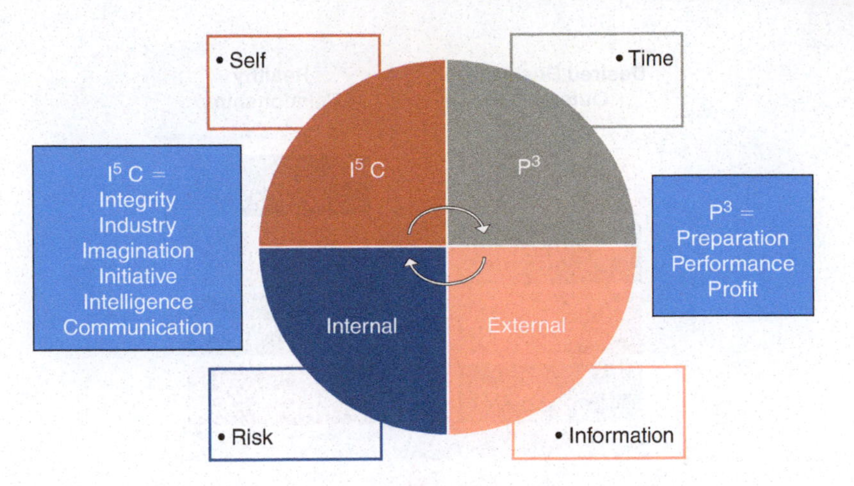

Figure 3.1
Development of professional managers.

estrous cycles and sex cells that result in a new calf crop. For the reproduction principles to be manifested, nutrition principles must also be utilized. Energy, protein, and other nutrients in essential amounts must be available at specific time periods or some of the reproduction principles will not be expressed. While reproduction and nutrition principles apply universally, how the nutrition (forage or feed) is provided—what plant species and amounts—can vary widely from enterprise to enterprise. Therefore, while reproduction and nutrition principles remain the same over wide geographical areas, the application of these principles through production practices can vary, even in enterprises only a few miles apart.

THE HUMAN RESOURCE

Without question, the human resource is of greatest importance. The foundation of the industry is built on human creativity and will power (Figure 3.2). Yet, in too many cases, managers do not have an effective approach to attracting, developing, and retaining talent.

Self-management, discussed later in the chapter, is an integral part of human resource management. It is human beings who develop mission statements, set goals, understand resource relationships, make decisions, and provide the labor that implements the written management plan. Successful managers understand that they

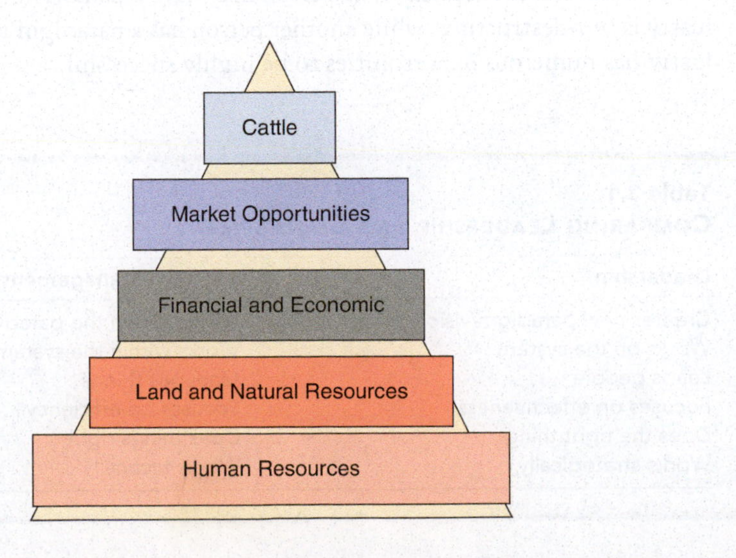

Figure 3.2
Management hierarchy is founded on human potential and creativity.

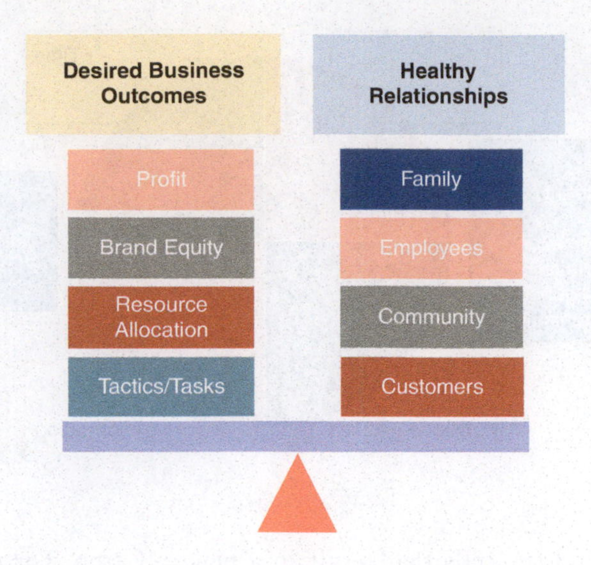

Figure 3.3
Long-term business success depends on the effective integration of business outcomes and relationships.

lead (not manage) people and manage the other resources (Figure 3.3). Developing an organization that effectively integrates business outcomes with relationships is critical to assuring sustained success.

Leadership and Management

Leadership and management are different and necessary elements that impact enterprise success in both the short and long term. Management is the process of taking an organization along an established route as smoothly and efficiently as is possible. Leadership, on the other hand, is the process of moving an organization into uncharted waters by effectively understanding and implementing change. Table 3.1 shows the differences between leadership and management. Figure 3.4 demonstrates the importance of combining leadership and management so that high levels of performance can be obtained.

It is interesting that in Table 3.1, the comparison of "doing the right things" before "doing things right" reinforces the long-time proven 80–20% rule of business (Figure 3.5). Another way of approaching this issue is to consider that a leader's primary task is to make doing the right thing easy and the wrong thing hard.

Leadership and management identify the need to create and implement paradigms. A paradigm is the map of our mind's perceptions, how a person sees the world or a particular situation. One person may have a paradigm that the current beef industry is self-destructing, while another person has a paradigm that the current beef industry has numerous opportunities to be highly successful.

Table 3.1
Comparing Leadership and Management

Leadership	Management
Creates new paradigm/vision	Works within the paradigm/vision
Works on the system	Works within the system
Leads people	Manages things
Focuses on effectiveness	Focuses on efficiency
Does the right thing	Does things right
Works strategically	Works tactically

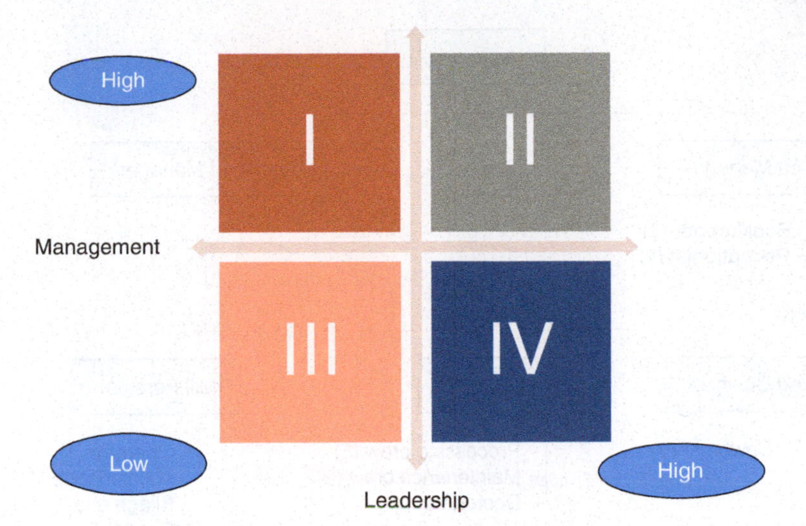

Figure 3.4
Intersection of leadership and management.

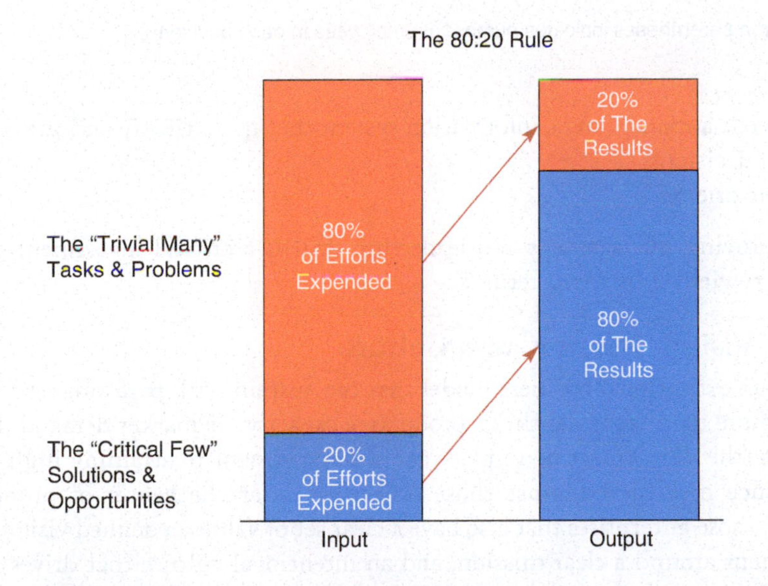

Figure 3.5
The 80–20% rule exists in each beef operation.

The manager of a beef cattle operation may be an owner-operator with only a minimum of additional labor or someone in charge of a complex organizational structure involving several employees (see Figure 3.6). An effective manager, whether involved in a one-person operation or a large complex one, needs to:

- Develop a written mission statement; identify and implement short-, intermediate-, and long-range goals as part of a written management plan.
- Set priorities and allocate resources accordingly.
- Execute the plan systematically.
- Keep abreast of current knowledge related to the enterprise and the beef industry.
- Use time effectively.
- Be self-motivated and incentivize team members appropriately; practice self-management to assure peak performance and to model effective behavior for the team; focus attention on the physical, emotional, and financial needs of oneself and those involved in the enterprise (employees and family).
- Communicate effectively to all employees and encourage a team approach to success. Development of an effective organizational culture involves accountability to business outcomes in a supportive and values-driven environment.
- Demonstrate integrity and conduct honest business dealings.
- Effectively implement risk management strategies.

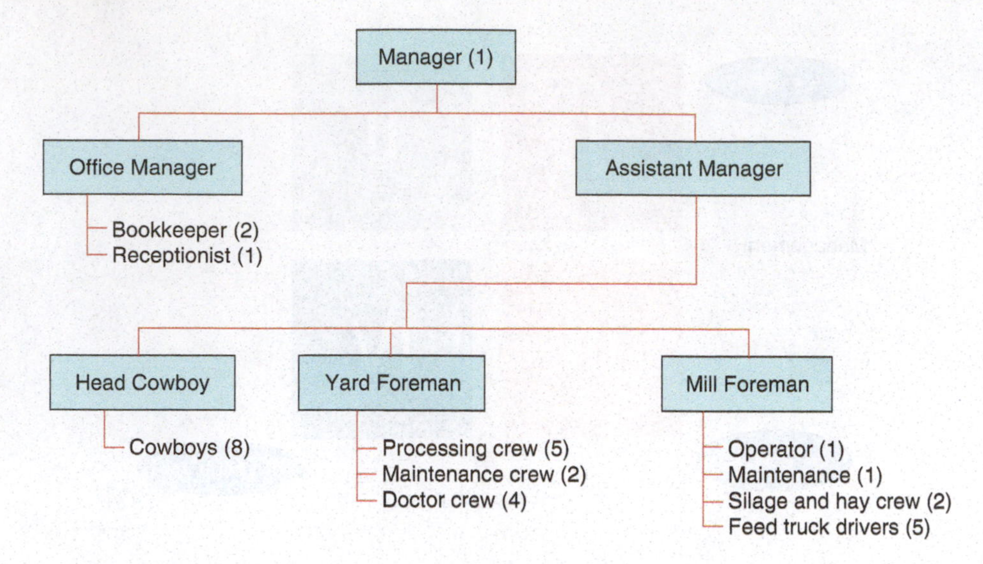

Figure 3.6
Organizational structure of a large commercial cattle feeding operation.

- Implement a management information system that gives timely and accurate feedback on decisions.
- Be profit-oriented.

Improving management and leadership skill is a critical investment with the potential to yield substantial returns.

Values, Vision, Mission, and Culture

Creating an exceptional business model that can sustain peak performance over time requires more than technical skills, productive capacity, or market demand. Business objectives (the what) must be in place and a commitment to attaining high levels of performance measured against those objectives is at the center of profitability. However, those enterprises that also have a clear set of values, a defined vision, unified commitment around a clear mission, and an intentional culture that drives how the business functions are the clear performance leaders.

Values are the non-negotiable standards upon which an organization is built. The values establish "true north" for the people in an organization. For example, the Engler Agribusiness Entrepreneurship Program at the University of Nebraska has six value pillars—aspire, courage, passion, grit, partner, and build. Identifying the core values provides expectations and standards upon which behavior and decision making can be framed at every level of the organization. Values provide the guiding principles for how team members interact with each other as well as with external constituents.

A vision statement establishes and defines the desired future state of being for an enterprise by clearly describing what success it wants to attain over time—5 years, 10 years, or even further into the future. The vision should be compelling, succinct, and memorable. For example, the vision of the beef industry long-range plan is to "responsibly produce the most trusted and preferred protein in the world."

A mission statement defines the current status or state of being for an organization and answers three critical questions—*what* does the organization do, *whom* does it serve, and *how* does it do what it does. A mission statement captures the organization's or team's purpose and identifies the path in making the vision a reality. A mission statement is often supported by a glossary that defines key words and phrases to avoid different interpretations.

An example of a mission statement applicable to a beef operation or the total beef industry follows.

> "Produce low-cost/high-profit cattle that consistently yield competitively priced, highly palatable and consistently uniform retail products."

Additional examples of mission statements from ranches and industry organizations are given below.

> "We are continually striving to improve the efficiency of converting God's forage into healthy, nutritious and great tasting beef to better feed His people."
>
> R. A. Brown Ranch, Throckmorton, TX

> "Our Mission is to provide our members and their customers with innovative programs and services, to continue advancing the quality, reliability and value of Red Angus and Red Angus-influenced seedstock used in the commercial beef industry."
>
> Red Angus Association of America, Denton, TX

Goals

Goals or strategies are needed to more specifically define the action to achieve the mission statement. They assure success based on the following definition: Success is the progressive realization of worthwhile goals. Goals should be few, specific, measurable, and with a reasonable time frame to accomplish them. A vague goal would be "to improve the quality of the cattle." *Quality* has several definitions and means different things to different producers. A more specific, measurable goal for a cow-calf producer might be: The breakeven price on weaned calves will be lowered from $1.75 to $1.50 per pound in three years.

Often times, goals may be arranged beneath a set of strategic objectives. For example, four strategic objectives were established by the beef industry long-range planning team in 2015:

1. Drive growth in beef exports.
2. Protect and enhance the business and political climate for beef producers.
3. Grow consumer trust in beef and beef production.
4. Promote and strengthen beef's value proposition.

Specific goals under each of these strategic objectives then drive resource decisions, allocation of time and talent, and provide a means to prioritize the order in which goals are to be accomplished.

Human

Human resources may include the manager, hired labor (seasonal or full-time), and labor supplied by family members. Successful managers know how to accomplish the following:

1. Assess the optimum labor needs for a business.
2. Identify prospective employees who can effectively contribute to the enterprise.
3. Motivate and adequately reward employees, not only monetarily but also through increased responsibility and opportunity for professional and personal growth.
4. Communicate goals, as well as the objectives and plans to accomplish the goals, so that each employee or family member understands his or her role in achieving the desired level of productivity and profitability.
5. Create a unified team approach by implementing a group-designed mission statement that effectively directs activities of the business.

Effective communication skills (including listening, writing, and speaking) are required of managers in order to utilize the human resource to the fullest extent. The importance of effective communication cannot be overemphasized in writing and communicating a plan and in understanding and motivating people in the desired direction. Listening to understand employees and other people is an important communication skill that is often overlooked by managers but is well understood and practiced by leaders.

Family Relationships

Family relationships in a family-owned and -operated beef operation can be enhanced or destroyed by effective or ineffective communication. The latter is tragic and usually is not perceived as a potential outcome when family members start working together. In addition to the five points mentioned earlier regarding successful management of the labor resource, the following items are pertinent to successful family operations.

1. Apply sound business principles rather than assuming things will work out simply because people are within the family. Involve all family members in financial decisions. At the same time, incorporate family values and goals into the strategic plan.
2. Evaluate other successful family operations. Determine why they are successful and how they resolve difficulties. Include all family members in the written plan of responsibilities (e.g., who will make the decisions, how each family member will be paid, and how vacation and other time away from the business will be handled). Develop a systematic approach to deal with potential conflict. Also, assure that compensation to family members aligns with contributions to the business.
3. Hold weekly family councils for additional planning, evaluating, and problem solving. Create an environment that encourages open communication.
4. Recognize that family relationships have a higher priority than profitability but that both can be compatible.
5. Have patience and tolerance with age differences in the family. Provide roles for family members and involve everyone (e.g., spouses and in-laws) in development of plans and goals.
6. Recognize that management changes can occur too fast or too slow in how they affect family relationships and profitability of the operation.
7. Assure that family members have the skills and abilities required to fulfill their job roles. Maintain the same performance standards for both family and nonfamily employees.

Family businesses will only be successful in the long term when the appropriate balance of planning, productivity, communication, trust, and respect are obtained.

Zimmerman and Fetsch (1994) offer the following steps to building a model that allows for consensus decision making in an environment that is both open and supportive.

1. Establish family rules and a shared vision.
2. Improve family communication and hold regular family meetings to enhance communication, delegation, and business effectiveness.
3. Create departments and appoint managers to spread responsibility among family members and to allow people a chance to develop their own levels of expertise. This process promotes shared responsibility, accountability, and training of people.

4. Develop job lists to allow the prioritization of tasks and the allocation of resources to assure their timely completion.
5. Establish monthly calendars to open communication between the departments of the business while helping family members share both their family and business needs. This step is critical in minimizing the surprises that originate from poor communication.
6. Resolve equality issues if they exist. Common equality issues include deskwork versus physical labor, generational/gender pay equity, and on- versus off-farm employment.

Employees

Perhaps the greatest challenge confronting the industry is the ability to attract and retain people to manage and work in the cattle and beef business. Without the ability to attract creative, motivated, and focused people no industry can remain sustainable over the long term.

Successful leadership and management of human resources revolve around communication of expectations, ongoing training and education, goal setting, and effective performance reviews. Just as is the case when dealing with family members, respect and trust are of critical importance.

The keys to excellent people management are:

1. Communicate expectations clearly.
2. Provide frequent feedback to employees with a goal of improvement. Seek input from employees.
3. Say thanks for a job well done and share credit.
4. Match a person's abilities to the job assigned.
5. Seek ways to help people grow and learn.
6. Avoid micromanagement.

Planning Process

The manager of a beef cattle enterprise is responsible for planning and decision making. In general, the management process involves (1) developing a written plan, (2) taking action, (3) evaluating results, and then (4) repeating the process. More detailed steps of the process are shown in Figure 3.7. The management process should not result in unproductive cycles but in meaningful progress toward written goals within required time limits (Figure 3.8).

Managers should develop an effective management plan in writing. A written plan makes it easier for everyone involved to identify the mission statement and goals and implement a plan of action of the operation. It also makes periodic evaluations more effective.

A written plan describes who will do what and when. The plan provides direction and motivation for the decision makers. Written goals and objective evaluations allow producers to measure improvements and weaknesses in the operation and prioritize needed management changes. An unwritten goal is not likely to be realized.

A written plan for a cattle operation can be determined in part by answering the following questions.

1. What do I want my cattle to do for my customers and me?
2. What are my cattle presently doing for me and my customers (e.g., what are the current levels of productivity and profitability)?
3. Are my cattle best matched to their environment? Is the biological type of cattle, emphasizing reproductive efficiency, well integrated with the most economical combination of forage/feed, markets, labor, and other resources?

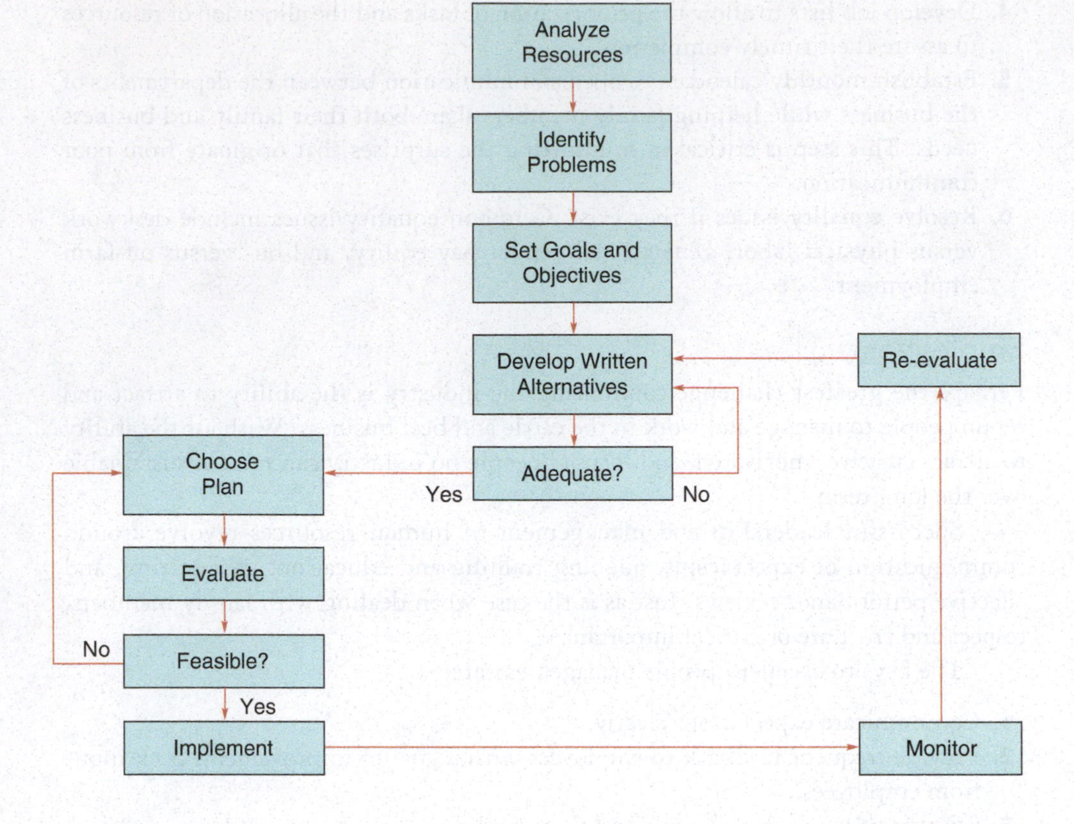

Figure 3.7
Major component parts of the planning process.

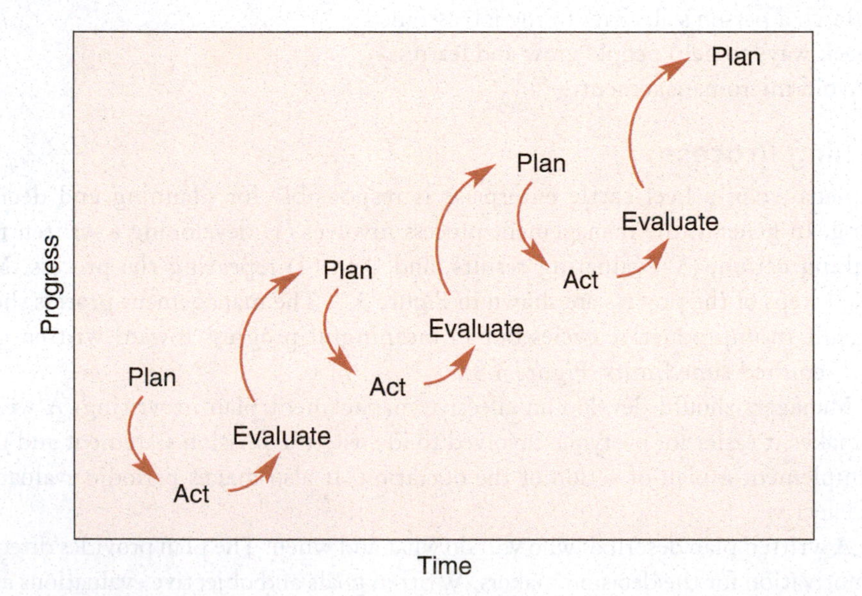

Figure 3.8
Progress is intended to be accomplished in the management decisions process.

A producer's written plan should take into account the biological and economic constraints imposed by resource variability of individual operations, yet be consistent with the two major beef industry goals: (1) achieving low-cost production, and (2) increasing consumer market share.

Family-owned businesses should extend the planning process into two additional areas—succession and estate transfer. A succession plan prepares the business for the transfer of power by identifying and training the individual designated as the "heir apparent." The estate plan anticipates the passing of the present owners and details the process of transferring assets from one generation to the next.

Time Management

Effective managers understand that end results are most important. To achieve the desired end results, priority must be given to time utilization. Every manager has the same resource—a fixed amount of time—but how effectively and efficiently that resource is used is the difference between a good manager and a poor manager. A successful manager knows that time must be allocated to working "in the business" as well as working "on the business."

All beef cattle operations have the "trivial many" and the "critical few" situations or problems. Managers should know the difference between these two categories and how each may affect the end results. An understanding of the 80–20% rule is important to the success of a manager (see Figure 3.5). Priorities can be established if managers are asking "Am I doing the right things?" rather than "Am I doing things right?" Once the right things are identified, then it is important to do them right. This gives direction to effective time management.

Establishing time priorities will help identify the 20% of the activities that will produce the 80% of the desired results. Figure 3.9 shows a method for managing time effectively. The demands on a producer's time can be separated into one of the four quadrants: A (Important and Urgent), B (Important and Not Urgent), C (Not Important and Urgent), and D (Not Important and Not Urgent). Obviously, the 20% time activities that count the most will be located in quadrants A and B. As management improves, increasingly more time will be spent in quadrant B.

To get into quadrant B, management should ask three questions: (1) Is it important? (2) Can I significantly influence it? and (3) Is it measurable? If the activity

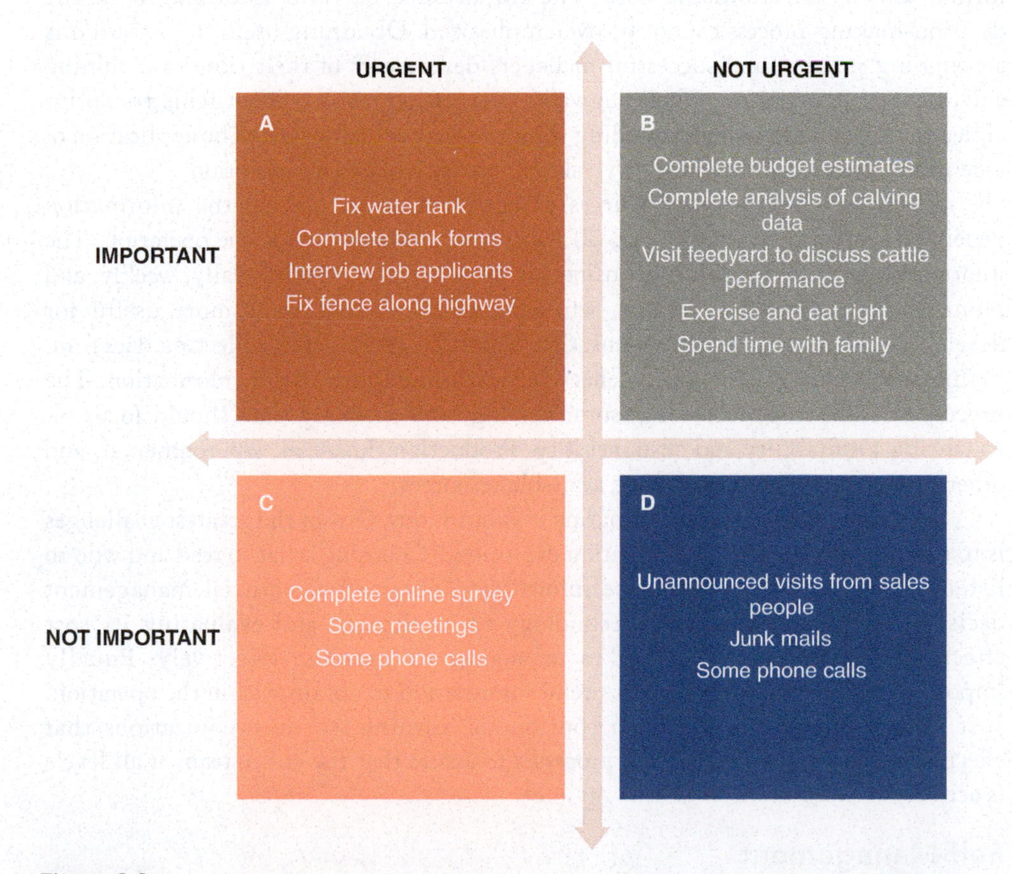

Figure 3.9
Time management priorities A, B, C, and D identify the four quadrants where activities can be categorized. Some example activities are shown.
Source: Adapted from Covey.

passes all three tests, then management should direct attention to it. Finally, management should deal with no more than three important items at any one time.

There is often not enough time to achieve all the activities that a manager wants to accomplish. So how can a manager make sure that the top 20% of priorities are accomplished? Available time comes from quadrant D by saying no to these activities. An example would be saying no to the time expected by a salesperson who calls or stops by. That salesperson will be available if the manager identifies the need to evaluate the product as part of quadrant A or B.

Most managers spend their time in quadrants A and C with urgent activities. Productive time can be accessed by spending less time in quadrant C. Most of crisis management comes in quadrant A. Crisis management is critical in most cases, but often can be prevented by spending more time in quadrant B. For example, water and forage development in quadrant B can alleviate much of the crisis drought management that will appear in quadrant A.

Information Management

Valid information is an essential part of making intelligent management decisions, for any decision is only as good as the information on which it is based. Lifelong learning is essential for success in the beef industry but with the volume of available information, people must also be able to discern fact from folly. Thus, a process to determine what information is needed as well as a means to filter it should be established. The manager begins by asking probing questions about the operation followed by consideration about where the answers can be found in the most useful format and at an affordable cost. The importance of valid information in the decision-making process cannot be overemphasized. Obtaining useful information is an ongoing process, and successful managers devote part of their time to obtaining and assessing information. Obtaining valid information involves identifying true principles or natural laws or understanding things as they actually exist. The application of true principles brings stability, survival, and profitability to an operation.

There are two primary sources of useful information: (1) the information generated within the business, and (2) that obtained from outside the operation. The information obtained within the business is most useful in making daily, weekly, and monthly management decisions, while outside information is more useful for developing future management plans. Care should be exercised in collecting data from within the operation that can be effectively translated into useful information. The process of collecting, recording, summarizing, and utilizing data should focus on enhancing profitability and sustainability. Production, financial, environmental, and other records are needed to manage available resources.

The amount of outside information is voluminous. One of the greatest challenges is to manage the so-called "information explosion." Choosing what to read and who to listen to and how to evaluate the information obtained are crucial management decisions. Choosing what new technology to implement and evaluating its cost effectiveness are also important to managing information effectively. Equally important is determining the most useful information to obtain within the operation. Just as individuals prosper from continuous learning, so do organizations that effectively implement policies and processes to assure that the entire team at all levels is actively seeking knowledge.

Self-Management

Most individuals are born with leadership and management talents and abilities that may or may not be developed later on in life. It is useful to reflect back on 5-year blocks of one's life to identify the personal characteristics and management skills that have been realized and acquired. This process provides motivation to develop other management abilities in the future.

Many outstanding managers continue to improve their personal management skills by following the examples of role models, attending professional development seminars, listening to motivational podcasts, and reading material on personal and human development. Weaknesses can be overcome and building self-esteem and consistently practicing the desired skills can enhance management skills. Achieving self-improvement goals will, in turn, enhance the accomplishment of goals identified for the beef operation. However, the promise of improvement to be translated into reality requires commitment and focus.

Included in self-management is managing or preventing stress. The inability to manage long-term stress can damage a manager's health and be destructive to family and employee relationships. Eventually, continued severe stress can cause the emotional and financial collapse of the human and physical resources in an operation. A manager's emotional responses to continued stress will appear in one or more of the following ways: denial of problems, depression, withdrawal from family and friends, blaming others, or blaming oneself for all the problems.

Even the best managers will have to make difficult decisions that may ultimately fail. Although stress cannot be totally eliminated, it can be reduced to manageable levels if the following guidelines are practiced.

1. Recognize personal limitations, both physical and financial.
2. Develop risk management plans consistent with one's physical, financial, and emotional limitations.
3. Work as a team with one's family and/or employees. Share disappointments and successes.
4. Improve one's problem-solving skills and increase the accuracy of one's decisions.
5. Accept the reality of this statement: "God grant me the serenity to accept the things I cannot change; courage to change the things I can; and the wisdom to know the difference."
6. Take time to relax each day by leaving behind the pressures of work and the challenging decisions yet to be made. Relaxation activities should be included in quadrant B activities (Figure 3.9).
7. Maintain one's physical well-being through a balanced diet and meaningful fitness training.
8. Develop a sense of humor. The ability to laugh at oneself or a situation can relieve tension and enhance communication.

THE FINANCIAL RESOURCE

As fulfilling as the agricultural lifestyle can be, only by effective business management can the desired lifestyle be assured. Thus, it is of critical importance that beef and cattle producers develop business management skills.

The costs, revenues, profitability, and net worth of a beef enterprise can only be critically assessed with a useful set of production and financial records. A record system is needed to (1) monitor cash flow and maintain financial control of the operation, (2) analyze the business so that effective management decisions can be made, (3) make loan applications, and (4) report and manage taxes. Managers should choose the record system that they can most effectively implement to make management decisions within the operation and to communicate financial information needed by themselves and other people, such as bankers or farm advisors. Although good records do not ensure success, successful managers usually have access to good records.

Financial and Economic Records

Table 3.2 identifies financial records used by most successful cattle managers. Examples of several of these records are shown in the Appendix and in later chapters. Production records that reflect cattle inventory and productivity are needed to generate financial records. These production records are discussed in detail in Chapters 4 through 7.

Prudent use of credit as a strategic financial tool can enable a cattle operation to grow more rapidly than it could through the use of reinvested earnings and savings, so long as borrowed funds return more over time than they cost. Financial records are used to understand operating expenses (cash and non-cash), interest for working capital, and term debt. Financial records do not evaluate opportunity costs or equity capital invested in the enterprise. Economic records deal with the opportunity costs associated with the enterprise as well as the expenses listed in the financial analyses.

Economic evaluations compare the potential returns from alternative enterprises to the beef operation. Economic analysis is typically used to evaluate longer-term decisions.

Cash versus Accrual Accounting

The **cash accounting** method recognizes income and expense items only at the time actual cash is received or paid. **Accrual accounting** recognizes income and expenses when they are effectively earned or incurred, rather than when cash changes hands.

Table 3.2
Financial Records for Beef Cattle Operations

Financial Record	Description and Purpose
Cash transactions	The recording of all cash receipts and expenditures is the simplest, yet most time-consuming of all financial records. This provides most of the information needed for completing cash flow statements, filing income tax returns, and making loan applications.
Balance sheet	Provides a financial picture of the operation at one point in time—usually on the last day of the year. It reflects the net worth of the operation. Net worth = assets (what is owned) minus liabilities (what is owed).
Income statement	A moving financial picture that describes most of the changes in net worth from one balance sheet to the next. Net income is calculated by subtracting the expenditures (cash, decrease in inventory, and depreciation) from income (cash receipts and increases in inventory).
Cash flow statement	Shows cash generated and cash needed on a periodic basis (usually monthly) throughout the year. It assesses times when money must be borrowed and times when money is available for additional purchases, investment, or retiring existing debts. A cash flow budget can be used to plan for the next calendar year. An active cash flow statement tracks what is actually happening and evaluates the accuracy of the cash flow budget.
Enterprise budget	Identifies costs and returns associated with a specific product or enterprise. It can aid in making financial decisions by identifying specific problem areas where management changes can be made. Enterprise budgets are also useful where operations have more than one enterprise or where additional enterprises are being considered. Components are production and marketing assumptions, operating receipts, direct costs, net receipts, and breakeven analysis. Estimates of market weight and price can be used in assessing risk in production decisions.
Partial budget	Involves only those income and expense items that would change when implementing a proposed management decision.
Income tax	Form 1040 Schedule F is the primary income tax form for sole proprietors and individual partners in a partnership. Producers who file Schedule F have basically completed a cash basis income statement. There are numerous other forms and 1040 schedules (e.g., Asset Sales, Asset Purchases, Self-employment Tax, Farm Rental Income and Expenses, Tax Withholding, and Depreciation) that may be completed, depending on the individual operation and circumstances.

Table 3.3
Financial Metrics Utilized to Assess Profitability

Metric	Definition
Net income	Return to unpaid family and operator labor, management, equity capital, and risk
Net income minus equity capital charge	Return to labor, management, and risk
Net income minus equity capital charge and labor	Return to management and risk
Net income minus equity capital charge, labor, and management	Return to risk
Net income minus equity capital charge, labor, management plus interest paid divided by capital investment	Return on investment capital

The cash flow statement is an important part of the financial picture of a beef operation; however, it is only one part of the financial picture. The record system of a beef operation should be maintained on the accrual accounting system or converted to the accrual system through inventory adjustments. The cash flow statement only measures cash inflow and cash outflow and says nothing about the profitability of the business. The accrual accounting method associated with the cash flow statement, income statement, and balance sheet will show profitability and overall financial strength and position.

Profit can be defined in several ways (Table 3.3). Most simply, profit is understood as total revenue (gross proceeds) less total expenses or costs. However, a straight cash basis assessment of profitability fails to account for a number of economic considerations. Profit is an economic term and should be evaluated from a strict accounting point of view using generally accepted accounting practices (GAAP). The only true measure of profit comes from accrual or accrual-adjusted income statements.

Profitability at the exclusion of all else is a doomed business model. However, without profitability an enterprise can neither employ people, reinvest in infrastructure, conduct wealth building activities within communities, nor provide investment into both "for profit" and "nonprofit" entities. With wealth creation, the conservation and advancement of culture, community, and natural resources become attainable.

Contingency, Succession, and Estate Planning

In addition to developing an effective business plan centered on financial performance, cattle producers can improve risk management by creating contingency, succession, and estate plans. Contingency planning involves determining appropriate chain of command, financial authorization, and other decision-making processes in the event that senior management are rendered unable to perform their duties as a result of injury, illness, or other unforeseen circumstance. Such a plan helps prevent an emergency from becoming a full-blown crisis.

Succession planning is the process of identifying the best candidates to assume leadership of a business when the senior generation retires. Given the financial scope of many beef cattle enterprises, identifying the best talent to assume leadership and then providing those individuals superior training and preparation is key to

sustained success. Estate planning is designed to provide an orderly transfer of assets from one generation to the next. Estate taxes and other considerations make decisions about property and wealth transfer critically important. Without a well-developed estate plan, assets may have to be sold to cover federal and state taxes that may in fact undermine the continuation of the enterprise.

LAND AND FEED RESOURCES

Land and feed are two important resources in beef cattle operations. They are obviously interrelated, as land (soil) is needed to produce the feed resource for cattle. In cow-calf operations, it is the largest component of investment cost on a per-cow basis. Feed costs are typically the highest (usually over 50%) of the annual operating costs in feedlot and cow-calf operations.

Much of the land cattle graze is not suitable for producing crops for human consumption. Even the higher-valued land produces crop by-products that cattle can utilize.

Sustaining a beef enterprise from generation to generation requires particular attention to natural resources, especially those related to grazing. Advocates of HRM argue that the interactions of animals with soil and plants have a significant effect on conserving and improving those resources. Their primary point is that it is the responsibility of cattle to adapt to what the land provides. The land, soil, and other renewable resources should not be over utilized and damaged in an attempt to meet the needs of the wrong biological type of cattle.

One of the most significant challenges in cow-calf management is to match the biological types of cattle (puberty, mature size, growth, and milk production) to the most economical feed resource. The "matching" implies economically producing the optimum number of pounds while sustaining or improving the feed resource. "Most economical" usually implies utilizing grazed forage rather than large amount of higher-priced harvested or supplemental feeds. This topic is covered in more detail in Chapters 5 and 15.

THE CATTLE RESOURCE

The typical business has production and market specifications that describe how it produces highly preferred consumer products. The beef business is no exception. The industry has struggled over the years not only to define such specifications but also to identify and communicate the general direction of the industry and the specific goals of each segment. The reasons for its struggles include the following.

1. Until recently, the beef industry has not had a well-defined concept of consumers and their preferences. Even today, much more refinement is needed in this area. Market-driven businesses serve very specific consumer markets where the goal is to cater to the specific and well-defined needs of targeted customers.
2. The industry is comprised of multiple segments (seedstock, commercial cow-calf, yearling-stocker, feeder, purveyor, packer, retailer, and consumer) where animal and product identity is typically lost as it passes from one segment to another. Also, communication between segments is difficult and in some cases almost impossible, especially between segments that are widely separated (e.g., seedstock and packer). Each segment is driven by different economic and value-determining parameters.
3. With a loss of animal and product identity between segments and a lack of useful production and financial records within certain segments, the needed specifications are often replaced with general descriptions and opinions. Some opinions are valid, while others are grossly incorrect. Even generalized opinions are not as useful in improving productivity and profitability as cost-effective records.

4. Both seedstock and commercial cow-calf producers produce cattle in a wide variety of environmental conditions. These variable conditions (such as climate and feed) dictate different production specifications if cattle are to be both productive and profitable.
5. Many, if not the majority of, individuals involved in cow-calf production (both seedstock and commercial cow-calf) do not solely derive their livelihood from their cattle operations. Their reasons for having cattle are oriented toward recreation, diversion from city life, or the availability of several acres with grass. Typically, the cattle are highly supplemented, sometimes with high-priced feed and usually with income from outside the cattle operation. Some "part-time" cattle producers manage costs and generate profits on their operations.

Cattle producers who are profit-oriented and earn their livelihood from their cattle want cattle that are both productive and profitable. The major components of productivity and profitability are shown in Table 3.4. Usually, cattle that are profitable are also productive. However, some highly productive cattle are not profitable.

Productivity and profitability, as presented in Table 3.4, need to be examined in more detail. Table 3.5 shows beef cattle productivity and market specifications that are applicable to the entire beef industry. The three broad classifications of these traits are reproduction, growth, and carcass. These areas identify the production part of the industry in producing numbers and pounds—and eventually producing pounds of consumer products, which are primarily proportions of palatable lean, fat, and bone. The specifications of retail beef products preferred by consumers were discussed in Chapter 2.

Cattle Identification

Cattle productivity and profitability can be measured on an individual animal, herd level, or per acre basis. As cattle move through the production, marketing, and processing stages of the industry, the reasons for individual animal identification may vary. However, when viewed from a total industry perspective, there are increasing benefits for the development of a coordinated identification system.

Table 3.4
FACTORS AFFECTING CATTLE PRODUCTIVITY AND PROFITABILITY

Productivity	Profitability
Numbers	Production costs
Cows	
Calves	
Pounds	Price and total value of products
Weaning weight	
Yearling weight	Weaned calves; yearlings
Carcass weight	
	Fed cattle; carcasses
	By-products; boxed beef
Product composition	
Fat	Retail products
Lean	
Bone	
Product palatability	Economic efficiency
Tenderness	Breakeven prices
Flavor	Cost-benefit ratios
Juiciness	Sustained net profit
Biological efficiency	

Table 3.5
Production and Market Specifications for Beef Cattle

Trait	Optimum Range[1]	Industry Target[2]
Reproduction		
Age at puberty (months)	12–16	14
Scrotal circumference (cm), yearling	32–40	36
Reproductive tract score at 14 mos of age	4–5	5
Weight at puberty (lb)		
Heifers	600–800	700
Bulls	950–1,200	1,100
Age at first calving (mos)	23–25	24
Body condition score (BCS) at calving	4–6	5
Postpartum interval (days)	55–95	75
Calving interval (days)	365–390	365
Calving season (days)	45–90	65
Calf crop weaned (% of cows exposed)	80–95	85
Cow longevity (years of age)	9–15	12
Growth		
Mature cow weight (lb) at BCS 5	900–1,300	1,100
Weaning weight (steer; lb at 7 mos)	450–600	525
Yearling weight (steer; lb at 365 days)		
Grazed and/or backgrounded	700–900	800
Weaning to feedlot	1,000–1,300	1,200
Feedlot gain (lb per day)	3.0–4.0	3.5
Feedlot feed efficiency (steers; lb feed/lb gain)	5–7[3]	6[3]
Carcass		
Carcass weight (lb)	750–999	850
Quality grade	Select$^+$–Choice$^+$	Choice$^-$
Yield grade	1.5–3.5	2.5
Fat thickness (in.)	0.10–0.60	0.30
Rib-eye area (sq. in.)	11–15	13
Palatability (% fat in retail cuts)	3–7	5
Frame score		
Steers	4–6	5
Cows	4–6	5
Bulls		
Maternal cross	4–6	5
Terminal cross	5–7	6

[1]Range will include most commercial beef operations where an optimum combination of productivity and profitability is desired.
[2]Target gives a central focus applicable to many commercial beef operations. Deviation from this target and optimum range is dependent on market, economic, and environmental conditions in specific commercial beef operations.
[3]High-energy ration.

A total quality management approach to meeting the needs of consumers calls for the implementation of a unified identification system for the following reasons:

1. Identification systems may be required to gain access to international markets.
2. To enhance consumer confidence in the production and processing of beef.
3. Facilitation of health monitoring and disease surveillance.
4. To provide a means of traceback in the case of food safety failures or disease outbreaks.

Furthermore, animal identification systems may be desirable from a production perspective as a means to (1) assist measurement of production parameters, (2) track financial performance, (3) facilitate monitoring of quality assurance efforts, and (4) track herd health performance.

Historically, hot iron brands have been used to mark the hides of cattle with the distinctive marks of individual owners. Branding has been considered an effective and affordable way to deter theft while identifying herd of origin. Tags, tattoos, earmarks, and ear clips are also frequently used in the industry. The use of electronic identification tags is becoming increasingly frequent in the industry and other high-tech approaches, such as retinal image scanning, have been adopted by some entities.

As alliances and other coordinated production systems become more widespread, the development of more sophisticated systems of identification and traceback will be developed independent of governmental mandates. Identification systems will be discussed in more detail in several of the following chapters (Chapters 4, 5, and 6, in particular).

Input resources

Most commercial cow-calf operations must manage production costs to be profitable. Sometimes it is difficult for the manager to differentiate the "needs" from the "wants" in a high-tech world of mechanization and other key inputs that can be used to enhance productivity. Although equipment and technology that can provide a lifestyle of convenience and labor saving is appealing, a careful economic analysis of cost should be made. One approach is that if the item to be purchased rusts, rots, depreciates, or runs on fossil fuels, then own as little as is possible. Table 3.6 provides a historical view of prices received and paid by producers for the two-decade period beginning in 1993. Clearly within this time frame, the costs of inputs increased more substantially than did cattle prices.

Table 3.7 provides a more recent evaluation of the ratio of cattle prices and key input costs with 2011 as the base year.

Ownership costs (sometimes called **fixed** or **overhead costs**) of equipment include depreciation, interest, taxes, insurance, and storage. Repairs, fuel, oil, and operating labor are the primary variable (operating) costs. Some of the equipment options that should be considered in terms of economics are new versus used purchases, leasing, custom hiring, and joint ownership. While controlling the price of inputs is

Table 3.6
INDICES[1] OF HISTORIC PRICES RECEIVED AND PRICES PAID BY BEEF PRODUCERS

	Year			
Item	1993	1998	2000	2004
Prices Received				
Beef cattle	100	85	104	116
Prices Paid				
Fuel	–	84	136	163
Fertilizer	97	112	110	141
Ag Chemicals	107	122	120	120
Supplies and repairs	107	119	124	137
Trucks	109	119	119	114
Machinery	106	132	137	162
Building materials	105	129	121	134

[1]Index based on 1990–1992 = 100
Source: Agricultural Outlook Statistical Indicators, ERS/USDA.

Table 3.7
INDICES[1] OF PRICES RECEIVED AND PAID BY BEEF PRODUCERS (2012–2015)

Item	2012	2014	2015
Prices received for beef cattle	107	137	133
Prices paid for:			
Fertilizer	101	95	89
Ag chemicals	105	110	106
Fuel	99	98	64
Supplies and repairs	103	106	106
Trucks	102	104	106
Machinery	105	112	115
Building materials	103	107	108
Feed grains	110	78	64
Hay and forages	118	113	100
Feed supplements	106	113	115

[1]Comparisons to 2011 as the benchmark.
Source: Agricultural Outlook Statistical Indicators, ERS/USDA.

beyond the reach of individual producers, cost-control management is critical to ongoing profitability in times when input costs are subject to high levels of volatility.

THE MARKET RESOURCE

Managing the market resource requires knowing consumer demand (discussed in Chapters 2 and 9) and the carcass specifications presented in Table 3.5. Producers are then concerned with the choice of various markets that will return the highest level of profitability to the enterprise. Too often, producers chose market highs without recognizing the costs associated with obtaining these premiums.

In addition, marketing involves the cost of goods and services needed to produce the cattle. Profitability, then, is the difference between costs and income. While this may seem simplistic, it is important to recognize that beef producers are price takers and not price makers. In other words, beef producers cannot pass on their higher production costs by increasing the price of their saleable products. Individual producers can only make minor changes in the market price for the cattle they sell. Good beef managers recognize that profitability can be enhanced by managing the costs of production and then utilizing a value-based marketing program such as retained ownership (marketing on a grid), or selling a branded product, or being part of an alliance, rather than attempting to increase market price of a commodity product.

More information on the marketing principles that affect management decisions is presented in Chapter 9.

MANAGEMENT SYSTEMS

Because the resource combinations for each operation are different, no single, fixed management approach for successful cattle production is applicable to all operations. The differences among beef cattle enterprises can include such variables as levels of forage production, marketing alternatives, energy costs, debt structure, biological type of cattle, environmental conditions (e.g., weather), costs of feed nutrients and labor, and management competence levels. These variables and their interactions

pose challenges to the producer, who must combine them into sound management decisions for a specific operation. There are, however, principles that apply to all producers.

A management systems approach provides a method of systematically organizing the information needed to make valid management decisions. It permits many variables to be critically assessed and analyzed in terms of their contributions to the desired end point. Management systems involve systems thinking rather than linear thinking. An example of linear thinking would be a producer implementing several management practices to increase weaning weight because he believes the more the pounds sold the higher the revenue and assumes that profit per cow is also higher. Systems thinking considers how increasing weaning weight might impact costs by increasing milk production, increasing mature cow weight, increasing forage consumption, and considers how body condition score (BCS) and reproductive performance might be changed, especially under a low-cost environment. The systems thinking example is consistent with actual studies that show herd weaning weight and net profit per cow to be independent of one another.

A systems thinker considers circles of influence and how each circle influences or interacts with other circles that are important in the system. These circles of influence are noted for a cow-calf focus (Figure 3.10). *Integrated Resource Management* is one of several management systems approaches being utilized to integrate the various resources of importance. Making the system even more complicated is the high degree of segmentation (Figure 3.11) within the beef supply chain as well as the impacts of decisions in one phase of production on outcomes in subsequent segments. The manager is challenged to make management decisions from a broad perspective, yet must also be aware of how each resource interacts with the others. Successful cow-calf managers use these management systems approaches today. However, some successful managers have used IRM/HRM principles and other management systems concepts for decades.

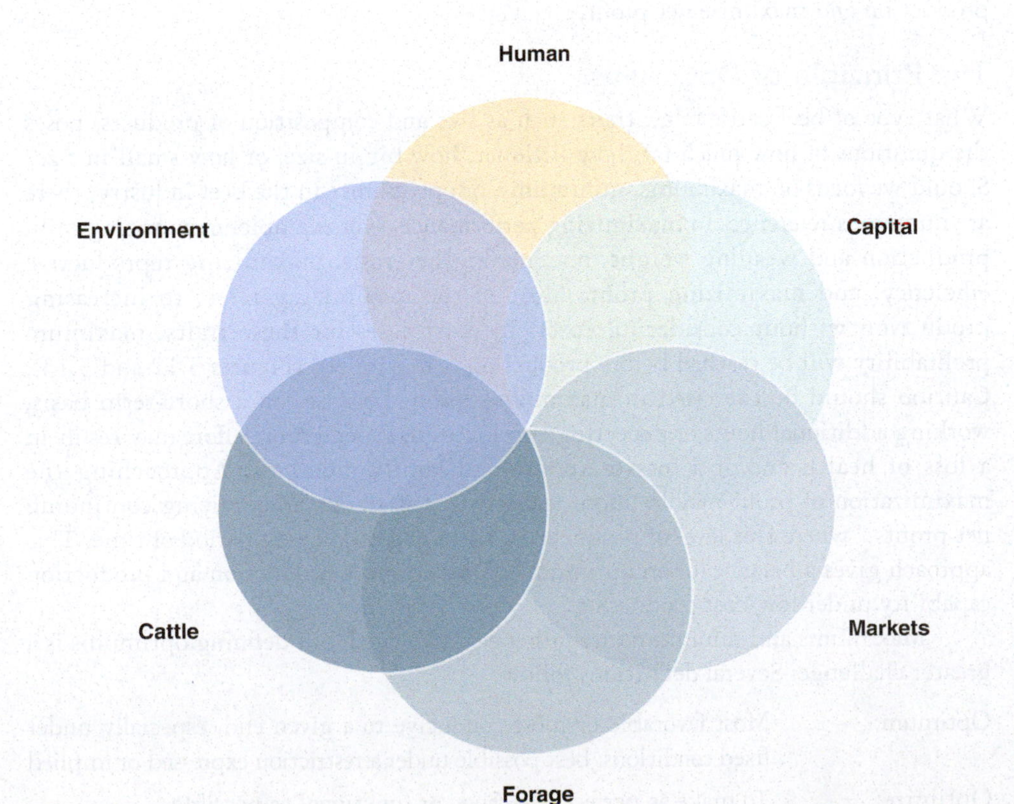

Figure 3.10
Circles of influence that impact enterprise outcomes and establish the need for systems management.

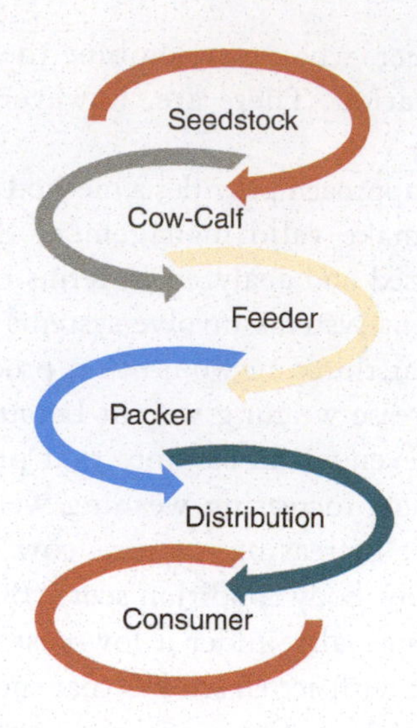

Figure 3.11
Decisions made at each industry segment impact performance and outcomes further down the supply chain.

Production is usually reflected in numbers and pounds. High and sustained profitability will allow managers to better enhance the resources under their care, to expand or diversify their business, and to create greater opportunities for themselves, their family, employees, and community. Cow-calf profitability in the short term must be balanced with the need to conserve and improve the resources associated with the enterprise. It has been a common practice to increase or maximize production of cattle by applying known biological relationships. Cattle production has typically been maximized without careful consideration of costs and how increased productivity relates to land, feed, and management resources. The need to optimize rather than maximize cattle productivity continues to receive more and more attention. Thus, there is an increased interest in a management systems approach that can optimize production and maximize net profit.

The Principle of Optimums

What type of beef cattle (e.g., traits such as size and composition of products) poses the questions of how much fat, how little fat, how big in size, or how small in size? Should we focus on maximums, minimums, or optimums? In the beef industry, there are numerous references to maximizing performance—for example, maximizing milk production and weaning weight, maximizing heterosis, maximizing reproductive efficiency, and maximizing profit. Most of the maximizing refers to increasing production without considering costs. In most cases for these traits, maximum profitability will be reached before production is maximized (Figures 3.12 and 3.13). Caution should be exercised in maximizing profit because, on a short-term basis, working additional hours or expecting employees to expend more effort may result in a loss of health and/or a loss of spouse and family members. Approaching the maximization of profit may be more realistic if it is stated "Maximizing continuing net profits," where this level of profit can be sustained over a long period of time. This approach gives a balance (or an optimum combination) of production and production capability under low-cost production.

Maximums and minimums are rather easily defined, but defining optimums is a greater challenge. Several definitions follow:

Optimum:	Most favorable or most conducive to a given end, especially under fixed conditions; best possible under a restriction expressed or implied
Optimize:	To make as perfect, effective, or functional as possible

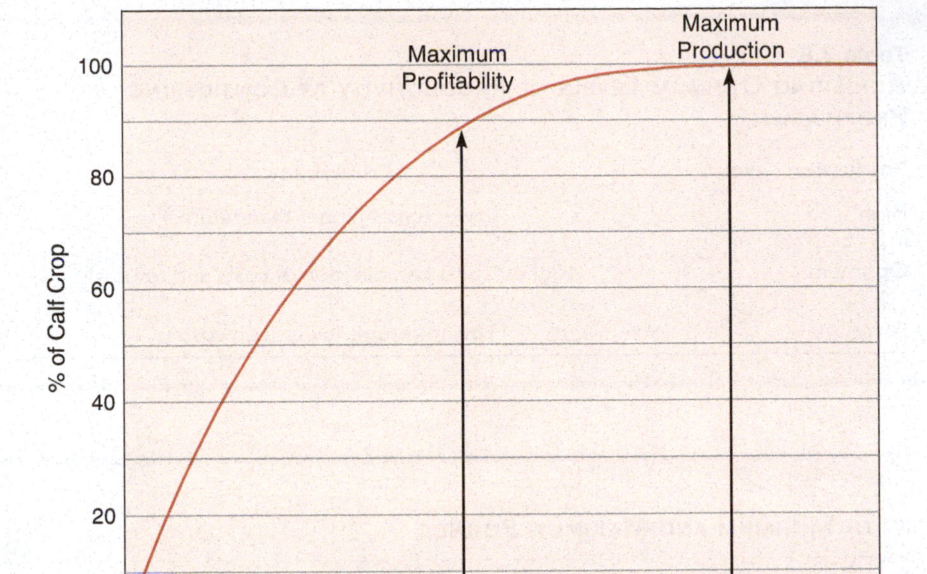

Figure 3.12
Maximum productivity versus maximum profitability.

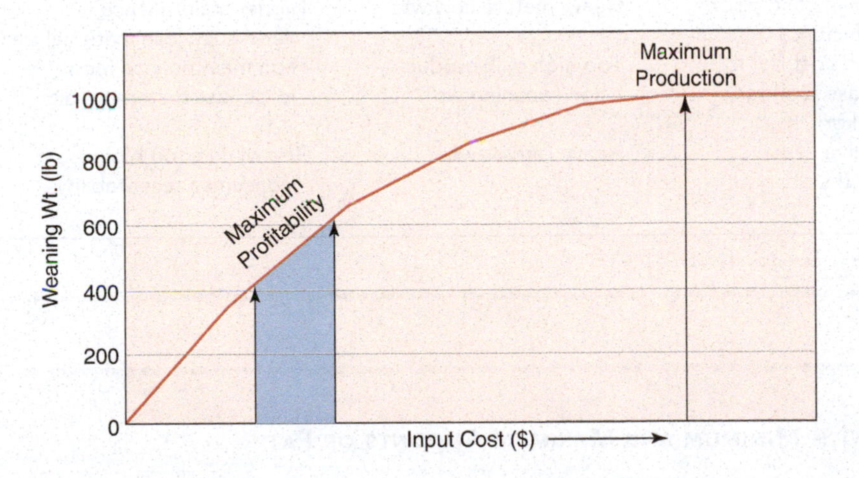

Figure 3.13
Maximum profitability and maximum production for average weaning weights of calves (5–9 months of age).

Optimal: Most desirable or satisfactory

Optimization: Process of orchestrating the effects of all components (resources) toward the achievement of the stated aim

Some of the key points in these definitions are "to a given end," "under fixed conditions," "under a restriction," and "achievement of a stated aim." These become defined with a mission/vision statement and goals.

Table 3.8 shows how productivity levels and profitability levels can be combined to achieve optimums in productivity. The arrows in Table 3.8 show that as high and low levels of productivity move toward an optimum level, profitability increases. With profitability, one would want to approach maximum levels as long as they were sustainable.

Producers sell pounds as weaned calves, yearlings, fed cattle, and carcasses. Maximums, minimums, and optimums can be considered for the traits that measure pounds (Table 3.9).

Consumers prefer highly palatable, lean beef. Maximums, minimums, and optimums can be considered for those traits that measure fatness in beef cattle (Table 3.10).

Table 3.8
ACHIEVING OPTIMUM LEVELS OF PRODUCTIVITY BY CONSIDERING PROFITABILITY

Production Level	Profitability
High	Lower (costs higher than returns)
↓	↓
Optimum	Highest (best combination of costs and returns)
↑	↑
Low	Lower (returns lower than costs)

Table 3.9
MANAGEMENT CHALLENGES WITH MINIMUM AND MAXIMUM POUNDS

Too Few Pounds	Problem	Too Many Pounds	Problem
Low sale wt.	Fewer dollars	Heavy birth wt.	Higher calf losses; longer rebreeding
More days to slaughter wt.	Higher gain cost; poor feed efficiency	Heavy mature cow wt.	Higher maintenance feed cost; lose maternal traits[1]
Light carcass wt.	Overhead cost per lb. of carcass too high for packer	Too high milk production (weaning wt.)	High maintenance feed cost; lose maternal traits
Light wt. replacement heifers	Won't calve at 24 months of age	Heavy carcass wt.	Size of cuts too big; lose consumer acceptability

[1]Maternal are primarily reproduction traits expressed by the cow.

Table 3.10
MANAGEMENT CHALLENGES WITH MINIMUM AND MAXIMUM AMOUNTS OF FAT

Too Little Fat	Problem	Too Much Fat	Problem
Less than 0.30 in. on carcass	Cold shortening (tenderness problem); difficult to grade choice	Yield grade 3.5 or higher	Low retail cutout; carcass discounts; poor consumer product; uneconomical feed use
Less than 3% intramuscular fat	Unacceptable taste to many consumers	Greater than 7% intramuscular fat	Fat getting too high to meet health guidelines
Body condition score (BCS) 3 and less at breeding	Cows—longer postpartum Interval Heifers— may not calve at 24 months of age Higher production costs	BCS 7, 8, and 9 at breeding	Poor reproduction; uneconomical feed use

Table 3.5 showed optimum levels of pounds (weaning weight, slaughter weight, carcass weight, and mature cow weight) and optimum levels of fatness (fat thickness, quality grade, % fat in retail cuts, and BCS) in breeding cattle.

It should not be surprising that some of the basic biological truths will not be useful or applicable to a specific cattle operation because an economic analysis will not permit them to be included in sound management decisions. For example, one well-known biological relationship is that calves born early in the calving season will have heavier weaning weights. Because of this relationship, some producers continue to move the calving season to an earlier time of the year. However, the focus on maximizing weaning weight may increase costs and reduce fertility. An evaluation of many beef cattle herds found that changing the calving season to later in the year better matched forage availability and increased profits significantly. On one ranch, changing calving season alone was estimated to increase the ranch's carrying capacity by 30–40% in terms of animal units. In addition, annual cow cost was reduced 18% per cow because of the reduction in winter feed requirements.

Sustainability

Any discussion of sustainability should be preceded by a conversation about core values and the culture of enterprises. Responsible beef production has always been and always will be dependent on the concepts of stockmanship and stewardship. Stockmanship is the application of skill, creativity, and knowledge to sound principles of livestock management founded on the model of the good shepherd. This concept integrates the results of sound science and practical experience. Stewardship is defined as the careful and responsible management of something entrusted to one's care. In the realm of beef production, this involves the thoughtful and wise management of livestock, people, natural resources, finances, and technology to create benefit for humanity. Decision making resulting in high-integrity outcomes is founded on four core considerations—values and ethics, community and people, economics and profit, and resource health.

Animal agriculture is a complex, multidimensional human activity influenced by cultural, environmental, political, economic, and social forces. Fundamentally, managers and leaders in the beef business have the opportunity to wrestle with systems questions and must make integrated systems decisions to be successful. Livestock management requires the ability to deal with the intertwined nature of humans, natural resources, animals, markets, communities, supply chains, and public policies. Such an environment is typified by the need to weigh competing interests, make thoughtful trade-offs, and balance the polarities of short- versus long-term needs as well as a host of cost–benefit relationships.

At the beginning of 2011, there were 7 billion people on earth. By 2050, that number will climb to 9 billion—essentially adding the equivalent of another 1.6 of China to the world population. Creating the food production, processing and infrastructure capacity to feed 9 billion people over the next four decades is daunting. However, there are substantial reasons that suggest such a feat can be accomplished. First of all, population growth is slowing. The global population growth rate over the next four decades is expected to be approximately 30% compared to a nearly 80% rate of expansion from 1970 to 2010. Secondly, agricultural yields have allowed a level of productivity sufficient to hold a global hunger crisis at bay. While concerns exist as to the ability to sustain rising yield rates, experts point to opportunities to increase food production by narrowing the gap between the low-yield producer and those with high production rates, intensification of production with the application of precision irrigation systems, enhanced housing systems for livestock and poultry coupled with improved nutritional management, genetic selection, and health management, and by harnessing the potential of new plant-breeding technologies and production systems. Given that somewhere

between 33% and half of the calories produced by the agricultural system never reach a human stomach due to losses from pests, spoilage, and waste; if half of this loss could be eliminated, it would equate to increasing food availability by 15–25%. Recapturing these losses requires enhanced production management, improved storage, handling and transportation of both raw and finished product, and changes in packaging, consumer behavior, and regulations that prevent left-over and minimal out-of-date food from reaching hungry people.

Nonetheless, reaching the goal of feeding 9 billion people by 2050 will not be without controversy. Too often, there is a tendency by individuals, organizations, and policy makers to seek popular and seemingly simple approaches to complex problems by focusing on one-size-fits-all approaches. History has shown that such an approach is not effective. Decision making almost always involves choices that require trade-offs. Careful analysis of agricultural issues clearly shows that there are very few black and white choices available. Good decision making requires acknowledgment of the assumptions that define both the problem at hand and the alternate solutions, careful consideration of both intended and unintended consequences, and rational assessment of cost–benefit relationships. In the end, the best decisions are those that are site- and situation-specific, based on knowledge and good information, and aligned with core values of the people involved.

Many problems are framed by polarities or competing interests. Managers of livestock systems must deal with a host of polarities. For example, profitability attainment involves having an enterprise that is sufficiently large to capture economies of size and scale. However, as more animals are added to the enterprise, the waste production of the business increases. If the waste management issue is not sufficiently addressed, then animal health declines as does the productive capacity of the natural resource base. Thus, long-term profits depend not only on productive capacity but also on the careful management of impacts on natural resources.

Sustainability has been defined as a method of harvesting or using a resource so that the resource is not depleted or permanently damaged. However, this definition is perhaps too narrow as it fails to consider the dynamics of other polarities, such as economy and culture. In addition to meeting ecosystem performance metrics, managers must also consider other factors. Assessing the viability of alternatives, defined as whether something is capable of growing, or something practical and able to be done; is foundational to good decision making. Livestock and poultry enterprises tend to be continuous in nature. Thus continuity—an uninterrupted succession or flow, a coherent whole—cannot be stopped and restarted in the same way as enterprises that generate storable products.

As people have struggled with the complexity of making effective natural resource decisions, a useful model has emerged called the triple bottom line approach to sustainability. The triple bottom line of sustainability is based on three columns of importance—ecosystems, economies, and communities (Figure 3.14). Sustainability

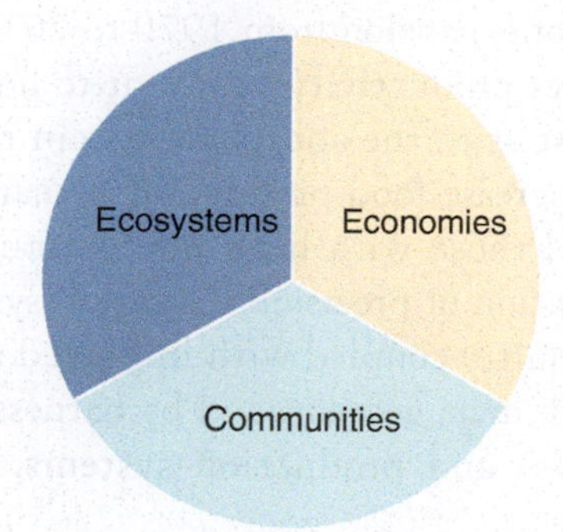

Figure 3.14
Triple bottom line components of sustainability.

is not simple, it is not a recipe, it is not a classification system by which industries, companies, or processes can be conveniently labeled as "sustainable" or "not sustainable," and it is not a task to be completed and then checked off the list. Rather, sustainability is a commitment to an ongoing journey that involves balancing decisions within the scope of the triple bottom line concept. The choice is to become *more* sustainable. Attempts to create black and white categories such as *sustainable* or *unsustainable* are misguided.

The beef industry led by the efforts of the Research, Education, and Innovation team at NCBA became the first agricultural supply chain to tackle the challenge of developing a holistic and comprehensive set of benchmarks coupled with a future-oriented plan to improve sustainability across the various segments. The groundbreaking lifecycle assessment study established 2005 as the baseline year. Analyses comparing the sustainability of the industry in 2011 to the baseline found that overall sustainability had increased 5% in that time period. A more detailed analysis can be found in Chapter 8.

Biological Efficiency versus Economic Efficiency

Tess (1995) summarized this key topic at the 1995 NCA Cattlemen's College:

- *Biological efficiency does not predict economic rankings very well.*
- *End users want economic comparisons.*
- *Production/management systems are ranked confidently on (1) breakeven prices and (2) net profit.*

Optimums can be effectively measured by having sustained, low breakeven prices and high profits that can be measured at endpoints such as weaning, stocker/yearling, feedlot, carcass, and retail cuts. Breakeven prices and consumer preferences determine the optimum production and market specifications applicable to each beef industry trait (Table 3.5).

The difference between biological efficiency and economic efficiency must be understood. **Biological efficiency** is measured in biological units—for example, the number of pounds of beef produced relative to the number of pounds of feed consumed. **Economic efficiency** is measured in economic units (dollars)—for example, the number of dollars returned for each dollar spent. While the two types of efficiencies are interrelated, they can be very different. A management practice that is biologically efficient may not be economically efficient or feasible. In fact, many measures of biological efficiency are of limited value in measuring the performance of the total integrated system. For example, producers may understand that increasing the feed supply to their cows would increase calf crop percent from 85 to 95. However, in some cases, this practice may not be economically efficient because the income from an increased number of calves would not cover the increased feed and other costs.

Another example of economic versus biological efficiency is feeding cattle to change from one carcass quality grade to another. It may be economically efficient to feed cattle a longer period of time in an attempt to move cattle from Select to Choice because of the price spread between the two grades. However, biological efficiency (pound of feed per pound of gain) is unfavorable during this period. Furthermore, once 100–120 days on feed have been surpassed, it is difficult to gain significant improvement in USDA Quality Grade.

Rather than contrast and compare biological efficiency and economic efficiency, it is best to combine them. The term bioeconomic efficiency reflects that combination. Management systems emphasize the best combination of biological and economic efficiencies.

Risk Management

Beef production involves making many decisions with uncertain outcomes. Risk management is a decision-making process that evaluates the chance or probability of adverse outcomes. Some of the risks that bring uncertainty into the management decision process are weather, changes in cattle prices, changes in input costs, equipment breakdown, changes in government regulations, variability in animal and crop performance, disease, and labor and human issues (e.g., loss of an employee or personal health).

Weather can reduce calf crop percent, such as by causing high death losses of young calves when they are exposed to wet and cold conditions. This risk can be reduced by providing additional shelter and avoiding a large concentration of calves when chances of severe storms are highest.

Drought can significantly affect feed supply and over time it has affected nearly every region of the United States often causing regional shifts in herd inventories (Figure 3.15). Producers in drought-prone areas manage drought risk by keeping herd size below the numbers that can be grazed in the best forage years. Also, some producers maintain a year's supply of stored forage to compensate for frequent years of drought.

Drought can reduce feed supply at critical times and thereby reduce next year's calf crop, particularly when cows are thin at calving time or when the feed supply is low from calving through rebreeding. The risk can be managed by providing supplemental feed to cows or calves. Calves can be creep fed during periods of low forage supply, which allows the cows to maintain a higher body condition and thus have good reproduction. Calves can also be weaned early to help assure a higher calf crop next year.

Market prices have a history of wide fluctuations on a yearly, a monthly, and even a weekly basis. Producers can study price trends, know current market prices, forward-contract their cattle, and hedge their cattle using the futures market (see Chapter 9) to manage risks of price changes.

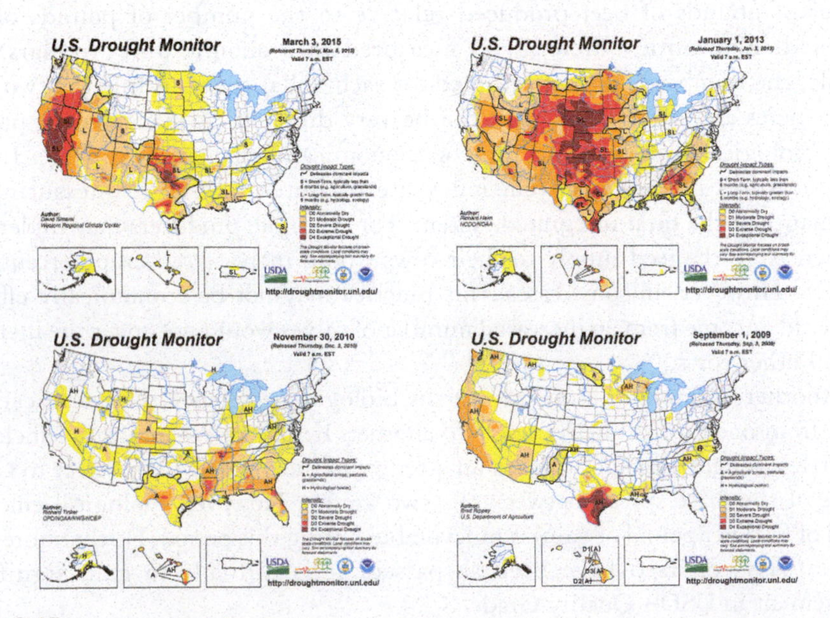

Figure 3.15

Drought status in the United States at one point in time. Droughts result in large annual fluctuation in feed supply. Drought occurs with greater frequency in certain areas of the United States and has to be managed accordingly.

Source: USDA.

SELECTED REFERENCES

Publications

Bourdon, R. 1986. *The systems concept of beef production.* Beef Improvement Federation (BIF) Fact Sheet FS8.

Covey, S.R. 1989. *The seven habits of highly effective people.* New York, NY: Simon & Schuster.

Dolezal, S.L., & Pollert, H. 1998. Cow-calf production record software. Oklahoma Cooperative Extension Service.

Dunn, B.H., Gates, R.N., Davis, J., & Arzeno, A. 2006. *Setting strategy and measuring performance.* South Dakota State University and Texas A&M Kingsville.

Farm Financial Standards Council. 2006. Management Accounting Guidelines for Agricultural Producers. Menomonee, WI.

Field, T.G. 2006. *Priorities first.* American Angus Association. Saint Joseph, MO.

Field, T.G., & Taylor, R.E. 1997. *Emotion, tradition and business.* Proc. BIF Research Symposium. Dickinson, ND.

Fowler, J.M., & Torell, L.A. 1987. Economic conditions affecting ranch profitability. *Rangelands.* 9:55.

Gersick, K.E., Davis, J.A., Hampton, M.M., & Lansberg, I. 1997. *Generation to generation: Life cycles of the family business.* Boston, MA: Harvard Business School Press.

Gutierrez, P.H. 1993. *Cost control using economic analysis and SPA.* Proc. The Range Beef Cow Symposium. Gering, NE.

Heifetz, R.A., & Laurie, D.L. 1997. The work of leadership. *Harvard Business Review.* 75:1.

Irvine, D. 1997. *Succeeding with succession: From parenting to partnership.* Proc. Alberta Hereford Association 106th Annual Meeting and Conference. Calgary, Alberta.

Kotter, J.P. 1996. *Leading change.* Boston, MA: Harvard Business School Press.

Lawrence, J.D., & Mintert, J.R. 2014. Fundamental forces affecting livestock producers. *Choices.* Agricultural & Applied Economics Association.

Maddux, J. 1981. *The man in management.* Proc. The Range Beef Cow: A Symposium on Production, VII. Rapid City, SD.

Miller, W.C., Brinks, J.S., & Greathouse, G.A. 1985. A systems analysis model for cattle ranch management. Proceedings—American Society of Animal Science, Western Section. 36:90.

Pratt, D.W. 2013. *Healthy land, happy families and profitable businesses.* Fairfield, CA: Ranch Management Consultants.

Savory, A. 1999. *Holistic resource management: A new framework for decision making.* Washington, DC: Island Press.

Swigert, S. 1996. *A new top hand: Your computer.* Proc. Beef Improvement Federation.

Tess, M.W. 1995. *Production systems and profit.* 1995 NCA Cattlemen's College.

United States Department of Agriculture: NRCS. 2015. *National range and pasture handbook.* Washington, DC.

White, L.D., Troxel, T.R., Pena, J.G., & Guynn, D.E. 1988. Total ranch management: Meet ranch goals. *Rangelands.* 10:3.

White, R.S., & Short, R.E. (eds.). 1987. *Achieving efficient use of rangeland resources.* Miles City, MT: Montana Agric. Expt. Sta, and USDA-ARS.

Zimmerman, T.S., & Fetsch, R.J. 1994. Family ranching and farming—a consensus management model to improve family functioning and decrease work stress. *Family Relations.* 43:125–131.

4

Management Decisions for Seedstock Breeders

The seedstock sector emerged over time as a specialized cow-calf enterprise focused on producing genetic resources and associated services for the larger cow-calf business. The early seedstock breeder would have been focused on the development of "pure" lines or breeds of stock sharing a common set of characteristics, such as color, the presence or absence of horns, body type, and other distinguishing traits. Eventually, like-minded breeders would form organizations dedicated to the creation of a herd book responsible for maintaining an accurate set of pedigrees for individual animals. These breed societies focused their attention on establishing guidelines for registration and transfer of pedigrees, on creating standards and preferred types of cattle, and utilized the show ring as a primary means to build relationships among breeders to promote their breed and to facilitate trade. As the science of genetics emerged, breed associations shifted their focus toward objective measurement of trait differences, the creation of large performance databases, and the development of national sire evaluation systems. Whether in the earliest days of the seedstock industry or in contemporary times, the seedstock business has been the one that has fascinated people and attracted not only those interested in the methodical improvement of cattle, but also those with a penchant for promotion and marketing.

Note: To better understand the technical nature of genetics, it is advised that materials from Chapters 12 and 13 should be evaluated in conjunction with this chapter.

Seedstock producers provide genetic inputs and related services that impact the entire beef industry (Figure 4.1). Some breeders develop the expertise, reputation, and marketing capacity to be viewed as elite breeders or those who have the highest level of influence on the genetic direction of a breed. Multiplier breeders are those who provide the invaluable service of propagating the influence of the superior seedstock produced by the elite herds for sale to commercial cow-calf customers. To be most effective, seedstock enterprise managers should understand how their decisions integrate with the management decisions made by other beef industry segments in producing cattle that meet production, market, and consumer specifications. Seedstock breeders must establish effective lines of communication with cow-calf producers to assure that the genetics produced meet customer needs. However, both breeders and cow-calf managers sometimes become focused on short-term trends that ultimately create negative cost–benefit outcomes. For example, in the 1980s, many cow-calf

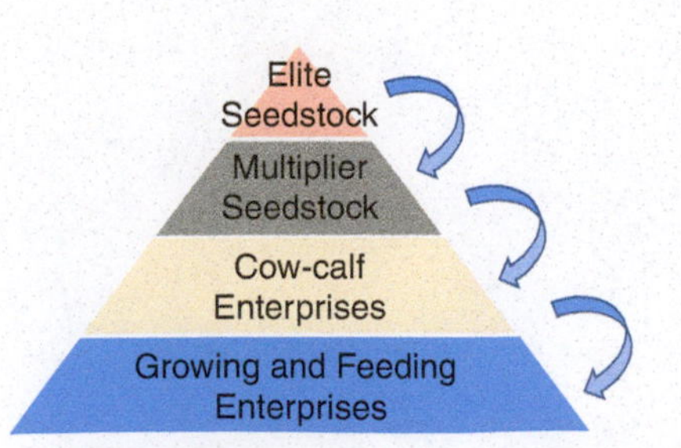

Figure 4.1
The flow of genetics in the beef industry.

producers wanted bulls that were extreme in frame size and milk production only to find out in later years that extreme types were not profitable in the long run.

BREEDING PROGRAM GOALS AND OBJECTIVES

Establishing credibility and market acceptance is critical to the sustainability of a seedstock enterprise. It has been estimated that the average life span of seedstock enterprises is 7 years—a time span that suggests how difficult it is to both produce quality seedstock and establish a profitable foothold in the marketplace.

The first step in establishing a seedstock enterprise is to develop a clear set of goals and objectives. As a result, seedstock breeders take diverse approaches to the business of producing genetics for the industry. Goals vary from the desire to produce award-winning show cattle to developing a breeding program that meets the needs of commercial cow-calf producers within a defined market territory to establishing a reputation as an elite breeder who supplies breeding stock, semen, and embryos to other seedstock breeders. For the majority of breeders, success depends on propagating seedstock that is well suited to the management system goals and objectives of commercial cattle producers where cost of production, live animal productivity, and producing beef products consistent with consumer demand are the focal areas.

Animal Identification

Breeders must establish individual animal identification systems to facilitate pedigree registration, performance recording, and marketing. Ear tattoos establish a permanent identification (Figure 4.2) and are usually paired with plastic ear tags to make it easy to identify individual animals without needing to restrain them to read the tattoo. An ear tattoo is typically required as permanent identification by the breed association,

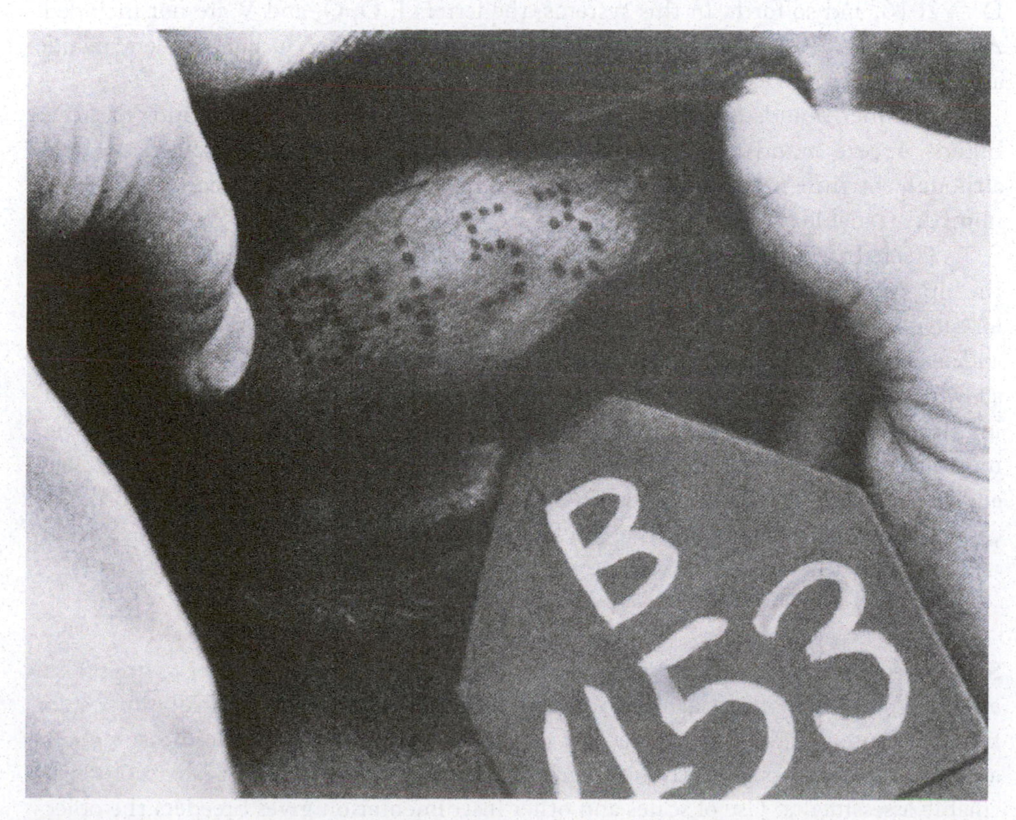

Figure 4.2
Double identification in the calf's ear using a tattoo and plastic ear tag. Note the same identification is used with both methods.
Source: American Angus Association.

and the same tattoo number is recommended for both ears because a single tattoo may be difficult to read.

Tattoos should be applied to animals prior to 3 months of age or younger to assure ideal readability. The process of applying the tattoo includes observing and verifying the calf's identification (usually by observing the dam's tag or identification brand), placing the appropriate ID into the tattoo applicator, double checking the accuracy of the number by applying the tattoo to a scrap piece of paper, cleaning the ear with a rag and rubbing alcohol, liberal application of ink to the ear, placing the tattoo in the upper lobe of the ear with the number horizontally aligned along the middle one-third, squeezing the applicator with moderate pressure, reapplying ink, and then rubbing the ink into the tattoo marks with a small brush.

Ear tags are a more convenient method of identification than tattoos because the tags can be read at a distance and when cattle are in a chute. The loss of ear tags is a problem (usually 2–10% per year are lost). When the ear tag is lost, the tattoo serves as the backup system for maintaining individual animal identity.

Establishing unique herd identities for each animal is key to maintaining appropriate and convenient production and pedigree records. Some producers use the last digit of the year as the first number for the ear tag. For example, the first calf born in 2015 would be given tag number 501 (5 for 2015 and 01 for being the first calf). However, this system may yield problems. For example, after 10 years, some females may have duplicate numbers. This is particularly challenging to computer records, which cannot distinguish between a 2005-born cow numbered 501 and a 2015 model with the same number. Some breeders solve this problem by changing ear tag colors every 10 years and including the tag color as part of the identification system.

A widely accepted and standardized system uses letters to represent the years in order to prevent the 10-year duplication of numbers: A = 2013, B = 2014, C = 2015, D = 2016, and so forth. In this systems, the letters I, O, Q, and V are not included. A full year listing of letter codes can be found in the Appendix. Other breeders include a sire code as part of each subsequent progeny's identification.

Number brands can be applied to the hide as either hot-iron brands or freeze brands. Freeze brands are usually most effective, especially on black-hided cattle, although the time needed to apply several numbers and to clip long hair to make the numbers readable are disadvantages.

Freeze brands are applied to an area where the hair has been closely clipped, usually the hip. A brass iron should be utilized: the appropriate size of the brand is 2 to 4 inches for calves and 4 to 6 inches for yearlings or mature cattle. The brands are cooled in dry ice and alcohol or liquid nitrogen. The hair where the brand was applied comes in white after a few weeks to several months. There are advantages of freeze brands over hot-iron branding: less damage to the hide, better readability if the application protocol is correct, and more mature cattle accept the process more readily. The disadvantages of freeze brands are additional costs, increased time required for application, poor results on white or light-colored cattle, and some numbers are difficult to read if the application process is flawed.

PRODUCTION RECORDS

Some breeding cattle currently are evaluated and selected solely on visual appraisal even though visual appraisal alone is a poor indicator of most economically important traits measured at the cow-calf feedlot or packer segments. Objective measurements almost always trump subjective evaluations and breeders are advised to focus on quantifiable assessments. Use of scales and other instrumentation gives breeders the objective measurements needed to identify beef cattle superiority or inferiority for traits affecting commercial profitability.

Most breed associations make performance record systems available to their members for a reasonable processing fee. The breed performance programs are variously named according to the breed: for example, Angus Herd Improvement Records (AHIR), Total Performance Records (TPR, for Hereford), Performance Registry System (for Simmental), and Charolais Herd Improvement Program (CHIP). Taken in total, these databases comprise perhaps the single highest valued asset of the beef industry.

A genetic evaluation system is only as good as the accuracy and volume of data submitted to the national association. For example, the American Angus Association has nearly 10 million individual animals that have contributed performance data to their national genetic evaluation system. In an attempt to increase participation in the breed performance program and as a result enhance the reliability of genetic estimates generated, many breed organizations instituted whole herd reporting. Whole herd reporting systems combine the fees for registry, pedigree transfer, and performance records calculation into one fee based on the cow herd inventory. Such an approach increases the amount of data submitted and allows for the development of new expected progeny differences (EPDs) for a variety of reproductive and convenience traits. In some situations, an animal cannot be registered without submission of data through the weaning or yearling phase of production.

The traits traditionally measured by most breed performance record programs included birth weight, weaning weight, yearling weight, and milk production (indirectly measured). Ratios of these weights are usually calculated for individual animals within the herd. However, the ratios are meaningful for within herd evaluation and valid only when the animals are compared with a contemporary group of animals exposed to a similar environment within the herd of origin. EPDs provide an accurate means to compare animals within a breed.

Table 4.1 outlines the relative value of available indicators of genetic merit for any specific trait.

Table 4.2 outlines a basic breed performance program showing both the input records and the final analysis records.

Once performance records are generated, it is important to put them into an appropriate format so they can be most effectively utilized. Some seedstock and commercial producers hand calculate their performance record data and maintain their own record forms. Most breeders and producers, however, submit their records to a breed association where the records are returned in a standardized and user-friendly form. Performance records that are used to calculate EPDs are especially useful to breeders.

Table 4.1
Relative Value of Indicators of Genetic Merit

Indicator	Relative Value as a Genetic Predictor
Raw data such as a scale weight or hip height	Lowest
Measure adjusted to a constant age and for age of dam differences	
Ratio comparing an individual's adjusted performance to the average of its contemporaries	
Expected progeny difference on a young animal	Moderate
Expected progeny difference on an animal with many progeny	
Genomically enhanced EPD	Highest

Table 4.2
Outline of a Breeding and Performance Program

I.		**INVENTORY**
	A.	Record input
	1.	Add cows and bulls to inventory and identify cows leaving the herd specifying reasons for disposal
	B.	Record analysis output
	1.	Current inventory; frequency of and reasons for involuntary culling
	C.	Selection decisions
	1.	Evaluate whether the genetic composition of the cattle is consistent with the management system and long-term goals
	2.	Look for specific problems traced to specific sire groups
II.		**BREEDING**
	A.	Record input
	1.	Sire
	2.	Dates bred or time exposed
	3.	Date and result of pregnancy exam
	B.	Record analysis output
	1.	Expected calving dates
	2.	Percentage bred of those exposed for each sire and for the entire group
III.		**CALVING**
	A.	Record input
	1.	Birth date
	2.	Birth weight
	3.	Calving ease score
	4.	Calf ID (identification)
	5.	Sire and dam ID
	B.	Record analysis output
	1.	Adjusted birth weight and ratio
	2.	Calving ease score (separately for first-calf heifers and for older cows)
	3.	Gestation length (if breeding dates are known) adjusted for cow age and sex of calf
	4.	Proportion calving in each 21 days of the calving season
	5.	Percent calf crop of those exposed
	6.	Birth weight and calving ease EPDs
	7.	Calving interval
IV.		**WEANING (CALF AGE RANGE OF 160–250 DAYS)**
	A.	Record input
	1.	Date weighed
	2.	Calf weight
	3.	Management code
	4.	Contemporary group code
	5.	Cow weight, hip height, and condition score
	6.	Cow pregnancy status
	B.	Record analysis output
	1.	Adjusted 205-day weaning weight and ratio
	2.	Repeat of calving information
	3.	Weaning-direct and weaning-milk EPDs for each calf
	4.	Weaning-direct and weaning-milk EPDs for each cow and bull
	5.	Mature size EPDs for cows and their sires
	C.	Selection decisions
	1.	Make initial selection/culling of heifers based on EPDs and structural problems
	2.	Cull cows based on soundness and EPDs
	3.	Consider culling open and very late bred cows
V.		**YEARLING (AGE RANGE OF 330–390 DAYS)**
	A.	Record input
	1.	Dates weighed and measured
	2.	Weight, hip height, and scrotal circumference

Table 4.2
(Continued) Outline of a Breeding and Performance Program

	3.	Ultrasonic measurement of composition
	4.	Management code
	5.	Contemporary group code
B.		Record analysis output
	1.	Adjusted yearling weight ratio, adjusted linear measures and frame score
	2.	Repeat of calving and weaning information
	3.	EPDs for weight and appropriate linear and compositional measures for each calf, dam, and sire
C.		Selection decisions
	1.	Select bulls by comparing the breeding values for each trait of relevance, both for sires in use and the yearlings
	2.	Cull any heifers that are structurally unsound and too extreme in frame (large or small)
	3.	Select replacement heifers based on EPDs
VI.		CONTRIBUTION TO COMMERCIAL COW-CALF, FEEDLOT, AND CARCASS (RETAIL PRODUCT) PRODUCTION
A.		Record input from sire progeny groups of herd sample from commercial herds
	1.	Production costs for weaned calves, feedlot cattle, and carcass weight
	2.	Carcass weight, quality grade, and yield grade
	3.	Warner–Bratzler shear or other measures of tenderness and total palatability
B.		Record output
	1.	Breakeven prices for weaned calves, feedlot cattle, and carcasses
	2.	EPDs for carcass weight, marbling, and percent retail product
	3.	EPDs for tenderness and other measures of palatability
C.		Selection decisions
	1.	Select bulls whose progeny excels in low breakeven prices
	2.	Select bulls whose progeny excels in optimum carcass weights, marbling, percent retail product, tenderness, and overall palatability.

Source: Adapted from BIF Guidelines for Uniform Beef Improvement Programs.

Useful Performance Records

Useful performance records have the following characteristics.

1. They are objectively, accurately, and honestly measured and reported.
2. They are adjusted for environmental effects (e.g., age of calf and age of dam) so that they are comparable.
3. They permit differences in performance records to be compared (*not* the absolute value of the records).
4. They can be compared within a contemporary group to minimize the environmental effects and improve the accuracy of estimating genetic differences.
5. They allow for the use of a national or international animal evaluation system of EPDs and other relevant genetic estimates.

SIRE SELECTION

Importance of the Sire

Effective bull selection should focus on two primary areas: (1) producing live calves and (2) making genetic improvements in economically important traits. This chapter is focused on genetic improvement; producing live calves is discussed in detail in Chapter 11.

Even though a bull contributes 50% of its genetic material to each calf, the magnitude of the bull's contribution is greater because of the increased number of

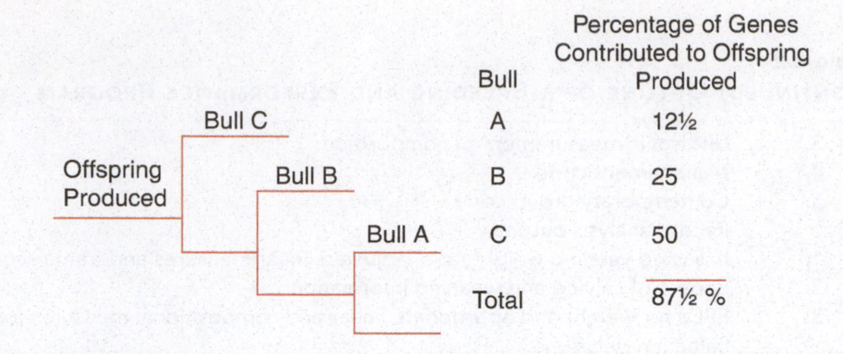

Figure 4.3
Genetic contribution of three bulls to the offspring currently produced in the herd.

Table 4.3
GENETIC CONTRIBUTION OF BULL SELECTION IN CATTLE SELECTED FOR WEANING OR YEARLING WEIGHT

	Selection Differential[1] (selection practiced)	
	Weaning Wt (lb)	Yearling Wt (lb)
Bulls	78	140
Heifers	19	18
Total	97	158
% from bull	78/97 = 80%	140/158 = 89%

[1]Calculation of selection differential is shown in Chapter 12.
Source: Adapted from USDA: U.S. Meat Animal Research Center, Clay Center, NE.

offspring he sires. The genetic importance of successive bulls used in a herd is shown in Figure 4.3. As observed in the figure, the three bulls (A, B, and C) contribute nearly 90% of the genetic material to the calves currently in the herd. Studies conducted by the U.S. Meat Animal Research Center, as shown in Table 4.3, clearly demonstrate that 80–90% of genetic improvement in a herd comes from bull selection.

Most genetic superiority or inferiority of cows, however, depends on the bulls previously used in the herd. Sire selection will be most effective when a thoughtful and thorough evaluation of the strengths and weaknesses of the herd and individuals within the herd has been conducted. An effective genetic improvement program also requires the application of selection pressure to the cow herd.

Before selecting a sire, a careful evaluation of marketing objectives, feed resources and environmental constraints, labor availability, and the gaps between current and desired levels of performance in critically important traits should be undertaken (Figure 4.4). The seedstock breeder must understand this for both the parent herd and the various customers being served.

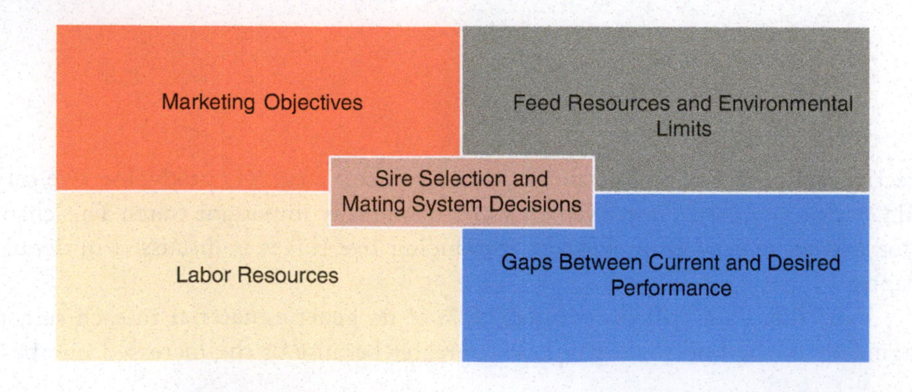

Figure 4.4
Factors affecting bull selection decisions and choice of mating system.

Contemporary Groups

A contemporary group is a set or group of calves that have been raised together and managed in a similar manner. Typically, contemporary groups are composed of animals of the same gender. Contemporary group comparisons help to minimize the effect of non-genetic influences on performance.

Contemporary groups provide the basis for genetic evaluation. A valid contemporary group is a set of cattle from a particular ranch that have been managed under comparable conditions. Comparisons between animals within a contemporary group are accomplished by calculating a ratio using the following formula:

Individual's adjusted measure/average adjusted measure of the contemporary group) × *100*

Animals with above average performance will have a ratio greater than 100, those with performance equal to the average of the group would be 100, and calves with better than average performance will have ratios greater than 100.

For example, if calf A and calf B are contemporary mates with adjusted 205-day weights of 610 and 655 pounds, respectively and the average adjusted 205-day weight of the contemporary group is 620 pounds, then the adjusted weaning weight ratios for A and B would be as follows:

$A = (610/620) \times 100 = 98.4 \qquad B = (655/620) \times 100 = 105.6$

Calf A has performance that is 1.6% below the average of the contemporary group while B has performance that is 5.6% higher than the mean.

Expected Progeny Differences

The development of EPD or expected progeny has made sire selection significantly more effective. Most breed associations make EPDs available from their respective websites while some publish a printed sire summary on an annual or semiannual basis.

Expected progeny difference and **accuracy** (ACC) are important terms used in understanding sire summaries. An EPD combines into one number a measurement of genetic potential based on the individual's performance and that of related animals (such as the sire, dam, siblings, progeny, and other relatives). EPD is expressed as a plus or minus value, reflecting the genetic transmitting ability of a sire for a particular trait. The EPDs reported by the various U.S. breed associations are described in Table 4.4.

Table 4.5 shows sire summary data for several Angus bulls. If bull B and bull C were used in the same herd (each on an equal group of cows), the expected performance of their calves would be as follows:

Bull B's calves would be 16 lb heavier at birth.

Bull B's calves would weigh 59 lb more at weaning.

Bull B's calves would weigh 78 lb more as yearlings.

Bull B's daughters would wean 7 lb more calf as the result of superiority in milk production.

As shown in Table 4.5, bull A and bull D have an optimum combination of all traits. Bull E is a promising young sire with an excellent combination of EPDs. However, the accuracy (ACC) is relatively low and could significantly change with more progeny and as daughters start producing. Accuracy is a measure of expected change in the EPD as additional progeny data become available. Lower accuracy estimates are expected to change more substantially as more information becomes available. High accuracy estimates are also subject to change as more data are added to the calculation, but the degree of possible change is significantly lower (Table 4.6).

Table 4.4
Description of the EPDs Reported by U.S. beef breed associations

PRODUCTION EPDs

Calving Ease Direct (CED) is expressed as a difference in percentage of unassisted births, with a higher value indicating greater calving ease in first-calf heifers. It predicts the average difference in ease with which a sire's calves will be born when he is bred to first-calf heifers.

Birth Weight EPD (BW), expressed in pounds, is a predictor of a sire's ability to transmit birth weight to his progeny compared to that of other sires.

Birth Weight Maternal EPD (BWM), expressed in pounds, is a predictor of the differences in birth weight due to maternal influence.

Weaning Weight EPD (WW), expressed in pounds, is a predictor of a sire's ability to transmit weaning growth to his progeny compared to that of other sires.

Yearling Weight EPD (YW), expressed in pounds, is a predictor of a sire's ability to transmit yearling growth to his progeny compared to that of other sires.

Residual Average Daily Gain (RADG), expressed in pounds per day, is a predictor of a sire's genetic ability for post-weaning gain in future progeny compared to that of other sires, given a constant amount of feed consumed.

Dry Matter Intake (DMI), expressed in pounds per day, is a predictor of difference in transmitting ability for feed intake during the post-weaning phase, compared to that of other sires.

Yearling Height EPD (YH) is a predictor of a sire's ability to transmit yearling height, expressed in inches, compared to that of other sires.

Scrotal Circumference EPD (SC), expressed in centimeters, is a predictor of the difference in transmitting ability for scrotal size compared to that of other sires.

Docility (Doc) is expressed as a difference in yearling cattle temperament, with a higher value indicating more favorable docility. It predicts the average difference of progeny from a sire in comparison with another sire's calves. In herds where temperament problems are not an issue, this expected difference would not be realized.

MATERNAL EPDs

Heifer Pregnancy (HP) is a selection tool to increase the probability or chance of a sire's daughters becoming pregnant as first-calf heifers during a normal breeding season. A higher EPD is the more favorable direction and the EPD is reported in percentage units.

30-month pregnancy (Pg30) predicts the probability that a bull's daughters will become pregnant and calve at 3 years of age, given that they calved as first-calf heifers. This EPD is expressed as a percentage with a higher number being more favorable meaning percentage of a sire's daughters will calve at 3 years of age, given they calved as first-calf heifers.

Sustained Cow Fertility (SCF) results, reported in percentage units, are oriented such that larger breeding values reflect sires whose daughters calve annually for more years. For example, if Bull A has a SCF = 110 and Bull B has a SCF of +95, Sire A's daughters are at 15% less risk of being open after a 60-day breeding season than Sire B's daughters.

Stayability (ST) predicts the genetic difference, in terms of percent probability, that a bull's daughters will stay productive within a herd to at least 6 years of age. The stayability EPD is one of the most effective measures to compare a bull's ability to produce daughters with reproductive longevity.

Calving Ease Maternal (CEM) is expressed as a difference in percentage of unassisted births with a higher value indicating greater calving ease in first-calf daughters. It predicts the average ease with which a sire's daughters will calve as first-calf heifers when compared to daughters of other sires.

Maternal Milk EPD (Milk) is a predictor of a sire's genetic merit for milk and mothering ability as expressed in his daughters compared to daughters of other sires. In other words, it is that part of a calf's weaning weight that is attributed to milk and mothering ability.

Total Maternal or maternal weaning weight (MWW) is measured in pounds of calf weaned by an animal's daughters. They account for average differences that can be expected from both weaning weight direct as well as from milk, and measure a sire's ability to transmit milk production and growth rate through his daughters. Milk EPD + ½ Milk EPD = MWW EPD.

Udder EPDs (UDDR) are reported on the scoring scale. Differences in sire EPDs predict the difference expected in the sires' daughters' udder characteristics when managed in the same environment. For example, if sire A has a UDDR EPD of 0.4, and sire B has a UDDR EPD of −0.1, the difference in the values is 0.5, or one-half of a score. If daughters of sires A and B are raised and managed in the same environment, you would expect half a score better udder suspension in daughters of sire A, compared to sire B. Scores range from 9 (very tight) to 1 (very pendulous) and represent assessments of udder support.

Teat size EPDs (TEAT) are reported on the scoring scale. Differences in sire EPDs predict the difference expected in the sires' daughters' udder characteristics when managed in the same environment. Scores range from 9 (very small) to 1 (very large, balloon shaped) and are subjective assessments of the teat length and circumference.

Table 4.4
(Continued) Description of the EPDs Reported by U.S. beef breed associations

Mature Weight EPD (MW), expressed in pounds, is a predictor of the difference in mature weight of daughters of a sire compared to the daughters of other sires.

Mature Height EPD (MH), expressed in inches, is a predictor of the difference in mature height of a sire's daughters compared to daughters of other sires.

Maintenance Energy (ME) predicts differences in daughters' maintenance energy requirements and is expressed in Mcal/Month. ME EPD allows for the selection of bulls whose daughters will require less feed; thus, reducing cow herd expenses.

CARCASS EPDs

Carcass Weight EPD (CW), expressed in pounds, is a predictor of the differences in hot carcass weight of a sire's progeny compared to progeny of other sires.

Marbling EPD (Marb) predicts differences in intramuscular fat and is expressed as a fraction of the difference in USDA marbling score of a sire's progeny compared to progeny of other sires.

Rib-Eye Area EPD (RE), expressed in square inches, is a predictor of the difference in rib-eye area of a sire's progeny compared to progeny of other sires.

Fat Thickness EPD (Fat), expressed in inches, is a predictor of the differences in external fat thickness at the 12th rib (as measured between the 12th and 13th ribs) of a sire's progeny compared to progeny of other sires.

Percent Retail Cuts (PRC) predict the average differences in cutability that can be expected between the progeny of animals at a given age endpoint. Expressed as a percentage.

Tenderness (WBSF) estimates genetic differences in pounds of force required to shear a rib-eye steak using the Warner–Bratzler shear force test.

$VALUE INDEXES

$Value indexes are multi-trait selection indexes, expressed in dollars per head, to assist beef producers by adding simplicity to genetic selection decisions.

Cow Energy Value ($EN), expressed in dollar savings per cow per year, assesses differences in cow energy requirements as an expected dollar savings difference in daughters of sires. A larger value is more favorable when comparing two animals (more dollars saved on feed energy expenses). Components for computing the cow $EN savings difference include lactation energy requirements and energy costs associated with differences in mature cow size.

Weaned Calf Value ($W), an index value expressed in dollars per head, is the expected average difference in future progeny performance for pre-weaning merit. $W includes both revenue and cost adjustments associated with differences in birth weight, weaning direct growth, maternal milk, and mature cow size.

Feedlot Value ($F), an index value expressed in dollars per head, is the expected average difference in future progeny performance for post-weaning merit compared to progeny of other sires.

Grid Value ($G), an index value expressed in dollars per head, is the expected average difference in future progeny performance for carcass grid merit compared to progeny of other sires. The values of quality grade (economic impact of differences due to marbling) and yield grade (economic impact of differences due to retail yield) can be distilled separately from the overall grid value estimate.

Beef Value ($B) an index value expressed in dollars per head, is the expected average difference in future progeny performance for post-weaning and carcass value compared to progeny of other sires.

Baldy Maternal Index (BMI$) is an index to maximize profit for commercial cow-calf producers who use Hereford bulls in rotational crossbreeding programs on Angus-based cows. Retained ownership of calves through the feedlot phase of production is maintained and the cattle are to be marketed on a CHB pricing grid.

Brahman Influence Index (BII$) utilizes Hereford bulls in a rotational crossbreeding system with Brahman. This index emphasizes fertility and age at puberty and less on growth. Because Brahman cattle are not used in the CHB program, a commodity pricing grid is used.

Certified Hereford Beef Index (CHB$) is a terminal sire index, where Hereford bulls are used on British-cross cows and all offspring are sold as fed cattle on a CHB pricing grid. There is no emphasis on milk or fertility since all cattle will be terminal. This index promotes growth and carcass.

Calving Ease Index (CEZ$) is used to select bulls that will be used in a heifer program. This index has increased emphasis on direct and maternal calving ease.

All-Purpose Index (API) evaluates sires for use on the entire cow herd (bred to both Angus first-calf heifers and mature cows) with the portion of their daughters required to maintain herd size retained and the remaining heifers and steers put on feed and sold on grade and yield.

Terminal Index (TI) evaluates sires for use on mature Angus cows with all offspring put on feed and sold on grade and yield.

(continued)

Table 4.4
(Continued) Description of the EPDs Reported by U.S. beef breed associations

$Cow represents the genetic value in dollars of profit of an animal when retained as a replacement female relative to other animals in the herd. A higher number represents more profitable genetics for maternal productivity. $Cow will serve producers in selecting bulls that will sire daughters with stayability and reproductive efficiency as well as other traits that lead to profitability in a production system, such as milk, calving ease, moderate mature weight, and the ability of calves to gain. A female's genetics also influence the performance of her calves in the feedlot and at slaughter, so traits such as feed efficiency and carcass value are also included in $Cow.

Efficiency Profit Index (EPI) is an economic selection index developed to aid producers in selecting for more feed-efficient cattle that still have acceptable amounts of gain. The EPI provides slight negative pressure on intake, while keeping gain at a constant value. By selecting on this index, producers will be able to find those animals that gain the same amount as their contemporaries while eating less.

Feeder Profit Index (FPI) is an economic selection index designed to aid producers in selecting sires whose progeny will perform in the feedlot and are sold on grade and yield. Well ranking sires for FPI have higher marbling and carcass weight than their contemporaries. As a terminal index, little emphasis is put on maternal traits such as stayability and calving ease.

HerdBuilder Index (HBI) will assist producers in building profitable herds. Significant influence is placed on stayability, heifer pregnancy, and calving ease. Additionally, marbling, yield grade, and growth EPDs impact the HerdBuilder index.

GridMaster Index (GMI) is designed to estimate differences between sires for maximizing profitability of progeny in the feedyard and on the rail.

Table 4.5
Example EPDs for Birth Weight, Weaning Weight, Milk, and Yearling Weight

	Birth Weight		Weaning Weight		Milk			Yearling Weight	
Sire	EPD	ACC	EPD	ACC	EPD	ACC	DTS[1]	EPD	ACC
A	–1	0.95	+20	0.95	+1	0.93	331	+39	0.94
B	+9	0.95	+47	0.95	–17	0.89	88	+74	0.90
C	–7	0.83	–12	0.85	–10	0.79	29	–4	0.85
D	+1	0.95	+22	0.95	+ 3	0.92	230	+54	0.94
E	–1	0.73	+15	0.70	+12	0.42	0	+45	0.66

[1]DTS = daughters.

Table 4.6
Confidence Level of Various Levels of Accuracy Associated with EPDs

Accuracy	Likelihood of EPD Changing with Addition of More Information	Confidence Level
Less than .4	High	Moderate if selecting groups of animals of comparable EPD, low for individuals
.4 to .6	Moderately high	
.6 to .8	Moderate	
Greater than .8	Low	High

Possible change adjustments for a number of EPDs at various levels of accuracy are listed in Table 4.7. For example, if two bulls have same yearling EPD estimates of +80 but bull A has an accuracy of .3 and bull B's estimate has an accuracy of .9, the true genetic estimate for bull B lies within a narrower range of possibilities than does bull A.

Table 4.7
Possible Change Estimates Associated with Various EPDs at Different Accuracies

ACC	CED	BW	WW	YW	MILK	ME	HPG	CEM	STAY	MARB	YG	CW	REA	FAT
0	9	3	16	26	12	3	11	9	7	0.26	0.2	19	0.43	0.04
0.1	8	2.7	15	23	11	2	10	8	6	0.23	0.18	17	0.39	0.04
0.2	7	2.4	13	21	10	2	9	7	5	0.21	0.16	15	0.34	0.03
0.3	6	2.1	11	18	8	2	8	6	5	0.18	0.14	14	0.3	0.03
0.4	5	1.8	10	15	7	2	7	5	4	0.16	0.12	12	0.26	0.02
0.5	4	1.5	8	13	6	1	6	4	3	0.13	0.1	10	0.22	0.02
0.6	3	1.2	7	10	5	1	4	4	3	0.1	0.08	8	0.17	0.02
0.7	3	0.9	5	8	4	1	3	3	2	0.08	0.06	6	0.13	0.01
0.8	2	0.6	3	5	2	1	2	2	1	0.05	0.04	4	0.09	0.01
0.9	1	0.3	2	3	1	0	1	1	1	0.03	0.02	2	0.04	0
1	0	0	0	0	0	0	0	0	0	0	0	0	0	0

Source: Adapted from Red Angus.

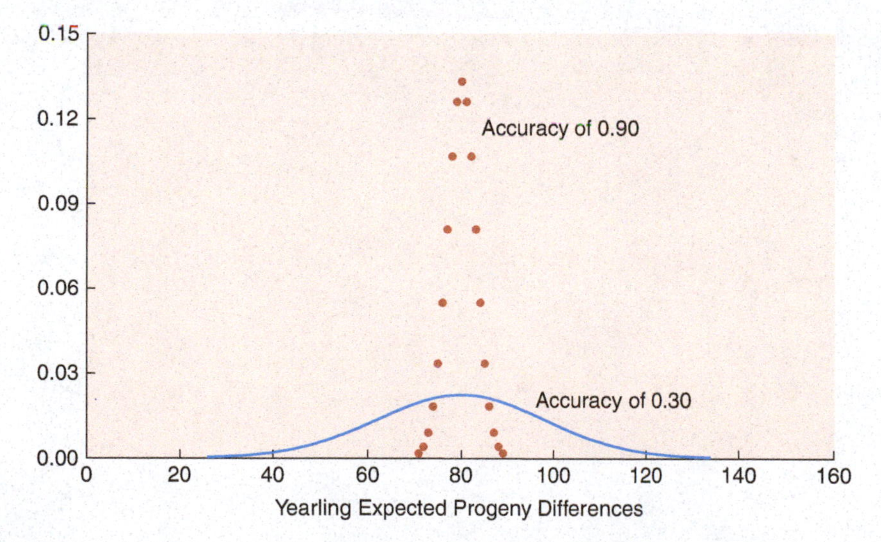

Figure 4.5
Possible change in EPDs where two bulls have the same yearling weight EPD (+80) but different accuracies (0.30 vs. 0.90).

Bull A has a possible change of plus or minus 18, which means that his true yearling value lies between +62 and +98 while B's true value (possible change of +9) lies within the range of +77 and +83 (Figure 4.5). The possibility of the EPD changing as more data are added is the same for both sires, but the magnitude of change is quite different.

Not all breed associations include the same traits for evaluation in their national genetic evaluation systems, although birth weight, weaning weight, milk, and yearling weight are common across most systems. Table 4.8 describes the various genetic estimates provided by breed associations. With the incorporation of genomic data, the estimates made on non-parent animals provide a boost in accuracy that is substantial.

One of the challenges with single-trait estimates is that profitability is affected by a multitude of traits. In response, economic multi-trait selection indexes, expressed in dollars per head, have been developed to assist beef producers by adding simplicity to genetic selection decisions. These indexes are developed by combining various traits, weighted for the level of their impact, on a breeding objective. A description of

Table 4.8
Various Genetic Estimates Calculated for the Major Beef Breeds

Trait	Angus	Beefmaster	Brahman	Brangus	Braford	Charolais	Gelbvieh	Hereford	Limousin	Red Angus	Shorthorn	Simmental
Production												
Birth weight	x	x	x	x	x	x	x	x	x	x	x	x
Calving ease direct	x			x		x	x	x	x	x	x	x
Docility	x		x						x			x
Dry matter intake	x						x	x				
Residual gain	x						x					
Scrotal circumference	x	x		x		x	x	x	x			
Weaning weight	x	x	x	x	x	x	x	x	x	x	x	x
Yearling height	x											
Yearling weight	x	x	x	x	x	x	x	x	x	x	x	x
Maternal												
30-month pregnancy							x					
Birth weight maternal			x		x							
Calving ease maternal	x			x		x	x	x	x	x	x	x
Heifer pregnancy	x							x		x		
Maintenance energy										x		
Mature height	x											x
Mature weight	x											x
Milk	x	x	x	x	x	x	x	x	x	x	x	x
Stayability							x		x	x		x
Sustained cow fertility								x				
Teat size								x				
Total maternal		x		x	x	x	x	x		x	x	x
Udder suspension								x				

(continued)

Table 4.8
(Continued) Various Genetic Estimates Calculated for the Major Beef Breeds

Trait	Angus	Beefmaster	Brahman	Brangus	Braford	Charolais	Gelbvieh	Hereford	Limousin	Red Angus	Shorthorn	Simmental
Carcass												
Carcass weight	x		x		x	x	x	x	x	x	x	x
Fat thickness	x	x	x	x	x	x		x		x	x	x
Marbling	x	x	x	x	x	x	x	x	x	x	x	x
Retail product			x									
Rib-eye area	x	x	x	x	x	x	x	x	x	x	x	x
Tenderness			x									x
Yield grade									x			x
Economic Indexes												
All-purpose $												x
Baldy maternal value $								x			x	
Beef value $	x											
Brahman influence profit $								x				
Calving ease profit $								x			x	
Certified Hereford Beef $								x				
Cow $							x					
Cow energy value $	x											
Efficiency profit $							x					
Feeder profit $							x					
Feedlot value $	x										x	
Grid master $										x		
Grid value $	x											
Herd builder $										x		
Terminal sire index $						x			x			x
Weaned calf value $	x											

various indices can be found in Table 4.4 while a comparison of which indexes are calculated by various breeds can be found in Table 4.8. The $Value is an estimate of how future progeny of each sire is expected to perform, on average, compared to progeny of other sires in the database if the sires were randomly mated to cows and if calves were exposed to the same environment.

Bull selection is complex because the desired genetic improvement is for a combination of several traits. Continuous selection for only one trait may result in problems in other traits. A good example is selecting for yearling weight alone, which results in increased birth weight because the two traits are genetically correlated. Birth weight is associated with calving difficulty, so increased birth weight might be a problem. Additionally, yearling weight and mature weight are positively correlated; thus, a singular focus on increasing yearling weight will likely lead to large mature sizes, which may be inappropriate for environments where feed resources are limited. The biggest challenge in bull selection is selecting bulls that will improve maternal traits. Frequently, too much emphasis is placed on yearling growth and frame size that often results in excessive mature size. These traits can be antagonistic to maternal traits, such as calving ease, early puberty, and maintaining a cow size consistent with an economical feed supply. Figure 4.6 shows an example of stacking a pedigree for maternal traits for yearling bull selection where the bulls

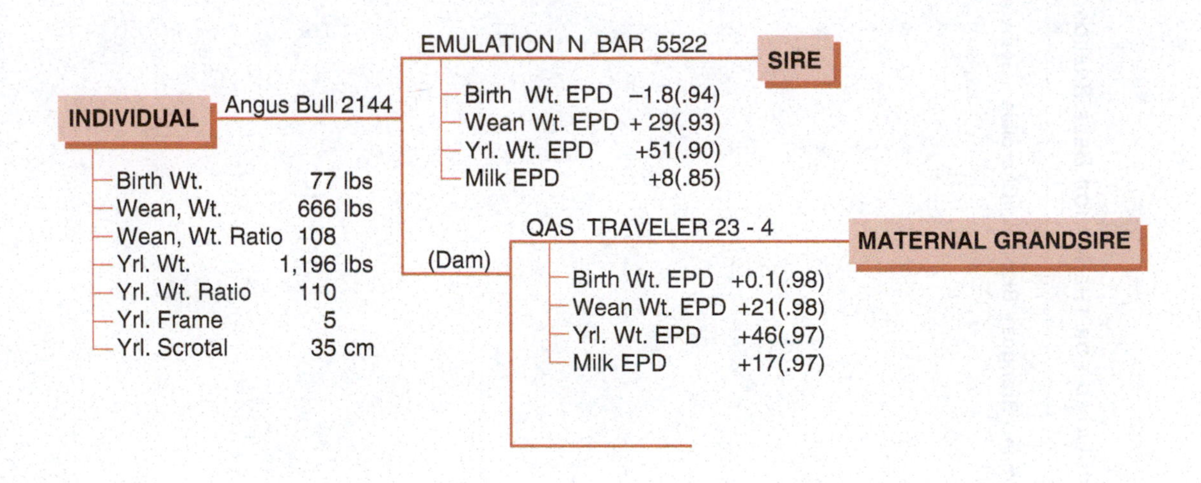

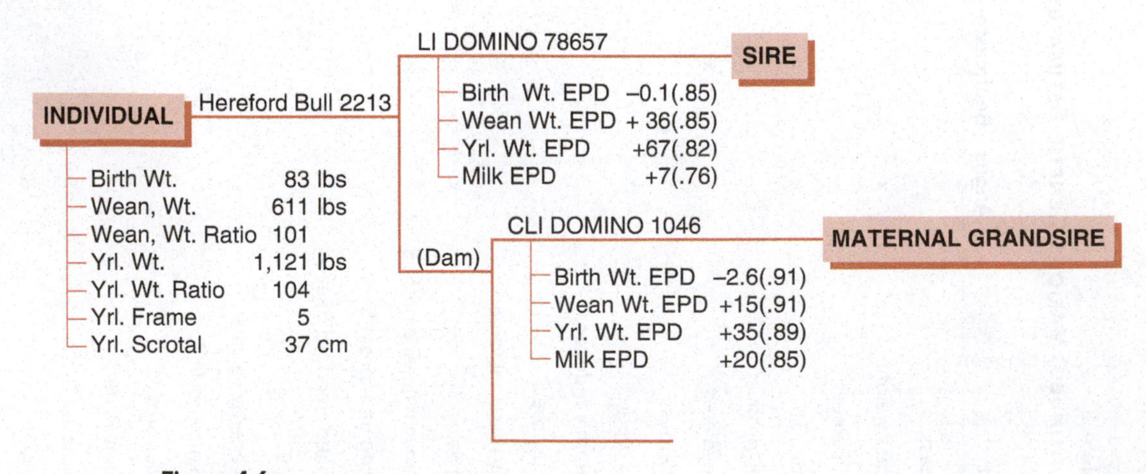

Figure 4.6
Yearling bull selection with a balanced trait emphasis. Bulls can be used naturally on replacement heifers.

are used natural service on heifers. If an older bull can be used AI, then the emphasis would be on accuracies higher than are available on young bulls. Examples of older bulls that could be used AI are the sires and maternal grandsires of the yearling bulls shown in Figure 4.6.

Table 4.9 identifies the bull selection criteria for commercial beef producers. These criteria should also help seedstock breeders focus on their selection criteria.

Table 4.9
BULL SELECTION CRITERIA FOR COMMERCIAL BEEF PRODUCERS AND BREEDERS

BREEDING PROGRAM GOALS

Selecting for an **optimum** combination of **maternal, growth, and carcass traits** to **maximize profitability**

(Avoiding genetic antagonisms and environmental conflicts that come with maximum production or single trait selection)

Provide genetic input so **cows** can be **matched with their environment**—cows that wean more lifetime pounds of calf without overtaxing the forage, labor or financial resources.

Stack pedigrees for maternal traits.

MATERNAL TRAITS

Early puberty/high conception:
Calve by 24 months of age

Calving ease: Moderate birth weights (65–80 lb hfrs; 75–90 lb cows). Calf shape (head, shoulders, hips) that relates to unassisted births.

Early rebreeding and longevity ("fleshing ability"): "5" body condition score at calving without high cost feeding.

Mature weight: Medium sized cows (1,000–1,250 lb; BCS 5).

Milk production: Moderate (wean 500–550 lb calves under average feed supply).

GROWTH AND CARCASS TRAITS—EARLY GROWTH AND COMPOSITION:

Rapid gains—relatively heavy weaning and yearling weights within medium (4–6) frame size.

Yield grade 2, grading ≥70% Choice (steers harvested at 1,250 lb).

FUNCTIONAL TRAITS (LONGEVITY):

Udders (shape, teats, pigment)
Eye pigment
Disposition
Structural soundness

Criteria for Maternal Traits[1] Birth weight EPD: Preferably under +1.0 lb. Evaluate calving ease of daughters.

Scrotal circumference: 32–40 cm at 365 days of age. EPD above +0.3 cm. Passed breeding soundness exam

Milk EPD: under +15[2]

Maternal EPD: 20–40 (prefer +25)

Weaning weight EPD: +20 lb

Yearling frame size: 4.0–6.0 (smaller frame size will adapt better to harsher environments—e.g., less feed, more severe weather, less intensive care).

Mature weight: Under 2,000 lb with BCS of 5 (preference for future). Currently, bulls under 2,400 lb will have to be considered.

Body condition: Backfat of approx. .20 inches at yearling weights of 1,100–1,250 lb. monitor reproduction of daughters BCS.

Yearling weight EPD: +40 lb

Accuracies: All EPDs of 0.90 and higher (older, progeny tested sires) for AI bulls. Young bulls without EPDs or with accuracies below 0.90—evaluate trait ratios and pedigree EPDs. Select sons of bulls that meet EPDs listed above.

Growth and carcass traits as noted in goals

Functional traits: Visual evaluation of bull and his daughters.

Visual:
Functional traits as noted in goals, sufficiently attractive to sire calves which would not be economically discriminated against in the marketplace.

Preference for "adequate middle" as medium frame size cattle need middle for feed capacity.

Gentle disposition.

Criteria for Terminal Traits[3] Birth weight EPD:
Preferably no higher than +5.0; want calf birth weights in 80–100 lb range (calves only from cows).

Scrotal circumference: 32–40 cm at 365 days of age. Passing score on the breeding soundness exam.

Milk EPD: Not considered

Maternal EPD: Not considered

Weaning weight EPD: +40 lb or higher

Yearling weight EPD: +60 lb or higher

Yearling frame size: 6.0–8.0 (should be evaluated with frame size of cows so slaughter progeny will average 5.0–6.0).

Mature weight: No upper limit as long as birth weight, frame size, and carcass weights are kept in desired range.

Body condition: Evaluate with body condition of cows so that carcass fat of 0.4 in. progeny at 1,100–1,300 lb slaughter weights.

Accuracies: All EPDs of 0.90 and higher (older, progeny tested sires) for AI bulls.

Young bulls without EPDs or with accuracies below 0.90—evaluate trait ratios and pedigree EPDs. Select sons of bulls that meet EPDs listed above.

Growth and Carcass Traits as noted in goals Visual:
Functional traits: as noted in goals sufficiently attractive to sire calves that would not be economically discriminated against in the marketplace.

[1]EPDs are Angus-based.
[2]Assumes extensive management under a grazed forage scenario.
[3]EPDs are for Continental breeds; assumes no breeding replacement female will be kept from these bulls.

First, the breeding program goals and the traits are identified. Second, the bull selection criteria are identified, emphasizing maternal traits, and then relatively high growth and carcass traits without causing a serious antagonism to maternal traits. Maternal, growth, and carcass traits would be balanced in selecting bulls of one biological type if a commercial producer was using only one breed, a rotational crossbreeding program, producing the cows to be used in a terminal crossbreeding program, or selecting bulls for a composite breed. Finally, Table 4.9 identifies the traits emphasized in the selection of terminal cross bulls. In selecting terminal cross sires, little emphasis is placed on maternal traits because replacement heifers from this cross are not kept in the herd.

Figure 4.7 shows a performance pedigree that includes the bull's pedigree and performance information on the bull, his sire, and his dam.

Birth Weight

Birth weight should be objectively measured and recorded on all calves (live and dead) within 24 hours of birth. Estimates of birth weights are inaccurate and are of limited value.

Figure 4.7
Performance pedigree of a Simm–Angus bull.
Source: Tom Field.

American Simmental Association
Certificate of Registration

3042663
Registration Number

SIMANGUS BULL

C580 LE
Tattoo/Brand - Loc

ASR SUPER BALDY C580

1/2 SM 1/2 AN	02-13-15	SINGLE	POLLED	PCS
	Birth Date	Birth Type	H/P/S Status	Genetic Status

Owner: 322240 THOMAS FIELD - RAYMOND, NE — 04-27-16 Reissue Date
Breeder: 008973 ALTENBURG SUPER BALDY RANCH LLC - FORT COLLINS, CO — 11-17-15 Date of Origin
Prev Ownr: 008973 ALTENBURG SUPER BALDY RANCH LLC FORT COLLINS CO — 03-19-16 Date of Sale

	Name	Identification		YR Born	H/P/S
	S A V BISMARCK 5682	AN15109865		2005	P
	S A V BRILLIANCE 8077	AN16107774		2008	P
	S A V BLACKCAP MAY 5270	AN15108286		2005	P
Sire	**PVF INSIGHT 0129**	AN16805884		2010	P
	P V F NEW HORIZON 001	AN13606093		2000	P
	PVF MISSIE 790	AN15746654		2007	P
	P V F MISS RAPTOR 024	AN13727121		2000	P
	SVF/NJC MO BETTER M217	PB SM	2,180,592	2002	P
	HSF BETTER THAN EVER	PB SM	2,399,920	2007	P
	HSF VICTORIA P30	PB SM	2,247,919	2004	P
Dam	**ASR MS DRAFT PICK X0171**	PB SM	2,554,457	2010	P
	ASR DRAFT PICK K062	PB SM	2,067,890	2000	P
	ASR MISS PICK M277	3/4 SM 1/4 AN	2,158,089	2002	P
	AGS MISS INCUMBENT C319	1/2 SM 1/2 AN	1,695,292	1993	P

Transfer Application: *Transfer of ownership is not effective unless recorded with the American Simmental Association*

Date of Sale: ______ ☐ *Check here if female sold bred or exposed* ☐ *Check here if female sold open*

Buyer/Ranch Name: ______ Member Number: ______

Address: ______ City: ______ ST: ______ Zip: ______

Please complete calf registration and transfer (on back of pedigree) if calf was sold and not registered prior to sale of dam with calf at side.

If female sold bred or exposed, complete the following information:

AI Sire Name and Registration Numbers	AI Dates	Pasture Sire Name and Registration Numbers	Pasture In Dates	Pasture Out Dates

Signature of seller or an authorized representative ___ Check here if representing seller as sales manager.

Personal Signature *Date*

Return Certificate to
___ Buyer
___ Seller
Other: ______

CORPORATE SEAL

Wade Shafer, Executive Vice President
American Simmental Association
1 Simmental Way, Bozeman MT 59715

The animal described in this registration certificate has been entered in the ASA's Herdbook under the rules and bylaws in effect at the time of registration and may be updated, corrected or cancelled as provided therein. The registration is based upon information provided to the ASA from outside sources and the ASA does not guarantee the genetic makeup, percentage, breeding, ownership or fitness of the animal for any purpose. The official record of any animal registered in the ASA's Herdbook or entered into the ASA's database is the electronic record maintained by the ASA. For the most current information on an animal, contact the ASA or visit the ASA's website: www.simmental.org.

Table 4.10
AGE OF DAM ADJUSTMENT FACTORS FOR BIRTH WEIGHT AND WEANING WEIGHT

Age of Dam (years)	Birth Weight[1] (lb)	Weaning Weight (lb)	
		Male	Female
2	8	60	54
3	5	40	36
4	2	20	18
5–10	0	0	0
11 and older	3	20	18

[1]Standard birth weight: male (75 lb) and female (70 lb). Standard birth weights for male calves are as high as 90–92 lb for some breeds.
Source: Beef Improvement (BIF) Guidelines for Uniform Beef Improvement Programs.

Age of cow and sex of calf both have an influence on the birth weight of the calf. Table 4.10 shows the BIF age of dam adjustments for birth weights so that birth weight records can be standardized for comparison. Some breeds use BIF adjustments in their record analysis while other breeds have their own specific adjustment factors. The birth weights of calves from 2-year-old heifers of different breeds are 4–8 lb lighter than the birth weights of mature cows (5–10 years of age). Bull calves are usually 5–8 lb heavier than heifer calves. A sex adjustment of birth weight is usually not done because performance records are compared within sex-of-calf categories.

Birth weight evaluations are more critical when calving 2-year-old heifers. However, to make mating decisions with the lowest likelihood of dystocia, the use of calving ease EPDs is advised.

Calving Ease

Birth weight is the best measure of calving ease or calving difficulty (dystocia). Although calving ease scores can be recorded and utilized, they are more subjective than birth weight. For example, a heifer may be manually assisted to give birth earlier than needed, but with additional time she could have calved without assistance. Furthermore, an easy pull to one producer may be a difficult pull to someone else. The recommended calving ease scores are shown in Table 4.11.

Weaning Weight

Pre-weaning growth is best measured as the adjusted 205-day weight. Calves should be weighed as near to 205 days of age as possible, with 160–250 days being the recommended acceptable range. Calves weighed outside this range will result in

Table 4.11
CALVING EASE SCORES

Score	Description
1	No difficulty; no assistance
2	Minor difficulty; some assistance
3	Major difficulty; usually mechanical assistance
4	Caesarean section or other surgery
5	Abnormal presentation

Source: BIF Guidelines for Uniform Beef Improvement Programs.

information that cannot be meaningfully compared to other performance data without risk of error.

Weaning weights are adjusted for age of calf and age of dam by using the following formula:

$$\text{Adjusted 205 day wt(lb)} = \frac{\text{Actual weaning wt} - \text{birth wt}}{\text{Age in days}} \times 205 + \text{birth weight} + \text{age of dam adjustment}$$

Weaning weights need to be adjusted for age of dam due to the variation in milk production by different aged females. For example, a 7-year-old cow typically produces more milk than a 2-year-old. The appropriate number of pounds must be added to the 205-day weight, previously adjusted for age of calf (see Table 4.10). For example, if the 205-day weight of a bull calf from a 2-year-old heifer is 465 lb, then 60 lb would be added (for a total of 525 lb) to adjust to the 5- to 10-year-old cow basis. Many breeds have their own weaning weight adjustments that differ from the BIF adjustment.

Weaning weight ratio is then computed by dividing each calf's adjusted 205-day weaning weight by the average of its contemporary group. The ratio is expressed as a percentage of the contemporary group. For example, a heifer calf with an adjusted weaning weight of 500 lb, where the contemporary group average was 550 lb, would have a weaning weight ratio of:

$$\frac{500}{550} \times 100 = 90.9$$

Thus, the example calf is 9.1% (100 – 90.9) lighter than the average of her contemporary group. A calf with a ratio of 105 would be 5% heavier than the average of its contemporary group.

Yearling Weight

After obtaining data on post-weaning gain over a 100–150-day feeding period, the adjusted yearling weight can be calculated as follows:

$$\text{Adjusted 365-day wt} = 160 \times \frac{\text{Actual final wt} - \text{Actual weaning wt}}{\text{Number of days between weights}} + \text{Adjusted 205-day wt}$$

Yearling weight ratios are computed separately for each contemporary group by dividing the individual animal's record by the contemporary group average and multiplying by 100.

Mature Cow Weight

Mature cow weight is an economically important trait because it is part of a biological-type description. It is related to reproductive performance, when cows are matched to their economical feed environment, and to the carcass weight of a cow's progeny.

Milk

Milk EPDs are used to compare genetic differences in weaning weight due to genes for milk production passed from parents to daughters. In essence, milk EPDs help breeders make comparisons between potential parents for their ability to contribute to weaning weight as a result of lactational differences. These estimates also help breeders match the desired levels of milk production to the available forage resources in both their own and their customer's production environment.

Total maternal EPDs combine the effects of weaning weight (direct) EPD and milk EPD to allow comparisons of weaning weight differences of calves from daughters. Total maternal EPDs are calculated by adding milk EPD to one-half of the weaning EPD.

Carcass Traits

The traits most commonly of interest are carcass weight, rib-eye area, marbling or intramuscular fat score, percent retail product, backfat, and tenderness. Carcass weight and tenderness (estimated by Warner–Bratzler shear force) are measured directly on progeny. Rib-eye area, marbling score, percent retail product, and backfat can be measured via direct progeny testing or estimated via ultrasound measures on live animals. The American Gelbvieh Association released a grid merit EPD in 2001 to allow for an economic comparison between sires for their progeny's ability to perform under the Gelbvieh Alliance grid pricing system. Since 2001, a number of breed associations have released EPD estimates for various profit indices. These will be discussed in more detail in Chapter 12.

Other Traits

Scrotal circumference, stayability, and docility are other traits included in some sire summaries. Scrotal circumference EPDs are utilized to ascertain differences in age of puberty. The desired level of performance in this trait will vary based on the breeds being used and their intended purpose.

Stayability is a measure of longevity that assesses the likelihood of a female remaining in the herd to at least 6 years of age. EPDs for this trait are expressed in percentages to allow comparisons of the percentage of a sire's daughters expressing longevity.

Docility is an important trait in some breeds where concerns have been expressed about the temperament of certain sire groups. Docility EPDs are calculated from subjective scoring of temperament as a 1 to 6 scale, with 1 being docile and 6 being very excitable.

Locating Genetically Superior Bulls

Commercial and seedstock producers should identify breeders who keep honest, comparative records on their cattle. Most performance-minded seedstock producers have useful weaning, yearling, and calving ease records on their bulls. Increasingly, seedstock producers obtain feedlot and carcass data on progeny from their own bulls or use bulls that have carcass EPDs. Seedstock breeders should provide meaningful performance data on their bulls and commercial producers should demand information that impacts profitability.

The best approach for selecting outstanding bulls involves (1) evaluating the objectives of the buyer, (2) scrutinizing performance records, and then (3) selecting several bulls whose records reflect the desired level of performance. From that group, bulls can be selected that meet the buyer's goals for skeletal size, freedom from predisposition to fat, scrotal size, and other soundness traits.

Bull testing stations provide a place where bulls from several sources can be compared for post-weaning growth. Some testing stations test maternal bulls (e.g., calving ease based on birth weights) in a separate contemporary group.

Bulls versus Steers

Seedstock and commercial producers should be aware of how sex differences affect the productivity and performance of economically important traits. Sexual maturity, as reflected in bulls, steers, and heifers, has a marked effect on growth and carcass characteristics (Chapter 17). Most feedlot and carcass goals are identified with steers as most cattle going to slaughter are steers.

Table 4.12 compares the expected productivity differences between steers and bulls. The data in the table demonstrate that selected bulls need to be higher in performance than their counterpart steers. Producers that select bulls at the same level of productivity expected in steers will generally be disappointed in the siring ability of the bulls.

Table 4.12
COMPARATIVE AVERAGE PRODUCTIVITY OF BULLS AND STEERS[1]

Trait	Bulls	Steers
205-day weight (lb)	525	500
Feedlot gain (lb/day)	3.25	3.00
365-day weight (lb)	1,050	980
Yearling hip height (in.)	51	49
Feed/lb gain	6.8	7.8
Backfat (in.)	0.2	0.4
Rib-eye area (sq. in.)	13.0	11.0
Cutability (%)	51%	48%
Quality grade	Select	Choice

[1]The growth data are compared at same age, while carcass data are compared at same slaughter weight.
Source: Adapted from multiple sources.

Seedstock breeders can maximize the progress they make by utilizing reproductive technologies, such as artificial insemination, embryo transfer, or the use of sexed semen. By virtue of these procedures, a breeder can access the very best genetics from a national, if not an international, gene pool. These techniques also make it possible for selection of proven parents with high associated accuracies.

SELECTING REPLACEMENT HEIFERS

Heifers, as replacement breeding females, can be selected for several traits at different stages of their productive life. The objective is to identify heifers that will conceive early in the breeding season, calve easily, give a flow of milk consistent with the most economical feed supply, wean a relatively heavy calf, and make a desirable genetic contribution to the calf's post-weaning growth and carcass merit while remaining free of structural defects to allow a high degree of longevity.

Beef producers have found it challenging to determine which young heifers will make the most productive cows. When a cow herd begins to reach an optimum level of performance, then producers should cull heifers at weaning and yearling ages that are in the low and high categories for the various traits. Heifers in the middle one-half of the group are then kept to evaluate their pregnancy status after breeding. Such an approach helps to hold performance of the cow herd at or near the current level.

Table 4.13 shows the process that some producers use to select the most productive replacement heifers. This selection process assumes that more heifers will be retained at each stage of production than the actual number of cows to be replaced in the herd. The number of replacement heifers that producers keep is based primarily on cost of production and market value at the various stages of production. More heifers than the number needed should be kept through pregnancy check time. Shortly after the end of the breeding season, a final selection of heifers is based on early pregnancy.

SELECTING COWS

Unless a producer is involved in a herd size reduction program, relatively few cows can be culled to enhance genetic progress in the herd. Most cows basically cull themselves because of poor reproductive performance, death loss, and age. Producers can practice some culling on performance records if the death loss is low, if cows over 9–10 years of age can sustain high levels of productivity for a few more years, or if a higher percentage

Table 4.13
REPLACEMENT HEIFER SELECTION GUIDELINES AT DIFFERENT PRODUCTIVE STAGES

Stage of Heifer's Productive Life	Emphasis on Productive Trait	
	Primary	Secondary
Weaning (7–10 months)	Cull only the heifers whose actual weight is too light to prevent them from showing estrus by 14 months of age. Also, consider the economics of the weight gains needed to have puberty expressed. Also, cull heifers that are too large in frame and birth weight.	Weaning weight ratio Weaning EPD Milk EPD Predisposition to fatness Adequate skeletal frame Skeletal soundness Cull for bad disposition Dam's performance summary
Yearling (12–15 months)	Cull heifers that have not reached the desired target breeding weight (e.g., minimum of 675–750 lb for medium-sized breeds or cross; minimum of 750–850 lb for large-sized breeds and crosses). Also, cull heifers that are extreme in size and weight.	Weaning weight EPD Milk EPD Yearling weight ratio Yearling EPD Predisposition to fatness Adequate skeletal frame Skeletal soundness
After breeding (19–21 months)	Cull heifers that are not pregnant and those that will calve in the latter one-third of the calving season.	Weaning weight EPD Milk EPD Yearling EPD Predisposition to fatness Adequate skeletal frame Skeletal soundness
After weaning first calf (31–34 months)	Cull to the number of first-calf heifers actually needed in the cow herd based on early pregnancy and weaning weight performance of the first calf. Preferably all the calves from these heifers have been sired by the same bull.	

of heifers are kept as replacements. The guideline "select heifers on EPDs and cull cows on actual performance" is a good approach to follow.

Cows can be culled on pregnancy status and calving interval. This basis for culling appears to be more economic than selection for genetic progress. Open cows can rebreed and produce calves regularly during the remainder of their productive life. The economic decision to cull open cows should be based more on replacement cost and the time value of money. It may appear obvious that the commercial producer should more frequently cull the open females; however, the decision is not as obvious or easy for a seedstock breeder to cull open females with desirable genetic merit. This decision is controversial in the minds of many breeders. The decision, however, should be made on the basis of known biological and economic relationships and not on opinions. Complicating the decision is the use of AI Failures in heat detection, semen thawing, and insemination technique that may result in lowered rates of fertility that should not be attributed to the female.

In most herds, more than 50% of culling is due to unsatisfactory reproductive performance. Cow longevity and profitability, for both seedstock and commercial producers, can be increased by matching the biological type of cow (mature weight, level of milk production, and body condition score) to the most economical feed conditions.

Table 4.14
PERFORMANCE DATA ON HIGH- AND LOW-PRODUCING COWS IN THE SAME HERD

Cow	No. Calves	Weaning Wt (lb)	Weaning Wt Ratio	Yearling Wt Ratio
1	9	583	112	108
2	9[1]	464	89	94

[1]One calf died before weaning (not computed in averages).

Culling cows to assist in making genetic change in a herd is relatively simple if the cows performed together under similar environmental conditions. One of the biggest problems in comparing cow records is that the cows likely have produced calves from different bulls. Relatively large genetic differences in the bulls used in the herd would make the cows' records less comparable than when all cows are bred to the same bull. This is usually not a serious problem when producers are culling the bottom 10–20% of their cows on performance records. Table 4.14 compares the performance records of high- and low-producing cows of the same herd. The lower weight ratios identify the low-producing cows.

Caution should be exercised in culling pregnant cows that are below herd average in performance. Heifers that replace these cows are usually high-cost inputs and the weaning weights of their calves can be less than the calves from culled cows. The latter is due to age-of-dam effect on weaning weight. Some cow-calf producers can market their less-productive cows to other producers who have better feed resources. These cows can be sold above slaughter price and still be profitable for the new owners. These cows can then be economically replaced with heifers that will improve profitable herd performance.

Genetic Trend

Evaluating the change in EPDs over the years can monitor the genetic trend for a seedstock herd or breed. Figure 4.8 shows the genetic trend for several traits in the Angus breed. Genetic trends result from the collective decisions made by breeders of a specific breed in response to both economic signals of their customers and their own

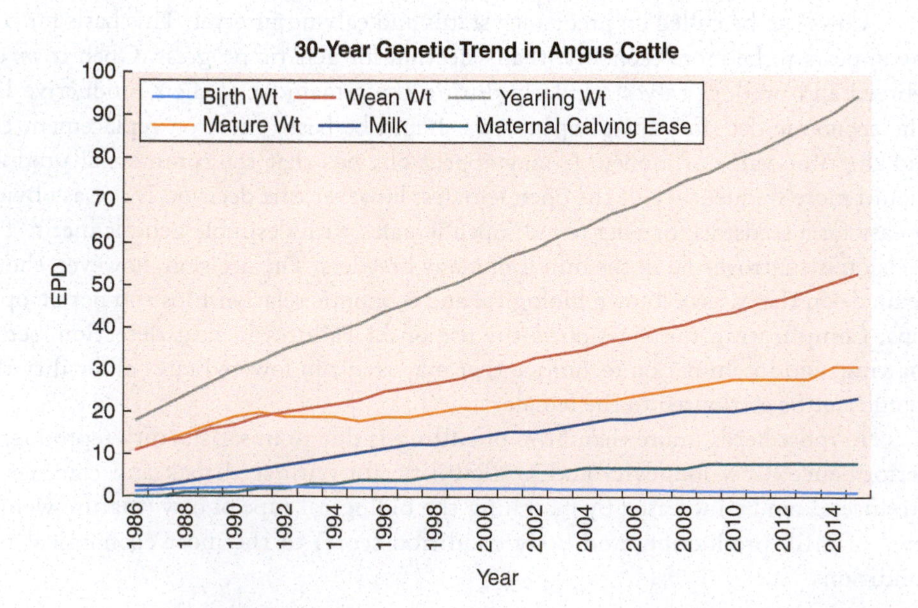

Figure 4.8
Genetic trends for multiple traits in Angus cattle.
Source: Based on American Angus Association Sire Evaluation Report.

perceptions about appropriate levels of performance in the traits that they deem to be important. Monitoring genetic trend for the breed is useful but even more so is comparing the genetic trend of a specific herd compared to changes in the breed. Such analysis provides a benchmarking system to help monitor progress toward an established set of goals and objectives.

Customer Service

Because of the dramatic impact from sire selection, it is important for cow-calf producers to carefully select not only their sires but also their breeders from whom they purchase genetics. The interface between seedstock and cow-calf producers is of critical importance. The communication of goals and needs and the ability to meet those needs are largely responsible for any progress made by individual producers to become more competitive and for the industry to make progress.

It is not enough to simply sell a bull or a breeding female, in fact given that AI and ET make elite genetics widely available to seedstock providers, it is difficult to create market differentiation on genetics alone. In light of the vast amount of performance data, multiple channels available for marketing customer cattle, and competitive nature of the seedstock business, creating a clear advantage with customers requires a higher degree of focus on service than ever before (Figure 4.9). Examples of value-added services include:

- Information packaged as a decision-support tool.
- Providing market outlets for calves from customers' herds.
- Facilitating the transfer of knowledge via customer seminars, newsletters, and discussion groups.
- Offering more than one breed or composite to customers.
- Assisting customers with the development of value-added management programs that enhance profits, such as participation in branded product initiatives and supply chain arrangements.
- Facilitating the collection, summarization, and transfer of useful information within the beef industry.

Customer orientation begins with developing a system to effectively interact with both existing and potential customers to understand their goals, challenges, and needs. Too often, seedstock breeders create cattle they like instead of developing a genetic program squarely focused on the needs of the customer.

The personal characteristics of a customer-oriented seedstock supplier include (1) a desire to actively listen to customers and the willingness to seek frequent opportunities for interactions with past and potential clients; (2) the willingness to make on-site visits to customers during key times when the performance of the genetics is on the line (calving, pregnancy checking, and weaning); (3) the ability to offer customers

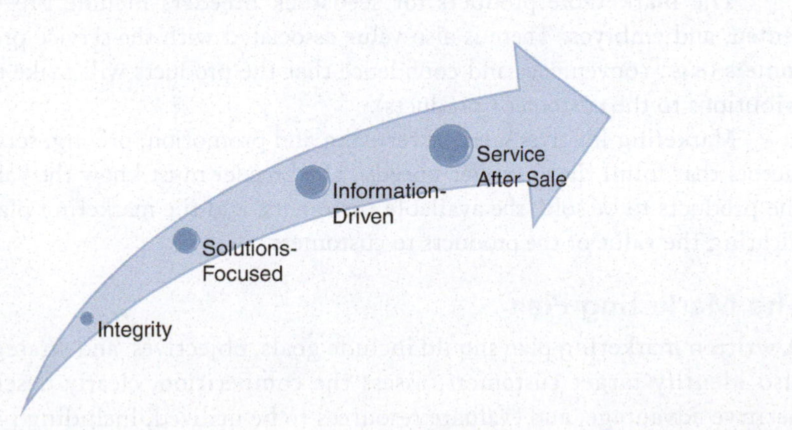

Figure 4.9
Characteristics of seedstock producers destined to succeed in the long run.

significant input into the sire selection process for the seedstock herd; and (4) a passion for the quality of the product and the long-term well-being of the client.

The dramatic structural changes occurring in the beef industry will not leave the seedstock sector unscathed. In fact, a number of macro trends will have a significant impact on how genetic suppliers function.

1. Selling cattle on an individual carcass basis is becoming the dominant method of merchandising fed cattle. Seedstock suppliers will be required to understand how their cattle perform throughout the beef chain, and provision of feedlot and carcass data will be a prerequisite for market entry.
2. Commercial cattle producers will not abandon heterosis. In an effort to simplify breeding programs at the commercial cow-calf level, the demand for hybrid bulls will grow.
3. The rapid advancement of genetic technologies will place more pressure on genetic suppliers to provide information interpretation, educational initiatives, and breeding system consultation services. Successful breeders will endeavor to direct change instead of being victimized by it.
4. The new business environment will nurture those who are partner, customer, and consumer driven. The creation of profitable alliances, partnerships, and supply chains should be at the forefront of a seedstock enterprise's business plan.
5. Adding value to seedstock cattle will be information driven. For this reason, genetic suppliers must develop a sincere commitment to the process of data collection, analysis, synthesis, and application. The professional seedstock breeder and the professional seedstock cattle organization will make whole-herd reporting of data a priority as well as the development of new genetic estimates of value to their customers.

Given the scope of these changes, the successful genetic supplier must undergo an ongoing evaluation process that will include a detailed and objective assessment of the business plan, breeding objectives, skills and knowledge management, products, and services.

MARKETING DECISIONS

The seedstock breeder who establishes long-term success develops a strong reputation or brand identity that creates marketing pull driven by strong demand from a customer base that may include various combinations of commercial cow-calf enterprises, other seedstock suppliers, and youth project participants (4.H, FFA). There is significant variation in each of these groups and it is critical that a seedstock provider develop a clear market focus.

The marketable products for seedstock breeders include breeding animals, semen, and embryos. There is also value associated with the service provided to customers (e.g., convenience and confidence that the products will make economic contributions to the customer's products).

Marketing involves sales, advertising and promotion, pricing, service, and other factors that fulfill the customer's needs. The breeder must know the value and cost of the products to be sold, the available customers, and the marketing plan for communicating the value of the products to customers.

The Marketing Plan

A written marketing plan should include goals, objectives, and strategies. It should also identify target customers, assess the competition, clearly describe the competitive advantage, and evaluate resources to be utilized, including production and marketing costs. The marketing plan should also be goal-driven. Goals might include identifying the number of customer contacts to be made either by phone or in person, reassessing the advertising copy and budget, assisting customers in the collection and use of feedlot and carcass data, or developing a social media presence.

Breeders marketing bulls to commercial cow-calf producers typically find that the majority of their customers reside within a 150–200 mile radius from their enterprise. Other breeders may sell semen to commercial producers or semen and breeding cattle to other breeders on a national or international scale. Thus, the merchandising reach can be broad or narrow, depending on the customers being targeted by the breeder.

An essential part of the marketing plan is to know production costs. Enterprise budgets should be kept so that breakeven prices can be calculated for weaned calves, yearling bulls and heifers, and semen (if it is produced for sale). The marketing plan should be developed based on a realistic assessment of the strengths and weaknesses of the genetic product being offered for sale. The following questions should be asked to accomplish this assessment.

1. Who are my customers?
2. What do my customers want and what constraints do they face?
3. How well does my program match the goals of my customers?
4. Upon what basis do my customers assess cattle (performance records, visual appraisal, etc.)?
5. Who are my competitors and how does my product/service compare with theirs?
6. Who are potential partners among my competitors with whom I might form an alliance?
7. How much are my cattle worth and how should they be valued?

Marketing Alternatives

The three primary marketing alternatives available to breeders are *private treaty sales*, *consignment sales*, and *production sales*. A breeder could use one or all of these marketing options, though the decision should be based on anticipated net returns and the method that will best fit the needs and personality of the breeder.

Private treaty sales involve selling cattle on the farm or ranch to one buyer at a time (Figure 4.10), whereas consignment sales involve cattle from several breeders' herds. In production sales, breeders host their own sale. Cattle in consignment and production sales are usually sold at auction. Table 4.15 shows the advantages and disadvantages of each marketing alternative.

Another marketing alternative for seedstock breeders is to participate in alliances. The alliance can be a simple agreement between two or more breeders or a more involved alliance where genetics are provided and integrated into a total industry plan.

One example of a cooperative effort between two seedstock breeders is combining their two production sales into one sale and having one advertising program instead of two. This effort reduces costs and provides buyers with larger numbers of genetics products.

In an industry alliance involving several production and processing segments, a seedstock breeder understands the needed genetic inputs. Genetics are supplied by contracts that meet biological and economics specifications. Costs are reduced because there are minimum advertising and sale costs. The genetic inputs are evaluated in the entire system with information feedback returning to the breeder. The breeders then adjust their breeding programs based on this input to improve future genetic inputs.

Figure 4.10
The selection decisions made by cattle breeders when purchasing herd bulls sets the genetic foundation of the commercial cow-calf industry. The use of private treaty sales techniques provides a means for seedstock suppliers to provide genetics and customer support most clearly aligned with the needs of the individual buyer.
Source: Coleman Locke.

Table 4.15
ADVANTAGES AND DISADVANTAGES OF PRIVATE TREATY SALES, CONSIGNMENT SALES, AND PRODUCTION SALES

Type of Sale	Advantages	Disadvantages
Private treaty	Breeder sets the price. Increased direct interaction with the customer. Lower sale costs.	Creating effective pricing structure. Breeder must be an effective communicator and salesperson. Time intensive. Determining order in which buyers get access.
Consignment	Several potential customers join forces to host sale. Sale costs are divided among consignors. Could lead to increased private treaty sales. Helps establish value of private treaty cattle. Opportunity to expand market area.	Cattle compared to other breeder's cattle. Sale management by committee may be inefficient. Cattle must be well displayed to be competitive. Consignor may not select the right cattle or plan far enough in advance. More price risk and higher costs. Difficult to share the full program when sale is held off-site.
Production	Buyers see total program. Breeder controls sale arrangements. Cattle not competing with those of other breeders. Encourages competitive bidding.	Need 40–50 lots to have a good sale and reduce sale costs. Encouraged to sell inferior cattle to boost numbers. May not attract enough buyers to meet expected sale average. More price risk and higher costs (may be as high as 20% of the sale gross).

Source: Adapted from various sources.

Advertising and Promotion Plan

Marketing is the act of furthering the acceptance and sale of merchandise through advertising and publicity. Promotion is not solely advertising, for breeders can promote their products without spending large amounts of money. Promotion communicates to customers where breeders are located and what they have to sell.

Some forms of cost-effective herd promotion include the following:

- Well-designed signage placed at the headquarters entrance and other highly visible sites on the farm or ranch; signage on company trucks and trailers.
- Maintaining an enterprise that reflects attention to detail and pride of ownership.
- One-on-one phone calls, herd visits, or other interactions with existing and potential clients.
- Developing an online presence with an excellent website and social media campaign.
- Developing and distributing effective brochures and posters to attract attention.
- Sponsoring youth activities, being involved in industry-related events, and being active in the community.

Marketing Effectiveness

The effectiveness of a marketing plan is best measured by profitability and customer satisfaction. Profitability should be achieved for both the buyer and seller. Breeders must sell higher than their breakeven price to realize profits. Commercial producers will realize profits when breeders' products help meet production and market targets that include matching biological types of cows to their most cost-effective feed environment.

In the end, customers want more than just a bull. They are seeking seedstock backed by excellent genetic performance data, economic information, and the sense of confidence that comes from dealing with breeders who guarantee their products.

SELECTED REFERENCES

Publications

Beef sire selection manual. 2010. National Beef Cattle Evaluation Consortium.

Boggs, D. 1992. *Understanding and using sire summaries*. Stillwater, OK: Beef Improvement Federation. BIF-FS3.

Bourdon, R.M. 1997. *Understanding animal breeding*. Upper Saddle River, NJ: Prentice-Hall, Inc.

Brinks, J.S., & Bourdon, R.M. 1989. Replacement heifer selection. *BEEF. 25:* 28.

Bull and Heifer Replacement Workshops. 1990. Proc. Fort Collins, CO: Colorado State University.

Bullock, D., Spangler, M., Van Eenennaam, A., & Weaber, R. 2011. Delivering Genomics Technology to the Beef Industry. National Beef Cattle Evaluation Consortium.

Cartwright, T.C. 1979. Size as a component of beef production efficiency: Cow-calf production. *Journal of Animal Science*. 48:974.

Beef sire selection manual. 2012. National Beef Cattle Evaluation Consortium.

Dekkers, J.C.M. 2004. Commercial application of marker- and gene-assisted selection in livestock: Strategies and lessons. *Journal of Animal Science*. 82(E. Suppl.):E313–E328.

Field, T.G. 2006. *Priorities First*. Saint Joseph, MO: American Angus Association.

Field, T.G. 1995. *Creating a revolution to enhance the competitive position of the beef industry*. Proc. Beef Improvement Association Conference. Armidale, NSW, Australia.

Field, T.G. 1999. *Meanwhile, down on the farm*. Proc. Australian Hereford Society Cattlemen's Conference. Brisbane, Australia.

Gelbvieh Seedstock Marketing Handbook. 1996. American Gelbvieh Association, Westminster, CO.

Gibb, J., Boggess, M.V., & Wagner, W. 1992. *Understanding performance pedigrees*. Stillwater, OK: Beef Improvement Federation. BIF-FS2.

Guidelines for Uniform Beef Improvement Programs (9th ed.). 2010. Manhattan, KS: Beef Improvement Federation.

Legates, J.E. 1990. *Breeding and improvement of farm animals*. New York, NY: McGraw-Hill. Performance record programs and national sire evaluation summaries (available from most beef breed associations).

Shike, J. 2000. *Who's who?* Angus Journal.

Silcox, R., & McGraw, R. 1992. *Commercial beef sire selection*. Stillwater, OK: Beef Improvement Federation. BIF-FS9.

5

Commercial Cow-Calf Management Decisions

Commercial cow-calf production is a highly diverse activity when compared across and within regions. Due to the dramatic differences in resources, motivation for the enterprise, and market targets, it is impossible to prescribe a fixed recipe for success. However, there are foundational principles that are utilized by most successful cow-calf managers that will be outlined in this chapter.

The motivation to initiate and sustain a cow-calf enterprise ranges from a profit orientation to the desire for a specific lifestyle. Several categories of cow-calf enterprises are listed below:

1. Professional cattleman—information driven—brand focused
2. Professional cattleman—tradition driven—commodity focused
3. Professional farmer—cattle as a by-product of land ownership
4. Professional in another industry—cattle as a secondary income
5. Lifestyle cattle producer—income not an issue
6. Lifestyle cattle producer—margin operator

The driving force behind the enterprise differs across these business categories and thus the management strategy, resourcing, staffing, and mission will vary considerably. The level of profit motivation will largely define the overall strategy. As Table 5.1 illustrates, herds with fewer than 200 cows are less likely to be the primary source of income for the owner-operator than are larger herds. For the majority of cow-calf enterprises across all regions of the United States, cattle are typically a secondary source of income.

Table 5.2 illustrates the time investment of owner-operators with various-sized cow herds. More than 50% of owners that have cow inventories in excess of 200 head

Table 5.1
Reasons for Engagement in Cow-Calf Enterprise by Herd Size and Region

	Size of Herd				Regional Location			
	1–49 (%)	50–99 (%)	100–199 (%)	200 (%)	West (%)	Central (%)	South Central (%)	East (%)
Produce primary income	5	24	43	65	25	21	11	9
Produce secondary income	78	68	51	32	55	70	74	76
Other reasons such as lifestyle	17	8	6	3	20	9	15	15

Source: USDA-NAHMS.

Table 5.2
PERCENT OF WORK TIME INVESTED BY OWNER-OPERATORS OF VARIOUS SIZED COW-CALF HERDS

Percent of time	1–49(%)	50–99(%)	100–199(%)	200+(%)
<25	58	27	19	11
25–49.9	19	27	23	15
50–74.9	11	20	22	19
75–99.9	4	7	14	19
100	8	19	22	35

Source: USDA–NAHMS.

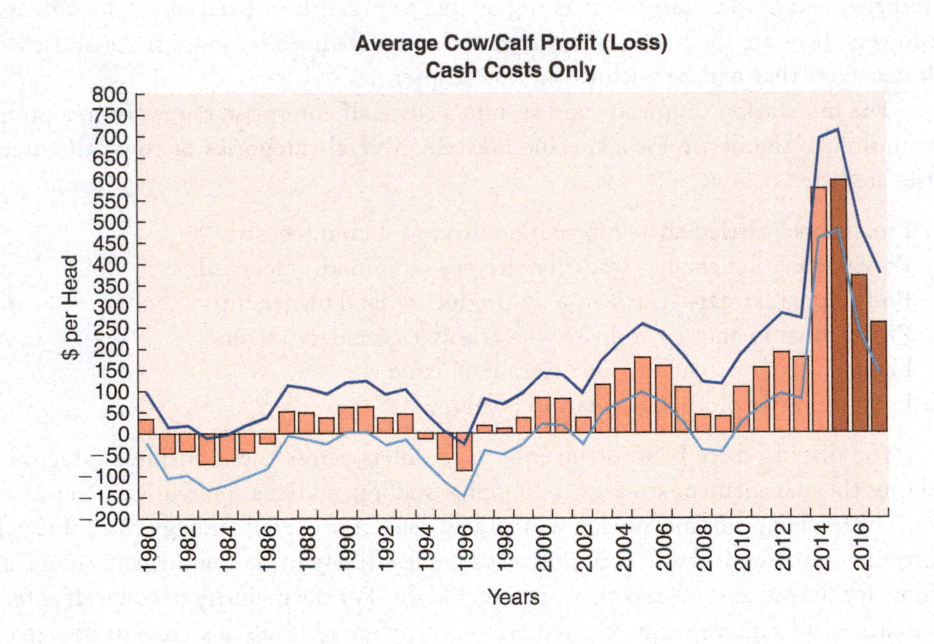

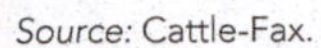
Figure 5.1
Estimated average cow-calf returns (revenue over cash costs including pasture rent).
Source: Cattle-Fax.

spend more than 75% of their time with the enterprise while more than half of herd owners with fewer than 100 cows allocate less than one-half of their time into the enterprise.

Each of these categories represents a valid reason for engaging in the cow-calf enterprise. This chapter however will focus on those enterprises that are focused on generating return on investment. Historical profitability in the cow-calf sector is highly cyclical (Figure 5.1). For example, in 1993, 70% of cow-calf enterprises were profitable but by 1996 fewer than 15% were generating profits due to the price of feeder calves dropping nearly $40/cwt. By 2000, 90% of cow-calf enterprises were profitable on a cash basis and high levels of profitability were sustained through 2015 as beef demand exceeded U.S. beef production, which was limited by drought, herd reductions, shifting lands from pasture to grain production during the commodity price run-up, and a number of other factors. However, as the national herd is rebuilt the cow-calf sector will experience declining prices and overall levels of profitability. Additionally, cash costs' association with beef production has risen over time challenging the ability of enterprise to generate sustainable profits (Figure 5.2).

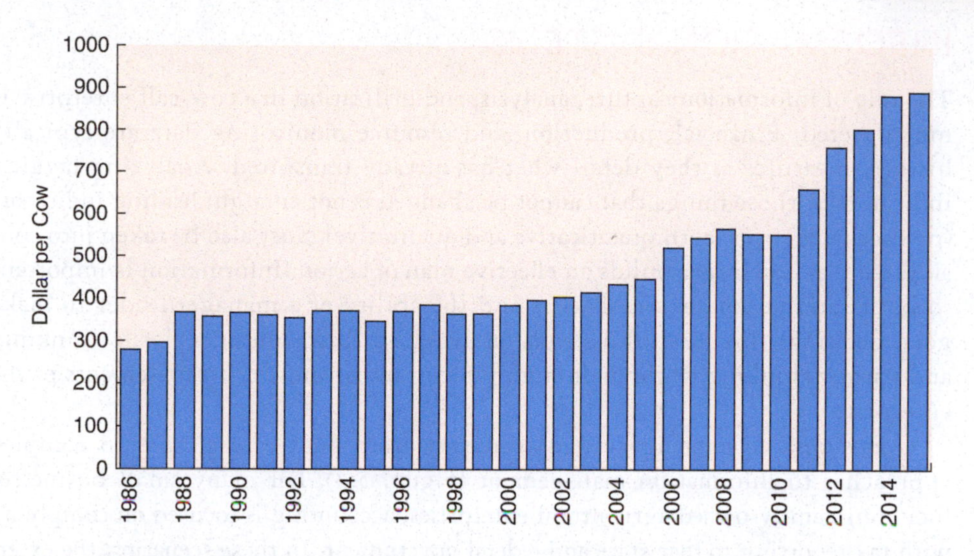

Figure 5.2
Historic trends in average annual cash costs per cow.
Source: USDA.

Profitable, low-cost producers are not successful only by virtue of their ability to manage cost of production. They also have the ability to optimize productivity, the flexibility to take advantage of opportunities in the marketplace, the ability to manage risk on a number of fronts, and they are effective in both the planning and implementation processes of their business. Successful cow-calf managers invest in profitable capital assets, maintain a focused business perspective, and proactively respond to the macro trends in the industry, marketplace, and society. Furthermore, they embrace the capture, synthesis, and utilization of meaningful information. Profitable cow-calf enterprises are always the result of the application of creativity, hard work, and excellent risk management.

CREATING THE VISION

Most important in profitable cow-calf management is the ability to see the enterprise as a system of interacting components and to understand the relationship between and among the components. Successful managers understand the impact of change in one component on other factors or resources. The system can be evaluated as being composed of six primary components: natural resources, the family, economic and finance factors, the production system, sociopolitical influences, and spiritual and cultural considerations (Dunn, 2002). To better understand a systems perspective, it is important to review Chapter 3 prior to continuing this chapter.

Many profitable producers have a written management plan that identifies their mission statement with specific goals to accomplish the plan. Two examples of mission statements are presented here.

1. Produce low-cost/high-profit cattle that yield competitively priced, highly palatable, beef products.
2. Manage the available resources (optimum, low-cost combination) for maximum continuing net profit, while conserving and improving the resources.

Specific, measurable goal might be to (1) Improve profitability $25 per acre over the next five years; or (2) Reduce the weaned calf breakeven price from $1.00/lb to $0.85/lb in 2 years.

A big challenge for most cow-calf producers in meeting these mission statements and goals is developing a cost-effective financial and production records system. Nationally, it is estimated that only 5–10% of cow-calf producers actually calculate cost of production. It is difficult to manage what is not measured.

INFORMATION SYSTEMS

The role of information capture, analysis, and utilization in a cow-calf enterprise is multifaceted. Financial, production, and resource monitoring data are typically historic measures as they detail what has already transpired. Analyzing lagging indicators or those things that cannot be changed is not enough; leading indicators (predictive metrics, both quantitative and qualitative) must also be taken into consideration as a manager builds an effective plan of action. Information is important to both internal and external users and the ability of a manager/leader to make good decisions (Figure 5.3) depends on avoiding "barn blindness," silo thinking, and narrow thinking while maintaining focus on the most critical elements of the enterprise.

Business accounting systems can vary from relatively simple to complex approaches to information management (Figure 5.4). For many small businesses including family-owned agricultural enterprises, accounting is focused on complying with tax reporting to that state and federal government. In these scenarios, the external user is the focal point while the managers use their own intimate knowledge of their business and years of experience upon which to make decisions. However, as the business grows adding family members, enterprise diversification, and new opportunities, the cash-only methodology is no longer appropriate. With rising complexity, the decision maker(s) requires more precise, meaningful, and focused information to assist in making correct choices for both the short- and long-term needs of the business. There is a need for an accurate and detailed accounting of business transactions plus the ability to integrate and systematize financial information with the information systems and data collection arising from the production, natural resource, and human capital aspects of the venture.

An effective accounting system should provide information for the following components of the business:

- Long-term planning and strategic direction
- Resource allocation decisions

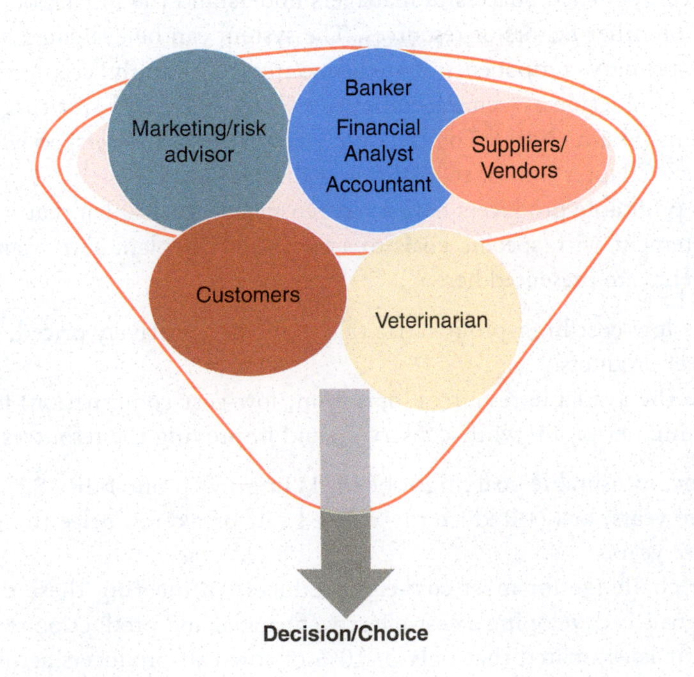

Figure 5.3
Effective decision making requires the ability to utilize input from a variety of sources.

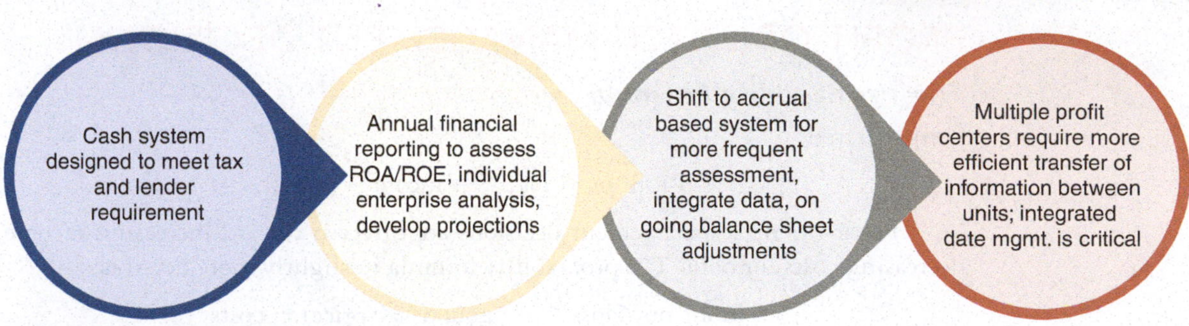

ROA- Return on Asset, ROE- Return on Equity

Figure 5.4
Progression of information systems requirements as a business increases in complexity.

- Cost control and monitoring critical control point metrics
- Performance measurement
- Conformance to legal and regulatory requirements

Given the complex nature of beef cattle production and the interlocking enterprises that comprise most cow-calf operations (cow-calf, replacement heifer development, stocker cattle, hay production, land and real estate), analytical evaluation is needed to determine cost assignments, resource allocation, and choosing which opportunities to exploit. Enterprise analysis allows managers to understand true cost of production, where to direct attention and exertion of control, and to better measure profitability. These analyses are critical under a number of scenarios:

- Determining whether to rent or own assets such as real estate, equipment, and facility infrastructure
- Deciding whether or not the business can afford to bring additional member of the family into the enterprise
- Creating staffing and compensation plans
- Assessing opportunities for diversification, partnership, or supply–chain relationships
- Determining which activities or enterprises to drop and which to expand

Incorporating an effective managerial accounting system requires an investment of effort, time, and treasure but once implemented the return on investment is substantial. Cow-calf managers will benefit from assuring that costs are organized logically with specific assignment to the appropriate enterprise, that metrics are shared with those who can execute control to make improvements, that the cost drivers are correctly determined (per cow or per acre for example), and that costs are aligned with revenue to determine whether an activity is a profit generator from the revenue side (sale of calves) or a cost reducer (sale of excess hay).

Development of a dashboard approach that allows an organization's leadership team to effectively monitor performance, track key trends, and effectively share information with both internal and external audiences is important to long-term profitability. One of the significant challenges faced by managers of cow-calf enterprises is to determine the key micro and macro indicators of highest value. As Chapter 3 pointed out, it is not enough to focus only on the cattle. Rather, the cow-calf enterprise is a complex set of relationships that connect natural, financial, human, livestock, and infrastructure resources.

PROFIT-ORIENTED MANAGEMENT DECISIONS

The Profitability Formula

Simply stated,

$$\text{Profit or} < \text{loss} > = \text{income} - \text{costs}$$

Profit-oriented management decisions are directed toward increasing income, decreasing costs, or both. The profitability formula in slightly more detail is:

$$\text{Profit or} < \text{loss} > = (\text{pounds} \times \text{price}) - \text{costs}$$

To increase profit (or minimize loss), cow-calf producers focus on three areas: (1) increase pounds (production), (2) increase price (revenue), and (3) decrease costs. Eventually, producers want to achieve an optimum combination of pounds, price, and costs so that profitability can be maximized.

Producers have the least management influence over prices because they are price takers—not price makers. Marketing management, however, should not be ignored. Good managers can influence price by a few cents a pound or several dollars a head depending on the timing of their decisions. Furthermore, by taking opportunity of retained ownership options, additional profits may be generated. Well-planned production and marketing through retained ownership and/or alliances can increase returns by several hundred dollars per head.

The profitability formula in more detail is:

$$\begin{aligned}\text{Profit or} < \text{loss} > = {} & [(\%\ \text{calf crop} \times \text{weaning wt}) \times \text{price}] \\ & + [(\text{lb market cows and bulls}) \times \text{price}] - \text{costs}\end{aligned}$$

There are more "pounds sold" in cow-calf operations than only "pounds of calves" (e.g., pounds of market cows and market bulls).

Breakeven Price Analysis

The profitability formula also can be expressed as a breakeven price analysis. Basically, it determines the price the calves must bring to cover the costs of production. Simply stated:

$$\text{Breakeven price} = \frac{\text{Annual cow cost}}{\text{Average weaning wt} \times \%\ \text{calf crop}}$$

An example for a producer with annual cow costs of \$600 per cow, average weaning weights of 525 pounds, and a 90% calf crop would have a breakeven as follows:

$$\frac{\$600}{525 \times .90}$$

$$\text{Breakeven price} = \$127.98/\text{cwt}$$

Obviously, if the sale price for this producer is above \$127.98/cwt, a profit is realized; if the calves sell for less than \$127.98, a loss is incurred.

Table 5.3 shows several breakeven prices for varying annual cow costs, weaning weights, and calf crop percentages. There is a range of approximately \$130 per hundredweight in the breakevens calculated in Table 5.3, which demonstrates the opportunity to lower breakeven price by focusing on the costs and productivity. These breakeven prices do not consider pounds of cull cows and cull bulls sold per cow, so the actual breakeven prices would be lower than those shown in the table. However, the breakeven price analysis is a simple way of looking at the profitability of a cow-calf operation. A producer could take the approach of breaking even on calf sales and then having sale of cull cows and bulls as a profit.

Table 5.3
BREAKEVEN PRICE WITH VARYING CALF CROP PERCENTAGES, ANNUAL COW COSTS, AND WEANING WEIGHTS (VALUE OF CULL BREEDING STOCK NOT ACCOUNTED)

		Average Calf Weight at Weaning		
Calf Crop (weaned, %)	Annual Cow Cost ($)	400 lb	500 lb $ per cwt	600 lb
95	600	157.89	126.32	105.26
95	500	131.58	105.26	87.72
95	400	105.26	84.21	70.18
85	600	176.47	141.18	117.65
85	500	147.06	117.65	98.04
85	400	117.65	94.12	78.43
75	600	200.00	160.00	133.33
75	500	166.67	133.33	111.11
75	400	133.33	106.67	88.89

The breakeven price formula that accounts for all cattle sales is as follows:

$$\text{Breakeven price} = \frac{(\text{annual cow cost} - \text{value of market cows and bulls sold})}{\text{Average weaning wt} \times \% \text{ calf crop}}$$

Breakeven price analysis can be used to evaluate the economics of different management alternatives.

Table 5.4 shows cow productivity, annual cow costs, and calf breakeven prices for average, low-cost, and high-cost producers in various regions of the United States.

It is evident that low-cost producers can be found in each region of the country and that the difference between low-cost and high-cost enterprises in net income per cow is significant. Table 5.5 shows that many of the factors contributing to these differences are largely due to cost per cow. While pounds weaned per cow are important, its contribution to net income is far outshadowed by cost per cow. This difference shows producers where management decisions should first be focused. Multiple research studies have shown that when comparing high profit to low profit cow-calf enterprises, 70% of the variation can be attributed to effective management of cost while the remainder results from realizing the maximum value of cattle production.

Table 5.4
COW-CALF PRODUCERS PROFITS BY COST GROUP

	Weaning (%)	WW (lb)	Lb/Wean Cow Exposed	Cost of Production ($/cow)	Cost of Production ($/cwt)	Net Income ($/cow)
North Dakota						
Top 20%	94	590	555	330		211
All	91	548	499	382		120
Difference between high versus low profit	+3	+42	+56	−52		+91

(continued)

Table 5.4
(Continued) Cow-Calf Producers Profits by Cost Group

	Weaning (%)	WW (lb)	Lb/Wean Cow Exposed	Cost of Production ($/cow)	Cost of Production ($/cwt)	Net Income ($/cow)
Beef Belt (1991–1999)						
Top 1/3	90	507	455	NA	61	51
Middle 1/3	87	525	455	NA	82	9
Low 1/3 (high cost)	83	493	413	NA	145	(57)
Difference between high versus low profit	+7	+14	+42	NA	−84	+108
Kansas—2000–2004						
Top 1/3	92	634	583	657		381
Middle 1/3	89	606	539	678		198
Low 1/3	90	592	534	817		11
Difference between high versus low profit	+2	+42	+51	−160		+370

Source: Multiple sources.

Table 5.5
Productivity Benchmarks for Traits Important to Determining Cow-Calf Profitability

	Southwest	Corn Belt	Northern Great Plains
Pregnancy percentage	89	95	NA
Calving percentage	85	92	96
Weaning percentage	82	87	91
Average weaning weight	525	496	548
Lbs. weaned per cow exposed	434	424	498

Source: Adapted from multiple enterprise analyses.

FACTORS AFFECTING POUNDS OF CALF WEANED

The number of pounds of calf weaned per cow in the breeding herd reflects both weaning weight and percent calf crop weaned. Table 5.6 demonstrates that both factors are important in calculating pounds of calf weaned per cow exposed. Producers should thus attempt to improve both traits when economically feasible. Further analysis of the table shows that a 1% change in percent calf crop is approximately equivalent to a 5-lb change in weaning weight.

Figure 5.5 shows the major factors that a cow-calf producer should critically evaluate in order to improve pounds of calf weaned per cow exposed. The two component parts (e.g., percent calf crop and weaning weight) are described in detail in this chapter. Each operation will differ in levels of productivity in each area. This also implies that priorities for improvement will usually differ for different operations. In past years, when cow-calf producers wanted to produce more beef, they

Table 5.6
Calculation of Pounds of Calf Weaned per Cow Exposed

% Calf Crop	Average Weaning Wt of All Calves (lb) 400	450	500	550	600
	Pounds of Calf Weaned Per Cow Exposed[1]				
75	300	338	375	412	450
80	320	360	400	440	480
85	340	382	425	467	510
90	360	405	450	495	540
95	380	427	475	522	570
100	400	450	500	550	600

[1]Calculated by multiplying % calf crop by average weaning weight.

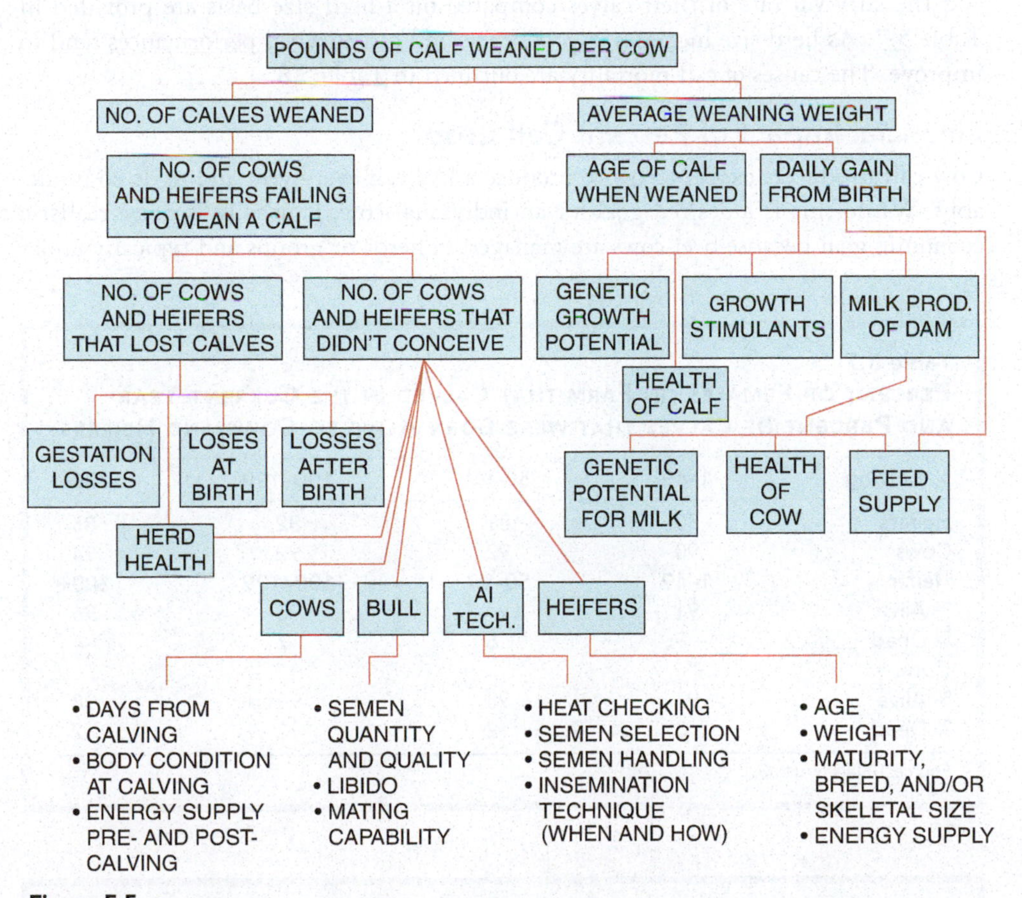

Figure 5.5
The major factors affecting pounds of calf weaned per cow bred.

typically purchased additional land and cows. Today, the emphasis has shifted to increasing the productivity and dollar return on productive units—the cow and the acre of land.

The remainder of this chapter emphasizes improving productivity and profitability for a cow, acre, and total operation. The various factors affecting productivity should be critically evaluated as to how they impact both costs and returns in the profitability formula.

MANAGING PERCENT CALF CROP

Calf crop percentages can be calculated in several ways. The calculation that best reflects production efficiency is as follows:

$$\%\text{ net calf crop} = \frac{\text{No. of calves weaned}}{\text{No. of cows in breeding herd (previous year)}} \times 100$$

This measure of percent calf crop, based on the number of cows exposed, is used throughout this book. If other measures of percent calf crop are used, they should account for all losses in reproductive efficiency.

Calf crop percentages for most cow herds in the United States will be between 70 and 95%. Extreme environmental conditions of drought, severe spring storms, and major disease problems may cause the calf crop percentage of individual herds to fall below 70%. Individual herds can document calf crop percentages over 95% for over 10 consecutive years; however, these are exceptional herds and there is evidence that this may not be optimum for many herds. The percent of females on farm that calved and the survival rate of their calves compared on a herd size basis are provided in Table 5.7. As herd size increases, reproductive and calf survival performances tend to improve. The causes of calf mortality are outlined in Table 5.8.

Financial Impact of Percent Calf Crop

Cow-calf producers expect a cow to produce a live calf every year and to do so profitably. While this is a desired goal for an individual cow, it usually is not a realistic economic goal because beef cows are managed in herds or groups and typically under

Table 5.7
Percent of Females on Farm that Calved in the Current Year and Percent of Calves that were Born Alive to Cows and Heifers

% Calving	1–49	50–99	100–199	200+
Heifers	80	81	82	85
Cows	90	92	92	94
Heifer	**1–49**	**50–99**	**100–199**	**200+**
% Alive	91	94	93	96
% Dead	9	6	7	4
Cows				
% Alive	97	98	98	98
% Dead	3	2	2	2

Source: USDA-NAHMS.

Table 5.8
Reasons for Calf Death Loss between Birth and Weaning (% of Losses)

Cause of Death	Prior to 3 Weeks of Age	3 Weeks and Older
Birth-related	17	0
Weather	28	12
Digestive	16	18
Respiratory	9	37

Source: USDA-NAHMS.

Table 5.9
Effect of Various Calf Crop Percentages on Breakeven Prices (Assumes a $500 Annual Cow Cost and 500-lb Weaning Weights)

% Calf Crop	Lb Calf Weaned (per cow exposed)	Breakeven Price ($ per cwt)	Change in Breakeven Price ($ per cwt) for Each 5% Change in Calf Crop %
100	500	100	
95	475	111	11
90	450	123	12
85	425	138	15
80	400	156	16
75	375	178	22
70	350	204	26

extensive conditions. However, forage-based management systems provide opportunities for beef cows to be managed under low-cost scenarios.

Reproductive performance, as reflected in calf crop percentages, is the most economically important production trait. It cannot be neglected or poorly managed without severe economic effects. Table 5.9 shows how changes in percent calf crop can affect the breakeven price. Note in the table that, as calf crop percentage improves, its effect on breakeven price is diminished.

Improving percent calf crop should be based on understanding the effect of management alternatives on costs and returns. Figure 3.12 (Chapter 3) shows that maximum profitability is reached before a 100% calf crop is obtained. The following evaluates the cost effectiveness of changing percent calf crop: A producer ($138/cwt breakeven with 500-lb weaning weights, 85% calf crop, and $500 annual cow costs) wants to raise calf crop from 85% to 90%. If weaning weight remains at 500 lb, what is the largest increase in annual cow cost that can be accepted and still remains at a $138/cwt breakeven? We can solve for x if 425 lb would increase to 450 lb (500×0.90) as a result of increasing percent calf crop:

$$\frac{500}{425} = \frac{x}{450}$$

$$425x = 225{,}000$$

$$x = \$529.41$$

If the producer could increase calf crop percentage from 85% to 90% at a cost per cow of less than $29.41, this management practice appears to be economically feasible.

Factors Affecting Calf Crop Percentage

Table 5.10 shows the relative importance of factors affecting the net calf crop in a herd where detailed records have been kept over several years. The most important factors affecting net calf crop in this example were failure of the female to become pregnant during the breeding season and calf losses at or shortly after birth. These two factors accounted for 82% of the reduction in net calf crop. Therefore, management to improve calf crop percentages must reduce losses from all four loss categories, but especially from these two major areas. Although the losses identified in Table 5.10 pertained to a herd of cattle in a research study, they are similar to the types of losses that occur in producers' herds. Bellows et al. reported in 2002 that reproductive failure in the U.S. beef herd equated to an average loss of $14.90 per cow. Reproductive losses were six

Table 5.10
Factors Affecting Net Calf Crop in a Disease-Free Beef Herd Bred by Natural Service (14-year summary)[1]

Factor	No.	% Reduction in Net Calf Crop
Females not pregnant at end of breeding season	2,232	17.4
Perinatal calf deaths	821	6.4
Calf deaths, birth to weaning	372	2.9
Calf deaths during gestation	295	2.3
Net calf crop weaned (%)	9,107	71.0
Total	12,827	100.0

[1]Includes females 14 mos to 10 years of age during breeding seasons of 45 or 60 days' duration.
Source: USDA Livestock and Range Research Station, Miles City, Montana.

Table 5.11
Trend of Calf Survival and Weaning Percent on a Large Western Ranch

Year	Cows Exposed[1] (N)	Calves Branded (N)	Calf Survival to One Month of Age (%)	Calves Died Branding/ Weaning (N)	Calves Weaned (N)	Calf Crop Weaned (%)
1	914	855	94	48	807	88
2	918	888	97	50	838	91
3	881	829	94	27	802	91
4	877	848	97	33	815	92
5	882	840	95	42	798	89
6	898	862	96	33	829	92
7[2]	917	851	93	24	827	89
8	908	850	94	17	833	91
9	900	876	97	28	848	94

[1]Only open cows sold.
[2]Improved vaccination program for respiratory disease was initiated at branding.

times higher than those for respiratory problems. The top three reproductive losses were female infertility (49.8%), dystocia (37%), and abortions/stillbirths (12.8%).

Factors affecting calf losses will vary in individual herds. Producers should maintain adequate records, so they will know where the greatest losses are occurring. A management program to improve calf crop percentage can be applied by producers being knowledgeable of losses and the cause and effect relationships affecting each loss area. Table 5.11 illustrates a nine-year trend in calf survival and weaning percentage on a large western ranch.

In this herd (Table 5.11), the average death loss from branding to weaning in years 1–6 was 4.5%. In the three years following the initiation of the vaccination protocol, the death loss was reduced to 2.7%. On average, 16 more calves were weaned following the management change. If an average weaning weight of 500 lb and an average price of $150/cwt are assumed, then an additional $12,000 was generated. Given a vaccine cost of $2/hd, and the average number of calves at branding was 850 head, then the total cost was $1,700. Thus, the management change generated a net return of $10,300 per year.

Recent research indicates that approximately two-thirds of the 20% reduction in calf crop percentage is due to fertilization failure and embryonic death. Part of this loss can be attributed to hormone imbalances, poor uterine environments, and sex cells lacking viability. It is not clear how to prevent these losses through management programs. Excellent herd health, nutrition, selection, and other current management practices should help prevent some fertilization failures and embryonic losses.

Mortality due to calf scours has historically affected a number of cow-calf enterprises regardless of regional location. The key to reducing the risk of a contagious outbreak of calf scours is to decrease the exposure of calves to pathogens and to diminish the concentration of pathogens. The Sandhills Calving System (SCS) was developed to attain these objectives by preventing direct contact between younger and older calves, by using uncontaminated calving pastures, and to prevent the youngest calves born onto the ranch from having to be exposed to an accumulated pathogen load. In short, the SCS is focused on moving pregnant cows to new calving pastures. There are several adaptations of the system including moving heavy cows to fresh pastures each week away from pairs. Another variation is to move pregnant cows once 100 calves have been born onto a pasture thus creating a progressive separation of females yet to calve from those that are already nursing a calf. One example of the effectiveness of this system comes from a 900 head cow-calf case study in Nebraska that had historically calved in designated calving lots where pathogens were concentrated and calf exposure was progressively increased throughout the calving season. Mortality rates were in excessive of 8% and veterinary care and supplies were costing the enterprise over $3,000 annually. Once the SCS was implemented, death loss from scours was reduced to zero and veterinary expenses were reduced to less than $200 per year. It was estimated that approximately $40,000 in savings were accumulated due to improved calf survival, reduced costs, and improved calf performance up to weaning.

Some of the major biological principles affecting calf crop percentage are summarized in Table 5.12.

MANAGING WEANING WEIGHTS

Calves are typically born in the spring and weaned in the fall. Most calves weaned are between 6 and 10 months of age. There are some producers developing management strategies that involve weaning cattle earlier than these traditional ages—even as early as 2 months of age. Some calves are born in the fall; in some operations, cows calve year round.

Weaning weights in individual herds will vary from approximately 300 lb to more than 700 lb. Average weaning weights at 207 days of age are approximately 530 lb in herds surveyed by the USDA. Average weights by gender, size of herd, and region of the country are provided in Table 5.13.

Financial Considerations

As noted earlier in the chapter, increasing the weaning weights of calves can lower the breakeven price required to cover production costs. Table 5.14 shows that at a given calf crop percentage and an annual cow cost, the increments in weaning weight have a decreasing economic advantage. Weaning weight, then, should be increased in a herd as long as it is cost effective. Average herd weaning weights appear to be approaching an optimum at around 500 lb to 550 lb for most cow-calf enterprises. The marketing goals of the enterprise will also have a significant impact on the targeted weaning weight.

Table 5.12
MAJOR MANAGEMENT PRINCIPLES AFFECTING PERCENT CALF CROP

1. AGE AND WEIGHT AT PUBERTY

a. Heifers calving at 2 years of age should reach puberty prior to 14 months of age and have one or two estrous cycles prior to breeding. Fertility of the first estrus is usually lower than those of subsequent estrous periods.
b. Heifers need adequate nutrition to reach a target weight at breeding that is consistent with biological type (target weight at breeding should be 60–65% of mature cow weight based on a body condition score of 5).
c. Body condition score (BCS) of replacement heifers at breeding should be a minimum of 5 and preferably 6.
d. Crossbred heifers will reach puberty earlier and at lighter weights when compared to their straightbred counterparts.

2. BREEDING HEIFERS

a. Synchronizing estrus of replacement heifers and breeding them 2 weeks prior to cow herd, if cost effective, are excellent management tools to prevent culling these females at 3–4 years of age because they calve late in the season.

3. FEEDING HEIFERS

a. Pregnant heifers should not be underfed or overfed during gestation. The best guide is to have a BCS of 5 or 6 at calving time. Heifers that are underfed will have a low BCS and a long postpartum interval. They also will produce calves that have a higher death loss and reduction in calf weight resulting from scours and other diseases.

4. CALVING EASE

a. Dystocia (calving difficulty) is a primary cause of calf deaths at birth. It costs $850 million annually. Additional losses can occur later due to the poor immune system of a stressed calf and a longer rebreeding period for the stressed dam.
b. Dystocia is influenced primarily by birth weight of the calf, and birth weight is controlled most effectively through sire selection (see Chapter 4).
c. The dam's pelvic area has a minor influence on dystocia; however, caution must be exercised in selecting for larger pelvic area. Larger-framed heifers typically have larger pelvic areas but in turn produce calves with larger birth weights. Selection of pelvic area within an optimum frame size may have merit. Growth implants will increase pelvic area at breeding, but there is little difference in pelvic area at calving time between implanted and nonimplanted heifers.
d. Two-year-old heifers experience more calving difficulty than older females. Therefore, bull selection and management at calving time is more crucial for these young females. Mature females can experience no calving difficulty with extremely large calves (over 100 lb). However, if heifers with heavy birth weights are saved as replacements they may experience calving difficulty, have heavier mature weight, and experience rebreeding problems, which reduces percent calf crop in later generations.

5. POSTPARTUM INTERVAL

a. Cost-effective BCS at calving, influenced primarily by nutrition, milk production level, and mature weight, is the best guide in managing the postpartum interval. Early weaning (5–6 months of age) of calves that is consistent with forage supply can help cows gain in BCS prior to the winter season. The cow has a maximum postpartum interval of 70–90 days if she produces a calf on a yearly basis. Postpartum intervals are most challenging on females being bred for a second calf. Calving seasons of approximately 70 days have been shown to be most profitable.

Table 5.13
AVERAGE WEANING WEIGHTS AND AGES BY SIZE OF ENTERPRISE AND REGION

	Size of Herd				Region of Operation			
Class	1–49	50–99	100–199	200+	West	Central	South Central	East
Replacement heifers	492	540	548	543	540	537	534	532
Other heifers	496	522	526	518	531	515	519	496
Bull & steers	532	565	572	564	572	565	560	531
Weaning age (days)	201	208	208	209	211	208	206	200

Source: USDA-NAHMS.

The following example shows one method of economically assessing a proposed management practice: A producer ($117.65/cwt breakeven with 500-lb weaning weights, 85% calf crop, and $500 annual cow costs) determines that by using a growth

Table 5.14
Changes in Breakeven Prices with 50-lb Increments in Weaning Weight (assumes an 85% calf crop and a $500 annual cow cost)

Weaning Wt (lb)	Lb of Calf Weaned	Breakeven Price ($/cwt)	Change in Breakeven Price ($/cwt)
350	297.5	168.07	
400	340	147.06	21.01
450	382.5	130.72	16.34
500	425	117.65	13.07
550	467.5	106.95	10.70
600	510	98.04	8.91

implant, pounds of calf weaned per cow exposed could be increased 10 lb with an increased cost of $2 per cow. What happens to the breakeven price?

$$\frac{502}{435} = 115.40$$

Breakeven price is lowered from $117.65/cwt to $115.40, so this management decision appears to be profitable.

Some research shows that herd average weaning weights between 500 and 700 lb are not related to profit per cow. Other studies evaluating the profitability of cow herds in the Great Plains have shown that weaning weight compared with cost of production and reproductive rate is a distant third in factors affecting profitability. These studies suggest that weaning weight has an optimum level within each cow-calf herd and that deviation from this optimum could decrease profits.

Factors Affecting Weaning Weight

Table 5.15 shows the major management principles affecting weaning weights. The application of these principles can be used to reach an optimum weaning weight for a herd. Commercial producers combining excellent cattle genetics with well-managed feed and health environments will have relatively heavy calves at weaning.

Genetic Factors Influencing Calf Gains

When an adequate nutrient supply is maintained, the most limiting factor influencing gain is the genetic makeup of the healthy calf. A major genetic influence on nutrient supply comes in the genetic ability of the cow to produce an optimum flow of milk. Growth potential of the calf and milk production of the cow is improved genetically through selection and crossbreeding.

The gene combination affecting growth can be changed through sire, heifer, and cow selection. It has been well demonstrated that effective selection can result in a genetic increase of 3–5 lb per year in the weaning weights of calves. Most of this change will come through bull selection based on evidence of genetic superiority in weaning and yearling weights. Some genetic improvement can be made through cow and heifer selection, but the amount of progress is limited due to the number of females that must be kept.

There is a breed effect on weaning weight where larger, faster-gaining breeds will have heavier weaning weights at the same age. This relationship is not only true when comparing breed averages, but is also noted when comparing biological types within a breed. Crossbreeding for the average cow-calf producer can produce a 20%

Table 5.15
MAJOR MANAGEMENT PRINCIPLES AFFECTING THE WEANING WEIGHTS OF CALVES

1. AGE OF CALF WHEN WEANED

a. Increasing calf age increases weaning weight. Age at weaning should be determined by the financial costs, especially feed costs. By assuring that a high percentage of calves are born early in the calving season, the age at weaning is increased. Figure 5.6 shows the calving distribution for a nearly 300 head cow herd with a 70-day calving season. To develop a concentrated time period in which calves are born then it is critical that a defined breeding season is established with some combination of specific duration of breeding season and the use of pregnancy testing. There continues to be relatively large numbers of enterprises that do not have a set breeding/calving season (Table 5.16).

2. CALVING SEASON

a. Calves born in summer and fall usually weigh less than calves born in spring. Exceptions to this occur in some geographical areas of the United States, where forage quantity and quality are high in the fall.

3. GROWTH IMPLANTS

a. Implanting suckling calves increases weight 10–25 lb. Consideration should be given to the market endpoint when establishing an implant program.
b. Caution should be exercised in implanting replacement heifers, as their reproduction may be affected.

4. FEED SUPPLY

a. Cost-effective management practices should focus on increasing the amount of grazed forage and its nutrient quality during the time when cows are milking the heaviest and calves are growing rapidly from dam's milk and the direct consumption of forage.
b. Weaning weights can be increased by feeding more harvested and purchased feed to cows and calves; however, this is usually not cost effective, especially over the long term.

5. SELECTION

a. Genetic selection, primarily through bulls, can improve weaning weights 3–5 lb per year. Weaning weight EPDs and milk EPDs should be emphasized if weaning weights need to be improved.
b. Selection for yearling weight EPD will increase weaning weight nearly as much as selecting only for weaning weight EPD. Selection for yearling weight EPD not only increases weaning weight but yearling weight and feedlot gain as well.
c. Caution should be exercised so that genetic selection for growth and milk does not exceed the most economical grazed forage program. Otherwise, BCS and reproduction will decline or a higher-cost feeding program (more harvested and purchased feeds) must be implemented to maintain reproduction at a relatively high level.

6. HETEROSIS

a. Utilization of heterosis (hybrid vigor) through well-planned crossbreeding can increase weaning weight by 10% and pounds of calf weaned per cow exposed by 20% over the average of the breeds in the crosses.
b. See section 5c in this table.

7. HEALTH

a. Calves that do not experience calving difficulty and that receive adequate levels of colostrum during the first 24 hours after birth usually weigh heavier at weaning time.
b. An excellent, cost-effective cow-calf health program assures an adequate milk flow from the cow and the ability of the calf to express its weaning weight potential.

increase in pounds of calf weaned per cow exposed as compared to a straight breeding system. Most of this increase occurs from improved reproductive performance; however, 25–40% of the increase is the effect of heterosis on (1) growth potential of the crossbred calf and (2) increased milk production of the crossbred cow. An economic evaluation of crossbreeding showed that a 16–20% increase in net income after taxes could result from effective use of crossbreeding. However, implementing high levels of heterosis and selection could yield weaning weights beyond the herd optimum, thus lowering profitability.

A cow-calf producer with an average weaning weight of 400 lb and an 80% calf crop in a herd of straightbred cows produces 320 lb of calf weaned per cow exposed.

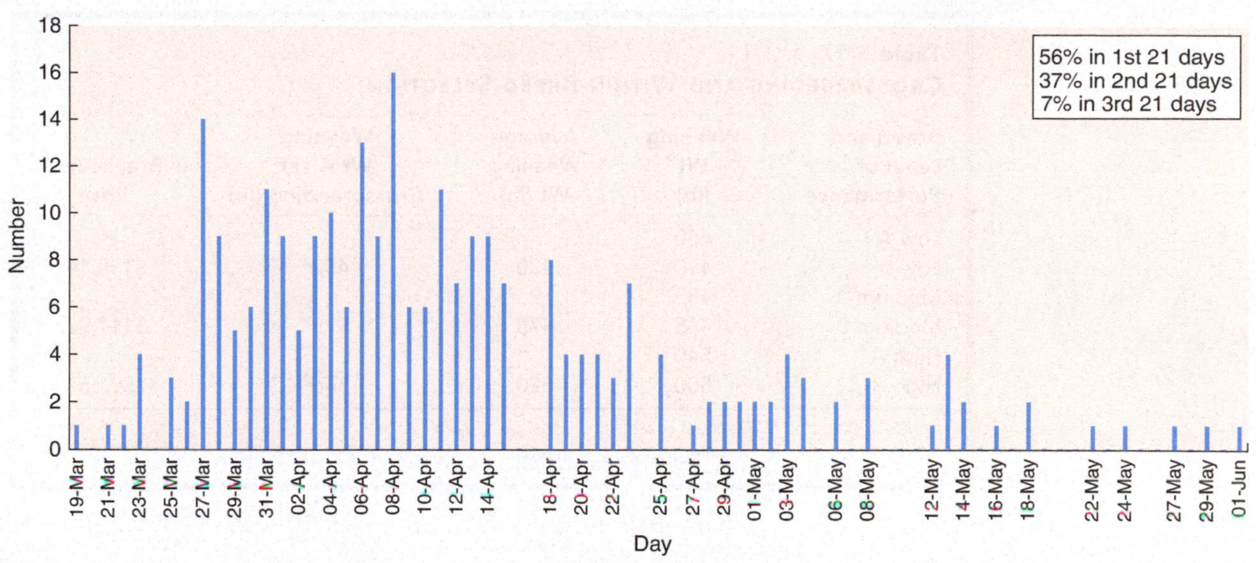

Figure 5.6
Calving distribution from a 300 head intermountain commercial cow herd.

Table 5.16
Number of Defined Breeding Season in Use by the U.S. Cow-Calf Sector and the Percent of Cows in Each and the Length of Breeding Seasons by Percent of Operations and Percent of Cows

Number of Defined Breeding Seasons	Operations (%)	Cows (%)
1	34	48
>2	12	18
No set season	54	34
Number of Estrous Cycles in Breeding Season		
≤3	26	23
4	13	16
5	22	23
6–7	17	18
>7	22	20

Source: USDA-NAHMS.

If this producer implemented an effective selection and crossbreeding program, the pounds of calf weaned per cow exposed would be 414 lb in approximately 6 years. Assuming the annual cow cost remained constant at $500, this would change the breakeven price from $156.25/lb to $120.77/lb.

Many commercial producers want to know what breeds to use in a crossbreeding program. Certain breeds complement one another more satisfactorily; however, producers need to pursue the genetic resource in more depth. It is extremely important to understand the variations that exist within each breed (see Chapter 13). In many cases, the biological types of cattle within the breeds have more economic significance than the heterosis obtained by crossing two or more breeds (see Table 5.17). However, it should not become a question of genetic superiority versus heterosis for commercial producers. They should take advantage of both to optimize productivity and maximize profitability.

Table 5.17
CROSSBREEDING AND WITHIN-BREED SELECTION

Breed and Level of Performance	Weaning Wt (lb)	Average Weaning Wt (lb)	Weaning Wt After Crossbreeding (lb)	Breakeven[1] Price
Low A	450			
Low B	410	430	452[2]	$130.14
Medium A	485			
Medium B	465	475	499[2]	$117.92
High A	540			
High B	500	520	546[2]	$91.57

[1]85% calf crop and $500 annual cow cost.
[2]5% heterosis above the average of the two breeds.

Growth Stimulants

The most common growth stimulants for calves are ear implants of Compudose®, Synovex®, and Ralgro®. Implants can be given to young calves when they are a few days or weeks old. (The average growth responses to various implants are shown in Chapter 7.) Implants should not be used on replacement heifers because of possible detrimental effects on reproduction.

The cost of a Ralgro® or a Synovex® implant is approximately $1. The approximate 10–20-pound increase in weaning weight well justifies the implant and labor cost. Proper implanting procedures as noted in Chapter 7 must be followed. The growth-promoting response of Compudose® lasts over a 200-day period. The implant costs approximately $2.50 and the expected response is a 5% increase in pre-weaning gain.

Feed Supply

Environment and the genetic potential of the calf determine the calf's weaning weight. Environmental effects such as health of cows and calves and feed supply are extremely important. It is estimated that approximately 60–70% of weaning weight is accounted for by milk production of the cow; the remaining 30–40% comes from grass and other forage that the calf consumes directly. Therefore, available forage to both cows and calves will significantly affect the weights of the calves at weaning time. Feed supply can be significantly affected by weather conditions, particularly precipitation. Amount and distribution of precipitation affects weaning weights through the amount and nutrient content of the forage.

Month of birth has some effect on weaning weights of calves. Highest daily gains appear to be realized when calves are born 2–3 months prior to green forage production. Calves born earlier and particularly calves born later (summer) usually have slower daily gains. Most cows have a peak in milk production approximately 2 months following calving. Milk production appears to be sustained longer if cows have lush, green pasture about the time they are peaking in milk production. Calf gains are more dependent on milk early in the calf's life as compared with the last couple of months prior to weaning.

The nursing calf usually has an adequate supply of nutrients when receiving cow's milk and an ample amount of forage. Salt and a phosphorus supplement should be provided to cows on a year-round basis. Most mature forages are low in phosphorus, so phosphorus supplementation should be considered.

Creep Feeding of Calves

Creep feeding can add an additional 20–50 lb to the weaning weights of calves. Caution should be exercised with creep feeding, however, because the added weight is often not economical and the gaining ability of the calves after weaning may be affected.

Replacement heifer calves should not be creep fed because future milk production is reduced. Apparently, fat accumulation in the udder decreases the amount of secretory tissue that develops. A producer should compute the expected cost of gain from creep feeding and assess the future production and utilization of the weaned calves before deciding to creep feed. Additional information on creep feeding is given in Chapter 14.

MANAGING ANNUAL COW COSTS AND RETURNS

Obviously, the breakeven price of calves is reduced if a producer can lower annual cow costs while maintaining percent calf crop and weaning weight at the same level (Table 5.18). Sometimes, it is profitable to lower annual cow costs even if weaning weight or percent calf crop is decreased.

Consider the following example: A producer ($117.65/cwt breakeven with 500-lb weaning weights, 85% calf crop, and $500 annual cow costs) determines that feed costs could be lowered $50 per cow, but that it would lower the calf crop from 85% to 80%. Would this be a feasible management decision if weaning weight remained the same?

$$\frac{450}{500 \times .80} = \frac{450}{400} = 112.50$$

This would be a profitable management decision, even though it may not seem logical to lower percent calf crop. The breakeven price is lowered from $117.65 per cwt to $112.50 per cwt, and the profit per cow would be higher at the $450 annual cow cost even though fewer calves would be sold.

Assessments of the costs of production are the most neglected area in many commercial cow-calf operations. Adequate expense records must be maintained so that cost areas can be carefully analyzed. Each producer should know where major costs occur and how the costs can be lowered or held in line with the level of productivity of the cattle (Figure 5.7). The distribution of cow herds by cost category has been estimated to be such that 60% have annual cash costs per cow less than $600, 30% within the range of $600–800, and 10% with annual cow costs greater than $800 per cow.

Table 5.18
EFFECT OF CHANGING ANNUAL COW COST ON BREAKEVEN PRICES (ASSUMES AN 85% CALF CROP AND 500-LB WEANING WEIGHTS)

Annual Cow Cost ($)	Pounds of Calf Weaned	Breakeven Price ($/cwt)
700	425	164.71
650	425	152.94
600	425	141.18
550	425	129.41
500	425	117.65
450	425	105.88
400	425	94.12
350	425	82.35

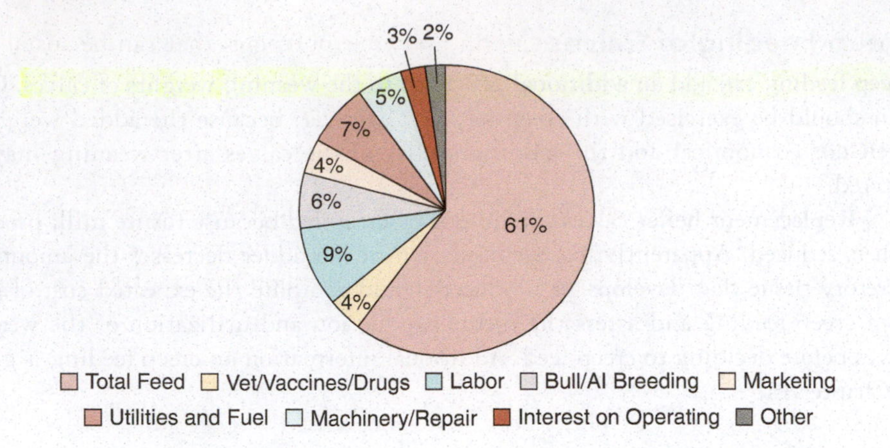

Figure 5.7
Cost categories in an average cow-calf enterprise budget.

Table 5.19
Comparative Regional 2015 Cash Basis Cow-Calf Budgets

	West	Central	Northern Plains	Southwest	Southeast	Mid-south
Variable costs			$ per cow			
Pasture/crop residue	221	222	135	168	222	256
Harvested forage	137	120	170	14	109	178
Supplements	20	19	30	90	92	28
Mineral	15	36	21	5	25	32
Total Feed	*393*	*397*	*356*	*277*	*448*	*494*
Vet/vaccines/drugs	28	28	25	32	19	30
Labor	34	56	25	45	60	80
Bull/AI breeding	35	43	10	45	45	51
Marketing	26	21	15	11	37	34
Utilities and Fuel	35	47	30	50	50	35
Machinery/repair	37	52	44	18	19	25
Interest on operating	10	28	14	15	29	19
Other	5	21	20	6	5	5
Total Variable costs	*603*	*699*	*539*	*499*	*712*	*773*
Revenue						
Calf sales	1,026	826	838	931	786	923
Cull breeding stock	163	150	183	199	200	201
Total Revenue	*1,139*	*976*	*1,021*	*1,130*	*986*	*1,124*
Cash profit	**536**	**277**	**482**	**631**	**274**	**351**

Source: Adapted from multiple sources.

Production and financial records are needed to make effective management decisions. An enterprise budget is a logical place to start. Table 5.19 provides the variable cash costs and returns for cow-calf enterprises in a variety of regions for 2015. Each enterprise should calculate a unique budget but these data provide a set of useful benchmarks for comparison. Figure 5.7 illustrates the contributions of each cost category when an average budget is created from Table 5.19.

Management Priorities

A study titled "Priorities First" (Field, 2006) evaluated more than 200 successful cow-calf producers and technical advisors to establish the rank order of their management priorities. Given the multifaceted nature of the decisions confronting cow-calf managers and the limitations of time and resources, established managers have learned where to invest energy to attain profitable outcomes that can be sustained through time (Table 5.20). This data provides a benchmark for startup cow herds or for those enterprises undergoing an evaluation process to develop more profitable strategies and protocols.

A comparison of data from high-return versus low-return producers also allows the identification of key areas requiring careful planning and effective plan implementation to achieve profitability (Table 5.21). In general, high-return enterprises had lower cow costs, lower feed costs, were less reliant on supplemental feed, had lower debt, and higher production per cow.

Reducing Feed Costs

Feed costs (purchased feed, harvested feed, and grazed forage) comprise approximately 60% of the annual cow costs in many cow-calf operations. Because feed cost is the largest component cost and different feeds have varying costs, this area must receive high priority in assessing production costs. Chapter 14 (Nutrition) and Chapter 15 (Managing Forage Resources) include principles that are applicable to assessing feed costs. USDA survey data analysis shows that 97% of all enterprises supplemented forage, 75% supplemented protein, and 51% supplemented energy on an annual basis. Tables 5.22 and 5.23 provide insight into the supplementation practices of different-sized cow-calf enterprises located in varying regions of the United States.

Table 5.20
RANKING OF THE FOUNDATIONAL AND IMPORTANT MANAGEMENT CATEGORIES AND SUBCATEGORIES WITHIN EACH

Management Categories Scored as "Foundational" or "Important"	Management Subcategories Scored as "Foundational" or "Important"					
Herd nutrition	Cow herd program	Replacement heifers	3rd trimester	Cows—calving to weaning	Bull program	
Pasture & range management	Stocking rate	Timing and duration of grazing				
Herd health	Calves—pre-weaning	Calves—post-weaning	Replacement heifers	Cow herd	Disease treatment	Herd bulls
Financial management	Cost accounting	Cash flow analysis	Enterprise analysis	Tax accounting	Annual balance sheet	
Marketing	Marketing calf crop	Selecting market channel	Marketing cull cows	Marketing replacement females		
Production management	Breeding management	Calving management	Selecting replacements	Weaning management	Herd bull management	
Genetics	Bull selection	Female selection	Use crossbred cows	Produce crossbred calves		
Labor management	Hired labor	Family labor				
Information management	Reproductive performance	Herd inventories	Production records	Health records	Weaning records	

Source: Adapted from *Priorities First* (Field, 2006).

Table 5.21
Comparison of Profitability Groups Over Multiple Regions of the United States

Central	Top 1/3 to Mid- 1/3	Top 1/3 to Low 1/3
Revenue		
Calf weight sold (lb)	+28	+41
Sale price ($/cwt)	+4.70	+3.9
Gross income ($/hd)	+162	+218
Expenses ($/head)		
Feed	−15	−72
Pasture	−3	+3
Interest	−8.9	−21.3
Health/vet	+5.2	−5.5
Machinery	−5.6	−29.8
Labor	−0.8	−11.9
Total variable costs	−3	−160.1
Return over variable costs	+183	+370

Plains/West/Corn Belt	Per Cow	Per Cow
Revenue		
Calf revenue	+75.53	+343.66
Non-calf revenue	+21.38	+97.30
Total	+96.91	+440.96
Expenses	*Top 1/3 to Mid-1/3*	*Top 1/3 to Low 1/3*
Purchased feed	−18.2	−82.81
Labor & management	−7.28	−33.124
Depreciation	−22.75	−103.51
Vet/health	−1.82	−8.28
Interest	−7.73	−35.19
Total	−97.82	−445.10
Net income	+194.28	+883.99

Northern Plains	Top 1/5 to Average (per cow)
Calving percentage (%)	+2
Weaning percentage (%)	+3
Average weaning weight (lb)	+42
Pounds weaned/cow exposed	+54
Expenses	
Feed	−13.5
Veterinary and supplies	−2.5
Interest	−4.2
Other direct costs	−9
Depreciation	−3.2
Other overhead (labor, etc.)	−1.4
Net return over direct costs	+90.65
Net return before operator labor	+96.51

Source: Adapted from multiple sources.

Table 5.22
PROTEIN AND ENERGY SUPPLEMENTATION PRACTICES ON COW-CALF ENTERPRISES—PERCENT OF ENTERPRISES (DURATION OF SUPPLEMENTATION IN DAYS)

	1–49	50–99	100–199	200+
Protein	72 (183)	84 (158)	75 (150)	79 (141)
Energy	50 (174)	55 (152)	47 (122)	61 (131)
	West	Central	Southeast	
Protein	65 (138)	47 (122)	61 (131)	

Source: USDA-NAHMS.

Table 5.23
ROUGHAGE FEEDING DURATION FOR ENTERPRISES (%) THAT FED AT LEAST SOME ROUGHAGE TO A PORTION OF THE COW HERD

Days (N)	1–49	50–99	100–199	200+	West	Central	Southeast
1–30	2	2	3	2	8	0	2
31–90	10	15	8	10	8	11	11
91–180	67	62	62	75	68	56	69
>180	21	21	27	13	16	33	18
Average	160	153	164	152	142	178	154

Source: USDA-NAHMS.

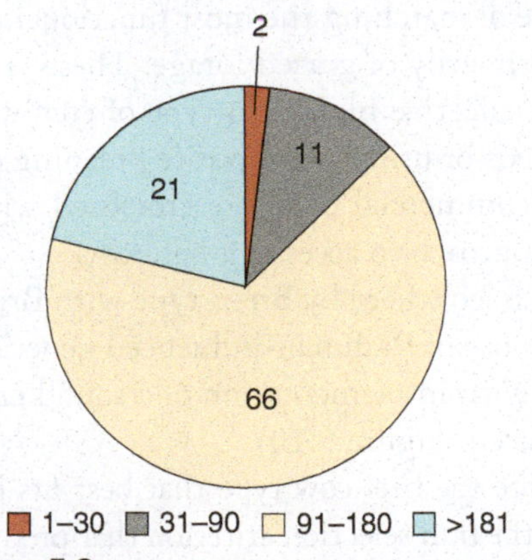

Figure 5.8
Duration (days) of roughage supplementation.
Source: USDA-NAHMS.

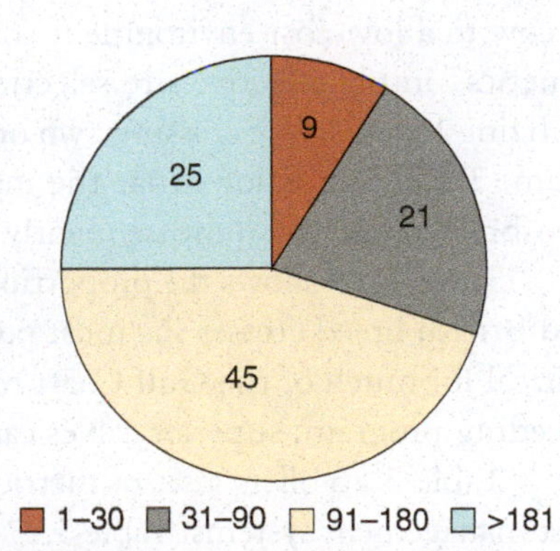

Figure 5.9
Duration (days) of energy supplementation.
Source: USDA-NAHMS.

Most low-cost producers have reduced their feed costs by increasing the number of days cows graze while decreasing the amount of harvested and purchased feed fed per cow. The duration of supplementation is variable as evidenced by Figures 5.8 and 5.9. Extending the grazing season and feeding less hay or harvested feed has been accomplished primarily by:

1. Matching calving season with green forage production. For the west and northern Great Plains, several low-cost producers are calving in April and May rather than February and March. Summer calving (June–August) shows substantial reductions in feed and labor costs (Lardy et al., 1998). High nutrient demands of the cow and calf are less costly if forage is grazed rather than feeding hay or other harvested or purchased feed.

2. Extending the green grazing through intensive and rotational grazing systems. This also can increase carrying capacity (more cows for the same land area), which reduces the fixed cost per cow.
3. Improving range/pasture forage production and utilization through burning, reseeding, interseeding, and grazing a mixture of grasses and legumes where possible.
4. Stockpiling forage for late fall, winter, and early spring grazing.
5. Weaning calves earlier (i.e., 5 months of age) so that cows can increase in body condition score (BCS 6 or 7) prior to the winter months. Less feeding of harvested feed occurs because cows in this condition can lose 75–150 lb prior to calving and still maintain excellent reproductive performance.

Feed costs can be reduced up to 15% by feeding replacement heifers to have low to moderate gains during most of the growing phase (Lynch et al., 1997). Heifers' growth is accelerated about 60 days prior to the breeding season with target weights achieved at breeding. The feeding of a high-concentrate diet following a period of feed restriction does not alter milk production like that experienced in creep feeding heifers prior to weaning.

Equipment and fuel costs are significantly reduced when cow-calf producers develop forage programs where cows graze more days during the year and less harvested feed is fed. While producers have no control over fuel prices, they can exert some control over substantial overusage and by developing buying strategies that take advantage of favorable price swings when they occur. Fossil fuel, machinery, and some supplies increase in cost due to inflation, while renewable resources such as water and sunlight needed to grow grazable forage are less inflationary.

Utilizing the "Right Genetics"

Profitable producers understand the importance of matching the most functional type of cow to a low-cost environment consisting primarily of grazed forage. These "right genetics" imply effective sire selection for cost-effective biological type of cow while utilizing heterosis by crossing two or more breeds or using a composite breeding program. Table 5.24 shows that the majority of commercial cows are crossbred with a two-breed cross, the most frequently utilized approach to accessing heterosis.

Table 5.25 shows the proportion of the U.S. cow herd by breed type with British and British breed crosses the most popular. The use of Brahman-influenced genetics is critical for much of the Gulf Coast region and with implementation of a sound crossbreeding program, superior calves can be produced (Figure 5.10).

Table 5.26 offers a set of metrics to describe the beef cow type that best fits low-cost management systems. Table 5.27 identifies the bull selection criterion that produces

Table 5.24
Description of the Mating System that Produced the Majority of Cows on Commercial Cow-Calf Enterprises

	Size of Herd				Regional Location			
	1–49 (%)	50–99 (%)	100–199 (%)	200+ (%)	West (%)	Central (%)	South Central (%)	East (%)
Straight-bred	16	21	18	24	28	19	15	16
Composite	14	13	7	8	9	9	15	16
2-breed cross	43	47	54	51	44	53	43	41
3-breed or more cross	27	18	20	16	19	19	27	28

Source: USDA-NAHMS.

Table 5.25
Proportion of Calf Crop of Differing Breed Types on Commercial Cow-Calf Herds

	British	Continental	Brahman-Influenced
0%	23	56	80
1–49%	9	15	7
50%	14	11	3
51–99%	15	6	3
100%	40	12	7

Source: USDA-NAHMS.

Figure 5.10
The use of Brahman-influenced genetics in a well-designed crossbreeding system produces both heat- and humidity-adapted females and high-quality calves.
Source: American Brahman Breeders Association.

Table 5.26
Biological Type of Cow for Low-Cost Production

Trait	Description
Mature cow size	Moderate (1,000–1,200 lb; BCS 5)
Milk	Moderate (can wean 475–550 lb calf [7 mos of age] with an average forage supply)
Muscling	Moderate (average thickness through the stifle area; progeny of same biological type have 11–14 sq. in. rib-eye)
Fatness	Fleshing ability (can increase BCS during late summer and early fall grazing); progeny of same biological type have yield grades from 2.0–3.4 at 700–750 lb carcass weights
Longevity/ Stayability	Cow produces a calf each year (365 days) beyond 9 years of age. This trait appears to be influenced primarily by cost-effective fleshing ability (BCS). Most cows leave the herd early because of poor reproduction. They are culled primarily as 2 to 4 years olds because they are open or late breeders—their cost-effective BCS is usually below 5.

Table 5.27
SIRE SELECTION CRITERIA CONTRIBUTING TO PROFITABILITY

1. Performed in one or more low-cost commercial cow-calf operations where the calves have been evaluated in the feedlot and carcass. Performance specifications in these low-cost herds:
   a. Feedlot performance of calves
      - Rate of gain of minimum 3.2 lb/day from 500–600 lb to slaughter weight (1,100–1,300 lb); slaughter age 13–15 months.
      - Feedlot cost of gain (<$.60/lb with $3.00/bu corn).
   b. Carcass performance of calves
      - >70% choice
      - Yield grade 2.0–3.5
      - Satisfactory tenderness (e.g., WBS < 8 lb/in)
2. Longevity and stayability of daughters: They remain in the herd beyond 10 years of age (highly related to fleshing ability and early rebreeding of 2- and 3-year-old cows under low-cost production).
3. Performance specifications for individual bulls (prefer to also see half-brothers to this bull having similar specifications): The following EPDs are Angus based (see across-breed EPDs for comparison):
   a. Birth Wt. EPD: Maximum of +1.5 (<0 for heifers); bull needs to sire actual birth weights as follows: heifers' calves (65–80 lb) and cows' calves (75–90 lb).
   b. Weaning Wt. EPD: +15 to +25 lb.
   c. Milk EPD: Under +15 lb.
   d. Yearling Wt. EPD: +30 to 50 lb.
   e. Scrotal Circumference EPD: Above +0.3 cm and/or minimum yearling scrotal circumference of 34 cm.
   f. Yearling Frame Score: Maximum 5.9 with most of the bull's calves in the 4.0–5.5 range (prefer bull's mature wt. near 2,000 lb and cow's mature wt. below 1,200 lb based on Body Condition Score of 5).
   g. Pedigree (parents and grandparents) stacked for specifications noted in a. through f. and many of those noted in h.
   h. Additional evaluations needed: (1) disposition, (2) Pulmonary Arterial Pressure (adaptability to high altitude), (3) adaptability to varying weather and feed conditions, (4) reduction of variation in a given trait (i.e., smaller standard deviation in birth weight of heifers' calves), and (5) other reproduction EPDs (e.g., fertility, calving ease, body condition score, etc.).

the biological type of female described in Table 5.26. It also identifies bulls that will sire low-cost progeny in the feedlot and superior carcasses in addition to low-cost cows.

Mature weight of cow and level of milk production are important factors affecting annual cow cost. Table 5.28 demonstrates how cow weight and level of milk production affect energy requirements. These requirements demonstrate that more feed is required when weight or milk production increases. A cow's stage of production has a significant impact on her feed requirements. For example, a 1,000 lb cow has higher requirements for energy and crude protein of 19% and 23% in late gestation as compared with the second trimester of pregnancy. The same size cow requires 29% more TDN and 48% more crude protein when she is milking as compared with her second trimester. Anticipating these differences and adapting nutritional management accordingly is key to profitability.

There is disagreement about the size of cow needed for commercial production. Many producers feel that cow size will and should vary according to environmental conditions, particularly feed supply. There are observations that large cows that wean heavier calves will not become pregnant in areas where the most economical feed supply comes primarily from grazed forage.

Table 5.28
Comparison of TDN Requirements for Cows of Various Weights and with Different Milk Production Levels to 1,000 lb Cow with an 8 lb Daily Lactation

Cow Wt (lb)	Annual TDN for Maintenance (lb)	TDN Required for Maintenance and Milk at Varying Production Levels (200 day lactation)		
		8 lb	16 lb	24 lb
800	−400	−200	300	800
1,000	par	par	500	1,000
1,200	400	200	700	1,260
1,400	800	400	900	1,480
1,600	1,200	600	1,200	1,680

Source: Adapted from multiple sources.

Table 5.29
Average Cow Weight by Herd Size and Region

	1–49	50–99	100–199	200+	West	Central	Southeast
Average mature cow weight	1,065	1,120	1,130	1,170	1,145	1,180	1,040

Weight range	Percent of cows		
	1–49	50–199	200+
<1,100	52	32	18
1,100–1,299	40	51	62
1,300+	18	17	20

Source: USDA-NAHMS.

When cow size is increased, the amount and nutrient quality of feed given to heifers are increased so they will reach puberty in sufficient time to calve at 2 years of age. The appropriate cow size is best determined by the amount and nutrient quality of available feed, cost of feed, realizing that grazed forage is usually lowest in cost, and the desired carcass weight and grades of the resulting progeny. In the latter category, carcass weights of 700–800 lb, Yield Grade 2, and Choice Quality Grade are reasonable targets for the majority of market steers and heifers. These carcass specifications strongly suggest a mature cow weight range of 1,000–1,200 lb (BCS 5). There are numerous examples of moderate-sized cows (1,000–1,200 lb in average body condition) that can reproduce regularly and wean 500–550-lb calves with excellent feedlot and carcass performance. Recent surveys show that the average mature weight of commercial cows is 1,050–1,170 lb with more than 20% of cows weighing in excess of 1,300 pounds (Table 5.29).

A terminal crossbreeding program can be used to achieve economic efficiency of smaller-sized cows that produce calves with high feedlot and carcass performance. Cow lines should be selected for moderate size and high maternal performance. Sire lines should receive selection emphasis for growth and cutability. The breed-cross of the cow has been shown to affect the amount of feed required for maintenance when mature cow weights are similar. Angus–Hereford crossbred cows required 19%, 14%, and 16% less feed than Jersey, Charolais, and Simmental cross cows, respectively (Ferrell and Jenkins, 1984). All cows were compared on a similar mature weight. These differences in maintenance requirements are related primarily to level of milk production. Similar differences in maintenance

requirements are observed when comparing cows of high versus low milk production within the same breed. These differences appear to persist, even during the dry period.

A research study (Van Oijen et al., 1993) compared the economic efficiency of three groups of cows of the same mature size but differing in level of milk production—high (>20 lb milk/day), medium (18 lb milk/day), and low (14 lb milk/day). The low milk group was more economically efficient (dollar value of output per $100 of total input cost) than the other higher milk groups. The low milk group would not be considered low by industry standards, as that amount of milk would produce 500 lb calves with good forage availability.

The emphasis on increased frame size, growth rate, and mature size during the past 25 years has significantly increased dystocia (calving difficulty). The risk of dystocia can be reduced by selecting bulls with desirable EPDs for direct and maternal calving ease. Calves with moderate birth weights can have excellent growth (weaning weights, feedlot gain, and yearling weights) by emphasizing the appropriate selection criteria. Attempting to maximize heterosis may increase production costs above optimal cost levels of birth weight, milk production, weaning weight, and mature cow weight. The primary value of heterosis in low-cost cow herds is (1) increased calf survival, and (2) increased cow longevity—best measured by lifetime performance of cows. One research study demonstrated that by 12 years of age, crossbred cows produced 1.6 more calves and 875 lb more calf weaning weight than straightbred cows (Sacco et al., 1989).

Bull costs can range from $1085 per cow as determined primarily by initial purchase price and bull-to-cow ratios (Table 5.30). There is ample evidence that the "right genetics" can be purchased without breaking the bank (Table 5.31) and that

Table 5.30
IMPACT OF BULL PURCHASE PRICE ON THE PER CALF COST OF GENETICS OVER THE PRODUCTIVE LIFE OF THE SIRE

Bull Purchase Price ($)	Total Bull Cost (4 yrs)[1] ($)	Bull-to-Cow Ratio[2] 1:20	1:30	1:40
		Bull Cost Per Cow ($)		
2,000	1,700	23.61	15.74	11.81
2,500	2,225	30.90	20.60	15.45
3,000	2,850	39.58	26.39	19.79
3,500	3,425	47.57	31.71	23.78
4,000	4,000	55.56	37.04	27.78
4,500	4,075	56.60	37.73	28.30
5,000	5,150	71.53	47.69	35.76

[1]Assumes $1,800 salvage value, $300/yr. bull cost (feed, maintenance, etc.), 10% risk (based on purchase price), 5% interest for 1 year.
[2]Assumes bull is used for 4 years; 90% conception rate each year.

Table 5.31
PRICES PAID FOR BULLS

Average Prices Paid for Bulls	% of Producers
<$3,500	38
$3,500–$7,000	46
>$7,000	16

Source: USDA-NAHMS

bull-to-cow ratios can be above 1 to 30 in many operations. However, genetic selection is not a decision that can be short-changed. One of the most important control points in the beef industry is the choice of both sires and mating systems.

Reducing Labor Costs

It is common to find cow-calf operations of similar size with varying numbers of employees. For example, a 300-cow-calf operation may have one to two employees, while another of the same size has three to four employees. In this example, labor cost per cow in one operation is two to four times higher than that of the other operation.

Some large, low-cost operations have reduced labor cost by having one employee responsible for 800–1,000 cows with some labor-intensive activities (e.g., branding and weaning) shared by several employees. Some smaller operations can improve their forage management and increase the number of cows by 20–30% on the same land area. This increased carrying capacity can decrease labor cost per cow.

The well-proven 80–20% rule of business tells labor/management to do the "right things" before doing "things right." Work activities should receive priority that answers the question, "How can my time be spent most effectively in lowering cost of production by concentrating on $100/hour activities before $5/hour activities?" Generally, this means more time spent on financial, forage, and human resource management prior to cattle management. Cattle management should not be overlooked, however, as cattle that are more problem-free (i.e., ease of calving, good dispositions, good mothering ability, etc.) require less labor and thus are more cost effective.

Improving grazing management and decreasing dependence on mechanical harvesting of forage usually require less labor. Some operations have reduced labor and equipment costs through contracting these services. At the same time, once the appropriate number of employees has been determined, it is critical to attract and retain good talent. Employee development is typically an area of weakness in most agricultural production settings.

MATCHING COWS TO THEIR ECONOMICAL ENVIRONMENT

Table 5.32 outlines the major management factors affecting annual cow costs. Producers who understand these factors and manage them effectively are likely to be high-profit commercial producers.

The management challenge for cow-calf producers is to match the biological type of cows to their most cost-effective environment (Figure 5.11). This is not an easy task, but a logical sequence can be found in the decision-making process.

1. Costs must be measured and recorded.
2. Feed costs should be carefully assessed and compared with those of other operations in the same area.
3. The most economical feed supply should be determined by answering the following questions:
   a. Can the grazing of forage be extended for the cows during the year and still meet nutrient requirements?
   b. Does the calving season best match the available green forage (usually calving a month prior to adequate levels of green forage)?
   c. Could the fall grazing season be extended by weaning calves earlier and allowing cows to add body condition prior to winter weather?
4. After identifying the most economical feed supply, determine the biological type of cattle that will be most profitable.
5. Integrate all management decisions by calculating the breakeven price on calves.

Table 5.32
MAJOR MANAGEMENT PRINCIPLES AFFECTING ANNUAL COW COSTS

1. ENTERPRISE BUDGET
   a. Annual cow costs can be determined and monitored through an enterprise budget. Producers must know annual cow costs to calculate breakeven prices and assess net returns.
   b. The individual cost components of the enterprise budget should be accurately recorded and then compared with other operations with similar environmental conditions. A specific cow-calf producer can then determine if his or her annual cow costs are high, low, or about average. Particular attention should be paid to feed, machinery, labor, interest, and health costs. Comparisons within the enterprise across time are also useful.
   c. Feed costs should be most critically assessed because they comprise from 40% to 70% of the annual cow cost. Producers should determine their most economical feed environment and then decide what biological types of cows are best matched to this feed environment. Grazed forage is usually the most economical feed environment, as harvested and purchased feeds are more expensive in most cases.

2. BIOLOGICAL TYPE OF COWS
   a. ***Puberty.*** Heifers that express early puberty and calve at 24 months of age under a low-cost feeding program will contribute to lower annual cow costs. This implies optimum weaning weights and frame size and using breeds (biological types within breeds) that are known for early puberty. An economical feeding program from weaning to a target breeding weight is essential.
   b. ***Calving ease.*** Calving ease, especially needed in 2- and 3-year-old females, reduces annual cow costs as less labor is required, health costs for cow and calf decrease, females will breed earlier and thus require less feed to produce a calf per cow per year.
   c. ***Milk production.*** If milk production is too low, the weaning weight of the calf is poor and the cow puts the nutrients "on her back" instead of "in her calf." Milk production that is too high results in a heavy calf but an open cow or late bred cow, which will increase annual cow cost. Research studies show that moderate levels of milk are more cost effective and profitable.
   d. ***Mature weight.*** Research studies demonstrate that moderate mature cow weights (1,000–1,250 lb under most cow-calf environments) are most profitable. Cows larger than these optimum weights can be biologically efficient but may not be economically efficient.
   e. ***Rebreeding.*** Rebreeding performance is best measured by body condition score (BCS) at calving time. Managing the BCS by utilizing primarily grazed forage with minimum amounts of harvested and purchased feeds will keep annual cow cost low.
   f. ***Longevity.*** Cows that reproduce regularly under a low-cost feed environment will remain in the herd for a long period of time. This situation requires a lower heifer replacement rate that usually lowers annual cow cost. Replacement heifer costs are high in most herds and high rates of heifer replacement are costly.

3. MANAGING FEED COSTS
   a. ***Calving season.*** The calving season for many operations should be approximately 30–45 days prior to grazed green forage. Earlier calving seasons will require large amounts of more expensive harvested and purchased feeds. Later calving seasons do not effectively utilize the peak of high-quality nutrients from grazed forage.
   b. See sections 1c, 2a, 2c, 2d, and 2e in this table. When BCS of cows is low at calving and breeding, producers should determine the most economical management decision—e.g., increase the supplemental feed to increase BCS or change biological type of cows to increase BCS on the original feed supply.
   c. ***Heifer development.*** Delaying the majority of weight gain until late in heifer development may decrease costs (up to 15%) without detrimental effects on reproductive performance.

4. INCREASING COW NUMBERS ON SAME LAND AREA
   a. Overhead costs (e.g., land payments, insurance, taxes, vehicle costs, and others) are expenses that remain approximately the same regardless of the number of beef cows in the operation. As cow numbers increase on the same land area, total overhead costs change little, if any; however, overhead costs per cow will decrease.
   b. Cow numbers can be increased on the same land area primarily by improving forage production and grazing management. This reduces the number of acres required per cow.

Figure 5.11 identifies the primary traits used in defining biological types. Breeds and breed types have been used to reflect biological types (refer to Chapters 4 and 13). Breeds can be used in a general sense, but for individual operations biological types must be described more specifically because several biological types exist within a breed.

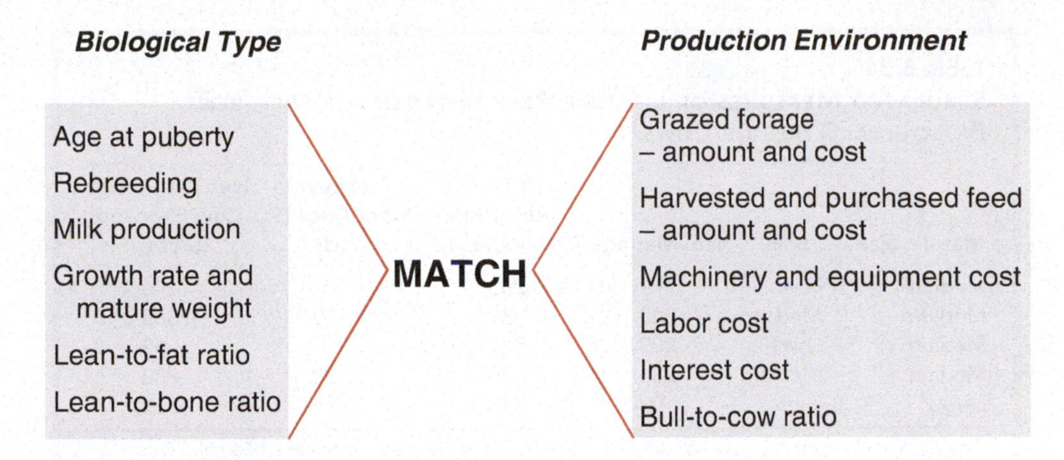

Figure 5.11
Matching biological types of cows to the most economical production environment is a management challenge that significantly affects profitability.

Table 5.33
PERFORMANCE OF DIFFERENT BREED TYPES (BIOLOGICAL TYPES) OVER SEVERAL CALF CROPS

	Biological Type			Milk Production per Day (lb)		Weaning Wt (lb) per Female Exposed[1]			
Breed Type of Dam[2]	Mature Size	Milk	Puberty/ Rebreeding	2-Year-Old Dams[3]	5–8-Year-Old Dams	2-Year-Old Dams[4]	5–8-Year-Old Dams	Mature Cow Wt (lb)	% Cows in Herd After 5 Years
HH	Medium	Low	Late	16.5	15.8	213	345	1,225	41
AH	Medium	Medium	Early	18.5	20.0	359	359	1,270	52
1S3H	Medium	Low+	Medium	18.4	18.7	260	394	1,240	53
SH	Medium+	Medium	Early	20.9	22.4	365	416	1,290	67
3S1H	Large	High	Late+	22.1	24.4	315	372	1,330	42

[1]Weaning weight at 180 days of age.
[2]Hereford (H), Angus (A), and Simmental (S).
[3]Taken 40 days into lactation.
[4]Taken 130 days into lactation.
Source: Based on Montana State University (*Journal of Animal Science* 68:54, 1910) and "Cows That Fit Montana" Video.

Tables 5.33 and 5.34 give the results of an excellent research study that focused on biological and economic assessments of different biological types of cows in a typical northern Great Plains environment. While the results may not apply to all other cow-calf environments, the importance of evaluating the economics of different biological types is emphasized. An analysis of biological types in several environments, however, indicates that moderate birth weight, moderate milk, moderate mature weight, and moderate frame size are realistic targets for many cow-calf producers.

The data in Tables 5.33 and 5.34 demonstrate that moderate performance in certain traits will result in maximum profits. However, it should be noted that the poor performance of the Herefords occurred because only 59% of the original group of heifers reached puberty at 14 months of age. There are different biological types of Herefords that perform differently for the traits shown. In addition, when Simmental bulls were crossed on Hereford cows to produce the S × H crossbreds, this mating group experienced the largest birth weight, most calving difficulty, and the lowest calf survival rate.

Table 5.34
SIMULATED MEASURES OF LIFETIME PERFORMANCE FOR DIFFERENT BIOLOGICAL TYPES OF COWS[1]

Mature Size	Milk	No. Matings	Lb Calf Produced per Cow Exposed	Breakeven Steer Price ($/cwt)	Net Return per Cow Exposed ($/cow)
Medium	Low	199	283	par	par
Medium	Medium	243	339	−$15	+$68
Medium	Low+	229	332	−12	+52
Medium+	Medium	246	370	−20	+88
Large	High	199	338	−13	+55

[1]Five replications were simulated for each breed type of dam × sire breed combination. Each replication simulated the life-cycle inputs and outputs for sixty replacement heifers in each biological type.
Source: Based on Montana State University (Proc. Western Section American Society of Animal Science 43:43).

Managing cow-calf resources implies matching the cows profitably to the available resources. Therefore, in most of the western and Great Plains states, and possibly other states as well, the cows should be of a medium-frame size, have medium milk, have early sexual maturity, and have a feed supply that originates from low-cost grazed forage.

Replacement Heifer Development

Generally, replacement heifers are weaned in the fall, are bred the following spring or summer, and calve the following spring at 2 years of age. From an economic standpoint, this is the best management scheme for many operations. It is not recommended that replacement heifers be creep fed because it leads to deposition of fat into the mammary system that can be detrimental to future milk production.

Puberty is primarily a function of age and weight. Most heifers are old enough to breed at 14–16 months of age. The critical limit is weight. A general management guideline has been to feed heifers to attain 65% of their mature weight by the start of the breeding season. To accomplish this, most heifers need to gain 0.5 to 1.5 lb per day depending on their weaning weight. There are two primary approaches to achieving target-breeding weights: feed heifers to gain at a consistent rate over the full development period or feed heifers to grow slowly with a ration increase to boost gains in the two months prior to mating. Research results suggest that heifers fed to gain 0.50 lb/day early with an accelerated gain of 2.5 lb/day two months prior to breeding are equal to, if not superior, in reproductive performance as compared with females fed to gain at a constant rate. Furthermore, under most feed price conditions, the slow–fast program was cheaper to implement. As Table 5.35 demonstrates, pregnancy rates are comparable for high, medium, and low–high dietary protocols. The same study found that these dietary regimes had no effect on birth weight, calving difficulty, or first-calf weaning weight. Furthermore, heifers consumed the least on the low–high treatment. However, there was an increase in calf death loss for the low–high protocol particularly when compared with the medium feeding level.

The use of a bypass protein in a heifer development ration may prove to be the key to developing heifers on pasture. Work at New Mexico State University shows that heifer pregnancy rates of 80% can be achieved on females grazing native range at a supplement cost of less than $30/head/year. With careful bull selection, heifers on this treatment experienced less than 5% calving difficulty. Rebreeding rates of 85% for 2-year-old females were also reported. The system depends on the ability to deliver

Table 5.35
PERFORMANCE OF HEIFERS RAISED ON THREE DIFFERENT NUTRITIONAL DEVELOPMENT PROGRAMS

					Summary of Female Culling (N)					
							Death Loss in First Calves			
Treatment	N	Weight at Breeding (lb)	Total ME Intake (med./high)	Pregnancy (%)	Open/ Aborted Yearling	Open as 2-Year-Olds	Birth	<72 hrs	>72 hrs	Total
High[1]	94	865	3,072	90.4	10	6	4	3	9	16
Medium[2]	94	845	2,854	92.6	9	7	2	1	6	9
Low–High[3]	94	843	2,652	91.4	10	3	5	6	9	20

[1]263 kcal ME/(BWkg)$^{0.75}$ for 205 days.
[2]238 kcal ME/(BWkg)$^{0.75}$ for 205 days.
[3]257 kcal ME/(BWkg)$^{0.75}$ for 83 days followed by 277 kcal ME/(BWkg)$^{0.75}$ for 122 days.
Source: Adapted from Freetly et al. (2001).

Table 5.36
HEIFER RETENTION STRATEGY/IMPACT ON REVENUE, RETURN OVER ECONOMIC COST AND RETURN OVER CASH COST (30-YEAR PERIOD)

Annual Revenue	Average ($1,000)	Minimum ($1,000)	Maximum ($1,000)	Ending Total Revenue ($1,000)
SS	44	27	65	40
CF	36	14	66	14
DCA	47	25	96	42
RAV	44	22	75	49
Return Over Total Economic Cost				
SS	−1.8	−16	19	0.5
CF	−0.9	−11	3	3
DCA	0.1	−21	37	2
RAV	−0.4	−18	28	3
Return Over Cash Cost				
SS	4.8	−8	27	6
CF	4.1	3	6	5
DCA	6.5	−15	48	8
RAV	5.6	−12	36	8

SS = Steady Size, CF = Cash Flow, DCA = Dollar-Cost Averaging, RAV = Rolling Average Value.
Source: Adapted from Iowa State University (2000).

a supplemental ration to the heifers and the ability of the producer to maintain adequate grazable forage during the development phase.

Determining the number of females to be retained each year is a critical decision in terms of its impact on both expenses and revenues. Iowa State researchers evaluated four heifer retention strategies: steady size (SS) to return the same number of heifers each year, cash flow (CF) to maintain income at a steady level (differing numbers of heifers sold to attain the goal), dollar-cost averaging (DCA) in which the same values of heifers are retained annually, and rolling average value (RAV) in which the 10-year average values of heifers are retained.

Table 5.36 shows that DCA and RAV strategies were the most profitable. This results from the effect of having lower-cost heifers entering production during the price peaks of the cattle cycle; thus, more calves are sold at higher prices. Also, when heifer prices are high, more females are sold.

However, there is a downside. Producers must accept more variation in year-to-year cash flow and producers must have flexibility in terms of access to land, as the herd size is variable over time.

WEANING MANAGEMENT

The process of weaning is potentially a period of high stress, especially for calves. In addition to the stress of removal from the cow, a calf faces the additional stressors of immunization, transportation, removal to new surroundings, and dietary change. If these stressors are not effectively managed, increased morbidity and mortality may result.

One approach to reducing weaning stress involves the use of a technique referred to as fence line weaning. This management strategy involves allowing calves to remain in close visual proximity to their dams following weaning. The success of this approach depends on a strong fence and is most effective when implemented on pasture conditions. If pasture is only available for cows or calves, it is usually best to dry lot the cows with calves in an adjacent pasture or paddock to avoid exposing the calves to dusty conditions. Furthermore, it is beneficial to vaccinate calves while still on their dams with the boosters given at the time of weaning.

The advantages of fence line weaning include reduced stress with up to 20–30% improvement in weight gain in the 10-week period following weaning and substantial reductions in morbidity. Notice that fence line weaned calves had better weight gains at both 2 and 10 weeks post weaning as well as increased eating times, time spent lying down, and reduced walking times.

Another management approach is called two-step weaning and utilizes a humane, anti-suckling device (resembles a plastic nose ring) that prevents calves from nursing but allows them to eat and drink while remaining in the presence of their dams. In one Canadian study, those calves that were prevented from nursing for four days but left with their mothers behaved identically to contemporaries that were allowed to nurse. However, once separated, the two-step weaned calves vocalized 85% less, walked 80% less, and spent 25% more time eating compared to their conventionally weaned contemporaries.

Cows prevented from being nursed via the use of anti-suckling devices for a period of 8 days but still allowed to have the calf at side were less stressed by weaning than if they had only been un-nursed for 4 days. Several studies have pointed to decreased weight gains in calves that carried the weaning rings for 14 days. These may have occurred as a result of nose irritation from the devices or reduced overall intake. Furthermore, as time progressed, retention of the devices declined.

Another weaning strategy that has been adopted by producers in times of limited feed availability (such as drought) or as a standard operating procedure is early weaning. In early weaning, calves younger than 150 days are removed from their dams. The benefits of this approach are:

1. Removing calves from pastures early makes about 10–15% more feed available to their dams.
2. Removing suckling calves from their mothers reduces cow feed requirements by about 1/3.
3. Improves rebreeding in young or thin cows.
4. Improves the opportunity to increase body condition of both calves and cows.
5. Offers more market flexibility.

One Nebraska ranch calves from February to April and then begins weaning calves in late May and June. Over a three-year period, this approach has yielded weaning weights of approximately 300 pounds and finished weights of 1,250 pounds sold in late March and April—ahead of the seasonal price break that typically occurs in May and June. Research at both the University of Illinois and Iowa State University demonstrates that early-weaned calves of both British and Continental descent have improved levels of marbling at harvest. One of the keys to successful weaning, either early or conventionally, is the use of low-stress animal-handling techniques. Also, early-weaned calves need to be transitioned to an appropriate and palatable concentrate ration during the weaning process.

Producers should be cautioned to carefully evaluate their facilities and time resources before implementing an early weaning program. Consultation with other producers or beef cattle experts with experience in early weaning protocols is advised to develop a successful approach.

MARKETING DECISIONS

Marketing begins with the decision about what will be produced. Thus, for cow-calf producers marketing actually starts at breeding time. However, too many producers do not begin thinking about marketing until a few weeks or days prior to weaning the calves.

Marketing decisions should be based on the profitability formula:

$$\text{Profitability} = (\text{product} \times \text{price}) - \text{costs}$$

Product is usually reflected in numbers and pounds and may include breeding bulls, weaned calves, background yearlings, slaughter cattle, carcass, or retail products. Price of product and cost of production are also important components of the marketing decision process. If cattle producers are to make intelligent and profitable marketing decisions, they must:

1. Know the cost of production and breakeven prices.
2. Know the value of the product they have produced, including how it will perform and be valued by other segments of the industry.
3. Understand prices (e.g., factors affecting prices, how to obtain current price information).
4. Know how to integrate the value of product, cost of production, and prices in evaluating different market alternatives.

Factors Affecting Market Value of Feeders

A number of variables impact the price received for feeder calves, including breed type (color), muscularity, frame size, body condition, and the degree of fill at time of sale (Table 5.37).

Interestingly, even though there have been long-standing market signals that discount horned or intact males in the feeder cattle market, the industry has not yet achieved conformance to these standards (Table 5.38). Cattle from the south central and eastern regions have the greatest levels of non-conformance to these management standards.

Producers can overcome some of these price differences by the careful sorting of market groups and the use of feeding strategies to conform to desired levels of performance in the marketplace. The feeder cattle supply is highly diverse and as those cattle move through the supply chain, the goal is to reduce variation through standardized management, sorting, and nutrition (Figure 5.12).

Table 5.37
SALE PRICE DIFFERENTIALS OF FEEDER CALVES OF VARYING TYPES

Factor	Price ($/cwt)	Price ($/cwt)
	Southeast	Midwest
Fill Condition		
Gaunt	+5.86	–3.60
Shrunk	+2.21	–0.99
Average	par	par
Full	–4.73	–0.72
Tanked	–11.10	–4.02
Muscle Score		
1	+13.40	+5.29
2	+8.68	+6.62
3	par	par
4	–9.25	–5.03
Frame Size		
Large	+0.96	+0.75
Medium	par	par
Small	–18.57	–5.98
Health Status/Other		
Polled	par	par
Horned	–1.49	–2.18
Lame	–27.45	NA
Sick	–25.45	–6.31
Body Condition		
Very thin	–7.69	–10.83
Thin	+2.43	–1.23
Average	par	par
Fleshy	–1.87	–0.86
Fat	–4.69	–4.87

Source: Adapted from multiple sources.

Table 5.38
PERCENT OF MALE CALVES CASTRATED AND PERCENT OF HORNED CALVES DEHORNED PRIOR TO LEAVING FARM OR RANCH OF ORIGIN BY REGION

	West	Central	South Central	East
Percent of male calves castrated	96	88	66	58
Percent of horned calves dehorned	74	67	36	39

West: OR, CA, MT, ID, WY, CO, NM.
Central: ND, SD, NE, KS, IA, MO.
South Central: OK, TX.
East: AR, LA, KY, TN, MS, AL, GA, FL, VA.
Source: USDA-NAHMS.

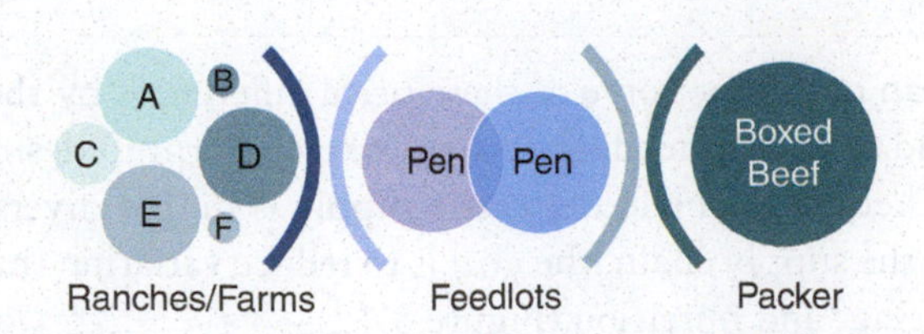

Figure 5.12
The process of reducing variation in the U.S. beef supply chain.

Marketing Alternatives

Cow-calf producers have several marketing alternatives. They may (1) sell calves at weaning, (2) sell bred heifers, (3) sell cull cows for slaughter or for breeding purposes to other producers who can make the cows profitable for a few more years, or (4) retain ownership via backgrounding or grazing yearlings or through the feedlot or retail stage. Any combination of these alternatives may also be used.

Selling calves at weaning is the most common marketing alternative because (1) it is a traditional marketing method, (2) the loan agency offer requires immediate payment rather than extending credit over a longer period of time, and (3) income tax considerations arising from shifting income from one year to another may restrict changes in the marketing strategy. Producers are also seeking opportunities for capturing value-added price differentials based on a number of management protocols (Table 5.39). Part of the development of value-added arrangements is the sharing of information on cattle as they move through the supply chain (Figure 5.13). Larger enterprises are more likely to share information with down-channel users and to participate in forward pricing that may require documentation.

When market conditions dictate the opportunity, more cow-calf producers are retaining ownership of their calves through the growing and feedlot stages. Table 5.40 lists the proportion of calves and the timing of their movement from the site of origin following weaning.

Table 5.39
Marketing Program and Value-Added Strategy Engagement by Herd Size for at Least Part of the Calf Crop

	1–49	50–99	100–199	200+
Breed influenced	12	16	16	29
Age and source verification	5	12	15	29
Natural	29	25	24	24
Organic	1	<1	<1	<1
Conventional	69	69	68	68

Source: USDA-NAHMS.

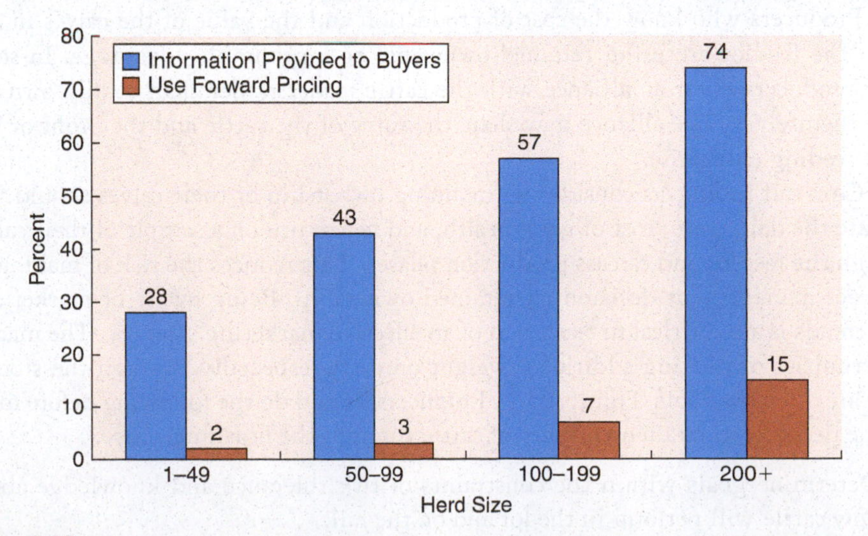

Figure 5.13
Percent of herds providing information to down-channel users and percentage participating in forward contracting pricing arrangements.
Source: USDA-NAHMS.

Table 5.40
Length of Time that Calves Remain on Site Following Weaning Prior to Transport for Sale or to Customer

	Percent of Calves			
Days (N)	1–49	50–99	100–199	200+
0	56	45	27	34
1–31	15	20	21	12
32–61	12	13	26	16
62–92	6	10	12	9
93–122	1	4	5	8
123+	10	8	9	20

Source: USDA-NAHMS.

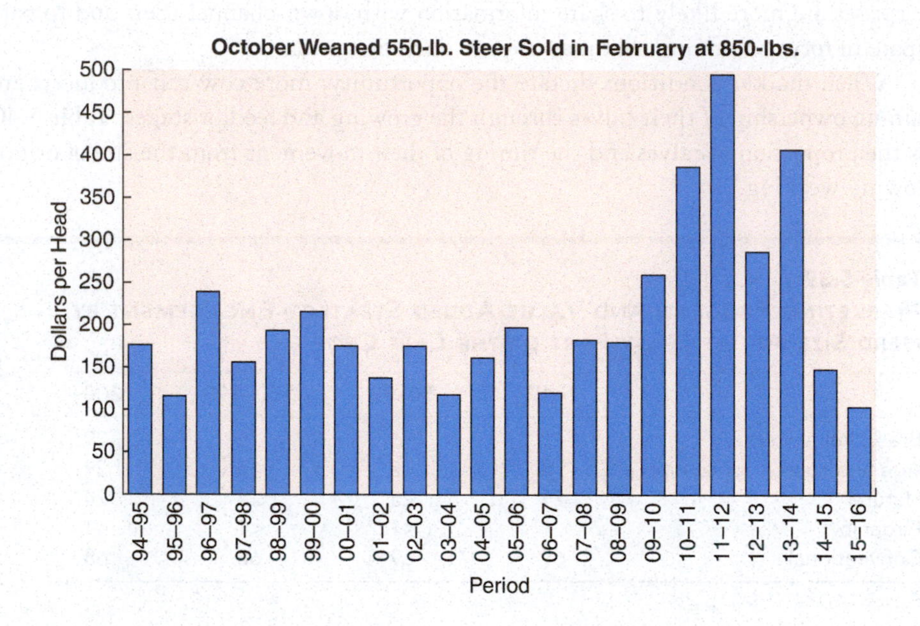

Figure 5.14
The market value of adding additional weight to calves beyond weaning.
Source: Cattle-Fax.

Producers who know the cost of production and the value of the calves in and out of the feedlot are using retained ownership to a profitable advantage. In some cases, producers enter an alliance with the cattle feeder rather than totally own the cattle themselves. The alliance may share the value of the cattle and the profit or loss in the feeding enterprise.

Cow-calf producers considering retaining ownership of their calves should first evaluate the daily gain, cost of gain, health, and net return on a sample of their calves through the feedlot and carcass production phases. This reduces the risk of making an incorrect management decision on retained ownership. Being aware of market and price trends is also critical to execution of an effective marketing strategy. The market opportunities of putting additional weight on cattle especially through the stocker phase are often available (Figure 5.14). Producers should do the following before making the decision to retain ownership of cattle through the finishing phase:

1. Determine goals within the constraints of risk tolerance and knowledge about how cattle will perform in the lot and on the rail.
2. Conduct a careful evaluation of market conditions at the anticipated dates of sale associated with each market alternative.
3. Seek professional consultation with regard to developing risk-management plans to protect equity against unforeseen shifts in the market.

Table 5.41
Reasons Cows were Culled from Cow-Calf Herds of Various Sizes (%)

Reason for Culling	1–49	50–99	100–199	200+
Pregnancy status	15	35	32	47
Other reproductive problem	5	2	7	3
Poor performance	3	3	5	4
Age	33	37	41	29
Unsoundness	1	4	4	3
Temperament	4	7	3	2

Source: USDA-NAHMS.

Table 5.42
Age at Which Cows are Culled on Commercial Cow-Calf Enterprises (%)

Age (years)	1–49	50–99	100–199	200+	West	Central	Southeast
<5	15	10	14	20	20	18	13
5–9	27	32	32	35	38	32	29
10+	58	58	54	45	42	50	58

Source: USDA-NAHMS.

4. Communicate with lenders to assure that the marketing plan is aligned with the debt-management plan.
5. Define and commit the necessary resources to implement the plan successfully.

Producers too often overlook the impact of marketing cows and bulls that have been culled due to reproductive failure, poor progeny performance, or age (Table 5.41)

Effective cow and bull marketing strategies are listed here.

1. Add weight to thin cows prior to sale.
2. Market older cows, non-producers, and problem cows (poor udders, disposition, etc.) in a timely manner to avoid production losses and market discounts. The age at which cows are culled compared by size of enterprise and regional location is listed in Table 5.42.
3. Sell cull cows outside of the seasonal fall run.
4. Evaluate opportunities to market cows/bulls directly to a packer.
5. Closely adhere to animal health-care product withdrawal dates.
6. Adhere to the marketing code of ethics.
7. Seek opportunities to add value to market cows and bulls.

ESTABLISHING A COMMERCIAL COW-CALF OPERATION

Many cow-calf operations have existed for decades, being inherited over several successive generations. During the past 10–20 years, land has become a major investment commodity primarily because of an expanding population with extensive available financial resources. The price of land has exceeded its ability to be profitable under many phases of agricultural production, including the production of beef cattle.

Cost of a commercial cow-calf operation can be evaluated on a per-cow basis. This would include the costs of land, cow, buildings, equipment, and other minor expenses. It is not uncommon to have investment costs of $1,500–5,000 on a per-cow basis.

Annual cow costs ranging from $400 to $700 are not uncommon. Even when calf prices are high, they will not cover all of these costs. Producers cannot plan on high calf prices to continue uninterrupted for long periods of time if beef continues as a commodity product subject to the typical beef cattle cycles.

Total capital investments for an economical cattle operation are extremely high. Several studies show that an economical unit is a minimum of 300 head while other industry experts suggest that an economic unit is an enterprise with a minimum of 800 cows. At a best price of $3,000 per cow, the total investment would be $900,000 for the 300 head outfit and $2,400,000 for the 800 head enterprise. It is evident that high total capital investment plus high annual interest costs prevent many potential beef producers from purchasing new beef operations. Some established cow-calf producers might be encountering a financial crisis because of an excessive debt load. They may solve this financial problem by selling some assets (e.g., land) or generating additional income from other resources (e.g., hunting, fishing, etc.). These solutions may allow them to keep an economic cow-calf unit intact.

Nearly half of the beef producers in the United States are 55 years of age or older. The primary reason for aging cow-calf producers is that younger people do not have the necessary capital or management experience. Some individuals are successfully leasing cow-calf operations rather than trying to purchase the land.

Land prices and interest costs fluctuate over time. There are certain times when cattle operations can be purchased on an economically sound basis. These purchases may be made by new producers or by established producers seeking to add to an existing operation. It is important to also evaluate the regional differences in land values. Note that cow-calf production is generally concentrated in those states with the lowest pastureland values.

One possibility to make entry into the cow-calf business more feasible is to incorporate a different business model. Several models to be considered as alternatives to the traditional cow-calf management calendar are listed below:

- Full confinement (FC)
- Pasture systems coupled with cornstalk grazing (PCS)
- Pasture systems with supplementation during grazing coupled with cornstalk grazing (PSCS)
- Semi-confinement systems with lot confinement for 6 months and corn residue grazing for 6 months (SC)

A 2-year study at the University of Nebraska found that full confinement systems were the most expensive to execute and systems incorporating semi-confinement with corn stalk grazing were the most economical (Table 5.43).

Table 5.43
Cost Comparison of Cow-Calf Management Systems

Year One	Traditional	FC	PCS	PSCS	SC
Total cost ($)	Par	+267	−17	+30	+11
Cost per lb of calf weaned ($/cwt)	Par	+77	+21	+19	−3
Weaning weight (lb)	Par	−77	−86	−48	+23
Year Two					
Total cost ($)	Par	0	−40	−39	−44
Cost per lb of calf weaned ($/cwt)	Par	+26	+11	+10	−15
Weaning weight (lb)	Par	−71	−57	−55	+23

Source: Adapted from Rasby (2015).

There are a number of considerations associated with confinement or semi-confinements systems. The minimum drylot space requirement for a pair is 500 square feet but up to 1,000 square feet may be required to assure cattle comfort across various seasonal conditions. Some systems incorporate either traditional roofed structures or hoop barn systems; in these situations 80 square feet per pair is the minimum space requirement with an additional 40 square feet needed during calving season. Bunk space can also be a limiting issue; thus facilities should be designed with at least 24 inches of linear bunk space for pregnant females, 36 inches for a pair, and 50 to 60 inches for bunks with access on both sides.

Pathogen concentration in confinement is particularly an issue for calves, so facility sanitation becomes a significant area of management focus. There is also evidence that calves under confinement may be slower to nurse, consume less colostrum, and experience weaker bonding to their dams as compared to pasture systems. Disease management and avoiding lameness are more critical in confinement where calfhood scours are more likely and hoof and foot problems are more prevalent.

Confinement or semi-confinement systems offer the opportunity to better manage diets by using a total mixed ration specifically formulated to the nutritional needs of differing stages of production, minimizing feed wastage, and taking advantage of cheaper feed sources such as grazing corn stalks.

Before moving to more intensive systems, a detailed financial and operational evaluation should be undertaken.

SUMMARY OF COW-CALF OPERATIONAL TYPES AND MANAGEMENT PRACTICES

The cow-calf industry is diverse in both organizational and operational approach. Tables 5.44 and 5.45 provide a summary of these differences. Regionally, cow-calf enterprises in the southeast and southern plains are the least likely to retain ownership beyond weaning and have lower average weaning weights. As precipitation levels decline, the acres per cow increase that results in larger land requirements per enterprise. Younger operators are also more likely to retain ownership beyond the weaning phase of production.

Table 5.44
DIVERSE STRUCTURE OF THE BEEF INDUSTRY BY ORGANIZATIONAL STRUCTURE

	Cow-Calf	Cow-Calf:Stocker	Cow-Calf:Feeder	All
Percent of calves:				
Sold at weaning	100	21	28	59
Backgrounded then sold	0	79	15	36
Retained through feedyard	0	0	57	5
Cows calving (N)	69	106	97	88
Average weaning weight (lb)	502	499	523	502
Percent farm income from cattle	36	43	34	39
Crops produced by farm (%):				
Corn	6	20	44	16
Soybeans	5	14	43	13
Small grains	10	17	21	14
Hay	77	78	79	78
Operator characteristics:				
Age	61	59	56	60
Percent over 65 years of age	42	34	21	36
Percent of total income from off-farm	41	34	20	36

Source: USDA-ERS.

Table 5.45
Diverse Structure of the Beef Industry by Regions

	North Central	Southeast	Northern Plains	Southern Plains	West
Percent of calves:					
Sold at weaning	44	70	41	69	53
Backgrounded then sold	45	28	49	29	39
Retained through feedyard	11	2	10	2	8
Cows calving (N)	61	63	118	82	186
Average weaning weight (lb)	501	480	543	493	538
Percent farm income from cattle	23	25	38	67	66
Acres operated					
Total	518	453	2,019	1,436	4,186
Pasture & range	208	246	1,359	1,272	4,028
Acres per cow	3	3	11	13	19
% grazing public lands	0	0	7	1	36

West: OR, CA, MT, WY, CO, NM.
Northern Plains: ND, SD, NE, KS.
Southern Plains: OK, TX.
North Central: IA, MO.
Southeast: AR, KY, TN, VA, AL, MS, GA, FL.
Source: USDA-ERS.

Table 5.46
Percent of Operations Using Various Beef Cattle Management Practices

Practices	Cow-Calf	Cow-Calf: Stocker	Cow-Calf: Feeder	All	North Central	Southeast	Northern Plains	Southern Plains	West
Defined calving season	54	66	79	61	82	45	92	42	85
Artificial insemination	4	11	19	8	11	4	17	6	14
Calf implants/ionophores	9	17	25	14	28	7	26	8	13
Regular veterinary services	17	28	32	23	27	12	41	18	32
Nutritional consulting service	4	6	18	7	11	2	15	5	9
Rotational grazing	59	62	56	60	54	60	58	62	71
Test forage quality	12	19	27	16	16	10	27	15	25
Individual cow records	40	50	56	46	52	35	59	45	52
On-farm computer records	17	22	29	20	20	13	23	22	35
On-farm access to internet	29	38	42	34	33	29	38	35	49
Forward purchase inputs	5	10	12	8	3	7	10	8	17
Negotiate input prices	14	23	24	19	14	16	21	19	31

West: OR, CA, MT, WY, CO, NM.
Northern Plains: ND, SD, NE, KS.
Southern Plains: OK, TX.
North Central: IA, MO.
Southeast: AR, KY, TN, VA, AL, MS, GA, FL.
Source: USDA-ERS.

Management practices incorporated by cow-calf enterprises change with both organizational structure and region of the country (Table 5.46). Generally as calves are owned further into the supply chain, the likelihood of technological adoption and use of more intensive management increases. Regional variation in management practices also exist and adoption of improved management protocols offer opportunities for increased productivity and profitability.

Integrating and Sustaining the Resources

The challenge for a producer's leadership and management decisions process is to integrate and sustain human, financial, forage, and cattle resources. Integration of resources reflects how the resources interact with one another. For example, if a producer decides to increase calf crop percentage, how much more forage or supplemental feed will it require or how much less forage will it take if mature weight and milk production are decreased, and how will these decisions affect the breakeven price of the weaned calves? Can the productivity goals of the cattle enterprise be accomplished while attaining natural resource management objectives? The answer is "yes" when managers take the long-term view and are open to new approaches, embrace measurement of key indicators, and retain decision flexibility as conditions both internal and external to the enterprise occur.

Managers should ask, "How do I best manage costs, produce cattle that are in demand by the marketplace, and build management systems that enhance the natural and human resources under my care and stewardship?" If a resource cannot be sustained or improved, it is likely being depleted. Short-term economics reflect resource depletion; however, long-term economics with sustained breakeven and sustained profits should be the goal.

Cow-calf producers with a vision of long-term profitability will have an optimum combination of cattle productivity (reproduction, growth, and end product) and cost management. Even with retained ownership programs, economic priorities for the most profitable cow/calf producers should be to:

1. Manage cost of production.
2. Attain optimal reproduction performance.
3. Produce market cattle with appropriate growth (moderate average weaning weights in the 475–550 lb range for 7-month-old calves, assuming an average forage availability; feedlot gain for calves in the 3.0–4.0 lb/day range) and replacement females with mature size and milk production matched to the environment.
4. Focus on end product (700–800 lb carcass weights; 2.0–3.4 yield grades, while avoiding discounts on yield grade 4s; quality grade of >70% choice).

Cost of production and maternal traits (primarily reproduction) are far more important than growth and end product in determining cow-calf profitability. Therefore, cow-calf producers should have vision and motivation to keep a focus on "first things first." Alliances among seedstock, cow-calf, and feedlot producers will encourage optimal combinations of cow, feedlot, and carcass traits for low-cost/high-profit cattle production.

SELECTED REFERENCES

Publications

Adams, D.C., et al. 1996. Matching the cow with forage resources. *Rangelands.* 18:57.

American Angus Association. 2004. *Sire evaluation.* Saint Joseph, MO.

Arthington, J.D., & Minton, J.E. 2004. The effect of early calf weaning on feed intake, growth, and postpartum interval in thin, Brahman-crossbred primiparous cows.*Professional Animal Scientist.* 20:34–38.

Azzam, S.M., et al. 1993. Environmental effects on neonatal mortality of beef calves. *Journal of Animal Science.* 71:282.

Beef Sire Selection Manual. 2012. 2nd Edition. National Beef Cattle Evaluation Consortium.

Bellows, R.A., Staigmiller, R.B., & Short, R.E. 1990. *Studies on calving difficulty.* Research for rangeland-based beef production. Bozeman, MT: Montana Agric. Expt. Station.

Bellows, D.S., Ott, L.S., & Bellows, R.A. 2002. Cost of reproductive diseases and conditions in cattle.*Professional Animal Scientist.* 18:26–32.

Beverly, J.R., & Spitzer, J.C. 1980. *Management of replacement heifers for a high reproductive and calving rate.* Texas A&M Ext. Public. B-1213.

Blasi, D., & Corah, L.R. 1993. *Assessment of forage resources to determine the ideal cow*. Proc. Cow Calf Conference III. Manhattan, KS: Kansas State University.

Buskirk, D. 2003. *Effect of a 2-step weaning system on performance of calves*. Michigan State University Beef Newsletter. 8:3.

Cattle industry reference guide. 2003. Cattle-Fax, Greenwood Village, CO 80111.

Corah, L.R., & Blanding, M.R. 1992. Use all four keys to higher profits in the cow herd. *Beef* (Spring).

Cow-calf focus (periodical). Cattle-Fax, P.O. Box 3947, Englewood, CO 80155.

Cow-calf management (Reducing production costs). 1991. *BEEF*. 27, no. 7A (Spring).

Cowley, J., Buskirk, D.D., & Black. J.R. 2001. *MSU-IRM-SPA analysis for cow-calf operations—1999 summary*. Michigan State University, East Lansing, MI. Research and Demonstration Report.

Davis, K.C., Tess, M.W., Kress, D.D., Boornbos, D.D., Anderson, D.C., & Greer, R.C. 1992. *Live cycle evaluation of five biological types of beef cattle in a range production system*. Proc. WSASAS 43:43.

Dunn, B.H. 2002. *Cow-calf profitability analysis – SPA update*. Proc. Nebraska Veterinary Medical Association Annual Meeting. Grand Island, NE.

Dunn, B.H., Gates, R.N., Davis, J., & Arzeno, A. 2006. *Setting strategy and measuring performance*. South Dakota State University and Texas A&M Kingsville.

Farm Financial Standards Council. 2006. *Management accounting guidelines for agricultural producers*. Menomonee, WI.

Ferrell, C.L., & Jenkins, T.G. 1984. Energy utilization by mature, nonpregnant, nonlactating cows of different types. *Journal of Animal Science*. 58:234.

Field, T.G. 2006. *Priorities first*. American Angus Association. Saint Joseph, MO.

Fields, M.J., & Sand, R.S. 1995. *Factors affecting calf crop*. Boca Raton, FL: CRC Press.

Fogelman, S.L., & Jones, R. 2003. *Beef cow-calf enterprise*. Farm Management Guide—MF-266. Manhattan, KS: Kansas State University.

Freetly, H.C., Ferrell, C.L., & Jenkins, T.G. 2001. Production performance of beef cows raised on three different nutritionally controlled heifer development programs. *Journal of Animal Science*. 79:819.

Grant, P. 2003. *Cow culling strategies*. Bottom Line. North American Limousin Foundation, Englewood, CO.

Gunn, P.J., Sellers, J., Clark, C., & Schulz, L. 2014. Considerations for managing beef cows in confinement. Proc. Driftless Regional Beef Conference. Dubuque, IA.

Hill, H.W. 2003. *Early and fence line weaning of calves*. Proc. Range Beef Cow Symposium XVII. Mitchell, NE.

Iowa Beef Center. 2015. *Summary of Iowa beef cow business record*. Iowa State University, Ames, IA.

Lardy, G., et al. 1998. *Spring versus summer calving for the Nebraska sandhills: Production characteristics*. Nebraska Beef Report.

Lawrence, J.D. 2000. *Profiting from the cattle cycle: Alternative cow herd investment strategies*. Presentation to the Southern Plains Beef Symposium. Ardmore, OK.

Lawrence, J.D., & Mintert, J.R. 2014. Fundamental forces affecting livestock producers. *Choices*. Agricultural Applied Economics Association.

Little, R.D., Williams, A.R., Lacy, R.C., & Forrest, C. S. 2002. Cull cow management and its implications for cow-calf profitability. *Journal of Range Management*. 55:112–116.

Loy, D., Maxwell, D., & Rouse, G. 1999. Effect of early weaning of beef calves on performance and carcass quality. *Beef Research Report*. A.S. Leaflet R1632. Iowa State Univ., Ames, IA.

Lynch, J.M., et al. 1997. Influence of timing of gain on growth and reproductive performance of beef replacement heifers. *Journal of Animal Science*. 75:1715.

McBride, W.D., and K. Mathews. 2011. The diverse structure and organization of U.S. beef cow-calf farms. United States Department of Agriculture: Economic Research Service Economic Information Bulletin Number 73. Washington, DC.

McGrann, J.M. 2000. *Cost effective decisions needed: Cow-calf SPA results for Texas—1991-99*. Texas A&M University, College Station, TX.

McGrann, J.M. 2011. Southwest Cow-Calf SPA Key Measures Summary. Texas A & M University, College Station, TX.

McGrann, J., NCA-IRM-SPA. 1992. *Workbook for the cow-calf enterprise*. College Station, TX. Dept. of Agric. Economics, Texas A&M University.

McGrann, J., & Walter, S. 1995. PE102, *Reducing costs with IRM-SPA data*. Cattlemen's College (NCA), Nashville, TN.

Metzger, S. 2015. Costs and returns for cow-calf producers. North Dakota State University. Carrington, ND.

Parker, R., Doye, D., Ward, C., Peel, D., McGrann, G.M., & Falconer, L. 2004. *Key factors contributing to cow/calf costs, profits and production*. Proc. Southern Agricultural Economics Assn. Annual Meeting, Tulsa, OK.

Pendell, D.L., Youngjune, K., & Herbel, K. 2015. Differences between high, medium, and low profit cow-calf producers. Kansas State University, Manhattan, KS.

Pratt, D.W. 2013. *Healthy land, happy families and profitable businesses*. Ranch Management Consultants. Fairfield, CA.

Price, E.O., Harris, J.E., Borgwardt, R.E., Sween, M.L., & Connor, J.M. 2003. Fenceline contact of beef calves with their dams at weaning reduces the negative effects of separation on behavior and growth rate. *Journal of Animal Science.* 81:116–121.

Rasby, R. 2015. Dry lot Beef Cow/Calf Enterprise. University of Nebraska Extension Bulletin. October.

Ritchie, H.D. The optimum cow: What criteria must she meet? *Feedstuffs.* Aug. 21, 1995.

Sacco, R.E., Baker, J.F., et al. 1989. Lifetime productivity of straightbred and F[1] cows of a five-breed diallel. *Journal of Animal Science.* 67:1964.

Simonds, G. 1995. *Matching cattle nutrient requirements to a ranch's forage resource*. Intermountain Cow Symposium (Jan. 4–5), Twin Falls, ID.

Smith, D.L., Wiggers, D.L., Wilson, L.L., Comerford, J.W., Harpster, H.W., & Cash, E.H. 2003. Postweaning behavior and growth performance of early and conventionally weaned beef calves. *Professional Animal Scientist*. 19:23–29.

Smith, D.R. 2007. *Basic principles used in the "Sandhills Calving System" and how they apply to other production environments*. Proc. Range Beef Cow Symposium XX. Fort Collins, CO.

Staigmiller, R.B., Short, R.E., & Bellows, R. -A. 1990. *Developing replacement heifers to enhance lifetime productivity*. Research for rangeland based beef production. Bozeman, MT: Montana Agric. Expt. Station.

Taylor, R.E., & Field, T.G. 1995. *Achieving cow-calf profitability through low-cost production.* Proc. Range Beef Cow Symposium XIII. Gearing, NE.

Troxel, T.R., Gadberry, M.S., Cline, S., Foley, J., Ford, G., Urell, D., & Wiedower, R. 2002. Factors affecting the selling price of feeder cattle sold at Arkansas livestock auctions. *Professional Animal Scientist*. 18:227–236.

USDA. 2015. *Agricultural land values*. National Agricultural Statistics Service. Washington, DC.

United States Department of Agriculture. 2008. Part I: Reference of Beef Cow-calf Management Practices in the United States, 2007–08. APHIS: NAHMS. Washington, D.C.

United States Department of Agriculture. 2009. Part II: Reference of Beef Cow-calf Management Practices in the United States, 2007–08. APHIS: NAHMS. Washington, D.C.

United States Department of Agriculture. 2009. Part III: Changes in the U.S. Beef Cow-calf Industry 1993–2008. APHIS: NAHMS. Washington, D.C.

United States Department of Agriculture. 2010. Part IV: Reference of Beef Cow-calf Management Practices in the United States, 2007–08. APHIS: NAHMS. Washington, D.C.

United States Department of Agriculture. 2010. Part V: Reference of Beef Cow-calf Management Practices in the United States, 2007–08. APHIS: NAHMS. Washington, D.C.

Van Oijen, M., Montano-Bermudez, M., and Nielsen, M.K. 1993. Economical and biological efficiencies of beef cattle differing in level of milk production. *Journal of Animal Science*. 71:44.

6 Stocker Management Decisions

The stocker phase of beef production bridges the gap between cow-calf enterprises and the feedyard phase of production (Figure 6.1). Stocker cattle are generally lightweight, post-weaned calves that are ideally suited to be grown on forage-based diets prior to the feedyard. The stocker phase of production is highly variable based on geographic region, forage conditions, desired weight for feeders entering the feedyard and price trends. Replacement heifers, intended to become brood cows, are often developed under stocker management systems.

The profitability of a stocker program depends on cost and availability of grass or forage, price of calves or yearlings, and successful health management. Stocker enterprises are located primarily in areas of the United States where there is an abundance of available forage for grazing and proximity to feedyards (Table 6.1). The top three states for stocker cattle production are Texas, Kansas, and Oklahoma. The choice of program depends mainly on the available feed, projected cost of gain, and weight and gain potential of calves. Calves that weigh more than 600 lb, are moderate to heavy muscled, and have a high gain potential are likely to go directly to the feedlot or to be backgrounded for a few weeks and then transitioned to a finishing ration. Spring born calves lighter than 500 lb with moderate gain potential are more likely to be grown through the winter, then pastured the following summer before entering the feedlot.

Based on a study of the stocker industry conducted by Kansas State University (2015), 25% of stocker operators are exclusively in stocker/backgrounding business, nearly half (46%) run stocker cattle in conjunction with a cow-calf enterprise, 13% manage stocker cattle in concert with their feedyard, and another 16% span the full spectrum of cow-calf, stocker, and feeder (Figure 6.2). Regardless of the structure, it is important to treat the stocker program as a separate enterprise. Nearly two-thirds of stocker enterprises market at least two turns of cattle per year. A 2008 Oklahoma State study found that the size distribution of stocker enterprises is broken into nearly equal one-thirds running fewer than 100 head, 100 to 500 head, and more than 500 head annually.

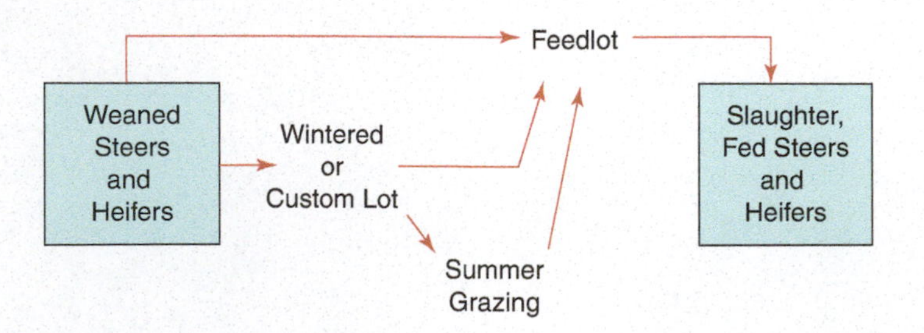

Figure 6.1
Alternative production pathways for weaned steers and heifers.

Table 6.1
Areas of Grazed Yearling-Stocker Production in the United States

Geographic Area/States	Description
Northwest	
Oregon and Washington	Grazing season is mid-April to early October; calves originate within the state and from surrounding states (ID, NV, and CA). Turnout weight is 500–650 lb with off-weights of 750–950 lb. Destination is feedlots within the states and to ID, CA, CO, NE, and KS.
West	
California	Primarily 3 regions: (1) northern—resembles OR, NV, and UT (see those states). (2) central—the winter/spring grazing season begins Dec. 15–Jan. 15. Turnout weight is 400–450 lb on calves originated within the states, NV, AZ, UT, and Mexico. Off weights are 750–900 lb in May and June. Cattle go primarily to feed yards in CA, WA, ID, and CO. 3) southern—comparable to the winter-spring grazing enterprises of AZ.
Colorado, Idaho, Montana, Nevada, Utah, Wyoming	Grazing can be mid-April through September. Turnout weights are typically 500–700 lb with market weights of 750–950 lb. Cattle go to feedlots in their own and bordering states and to CA, NE, and KS.
Southwest	
Arizona	Arizona has two main grazing seasons: (1) April to October and (2) mid-December to April 1. During summer grazing, a few stockers run in the higher elevations with turnout weights of 500–600 lb and off-weights of 600–800 lb. The winter grazing period in the desert utilizes light calves weighing 350–550 lb with off-weights of 600–750 lb. These lightweight calves originate in AZ, TX, Gulf Coast states, and Mexico. They are marketed to feedlots primarily in AZ, CA, TX, CO, and KS.
New Mexico	New Mexico is similar to summer grazing in Arizona. They are marketed to feedlots similar to Arizona with the Oklahoma Panhandle being an additional marketing outlet.
Great Plains	
North and South Dakota	Grazing primarily west of Missouri River during May to mid-October. Turnout weights are 525–600 lb; off-grass weights 750–850 lb. Cattle go primarily to NE, KS, and CO feedlots.
Nebraska	Grazing season is mid-May to mid-October. Turnout weights are 550–650 lb; off-grass weights are 800–900 lb. Majority of yearlings will go to Nebraska feedlots.
Kansas	Summer grazing from mid-April to mid-October. Intensively grazed cattle (70% of total) may be marketed mid-July to mid-August. Turnout weights are 500–600 lb; off-grass weights are 700–900 lb. Winter wheat grazing is mid-November to Mid-March with some limited graze-out until mid-May. Most wheat-grazed cattle are marketed in March. Most cattle go to feedlots in the Great Plains area.
Oklahoma	Grazing is from mid-April to mid-September that occurs primarily in Western Oklahoma Panhandle and extreme Northeast corner of the state (Osage country). Cattle originate from within the state, MO, TX, or Southeast. Turnout weights are 575–625 lb; off-grass weights are 800–900 lb. Fall wheat grazing is mid-November to mid-March involving 400–450 lb calves. Five-month gains of 250–270 lb are expected. Yearlings are marketed primarily to Texas/Oklahoma Panhandle and Kansas feedlots.
Texas	Summer grazing season is mid-March to mid-August. Light calves (450–550 lb) are grazed on native grasses in western part of the state and a combination of native grasses and improved grasses (e.g., Bermuda grass) in the eastern and southern regions of the state. Cattle are sold to feedlots weighing 675–775 lb. Fall grazing programs involve wheat, oats, and some rye grass during mid-October through March. In-weights and out-weights are similar to those on summer grazing program.

(continued)

Table 6.1
(Continued) Areas of Grazed Yearling-Stocker Production in the United States

Geographic Area/States	Description
Midwest	
Iowa, Minnesota, Illinois, Wisconsin, Indiana, Ohio, Michigan Missouri	Southern Iowa produces a limited number of yearlings, but the Midwest is typically not a stocker-yearling area. Missouri is a diverse cattle state with numerous grazing seasons and various combinations of cow-calf and stocker operations. The summer grazing season generally begins in late April and runs through September. Turnout weights are 500–600 lb; off-grass weights are 650–850 lb. Most of the stockers will be sold to feedlots in MO, KS, TX, NE, IA, and IL.
Northeast	
	There are limited numbers of stocker-yearling cattle produced in the Northeast area of the United States.
Southeast	
Kentucky and Tennessee	Due to widespread spring calving in these states, grazing programs are somewhat limited to fall through early spring. Improved pastures consist of rye grass, blue grass, and clover. The calves are raised to 700–850 lb, and then sold to feedlots primarily in the Central and Southern Plains and Corn Belt states.
Louisiana	Summer stocker programs are limited due to heat and humidity. Winter grazing operations using primarily native calves consist of rye grass, oats, and some wheat in the north. Yearling cattle weighing 650–850 lb will be sold primarily to Texas Panhandle feedlots in early spring.
Florida and Georgia	Florida produces few stocker-yearling cattle due to weather conditions and grazing programs that focus primarily on cow-calf. Grazing programs are more indicative of typical Gulf Coast stocker operations. Calves in these states and areas are marketed in April and June. This marketing time coincides with KS, TX, and OK stocker demand because of the available grass at this time.

Source: Based on multiple resources including Cattle-Fax Resources, Inc.®

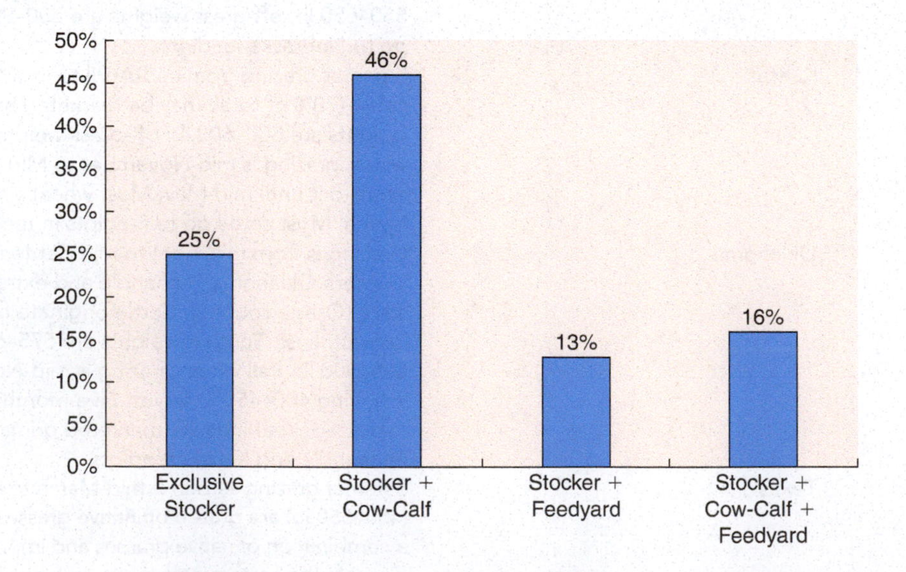

Figure 6.2
Types of stocker enterprises. Adapted from KSU.

The primary function of the yearling-stocker operation is to market available forage and high roughage feeds such as grass, crop residues (corn stalks, grain stubble, and beet tops), wheat pasture, and silage while achieving targeted rates of gain (Figure 6.3). Table 6.2 lists the utilization of various forages by stocker cattle.

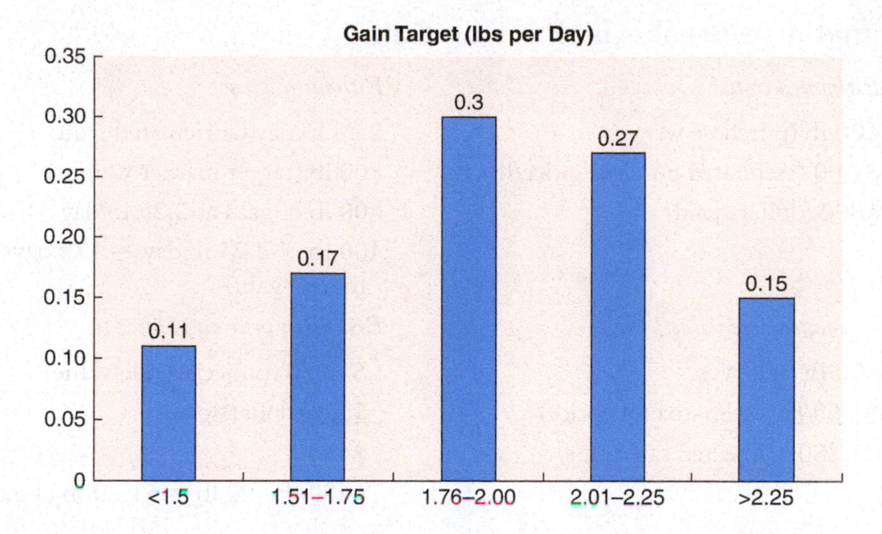

Figure 6.3
Targeted rates of gain by stocker operators (%).

Table 6.2
PERCENTAGE OF STOCKER ENTERPRISES UTILIZING VARIOUS FEED SOURCES

Feed Source	Percentage of Enterprises
Harvested forages/by-products fed in drylot	32
Cool season pastures	29
Warm season pastures	26
Fall small grain pastures	17
Stockpiled forages and crop residues	12
Other	7

Note: Does not total to 100 due to multiple feed sources in some enterprises
Source: Adapted from Tonsor, Hill, and Blasi, 2015.

A distinction is sometimes made between wintering and **backgrounding**. A wintering program usually emphasizes slower gains on high roughage diets with the intent that the calves go to a pasture program in the following spring and summer. Backgrounding emphasizes a faster rate of gain, using more grain and less roughage to prepare the calf for a more immediate feedlot placement. Purchases of stocker cattle is greatest in the fall months (October-30%, November-15%, and September-10%) with another peak of activity in the spring (March, April, and May-6% each). A vast majority of stocker cattle weigh between 400–600 lb at the time of purchase.

COMPUTING BREAKEVEN PRICES

Full time yearling-stocker operators have the greatest decision flexibility and so each grazing season have the opportunity to carefully evaluate their options—buy stocker cattle to graze available forage, mechanically harvest forage for sale, or rent the grazable forage to someone else. Backgrounding enterprise managers have somewhat less flexibility although if they choose not to own cattle, they can provide custom backgrounding service to others. Regardless of the situation, managers must assess breakeven prices as part of their decision-making process.

Estimating a Breakeven Price on Gain

Estimated cost:

400 lb (purchase wt)
$2.00 (estimated purchase price/lb)
$800 (dollars paid)

Estimated gain:

2.25 lb/day (anticipated gain)
800 lb (target market weight)
400 lb of gain at 2.25 lb/day
400 lb ÷ 2.25 lb/day = 178 days
lb day (gain)

Estimated sale value:

800 lb (sale wt)
$1.60/lb (estimated sale price)
$1,280 (projected sale value)

Breakeven price on gain:

$1,280 (projected sale value)
−$800 (dollars paid)
$480
$480 ÷ 400 lb = $1.20/lb of gain

If the anticipated cost of gain is less than $1.20 per pound or $120/cwt, the scenario appears profitable (assuming the projected gain and projected sale price are accurate estimates).

Estimating the Breakeven Sale Price

When historic costs of gain are well established, breakeven analysis can be conducted as follows: Anticipated cost of gain is $.80 per lb, anticipated calf purchase weight of 425 lb and a sale weight of 700 lb.

Total cost of gain:

700 lb (sale wt)
− 425 lb (purchase wt)
275 lb (total gain)
275 lb × $0.80 = $220 (total cost of gain)

Estimated purchase cost:

425 lb (purchase wt)
× $2.00 (price per lb)
$ 850 (dollars paid)

Necessary selling price:

$ 850 (dollars paid)
+$220 (cost of gain)
$ 1,070 (total cost)
$1,070 ÷ 700 lb = $1.53/lb

A selling price over $153 per cwt would be profitable, assuming the estimated cost of gain is valid.

After calculating the projected cost of gain and the breakeven selling price, the producer can prepare budgets based on different rates of gain and potential management options such as wintering, short-term backgrounding, summer grazing, etc.

THE BUDGETING PROCESS

Examples of the budgeting process for wintering and backgrounding operations are shown in Table 6.3. The faster-gain programs show a higher return potential than the slower-gain programs. Even though the total costs are higher for the faster-gain programs, their cost per pound of gain is less. This occurs because a higher percentage of feed is used for growth rather than for maintaining body weight.

In some cases, slower-gain programs may provide a useful alternative to selling weaned calves in the fall because there is a market for feed that might be difficult to sell.

Table 6.3
The Budgeting Process for Alternative Management Programs for Weaned Calves

	Wintering Rate of Gain		Backgrounding Rate of Gain	
	Low	High	Low	High
Initial wt (lb)	425	425	425	425
Average daily gain	1.0	1.25	1.5	2.00
No. days feeding	150	150	150	150
Sale wt (after 2% shrink)	563	600	637	710
Purchase cost (425 lb at $2.80)	$1,190	$1,100	$1,190	$1,190
Feed costs	215	250	270	300
Other variable costs	65	65	65	65
Total costs	$1,470	$1,505	$1,525	$1,555
Breakeven sales price (total cost/sale wt) × 100 =	$261.10	$250.83	$239.40	$219.01
Estimated selling price/cwt	255.00	248.00	240.00	230.00
Estimated returns/cwt	−$6.10	−$2.83	$0.60	$10.98
Estimated returns/head	−$34.35	−$17.00	$3.80	$78.00

In this example, the cattle represented in the first three columns are strong candidates to be retained through a summer grazing program.

Table 6.4
Example of Stocker Summer Grazing Budget

Item	$ Per Head	Percentage of Cash Costs (%)
Feeder cattle (550 lbs. × $2.00/lb)	1,100.00	88
Pasture cost ($18/acre/month–4 mos)	72.00	6
Mineral and salt	10.00	1
Labor	20.00	2
Vet/medicine/death loss	8.50	1
Marketing costs	10.00	1
Interest on cattle loan	2.00	<1
Fuel and energy	5.00	<1
Machinery/facilities/repairs	10.00	1
Miscellaneous cash variable costs	2.50	<1
Total variable costs	1,240.00	99
Cash fixed costs	12.50	1
Total costs	1,252.50	100

Also, if calves are retained through the winter, they could command a higher price in the spring, or pasture gains may compensate for the reduced returns of the wintering program. In this case, producers with a wintering program and a summer pasture program should compute a separate budget for each program.

Table 6.3 is only one example of the budgeting process. Producers should prepare their own individual budgets using data specific to their own situation. Valid historic cost and return data can increase the accuracy of budget estimates for the current year. Stocker enterprises have historically been profitable year to year. The KSU (2015) study reported that stocker operators had experienced an average net return of $77 per head in each of the previous 10 years. The average worst year was a loss of $14/hd while the best year averaged $193/hd profit for all enterprises studied.

Table 6.4 shows an enterprise budget for summer grazing yearling cattle. Clearly, the purchase of calves at the front end of the grazing period accounts for the

greatest percentage of the costs; a fact that has led to the adage that "bought right is half sold". Purchasing healthy and productive calves at the right price along with having access to forage largely determines success of the summer grazing enterprise.

MANAGEMENT CONSIDERATIONS

Market Prices

The marketing decisions that affect market prices often have the greatest effect on the profit potential of yearling operations. Stocker operators typically deal with both sides of the marketing equation—purchase and sale. Purchase weight and price, along with sale weight and price, have a major influence on total dollars returned to the operation. Cattle prices may be discounted or enhanced based on a number of factors (Table 6.5).

The buy-sell margin—that is, the difference between calf purchase price and selling price—should be carefully evaluated by a breakeven price analysis. High buy-sell margins limit profitability. As a rule of thumb, the calf-buying price should not exceed the selling price by more than 15%.

Stocker-yearling producers prefer to purchase shrunk cattle at average prices and cattle that are low to moderate in condition, healthy, relatively light for their age, and moderate or better in both muscle and frame. Although the cow-calf producer might identify these animals as somewhat "mismanaged cattle," they are potentially profitable to the yearling operator. Calves heavy at weaning and that possess the ability for rapid gains need to be utilized in the finishing phase of the feedlot, not in the yearling operation. An exception to this might occur when yearling feed costs and interest rates are very low.

Feed prices, forage supplies, and projected calf and yearling prices are needed to determine the best time for marketing. Seasonal price patterns suggest that fall calf prices are usually the lowest for the marketing year. However, even though spring

Table 6.5
Premiums and Discounts Paid for Feeder Cattle

Factor	Impact on Price
Breed	Herefords sell at premium to Angus (1980s) Angus sell at premium to Herefords (1990s) Black hides sell at a premium (2000s) >25% Brahman discounted more than those with <25% Brahman especially in northern markets
Weight, condition, muscle	Cattle <600 lb and light muscled discounted $18–26/cwt compared to heavy muscled Lighter weight cattle that are healthy, ≥ mod. framed, ≥ mod muscle sell at a premium to heavier weight calves (premium declines as feed prices rise). Thin cattle more likely to be discounted in fall compared to spring Over-conditioned feeders discounted $1–3/cwt
Fill	Over-filled cattle heavily discounted
Health status	Sick calves discounted $10–25/cwt
Lot size	Cattle sold in truck-sized lots bring a premium compared to small lots ($7.50/cwt for feeders; $4.50/cwt for yearlings)

Source: Adapted from multiple sources.

prices are often higher than in the fall, net returns may be maximized by waiting till spring to buy calves for summer grazing to avoid the costs of wintering. This is not always the case so using a simple budget should be the deciding factor. Producers can obtain some flexibility by considering the purchase of lighter weight calves or heifers when market conditions are favorable.

Conditions can change, especially prices, from the time a budget is constructed. The conditions that affect costs and returns should be periodically evaluated to determine optimal marketing alternatives. The method used in the previous year may not be the best approach for the current year. In addition, when large numbers of cattle are held over the winter, heavy marketing of yearlings in the spring can cause reduced market prices to be depressed. Prudent producers may want to take a hard look at forward-pricing opportunities to lock in profitable returns.

Utilizing the concepts of arbitrage and asset turnover offer additional profit opportunities. Arbitrage is the simultaneous purchase and sale of a commodity in different markets to capture profit from price and value discrepancies. Stocker operators who opportunistically buy calves at favorable prices, take advantage of relatively cheap gains during the grazing cycle, and create higher valued calves over and over again are practicing arbitrage. Asset turnover is a financial concept that is particularly important in low margin commodity enterprises. For example, a cow-calf producer may attain maximum profitability by selling their own calves and then buying back lighter-weight calves with high compensatory gain potential for use in the stocker enterprise. When forage and market conditions are favorable, making multiple turns of stocker calves over a period of time is often more profitable and less risky than depending on a one-turn retained ownership approach.

PASTURE LEASES

Leasing or renting agreements for pasture are influenced by a variety of factors including the costs associated with land ownership, the tenants expected gross or net income, demand for use of the land, federal farm programs, and land productivity. Rental rates are typically determined by a fixed rate per head per season, per acre rate, or a rate per pound of gain (Table 6.6).

As Figure 6.4 illustrates, more productive the land and lower the likelihood of drought, the higher are pasture lease fees. The pasture lease agreement typically allocates responsibility to either the land owner, the stocker operator, or both for activities such as checking cattle, maintaining fences and water sources, and providing salt and mineral supplementation (Table 6.7).

Table 6.6
Pasture Rental Rates ($ Per Acre Per Month)

	Kansas	Oklahoma	Texas	U.S.
2003	12.60	8.50	7.80	9.40
2005	13.40	9.00	6.20	10.40
2007	14.50	9.50	6.20	10.60
2009	15.50	10.50	6.20	10.50
2011	16.00	11.50	7.50	11.50
2013	17.50	12.00	6.50	12.00
2014	17.50	12.00	6.50	12.00

Source: USDA.

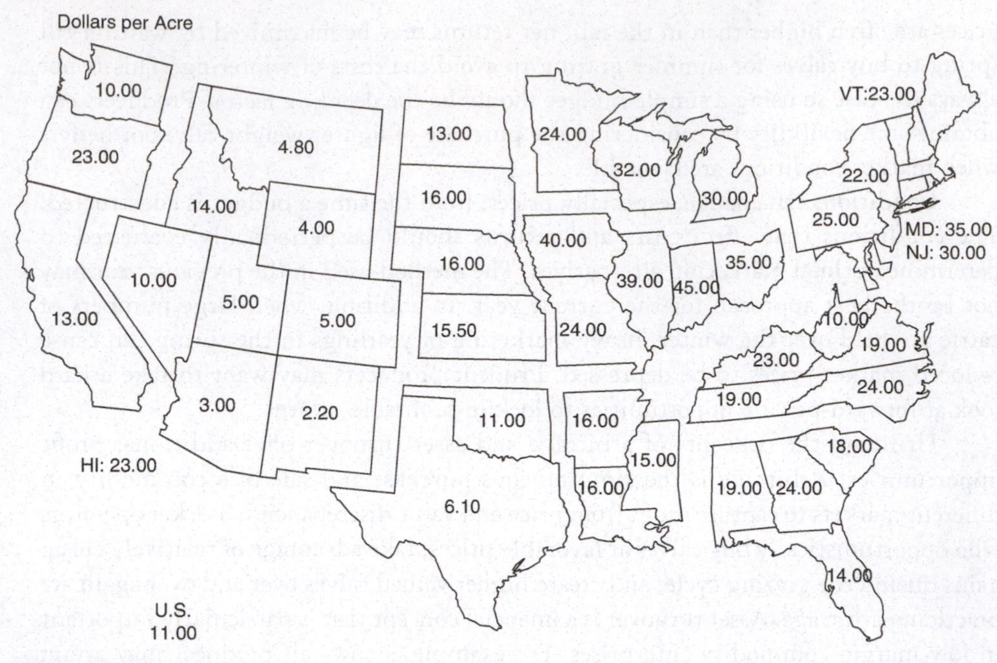

Figure 6.4
Pasture rent costs.
Source: USDA.

Table 6.7
Assigned Responsibilities in Pasture Lease Agreements (%)

	Small Grain Winter Grazing			Winter grazing and graze out		
Task	Land Owner	Cattle Owner	Both	Land Owner	Cattle Owner	Both
Checking cattle	15	58	27	16	70	14
Mineral/salt	16	60	23	18	70	12
Fencing materials	29	44	28	28	54	18
Fencing labor	23	48	30	24	62	17
Fertilizer	35	44	21	23	62	15
Supplemental feeding	17	62	21	16	72	11

Source: Adapted from Doye and Sahs (2015).

Cattle Health

The first priority of a stocker enterprise manager is to deal with the multitude of factors that contribute to calf stress. These influences include weaning, sorting and commingling, transportation, handling, processing and vaccination, dietary changes, weather, and introduction to new surroundings. Any one of these factors alone may cause sickness, and taken in total, these stressors can yield significant health-related economic losses. The highest risk calves are light weight, freshly weaned, and have been subjected to several of the aforementioned stressors.

Associated financial losses related to health issues account to 62% due to death loss, 21% due to performance losses in sick cattle, and 17% for the expense of treatment. While the costs associated with cattle mortality are obvious, cattle operators often do not recognize all of the costs associated with sick cattle. The average sick animal shrinks 10–20%. Considerable additional labor is required per sick animal. Medicine, treatment programs, and death losses are expensive. An average 3-day medication cost for a 400-lb calf can cost approximately $10. A 5% death loss increases

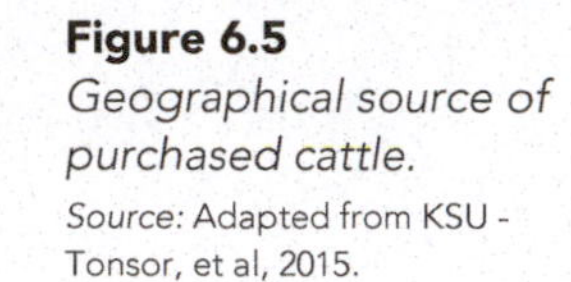

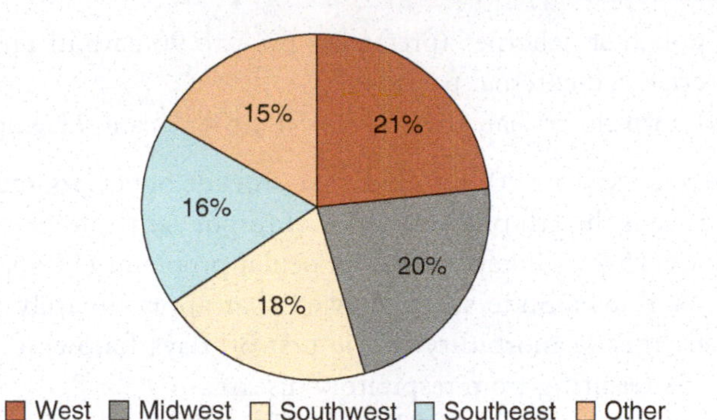

Figure 6.5
Geographical source of purchased cattle.
Source: Adapted from KSU - Tonsor, et al, 2015.

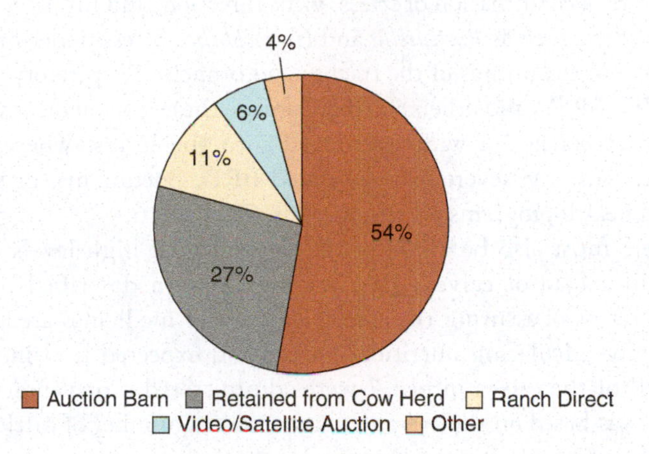

Figure 6.6
Market source of purchased cattle by stocker operators (%).
Source: Adapted from KSU - Tonsor, et al, 2015.

the cost of each calf $10–20, or 5–8%. Since the cost of "one time through the chute" is considered equivalent to a 7-day-feeding period, it is cheaper to prevent disease than it is to treat it. The best way to prevent disease in a yearling-stocker operation is to know the origin of the cattle and their previous health program and work to minimize stress levels. Stocker cattle are sourced from a diverse set of locations (Figure 6.5) which can increase management challenges. Purchasing "fresh cattle" involves less risk than buying feeder calves of unknown origin that have likely been transported long distances and exposed to infected cattle. Shipping stress can significantly decrease the profitability and production efficiency of native-range-based stocker cattle enterprises. A Texas shipping stress study showed that morbid heifers returned $20–52 per head less than healthy heifers. Morbid replacement heifers had a conception rate that was 9% lower and a 1-month later pregnancy date than healthy heifers (Pinchak et al., 1995).

Increasingly, both stocker and feedlot buyers are recognizing the value of preconditioning (calves fully weaned and vaccinated before leaving the farm of origin) to assure future performance of the cattle they purchase. It is generally accepted that there should be a $100 per head advantage in price for preconditioned calves as compare to their non-preconditioned contemporaries. An analysis of Oklahoma markets demonstrates that from 2012 to 2014, buyers increased the premium paid for preconditioned calves from $9 per cwt to $19 per cwt. Results from Superior Livestock Video sales showed that buyers increased the preconditioning premium from 2013 to 2014 by nearly $7 per cwt (4.78/cwt to $12.06/cwt). The markets used to acquire stocker cattle are provided in Figure 6.6.

Producers must also have an excellent health and nutrition program upon receiving the cattle. How the cattle are handled and fed during the first 5 weeks will have a significant effect on their health and profitability. Cattle that meet the following requirements are the least likely to have a high degree of health related risk:

1. Weaned for 30–45 days prior to shipment
2. Males castrated (preferably prior to 90 days in age)

3. All calves polled or dehorned (preferably prior to 90 days in age)
4. Free of internal and external parasites
5. Vaccinated with clostridial, IBR, PI3, BVD, BRSV, and Haemophilus somnus

A Drovers' survey in 2001 found that stocker operators reported respiratory disease as the greatest health problem (40% of respondents), followed by internal and external parasites (15%), pinkeye and other ocular problems (11%), and lameness and foot rot (7%). A Kansas State study showed that approximately 9% of all stocker cattle exhibited signs of morbidity in the first 30 days following arrival with two-thirds of the cases resulting from respiratory disease.

Bovine respiratory disease (BRD) is seldom the result of a single factor. It is usually caused by a combination of stress, virus infection, and invasion of the lungs by pathogenic bacteria such as *Pasteurella* and *Haemophilus*. Stress undermines the natural defenses built into the linings of the trachea and bronchi. Respiratory viruses (such as IBR, PI3, BVD, BRSV, and others) further damage these natural defenses. Ultimately, pathogenic bacteria find a wide-open road into the lungs, where they localize, multiply, and cause the severe damage called BRD, pneumonia, or shipping fever. More detailed health programs are discussed in Chapter 16.

Managers must also be aware of the potential for high levels of variation in weight within a load of calves. The weight variation described in Table 6.8 is important in terms of assuring the correct amount of medicines are administered as well as correctly calculating nutritional needs and expected feed intake levels. For example, if all of the calves in pen 7 were administered a preventative medication where dosage was based on the average weight, then a number of cattle would receive too high a dose while others would receive an insufficient amount.

Nutrition

Successful nutritional programs for stocker-yearling cattle must be carefully coordinated with a well-planned health program. Many newly received cattle are stressed; they have been sold through several markets, exposed to diseases, and shipped long distances through varying kinds of weather without feed and water. Complementary nutritional and health programs will largely determine how these cattle adjust to their new environment.

Low feed intake and water consumption are the two biggest nutritional challenges for newly arrived stocker-yearling cattle. These challenges exist for the first 3–4 weeks. It is important that the cattle consume feed before filling up with water. The physical environment for receiving calves is of importance, with dust minimization

Table 6.8
Variation within Seven Loads of Calves Received by a Stocker Enterprise

Load	Average Weight (lb)	Standard Deviation (lb)	Weight Range (lb)	Difference from High to Low (lb)
1	471	43	342–600	258
2	439	32	343–535	192
3	428	36	320–536	216
4	456	34	354–558	204
5	447	31	354–540	186
6	443	25	368–518	150
7	443	33	344–542	198

Source: Based on Blasi, 2004.

being a critical factor. If possible, newly received stockers should be managed in a grass lot instead of a drylot. Morbidity rates may be reduced by as much as 40% as a result of this practice. Calves should receive 2 or more lb of high quality, long stemmed hay (brome, Bermuda grass, or native grass hays) in addition to a starter ration (chopped feeds). The starter ration should minimize dust to prevent respiratory stress, minimize fermented feeds (silage, high moisture grain) to enhance intake, and avoid excessive starch to prevent acidosis. Percent dry matter intake for stressed, freshly weaned calves in the first, second, and third weeks after weaning should range from .5–1.5, 1.5–2.5, and 2.5–3.5, respectively.

Table 6.9 shows other nutritional recommendations for 450–600 lb stocker-yearling cattle that can increase feed intake and contribute to the nutritional well-being of cattle during the initial adjustment period. Protein should be supplemented on a pounds-per-head basis rather than a percent-of-ration basis because of the variability in feed consumption. Natural protein is recommended in the ration, with only minimal amounts of non-protein nitrogen. Bypass protein may also be beneficial. Potassium additions tend to improve the performance and health of cattle.

Water should be provided 3–4 hours after the cattle's arrival. Water tanks should be clean and the cattle (especially small cattle) should be able to reach the water without being frightened. Water that is allowed to run continually into the tank or trough may attract the cattle to the water.

Calves may be fed roughage on the ground or in feed bunks. Feeding cattle in feed bunks allows more flexibility in moving from a wintering to a backgrounding program in that the grain can be more conveniently added to the feeding program.

Some yearling-stocker cattle are not fed harvested feeds through most of the wintering program, particularly when the climate, in combination with certain forage systems, permits green forage to be grazed on a nearly year-round basis. In certain areas of the Southeast, for example, winter annuals such as rye, oats, ryegrass, wheat, and barley provide a forage base for winter grazing. Also, some types of clover planted in the fall will allow grazing by late January in certain areas. Table 6.10 shows several rations for wintering and backgrounding calves at different daily gains.

Rate of Gain

Rate of gain is affected by the health, body condition, and nutrition of cattle. Most producers prefer the daily rate of gain of backgrounded and grazed cattle to be 1.5 lb per

Table 6.9
GENERAL RATION GUIDELINES FOR RECEIVING STOCKER CALVES (450–600 LB)

Ration Component	Recommendation
Dry matter	70–85%
Crude protein	14%[1]
Crude fiber	20–25%
Potassium	1.2–1.4%[2]
Copper	20 ppm[3]
Zinc	80–100 ppm[3]
High quality hay	2+ lb/hd/day

[1]Highly stressed light weight cattle should receive 16%.
[2]Electrolyte responsible for fluid replenishment of tissue.
[3]Enhance immune response.

Table 6.10
Rations for Different Rates of Gain for Growing Calves

	300–400 lb			400–500 lb			500–700 lb		
Ration	.5–1 lb Day Gain	1–1.5 lb Day Gain	1.5–2 lb Day Gain	.5–1 lb Day Gain	1–1.5 lb Day Gain	1.5–2 lb Day Gain	.5–1 lb Day Gain	1–1.5 lb Day Gain	1.5–2 lb Day Gain
1. Corn silage	6.5	11.5	—	6.5	21.0	11.0	—	19.0	21.0
Alfalfa hay, mid-bloom	7.5	4.0	5.0	10.5	5.5	5.0	17.0	9.5	5.5
Corn	—	1.5	5.0	—	—	4.5	—	—	3.0
Dicalcium phosphate	0.02	0.04	0.04	0.03	0.03	0.04	0.04	0.04	0.02
2. Corn silage	—	—	17.5	—	—	30.5	—	—	38.5
Corn	—	—	1.5	—	—	—	—	—	—
32% protein supplement	—	—	1.7	—	—	1.6	—	—	1.8
Dicalcium phosphate	—	—	0.01	—	—	0.01	—	—	—
3. Sorghum silage	12.0	11.5	—	12.5	20.5	11.0	5.5	34.0	21.5
Alfalfa hay, mid-bloom	6.0	4.0	5.0	9.0	4.5	4.5	15.0	5.0	5.0
Corn	—	2.5	5.0	—	2.0	5.0	—	—	4.5
Dicalcium phosphate	0.02	0.03	0.04	0.03	0.02	0.03	0.04	0.04	0.02
4. Prairie hay, early bloom	8.0	6.5	4.5	11.5	9.0	7.0	15.5	13.5	10.0
Corn	—	3.0	4.5	—	3.5	5.5	—	2.5	6.0
32% plant protein suppl.	2.0	1.0	1.5	1.5	1.0	1.0	1.0	1.0	1.0
Dicalcium phosphate	—	0.01	0.08	—	—	—	0.02	—	—
5. Alfalfa hay, mid-bloom	10.0	7.0	5.0	13.5	10.0	7.5	17.0	14.0	10.5
Corn	—	3.0	5.0	—	4.0	6.0	—	3.0	6.5
Dicalcium phosphate	0.02	0.03	0.04	0.03	0.01	0.03	0.03	0.02	—
6. Alfalfa hay, early bloom	10.0	7.5	5.5	13.5	11.0	8.0	16.5	15.5	11.5
Corn	—	2.5	4.5	—	2.5	5.5	—	1.5	5.0
Dicalcium phosphate	0.01	0.02	0.04	0.01	—	0.02	0.01	0.01	—

day or higher. Cattle that gain faster usually are more profitable. In some cases, cattle with a high rate of gain lose money, but they lose less than slower-gaining cattle. Fast-gaining cattle may not be profitable if production costs are too high or when the buy-sell price margin is too low. Slow-gaining cattle usually do not produce enough pounds to cover maintenance feed and interest costs.

Growth stimulants, such as ear implants or ionophores, are included in the management programs of many yearling-stocker programs(Chapter 7 identifies the available implants). The preferred implant for most stocker enterprises utilizing forage pastures is Ralgro (70%) with Synovex-S and H used on about 20% of the enterprises of this type. For those stocker enterprises utilizing a higher concentrate feed, Ralgro is

Table 6.11
EFFECTS OF IONOPHORES ON DIFFERENT TYPES OF PASTURES

	Pasture Types			
	Native, Growing	Native, Dormant	Improved	Small Grains
No. trials	54	9	10	7
No. cattle	3,880	649	504	470
Average daily gain (lb/day)	1.41	1.02	1.58	1.58
Ionophore response (lb/day)	0.18	0.12	0.14	0.11

Source: Michigan State University (summary of eighty research trials, 1988).

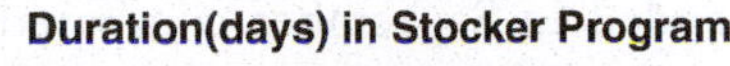

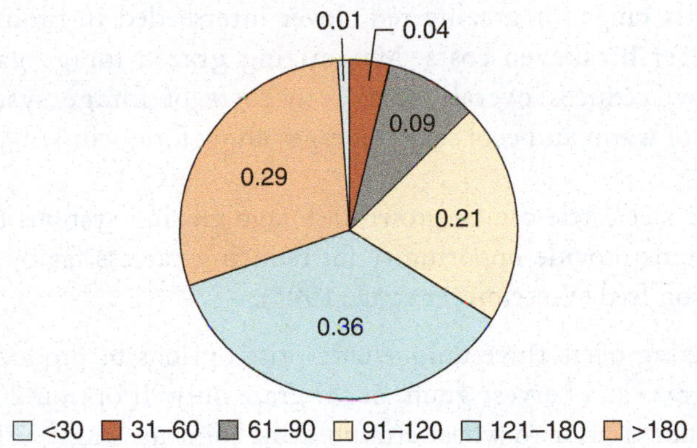

Figure 6.7
Period of time that cattle are retained in the stocker enterprise.
Source: Adapted from KSU - Tonsor, et al, 2015

the preferred implant (60%), but the use of Synovex-S and H is higher at 35%. Monensin (which has the trade name Rumensin) and lasalocid (which has the trade name Bovatec) are ionophores—antibiotics that alter rumen metabolism—used for wintering, backgrounding, and grazing programs. The response of grazing cattle to the two types of ionophores is similar (Table 6.11).

Producers identify gaining ability in healthy cattle by associating gain with a relatively low body condition score. Genetics also affects gaining ability; however, it is difficult to assess unless the source and breeding of the cattle are known.

Production Systems

Numerous alternatives are available for managing weaned calves to harvest weights. Calves placed into a stocker enterprise will be managed for variable periods of time depending on feed conditions, growth rate of cattle, and market conditions (Figure 6.7). Efficiency of production includes the growing of calves prior to receiving finishing diets in the feedlot. Frequently, the economics of production only considers a single phase of the total production system. As a consequence, one segment of the beef industry may make management decisions based on maximum profit while they manage the animals that may adversely affect the profit of the next owner. Economic efficiency of the total production system could be reduced. For example, cost per pound of gain is usually lower when calves are wintered at a relatively fast rate of gain and feedlot operators also want high daily gains so cost of gain will be relatively low. However, high feedlot gains may not be realized if calves wintered at high rates of gain are grazed during the summer months.

The University of Nebraska has made excellent economic assessments of several production systems evaluating weaned calves through the feedlot. The following are given for some of these systems.

- Cattle grazing corn stalks that had a relatively low winter gain (0.79 lb/day) compensated during the summer and experienced faster summer gains than those wintered at a higher rate (2.0 lb/day). Most of the compensatory gain was achieved during the first 62 days of grazing. Compensatory growth is a phenomenon of rapid growth by cattle that usually follows a period of growth restriction.
- Cattle that were on grass for the short grazing period (62 days) had faster finishing gains and tended to be more efficient. The cattle that were wintered at a fast rate (2.0 lb/day) and pastured for the full summer period (120 days) had a higher breakeven. Cattle wintered at a fast rate of gain (2.0 lb/day) should only be grazed on high quality forage in the spring and early summer to be competitive with systems that have lower winter input costs (Morris, 1996).
- Systems where grazing ended in September compared with November, using native Sandhills range, or grazing red clover interseeded in bromegrass had the lowest slaughter breakeven costs. Maximizing grazed forage gain, while cost of gain is low, reduces overall breakeven costs of forage systems. Grazing combinations of warm and cool season forages allow for optimizing forage quality (Shain, 1996).
- Medium frame size cattle can be grown in forage grazing systems before finishing and these systems provide opportunity for reducing carcass fat by shortening the finishing days on feed (Vieselmeyer et al., 1995).

Wheat grazing offers three unique enterprise options to producers: (1) harvest grain only, (2) graze and harvest grain, or (3) graze only. If option 2 is chosen, then grazing is terminated prior to wheat attaining the jointing stage. Wheat grazing in the midwest is typically initiated in the late fall months. Supplementation is provided while the plant is dormant. As the pasture becomes lush, it is critical to manage cattle to avoid losses associated with bloat.

Yearling-stocker production systems can be evaluated on how they contribute to a carcass breakeven price. For example:

$$\text{Carcass breakeven } (\$/\text{lb}) = \frac{\text{Weaned calf cost} + \text{yearling/stocker cost} + \text{feedlot cost}}{\text{Carcass weight}}$$

$$= \frac{\text{Total cost}}{\text{Carcass weight}}$$

Because cost of gain in each production phase affects the other two phases, it is important to achieve an optimum combination of all three phases of production.

Management Differences Due to Enterprise Size

The size and scale of the stocker enterprise impacts the management and marketing decisions made by stocker enterprises (Table 6.12). Relative to grazing and nutrition, large enterprises that manage more than 500 stockers annually are more likely to utilize small grains grazing systems, to forage test and provide mineral supplementation as compared to those enterprises managing fewer than 100 head of stockers per year. Larger enterprises are also more likely to use modified live vaccines and preventative approaches to ticks and internal parasites. Larger enterprises have significantly higher utilization rates of growth promoting implants and more sophisticated approaches to risk management.

Managers of smaller enterprises are advised to carefully evaluate the profitability of incorporating some, if not all, of the strategies utilized by their larger competitors.

Table 6.12
Management Practices Incorporated by Stocker Operators (%)

Nutrition/Grazing	All Enterprises	Small (<100 hd)	Large (500+ hd)
Stockpile fescue or Bermuda grass for fall/winter grazing	46	43	41
Winter small grains grazing	34	30	56
Summer grazing	58	56	61
Warm season species grazed	61	62	68
Cool season species grazed	16	24	8
Year round grazing	49	57	52
Forage test	31	23	44
Soil test (every 3–4 years)	52	52	54
Mineral supplementation	79	71	87
Animal Health			
MLV vaccines	61	44	88
Killed vaccines	25	27	18
Dewormed	93	88	98
Tick control	84	81	90
Individual identification	89	82	79
Growth promotants			
Implant steers	59	38	78
Implant heifers	43	26	63
Risk management			
Uses futures contracts	34	21	65
Uses options contracts	29	10	55
Uses cash contracts	26	10	43

Source: Adapted from OSU (2008).

SELECTED REFERENCES

Publications

Blasi, D.A. 2004. *Tools and technologies for the beef stocker producer.* Presented at the Colorado State University IRM Shortcourse, Fort Collins, CO.

Doye, D., & Sahs, R. 2015. Oklahoma pasture rental rates: 2014–15. Oklahoma State University Extension Service. Stillwater, OK.

Drovers 2000 Stocker Survey. *Drovers*. Vol. 129, No. 4, pg. 55.

Dumler, T.J., Jones, R., & O'Brien, D.M. 2003. Summer grazing of steers in Western Kansas. *KSU Farm Mgmt. Guide* MF-1007. Kansas State University. Manhattan, KS.

Jones, R., & Dumler, T.J. 2003. Winter wheat grazing. *KSU Farm Mgmt. Guide* MF-1009. Kansas State University. Manhattan, KS.

Johnson, R.J., Doye, D., Lalman, D.L., Peel, D.S., & Raper, K.C. 2014. Stocker cattle production and management practices in Oklahoma. Oklahoma State University Extension Service. Stillwater, OK.

Morris, C., et al. 1996. Beef production systems from weaning to slaughter in Western Nebraska. *Nebraska Beef Report*, University of Nebraska, Lincoln, NE.

Pinchak, W.E., et al. 1995. *Shipping stress impacts on production efficiency and profitability of stocker cattle.* Beef Cattle Research in Texas. Texas A&M University.

Proc. Light Cattle Management Seminars. 1982. Colorado State University, Fort Collins, CO.

Roth, L., Klopfenstein, T., & Sahs, W. 1988. *Corn stalklage for growth calves—A review.* Beef Cattle Report MP-53: 51, University of Nebraska, Lincoln, NE.

Rust, S.R. 1988. *Ionophores for Grazing Cattle.* Extension Bulletin E-2100, Michigan State University, East Lansing, MI.

Shain, D., et al. 1996. *Grazing systems utilizing forage combinations.* Nebraska Beef Report, Lincoln, NE.

Sindt, M., Klopfenstein, T., & Stock, R. 1990. *Production systems to increase summer gain.* Nebraska Beef Cattle Report, University of Nebraska, Lincoln, NE.

Tonsor, G. T., Hill, S., & Blasi, D. 2015. Benchmarking situation and practices of U.S. stocker operations. Kansas State University Cooperative Extension.

Vieselmeyer, B., Klopfenstein, T., et al. 1995. *Physiological and economic changes of beef cattle during finishing.* Nebraska Beef Report, University of Nebraska, Lincoln, NE.

7 Feedlot Management Decisions

As settlement reached across the Appalachians and into a region that would become the Corn Belt, the stage was set for the development of the cattle feeding industry. The production of surplus feed grains is the foundation of cattle feeding as a distinct sector of the industry. In addition to high-energy feeds, both grains and by-products, the feedyard sector requires access to a steady supply of feeder cattle as well as beef packing and processing capacity.

The term "**fed cattle**" refers to cattle that have been fed for high-energy rations for a period of time following weaning or a backgrounding/stocker phase. The duration of feeding depends on several factors—targeted finished weight and composition goals, genetic potential of cattle to gain and finish, as well as the price of inputs and outputs. The major cattle feeding areas in the United States are generally concentrated in the Great Plains region where access to feed, feeder cattle, and packing capacity is highly available. The concentration of cattle feeding is illustrated in Figure 1.10. The major growth of the modern fed cattle industry occurred between 1945 and 1972 driven by the economic energy that occurred following WWII, an abundance of affordable feed grains, and relatively low cost of fuel. As the farming sector ramped up its capacity to increase grain yields, the growth of the cattle feeding sector, the integrated swine business, and commercialization of poultry production followed.

TYPES OF CATTLE FEEDING OPERATIONS

There are two basic types of cattle feeding operations: the commercial feeder (Figure 7.1) and the farmer-feeder. The two types are generally distinguished by ownership structure and size of feedlot. Commercial feedlots are usually defined as having more than 1,000 head capacity and farmer-feeder feedlots as having less than 1,000 head one-time capacity.

The farmer-feeder operation is usually owned and operated by an individual or family. The commercial feedlot may be owned by an individual, partnership, or corporation—the last type is most common, especially as feedlot size increases. Commercial feedlots may own the cattle, feed cattle owned by someone else (**custom cattle feeding**), or engage in some combination of the two. Custom-fed cattle are owned either separately or in partnerships by other cattle feeders, investors, cattle producers, or packers and are fed on a contractual fee for service basis. The feeding capacity of various-sized feeding enterprises as well as the number of feedyards has changed over time (Tables 7.1 and 7.2).

Approximately 80% of cattle feeding is conducted by commercial feedyards that account for less than 5% of all feedyards. Due to economies of scale, the large feedyards increased the percent of total cattle fed in the United States from 34% in 1999 to nearly 45% in 2014. Even though the feeding capacity and number of farmer-feeders

Figure 7.1
A commercial yard depends on the capacity to deliver high quantities of formulated rations year-round.

Table 7.1
Cattle on Feed Numbers (1,000 hd) by Differing Sizes of Enterprises

Capacity of Feedyard (N)	1999	2011	2014
<1,000	2,616	2,509	2,399
1,000–7,999	2,212	2,284	2,016
8,000–15,999	1,424	1,250	1,060
16,000–31,999	2,546	2,180	1,750
32,000+	4,485	5,800	5,800

Source: USDA.

Table 7.2
Numbers of Various-Sized Feedyards From 1999 to 2011

Capacity of Feedyard (N)	1999	2011
<1,000	100,000	75,000
1,000–7,999	1,674	1,675
8,000–15,999	193	170
16,000–31,999	141	138
32,000+	111	137

Source: USDA.

has diminished over the past two decades, it is important to understand the inherent advantages held by farmer-feeder enterprises:

1. The farmer-feeder utilizes cattle as a market for homegrown feeds.
2. The farmer-feeder effectively utilizes high roughage feeds in a backgrounding operation.
3. The farmer-feeder distributes available labor over several different enterprises with cattle feeding being only one of those enterprises.
4. The farmer-feeder may have advantages in flexibility. When cattle feeding is unprofitable, the farmer-feeder can totally close out the cattle feeding operation and divert time and dollars into other phases of the farming operation. The large commercial feedlot has much higher overhead costs, which are fixed regardless of whether the pens are filled or empty.

However, the commercial feedyard holds distinct advantages as described below:

1. Commercial feedlots usually obtain and analyze more records and information and make more effective management decisions. Commercial feedlots typically utilize more professional expertise (consultants) in managing nutrition, health, and marketing.
2. Custom cattle feeding greatly reduces the operating capital requirement for the commercial feedlot and shifts some of the risk to the customer. However, one of the greatest risks for commercial feedlots is keeping the lots full of cattle. Customers will stop putting cattle into custom lots when financial losses occur over a period of time. Farmer-feeders usually feed just one group of cattle per year, which requires leaving the facilities vacant for several months. Commercial feedlots attempt to keep their lots full year-round with a goal of attaining a yard capacity turnover rate of 2–2.5 times per year.

Feeding Dairy Calves

Most of the bull calves from the 9.1 million dairy cows become feedlot steers with a predominance of them being Holstein. These young dairy calves are managed differently than calves from the beef breeds because they are removed from the dairy cows shortly after calving and thus are dependent on nutrient sources beyond their dams earlier in life and thus have a distinctly different growth curve. Following weaning at an early age, dairy calves are grown on milk replacer, transitioned to grazed forages or complete feeds and eventually to a finishing ration. Dairy-type cattle have an extended time on feed as compared to their beef-type contemporaries. Holstein cattle consume relatively large amounts of feed and gain rapidly and efficiently up to 1,100 lb; they become quite inefficient at heavier weights primarily due to their increased requirements for maintenance. Net energy requirements for maintenance are about 12% higher than those established for beef cattle. From a carcass standpoint, Holsteins have a lower dressing percentage and lower muscle-to-bone ratio. However, Holstein cattle marble quite well and have high cutability carcasses. They are more uniform in carcass characteristics than most beef breeds.

NON-FINISHING FEEDING

Feedyard enterprises may seek to diversify income or to take advantage of unique market opportunities to allocate a portion of their feeding capacity to non-finishing activities. For example, developing beef breeding stock, dairy replacement heifers, or backgrounding calves as a transition to either a stocker grazing program or to the finishing phase are potential options (Figure 7.2). Smaller yards are more likely to engage in these activities as compared to larger enterprises. However, a relatively small percentage of total cattle throughput is accounted for by these alternative management programs—less than 10% of total cattle fed by small yards and less than 1% of activity by the largest commercial feedyards.

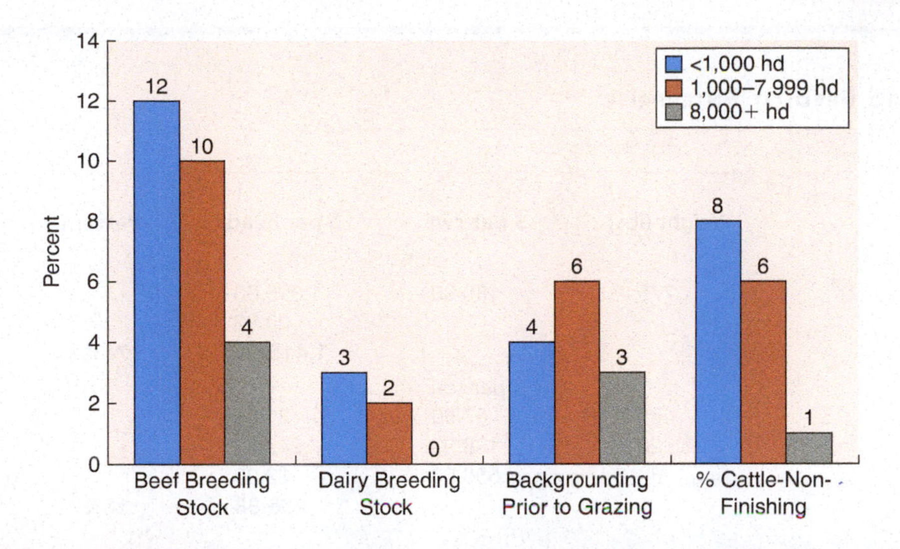

Figure 7.2
Non-finishing feeding activities engaged in by feedyards of various sizes.
Source: USDA-NAHMS.

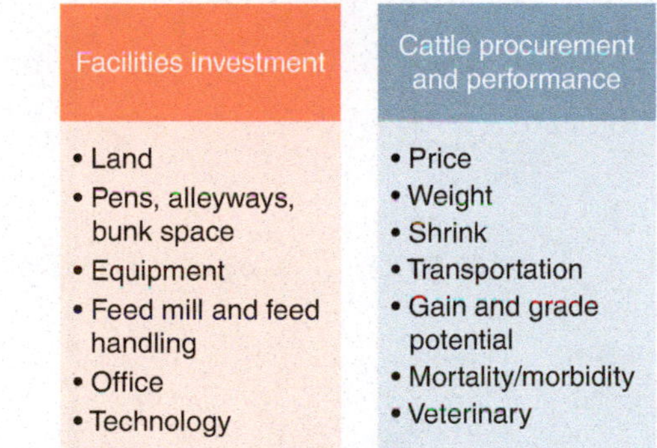
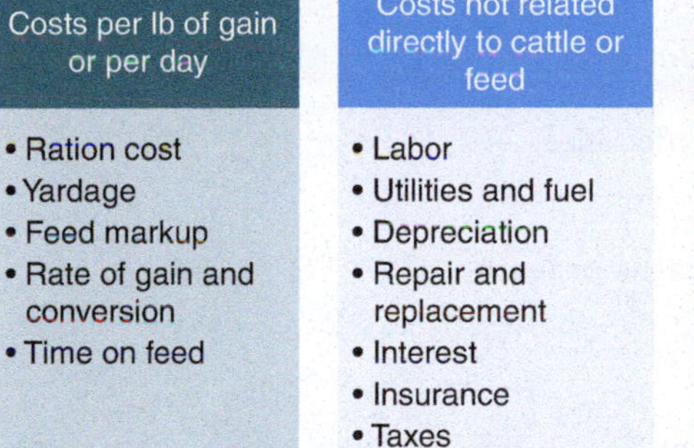

Figure 7.3
Major categories of assessing costs and returns associated with cattle feeding.

MANAGING A FEEDLOT OPERATION

The five primary factors needed to assess the performance of a feedlot operation are shown in Figure 7.3.

All five factors are included in the enterprise budget shown in Table 7.3. As Table 7.3 illustrates, feeder cattle and feed accounted for 76% and 15% of the total costs, respectively. By comparison, in 2004, a yearling steer entering the feedlot was valued at $740 (71%) and the cost of feed was $235 (23%) while in 1996, feeder calves were worth $420 (57%) and feeding cost was approximately $250 (34%).

Without effective risk management and exceptional operational management, feedyards are vulnerable to the highly volatile nature of commodity markets. While profitability is certainly within reach of well-run enterprises, significant losses may also be incurred.

Facilities Investment

Facility planning efforts should take into account the following objectives:

1. Minimize animal and employee stress while assuring safety for each.
2. Contribute to profitability and the production of safe, wholesome products.
3. Protect the surrounding environment.
4. Allow for the effective delivery of feed and care to cattle.

Most commercial feedlot facilities throughout the United States are similar in terms of layout and equipment. The typical feedlot has an open lot with dirt pens that

Table 7.3
EXAMPLE BUDGET FOR A 2015 FEEDLOT ENTERPRISE

Assumptions				
Mortality rate of 1.5% Time on feed—160 days	**Weight (lbs)**	**$ per cwt**	**$ per head**	**Percent**
Cattle costs				
Feeder cattle in weight	775	180.00	1,395.00	
Death loss			20.92	
Total cattle expenses			**1,415.92**	**76%**
Feed Costs		$ per ton		
Harvested forages	480	87.50	21.00	
Grain/supplements	3,520	138.00	242.88	
Mineral	80	550.00	22.00	
Total feed expense			**285.88**	**15%**
Other Variable costs				
Labor			16.00	
Veterinary/medicine			20.00	
Marketing			11.00	
Utilities/fuel/lubricants			13.00	
Repairs (equipment/machinery/facilities)			16.00	
Interest on cash			17.00	
Others			8.00	
Total variable expense (not cattle or feed)			**101.00**	**6%**
Fixed Costs				
Depreciation			13.00	
Taxes and insurance			5.75	
Opportunity cost on investment			51.00	
Total fixed expense			**49.75**	**3%**
Total cost per head			1,852.55	
Fed cattle revenue	1,375	126.00	1,757.70	
Profit (loss) over total costs			***−$95.05***	
Profits (loss) over all variable costs			***−$45.30***	

Figure 7.4
Feed is delivered into a concrete bunk at most feedyards. The bunk is then read by trained personnel several times a day to assess intake.

can hold 100–500 head of cattle. The pens are usually mounded in the center so that the cattle have a dry resting area. The fences are made of pole, cable, or pipe. There are fence-lined feed bunks (Figure 7.4) with concrete aprons inside the pens, where the cattle stand while eating. A feed mill (Figure 7.5) to process grains and other feeds is usually a part of the facility. Feed is distributed to the fence-lined bunks with specialized feed trucks (Figure 7.6) that mix the feed while traveling to the feed bunks. Bunker trench silos hold corn silage and other roughages. Grains may also be stored in

Figure 7.5
Rations are formulated and then mixed in high-tech mills located on-site for most commercial feedyards.

Figure 7.6
Feed is delivered to the bunks by a specialized feed truck that is able to evenly distribute feed down the feed bunk. These trucks are equipment with a scale to assure proper delivery of sufficient amounts. Feed is sometimes mixed on the truck when complete mill capacity is not available.

these silos; however, they are often stored in steel bins located above the ground. Specifications for a feedyard are provided in Table 7.4.

Facilities for farmer-feeder operations vary from unpaved, wood-fenced pens to paved lots with shelter (sheds) to partial confinement or total confinement feedlots. The latter may have manure collection pits under the cattle with feed stored in airtight facilities. Most feeds are stored in upright silos and grain bins. Feeds are typically processed on the farm and distributed to feed bunks (located inside or outside the pens) with tractor-powered equipment. The costs of such facilities are shown in Table 7.5. The costs are computed for farmer-feeders under Midwest feeding conditions, which is where the majority of farmer-feeders are located.

Table 7.4
SITE SPECIFICATIONS FOR FEEDYARDS

Factor	Specification
Land slope	3–6%
Soil type	>25% clay
Land mass	1 acre/100 head for pens, alleys, and feed roads
Home pen size—open lot	300 ft^2/hd for wetter conditions and 250 ft^2/hd for drier conditions
Home pen size—confinement	25 ft^2
Home pen size—confinement deep bedded	40 ft^2
Holding pens (feeder cattle)	16 ft^2 per head
Holding pens (fed cattle)	20 ft^2 per head
Hospital pens	75 ft^2 per head
Bunk space (receiving and hospital)	16–24 inches per head
Bunk space (finishing ration)	10–12 inches per head
Bunk pad	12–20 feet depth
Mounds—slope	3–4%
Mounds—top dimension	30 ft^2 per head
Mounds—side dimension	90 ft^2 per head
Water tank (center of pen)	At least 4 inches per head
Water tank flow rate	75 gallons per minute for a 100-head pen
Water access	Every animal within a pen should be able to access water within a 30-minute period

Source: Adapted from multiple sources.

Table 7.5
INITIAL INVESTMENT FOR VARIOUS TYPES OF FEEDYARDS, $ PER HEAD

	Open Lot—Windbreak	Open Lot—Shed	Covered—Solid Floor	Covered—Slatted Floor
Lot, fences, feed bunks, and building	196	513	651	1,121
Feed storage and handling plus animal handling	65	65	65	65
Environmental mitigation structures and engineering (<1,000 hd)	56	56	0	0
Environmental mitigation structures and engineering (1,000+ hd)	138	138	4	4

Source: Based on Iowa Feedyard Manual.

Most total confinement feedlots are farmer-feeder feedlots located in the Midwest with fewer than 1,000-head capacity. The world's largest confinement feedlot—a commercial feedlot owned by J.R. Simplot Company and located in eastern Oregon—has a capacity of 32,000 head and was built at an estimated cost of $6.2 million.

Environmental Management

The four primary environmental management issues for feedyards are dust, odor, flies, and water quality. Dust management is best accomplished via regular pen maintenance with timely manure removal in late spring. The regular use of box scrapers reduces manure accumulation without cratering the pen hardpan. The use of overhead sprinklers or water truck sprayers can be utilized to maintain the recommended moisture rate of 25–35% in a loose soil layer of 1 inch or less. Water delivery systems may be cost prohibitive. For example, to increase moisture from 10% to 35% in 1 inch of manure requires approximately 14 gallons of water per head of pen capacity. Achieving the same moisture increase in 2 inches of manure would require roughly 28 gallons per head of pen capacity.

Control of odor requires regular pen maintenance, the use of correctly constructed and maintained runoff holding ponds, and proper nutritional management. The use of a more precisely balanced ration to avoid overfeeding of phosphorus is a key strategy.

Research is being conducted to use plant-oil extracts or fat extracts as a treatment for pen surfaces to control both odor and dust. Plant-oil extracts may be useful in inhibition of manure fermentation as a means to preserve nutrient value and suppress odor. These extracts may also have antimicrobial properties that help to inhibit the growth of organisms that contribute to food-borne illness.

Another strategy is to compost manure. Composting reduces volume and weight, kills weed seeds, concentrates nutrients, reduces odor, and reduces fly populations. The disadvantages to composting are labor and equipment costs, storage space, and nutrient loss. The nutrients in manure vary by type of lot construction (Table 7.6).

Fly control can be achieved via chemical, biological, or combination strategies. Chemical controls used alone are typically costly, short-term in effect, and increase risk of human or environmental chemical exposure. The use of biological controls via fly parasites has been successfully adopted in the industry.

Fly parasites, also known as parasitic wasps, lay their eggs inside fly pupae. The fly pupae then become a food source for the fly parasites. The use of early releases of fly parasites before fly season followed by scheduled weekly releases is recommended. The cost of biological control ranges from $.20 to 1.00 per head of cattle.

Table 7.6
Nutrients in Manure from Various Styles of Feedlot Design (lb/hd/yr)

Source	N	P_2O_5	K_2O
Solid manure from open lots	45	24	33
Solid manure from deep-bedded building	90	55	70
Liquid runoff from open lots	5	2	11
Liquid manure from deep pit	113	60	90

Source: Based on Iowa Feedyard Manual.

Water-quality issues are likely to have a significant impact on feedlots and other intensive animal management facilities. Waste management is a primary concern as the decomposition of manure can negatively affect water quality via pathogens, nitrate, ammonia, phosphorus, salts, and organic salts. Incorrect handling, storage, or land application of manure can result in contamination of ground water or surface water. The revised Clean Water Act regulations administered by the Environmental Protection Agency apply to animal feeding operations (AFOs) and confined animal feeding operations (CAFOs).

An AFO is defined as an operation where animals are confined for at least 45 days during any 12-month period and where crops, forage, and other vegetation are not grown in the area where the animal is confined. A CAFO meets the description of an AFO and has a certain number of animals. CAFOs are classified as large (at least 1,000 cattle or cow-calf pairs), medium (300–999 head of cattle or cow-calf pairs and meets one of the medium category discharge criteria or is designated by the permitting authority as such), and small (fewer than 300 head of cattle or pairs and has been designated by the permitting authority). CAFOs must secure a National Pollutant Discharge Elimination System (NPDES) permit.

A NPDES permit will address four basic requirements: effluent limitations, special conditions, standard conditions, and monitoring, record keeping, and reporting requirements. More specific details can be found at www.epa.gov. However, one of the most important steps is to design and utilize a plan that specifies nutrient and waste management protocols. The plan must contain the following sections:

1. Adequate storage capacity
2. Proper disposal of dead animals
3. Clean water management
4. Protection of U.S. waters from direct animal contamination
5. Correct chemical handling
6. Conservation practices to control nutrient loss
7. Testing provisions for manure, soil, and process wastewater
8. Protocols for land application of manure and process wastewater
9. Record keeping systems

In an active feedyard, a layer of soil and manure becomes sufficiently compacted to create a seal that serves as a barrier to seepage. If the integrity of this layer is ensured, then water infiltration can be kept to less than 0.05 inches per day. It is critical that this layer be left undisturbed during pen cleaning.

Applying manure or wastewater directly to agricultural lands requires thorough knowledge about the following factors associated with a particular site: soil type, slope, irrigation practices, precipitation levels, crop nutrient requirements, nutrient levels of manure, and proximity to waterways or wells. The goal of manure management is essentially to collect, store, and apply wastes to lands at appropriate agronomic rates with the objectives of optimizing crop growth rates, economic returns, and protecting water quality.

Runoff problems can be minimized via employment of Best Management Practices (BMP) such as building up-gradient ditches, dams, grass filter strips, filter fences, and other appropriate controlled drainage systems. Lagoons or ponds may be required to contain wastewater (Figure 7.7) and runoff. BMP for manure handling are provided in Table 7.7.

Adoption of innovations in environmental management by feedyards of various sizes is provided in Table 7.8. The practices utilized vary between smaller and larger lots driven not only by regulatory compliance, but also by differences in scale of the enterprise as well as business structure.

Figure 7.7
Waste water lagoons are designed to prevent contamination to streams and waterways from feedyard runoff. Some of these are aerated to enhance aerobic breakdown of manure.
Source: Justin Field.

Table 7.7
Summary of Best Management Practices for Manure Handling, Storage, and Application

Periodic analysis for nutrient content.
Account for available N from the total system.
Apply to land areas large enough to accommodate manure volume.
Calculate long-term manure loading rates.
Maintain records of manure and soil analysis; application volume, timing and methodology, additional fertilizer applications, and plant yields.
Manure application rates in accordance with site-specific nutrient plans.
Incorporate manure soon after application to avoid runoff.
Determine application protocol based on soil composition and risk of aquifer contamination.
Apply manure uniformly, utilizing correct calibration.
Utilize buffer zones to prevent water contamination.
Use grass strips to catch and filter nutrients and sediments from runoff.
Use rotational application schemes when planting high-N-use crops or forages.
Locate manure stockpiles away from wells.
Divert runoff via correctly engineered ditches, terraces, etc.
Maintain integrity of manure-soil seal when cleaning feedlot pens.

Source: Based on Colorado State University, 1994.

Breakeven Prices

Computation of breakeven prices is the central focus of a feedlot risk assessment and management program. Breakeven prices can be calculated on the price that can be paid for feeder cattle, cost of gain, and (or) sale price on fed cattle. Breakeven prices are usually calculated on an individual lot of cattle that are fed as a unit. Costs are estimated based on current and future market prices and gain costs that have been previously experienced.

Table 7.8
Environmental Management Protocols Employed by Feedyards (%)

	1,000–7,999	8,000+
Routine environmental testing:		
Ground water	52	78
Surface water	19	48
Nutrient content of manure	71	87
Air quality	5	12
Manure management protocols:		
Land application on land owned/managed by feedyard	91	46
Manure sold	10	46
Manure given away	15	52
Manure hauled away by third-party	2	13
Runoff control protocols:		
Lagoons to capture runoff	65	95
Berms to control runoff	74	77
Dust control protocols:		
Permanent sprinklers	14	22
Water truck	29	54
Mechanized pen scraper	56	82
Increased cattle density	13	27

Source: USDA-NAHMS.

The breakeven price for final live weight of the fed cattle is calculated as follows:

$$\text{Breakeven price} = \frac{\text{Cost of feeder cattle} + \text{cost of gain (feed and non-feed)}}{\text{Estimated final weight of fed cattle}}$$

For example:

$$\text{Breakeven price} = \frac{(750 \times \$1.70) + (600 \times \$0.80)}{1{,}350\text{ lb}} = \frac{\$1{,}755}{1{,}350\text{ lb}} = \$130\text{ per lb}$$

Breakeven price calculations for cost of feeder cattle and cost per pound of gain are presented later in the chapter.

As alliances and other retained ownership programs are implemented, a total production breakeven price should be calculated and utilized. A carcass weight breakeven price combine s the costs of the cow-calf, yearling-stocker, and feedlot production phases. The calculations are:

$$\text{Breakeven price (cost/lb of carcass)} = \frac{\text{Cost of weaned calf} + \text{yearling} - \text{stocker cost} + \text{feedlot cost}}{\text{Carcass weight (lb)}}$$

Using an example of current industry costs and weights for steers:

$$\text{Breakeven price (cost/lb of carcass wt)} = \frac{\$900 + \$200 + \$325}{875\text{ lb carcass wt}} = \$1.62/\text{lb}$$

A variety of factors can change the profitability of a pen of cattle. Poor health management resulting in high levels of morbidity, mortality, and realizers, poor rates of gain, and carcass discounts are likely contributors to financial losses. Table 7.9 summarizes data from the Texas Ranch to Rail Program as to the causes of economic losses. For the cattle fed in 1999–2000, it took the gains of 335 cattle (26%) with the lowest profits to offset the losses incurred by the 10% of the cattle that lost money.

Table 7.9
Causes of Economic Losses at the Feedyard

Cause	N	Average Loss ($/hd)
Death	18	608.54
Realizers	17	281.95
Dark cutters	23	86.53
Poor gain[1]	47	47.09
Poor gain and grade discount[2]	18	89.54
Yield Grade 4 and 5 discount	11	26.93

[1]Gain was 24% below average; medicine cost 3 times greater than average.
[2]Gain was 27% below average; medicine cost 5 times greater than average.
Source: Based on Ranch to Rail Data Texas A&M University, 2001.

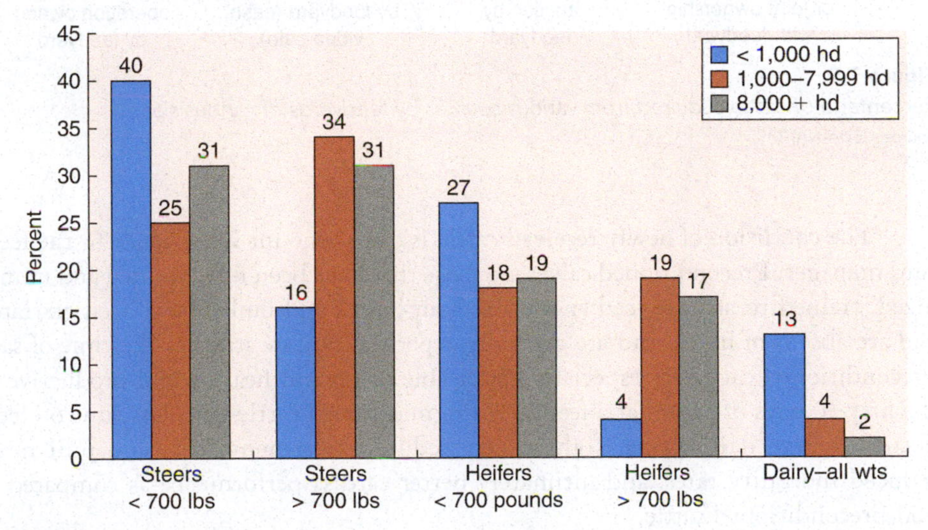

Figure 7.8
Weight, gender, and type differences of feeder cattle sources by feedyards of varying sizes.
Source: USDA-NAHMS.

Cost of Feeder Cattle

Cattle feeders source cattle of various weights, genders, and types based on market conditions, feed sources, and weather conditions (Figure 7.8). Larger yards are more likely to source cattle as yearlings and at weights over 700 pounds and less likely to place dairy-type cattle as compared to their smaller capacity contemporaries.

Cattle feeders usually calculate what they can pay for feeder cattle based on projected feeding costs (feed and yardage) and anticipated selling price of slaughter cattle. The price spread between feeders and fed steers is an important consideration for cattle feeders; however, it should be noted the cattle feeder is projecting the price of fed cattle 4–6 months from the time the feeder cattle are purchased. The prices of both feeder cattle and fed cattle can vary considerably during this short time period. The use of risk management tools to offset volatility in cash prices is most prevalent in the cattle feeding industry and its use rises with size of the feedyard.

Feeder cattle are typically acquired from cow-calf producers on stocker operators via direct purchase or by virtue of a custom-feeding arrangement. The sources of feeder cattle and ownership mix are described in Figure 7.9. The large yards custom feed 40% of their cattle with an additional 30% purchased direct from the ranch of origin. The mid-sized enterprises feed cattle that were acquired in custom arrangements, auction barns, and direct sales in nearly equal thirds while the smaller yards buy 40% of their cattle from auction barns and feed 30% as part of a wholly owned integrated operation.

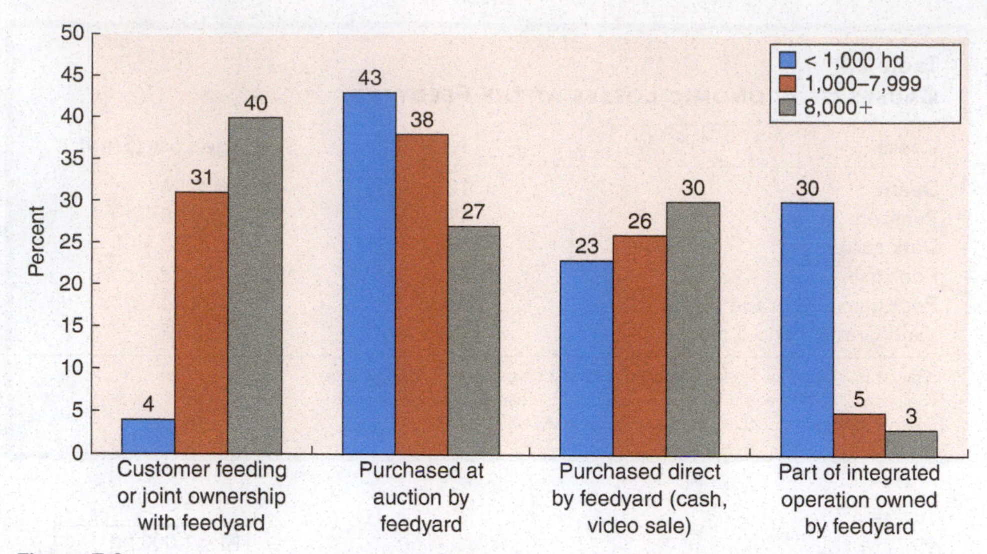

Figure 7.9

Percentage of cattle acquired from various sources by feedyards of various sizes.

Source: USDA-NAHMS.

The condition of newly received cattle is of paramount importance to the feedyard manager. Preconditioned calves are those that have been fully weaned and immunized, trained to acquire feed and water from bunks and tanks, have been castrated and are absent of horns, and are generally experiencing low stress at the time of sale. Preconditioned cattle are especially better able to remain healthy and productive in the first 30 days of arrival at a feedyard. Preconditioned cattle have less time on feed, better average daily gains, significantly reduced treatment costs and pull rates, reduced mortality rates, and ultimately better carcass performance as compared to non-preconditioned cattle.

The benchmark goal for death loss is to hold mortality rates to less than 1.5%. The National Beef Quality Audit in 2011 found that the average death loss for feedyards was 1.7%. The NAHMS study found that feedlots with less than 1,000 head capacity, between 1,000 and 8,000 size, and greater than 8,000 head had mortality rates of 1.2, 1.4, and 1.6%. Yearling cattle typically will have lower morbidity and mortality rates than weaned calves. However, in either case, poor health management can lead to severe financial losses. To help mitigate, the risk of lighter calves contracting shipping fever (respiratory disease complex), about 40% of calves less than 700 pounds arrival weight receive a prophylactic antibiotic treatment while only 5% of cattle arriving at weights in excess of 700 pounds receive the protocol.

Table 7.10 shows both the origin and destination of cattle entering and leaving the feedyard as well as the average distance that cattle will travel. Effective management requires minimization of stress during transport as well as assuring that receiving management and pre-shipment protocols are designed to enhance cattle comfort and performance.

Non-weaned calves arriving at the feedyard are 3–4 times more likely to contract bovine respiratory disease (BRD) than weaned calves, independent of age or vaccine status. Steers treated for BRD have net profits that are $57.48 per head lower than non-treated contemporaries. Forty percent of BRD occurs in the first 14 days of arrival; 81% of high morbidity cases expressed BRD in the first 42 days in the feedyard. Research consistently demonstrates that as the number of times cattle are pulled due to morbidity, average daily gain declines, percent USDA Select and Standards increase, and profits decline.

Table 7.10
SHIPMENTS TO AND FROM FEEDYARD BY ORIGINATION/DESTINATION (%) AND DISTANCE SHIPPED (MILES)

Feeder Cattle Shipments to Yards	<1,000 Head Capacity	1,000–7,999 Head Capacity	8,000+ Head Capacity
Auction barn	38 (142)	65 (280)	68 (356)
Direct from other beef producer (ranch, stocker)	45 (61)	25 (334)	25 (370)
Another feedyard	3 (88)	1 (160)	1 (158)
Other	14 (43)	9 (351)	6 (518)
Fed cattle shipments from feedyard			
Direct to packer	67 (87)	94 (184)	99 (119)
Auction	32 (38)	3 (74)	0
Another feedyard	0	3 (177)	1 (152)

Source: USDA-NAHMS.

Cost of Gain

Once feeder cattle are placed in a feedlot, cost of gain becomes a major economic concern for the cattle feeder. Total cost of gain typically is divided into two parts: (1) feed cost per pound (or cwt) gain, and (2) non-feed cost per pound (or cwt) gain.

Breakeven cost per pound of gain is calculated as follows:

$$\text{Breakeven cost (price per lb of gain)} = \frac{\text{Value (\$) of fed cattle} - \text{cost (\$) of feeder cattle}}{\text{Total pounds of gain}}$$

Grain prices have a tremendous effect on cost of gains for the cattle feeder. The price of feed comprises 70–80% of the total gain cost. Other feed grains (e.g., sorghum, barley, and oats) are important grains in certain cattle feeding areas; however, their total contribution to feed availability is only about 20% of corn production and their price trends often follow the price of corn. Historically, corn prices remained relatively constant from 1997 until 2003. In 2003, the price of corn was 15–20 cents higher per bushel than in the previous year. Prices in 2004 were the highest in nearly a decade but declined to 20-year lows in 2005. While demand from the livestock and poultry industries is an important factor influencing grain prices, the increasing focus on ethanol production is a substantial demand driver. The price run-up that led to historically high corn prices (Figure 7.10) from 2009 through 2013 generated tremendous wealth production for grain farmers and at the same time was very costly to the livestock industry. As corn prices skyrocketed, production capability was expanded and predictably the price of corn retreated significantly beginning in 2014. Most feedyards incorporate risk management strategies to offset the volatility of feed grain markets.

Factors Affecting Feed Cost per Pound of Gain

Next to the cost of feeder cattle, feed-related expenses are the largest category of expense for feedlots. Feedyards utilize feed grains, harvested forages and by-products from grain milling, ethanol production, food manufacturing, and brewing as their primary ration ingredients. Formulating rations that are optimized for cost, performance, and ease of handling are typically handled by nutritional consultants. The percentage of concentrate feeds high in energy fed to cattle in the receiving and finishing phase are listed in Table 7.11.

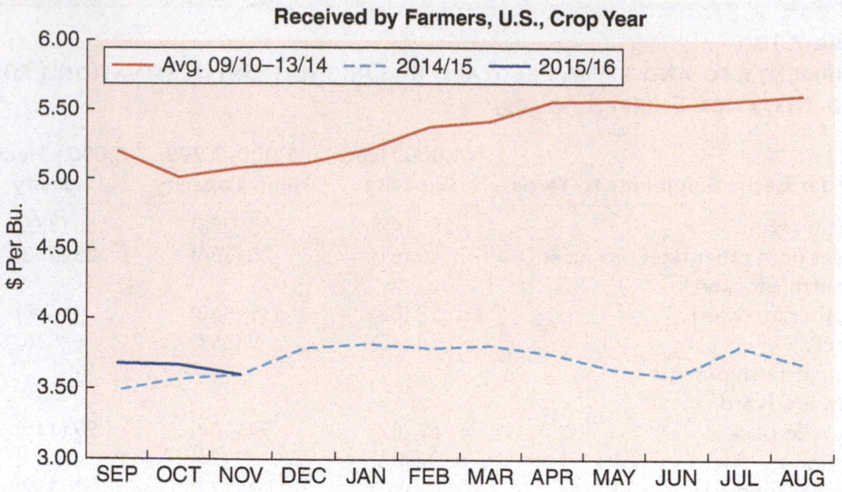

Figure 7.10
Average monthly corn price trends from 2010 to 2016.
Data source: USDA National Agricultural Statistics Service; compiled by Livestock Marketing Information Center.

Table 7.11
Percent Concentrate Feeds in Receiving and Finishing Rations Used by Various-Sized Feedyards

% Concentrate in Receiving Ration	1,000–7,999 hd	8,000+ hd
1–25	29	13
26–50	27	27
51–75	37	52
76–100	7	8
% concentrate in finishing ration		
1–25	6	2
26–50	22	13
51–75	43	34
76–100	28	51

Source: USDA-NAHMS.

Feed efficiency is an economically important trait for the cattle feeder. It has been determined that a 5% improvement in feed efficiency is equivalent to:

1. Reducing ration costs on a dry matter basis
2. Reducing purchase costs of the feeder animal
3. Increasing daily gain
4. Reducing the interest rate on operating capital

Daily gain and feed efficiency have a relatively high relationship (more rapid gains result in less feed required per pound of gain) where cattle are fed from the same in-weights to the same slaughter weights or to the same carcass compositional endpoint. The primary reason for this relationship is that cattle that gain more rapidly require less feed to maintain their body weight as they are fed for shorter periods of time. Cattle that gain 3 lbs per day as compared to those that gain 2 lbs daily utilize 10% less feed for maintenance.

Because feed efficiency has such a significant relationship to cost of gain, a discussion of major factors affecting it is warranted. The three most important variables affecting feed efficiency are: (1) energy density of the feed, (2) amount of dry matter consumed, and (3) rate of gain. Rations high in roughage fill the rumen with bulk, thus limiting energy intake and gain. Physical distention of the rumen is one of two main factors controlling dry matter intake. Adding grain to a high roughage ration reduces fill, thereby increasing energy intake and improving gain.

Feed intake under high-energy ration management is controlled by the products of digestion curtailing the animals' appetite (called chemostatic regulation) rather than rumen fill. Thus, increasing the energy density in an already high-energy ration usually has little effect on gain. Simply stated, improvements in feed efficiency are reflected mainly as increases in gain in low-energy rations and as reductions in feed intake in high-energy rations.

There are several ways the cattle feeder can improve feed efficiency in a feedlot operation:

1. Increase the energy density of the feed by feeding less roughage. The highest practical energy level for a finishing ration is in the range of 0.64–0.68 Mcal/lb net energy for gain. Rations containing higher energy levels increase the risk of serious problems, such as acidosis. Some farmer-feeders, raising their own silage, may feed high roughage levels to optimize the amount of feed produced per acre with cost of gain.
2. Processing grain is another method of improving feed efficiency. Steam flaking and reconstituting milo improves feed efficiency 8–17% depending on the processing method used and the type of grain. However, cost of feed is also increased.
3. Feeding additives and growth stimulants can improve feed efficiency from 3% to 12%. Several of these stimulate rate of gain but also indirectly improve feed efficiency as a result of the relationship between gain and feed efficiency. A more detailed discussion follows later in the chapter.
4. Feed cattle to the proper composition of gain. It takes more feed to deposit a pound of fat than a pound of lean. Cattle that mature too early and cattle fed beyond their logical slaughter potential will be very inefficient and gain will be costly.
5. Keeping cattle free of climatic stresses such as muddy pens and minimizing their exposure to extremely cold or hot temperatures will contribute marked changes in feed efficiency. Cattle exposed to these stresses usually have slower gains through reduced feed consumption (particularly in times of extreme heat stress) and higher maintenance requirements with resulting poor feed conversions. Black cattle are more prone to heat stress than their lighter-colored contemporaries. Mounds in the feed pens or well-drained pens will help keep cattle dry and free of muddy conditions. Cattle fed in pens with 4 inches of mud will have feed efficiency lowered by approximately 10%. Table 7.12 provides management approaches for stress reduction.

   The management of heat stress is of particular importance to cattle feeders in those seasons of the year when temperature and humidity rise to critical levels. The temperature humidity index categorizes livestock safety situations as alert, danger, and emergency based on the relationship of temperature and humidity (Figure 7.11). Additional steps that can help avoid heat stress include not allowing cattle to become overly fat prior to marketing; transporting and processing cattle in the coolest time of the day; keeping fresh water available; wetting down pens to allow cattle an opportunity to dissipate heat; spraying cattle with large water droplets at 5–10-minute intervals several times per hour (be sure that the entire hide becomes wet, not just the hair, and avoid cooling down cattle that are already hot too rapidly); keeping air moving at 5-10 mph; and feeding the bulk of the ration (65–70%) in the early evening, after temperatures have declined.

Table 7.12
Factors Initiating Stress and Management Protocols to Reduce the Impact of Stress

Factor	Effect of Stress	Method(s) to Reduce Stress
Behavior		
Novel experiences[1]	Reduced gain, increased dark cutters	Standardize handling, processing, and transport to minimize stress.
Cattle handling	Reduced gain, increased dark cutters	Minimize noise, hotshots, animal discomfort during processing and transport; do not overcrowd tubs and alleyways.
Environment		
Pen condition	Reduced gain, increased morbidity	Keep pens dry, well drained, and mounded; minimize dust in pens and feed.
Shade and windbreaks	Increased maintenance cost, poorer feed efficiency and gain	Provide shade and sprinklers to minimize effects of extreme heat, and windbreaks to minimize wind chill; do not ship or process in the hottest part of the day.
Insects	Increased eye problems, disease, loss of performance	Control flies via biological or chemical control.
Management	Reduced performance	
Feed and water		Make water available at arrival, during feeding periods, and prior to shipment; feed consistently to allow cattle to settle into a routine of habit.
Bullers	Reduced gain plus increased injury and bruising	Relocate bullers immediately.
Processing and transport	Reduced gain, increased injury; increased bruises and dark cutters	Wait 48 hours following arrival for processing.

[1]Novel experiences include new pens, handling by people on horseback versus on foot, etc.
Source: Adapted from multiple sources.

Relative Humidity \ Temp (F)	82	83	84	85	86	87	88	89	90	91	92	93	94	95	96	97	98	99	100
	Temperature Humidity Index																		
90	81	82	83	84	84	85	86	87	88	89	90	91	92	93	94	95	96	97	96
85	80	81	82	83	84	85	86	86	87	88	89	90	91	92	93	94	95	96	95
80	79	80	81	82	83	84	85	86	86	87	88	89	90	91	92	93	94	94	94
75	79	80	80	81	82	83	84	85	86	86	87	88	89	90	91	92	93	93	93
70	78	79	80	81	81	82	83	84	85	86	86	87	88	89	90	91	91	92	92
65	77	78	79	80	81	81	82	83	84	85	85	86	87	88	89	89	90	91	91
60	77	78	78	79	80	81	81	82	83	84	85	85	86	87	88	88	89	90	90
55	76	77	78	78	79	80	81	81	82	83	84	84	85	86	87	87	88	89	89
50	75	76	77	78	78	79	80	80	81	82	83	83	84	85	86	86	87	88	88
45	75	75	76	77	78	78	79	80	80	81	82	82	83	84	85	85	86	87	87
40	74	75	75	76	77	77	78	79	80	80	81	81	82	83	83	84	85	85	86
35	73	75	75	75	76	77	77	78	79	79	80	80	81	82	82	83	84	84	85
30	73	74	74	75	75	76	76	77	78	78	79	80	80	81	81	82	83	83	84
25	72	73	73	74	74	75	76	76	77	77	78	79	79	80	80	81	82	82	83

Normal | Alert | Danger | Emergency

Figure 7.11
Various risk categories associated with differing temperature and humidity indices (THI).

6. Cattle that have the genetic ability to gain rapidly will produce efficient gains. These cattle have been genetically selected for appetite, which significantly increases their consumption of feed. Most cattle feeders have not tapped this genetic resource because compensatory gain has been available. Cattle with compensatory gain are typically light for their age, which means they have been given somewhat restricted nutrition during part or all of their life. When they are fed finishing rations in the feedlot, they compensate for the earlier restriction of feed by gaining very rapidly. Cattle with compensatory gain are not as numerous as in past years, primarily because those producing them have found it unprofitable. Also, feeders have been more competitive in the marketplace for these cattle, so their economic advantage has been reduced compared with several years ago. Cattle feeders who can identify and purchase these genetically superior cattle can usually expect low gain costs because of excellent feed conversion. It has been estimated that cost of gain can be lowered $4–8 per hundredweight by feeding cattle with high genetic gaining ability.
7. Sex of feedlot cattle also affects feed conversion, though differences in feed efficiency vary depending on the slaughter endpoint. It is generally accepted that heifers are less efficient than steers and bulls and that bulls are more efficient than steers. These relationships are valid when the different sexes are slaughtered at similar slaughter weights. Heifers typically have 10–15% lower average daily gains, 6–8% poorer feed efficiency, and slightly higher death losses than steers. Heifers that come into heat during the feeding phase have significantly diminished performance, so the feeding of MGA (melengestrol acetate) to suppress heat is often utilized in the industry.

Table 7.13 outlines the nutritional protocols implemented by various-sized feedyards to enhance animal performance. Feedyards with greater than 1,000 head capacity are more likely to use ionophores to enhance rumen health, distiller's by-products (predominately from the ethanol industry), probiotics to enhance rumen function, and MGA to suppress reproductive behavior in heifers. The large yards are also more likely to incorporate repartitioning agents such as Optaflex into the nutritional management plan.

Factors Affecting Non-Feed Costs

Among non-feed costs per pound of gain, gaining ability, health, yardage, and interest rates are the most important. Cattle with more rapid daily gains have less maintenance feed cost and less overhead cost because of the fewer days required to accumulate a given amount of weight gain. A 700-pound steer that gains 3.75

Table 7.13
Percentage of Cattle Receiving Various Nutritional Protocols

	<1,000 Head Capacity	1,000–7,999 Head Capacity	8,000+ Head Capacity
Ionophore	49	93	89
Coccidiostat (not ionophore)	21	19	21
Distillers grains	53	90	87
Probiotic	8	19	60
MGA to suppress cycling in heifers	NA	63	89
Optaflex	13	23	49

Source: USDA-NAHMS.

pounds per day in the feedyard attains a target weight of 1,350 pounds 43 days faster than one that gains 3 pounds per day. Assuming a non-feed yardage cost of 0.40 per day, the faster gaining animal has $17.20 less cost over the feeding period.

Healthy cattle are less costly because they do not require additional labor for treatment and the cost of medication is eliminated. Also, healthy cattle will have a lower death loss and a higher rate of gain because of a consistent high-feed consumption pattern.

Respiratory diseases (e.g., BRD) are consistently the major health problem, usually accounting for approximately 75% of morbidity and 50% of mortality during the feeding phase.

Feeder cattle from one source, arriving from backgrounding operations and low-stressed, weaned calves, are usually lower risk cattle for respiratory and other disease problems.

More than 70% of feedyards ride or walk pens at least two or more times a day for cattle received within 14 days. For cattle that have been at the yard for at least 1 month, pen observation is typically reduced to once daily in larger yards while smaller yards are more likely to maintain a 2× per day schedule (Table 7.14). The majority of cattle in the United States are processed in the first 24 hours following arrival at a feedyard with larger yards more likely to incorporate this practice. Smaller yards may be slower to conduct vaccinations and processing as they are more likely to be handling cattle that are part of an integrated operation where cattle are not changing ownership and have been fully processed prior to reaching the yard (Table 7.15).

Standard processing protocols vary somewhat by feedlot size (Table 7.16). Most cattle receive two processings to facilitate the administration of a second vaccination that substantially increases immune response. Feedyards with capacity greater than 8,000 head are more likely to utilize implants at both first and second processing but are less likely to use prophylactic antibiotic protocols in the second round of processing, and less likely to administer a second round of clostridial vaccines or parasite prevention protocols. The inclusion of nonlactating, mature cows with newly received

Table 7.14
Frequency of Pen Riding/Pen (times per day) Walking to Check Cattle by Size of Feedyard

# Days at Yard	>2×/day %	2×/day %	1×/day %	<1×/day %	No Standard Protocol %
<1,000 hd. capacity					
<15 days	19	46	21	0	14
15–29 days	8	51	25	0	15
30+ days	6	48	31	1	14
1,000–7,999 hd capacity					
<15 days	20	49	27	1	3
15–29 days	6	35	51	5	3
30+ days	5	18	62	10	5
8,000+ capacity					
<15 days	20	34	45	1	0
15–29 days	7	25	68	0	0
30+ days	7	2	90	0	0

Source: USDA-NAHMS.

Table 7.15
Timing of Initial Processing Following Arrival at Feedyard (% of Cattle)

Hours (N) Post Arrival	<1,000 Head Capacity	1,000–7,999 Head Capacity	8,000+ Head Capacity
<24	34	49	63
25–72	24	35	32
>72	26	14	5
Not processed	17	2	0
% of yards using second processing	58	60	79

Source: USDA-NAHMS.

Table 7.16
Percent of Cattle (Processed in Groups) Receiving Various Protocols at 1st and 2nd Processing

	<1,000 Head Capacity	1,000–7,999 Head Capacity	8,000+ Head Capacity
1st processing			
Respiratory diseases vaccinations	93	95	96
Clostridial vaccinations	73	71	72
Injectable prophylactic antibiotic	31	20	27
Implanted	56	74	86
Treated for parasites	84	96	91
2nd processing			
Respiratory diseases vaccinations	64	67	67
Clostridial vaccinations	34	46	15
Injectable prophylactic antibiotic	14	8	4
Implanted	42	63	72
Treated for parasites	49	37	9

Source: USDA-NAHMS.

calves may reduce morbidity, improve average daily gain, and increase the speed by which calves learn to eat at a bunk.

Table 7.17 outlines the specific preventative protocols administered by feedyards of greater than 1,000 head capacity and the percent of cattle receiving each. However, protocols vary based on the source of cattle, weather conditions, level of known history about previous management, transportation distance, and indications of sickness at arrival. The integrated health management plan at a feedyard is developed by incorporating the expertise of the feedyard veterinarian, yard manager, and key personnel.

Labor is another cost center in a feedyard enterprise. Yards with less than 8,000 head capacity employ an average of 4.2 people per 1,000 head of cattle on feed with about half of those employees directly responsible for daily cattle care. Larger feedlots employ 1.4 people per 1,000 head of cattle on feed, but due to efficiencies of scale and operational management maintain performance comparable or superior to smaller yards and one-half of the staff is dedicated directly to cattle care.

Table 7.17
Percent of Cattle Receiving Various Preventative Disease Vaccinations or Protocols in Feedyards of Greater Than 1,000 Head Capacity

	% of Cattle
BVD	95
Injectable IBR	93
Intranasal IBR	13
PI3	55
BRSV	61
Haemophilus Somnus	28
Pasteurella	29
Leptospirosis	12
Clostridial	62
Anthelmintic	76
Prophylactic antibiotic to minimize shipping fever (calves <700 lbs)	39
Prophylactic antibiotic to minimize shipping fever (calves >700 lbs)	5

Source: USDA-NAHMS.

Table 7.18
Variation in Feedlot Performance of 1,311 Cattle from 111 Ranches

A.D.G. (lb/day)	Cattle (%)	Cost of Gain Feed ($/cwt)	Cattle (%)	Medicine Cost ($/hd)	Cattle (%)
≤2.5	7	≤40	15	0	30
2.51–2.75	15	41–45	30	≤5	37
2.76–3.00	23	46–50	23	5.01–10	18
>3.0	55	51–55	22	10.01–15	6
		>55	10	>15	9

Source: Based on Texas A&M Ranch to Rail, 1999–2000.

Total Dollars Received

Cattle feeding is a high-risk business because of fluctuating market prices, the biological variability of the cattle being fed, and the complexity of the many other factors that require careful management. The variation in cattle performance in the feedyard and on the rail can have a dramatic effect on feedyard profitability (Tables 7.18 and 7.19). Research has shown that approximately three-quarters of the variation in profitability is due to difference in the price of feeders and fed cattle.

Sale price of the market steers and heifers has the most significant effect on total dollars received in the cattle feeding operation. Market prices can fluctuate widely between years and within a year, with both having a major influence on profits. Week-to-week variation over a 4-month period (Figure 7.12) swung more than $21 per cwt between the high and low. This price variability drives the need for an effective risk management strategy.

Historically, most fed cattle were sold on a live weight basis where packer buyers bid on pens of cattle either in person or over the phone with the final price the

Table 7.19
VARIATION IN CARCASS PERFORMANCE AND PROFITABILITY OF 1,311 CATTLE FROM 111 RANCHES

Carcass Weight (lb)	Cattle (%)	Fat Thickness (in.)	Cattle (%)	Rib-Eye Area (in.^{2})	Cattle (%)	Yield Grade	Cattle (%)	Quality Grade	Cattle (%)	Profit ($/hd)	Cattle (%)
≤650	2	<.20	22	≤11	2	1	15	Prime	0	<0	6
651–750	22	.20–.29	14	11.1–13	31	2	48	Choice	51	1–50	9
751–850	50	.30–.39	20	13.1–15	41	3	32	Select	45	51–100	17
>850	26	.40–.49	13	15.1–17	21	4	10	Standard	4	101–150	37
		.50–.59	13	>17	5					>150	31
		.60–.69	9								
		≥.70	9								

Source: Based on Texas A&M Ranch to Rail, 1999–2000.

Figure 7.12
Variation in fed cattle prices.
Source: USDA.

result of negotiating with the feedyard. Nearly 60% of fed cattle are sold on a formula basis with some 18% sold via negotiated cash pricing and forward contract. The remainder is sold on a grid basis where premiums and discounts are applied based on a number of carcass criteria. Formula pricing is based on adjustments determined by changes in futures markets or boxed beef markets.

Manure, as a by-product of the feedlot, can be a minor source of income. Some feedlots will compost the manure that is later sold primarily for home lawns and gardens. Other feedlots will sell or give the manure to farmers or to contract haulers who transport the manure from the feedlot. Some manure is not acceptable to farmers if the salt content is high.

Implants and Additives

Rate of gain and feed efficiency can be improved by using growth-promoting implants and feed additives. Some commonly used implants and feed additives for calves, yearlings/stockers, and growing–finishing cattle are shown in Tables 7.20 and 7.21.

Implants can be categorized into several classifications: medium potency estrogens, lower potency estrogens, androgens, higher potency combinations, and lower

Table 7.20
Approved Growth-Promotant Implants

Implant[1]	Ingredient, Level (mg), and Use	Class[2]
Calfoid	10 estradiol benzoate/100 progesterone Calves < 400 lb	LPE
Component E-C	10 estradiol benzoate/100 progesterone Calves < 400 lb	LPE
Component E-S	20 estradiol benzoate/200 progesterone Steers > 400 lb	MPE
Component E-H	20 estradiol benzoate/200 testosterone Heifers > 400 lb	MPE
Component T-S	140 trenbolone acetate Finishing steers	A
Component T-H	200 trenbolone acetate Finishing heifers	A
Component TE-S	24 estradiol 17 beta/120 trenbolone acetate Finishing steers	CHP
Compudose 200	24 estradiol 17 beta All classes	MPE
Encore	48 estradiol 17 beta All classes	MPE
Finaplix-S	140 trenbolone acetate Finishing steers	A
Finaplix-H	200 trenbolone acetate Finishing heifers	A
Implus-S	20 estradiol benzoate/200 progesterone Steers > 400 lb	MPE
Implus-H	20 estradiol benzoate/200 testosterone Heifers > 400 lb	MPE
Ralgro	36 zeranol All classes	LPE
Ralgro Magnum	72 zeranol Finishing steers	MPE
Revalor-S	24 estradiol 17 beta/120 trenbolone acetate Finishing steers	CHP
Revalor-H	14 estradiol 17 beta/140 trenbolone acetate Finishing heifers	CHP
Revalor-G	8 estradiol 17 beta/40 trenbolone acetate Pasture steers and heifers	CLP
Revalor 200	20 estradiol 17 beta/200 trenbolone acetate Finishing steers	CHP
Synovex-C	10 estradiol benzoate/100 progesterone Calves < 400 lb	LPE
Synovex-S	20 estradiol benzoate/200 progesterone Steers > 400 lb	MPE
Synovex-H	20 estradiol benzoate/200 testosterone Heifers > 400 lb	MPE
Synovex-Plus	28 estradiol benzoate/200 trenbolone acetate Finishing steers	CHP

[1] Follow manufacturer's directions for implanting method and withdrawal time.
[2] A = Androgen, CHP = combination high potency, CLP = combination low potency, LPE = low potency estrogen, MPE = medium potency estrogen.

Table 7.21
Selected Feed Additives for Growing–Finishing Cattle[1]

Common or Trade Name (Additive)	Claims	Recommended Level	% Improvement Gain	% Improvement Efficiency
Bovatec (Lasalocid Sodium)	An ionophore that increases the proportion of propionic acid in the rumen and decreases occurrences of acidosis and bloat.	10–30 g per ton of feed	2	8
MGA (Melengestrol Acetate)	A synthetic progesterone that inhibits estrus.	0.25–0.50 mg per day	5–11	5
Rumensin (Monensin)	An ionophore that increases the proportion of propionic acid in the rumen and decreases occurrences of acidosis.	5–30 g/ton of complete feed; do not feed more than 360/mg/hd/day	0	5–12

[1] Follow manufacturer's directions for proper feeding and withdrawal times.

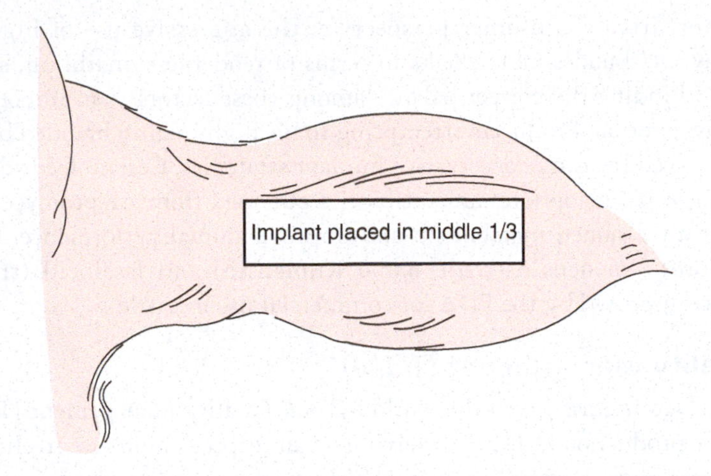

Figure 7.13
Proper location of implant placement.
Source: Sean Field.

potency combinations. When choosing an implant program, consider cost of implant, expected response, type of cattle, and convenience. The proper combination of implants can yield improvements in daily gain, feed efficiency, and lowered cost of gain of up to 20%, 15%, and 10%, respectively.

Implants containing trenbolone acetate (TBA) are typically considered terminal implants or the last implant given. The general recommendation is that these products should be administered in the last 100–140 days on feed. Early implants tend to be estrogenic in nature.

For implants to be effective, they must be implanted properly. First, the animal must be restrained so that movement of the head is minimal during the implanting process. Determine where the implant will be located (Figure 7.13), then insert the needle the proper distance away from that point, push it to the hilt, and back the needle off slightly before commencing to squeeze the trigger and withdraw the needle. This prevents possible crushing of the implant, which might lead to decreased effectiveness or possible problems due to too-rapid absorption. The needle must be sharp; if a bigger hole is put into the skin, it invites bacteria into the wound and may cause infection and poor absorption of the pellets.

Once the needle has been removed from the ear, it is a good practice to press the injection opening between the thumb and finger so the opening tends to seal. This practice is done not to prevent the implants from coming out, but to prevent infection of an open wound.

Assuring proper implanting protocols is critical. The ear should be cleaned with a brush and then disinfected with a chlorohexadine solution prior to placement of the implant. This procedure reduces the possibility of an abscessed ear. Cattle with abscessed implant sites will experience as much as a 0.2 lb per day reduction in gain. The net economic effect of an infected ear can approach $20.00 per head.

According to an Oklahoma State University study, the economic returns from implanting were $18.32 per implant investment if cattle were sold live, and $13.53 per implant investment if cattle were sold grade and yield (Duckett et al., 1996). The difference in profit was a result of the decreased percent of prime and choice carcasses from the implanted cattle. Producers should be cautious in the use of aggressive implant protocols because of potential negative effects in quality.

The trade-off is between the increased growth of fed cattle versus the impact on product quality. A comparison of cattle receiving TBA implants versus those not receiving TBA showed a decline of 11% in USDA Prime and Choice grades, but a gain of more than 40 lb in carcass weight in the treated animals. The advantage in weight is typically more profitable except in those cases when the Choice–Select spread is exceptionally high.

However, from a consumer perspective, the aggressive use of high potency implants may yield undesirable effects in terms of tenderness, marbling, and overall acceptability. Finding the correct balance among these factors is an important management consideration. Producers attempting to hit high-quality branded beef targets may be best served by using conservative implant strategies. Certain feed additives are known as *ionophores*. Ionophores are antibiotics obtained from streptomyces microorganisms that alter rumen metabolism and improve animal performance. Currently, two ionophores—monensin (trade name Rumensin) and lasalocid (trade name Bovatec)—are approved by the FDA for commercial use in cattle.

Total Quality Management (TQM)

W. Edwards Deming has given the world a Total Quality Management (TQM) process of quality production rather than relying on an inspection process to eliminate the creation of mistakes. Deming defines *quality* as meeting or exceeding customers' expectations at a cost that represents value to them every time.

Mies (1993) reviews how Deming's 14-point TQM plan applies to feedyard management. The 14 points are:

1. Create a constancy of purpose toward improvement of the product and service, with a plan to become competitive, stay in business, and provide jobs.
2. Work to eradicate delays, mistakes, defective materials, and defective workmanship.
3. Cease dependence on inspection to achieve quality. Eliminate the need for inspection on a mass basis by building quality into the production in the first place.
4. End the practice of awarding business on the basis of price. Instead, depend on meaningful measures of quality, along with price. Move toward a single supplier for any one item in a long-term relationship based on loyalty and trust.
5. Improve constantly and forever the system of production and service to increase quality of the product and constantly decrease costs.
6. Institute modern methods of training for all employees.
7. Institute leadership. Focus supervisors on helping people do a better job. Ensure that immediate action is taken on conditions detrimental to quality.
8. Drive out fear so everyone may work efficiently for the company.
9. Break down barriers between departments and encourage problem solving through teamwork.
10. Eliminate arbitrary goals, posters, and slogans for the workforce that seek new levels of productivity without providing methods.
11. Use statistical methods for continuing improvement of quality and productivity. Eliminate work standards that prescribe numerical goals.
12. Remove barriers that rob workers of their pride of workmanship.
13. Institute a vigorous program of education and training.
14. Take action to accomplish transformation. Transformation is everybody's job.

SELECTED REFERENCES

Publications

Brink, T. 2000. The impact of TBA implants on Gelbvieh-sired cattle. *Gelbvieh World* (Feb.).

Colorado State University and Texas A&M University. 1999. Impact of feedlot growth-promotant implant strategies on carcass grade characteristics and subsequent cooked beef palatability traits when applied to small/medium framed, 3-way, British crossbred steers. Final report to NCBA.

Colorado State University and Texas A&M University. 2012. National Beef Quality Audit 2011–Phase III: Quality enhancement by the seedstock, cow-calf, and stocker sectors. Final report to the National Cattlemen's Beef Association.

Duckett, S.K., Wagner, D.G., Owens, F.N., Dolezal, H.G., & Gill, D.R. 1996. Effects of estrogenic and androgenic implants on performance, carcass traits and meat tenderness in feedlot steers: A review. *The Professional Animal Scientist.* 12:205–214.

Environmental Protection Agency. 2003. *Producers' compliance guide for CAFOs.* EPA 821-R-03-010. Washington, DC.

Great Plains beef cattle feeding handbook. Cooperative Extension Services of CO, KS, MT, NE, NM, ND, OK, SD, TX, and WY.

Harner, J.P., & Murphy, J.P. 1998. *Planning cattle feedlots*. Kansas State University Extension Publication MF-2316. Manhattan, KS.

Iowa State University. 2013. *Beef feedlot systems manual.* ISU Extension Publication 1867. Ames, IA.

Loy, D. 2000. *Growth promotants in cattle.* Iowa State University Fall Extension Newsletter.

Livestock and poultry situation and outlook report. 1970–2001. Washington, DC: USDA.

Lorimor, J.C., Shouse, S., & Miller, W. 2003. *Best environmental management practices for open feedlots.* ISU Extension Report 1946. Iowa State University. Ames, IA.

Mader, T.L., Davis, M.S., & Halt, S.M. 2000. Management of feedlot cattle exposed to heat is important. *Feedstuffs.* July 17, 11–13.

McNeill, J.W. 2000. Where the profits came from . . . and what caused the losses. *Ranch to Rail Summary.* College Station, TX: Texas A & M University.

Mies, W.L. 1993. *Adapting total quality management to feedyard management.* Englewood, CO: The National Cattlemen's Association.

Owens, F.N., et al. 1997. The effect of grain source and grain processing on performance of feedlot cattle: A review. *Journal of Animal Science.* 75:868.

Palo Dura Consultation Research and Feedlot and Global Animal Products, Inc. 2002. Beef Production Management. Canyon, TX.

Roeber, D.L., Cannell, R.C., Belk, K.E., Miller, R.K., Tatum, J.D., & Smith, G.C. 2000. Implant strategies during feeding and impact on carcass grades and consumer acceptability. *Animal Science.* 78:1867–1874.

Texas A&M University. 2000. *1999–2000 Ranch to Rail summary report.* College Station, TX.

Texas A&M University. 2011. Assessing beef quality and BQA adoption at the feedyard receiving interface: Final report to the National Cattlemen's Beef Association. College Station, TX.

USDA. 2013. Feedlot 2011 Part I: Management practices on U.S. feedlots with a capacity of 1,000 or more head. National Animal Health Monitoring System. APHIS-VS.

USDA. 2013. Feedlot 2011 Part II: Management practices on U.S. feedlots with a capacity of 1,000 or fewer head. National Animal Health Monitoring System. APHIS-VS.

USDA. 2013. Feedlot 2011 Part III: Trends in health and management practices on U.S. feedyards: 1994–2011. National Animal Health Monitoring System. APHIS-VS.

USDA. 2013. Feedlot 2011 Part IV: Health and health management practices on U.S. feedlots with a capacity of more than 1,000 head. National Animal Health Monitoring System. APHIS-VS.

8 The Beef Supply Chain

For most of its history, the U.S. beef industry was composed of a set of loosely coordinated segments populated by competing enterprises with an associated purveyor or "middleman" handling transactions and transfer between each of these segments. As a result, a number of inefficiencies became embedded in the overall system. With each of the entities engaged in the business operating from a silo mentality, the transfer of information and clear market signals was difficult, if not impossible to achieve. As a result, a commodity mentality became the driving philosophy of the industry which impaired the creation of value aligned with the shifting needs and desires of downstream buyers. However, under the leadership of Chuck Schroeder, who became CEO of the National Cattlemen's Beef Association that was created with the merger of the National Cattlemen's Association and the National Livestock and Meat Board, the industry began to develop its strategies on the basis of becoming a customer driven industry. As a result, the organization began to build stronger relationships across the supply chain to create the momentum to remove old barriers that inhibited the ability of the beef business to capture market share.

In 2015, the Industry Long Range Plan was updated with the following as the stated mission, vision, and strategic objective:

Mission—*"A beef community dedicated to growing beef demand by producing and marketing the safest, healthiest, most delicious beef that satisfies the desires of an increasing global population while responsibly managing our livestock and natural resources."*

Vision—*"To responsibly produce the most trusted and preferred protein in the world."*

Strategic Objective—*"Increase the wholesale beef demand index by 2 percent annually over the next five years."*

While independent production and processing decisions, local and regional marketing campaigns, and segment-driven initiatives can impact the mission, vision, and strategic objective stated above, substantial progress requires system-wide commitment and engagement. In recognition, the task force responsible for writing the Industry Long Range Plan in 2015 created four core strategies for the industry (Table 8.1).

The Beef Industry Long Range Plan task force concluded that there were a series of questions that must be addressed by industry leadership (Table 8.2). These questions align with the core strategies of the 2011 National Beef Quality Audit that focused on protecting the integrity of beef production and beef products, enhancing eating satisfaction, and telling the beef story. These questions provide the strategic framework for organizational and individual participants in the beef industry to determine the most impactful tactics to move forward. These questions open the door for significant innovation and entrepreneurship in search of solutions to pain points that if resolved will enhance opportunity, profitability, and sustainability.

Table 8.1
Beef Industry Long Range Plan

Core Strategy 1: Drive growth in beef exports

Rationale

- The U.S. beef industry generates $350+ per head of additional value by exporting beef and beef products.
- The global middle class is expected to grow from 2 billion people to 4.9 billion people by 2030.
- China's middle class is expected to grow from 300 million to 640 million by 2020.
- Asia's middle class spent $4.8 billion in 2010 and is expected to spend $32.6 billion by 2030.
- Global beef production has remained flat for the past 10 years compared to pork and poultry supplies which have been rising.
- At an average export price of $3.27/pounds, in 2014, the United States is garnering a $.75 per pound premium compared to its next closest competitor (Australia) indicating the "premium" position U.S. beef holds in the global marketplace.
- The United States is uniquely qualified to produce large volumes of the high-quality, grain-fed beef demanded by global consumers.

Goal

By 2020, grow the value of U.S. beef exports as a percent of total beef value from EOY 2015 benchmark to at least 16%.

Initiatives

Increase Market Access
Advocate for international trade and ultimately gain unfettered access to key export markets.
Adopt Animal I.D. Traceability Systems
Secure the broad adoption of individual animal I.D. traceability system(s) across the beef community to equip the industry to effectively manage a disease outbreak while enhancing both domestic and global trust in U.S. beef and ensuring greater access to export markets.
Promote Unique Attributes of U.S. Beef
Promote the unique attributes of U.S. beef in foreign markets (quality, safety, sustainability, and nutritional value). (Chapter 10 covers beef exports in detail.)

Core Strategy 2: Protect and enhance the business and political climate for beef

Rationale

- Local, state, and federal regulations continue to increase in number and scope.
- Politicians, Non-governmental Organizations (NGOs) and activists are using state ballot initiatives, regulatory agencies, and public relations campaigns to change agricultural production practices and limit access to production inputs and technologies.
- Large retail, restaurant, and food service organizations are beginning to mandate production protocols—sometimes with limited input/involvement from producers.
- Increases in regulations can create "barriers to entry" and make it more difficult for young producers to enter the beef business.
- Tax laws can make it difficult, if not impossible, to pass ranches across generations.

Goal

By 2020, decrease the percentage of producers saying regulations imposed on their business make it more difficult to operate freely from 72% to 62%.

Initiatives

Manage the Political and Regulatory Environment
Protect the business climate for beef against legislative policies and/or agency regulations that have a negative impact on the economic health of the beef community while supporting public policy that can improve the overall business climate for the beef community.
Ensure Beef's Inclusion in Dietary Recommendations
Develop a comprehensive strategy for effectively positioning beef as part of a healthy diet in future dietary guidelines.
Motivate Producers and Stakeholders to Engage in Policy Issues
Secure expertise to activate and energize beef, dairy, and veal producers and other industry stakeholders to become more engaged in beef industry policies and issues.

(continued)

Table 8.1
(Continued) Beef Industry Long Range Plan

Core Strategy 3: Grow consumer trust in beef and beef production

Rationale

- Consumers are increasingly interested in where food comes from and how it is grown/raised.
- There are a growing number of well-funded organizations focused on cultivating mistrust in modern agriculture production tools and practices.
- Consumers receive mixed messages regarding the healthfulness of beef.
- Only 38% of consumers "strongly" or "somewhat" trust that producers are committed to the welfare and well-being of livestock.

Goal

Benchmark consumer trust using the July 2015 Consumer Beef Index and set a goal for 2020.

Initiatives

Ensure Antibiotic Stewardship
Aggressively invest in initiatives and research that ensure the responsible stewardship of antibiotics to safeguard human as well as animal health and well-being, while committing to the development and use of alternative technologies and practices.
Certify and Verify Production Practices
Facilitate the creation of a standard to certify and verify beef production and management practices to address consumer concerns.
Ensure Beef Safety
Increase industry investment in beef safety and production technology research and communicate the beef community's commitment to safe and healthy beef.

Core Strategy 4: Promote and strengthen beef's value proposition

Rationale

- 50% of Americans believe animal protein is the best source of protein.
- 60% of millennials believe they can get their daily allowance of protein without eating meat.
- The primary and secondary reasons why people eat less beef are "health reasons" (38%) and "other meats seem healthier" (35%).
- Competitive meats are expected to increase production.
- Beef prices are 50% higher than 2009.
- In a recent Consumer Beef Index study, 34% of millennial parents "strongly" or "somewhat" preferred chicken compared to 31% who preferred beef.
- There is a growing interest from consumers to understand how animal protein production impacts the environment.
- There is an increased interest in food and food preparation.

(Chapter 2 provides a detailed evaluation of beef demand and merchandising.)

Goal

By 2020, increase the number of consumers who agree that the positives of beef "strongly" or "somewhat" outweigh the negatives of beef by five percentage points.

Initiatives

Revolutionize Beef Marketing and Merchandising
Invest in efforts to revolutionize and rapidly implement innovations in the presentation of beef as a convenient and essential staple of a healthy lifestyle (new products for domestic and global markets, packaging innovations, processing innovations, merchandising innovations, preparation methods, etc.).
Research and Communicate Beef's Nutritional Benefits
Invest in credible research to assess, document, and communicate the value of beef's nutrition and health benefits.
Connect and Communicate Directly with Consumers
Capitalize on flexible new media technologies and innovative communication tactics to design and deliver memorable messages directly to consumers, regarding beef's quality, safety, and nutritional value and the beef community's commitment to animal well-being and the sustainable use of natural resources.
Measure and Improve Sustainability
Collaborate with beef industry stakeholders to conduct additional sustainability research, demonstrate continuous improvement, and communicate beef's sustainability efforts and progress to key target audiences.

Source: Adapted from Beef Industry Long Range Plan.

Table 8.2
CRITICAL STRATEGIC QUESTIONS TO BE ADDRESSED BY BEEF INDUSTRY PARTICIPANTS

Product Integrity
How can the beef industry become the "protein of choice"?
What is the best way to measure, demonstrate, and communicate the nutritional benefits of beef?
How to verify the best practices of production to increase consumer confidence?
How can industry proactively address meat/food safety and animal health issues?
How can the industry help consumers trust the way we produce beef?

Eating Satisfaction
What are most effective strategies take advantage of the increasing demand for high-quality beef in the global market?

Telling the Beef Story
How do industry participants better listen and respond to customers?
How to effectively communicate with the millennial generation in a way that increases their long-term commitment to include beef as a significant part of their family's diet?
What should the U.S. beef industry do to maximize the use of social media in a positive and proactive manner?
How can the beef industry be more transparent in telling our story to the consumer, especially millennials, who wants more information?
What are the strategies to increase our influence in media and news coverage?

Mobilizing Action
How to effectively communicate the changes that need to take place to grow world-wide demand for U.S. beef through all facets of our industry?
Is the industry set up in an appropriate manner to execute and capitalize on whatever plan is developed?
How can influence in Washington, D.C. be increased?
How to motivate industry participants across the supply chain to be proactive and innovative?

TRUST

Agriculture in its entirety is a complex and multifaceted production, distribution, and merchandising series of business models that operates on both the domestic and international market stage. Humanity and the very foundations of civilization depend on the success of the agricultural infrastructure to nourish people. When famine and chronic persistent hunger were more prevalent and widespread, the goal was clear—produce more calories, improve distribution, and as economies improved to then enhance the nutritional choices available to consumers. Agriculture and the beef industry were successful in that environment. In today's market, another element has been introduced that significantly influences consumer demand—an interest that extends beyond *what* is produced to *how* it is produced. Thus, agriculture must thoughtfully and consistently enhance society's trust level.

Trust can be thought of as the ability to rely on the integrity and dependability of a person, organization, or thing. Developing trust in a business and its products requires three basic elements—an intentional strategy to build quality and value into products and services from the very beginning that are aligned with specific market demands, an ongoing process of internal checks, balances, and review processes to assure that quality standards are being met, and finally implementation of a third-party verification system that demonstrates the level of conformance to quality standards from the perspective of an impartial third-party. In an age when most consumers were only a generation or two away from farming, trust in the agricultural system was high. Today, fewer than 5% of American consumers consider themselves to have linkage to farming and ranching. This creates a space of uncertainty and low knowledge levels which if left unbridged yield a low trust environment.

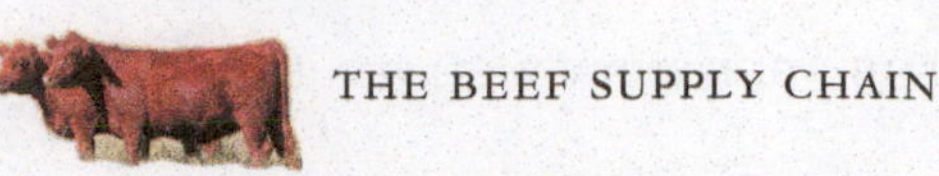

The trust gap has given way to the creation of products and food brands focused on credibility attributes, such as source and method of production. These credibility attributes relative to the beef industry tend to focus on characteristics, such as the location of production, direct identification of the producer, animal well-being, and environmental considerations. While the core drivers of beef demand—taste, price, and convenience—continue to be important, the industry must address the new layer of demand drivers that include sustainability and the credibility of the story of how beef is produced.

SUSTAINABILITY

An overview of sustainability is provided in Chapter 5. The discussion here will focus on the measurement of sustainability and mechanisms for improvement throughout the beef chain. Despite the politicized rhetoric surrounding sustainability fueled by the illogical view that broad generalizations of good and bad are an appropriate approach to complex problems, the central issue for all enterprises, companies, and industries is to develop innovative ways to become more sustainable over time. This process is not a sprint and quick fix approaches will be expensive and unproductive; rather the process of improving sustainability is a long-term approach.

A number of factors make lifecycle assessment of a product(s) difficult to conduct and also difficult to compare results across studies—geographic location, system boundaries (farm gate vs. full supply chain), allocation of resources and impacts to inter-related sectors (feed grains, dairy, and imported cattle, for example), functional units selected as the basis of measurement and comparison, plus the inherent assumptions associated with a diverse set of methodologies. Given these factors, it is critical to thoughtfully evaluate research summaries in light of their individual and collective impact on interpretation.

Funded by an investment of checkoff dollars, the U.S. beef industry has undertaken the most comprehensive study by any agricultural entity to develop benchmarks and metrics that will allow the industry to approach sustainability from the perspective of a continuous improvement process that will reap value far into the future. While the ideal benchmark for the industry would have been at its peak inventory numbers in the mid-1970s, a lack of critical data prevented that opportunity. Thus, 2005 was chosen as the benchmark for the beginning of the industry assessment. The results of the initial assessment are provided in Table 8.3. It is very important to understand that this research was certified by the National Standards Foundation, the largest accredited third-party body that certifies national conformity standards to measure sustainability of a product. On virtually every measure of sustainability, the industry had made progress.

Table 8.3
Improvements in Beef Industry Sustainability from 2005 to 2011

Sustainability Metric	Percent Change	Primary Contributors to Improvement
Emissions to water	10% decrease	Precision farming techniques
Emissions to soil	7% decrease	Improved crop yields
Greenhouse gas emissions	2% decrease	Better cattle genetics, herd health, and nutrition
Occupational illnesses and accidents	32% decrease	Improved training and process control
Energy use	2% decrease	Increased use of biogas capture and conversion
Resource consumption	2% decrease	Improvements in right-sized packaging
Water use	3% decrease	Innovations in irrigation technology

Source: Adapted from Beef Sustainability Report.

Perhaps the most startling finding of the study was that approximately 4% of beef was lost to wastage at the retail level while 20% of edible beef is disposed of and thus wasted by consumers. Food waste is one of the most significant contributors to the industry's footprint and is a problem shared by global agriculture (Figure 8.1). Wastage is often the result of inadequate storage and refrigeration infrastructure in emerging economies. In established economies, the problem of food waste is driven by plate waste and failure to consume foods before the expiration date. Attacking food waste is an important strategic initiative with high potential for immediate impact.

Attaining the goal of continuous improvement in sustainability requires the application of creativity and innovation to each of the metrics in Table 8.3 will be required at critical steps in the beef supply chain. For example, the water footprint of the industry can be improved through a number of tactical approaches (Figure 8.2). To gain meaningful improvements in water efficiency will require multidisciplinary research, collaboration, and innovation involving soil scientists, plant genomics specialists, irrigation system engineers, processing plant design engineers, food scientists, and technical expertise spanning the width of the beef supply chain. Coupled with meaningful partnerships with industry professionals, a systems approach will yield improvements.

Table 8.4 provides an evaluation of seven separate studies that evaluated the carbon footprint of beef production across a number of geographic regions. Thus, caution is advised in making direct comparisons. However, the trend of these analyses is that more intensive beef production systems that utilize feedyards systems as part of the production chain have lower footprints than those systems that depend exclusively on grasslands. In regard to greenhouse gas production, according to the

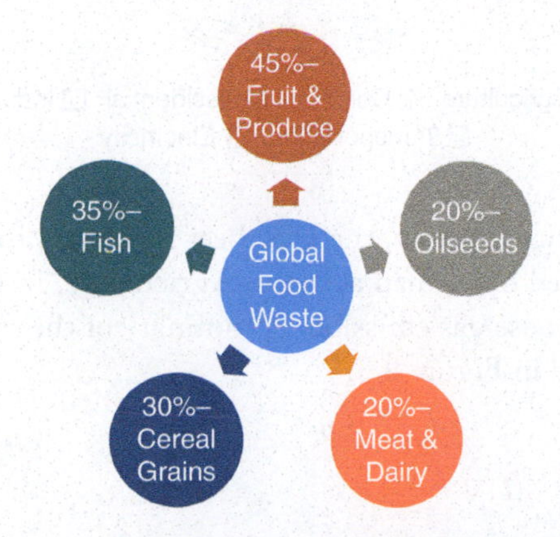

Figure 8.1 *Percentage of various foodstuffs lost to wastage on a global scale.*

Source: Adapted from FAO data.

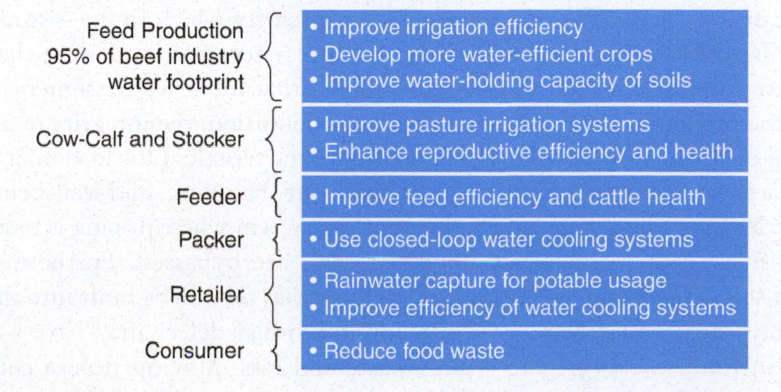

Figure 8.2 *Approaches to improving industry water footprint.*

Table 8.4
Estimates of Carbon Footprint Across Nations and Production Systems

Country and Production System	Kg CO_2–Equivalent per Kg Carcass Weight
Australia w/ feedlot finishing	9.9
Australia w/ grass finishing	12.0
Brazil–without deforestation	28.0
Brazil–with deforestation	44.0
Canada	21.7
Japan	36.4
Sweden	19.8
United States w/ feedlot finishing–study 1	17.9
United States w/ feedlot finishing–study 2	14.8
United States w/ grass finishing	19.2

Source: Adapted from seven different studies, so the figures are not directly comparable.

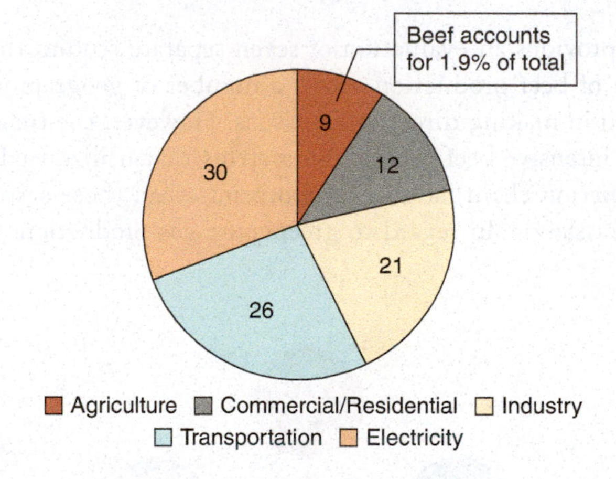

Figure 8.3
Sources of greenhouse gas emissions in the United States.
Source: EPA.

Environmental Protection Agency, the U.S. beef industry produces about 2% of the total emissions produced by human activity. Agriculture, in total, accounts for only 9% of the total green house gas emission. A summary of the role of industry in GHG production is described in Figure 8.3.

VALUE CREATION

Ultimately, the goal of a supply chain is to create value for downstream users, especially consumers. Value chains develop high levels of buyer and consumer trust. At the forefront of this trust is the concept of product integrity which in the case of the beef industry is equal to the combined effects of all processes used to produce the product along with the resulting attributes of the product that influence consumer confidence in both the product and its suppliers. Given this definition, the integrity of beef is the sum total of all the nutritional and palatability characteristics plus food safety, source, and the health of the cattle resulting from their care, handling, and well-being.

Under a commodity pricing and marketing system where pricing is based on the average, many value differences among animals are bypassed. Furthermore, in a multiple-segmented commodity system, inefficiencies tend to be built into the system in the form of product defects. While any individual defect may have a minimal impact, in total, the aggregate creates waste and loss. Moving from a commodity approach to value creation requires that an industry or enterprise carefully analyzes its products and processes to find the inefficiencies and value losses, build a strategy to rectify them, and to then execute strategy to move the bar from mediocre toward measurably differentiated. As a result, beginning in 1991, the beef industry undertook a benchmarking study called the National Beef Quality Audit (NBQA). Repeated approximately every 5 years since, the NBQA has provided a measurable assessment that provides a scorecard and systematic approach to improving the value proposition of the beef chain.

For example, one of the strategic objectives of the 2000 NBQA was to change the Quality Grade and Yield Grade mix to 6% Prime, 60% Choice, and 35% Select, and to 15% Yield Grade 1, 50% Yield Grade 2, and 35% Yield Grade 3. After a 15-year-period, the industry grade performance was as follows—3% Prime, 60% Choice, and 32% Select; 12% Yield Grade 1, 41% Yield Grade 2, and 37% Yield Grade 3 (Table 8.5). This provides evidence that the industry is more than capable of attaining a specific set of goals despite the hardships of moving the performance of populations that have a slow generation turnover. However, there is still work to be done to optimize the yield and quality mix by reducing the percent of Yield Grade 4 and 5 carcasses and keeping the pressure on eliminating carcasses with Quality Grades of Standard or poorer.

Specific changes in the various grade factors over the 20-year-period of the NBQA are listed in Table 8.6 and demonstrate that the industry was able to optimize yield and quality grade performance by increasing marbling, rib-eye area, and carcass weight while stabilizing backfat thickness. A challenge for purveyors, food service operators, and retailers is dealing with the variability in rib-eye size. While the industry has made cattle more muscular, which has been valuable in terms of

Table 8.5
USDA Yield and Quality Grade Distribution Observed in the 2011 National Beef Quality Audit

	USDA Quality Grade (%)			
USDA Yield Grade (%)	Prime	Choice	Select	Other
1	0.0	3.6	7.3	1.4
2	0.4	22.8	15.3	2.4
3	1.8	25.9	8.0	1.5
4	0.5	6.3	1.4	0.4
5	0.1	1.3	0.1	0.1

Source: Adapted from NBQA and USDA.

Table 8.6
Average Carcass Performance Over 20 Years as Measured by the NBQA

Factor	1991	1995	2000	2005	2011	20-yr. Change
YG	3.2	2.8	3.0	2.9	2.6	0.6 improvement
Choice and Prime (%)	55	49	51	55	61	6% improvement
Fat thickness	0.6	0.5	0.5	0.5	0.5	0.1 in. improvement
Rib-eye area	12.9	12.8	13.1	13.4	13.8	0.9 in.^{2} improvement
Hot carcass weight	761	748	787	793	825	64 pound increase

Source: Based on the various NBQA final reports.

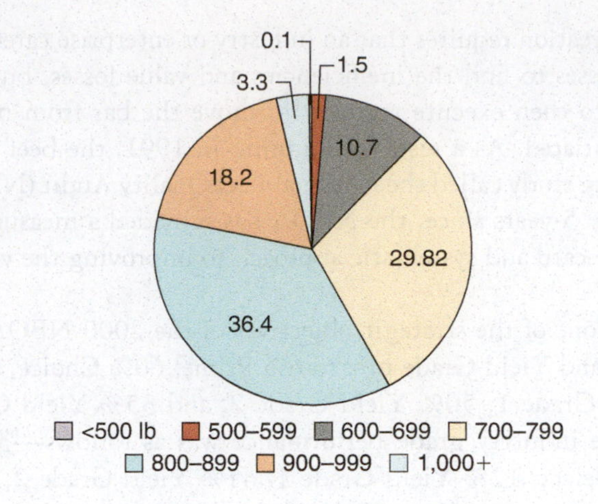

Figure 8.4
Percentage of beef carcasses by weight category.
Source: USDA.

Table 8.7
Incidence of Polled Cattle and Bruising Observed in Fed Cattle and Carcasses

Year	Polled (%)	Carcasses with zero bruising (%)
1991	69	61
1995	68	52
2000	77	53
2005	78	65
2011	76	77
20-year improvement	*7%*	*16%*

Source: Based on the various NBQA final reports.

improving sustainability and per animal productivity, more than 25% of beef rib eyes fall outside the 12–16 square inch preferred range. Part of the driver of increased rib-eye size is the increase in beef carcass weights over time (Figure 8.4). The 2011 NBQA cooler audit found that nearly 20% of beef carcasses were in excess of 900 pounds, a number that prior to 2000 was viewed as the upper acceptable limit on carcass weight.

Other examples of defects uncovered by the NBQA included the problems associated with horned cattle especially as it relates to carcass bruising rates. Table 8.7 shows the progress made and quantifies the opportunities to reduce defect rates even further.

The 2011 NBQA provided an innovative approach to better understanding the factors that influenced buying decisions by various sectors of the supply chain while establishing measures of price sensitivity to premiums and discounts associated with those decision influencers. The study also determined that not all sectors interpret opportunities or requirements in the same way. Study respondents representing the various sectors clearly have differing perceptions about palatability (Table 8.8). Retailers were more likely to mention multiple factors related to palatability. Especially relative to food service operators, retailers are more sensitive to factors that influence palatability because their customers control the crucial steps of preparation and cooking that ultimately impact the eating experience. Cattle feeders were less aligned with downstream participants about palatability perhaps suggesting that due to the gap between their stage of production and consumers, producers are less aware and sensitive to the consumer's eating experience.

Table 8.8
DESCRIPTORS OF EATING SATISFACTION AS REPORTED BY RESPONDENTS (%) IN THE 2011 NBQA BY SECTOR

	Feeders	Packers	Food Service, Further Processors, and Distributors	Retailers
Tenderness	44	65	52	67
Flavor	20	54	62	70
Customer satisfaction	25	15	30	37
Juiciness	8	11	12	23
Consistency	10	11	19	40
Marbling	25	31	8	13

Source: Based on 2011 NBQA.

When asked what the concept of traceback meant to them, various sector representatives responded differently. Feeders (29%) and packers (23%) responded that traceback systems enabled the ability to trace an animal from farm to fork while an additional 25% and 13% of feeders and packers respectively considered traceback to be the equivalent of source and age verification. Meanwhile 56% of food service operators, further processors, and beef distributors as well as 37% of retailers considered traceback systems to be the capacity to quickly trace an outbreak or recall to its source.

In regards to sustainability, 25% of food service, further processors, and distributors considered the concept to mean a triple-bottom line approach (environment, economic, and community). Twenty-five percent of food service operators and 37% of retailers equated sustainability with being environmentally friendly. Assuring the transfer of an enterprise to the next generation and maintaining profitability were the key descriptors of sustainability for 20% and 35% of feeders.

Relative to animal well-being, 30, 42, 58, and 30% of feeders, packers, food service operators, and retailers equated this attribute with humane handling and management practices. In regards to food safety, all sector participants were in agreement that food safety is a table stake for the industry and anything less than safe is unacceptable.

The 2011 NBQA also found that there was variability on the importance of purchase price to the various sectors of the industry. Feeders and food service operators viewed purchase price as highly important while packers and retailers were somewhat less focused on price as a primary element of their purchase decisions. This variation may have been due to the timing of the 2011 study where the industry was in a situation of declining numbers and both packers and retailers were in serious competition to find sufficient sources of cattle and beef to meet market demands.

Beginning in 2000, strategies focused on genetics emerged from the NBQA process. The industry has struggled, however, with meaningful assessment of genetic differences for performance in the supply chain beyond the seedstock purchase level. Without a cost effective, simple system to assess genetic potential for carcass performance, the industry has depended on a phenotypic measure—hide color—as its preferred proxy. Figure 8.5 shows the trend in the industry to increase the percentage of

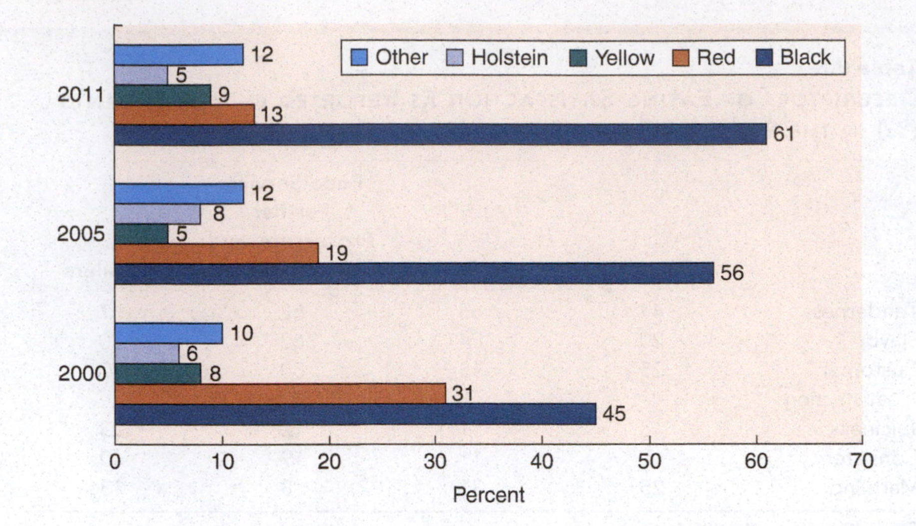

Figure 8.5
Hide color trends in fed cattle.
Source: Adapted from NBQA, 2011.

cattle with black hides driven by brands that use hide color as an indicator of breed type, in this case Angus.

The 2011 NBQA strategy team outlined the following as key industry priorities:

Food safety–Animal Health: develop and implement an effective identification sharing system, execute full supply chain safety interventions, implement and document BQA practices industry wide, and continuously improve cattle health.

Maximize and reduce variation in eating quality: develop information systems to enhance supply chain coordination, develop strategies for precision management of production technologies, and match growth promotion protocols to specific cattle focused on specific market targets, use genetics to optimize cutability and palatability.

Optimize beef's value and reduce waste: develop and implement an effective animal identification sharing system, enhance market signal communication between sectors, reduce extremes in rib-eye area, carcass weight and back fat, precisely define product and weight inconsistencies, and document the economic value of beef quality assurance for the chain.

SOURCE VERIFICATION, TRACEBACK, AND IDENTIFICATION SYSTEMS

As consumers, federal regulatory agencies, and international markets continue to increase demand for source verification and traceback systems as part of a total quality assurance effort, the pressure to develop supply chain or national identification systems grows. There are a variety of reasons for adoption of a source verification system. Perhaps the most important of these is the ability to better capture and utilize data from conception to consumption on a host of factors influencing profitability. Furthermore, the ability to source verify may provide sufficient market differentiation to enhance profitability, at least in the short term. Finally, the development of a beef industry identification system would provide some degree of ability to track a food safety problem to its root cause.

The challenge in the design of source verification and identification systems is finding an approach that is both functional at each industry segment and cost effective.

For these reasons, producers are understandably hesitant to accept federally mandated schemes. The most likely scenario is the development of voluntary systems tied to alliance on branded beef production systems. It is absolutely critical to note that identification systems will not prevent problems and, as such, should be considered as one piece of a broader quality program.

An ideal identification system would be able to provide unique identification of individual animals tied to their source of origin, and provide transaction tracking as the animal moves through the beef production and processing steps. Furthermore, this system would be tied to a highly user-friendly database system to record meaningful data on an animal's management and performance.

There are a variety of identification methods that can be utilized individually or in combination. These include visual identification (ear tags—plastic, metal clips; back tags; brands, tattoos; ear notches, and color patterns) and non-visual identification (electronic and biometric). Electronic identification systems include bar codes, two-dimensional symbols, and optical character recognition (scanners to record information from an existing tag). Biometric approaches include retinal or iris scanning, retinal imaging, DNA sequencing, and antibody fingerprinting. The advent of cloud-based computing, real-time electronic devices that are both durable and affordable, and the integration of technologies speeds the process. For example, Quantified Ag, an innovations company in Lincoln, NE has developed an ear tag that is both an identifier but also serves as a means to track body temperature, animal activity, and other wellness attributes. With active data transmission, the system allows managers a real time evaluation of cattle health.

The most likely technology to be of value to the beef industry is the use of RFID devices in the form of ear tags. In many cases, supply chains, branded product companies, and other users will likely merge several of the available identification systems into their protocols. The advent of relatively cheap electronic Identification systems in the industry were quantified in both the 2005 and 2011 National Beef Quality Audits (NBQA) (Table 8.9).

Certainly, technology will bring new alternatives and improvements. However, it is no longer a question of *if* source verification will occur but *when*. The integration of identification systems with data management, quality assurance, and branded-product initiatives will clearly bring benefits to producers, processors, and consumers. These benefits will not be derived without the challenges associated with implementation of a new approach. While an integrated national identification system is not yet in place, voluntary implementation by various stages of the beef production system to tie animal identification to both individual and group records is reasonably well developed (Table 8.10).

Table 8.9
Forms of Identification Observed at Harvest

Identification Device	2005 (%)	2011 (%)	Percent Change
Identified by any form	93.3	97.5	+4.2
Lot visual tag	63.2	85.7	+22.5
Individual visual tag	38.7	50.6	+11.9
Electronic tag	3.5	20.1	+16.6
Metal clip tag	11.8	15.7	+3.9

Note: N exceeds 100 due to multiple forms of identification used in some cases.
Source: Based on National Beef Quality Audit, 2011.

Table 8.10
IDENTIFICATION SYSTEMS IN PLACE TO TRACK INDIVIDUAL ANIMALS HEALTH TREATMENTS AT THE ENTERPRISE LEVEL

Sector	Individual ID[1]	Animal Tied to Group[2]	Tracking Group[3]	More than One Form used
Overall	78	11	9	2
Seedstock	89	4	7	<1
Cow-calf	77	11	11	2
Backgrounder	73	13	10	4
Stocker	62	22	15	2
Feeder	78	15	5	1
Dairy	83	12	4	2

[1]Records individual identification on all cattle.
[2]Identifies only treated animals.
[3]Tracks groups where individuals within the group were treated.
Source: Based on NBQA, 2011.

STRUCTURAL CHANGES IN THE BEEF INDUSTRY

Consolidation of agricultural enterprises into larger enterprises capable of capturing economies of scale, adopting and executing innovation and new technologies, and having the ability to adapt to the rapid rise in government regulations has decidedly changed the business landscape. For example, mid-aggregate enterprise size has increased dramatically over the past three decades. Mid-aggregate enterprise size is that level where one-half of all production on all farms producing a specific commodity are more than the average while half of production is on farms producing less than the average. Mid-aggregate enterprise production across a range of categories is provided in Table 8.11. Interestingly, between 1987 and 2007, U.S. agriculture reduced its total land use by 10% at the same time that consolidation was occurring.

Consolidation in the industry result from a variety of causative factors. Among these are the advantages of economies of scale. The benefits that arise from capturing economies of scale include the opportunity to lower costs via the advantages of volume purchasing power that lowers per unit input cost, capability to access useful new technology, the ability to spread fixed costs over higher levels of production, and gaining access to markets as a result of scale.

In a commodity industry, where by definition the average producer breaks even over time, one strategic opportunity to capture profits is via the process of lowering costs. An analysis of cow-calf enterprise cost structures from 1991 through 1999 in Texas, Oklahoma, and New Mexico revealed that the cash production costs per cow for herds with 1,000 or more beef cows was 20% lower than that of herds with 300–499 head of beef cows and nearly 40% lower than those enterprises with fewer than 50 cows. The mid-sized group also had an approximately 20% advantage in production costs as compared to the smallest herds. Relative to the packing industry, the largest packers have a 3% advantage in cost of production—a significant advantage in a low-margin industry. Net profit as a percentage of sales in the supermarket business hovers in the 1–1.5% range demonstrating the relatively thin margins in retailing. Five of the top ten food retailers in 1982 no longer exist or have fallen from the list. In almost every case, these firms were unable to compete because they couldn't find inventory management opportunities or cost savings—two skill sets that propelled Wal-Mart into a global food retailing giant.

However, concerns arise when the forces of consolidation concentrate either market power or processing capacity into fewer but larger firms. For example, the largest

Table 8.11
Mid-Aggregate Enterprise Production of Agricultural Commodities

Enterprise	1987	1997	2007
Livestock (head/farm)			
Chickens	300,000	480,000	681,600
Dairy cattle	80	140	570
Fed cattle	17,532	38,000	35,000
Hogs	1,200	11,000	30,000
Farming (harvested acres per farm)			
Corn	200	350	600
Cotton	450	800	1,090
Rice	295	494	700
Soybeans	243	380	490
Wheat	404	693	910

Source: USDA ERS.

Table 8.12
Beef Harvest Concentration by the Top Four Firms

Year	Fed Cattle (%)	Market Cows and Bulls (%)
2003	80	44
2005	80	48
2007	80	55
2009	81	54
2011	84	53
2012	85	56

Source: Packers and Stockyards Administration-USDA, 2015.

four packers have accounted for approximately 80% of fed cattle harvest and 50% of market cow harvest since about 2000 (Table 8.12). This level of consolidation has caused some to question whether or not there are issues with maintaining competitive pricing in the industry.

Consolidation in the agricultural sector has also been driven by consumer expectations for plentiful and relatively cheap foods. The net result for U.S. consumers has been that instead of paying 21 cents of each earned dollar for food, as was the case 50 years ago, they now pay 11 cents. Consolidation resulting in large food retailers and nationally franchised food service offerings are part of the economic landscape because they fulfill consumer demand. Consumers want to do business in an environment where access to a variety of brand names is readily available at affordable prices.

The consequences of industry structural change have been summarized by Boehlje (1995) and Field (2000) as follows:

1. The deadwood gets trimmed as the low-profit, inflexible players are removed from the system.
2. As the focus shifts from segment to system, total cost reduction in the system becomes important. To assure critical communication of information within the system, the attention given to data capture, transfer, and utilization is magnified.

3. Inputs (health program, genetics, and nutrition) become a contractual obligation instead of a free-agent decision. This is necessary to assure that customers receive a very specialized product that meets detailed specifications.
4. Value-added, specification-oriented decisions replace mass quantity production goals.
5. As product specifications become more clearly defined, the need for standardization at critical control points becomes very important.
6. Spot markets are replaced by contracts, formulas, and performance-oriented pricing systems.
7. Risk management is expanded to include pricing, environmental impact, animal welfare, and partnerships with suppliers and customers.
8. Information becomes the cake—not the frosting. Access to, sharing, and integration of data become the center of attention.
9. Control measures at the interface with consumers and the point of genetic decision making become increasingly critical.
10. Partnerships and alliances take the place of isolated, free agent decision making.

Consolidation results from consumer demand, public policy, and survival of the fittest in an environment of increasing competition, and the loss of producers who choose to retire rather than adapt. The benefits of consolidation include lower production costs at the farm and ranch level, lower food prices for consumers, and a more competitive position for the United States in global markets. The drawbacks include the loss of some people and firms from the agricultural sector and dramatic changes in rural communities, both in terms of culture and infrastructure.

The structural changes that have occurred in the beef industry have forced managers and industry leaders to look for unique and innovative approaches to improve profitability in the new landscape. One option that exists in the industry is to realign ownership or contractual arrangements via integration.

SUPPLY CHAIN COORDINATION/BUSINESS RELATIONSHIPS

The business model and operational structure of the beef industry is very complex with highly variable resources required as the product moves from farm and ranch to the consumer's table. Because of its regional and structural diversity, inefficiencies are inherent. Solving these challenges is beyond the influence and reach of any one entity or segment within the industry. However, efficiencies and improvements in product value can be obtained via enhanced supply chain coordination and creation of business relationships that clarify demand signals, reduce risk, and assure a continuous flow of product into the various market channels served by the beef business. W. Edwards Deming defined the practice of sub-optimization (opposite of optimization) as the lack of coordination between business units. Managers of these business units practice sub-optimization when they pursue goals of lowest possible cost without receiving feedback on how their decisions affect the results of the operation (system) as a whole. Sub-optimization may result in one enterprise or segment of an industry thriving while others struggle. Furthermore, under sub-optimization, the consistency and quality of finished products is often reduced.

In the mid-to-late 1990s, a number of innovative approaches began to take shape to better meet specific market demands for value and consistency. These efforts have been referred to as industry alliances.

Part of the stimulus in developing alliances came from a strategic alliance pilot project coordinated by the National Cattlemen's Association (now NCBA) in 1993.

The project was sponsored by several beef organizations and involved a partnership of producers, feedlots, packers, and retailers as a group. The strategic alliance cattle were superior to other groups of typical feedlot cattle in both cost of production and retail product. Among the strategic alliance cattle, there were large individual differences in productivity and profitability. This demonstrated that future alliances could select and manage cattle that would be superior in cost of production and retail product.

Alliances usually involve a contractual arrangement between different segments of the beef industry. These alliances are a form of integration where different individuals and companies can operate somewhat independently of one another but still share in risks and profits when cattle and beef products meet certain specifications. Success of alliances is highly dependent on participants having common paradigms, effective communication, trust, and "win-win" attitudes. Alliances can be simple, involving only two or more producers or organizations, or they can be very complex, where all segments of the beef industry are in alliance. These alliances can be organized and driven by producers, packers, breed associations, or the total industry. The net result of each alliance is that bulls, feeder cattle, carcasses, retail products, or all of these are produced for targeted markets.

While alliances can differ widely from one another, most current alliances are marketing alliances, breed alliances, and closed cooperatives. Marketing alliances are coordinated approaches designed to meet specific market targets while better managing price risk. Breed alliances represent efforts to increase markets for specific breeds. Closed cooperatives are member-owned organizations that attempt to control the product from gate to plate. Most alliances typically target carcass trait conformance, specify management protocols, or specify adherence to brand label characteristics.

Alliances are typically formed to better meet consumer expectations by improving the process of production and processing via the application of Total Quality Management (TQM), Hazard Analysis Critical Control Point (HACCP), or Best Manufacturing Practice (BMP) principles to improve profitability via a combination of cost-control and value-added efforts, and to better manage total risk in the system.

However, several factors stand in the way of creating successful alliances. These include lack of trust, availability of capital, poor planning, lack of management expertise, inflexibility, poor decision making, and lack of commitment to the alliance when economic conditions shift.

Producers who desire the opportunity to participate in alliances should carefully research the following questions:

1. Which management practices will I have to change?
2. Do I prefer collaborative or independent decision making?
3. How much financial risk can I undertake in both the short and long term?
4. Do I have sufficient borrowing power?
5. Do the goals of the alliance align with those of my enterprise?
6. What are the costs and benefits?
7. How much experience does the management of the alliance have in critical areas?

Consumer-based alliances that focus on breed specifications were dominated in 2014 by requirements for Angus influence or black hide color with more than 4 million head marketed. The closest competitor is a Hereford influenced specification program with 250,000 in annual marketings. An additional 2.2 million head were marketed through alliance structures without a breed requirement. There are also a number of feeder cattle marketing alliances with 185,000, 135,000, and 6,000 traded through Angus, Red Angus, and Gelbvieh specified programs in 2014.

Alliances specifications for fed cattle lot size range from 1 to 60 head and from 1 head to a semi-load for feeder cattle programs. Premiums received for fed cattle in

2014 were highly variable ranging from $15 to $250 per head while feeder cattle premiums ranged from $20 to $180 per head. Costs for participation were highly variable ranging from one time participation fees to per head charges.

Alliances function as part of a free enterprise system and, as such, are not guaranteed success. Start-up efforts will likely require a minimum of 3 years and typically 5 years before profits are realized on the investment. However, when the partners and conditions are aligned within an alliance structure, partners can achieve profitability while experiencing the meaningful sense of fulfillment that comes from successful innovation.

Even if highly palatable retail products can be consistently produced and effectively marketed, the beef industry could have large numbers of producers exiting the business and cow numbers decreasing because of high production costs. It is possible to make money with superior feedlot and carcass performance, yet lose money in the cow-calf segment. If both cost of production and retail product is emphasized, there are several hundred dollars profit that can be captured in both the live cattle production and processing (carcass/retail cut) phases. The combined total is greater than emphasizing only one of the segments. Table 8.13 shows how a carcass breakeven price can be calculated for beef. This emphasizes the need for an optimum combination of costs rather than low costs in one of the production phases that may result in high cost in another production phase. For example, a group of yearling cattle may have high gains with a low breakeven price, but these high yearling gains may cause lower feedlot gains and a higher breakeven price.

Table 8.13
COMPONENTS OF A TOTAL BEEF INDUSTRY ALLIANCE

Low Cost of Production and Higher Net Profit	Retail Products and Improved Market Share
• Low weaned calf breakeven price ($/cwt)	• Consistently uniform and highly palatable retail products (e.g., tenderness, flavor, and juiciness)
• Low yearling breakeven ($/cwt) • Low feedlot breakeven ($/cwt) • Low carcass weight breakeven, e.g., $\frac{\text{Weaning cost + yearling cost + feedlot cost}}{\text{Carcass weight}} = \text{Cost per lb of carcass}$	• Highly tender –breed and sire selection within breed –low Warner–Bratzler Shear –carcasses electrically stimulated –carcasses aged 14–21 days
• Resources (e.g., forage) sustained or improved to optimum levels	• Carcass quality grade
• Continuing high net margin (profit) each year	• Carcass yield grade • Optimum carcass weights • Carcasses free of defects (e.g., darker cutters, injection- site blemishes, bruises, etc.) • Free of microorganism contamination and harmful residues • Environmentally-safe image • Cost competitive with other meats • Regain market share to 50% or higher • Value-added by-products (e.g., hides free of brands or butt branded, large choice of convenient, and highly palatable retail products)

Source: Adapted from multiple sources.

SELECTED REFERENCES

Publications

Alliance yellow pages. August, 2015. Beef. Overland Park, KS.

Alternative marketing programs. 1999. (7th ed.) Greenwood Village, CO: Cattle-Fax.

Barkema, A., & Drabenstott, M. 1996. *Consolidation and change in heartland agriculture*. Federal Reserve Bank of Kansas City.

Boehlje, K. 1995. *Industrialization of agriculture: What are the consequences?* Proc. Industrialization of Heartland Agriculture Conference. Minneapolis, MN.

Cattle-Fax. 2015. *Outlook and strategies*. Centennial, CO.

Council for Agricultural Science and Technology. 2001. *Vertical coordination of agriculture*. Ames, IA.

Corah, L. 1997. *An overview of alliances and packers interest in genetic control*. Proc. Virginia Beef Cow-Calf Conference. Blacksburg, VA: Virginia Cooperative Extension.

Drabenstot, M. 1999. Consolidation in U.S. agriculture: The new rural landscape and rural policy. *Economic Review* (1st Quarter, p. 63).

Field, T.G. 2000. *Balancing the economical and social importance of ruminants with their environmental impact*. Proc. of World Buiatrics Congress. Punta del Estee, Uruguay.

National Cattlemen. 2015. *Directions*. Centennial, CO.

National Cattlemen's Beef Association. 2015. 2016–2020 Beef Industry Long Range Plan: Final Report. Centennial, CO.

Lawrence, J.D., & Mintert, J.R. 2014. Fundamental Forces Affecting Livestock Producers. *Choices*. Agricultural & Applied Economics Association.

NCA. 1989, Oct. 25. *Beef in a competitive world*. NCA Beef Industry Concentration/Integration Task Force Report. NCA: Englewood, CO.

Richie, H.D. 2001. *Understanding alliances: What they are, how they function, and questions to ask*. Michigan State University. Research and Demonstration Report.

Schroeder, T.C., & Kovanda, J. 2003. Beef alliances: Motivations, extent, and future prospects. *Food Animal Practice*, 19: 397–417.

Smith, G.C. 2001. *Increasing value in the supply chain*. Proc. Conference of the Canadian Meat Council. Vancouver, BC.

USDA. 2015. *Packers and stockyard statistical report*. Washington, DC: Grain Inspection, Packers, and Stockyard Administration.

Ward, C.E. 2000. *Characteristics and dynamics of alliances in the beef industry*. Stillwater, OK: Oklahoma State University.

9 The Marketing System

The process of selling and distributing breeding cattle, feeder cattle, fed and non-fed cattle, carcass beef, and retail cuts for the consumer in the United States is accomplished through a dynamic and complex marketing system. Marketing is the physical movement, transformation, and pricing of goods and services, with numerous buyers and sellers working to move cattle and beef products from the point of production to the point of consumption. Price determination is the result of interactions of supply and demand to establish market price levels.

Producers need to understand marketing to produce products preferred by other beef industry segments, including the consumer. They also need to decide intelligently among various marketing alternatives and understand how animals and products are priced if profitable cattle are produced and effectively marketed. The material in this chapter, which covers marketing principles, should be integrated with the marketing information discussed in Chapters 1–8 and 10.

MARKET CLASSES AND GRADES

Market classes and grades have been established to segregate cattle, carcasses, and products into uniform groups based on the preferences of buyers and sellers. The USDA has established market classes and grades to make the marketing process simpler and more easily communicated. Use of USDA grades is voluntary. Some packers and other organizations, including alliances, have their own private grades and brand names that they use in combination with USDA grades. An understanding of market classes and grades helps producers recognize the quantity and quality of products they supply to consumers.

Several national surveys have shown that consumers are not familiar with most USDA grades for beef (e.g., Prime, Choice, Select, and so on). Consumers typically confuse these grades with the intent of the inspection mark. They often assume that all beef is graded and frequently report certain grades when no such grade exists for beef (e.g., "Grade A," "First Cut," and "Grade AA"). Many consumers are familiar with Choice beef, associating it with a desirable product, but usually they do not understand the grading criteria.

Markets for Fed Cattle

Fed cattle are separated into classes based primarily on age and sex (Table 9.1). Age of the animal has a significant effect on tenderness, with younger animals typically producing more tender meat than older animals. Age classifications for meat from cattle are **veal**, **calf**, and **beef**. Veal is from young calves, 1–3 months of age, with carcasses weighing less than 150 lb. Calf is from animals ranging in age from 3 months to 10 months with carcass weights between 150 lb and 300 lb. Beef comes from cattle over 12 months of age with carcass weights higher than 300 lb.

Table 9.1
Official USDA Grade Standards for Fed Cattle and Their Carcasses

Class or Kind	Quality Grades (Highest to Lowest)	Yield Grades (Highest to Lowest)
Beef		
Steer and heifer	Prime, Choice, Select, Standard, Commercial, Utility, Cutter, Canner	1, 2, 3, 4, 5
Cow	Choice, Select, Standard, Commercial, Utility, Cutter, Canner	1, 2, 3, 4, 5
Bullock		1, 2, 3, 4, 5
Bull	No designated quality grades	1, 2, 3, 4, 5
Veal	Prime, Choice, Select, Standard, Utility	Not applicable
Calf	Prime, Choice, Select, Standard, Utility	Not applicable

Source: USDA.

The classes and grades established by the USDA for cattle are based on sex, quality grade, and yield grade, all of which are used in the classification of both live cattle and their carcasses.

The sex classes for cattle are **heifer, cow, steer, bull,** and **bullock**. Sex classes separate cattle and carcasses into more uniform carcass weights and tenderness groups and identify how carcasses are processed. Occasionally, the sex class of **stag** is used by the livestock industry to refer to males that have been castrated after their secondary sex characteristics have developed. **Heiferettes** are those females that are approximately 23–36 months of age. Heiferettes are typically first-calf heifers that have lost a calf or were weaned early. The female is put into the feedlot and typically produces a youthful carcass that has the potential to grade Choice.

Quality Grades

Quality grades are intended to measure certain consumer palatability characteristics, whereas yield grades measure amount of fat, lean, and bone in the carcass. Fed steers representing some of the eight quality grades and five yield grades are shown in Figures 9.1 and 9.2. To a large degree, slaughter cattle of different quality and yield grades reflect visual differences primarily in fatness and to a lesser degree in muscling.

The factors used to determine quality grades are (1) bone maturity, (2) marbling, (3) lean color, and (4) firmness and texture of lean tissue. Marbling and bone maturity are the two most important factors. Marbling—flecks of fat within the rib-eye muscle or intramuscular fat—is visually evaluated at the twelfth and thirteenth rib interface. The nine degrees of marbling vary from abundant to practically devoid.

Maturity classifications are based on the relationship that as animals increase in age and deposit higher levels of connective tissue, the tenderness of the beef from their carcasses is less tender than younger animals. Bone maturity estimates the physiological age of the animal and maturity is determined primarily by the degree of bone ossification. After the carcass has been split into two halves, the dorsal processes of the vertebrae are exposed and then evaluated for degree of ossification. Carcasses classified as A-maturity have the largest amount of cartilage on the tips of the vertebrae processes, while complete ossification of cartilage into bone has occurred in carcasses with E maturity. Carcasses with C, D, or E maturity are eligible for only the Commercial, Utility, Cutter, and Canner grades.

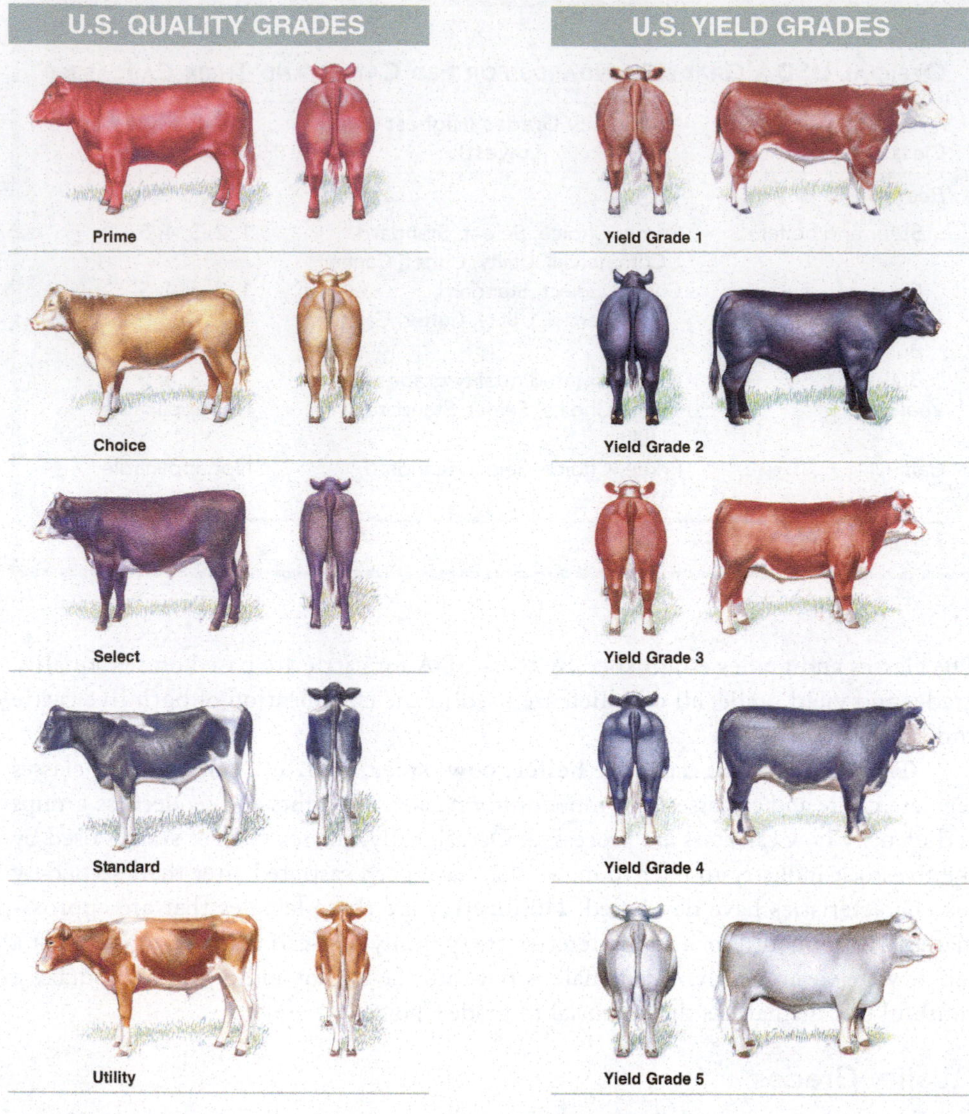

Figure 9.1
Fed cattle types associated with USDA quality grades.
Source: USDA.

Figure 9.2
Fed cattle types associated with USDA yield grades.
Source: USDA.

Figure 9.3 shows how the final carcass quality grade is determined by combining maturity and marbling; Figure 9.4 visually depicts several quality grades. Note in Figure 9.3 that as maturity increases, a carcass usually has to have more marbling to stay in the same quality grade. The vast majority of fed steers and heifers are harvested before reaching 30 months of age. B maturity carcasses are only graded Choice or higher if they have marbling scores of modest or higher. Note that C maturity starts at 42 months of age. Most cows are harvested after 4 years of age and graded Commercial, Utility, Cutter, or Canner regardless of the amount of marbling.

Color of lean is evaluated along with bone maturity to determine the final maturity classification. Color of lean becomes darker with advancing age of the animal. An A maturity carcass is usually a bright, cherry red, whereas an older (C, D, or E) maturity carcass is usually a darker, coarser-textured lean. Occasionally, an A or B maturity has dark lean ("dark cutter"), typically resulting from the animal being stressed shortly before slaughter. Dark cutters, with A or B maturity, are often given a quality grade lower than what the marbling score would indicate.

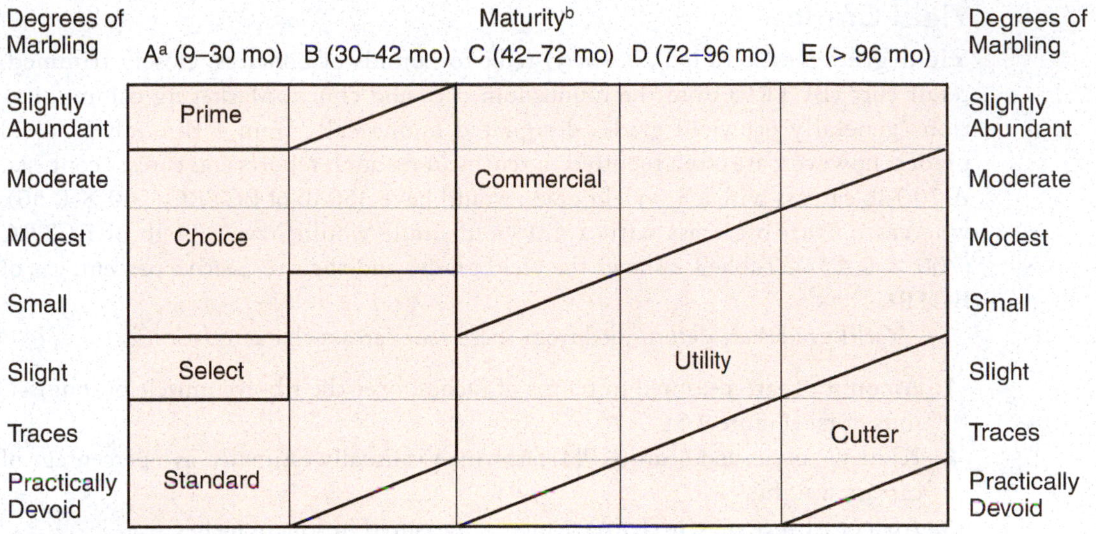

Degrees of Marbling
Maturity[b]
A[a] (9–30 mo)
B (30–42 mo)
C (42–72 mo)
D (72–96 mo)
E (> 96 mo)
Degrees of Marbling
Slightly Abundant
Moderate
Modest
Small
Slight
Traces
Practically Devoid
Prime
Choice
Select
Standard
Commercial
Utility
Cutter

[a] Assumes that firmness of lean is comparably developed with the degree of marbling and that the carcass is not a "dark cutter."

[b] Maturity increases from left to right (A through E).

[c] The A maturity portion of the Figure is the only portion applicable to bullock carcasses.

Figure 9.3
Relationship between marbling, maturity, and carcass quality grade.
Source: USDA.

Moderately Abundant (MdA°)
Slightly Abundant (SlA°)
Moderate (Md°)
Modest (Mt°)
Small (Sm°)
Slight (Sl°)

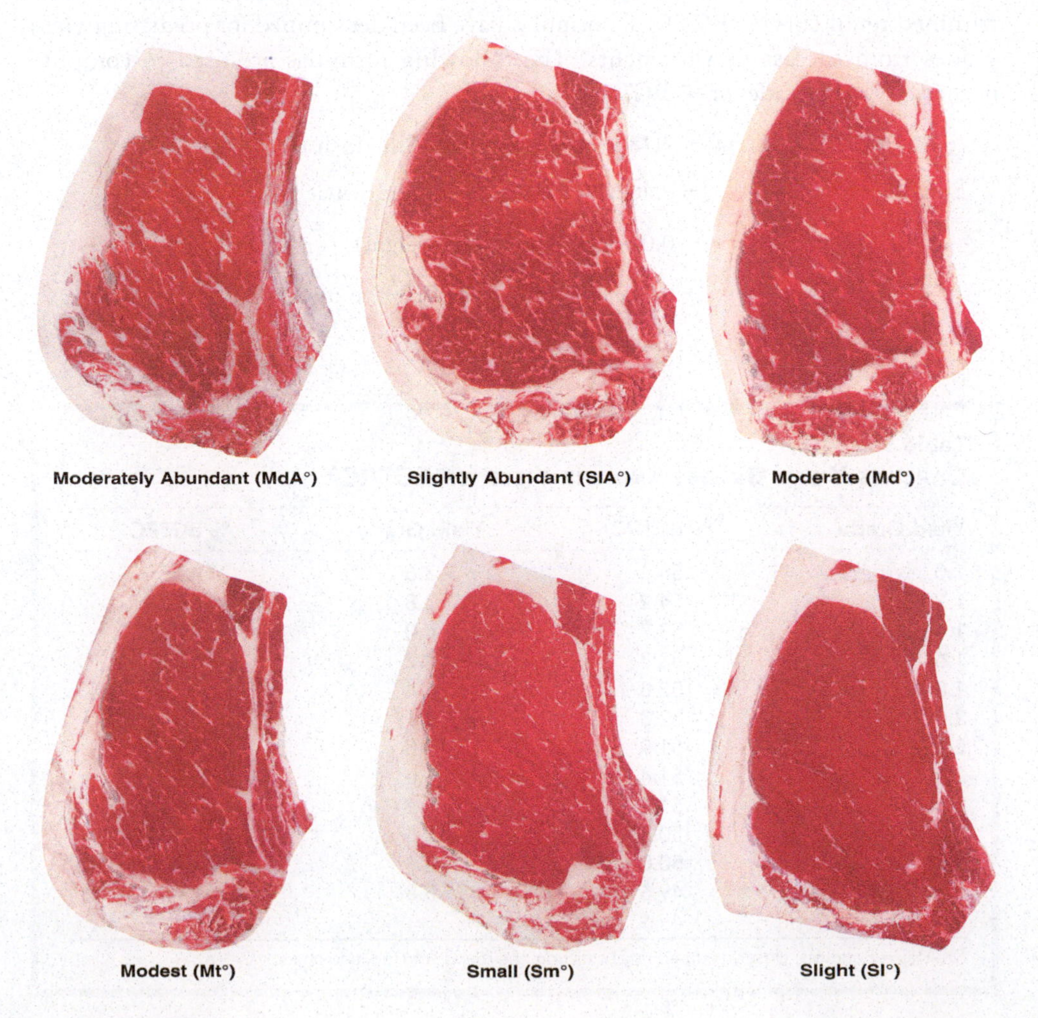

Figure 9.4
Exposed rib-eye muscles (between twelfth and thirteenth ribs) showing several carcass quality grades resulting from combinations of marbling and maturity.
Source: USDA.

Yield Grades

Yield grades (also called *cutability*) refer to pounds of boneless, closely trimmed, retail cuts (BCTRC) from the round, loin, rib, and chuck. Marketing communications generally use yield grades designated numerically from 1 through 5. Yield grades, however, are often reported in tenths in research reports and carcass contests. A 700-lb carcass with a 3.0 yield grade would have 350 lb of BCTRC (700 × 0.50), whereas a 700-lb carcass with a 4.0 yield grade would have 334 lb of BCTRC (700 × 0.477). Table 9.2 shows the yield grades and their respective percentages of BCTRC.

Yield grades are determined from these four carcass characteristics:

1. Amount of fat, measured in tenths of inches, over the rib-eye muscle or longissimus dorsi (Figure 9.5).
2. Kidney, pelvic, and heart (KPH) fat, which is usually estimated as a percentage of carcass weight.
3. Area of rib-eye muscle (REA), which is measured in square inches.
4. Hot carcass weight. Carcass weight reflects amount of intermuscular fat. Generally, as the carcass increases in weight, the amount of intermuscular fat increases as well.

Fatness is the primary factor in determining yield grades. The fat measurement, over the rib-eye muscle, measures most differences in fatness in the carcass (Figure 9.6).

Yield grades were developed to estimate carcass yields of boneless, closely trimmed retail cuts (% BCTRC). Formulas have been determined for predicting yield grades from carcass measurements. The following formulas are used to compute numerical yield grades or % BCTRC.

$$
\begin{aligned}
\%\ \text{BCTRC} = 51.54 &- 5.784\ (\text{inches of fat at 12th–13th rib}) \\
&- 0.462\ (\%\ \text{kidney, heart, and pelvic fat}) \\
&- 0.0093\ (\text{lb hot carcass weight}) \\
&+ 0.740\ (\text{square inches of rib-eye muscle})
\end{aligned}
$$

Table 9.2
Carcass Yield Grades and the Yield of BCTRC[1]

Yield Grade	% BCTRC	Yield Grade	% BCTRC
1.0	54.6	3.6	48.7
1.2	54.2	3.8	48.2
1.4	53.7	4.0	47.7
1.6	53.3	4.2	47.3
1.8	52.8	4.4	46.8
2.0	52.3	4.6	46.4
2.2	51.9	4.8	45.9
2.4	51.4	5.0	45.4
2.6	51.0	5.2	45.0
2.8	50.5	5.4	44.5
3.0	50.0	5.6	44.1
3.2	49.6	5.8	43.6
3.4	49.1		

[1]BCTRC = Boneless, closely trimmed retail cuts from the round, loin, rib, and chuck.

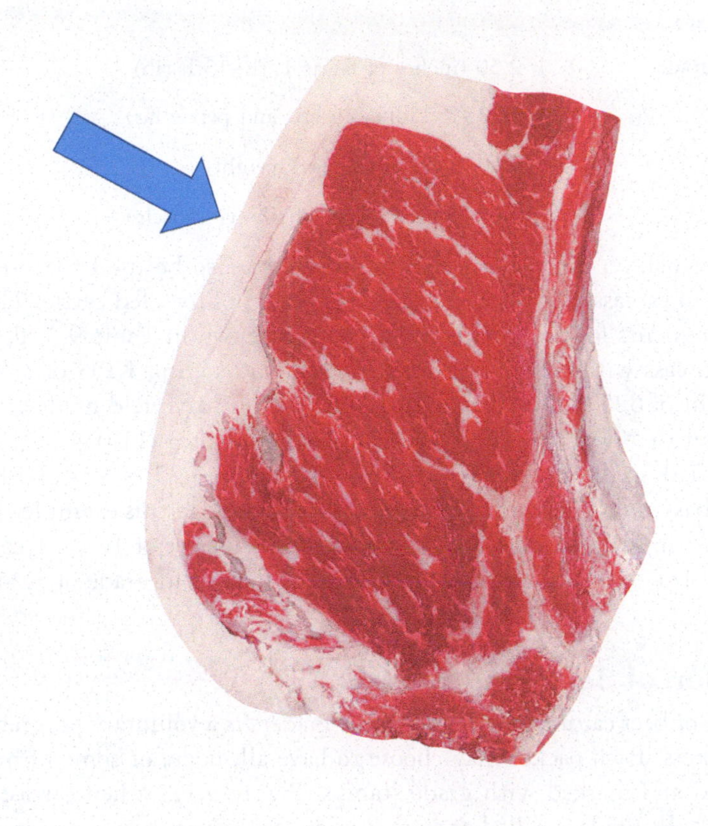

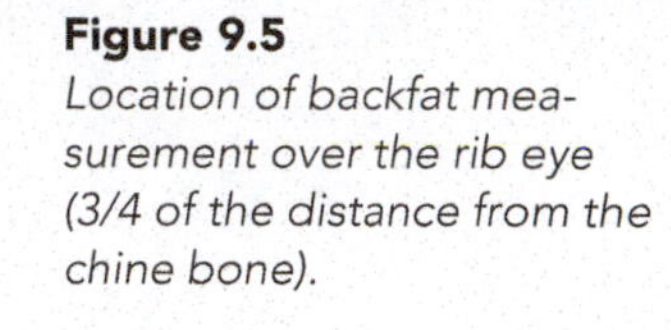

Figure 9.5
Location of backfat measurement over the rib eye (3/4 of the distance from the chine bone).

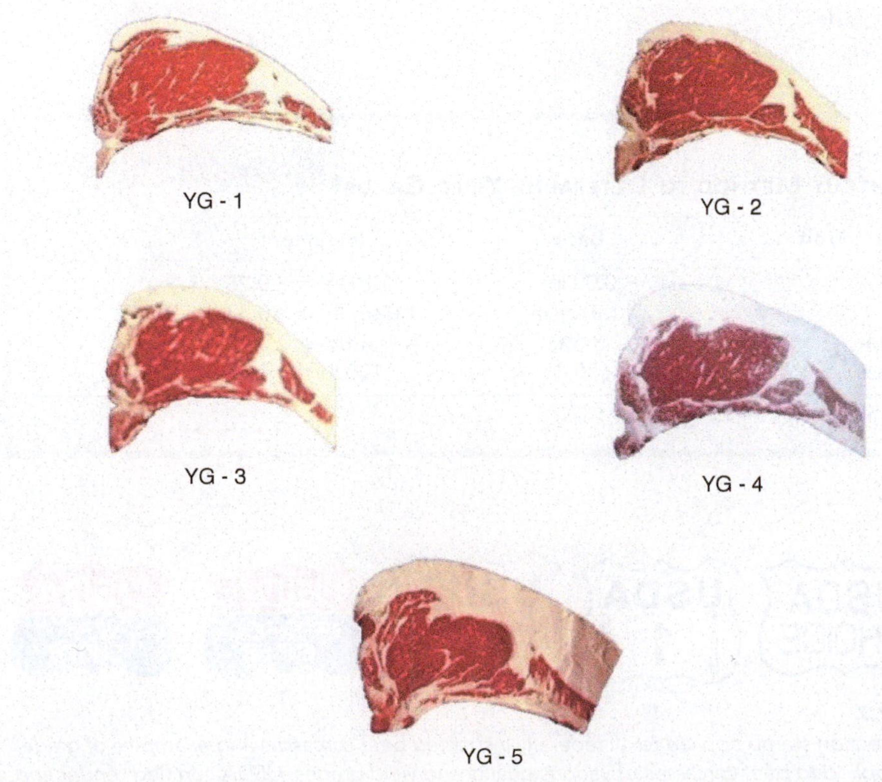

Figure 9.6
The five yield grades of beef.
Source: USDA.

$$\begin{aligned}\text{Yield grade} = 2.50 &+ 2.50 \text{ (inches of fat at 12th–13th rib)}\\ &+ 0.20 \text{ (\% kidney, heart, and pelvic fat)}\\ &+ 0.0038 \text{ (lb hot carcass weight)}\\ &- 0.32 \text{ (square inches of rib-eye muscle)}\end{aligned}$$

An alternative, more simplified method may also be used to determine the yield grades of carcasses or the potential yield grades of live, fed cattle. This method uses the increments for the carcass characteristics shown in Table 9.3. For example, consider a carcass with 0.50 in. of fat, REA of 11.30 sq. in., KPH of 3.50%, and a carcass weight of 625 lb. The starting base is a preliminary yield grade of 2.00, with 0.50 in. of fat (5 × 0.25 + 2.00 = 3.25). A REA of 11.3 (11.0 – 11.3 = 0.30) is one-third of 1.0, so the increment would be –0.11 from 3.25 = 3.14. KPH fat is the same as the base, so KPH has no influence on yield grade in this example. The carcass weight is 625 lb (625 – 600 = +25, which is one-quarter of 100 lb), so the increment is 4 × 1/4 = +0.1, added to 3.14 equals a final yield grade of 3.24 or a yield grade of 3.2.

Distribution of Beef Grades

The grading of beef carcasses by USDA meat graders is a voluntary program available to meat packers. Each packer may choose to have all, none, or some of his beef carcasses graded and marked with grade stamps (Figure 9.7). When carcasses are officially graded, the grade is rolled or stamped onto the carcass.

Carcass quality and yield grade stamps should not be confused with the inspection stamp also shown in Figure 9.7. The inspection stamp verifies the wholesomeness of the meat from a food safety standpoint. Most packers have at least some of their beef marked with grade stamps, usually the carcasses that are most desirable in yield and quality grades.

Table 9.3
Shortcut Method to Determine Yield Grade[1]

Carcass Trait	Base	Increment
Fat	0.00 in.	0.10 in. = ±0.25
REA	11.0 sq. in.	1.0 sq. in. = ±0.33
KPH fat	3.5%	1.0% = ±0.20
Carcass weight	600 lb	100 lb = ±0.40

[1]Assumes a starting yield grade base of 2.00.

Figure 9.7
The inspection stamp applied to all federally inspected beef carcasses and examples of grade marks applied to beef carcasses of various quality and yield grades. USDA certified tender and very tender stamps are also available to those carcasses that meet the applicable American Society for Testing and Materials (ASTM) International tenderness standards.
Source: USDA.

Table 9.4
QUALITY AND YIELD GRADING OF BEEF CARCASSES, 1980–2014

	Quality Grades (%)[1,3]			Yield Grades (%)[1]				
Year	Prime	Choice	Select[2]	1	2	3	4	5
1980	5.9	89	4	2	29	58	10	1.4
1985	3.6	93	3	4	41	50	5	0.6
1990	2.2	82	15	7	46	44	3	0.2
1995	2.4	63	34	11	47	40	2	0.2
2000	3.0	52	36	10	41	36	2	0.2
2005	3.1	57	39	1	40	40	8	1
2010	3.3	64	31	12	40	40	7	1
2014	4.4	69	26	8	36	45	10	1

[1]Percent of cattle graded (~1/3 ungraded up to 1992, ~15% ungraded 1993–97, ~8% ungraded 1998–2004) and 6% ungraded from 2005 to 2015.
[2]Name of grade changed from Good to Select in 1987.
[3]Less than 1% of graded steers and heifers are identified as Standard.
Source: USDA.

Table 9.4 shows the distribution of grades for beef actually graded and stamped during recent years. In 2014, for example, 94% of steer and heifer carcasses were graded and stamped for both quality grade and yield grade; the remaining 6% were deemed to be undesirable for quality or cutability attributes and thus, were either left entirely ungraded or received only a quality or a yield grade stamp. There are regional differences in USDA Quality Grade performance with Nebraska and Kansas fed cattle carcasses having a 20 and 10% advantage over those fed in Texas. Cattle fed in the southern half of the United States, however, have a distinct advantage in yield grade nearly equal to the differences observed in quality grades.

Instrument Grading

USDA grading systems have depended on subjective evaluations by trained graders. As chain speeds increased, the potential for human error was also heightened. Given the significant economic impact of incorrect USDA quality and yield grades on the total return from a pen of fed cattle and the concerns about grader error, the industry invested considerable dollars into the development of instrument grading systems. Ultimately video image analysis became the preferred technology for assessing rib-eye area, USDA yield grade, and marbling score assessment. Using differences in color and texture, these systems are able to distinguish differences in tissue types and amounts (Figure 9.8).

These systems whether used to augment human graders or to replace them entirely provide increased confidence that differences between carcasses are accurately and consistently determined. The use of objective assessment of carcasses for both eating quality and yield offers a powerful tool in the delivery of beef products more clearly matched to the demand of various market targets and allows for the transfer of highly accurate and meaningful data to producers and feeders.

Feeder Cattle Grades

The USDA feeder cattle grades are intended to predict feedlot weight gain and the slaughter weight end point of cattle fed to a desirable fat-to-lean composition. The three criteria used to determine feeder grade are frame size, thickness, and thriftiness.

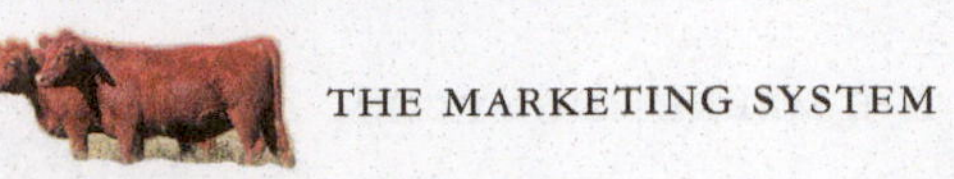

Table 9.5
Slaughter Weights of Large-, Medium-, and Small-Framed Slaughter Cattle at 0.50 Inches of Fat

	Slaughter Weight	
Frame Size	Steers (lb)	Heifers (lb)
Large	> 1,250	> 1,150
Medium	1,100–1,250	1,000–1,150
Small	< 1,000	< 1,000

Frame sizes are used to predict compositional variation and are differentiated into large, medium, and small. Cattle in these categories would be expected to reach U.S. Choice at about 0.50 in. of backfat at different weights. For example, large-, medium-, and small-framed steers would reach this market target at weights of greater than 1,250, 1,100–1,250, and less than 1,100 pounds, respectively (Table 9.5).

Thickness scores are used to help distinguish between cattle of differing levels of muscularity and are categorized into four classifications (No. 1, No. 2, No. 3, and No. 4). No. 1 feeders are typically from beef-type breeds, are moderately thick, and show particular thickness through the forearm, stifle, and gaskin. No. 2 feeders are considered slightly thick, while No. 3 feeders are thinly muscled. No. 4 feeder cattle are still thrifty, but less muscular than No. 3s. Dairy cattle typically fall into categories 3 and 4. Any feeders that are unthrifty or double-muscled are classified as USDA Inferior. These are cattle not expected to perform normally and may be of any combination of thickness and frame size. Frame and muscle classifications are combined to create 12 grades (in addition to Inferior). These grades are reported as Large No. 1, Large No. 2, and so forth.

Although frame size and ability to gain weight in the feedlot are apparently related in the sense that large-framed cattle usually gain more rapidly than other frame-size cattle, frame size appears to more accurately predict carcass composition or yield grade at different slaughter weights than gaining ability.

MARKETING COWS AND BULLS

A frequently overlooked source of cash flow is market cows and bulls that have been deemed to be unacceptable as breeding animals. The profitable marketing of these cattle can make the difference between profit and loss for a cow-calf enterprise. In fact, 15–25% of gross revenue to the average cow-calf enterprise originates from the sale of market cows and bulls. This value is particularly important in low margin years.

Once a producer understands the potential value of market cows and bulls, it is important to implement a proactive management plan to ensure that this value is maximized. By utilizing appropriate husbandry, handling practices, and management protocols, beef producers can prevent quality defects. One of the core strategies in this process is to closely monitor herd health and to market in a timely fashion.

The following recommendations can be useful in developing an effective merchandising plan for market cows and bulls.

1. Recognize and maximize the value of market cows and bulls.
2. Be proactive to assure the safety and integrity of beef products.
3. Use appropriate management and handling practices to prevent quality defects.
4. Closely monitor herd health and market cull cattle in a timely manner.

To facilitate the implementation of these four directives, a Quality Assurance Marketing Code of Ethics has been developed for the beef industry. The tenets of the Marketing Code of Ethics are as follows:

Only market cattle that:

- Do not pose a known public health threat
- Have cleared proper withdrawal times
- Do not have a terminal condition
- Are not severely disabled
- Are not severely emaciated
- Do not have uterine/vaginal prolapses with visible fetal membranes
- Do not have advanced eye lesions
- Do not have advanced lumpy jaw

Cattle will be humanely gathered, handled, and transported in accordance with accepted animal husbandry practices. When appropriate, cattle will be humanely euthanized to prevent suffering and to protect public health.

Several strategies can be useful in implementing a marketing plan for non-fed cows and bulls, including adding weight to thin cows prior to selling (optimal BCS at sale is a 6), marketing older cows before they begin to lose body condition and fail to rebreed, using annual pregnancy checks to identify non-productive females, and evaluating opportunities to price cows directly to a packer.

Markets for cows and bulls typically get stronger in the spring months following the large run of cows that are sold in October through December. As such, producers need to consider a short retained ownership period. If cheap feed sources are available, a 2- to 3-month retained ownership may well be advisable (Table 9.6). Retained ownership of cows from November to February returned a profit in almost every year from 1980 to 2015.

A number of factors impact the value of market cows including pregnancy status, fill condition, muscle score, frame size, age, and health status. Table 9.7 highlights the price differences of market cows (n = 13,104) sold in Arkansas auction markets during 2001.

Table 9.6
Effect of Retained Ownership on Carcass Traits and Value of Cull Cows

Trait	0 Days	42 Days	87 Days
Live weight (lb)	1,164	1,253	1,305
Carcass weight (lb)	546	631	741
Dressing percent (%)	47.0	54.9	56.8
Fat thickness (in)	0.11	0.24	0.49
Carcass value ($)	477.53	585.50	656.42

Source: Adapted from Lardy (2002).

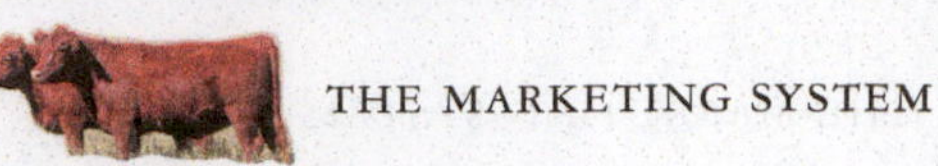

Table 9.7
FACTORS AFFECTING SALE PRICE OF MARKET COWS

Factor	Price ($/cwt)	% of Cows
Pregnancy Status		
3rd trimester	+0.86	NA
2nd trimester	+0.68	NA
1st trimester	+0.69	NA
Open	par	NA
Not pregnancy checked	−0.52	NA
Fill Condition		
Gaunt	−0.95	11.3
Shrunk	+0.26	35.1
Average	par	38.1
Full	−1.82	13.2
Muscle Score		
1	−1.47	46.7
2	par	42.1
3	−1.99	11.2
Frame Size		
Large	+1.12	44.6
Medium	par	43.0
Small	−3.79	12.4
Health Status		
Healthy	par	92.4
Lumpy jaw	−5.39	2.1
Lame	−6.97	1.8
Sick	−14.36	1.6
Bad eyes	−14.55	2.1
Age		
2	par	0.3
3	−1.50	0.7
4	−2.17	1.9
5	+6.60	8.7
6	−3.47	9.3
7	−3.83	14.1
8	−4.91	29.5
9+	−7.65	35.6

Source: Based on Troxel et al. (2002).

MARKET CHANNELS

Market channels are the pathways through which cattle move from farm or ranch to feedlots, then to packing plants, and eventually to consumers. Figure 9.9 shows the marketing channels currently being used for cattle. The market channels available to cattle producers have evolved through a series of significant transitions, from the cattle drives of yesteryear to the utilization of informational technologies available today.

The auction market is most important in the marketing of feeder cattle and cull breeding stock. Direct selling (also referred to as *country selling*) is most important for fed steers and heifers. In this situation, finished cattle are being marketed directly from the feedlot to the packer. Figure 9.9 does not show the market channels for breeding cattle that are marketed primarily through auctions (private or consignment) and that are sold directly to commercial producers through private treaty. Producers should understand the various market channels and how they differ in terms, availability, marketing costs, and buyer competition.

Table 9.8 illustrates the market channels utilized by various cow-calf enterprises. Small and midsize herds rely more heavily on auction markets than do large herds that are more likely to utilize direct, forward contracting, and carcass basis sales. In one southeastern study of feeder cattle markets, 81.2% of the feeders sold at local or regional auction markets were marketed as single animal lots. Only 1% of the feeders were sold in lots of six or more. This is of significance given that approximately 53%

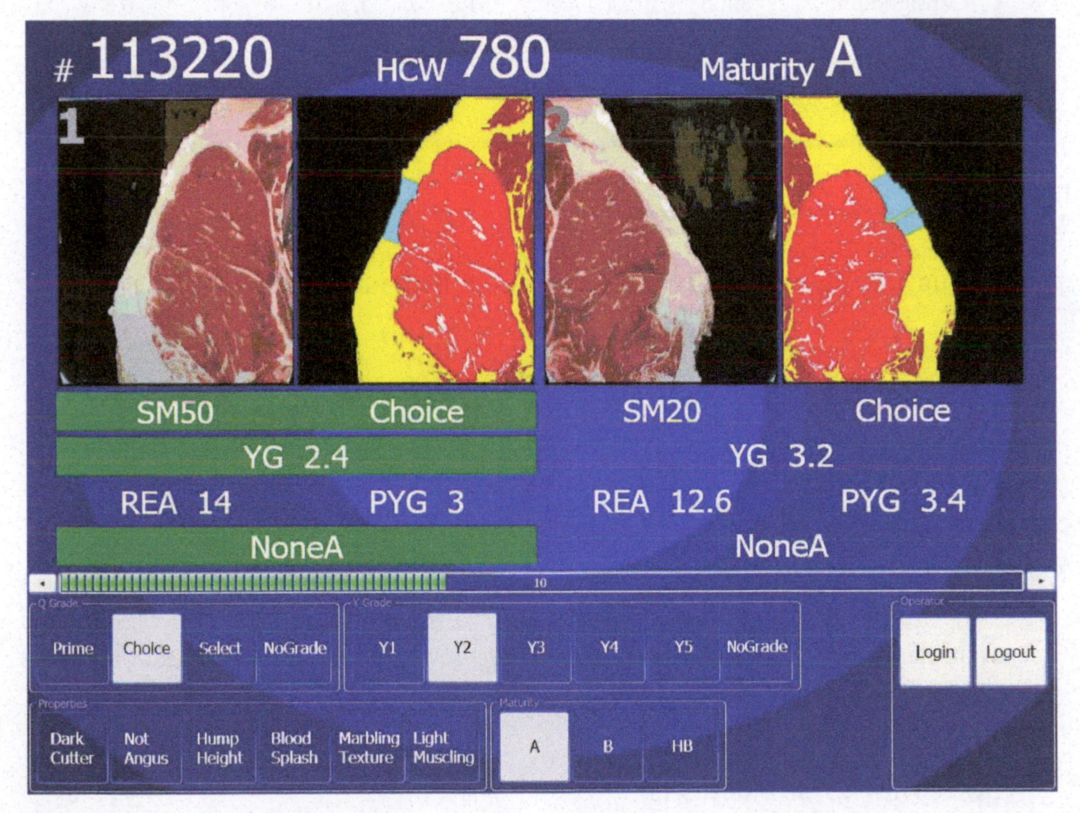

Figure 9.8
Instrument grading provides a consistent evaluation of both yield and quality attributes.

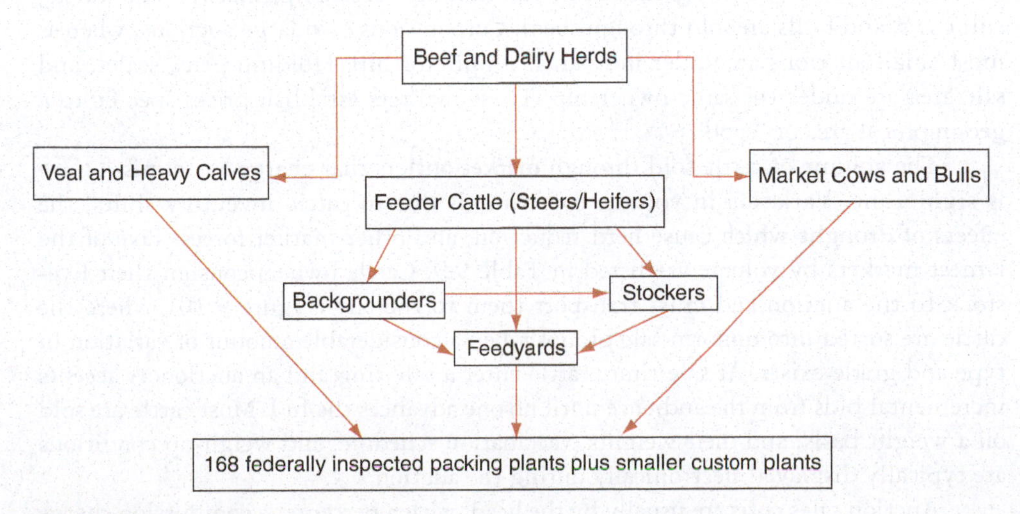

Figure 9.9
Flow of cattle through marketing channels.

Table 9.8
PERCENTAGE OF OPERATIONS MARKETING CATTLE BY VARIOUS METHODS

	All Operations		Herd Size (2007–08)			
Method	1997	2007–08	<50	50–99	100–299	≥ 300
Auction	85	82	86.9	84.8	78.9	49.9
Direct-video	0.7	1.1	0.4	0.4	1.4	7.0
Direct-private treaty	10.4	11.6	9.5	10.1	13.8	28.7
Consignment	1.2	1.1	1.2	1.3	1.1	0.5
Forward contract	0.5	0.3	0.2	0.3	2.6	5.6
Carcass basis	1.3	1.0	1.0	1.9	1.3	6.8

Source: NAHMS, 1997 and 2008.

Table 9.9
AUCTION MARKETS—CATTLE RECEIPTS—1,000 HEAD—(1996–2013)

	1996	2003	2007	2011	2013
Oklahoma City, OK	554	573	422	468	409
Sioux Falls, SD	239	221	82	30	46
South St. Paul, MN	172	146	65	60	35
South St. Joseph, MO	118	109	52	94	90
West Fargo, ND	108	49	115	104	156

Source: USDA.

of the beef cattle enterprises and 47% of the beef cows are located in the 13 southeastern states. Furthermore, 83% of southeastern beef cattle enterprises had fewer than 50 head of cattle.

Spot Markets (Auctions)

The **auction market** is made up of approximately 2,000 livestock auctions, sometimes called sale barns, located throughout the United States. Cattle auctions are more numerous in areas with the greatest cow-calf numbers because primarily feeder cattle, cull cows, and bulls are sold through them. Corporations own large auctions, whereas most small auctions are under individual proprietorship. Holding pens, scales, and sale area are under the same ownership. These markets establish prices specific to a geographical area or "spot".

The volume of cattle sold through market outlets that charge a commission fee is significant. Variation in volume over time is due to cattle inventory shifts, the effects of drought which cause herd reduction, and other market forces. Five of the largest markets by volume are listed in Table 9.9. Cattle owners consign their livestock to the auction and must transport them to the site (Figure 9.10), where the cattle are sorted into uniform sale groups when a considerable amount of variation in type and grade exists. At sale time, cattle enter a sale ring and an auctioneer accepts incremental bids from the audience until no one advances the bid. Most cattle are sold on a weight basis, and their weights, vaccination schedule, and weigh-up conditions are typically displayed electronically during the auction.

Auction sales costs are usually by the head, with a percentage commission charge on the gross revenue ranging from 1.5% to 4% depending on the market and the class

Figure 9.10
Traditionally, cow-calf producers have utilized local auction markets as means to sell feeder cattle and breeding stock that have been culled from the herd. This form of marketing requires transport of cattle from the ranch or farm to a local or regional auction adding a layer of cost.

of livestock sold. Additional representative costs per head are feed (cost plus 25%), brand and health inspection ($0.64), insurance ($0.10), and a beef checkoff program fee ($1). Auction markets, which handle livestock involved in interstate movement, are considered public markets and as such are subject to regulations specified by Packers and Stockyards (P&S), an agency of the Agricultural Marketing Service of the USDA. P&S ensures that markets are properly bonded, prescribes rules for fair trade and competition, and periodically verifies accuracy of scales used to establish sale weight.

The advantages and disadvantages of this market outlet are listed below:

Advantages:	Disadvantages:
• competitive bidding • available 48–50 weeks out of the year • easy access to willing sellers and buyers. • prompt cash payment by bonded agent • public and open pricing • regulated by the U.S. government • no minimum lot size	• seller has no control over prices • overhead costs in commission and yardage are incurred • shrink losses may occur • producers may find it hard to establish a reputation • prices are uncertain • biosecurity risk is higher • competitiveness of bidding depends on the number of buyers

Cooperative Marketing

Small producers often find themselves in the most difficult position in regard to marketing because an individual producer cannot attract on-farm buyers due to the small number of cattle that can be offered for sale at any one time. Therefore, cooperative efforts become essential to broaden marketing opportunities.

The formation of marketing associations allows groups of small producers to join together to function as a larger supplier of feeder cattle or replacement females. These producers agree to a standardized health, genetic, and management practice as a means to create larger pools of feeder cattle. As a result, they are able to attract not only more attention from buyers, but price premiums as well. Compared with conventional marketing of small groups of calves via local auctions, pooled marketing groups have been able to capture premiums in excess of $8, $6, and $5/cwt for 500, 600, and 700 lb steers, respectively. The advantages and disadvantages are summarized below:

Advantages:	Disadvantages:
• Create large and uniform lots of cattle • Cost savings for buyers are passed along to sellers • Larger numbers consigned attract more competition	• Grading, sorting, weighing, and penning before sale creates logistics challenges • Individual producers lose their identity • May be difficult to get multiple producers to agree on terms of sale

Direct Marketing

Historically, direct markets were structured around order buyers and commission representatives who functioned as intermediaries between the seller and the ultimate buyer. This created another cost in the marketing process and these systems were eventually eroded by sellers wanting to avoid these costs and by the speed of information transfer facilitated by the spread of high speed computing.

Examples of direct marketing are commercial producers who sell their feeder cattle directly to cattle feeders and packer buyers who purchase slaughter cattle directly from the feedlot. Direct marketing of fed steers and heifers is accomplished in three ways: (1) cattle are sold on a live weight basis, where buyers base their live price on estimated carcass weight, quality grade, and yield grade; (2) dressed beef basis(flat price on carcass weight regardless of quality grade or yield grade); and (3) grade and yield (priced on quality grade and carcass weight). Examples of these three methods of selling fed cattle are shown in Table 9.10. In live weight and dressed weight pricing, sellers are paid for the average performance of their cattle. As a result, the highest performing cattle receive a sub par price while the poorest performing animals receive a premium.

The feedlot develops a show list of cattle to be presented to order buyers at the beginning of each week. Buyers evaluate the cattle, estimate their value, and then either accept or reject the sellers' asking price. Date, time, and cost of delivery are typically a part of the negotiation. It is becoming more common for a majority of the transactions in any given week to be completed within a 1- to 3-day range.

Approximately two-thirds of fed cattle are currently sold on a carcass basis, an increase from the approximately 20% in 1971. Nebraska, Iowa, and Colorado account for about two-thirds of all carcass grade and weight marketings of steers and heifers. Packers in Texas, Iowa, Nebraska, Minnesota, and Wisconsin account for approximately three-fifths of all cows and bulls purchased on a carcass basis.

Forward Contracts and Formula Pricing

Transactions may also occur when cattle are priced based on future performance and tied to a running average price or some other system of establishing a future value.

Table 9.10
Methods of Selling Fed Cattle[1]

Method Example			
Live Cash Sale			
Cash bid of $74			
Live weight	1,350 lb		
	0.04% shrink		
	54 lb		
	1,350 lb		
	−54 lb		
Pay weight	1,296 lb		
	× $130 per cwt		
	$1,684.80		
Freight	(generally, when cattle are sold live, packer pays the freight)		
	$1,684.80 net		
Dressed Beef Sale			
Beef bid of $116/cwt of carcass			
Live weight	1,350 lb	1,350 lb	
	× 64% actual dressing percentage	× 62% actual dressing percentage	
	864 lb carcass weight	837 lb carcass weight	
	× $200 per cwt	× $200 per cwt	
Freight	$1,728.00	$1,674.00	
	−7.30 (transport–120 mi.)	−7.30 (transport–120 mi.)	
	$1,720.70 net received	$1,666.70	
	$54.00 advantage for higher dressing percentage.		
Grade and Yield	Choice	Select	Standard
Prices/cwt: Choice ($200), Select ($190), Standard ($180)	1,350	1,350	1,350
Live weight	× 0.63	× 0.63	× 0.63
Dressing percentage	850 lb	850 lb	850 lb
Carcass weight	× $ 200	× $ 190	× $ 180
Freight to packing plant	$1,700	$1,615	$1,530
Net received	−7.30	−7.30	−7.30
	$1,692.70	$1,607.70	$1,522.70
	× 0.60	× 0.35	× 0.05
% of total head in each quality grade	$1,015.62	$562.69	$76.13
Average net received (per head for the 100 head)	= $1,654.44		

Source: USDA.

Marketing agreements involve a longer-term relationship for the ongoing delivery of cattle where the number of animals, the date and conditions for delivery, performance specifications, and pricing method are predetermined. These arrangements are typical of some alliances.

Forward contracting may be part of a marketing agreement or a one-time transaction. In this system, price is either fixed or based on some publicly reported future price. The base price is often determined from the futures market. In this case, a buyer and seller agree to a differential, or basis, from the futures market for a specified contract month. Premiums and discounts are then applied. Packer feeding is the situation where a packer owns outright or in partnership part of the cattle scheduled for delivery to the plant. Packer-owned fed cattle account for approximately 5% of all fed cattle marketings.

Electronic Marketing

Electronic marketing and market information transfer have significantly changed the process of cattle marketing. By utilizing the various forms of electronic transmission of information, buyers and sellers can make decisions based on a host of relevant data. Unlike the comparatively long and slow marketing process of the past, today large numbers of cattle can be sold within only a few hours and without the cost of transporting them to a centralized marketing facility.

Most electronic marketing systems share the following characteristics: (1) pricing is determined at a single location or over a single communication system, (2) the cattle are not moved from the seller's farm until they are sold, and (3) buyers can participate without being at the location where pricing occurs. To date, the most successful form of marketing cattle by electronic means has been the video auction. Video auctions are conducted by showing buyers a videotape (2–4 minutes) taken at the farm or ranch of the cattle being offered for sale. The audio portion of the tape describes weight, location, and background information on the cattle. The seller also states the weighing location and various options for delivery dates. A satellite hook-up allows the videotape to be shown to buyers gathered in several locations, and then an auctioneer sells the cattle to the highest bidder. Some individuals or companies can receive the video auction by tuning their own satellite dish to the correct channel. After the sale, buyer and seller arrange for the shipping of the cattle. This step is usually handled by the agent who represents the buyer.

Web-based livestock marketing is another innovative alternative used by some in the industry. While some sites are simply listings of cattle for sale, others conduct real-time Internet auctions where several hundred buyers might be logged on to a site to bid for the available cattle. In some sale situations, buyers have the option of attending an event live, bidding via the web, or bidding over the phone. This level of flexibility extends market reach and provides additional opportunities to assure competitiveness among buyers.

Advantages and disadvantages are summarized below:

Advantages:	Disadvantages:
• exposure to the largest number of potential buyers • reduced buyer cost passed along to seller • allows for direct buyer-to-seller transportation • flexible delivery coordination; for instance, cattle can be sold in spring and summer months for delivery in fall • seller can reject the price without having incurred the cost of moving cattle to a sale facility	• requires truckload or larger consignments of relatively uniform cattle • requires some investment in setting up the consignment paperwork, arranging for videoing of cattle, and assuring loadout conditions meet the contract • incorrectly estimating calf delivery weight may have economic consequences

GRID PRICING

Another pricing mechanism that is widely used in the marketing of fed cattle is known as grid pricing. Grid pricing is a more complex approach that applies a series of premiums and discounts based on Quality Grade, Yield Grade, carcass weight, and level of carcass defects to a pre-established base price. While similar to Grade and Yield pricing, grid pricing typically determines base price from some average price in the week prior to delivery. The Choice–Select mix and other premiums and discounts are calculated from the plant average.

In grid pricing, the individual performance of each animal in a particular lot is evaluated and rewarded accordingly. The goal of most grids is to reward cattle that grade Choice or better, have USDA Yield Grades of 1 or 2, have carcass weights between 600 and 900 pounds, and are free from any defects such as dark cutter. Producers must be careful to balance the pursuit of premiums with other factors that influence profitability. Producers who enjoy success in marketing cattle via a grid system are able to avoid the discounts (< 3%), optimize pounds sold, and produce cattle that are capable of being 70% Choice and 70% Yield Grades 1 and 2.

In most grid-pricing systems, premiums are paid for cattle that have USDA Quality Grades in excess of a plant or regional average, and that have a higher percentage of USDA Yield Grades 1 and 2. The specific premiums paid depend on the emphasis in the product line—superior palatability, superior cutability, or a combination of the two. A series of discounts are also applied to carcasses that fall outside of a pre-determined carcass weight range, that have excessively large or small rib-eye area, that fail to make A-maturity, or that have undesirable lean color or other defects. The discounts associated with are typically larger on a per-weight basis as compared with the premiums (Table 9.11). As a result,

Table 9.11
National Carcass Premiums and Discounts for Slaughtered Steer and Heifers for the Week of April 25, 2016

Value Adjustments	$/cwt	
Quality	**Range**	**Simple Avg.**
Prime	0.00–22.00	12.43
Choice	0.00–0.00	0.00
Select	(2.00)–(12.00)	−9.70
Standard	(43.00)–(10.00)	−26.54
Certified programs		
Avg. Choice/Higher	2.00–8.00	3.69
	—	
Bullock/Stag	(55.00)–(15.00)	−32.78
Hardbone	(55.00)–(20.00)	−34.08
Dark cutter	(55.00)–(20.00)	−34.27
Over 30 months	(40.00)–0.00	−16.17
Cutability[1]		
Yield Grade/Fat/Inches		
1.0–2.0, <0.1″	0.00–8.00	3.69
2.0–2.5, <0.2″	0.00–5.00	1.63
2.5–3.0, <0.4″	0.00–5.00	1.21
3.0–3.5, <0.6″	0.00–0.00	0.0
3.5–4.0, <0.8″	0.00–0.00	0.0
4.0–5.0, <1.2″	(15.00)–(8.00)	−11.67
5.0/up, > 1.2″	(20.00)–(10.00)	−16.58
Weight		
400–500 lbs	(40.00)–0.00	−28.46
500–550 lbs	(40.00)–0.00	−20.50
550–600 lbs	(20.00)–0.00	−8.00
600–900 lbs	0.00–0.00	0.00
900–1,000 lbs	(15.00)–0.00	−1.79
1000–1,050 lbs	(25.00)–0.00	−7.50
Over 1,050 lbs	(50.00)–(5.00)	−23.93

Based on individual packer's quality, cutability, and weight buying programs. Values reflect adjustments to base prices, dollars per cwt, on a carcass basis.
[1]If yield grades are not available, yield differentials may be based on fat at 12th rib using a constant of average rib-eye area and muscling for carcass weight and KPH. Superior or inferior muscling may adjust lean yield.
Source: USDA Livestock and Grain Market News.

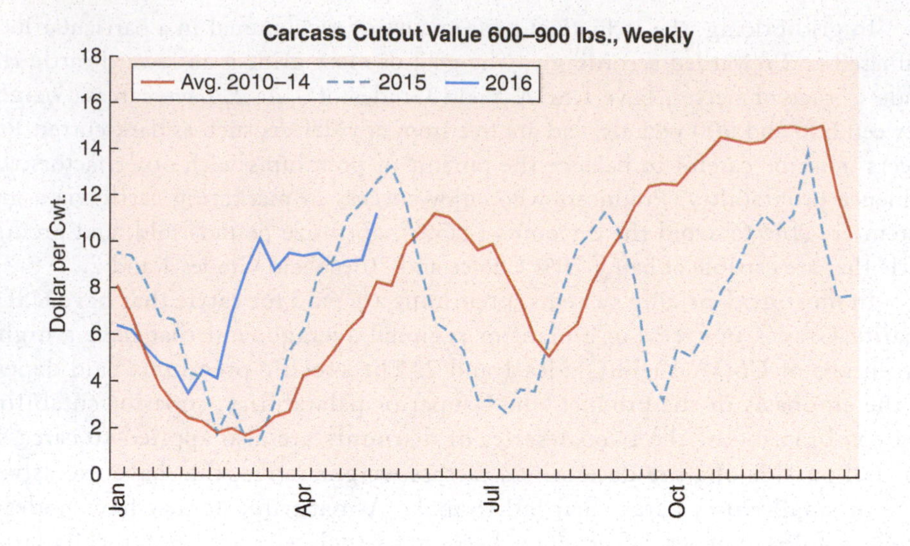

Figure 9.11
Choice minus Select beef prices (seasonal).
Data source: USDA Agricultural Marketing Service (Market News); compiled by Livestock Marketing Information Center.

cattle producers are advised to focus on avoiding the discounts as their first priority, and to then focus on the fine-tuning of their programs to better capture premiums.

In a grid-pricing formula, the Yield Grade price spreads tend to stay reasonably stable over time. The greatest value shortfall comes with Yield Grades 4 and 5 carcasses that may incur a \$10–20/cwt carcass discount. The Choice–Select price spread is much more volatile and may move \$15 to \$20 per cwt over the course of a single year. This spread is typically highest in the late fall and lowest in the early spring, although market conditions may shift these trends in some years (Figure 9.11). Understanding the seasonal pattern of premiums and discounts is important in determining market strategy.

Monitoring the Choice–Select price differential is a key to making timely marketing decisions based on the genetic marbling potential of cattle. If the variation in this price spread ranges from \$0–20 per cwt over a period of time, then the difference between a Choice 800 lb carcass versus one that is Select ranges from \$0 to \$160 per head.

Because premiums and discounts are applied to a base price, the negotiation to determine base price is of importance. Understanding how the grade base is calculated is also important because it determines how the Choice–Select spread will be applied to price calculation. Packing plants in the northern Great Plains tend to run a higher percent Choice than those in the southern Great Plains. The Select discount is calculated by multiplying the grade-base percentage by the Choice–Select spread.

For example, if the plant average is 60% choice and the Choice–Select spread is \$5.00/cwt, the Select discount is $0.60 \times \$5.00 = \$3/\text{cwt}$. The difference between the spread and the Select discount equals the Choice premium ($\$5.00 - \$3.00 = \$2.00$). As the grade-base increases, and/or as the Choice–Select spread widens, the discounts associated with Select grading carcasses become more severe. Figure 9.12 illustrates this relationship.

Table 9.12 shows the performance of two pens of cattle under a standard grid-pricing structure. Pen 1 receives a \$37.74 advantage because they avoided discounts and had more desirable Quality and Yield Grades.

Figure 9.13 documents the trend in fed cattle marketing strategies shifting toward formula and forward pricing. In addition, packers own about 5% of their total harvest—a number that has been nearly unchanged for the past decade. The growth of

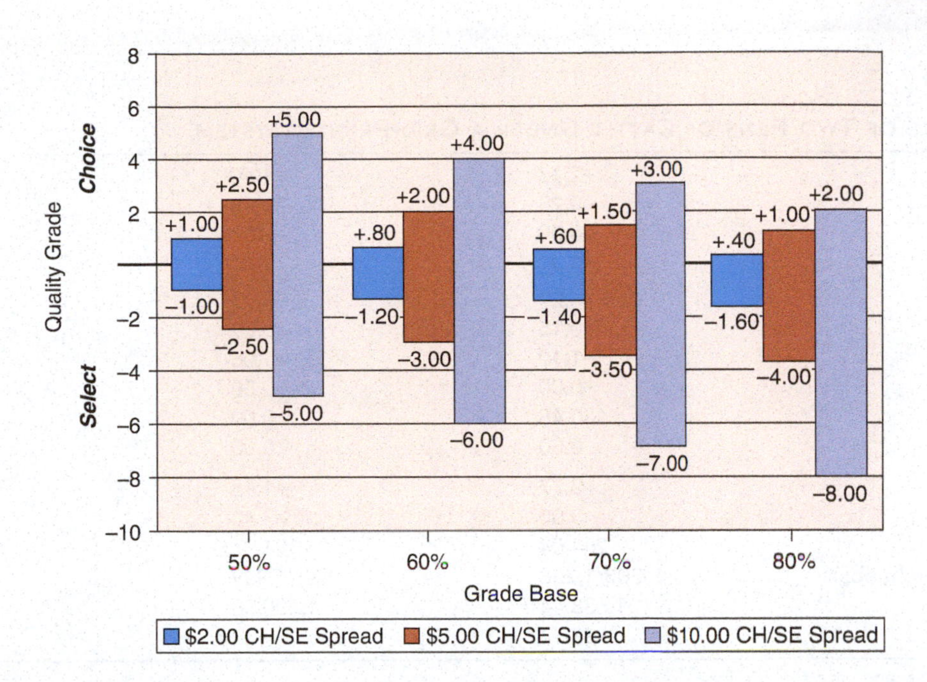

Figure 9.12
Effect of grade base and Choice–Select spread on the premiums and discounts associated with quality grade.
Source: Based on Field, Tatum, and Kimsey (1998).

Table 9.12
PERFORMANCE OF TWO PENS OF CATTLE UNDER A GRID-PRICING SYSTEM

		Premiums and Discounts		
Example Program	Marketing Specifications	Quality Grade	Yield Grade	
Live price	$127/cwt	Prime +$12.00 over Choice	YG1	+$3.00
Base carcass price	$200/cwt	Certified program +$3.50 over Choice	YG2	+$2.00
Base dressing %	63.5	Choice +$2.80	YG3	$0.00
Out discount[1]	$20.00/cwt	Select −$4.20	YG4	−$12.00
		Standard −$25.00	YG5	−$16.00
Grade base	60% Choice			

Carcass Performance of Two Pens of Cattle

	Pen 1	Pen 2
% Prime	9	0
% Choice	76	50
% Select	6	40
% No roll	0	10
% of Choice for certified program	9	0
% YG 1	5	3
% YG 2	55	53
% YG 3	36	44
% YG 4	4	0
% YG 5	0	0
% Outs[2]	0	17
Dressing percentage	64.8	62.0

Computation of Price Under a Grid System

	Pen 1	Pen 2
Base Price	$200.00	$200.00
Out allowance	0.50	0.50
	200.50	200.50
Prime	+1.08	0.00

(continued)

Table 9.12
(Continued) Performance of Two Pens of Cattle Under a Grid-Pricing System

Choice	+2.13	+1.40
Select	−0.25	−1.68
No roll	0.00	−2.50
Certified	+0.31	0.00
Quality Adjustment	+3.27	−2.78
YG 1	+0.15	+0.09
YG 2	+1.10	+1.06
YG 3	0.00	0.00
YG 4	−0.48	0.00
YG 5	0.00	0.00
Yield Adjustment	+0.77	+1.15
Outs Adj.	0.00	−3.40
Adjusted price ($/cwt)	204.04	194.97
Ave. carcass weight (lbs)/live weight (lbs)	803/1,238	821/1,329
Gross return ($/hd)	1,638.44	1,600.70
Difference ($/hd)	+37.74	

[1]Dark cutters, carcass weights outside the range of 600 to 950 lbs.
[2]Weight nonconformance.

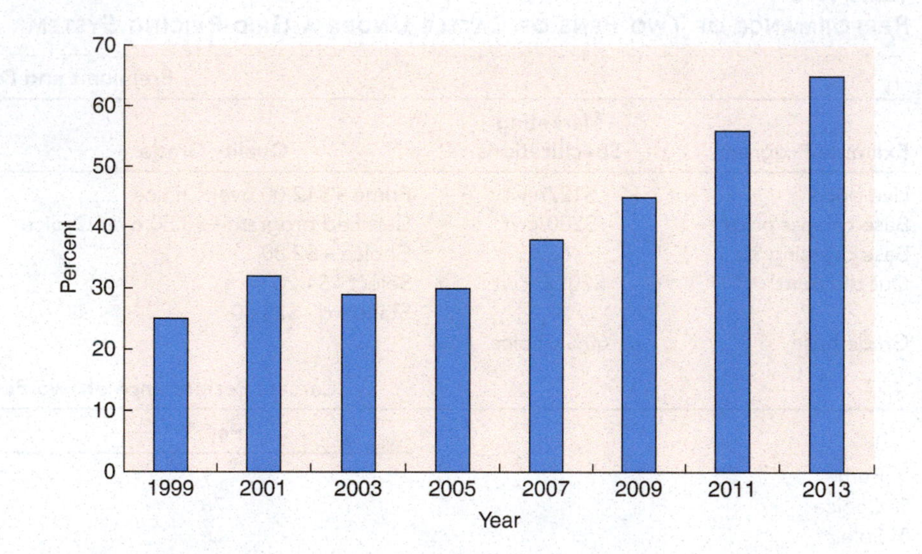

Figure 9.13
Percentage of fed cattle marketing via formula and contract pricing.
Source: USDA.

forward pricing and formula pricing is due to the desire of industry participants to reduce risk in the midst of volatility. As cow-calf producers and feeders evaluate marketing alternatives such as grid pricing, several factors must be considered. An entire set of cattle should not be subjected to a new pricing system without the knowledge of how they might perform. Therefore, data should be collected on a sample of cattle to determine the risks that may be incurred. Producers should also determine if there are any up-front fees or membership charges that are associated with participation in the grid.

Capturing premiums must be balanced with assuring desirable levels of feedlot performance and weight to achieve maximum profitability. As Table 9.13 illustrates, heavier carcasses that receive lower prices on a per pound basis may yield more gross revenue than lighter weight carcasses that receive significant premiums.

Table 9.13
GROSS REVENUES FOR VARIOUS CARCASS WEIGHTS AT DIFFERING PRICES

	Carcass Price ($/cwt)					
Carcass Weight	200	195	190	185	180	175
650	1,300	1,267.5	1,235	1,202.5	1,170	1,137.5
700	1,400	1,365	1,330	1,295	1,260	1,225
750	1,500	1,462.5	1,425	1,387.5	1,350	1,312.5
800	1,600	1,560	1,520	1,480	1,440	1,400
850	1,700	1,657.5	1,615	1,572.5	1,530	1,487.5
900	1,800	1,755	1,710	1,665	1,620	1,575

Note: A heavy carcass can receive a lower price and still generate more gross revenue than a higher-priced, but lighter carcass.

MAJOR FACTORS AFFECTING CATTLE PRICES

The major goal of the beef industry—efficient production of a highly palatable product with a profitable return to those who produce it—implies that two primary factors affect profitable beef production: (1) efficiency of production (or cost of production) and (2) the price received for the product. Although there are limits to improving efficiency of production, the current beef industry has the opportunity to make tremendous strides in cost-effective productivity, especially the cow-calf segment. Even at optimum levels of production efficiency (when numbers and weights are at the least cost), if market price is not high enough to cover costs, profitability will not be realized. Producers need to understand the advantages and disadvantages of different markets as well as the factors affecting the supply and demand for beef. These factors determine the price structure for beef, allowing producers to identify ways of predicting and possibly changing beef prices.

Cattle prices are the most widely discussed topic in the beef industry. The factors affecting cattle prices are complex, influenced by supply, demand, psychology, and several other factors. Because of this complexity, many misunderstandings are generated among various beef industry segments, some of which have resulted in serious accusations, lawsuits, price freezes, and consumer boycotts.

Cattle prices result from a free market system in which supply and demand factors determine the price of cattle. Producers find themselves "price-takers" rather than "price-makers," yet they purchase most of their goods and services in a reverse pricing structure. Even within the beef industry, the pricing structure changes from the producer who says, "What will you pay me?" to the retailer who establishes a price for beef.

Supply and Demand

The various factors affecting the supply and demand of beef and ultimately its price are shown in Figure 9.14. Numbers of animals and pounds per animal eventually reflect the tonnage of beef produced. Poultry and pork are the most significant competitors to beef.

Psychology (Market Perception)

The supply of and demand for beef have the most significant influence on cattle prices; however, how buyers and sellers perceive the market can also influence prices. The perceptions of a few individuals can be communicated rapidly to large numbers of people who impact the market. Significant and unexpected political, social, or

international trade issues can alter the marketplace strictly due to the level of uncertainty. The markets should be viewed as fluid processes where trends and unpredictability are hallmarks.

Beef Cattle Cycles and Prices

During the past 100 years, there have been several beef cattle cycles with characteristic peaks and valleys in cattle numbers and prices. These periodic changes in cattle numbers have occurred primarily in response to changes in profitability and weather (drought). During the peaks in cattle numbers, the supply of beef was excessive and beef prices dropped; whereas during the low points in cattle numbers, the supply of beef was limited and the price of beef increased. These cycles for the past years are shown in Figure 9.15. The cyclic nature of prices is shown in Figure 9.16.

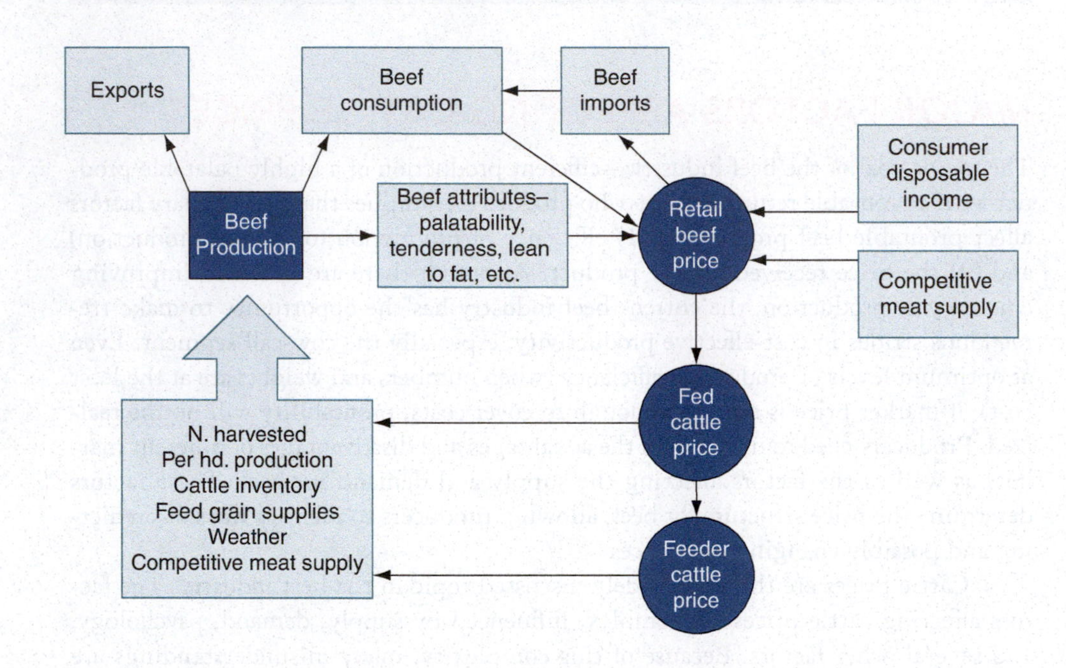

Figure 9.14
Factors affecting beef supply and demand.

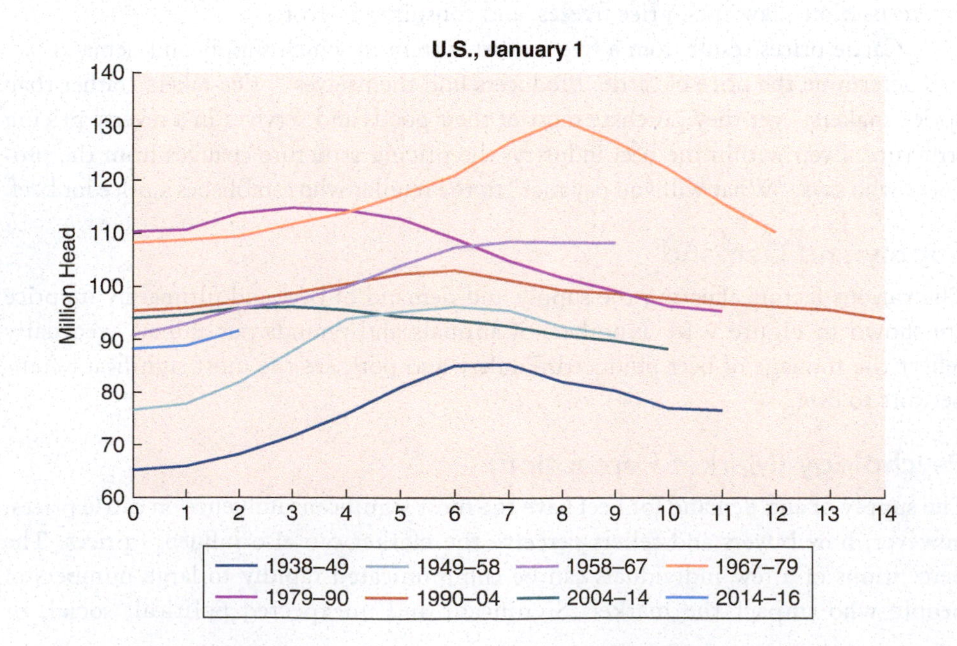

Figure 9.15
Total U.S. cattle inventory by cycle.
Data source: USDA National Agricultural Statistics Service; analysis and compilation by Livestock Marketing Information Center.

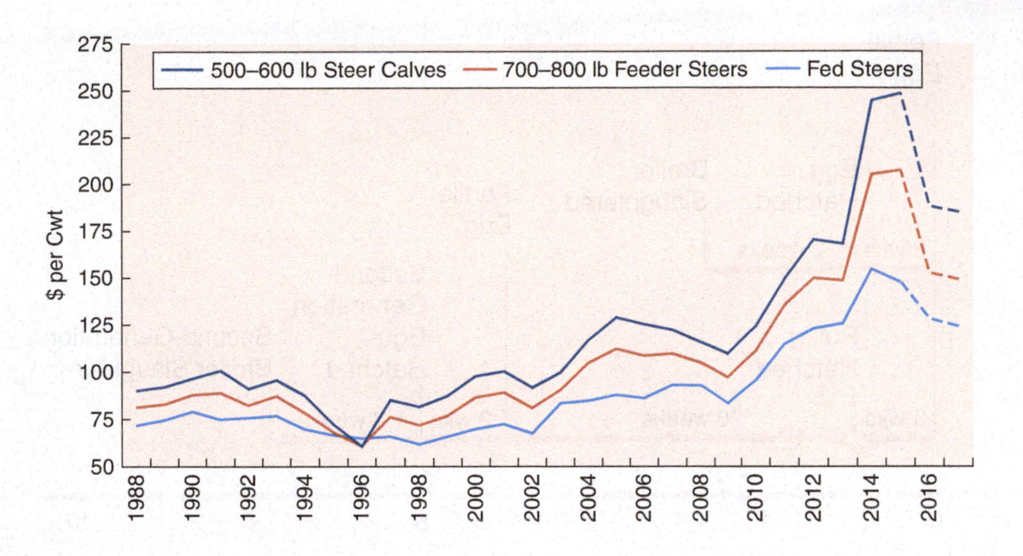

Figure 9.16
Average annual cattle prices, 1988–2016.
Data source: USDA Agricultural Marketing Service (Market News); compiled by Livestock Marketing Information Center.

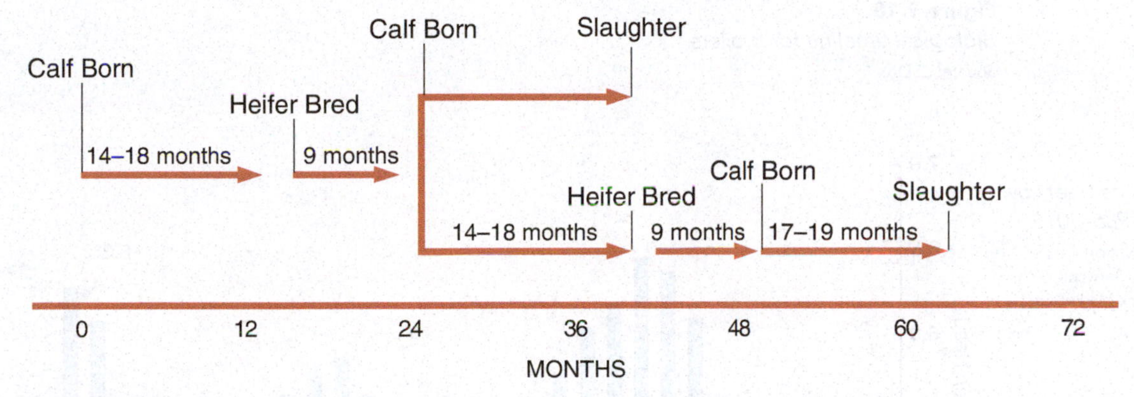

Figure 9.17
Biological timeline for cattle supply.
Source: USDA.

Typically, price cycles lag behind inventory number cycles due to the extended production cycle of cattle production (Figure 9.17) that prevents a quick turnaround in inventory when price signals that changes should be made. The production cycle is much shorter for poultry (Figure 9.18) than for beef cattle which allows poultry producers a faster response to demand signals.

Some market analysts argue that the cattle cycles of the past will not repeat themselves in the same way. They say that rapid increases in cattle numbers have occurred in the past because of the availability of cheap cattle, cheap land, cheap feed, and/or cheap money. From 1980 to 1992, none of these factors existed. Yet high cattle prices increased cattle numbers, and the resulting large beef supply depressed cattle prices in 1993–1996, resulting in a reduction in herd inventories. This implies that even in a mature beef industry there can be changes in cattle numbers that result in years of unprofitable beef prices as well as years in which beef prices can be profitable. Note that the cattle inventory in 1990–2004 was less changed than in previous cycles, but that numbers did decline following the price trough of 1996–1997. In the period of 2004–2014, the cow-calf business had the most profitable run in the history of the industry. However, cow numbers declined for a good portion of the cycle due to the run-up in corn price, the ethanol boom, and the resultant rise in land prices.

Relatively high prices for fed steers and heifers usually result in an increased number of cows and heifers retained for breeding and thus fewer being slaughtered.

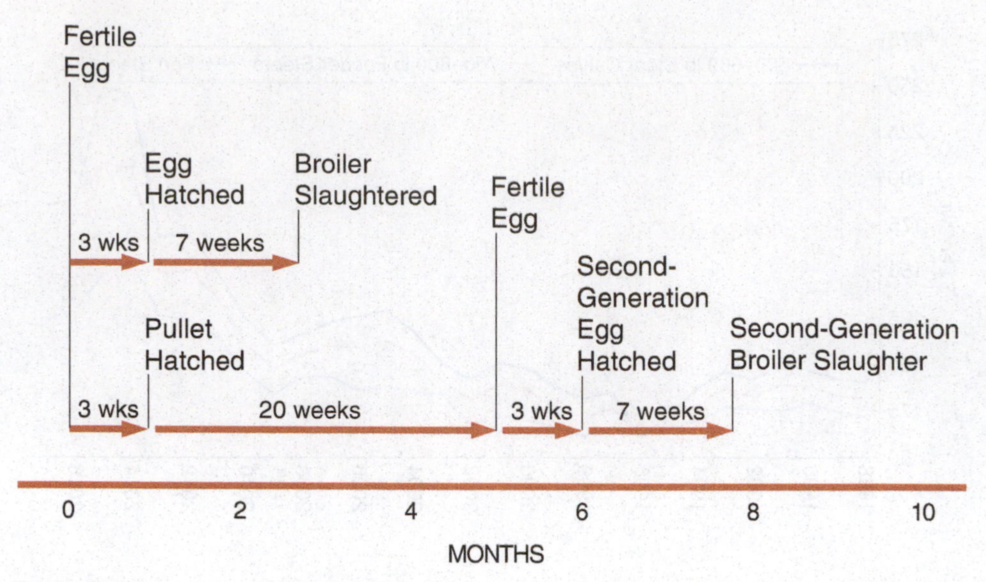

Figure 9.18
Biological timeline for broilers.
Source: USDA.

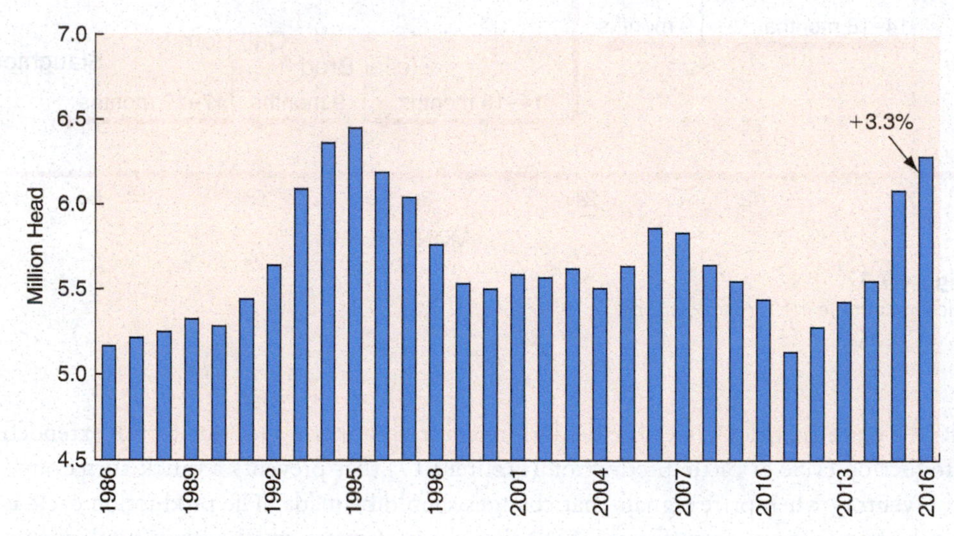

Figure 9.19
Heifers retained as beef cow replacements, 1986–2016.
Data source: USDA National Agricultural Statistics Service; compiled by Livestock Marketing Information Center.

Conversely, as prices decline, fewer females are kept as replacements. This relationship is demonstrated in Figure 9.19. Replacement retention rate impacts the size of the U.S. cow herd (Figure 9.20) but the response of retention on cow numbers is slowed by the extended and relatively slow production cycle of beef cows.

The relationship of gross returns and cattle inventory is illustrated in Figure 9.21. Price, and to some extent profit, responds inversely to supply. For example, the declining inventory from 1985 to 1991 was accompanied by relatively high cow-calf profits. Rising inventories in 1980–1983 and 1993–1996 resulted in declining returns. From 1997 on, the commercial cattle producer has experienced a price run unlike any in previous history. However, the reduction of the cow herd to historic low inventories is a double edged sword that while it supports high prices, it has left substantial levels of industry infrastructure in feeding and packing underutilized which has led to reduced total productive capacity.

Factors such as drought, calf crop percentages, competitive meat prices, grain prices, and global and national economies also significantly affect cattle prices.

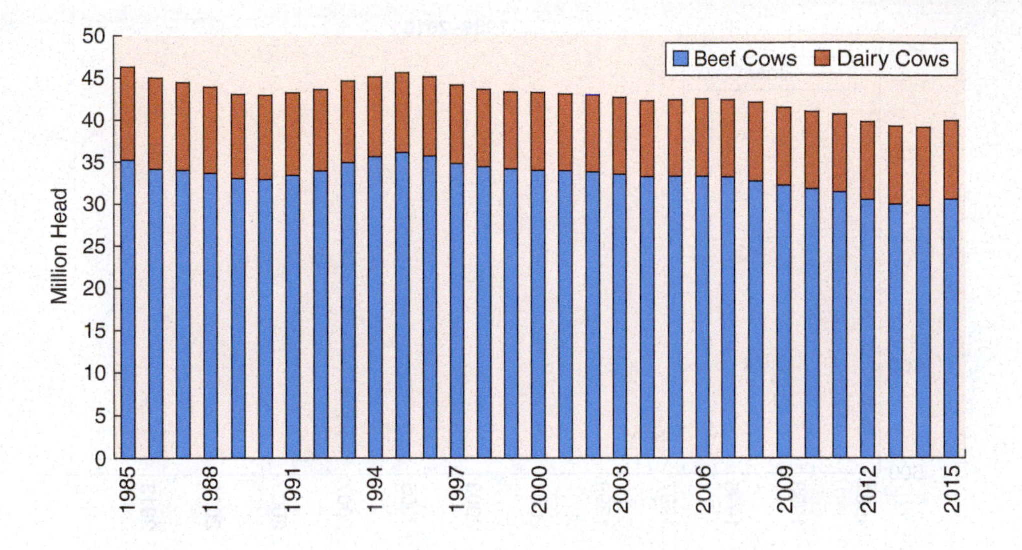

Figure 9.20
Historical changes in beef and dairy cow inventories, 1985–2015.

Data source: USDA National Agricultural Statistics Service; compiled by Livestock Marketing Information Center.

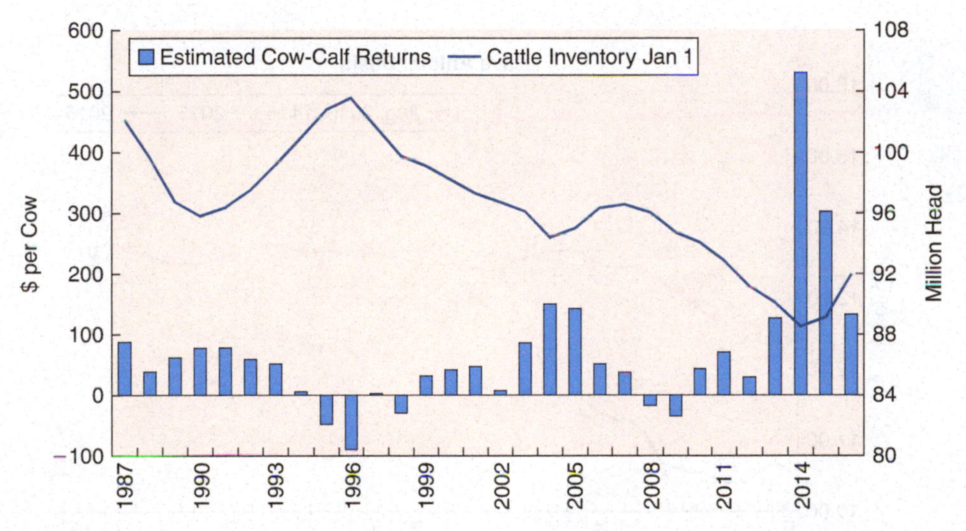

Figure 9.21
The relationship between cow-calf enterprise gross returns and cattle inventory, 1987–2016.

Major data sources include: USDA Agricultural Marketing Service (Market News) and National Agricultural Statistics Service; analysis and compilation by Livestock Marketing Information Center.

Carcass Weight

Tonnage of beef produced is a function of the number of head harvested and average carcass weight. Individual animal performance reflected in carcass weights have been increasing over the past several decades, resulting in much higher production per cow (Figure 9.22). Note that despite the dramatic increases in individual animal productivity, profits were variable over the same time period. The increase in average carcass weights has allowed the beef industry to maintain or even increase beef production while the total cattle inventory has declined.

There is a relationship between carcass weight and cattle prices. Heavy carcass weights do not necessarily directly cause lower beef prices, but they increase the total beef supply, which, in turn, tends to lower beef prices. When market prices begin to drop below producers' breakeven price, feeders typically feed their cattle longer, anticipating a price increase. This usually has an antagonistic effect on prices by increasing the tonnage of beef and the number of Yield Grade 4–5 slaughter cattle. As cattle prices increase and fed cattle become profitable, feeders usually market their cattle at lighter weights, desiring to market as many cattle as possible before a price break occurs.

By-product Value

In addition to the value of the carcass, by-product value (e.g., hide and offal) contributes to the total value of slaughter cattle. The value of by-products ranged from

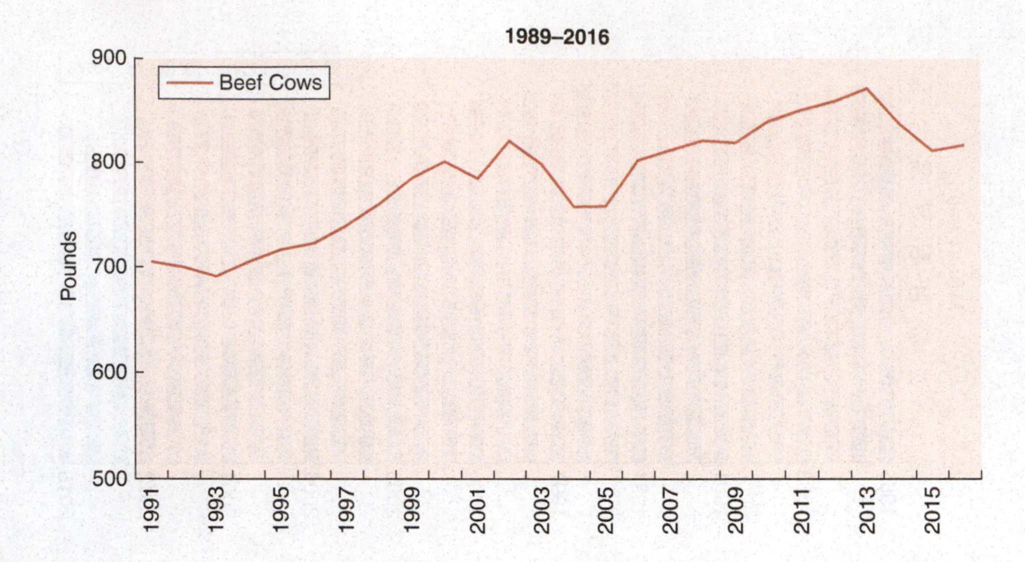

Figure 9.22
U.S. beef production per cow (carcass weight), 1991–2016.
Data source: USDA National Agricultural Statistics Service; analysis and compilation by Livestock Marketing Information Center.

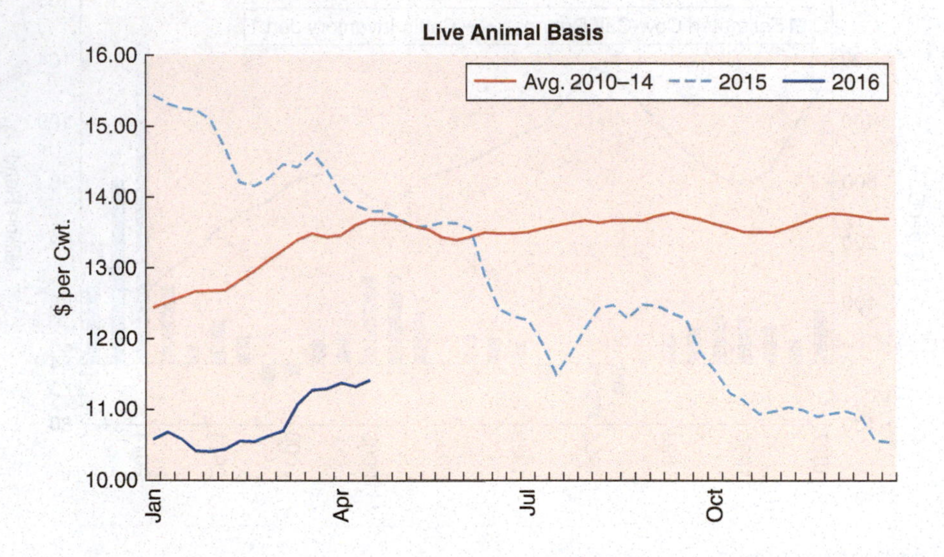

Figure 9.23
Steer hide and offal value—live animal basis.
Data source: USDA Agricultural Marketing Service (Market News); compiled by Livestock Marketing Information Center.

$70 to $190 per head from 2000 to 2015. The hide accounts for approximately two-thirds of total by-product value. Hide values can fluctuate dramatically over a short time period and thus affect live cattle prices (Figure 9.23).

Seasonal Prices

Seasonal differences in cattle prices have existed over time. Seasonality affects the prices of feeders, fed cattle and market cows. While seasonality tends to follow specific patterns, cattle producers should not depend on these seasonal price cycles for market steers to hold true for all years. Feeder cattle tend to come to the market in highest numbers during the late fall months and into the first quarter of the new year which creates the price drop typically observed in this time frame (Figure 9.24). Fed cattle prices are also variable and are impacted by seasonality of consumer demand (increased demand for beef during grilling season) as well as inventory trends where low availability of cattle tends to even out seasonal variation while excess inventory tends to enhance seasonal movement (Figure 9.25). Cow prices are consistently higher in the spring months than in the fall months (Figure 9.26). This is true for most years because the heaviest culling time for cows is in the fall when cows are pregnancy-checked and calves are weaned.

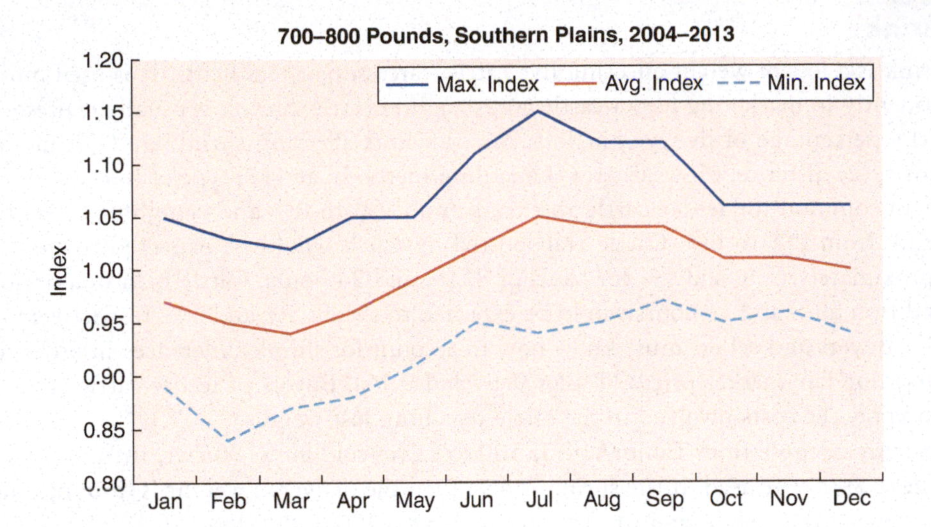

Figure 9.24
Seasonal variation in the feeder cattle price index.
Data source: USDA Agricultural Marketing Service (Market News); compiled by Livestock Marketing Information Center.

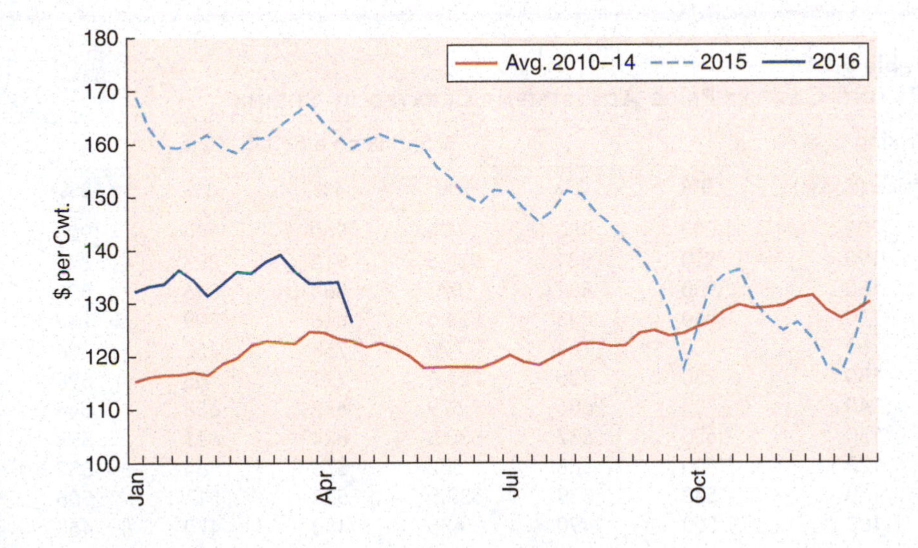

Figure 9.25
Seasonal variation in the fed steer prices.
Data source: USDA Agricultural Marketing Service (Market News); analysis and compilation by Livestock Marketing Information Center.

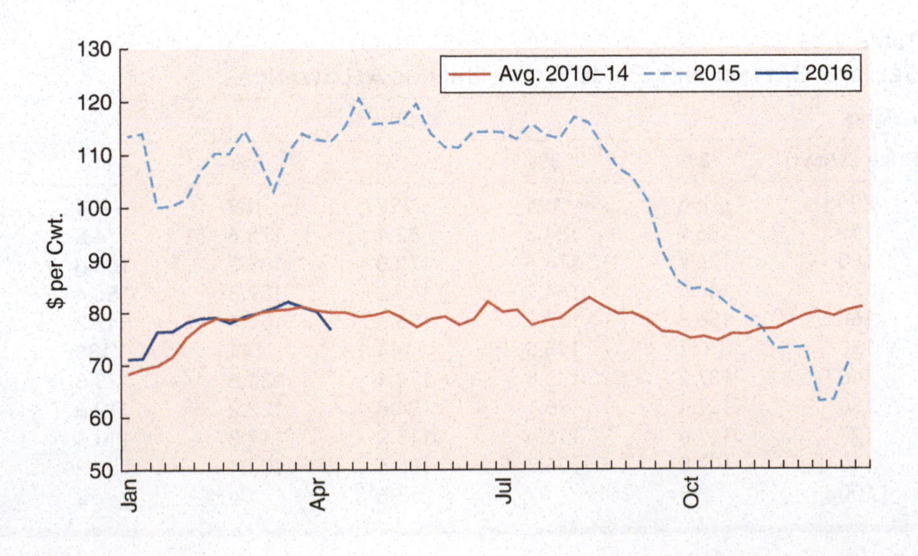

Figure 9.26
Seasonal variation in market cow prices.
Data source: USDA Agricultural Marketing Service (Market News); analysis and compilation by Livestock Marketing Information Center.

Shrink

Shrink is a loss of weight affecting live cattle, carcasses, or retail cuts. It is used most frequently in marketing feeder cattle and slaughter cattle. Shrink is usually expressed as the percentage of live weight loss. Major factors affecting shrink are (1) time in transit, (2) distance, (3) weather, (4) handling methods, and (5) type of feed fed. It is not uncommon for feeder cattle to shrink from 2% to 8% and slaughter steers to shrink from 2% to 6%. Cattle transported by truck would be expected to shrink approximately 6, 8, and 9% for hauls of 8, 16, and 24 hours. Cattle held in a drylot condition for 8 and 16 hours would be expected to shrink 3% and 6%, respectively.

Buyers and sellers must know how to account for shrink differences in order to determine fair market prices (Tables 9.14 and 9.15). Buyers of feeder cattle should also know the costs involved in the cattle regaining lost weight.

An example from Table 9.14: If 500 lb calves sold for $180/cwt, then the total value is $900 per head. If the shrink is 4%, then the delivered weight is 480 lb; thus the value of the weight delivered is $187.50/cwt ($400 ÷ 480 lb).

An example from Table 9.15: A cattle feeder hears a price quote of $170.00/cwt for 750 lb feeder heifers. The first impression is a total value of $1,275/head. However,

Table 9.14
BUYING CATTLE: PRICE ADJUSTMENT CREATED BY SHRINK

Asking	% Shrink on a 500 lb. Calf					
Price ($/cwt)	0%	2%	3%	4%	6%	8%
200	1,000	980	970	960	940	920
190	950	931	921.5	912	893	874
180	900	882	873	864	846	828
170	850	833	824.5	816	799	782
160	800	784	776	768	752	736
150	750	735	727.5	720	705	690
140	700	686	679	672	658	644
130	650	637	630.5	624	611	598
120	600	588	582	576	564	552
110	550	539	533.5	528	517	506
100	500	490	485	480	470	460

Table 9.15
SELLING CATTLE: NET PRICE AFTER SHRINK ALLOWANCE

Asking	% Shrink				
Price ($/cwt)	2%	3%	4%	6%	8%
200	196	194	192	188	184
190	186.2	184.3	182.4	178.6	174.8
180	176.4	174.6	172.8	169.2	165.6
170	166.6	164.9	163.2	159.8	156.4
160	156.8	155.2	153.6	150.4	147.2
150	147	145.5	144	141	138
140	137.2	135.8	134.4	131.6	128.8
130	127.4	126.1	124.8	122.2	119.6
120	117.6	116.4	115.2	112.8	110.4
110	107.8	106.7	105.6	103.4	101.2
100	98	97	96	94	92

in this transaction a 4% shrink is stipulated. Thus, the total dollars received would be $1,224 (720 lb × $170/cwt or 750 lb × $163.20/cwt). The $163.20/cwt is derived from a 4% shrink on the asking price.

Managing the factors that contribute to shrink can bolster profits. A few guidelines follow:

1. Cattle on green pasture shrink more than those consuming a drier ration. Also, feed changes near time of transport can increase shrink loss.
2. Weaning is a stressful time for calves and adding transport to the stress can increase shrink losses and morbidity/mortality rates. Pre-conditioned calves typically fare better than do those calves that are weaned and marketed back to back.
3. Gathering cattle gently and minimizing their stress is a key to minimizing shrink.
4. Cattle should not be overcrowded into trucks and trailers, and the footing should be good.

ASSESSING MARKETING COSTS

A good marketing plan includes knowing the costs of production and the break-even prices. Part of cost assessment involves marketing costs and costs of different marketing alternatives. Direct markets and public markets as marketing alternatives were discussed earlier in the chapter. Costs for these various markets can be determined.

Direct marketing costs are primarily the costs associated with selling at the point of production. Examples include the sale of feeder cattle at the farm or ranch and the sale of fed cattle at the feedlot to the packer. Direct sale costs are mostly tied to shrink adjustments and the buyer's ability to sort off calves deemed undesirable.

Public marketing costs are those associated with auctions and **terminal markets**. Transport costs can have a significant impact on net price received. Most cattle are transported via large cattle trailers known as "pots," which can carry 48,000 to 52,000 pounds of loaded weight. Most livestock hauling companies charge based on mileage rate or a mileage rate plus fuel surcharge. Yardage expenses are also incurred at auction markets and are charged on a per head per day boarding fee. In addition to yardage, livestock commission companies collect commission charges, insurance, and checkoff assessments.

Cattle-Fax

Cattle-Fax is a member-owned market information, analysis, and research organization that serves cattle operators in all segments of the U.S. beef industry. It maintains the largest private database on the U.S. cattle industry. The mission of Cattle-Fax is to collect, analyze, and disseminate information so that its members and clients can make more profitable market and management decisions (see address in the Appendix).

Cattle-Fax members receive up-to-the-minute cattle marketing information from regional analysts. The organization also provides written report updates; these are mailed to members weekly and are current, intermediate, and long-term market evaluations.

FORECASTING BEEF PRICES AND MANAGING PRICE RISKS

One of the most significant variables affecting profitable beef production is the ability to project future market prices. For example, the cattle feeder currently purchasing a pen of feeder cattle will be selling them at an unknown price some 100–150 days in the future.

Price forecasting is formulating an outlook (or opinion) on where the market will be at some point in the future. The most important factors analyzed by price

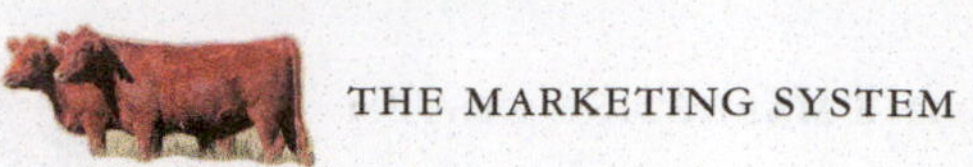

Table 9.16
METHODS OF MANAGING PRICE RISKS

Method	Comments
Sound management practices	Attain low costs and high productivity to maintain competitive breakeven prices. Enterprises with lower breakevens are more profitable when prices decline.
Continuous buying and selling during the year	Producer obtains close-to-average price for the year. Avoids the risk of one time selling.
Arbitrage	Producer has a better knowledge of marketing costs and returns. Simultaneous buying and selling of the same commodity in different markets in order to profit from price discrepancies.
Distributing cattle geographically	Market prices vary in different regions of the United States. The high and low markets average out. This also reduces weather risks.
Partnerships	Spreads the risk over two or more production segments of the beef industry coupled with profit sharing.
Retained ownership	Producer can take advantage of genetic superiority for added productivity beyond usual point of sale.
Contractual arrangements	Provides a market price at a future date so that net return can be projected prior to delivery of the cattle.
Combination of several of the above	

Source: Based on Cattle-Fax, CME, Dunn.

forecasting are supply, demand, psychology (especially in the short term), and current data (how fed cattle are moving out of the feedlots).

No person or group is right all the time in forecasting prices. The objective is to be accurate most of the time by recognizing changes as they occur and adjusting forecasts as necessary.

Table 9.16 identifies how some price change risk can be managed.

THE FUTURES MARKET

The futures market is a highly standardized, regulated method of exchanging forward contracts. In the cash cattle market, it is not uncommon to forward-contract a prescribed number of cattle at a mutually agreed-upon price for delivery at a pre-determined date in the future. The language of a futures and options market contract specifies the number and kind of cattle and the month of delivery. Price is the only variable that is determined when a contract is bought or sold.

Trading of futures and options contracts is regulated by the exchange trading the contract and by the Commodity Futures Trading Commission, the government agency overseeing all futures activities. The Chicago Mercantile Exchange (CME) involves both fed cattle and feeder cattle contracts. The CME notes the following advantages of the futures and options markets in helping cattle producers manage market risks: the ability to (1) lock in profits, (2) enhance business planning, and (3) facilitate financing. Self-study guides, videos, and computer programs are available from the CME for those interested in a more detailed study of the futures and options markets (see address in the Appendix).

Those people who use the futures markets fall into two general groups: hedgers and speculators. Hedgers are typically beef producers who produce, feed, and market cattle and who seek to transfer the price risk. Speculators may or may not own or produce cattle, but they buy or sell contracts in the hope of making a profit on price changes. Commodity prices can be very volatile, so the potential for profit and loss is greater in commodity trading than in many other kinds of investments.

What Is a Futures Contract?

A futures contract is a standardized agreement to buy or sell a commodity at a predetermined date in the future. The contract specifies the following information:

Commodity—live cattle or feeder cattle

Quantity of the commodity—pounds of livestock as well as the range or weight for individual animals

Quality of the commodity—specific U.S. grades

Delivery point—location at which to deliver the commodity, or a cash settlement in the case of feeder cattle

Delivery date—within a month of the contract termination date

A futures contract does not specify the price at which the commodity will be bought or sold. Rather, the price is determined on the floor of the exchange as floor brokers execute buy and sell orders from all over the country. The prices they bid reflect the supply and demand for the commodity as well as expectations of whether the price will increase or decrease. Contracts traded on the CME are shown in Table 9.17. The specifications for such contracts change periodically, thus a broker should be consulted.

Settlement of open live cattle contracts require physical delivery of equivalent loads of either all steers weighing between 1,050 and 1,500 pounds or all heifers weighting between 1,050 and 1,350 pounds that meet the carcass specifications described in Table 9.17. CME rules specify par delivery of live beef cattle from approved livestock yards in Wray, Colorado; Worthing, South Dakota; Syracuse, Kansas; Tulia, Texas; Columbus, Nebraska; Dodge City, Kansas; Amarillo, Texas; Norfolk, Nebraska; North Platte, Nebraska; Ogallala, Nebraska; Pratt, Kansas; Texhoma, Oklahoma; and Clovis, New Mexico. Delivery can also be made to specified packing plants. Full specifications and contract rules are available at www.cmegroup.com.

As of November, 2016, delivery of live feeder cattle as settlement will no longer be required. All open contracts will be cash settled based upon the CME Feeder Cattle Index™ for the seven calendar days ending on the day on which trading terminates. The CME Feeder Cattle Index™ is based upon a sample of transactions from these weight/frame score categories: 700 to 899 pound Medium and Large Frame #1 feeder

Table 9.17
SPECIFICATIONS FOR LIVE AND FEEDER CATTLE CONTRACTS

	Contract Type	
Specification	Live (Fed) Cattle	Feeder Cattle
Size	40,000 lb of USDA 55% Choice, 45% Select, Yield Grade 3 USDA grade live steers or heifers	50,000 lb of 700–849 lb medium and large frame #1 plus #1 and 2 of the same weight and frame category
Contract months	Feb, April, June, Aug, Oct, Dec	Jan., Mar., Apr., May, Aug., Sep., Oct., Nov.

steers, and 700 to 899 pound Medium and Large Frame #1–2 feeder steers. The sample consists of all feeder cattle auction, direct trade, video sale, and Internet sale transactions within the 12-state region of Colorado, Iowa, Kansas, Missouri, Montana, Nebraska, New Mexico, North Dakota, Oklahoma, South Dakota, Texas, and Wyoming for which the number of head, weighted average price, and weighted average weight are reported by the USDA.

What Is an Option?

An option is a choice—that is, the *right* but *not* the obligation to buy or sell a futures contract at a specific price on or before a certain expiration date. There are two different types of options: puts and calls. Each type offers opposite pricing alternatives and the opportunity to take advantage of futures price moves without actually having a futures position.

Hedging

Hedging is a risk management tool that permits producers to establish a buying or selling price for their livestock months before they are ready for sale or purchase. In hedging, a producer is taking a position opposite the cash position or what the producer owns. The cattle feeder who owns cattle, for example, sells a futures contract at a favorable price to cover the animals on feed. Packers wishing to cover future purchases buy futures contracts, as do cattle feeders planning to buy yearlings. The contract runs to some specified maturity date, usually months in advance. At that point, the producer can either (1) buy back the futures contracts to offset the contracts originally sold or (2) deliver the cattle to one of the specified delivery points.

Hedging guarantees the selling price but it does not necessarily guarantee a profit. Profitability depends on placing the hedge at a price that exceeds all costs. It is extremely important for a person to know costs before becoming involved in hedging. The two major types of costs are (1) production and (2) basis. The latter is the difference between the futures market price and the producer's local cash price. The spread between these prices includes transportation, shrink, and other marketing costs as well as differences in supply and demand. Producers should know their basis and watch future quotations daily, even if they do not have a hedged position, because this tells them what they could lock in through the use of a hedge.

There are times when futures are favorable for hedging and other times when they are not. The management rule is this: If producers cannot figure a satisfactory profit, they should not hedge, but the producer should take a chance on the cash market instead.

The decision to hedge depends on the market situation and the degree of risk the producer is willing to assume. Each producer has a different degree of risk-carrying ability. Some cattle producers do not like hedging because they would rather gamble for higher profits (but they are also gambling for higher losses).

Choosing the timing to place a hedge is an extremely difficult management decision. The use of a professional broker who can provide advice and account service is advisable. Before placing the hedge, a producer needs to consider all the market facts on supply and demand and how prices might be affected. After all of the facts have been examined, the producer will come to one of two possible conclusions:

1. Higher prices are coming, so the producer will wait to place the hedge at a higher price.
2. Lower prices are likely, so the producer will go ahead and "lock in" a price on the futures market.

Branded Beef Products

Beef was traditionally marketed as a commodity. When other competing products began to sell under brand names, the beef industry was initially slow to respond. However, the industry soon realized that branding approaches would indeed be required, especially to carve out market share in higher end, premium market niches.

During the 1980s, branded beef products made their appearance primarily because consumer demand for beef appeared to be slipping and the beef industry was emphasizing the need for improvements in marketing. Several branded products were marketed to consumers who were looking for: (1) "lite beef" with less fat and cholesterol, (2) organic or natural beef from cattle not fed antibiotics or growth stimulants, and (3) high palatability beef. In 1988, when the major packers and retailers began trimming fat, many of the "lite beef" markets were lost. Thus, most name brands featuring less fat were abandoned or became less significant as their distinctiveness was lost compared with commodity beef. However, some of these companies pivoted and added label claims such as natural or grass-fed to establish differentiation.

The current market appears to have four primary target segments: "white table cloth," where superior marbling and tenderness are critical attributes; "calorie conscious—lifestyle," characterized by demand for tenderness coupled with specific lean yield parameters, "retail," where a balance of quality and yield must be managed; and attribute-based demand such as natural, organic, or grass-fed. To create standardization, verification, and certification of beef brand claims, the USDA has instituted a number of programs to provide clarity in the market place. There programs include certification services, process verification, quality systems assessment, and export enhancement.

Certification of beef brands is provided by Agricultural Marketing Service (USDA) and focuses on marketing programs that make breed and carcass characteristic claims. These specified attributes go beyond the requirements of the USDA grading system. Certification is often the basis for approval of meat product labels making a variety of marketing claims. The attributes include live characteristics such as phenotype (hide color predominately) or breed composition as well as carcass traits, such as specified marbling levels, USDA quality and yield grades, rib-eye size, degree of fatness, marbling texture, and freedom from defects such as dark cutting and blood splash. In 2016, there were 107 certified beef programs, with 71 of those specifying black or red hide color and using the name Angus. However, within those 71, there were programs specifying minimum marbling levels ranging from Select to Prime as well as one that specified marbling standards of Standard and lower. This diversity clearly points out the fragmentation of the beef market into a number of unique niches that open the door for a number of products to compete for market share while providing a high level of responsiveness to market demand. Consumers are increasingly mindful of the management and production practices employed in the production of food, some beef companies are moving toward process-verified programs that document production practices, such as health history, breed of origin, feeding protocols, unique fabrication processes, and proof of not using antibiotics or hormonal growth promotants. The use of electronic identification systems and intensive data management are critical components of these systems.

The process-verified approval protocol involves the following seven steps: document program, operate and test, conduct an adequacy audit, conduct a compliance audit, receive product approval, begin marketing, and submit to USDA for monitoring. A documented program supported by a management system allows the opportunity to assure customers that consistent quality products can be provided. Process points may include company conformance to internal, industry, international, or customer defined requirements. As such, process points must be verifiable, repeatable, feasible, and factual.

Approved USDA Process Verified Programs may develop promotional materials associated with their process verified points, use the USDA PVP shield in accordance with Program requirements and market products as "USDA Process Verified".

The Quality Systems Assessment (QSA) Program, like the process verified system uses the requirements of the ISO 9001, but some requirements are excluded, thus making programs less comprehensive. The QSA Program is designed to aid in the marketing of products that have specific claims, and are limited in scope to those specific items associated with the product or process claim. Because there are fewer requirements for a comprehensive quality management program, companies operating under a QSA Program are audited a minimum of two times per year. Table 9.18 illustrates the differences between these two programs.

Export oriented programs offered via USDA include the non-hormone treated cattle (NHTC) and under 30-month certification program. Started in 1999, when the European Union (EU) and the United States agreed to control measures to facilitate the trade of non-hormone treated beef, including veal, the NHTC has three basis components:

1. Cattle are grown on approved farms/feedlots and delivered to the packer with shipping documentation that includes the statement "Cattle Meet EV Program Requirements for the EU" and clearly identifies the animals and the quantity delivered.
2. Non-hormone treated cattle and beef are segregated at the packer and are not commingled with other animals or meat.
3. Tissue samples from NHTC are collected at harvest and analyzed in order for FSIS to offer export certification.

Additionally, USDA offers a program to assure that carcasses have originated from cattle 30 months of age or younger at the time of harvest. In 2016, there were 86 packers, fabricators, or further processors participating in the program. This program was initiated to alleviate international concerns about bovine spongiform encephalopathy.

Table 9.18
COMPONENTS OF THE USDA QUALITY SYSTEMS ASSESSMENT AND PROCESS VERIFIED PROGRAMS

	Quality Systems Assessment Program	Process Verified Program
Audit Criteria	ISO 9001 Requirements for Quality Management Systems (for applicable product or process claims only [predetermined])	ISO 9001 Requirements for Quality Management Systems (comprehensive with company determining claims or standards to use)
Approval Process	Submit quality manual (desk audit); on-site audit; bi-annual audit	Submit quality manual (desk audit); on-site audit; annual audit including companies associated with claim
Marketing	Official listing on QAD website, no USDA shield available	Official listing on QAD PVP website, USDA PVP shield is allowed for use on marketing materials with prior AMS approval
Types of Claims (examples)	Non-hormone treated beef and pork for the European Union; age verification; removal of specified risk materials (for example, spinal cord); beta agonist free	Source verification of animals; genetic traceability; never fed or given hormones, antibiotics, or animal protein in feeds; animal raising claims; feeding claims
Specific Programs (examples)	Poultry, Egg, and Egg Product Export Verification Program; Less than 30-month-old cattle (LT30) Program for export of products to the Pacific Rim	USDA Certified Tender/Very Tender; Never Fed Beta Agonists; service claims for livestock and poultry producers

Source: USDA.

Table 9.19
Certified Angus Beef Percent Acceptance: Percent Cattle Certified and Volume Pounds Sold

Year	Acceptance[1]	Sales (mil lb)
1987	24	43
1991	16	80
1995	17	226
1999	18	495
2003	17	584
2007[2]	16	544
2011	24	778
2015	27	896

[1]Percent of cattle that meet the live specifications also meeting the carcass specifications.
[2]Decline due to closure of international borders, as well as short-term supply issues.
Source: Adapted from Certified Angus Beef LLC.

The branded product leader is Certified Angus Beef (CAB), which sold 882 million pounds in 2014 that equates to an increase of nearly 300 million pounds over the previous decade. The CAB program is based on a set of live animal and carcass specifications designed to assure palatability. Trends in acceptance percentage and total sales volume are outlined in Table 9.19. To expand its appeal to a broader market, a natural product line was added to the CAB family of products in 2004. The new product line will meet the original carcass standards established by the company and will be process-verified under USDA guidelines for adherence to production practices that eliminate the use of hormone-based growth promotants, antibiotics, and animal by-products in the ration. The licensed packer for the CAB Natural product pays a premium for cattle managed under the all-natural product criteria of no implants, antibiotics, or feeding of animal by-products.

It is interesting to note that acceptance rates have been relatively stable over the past decade, while the tonnage of sales has increased dramatically. The ability to effectively market the product while building a supply chain that consists of 80 licensed feedlots has given the CAB program tremendous momentum. Furthermore, CAB has implemented a genetic research and education effort with a goal to increase adoption of relevant technologies, to substantially enhance carcass data delivery and interpretation, and to provide industry-wide communication and education about the value of Certified Angus Beef®. In the end, their goal was to achieve 80% Angus-based beef cattle population by 2005, followed by CAB acceptance rates of 30% by 2007.

Through these efforts, prices for Angus and Angus-cross feeder calves have risen relative to the total supply. Premiums paid for CAB carcasses top $50 million annually with accumulated premiums of more than $450 million. Angus steer and heifer calves typically receive a premium as do black hided cattle regardless of breed. The success of the program has contributed to the rise in bull prices paid for Angus bulls. Certified Hereford Beef (CHB) sales increased from 34.7 million pounds of volume in 2003 to 50.2 million pounds in 2014 when it was featured in 304 supermarkets and 40 foodservice distributors spread across 35 states. As a breed-based program, CHB is designed to grow demand for Hereford and Hereford-cross feeder cattle. The certification rate for cattle deemed eligible for receiving the CHB label is approximately 75%. One of the keys in branded beef markets is to assure delivery of a product that meets or exceeds consumer expectations while connecting the product and producers to the needs of consumers. The family ranchers who produce Country Natural Beef describe their product as follows:

> "Our product is more than beef. It's the smell of sage after a summer thunderstorm, the cool shade of a Ponderosa Pine forest. It's 80-year-old weathered hands saddling a horse in the Blue Mountains, the future of a 6-year-old in a one-room school in the high desert. It's a trout in a beaver built pond; haystacks in an Aspen-framed meadow. It's the hardy quail running to join the cattle for a meal, the welcome ring of a dinner bell at dusk."

Shifting beef products from commodities to uniquely identified products that resonate with the deepest needs of consumers is the key to the future of the beef industry.

Natural Beef

Natural beef is produced to fit a specific branded beef program where the requirements are defined by the brand's business model. Because the brand entity establishes the requirements and is accountable for compliance to the requirements, natural claims are largely undefined and unregulated in the marketplace. A uniform definition of natural is not available but typically these programs do not allow the use of antibiotics or growth promotants.

Organic Beef

Beef sold under an organic label must originate from cattle produced and processed by a USDA-certified organic farm and processor. The organic label is regulated by USDA under a strict set of standards with the government providing both the certification and the auditing authority to verify that the standards have been met. A selection of the requirements for organic production is listed below:

- The cattle must be free of any antibiotics or growth hormones.
- They must not be fed mammalian or poultry protein or by-products. Producers are required to feed livestock agricultural feed products that are 100% organic but may also provide allowed vitamin and mineral supplements. Feed must not have been exposed to pesticides, fertilizers made from synthetic ingredients, or bioengineering. In order to produce 100% organic feed, the land will have no prohibited substance applied to it for at least 3 years before the harvest of an organic crop.
- Cattle for harvest must be raised under organic management from the last third of gestation.
- Soil fertility and crop nutrients will be managed through tillage and cultivation practices, crop rotation and cover crops, and the application of plant and animal materials.
- Preference will be given to the use of organic seeds and other planting stock, but a producer may use nonorganic seeds and planting stock under specified conditions.
- Crop pests, weeds, and disease will be controlled primarily through management practices including physical, mechanical, biological controls and grazing. When these practices are not sufficient, a biological, botanical or synthetic substance approved for use on the National List may be used.
- Preventive management practices, including the use of vaccines, will be used to keep animals healthy. Producers are prohibited from withholding treatment from a sick or injured animal; however, cattle treated with a prohibited medication may not be sold as organic.
- All organically raised cattle must have access to the outdoors, including access to pasture for ruminants. They may be temporarily confined only for reasons of health, safety, the animal's stage of production, inclement weather or to protect soil or water quality. Continuous total confinement of any animal indoors is prohibited.

- The producer of an organic beef operation must maintain records sufficient to preserve the identity of all organically managed animals and edible and nonedible animal products produced on the operation.

Grass Fed

As is the case with natural beef marketing, USDA has chosen not to provide certification or verification services. The Grassfed Beef Association, an industry group, however, has established a set of guidelines that focus in four areas:

Diet—animals are fed only grass and forage from weaning to harvest.

Confinement— animals are raised on pasture without confinement on feedlots.

Antibiotics and hormones—animals are never treated with antibiotics or growth hormones.

Origin—all animals are born and raised on American family farms.

For the foreseeable future, it is most likely that grassfed marketing systems will be merchandised under a private labeling model.

Local

Beef marketed as 'local' is in vogue with some consumers who prefer to purchase food via farmer's markets or other venues that feature small scale agricultural production (Figure 9.27). These ventures are potentially profitable but face a number of challenges that limit the sustainability of the business model. These challenges include access to packing capacity, effectively merchandising the entire beef carcass, including middle meats (easier to sell) plus end meats, ground beef, and dress off items; and developing the logistics capability to effectively and safely harvest, process, package,

Figure 9.27
Local beef supply chains meet the needs of a specific market niche but often must sell at a substantial premium to allow complete cost recovery as well as to recoup the lost opportunities due to lack of economies of size and scale.

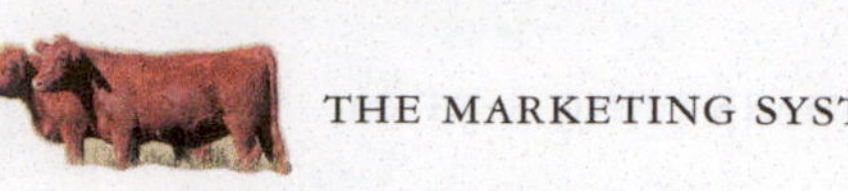

store, and merchandise beef direct to consumers. As is the case with natural, organic, and grass-fed products, "local' beef products must be sold at a significant premium to offset the costs of leaving management options on the table plus the costs of not having access to the economies of scale that come with regional or national supply chains.

ADVERTISING AND PROMOTING RETAIL BEEF

It is difficult to develop meaningful advertising and promotion strategies for commodities. In a market space dominated by brands, generic campaigns are oftentimes underfunded and struggle to compete for attention in a communications environment that favors clear messages, attributes, and messaging. Nonetheless, in 1986, the beef industry became the last major agricultural commodity group to initiate a checkoff program to fund a marketing program. The Cattlemen's Beef Promotion and Research Board (Beef Board) was created to manage the checkoff dollars originating from the $1 per head collected each time an animal was sold.

The National Beef Checkoff is designed to stimulate the supply chain to sell more beef and to grow consumer demand (Figure 9.28). All producers and importers pay the equivalent of $1 per head every time a bovine animal is sold. U.S. Customs collects $1-per-head or equivalent on all imported live cattle, beef, and beef products and forwards the full dollar to the Beef Board. Qualified State Beef Councils (QSBCs) collect the dollar retaining half for state programs and sending 50 cents to Cattlemen's Beef Board for investment into national checkoff programs. QSBCs may invest their 50 cents into state programs and/or invest an additional portion into national efforts through the Federation of State Beef Councils or the Beef Board.

All national checkoff-funded programs are budgeted and evaluated by the Cattlemen's Beef Board, comprising 103 checkoff-paying producer volunteers who have been nominated by producer organizations in their states and appointed by the U.S. Secretary of Agriculture. By law, checkoff funds cannot be used to influence government policy or action, including lobbying; for any unfair or deceptive practices; or to reference any particular brand or trade name without approval by CBB and USDA.

Total collections are very dependent on the growth of the industry as evidenced by the loss of nearly 8.9 million dollars in collections between 2000 and 2015, a time when the U.S. cow herd was declining.

Checkoff dollars are used to achieve results in six key performance categories—promotion, research, consumer information, industry information, foreign marketing,

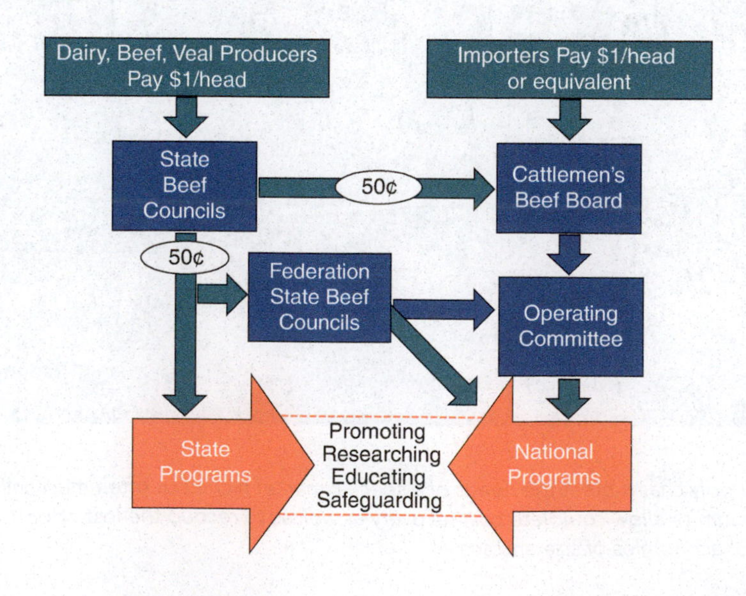

Figure 9.28
Structure of the beef checkoff program in the United States.

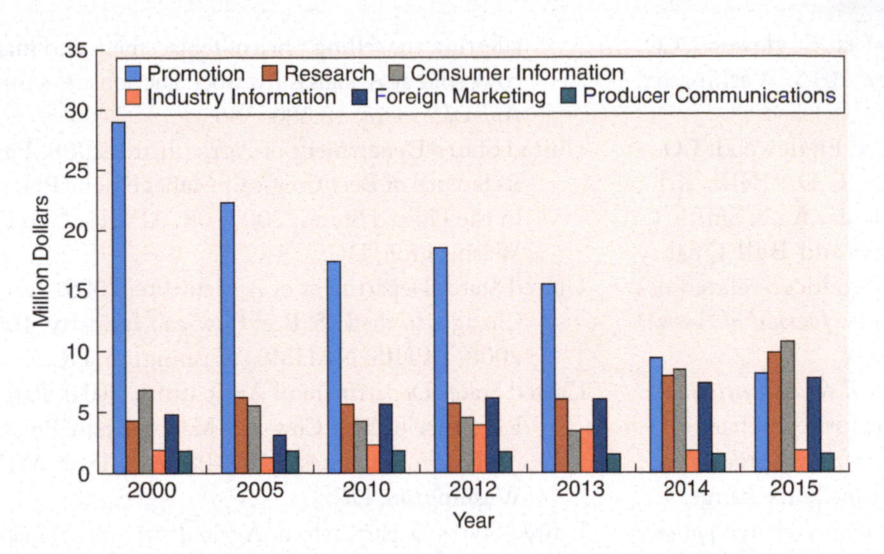

Figure 9.29
Trends in checkoff funded expenditures.

and producer communications (Figure 9.29). In the early days of the checkoff, most of the investment was in print and broadcast advertising campaigns but with the advent of social media and other mechanisms to reach both consumers and influencers, the industry pivoted from the very expensive and moderately effective advertising approach to a more targeted effort that better communicated beef's advantages in terms of health, nutrition, and versatility. Many in the industry are advocating for additional investments into either current checkoff program or some modified version designed to better connect the beef supply chain to food service operators, retailers, and consumers both domestically and abroad.

SELECTED REFERENCES

Publications

Apple, J.K. 1999. Influence of body condition score on live and carcass value of cull beef cows. *Journal of Animal Science.* 77:2610–2620.

Barnes, K., Smith, S., & Lalman, D. 2000. *Managing shrink and weighing conditions in beef cattle.* Fact sheet F-327. Stillwater, OK: Oklahoma State University Extension Service.

Beef Board Annual Report. 2015. Cattlemen's Beef Promotion and Research Board. Centennial, CO.

Cattle industry reference guide. 2015. Cattle-Fax. Englewood, CO.

Field, T.G., Tatum, J.D., & Kimsey, K.M. 1998. *Grid-pricing: A workbook.* Ft. Collins, CO and Englewood, CO: Colorado State University and NCBA.

Food Marketing Institute. 2004. *The law of unintended consequences.* White paper. Washington, DC.

Gill, D., Barnes K., & Lalman, D. 2002. *Rancher's guide to custom cattle feeding.* Ext. Fact Sheet F-3022. Stillwater, OK: Oklahoma State University.

Gill, R. 2002. Marketing cull cows. *American Red Angus Magazine.* October, p. 42–44.

Improving the consistency and competitiveness of market cow and bull beef and increasing the value of market cows and bulls. 1999. Ft. Collins, CO and Englewood, CO: Colorado State University and NCBA.

Improving the quality, consistency, competitiveness, and market share of beef. 2000. NCBA, Colorado State University, Texas A&M University, Oklahoma State University, and West Texas State University.

Lardy, G. 2002. Adding value to cull cows. *SimTalk*, March. p. 40–41.

Little, R.D., Williams, A.R., Lacy, R.C., & Forrest, C.S. 2002. Cull cow management and its implications for cow-calf profitability. *Journal of Range Management.* 55:112–116.

Marousek, G.E., Stodick, L.D., Carlson, P., & Gibson, C.C. 1994. Economics of value-adding rangeland beef cattle enterprises. *Rangelands.* 16, p.1.

McBride, W.D. & Mathews, K. 2011. The Diverse Structure and Organization of U.S. Beef Cow-Calf Farms. United States Department of Agriculture: Economic Research Service Economic Information Bulletin Number 73. Washington, DC.

Outlook and strategies. 2015. Cattle-Fax. Englewood, CO.

Packers and stockyards statistical report. 2015. Washington, DC: USDA.

Retained ownership. 2010. Cattle-Fax. Englewood, CO.

Roeber, D.L., Mies, P.D., Smith, C.D., Belk, K.E., Field, T.G., Tatum, J.D., Scanga, A., & Smith. C. 2001. National Market Cow and Bull Quality Audit—1999: A survey of producer-related defects in market cows and bulls. *Journal of Animal Science.* 79:658–665.

Self-study guide to calculation of CME feeder cattle index. 2004. Chicago: Chicago Mercantile Exchange.

Self-study guide to final settlement price of a live cattle delivery. 2004. Chicago: Chicago Mercantile Exchange.

Self-study guide to forward pricing with livestock options. 2004. Chicago: Chicago Mercantile Exchange.

Self-study guide to hedging with livestock futures. 2004. Chicago: Chicago Mercantile Exchange.

Self-study guide to introduction to livestock and meat fundamentals. 2004. Chicago: Chicago Mercantile Exchange.

Steiner, R., Wyle, A.M., Vote, D.J., Cannell, R.C., Belk, K.E., Scanga, J.A., Wise, J.W., O'Connor, M.E., Tatum, J.D., & Smith, G.C., 2001. *Video image analysis determination of percentage subprimal yield of beef carcasses.* Animal Science Research Report. Fort Collins, CO: Colorado State University.

Troxel, T.R., Gadberry, M.S., Cline, S., Foley, J., Ford, G., Urell, D., & Wiedower, R. 2002. Factors affecting the selling price of replacement and market cows sold at Arkansas livestock auctions. *Professional Animal Scientist,* 18:380–386.

United States Department of Agriculture. 2008. Part I: Reference of Beef Cow-calf Management Practices in the United States, 2007–08. APHIS: NAHMS. Washington, DC.

United States Department of Agriculture. 2009. Part III: Changes in the U.S. Beef Cow-calf Industry, 1993–2008. APHIS: NAHMS. Washington, DC.

United States Department of Agriculture. 2010. Part IV: Reference of Beef Cow-calf Management Practices in the United States, 2007–08. APHIS: NAHMS. Washington, DC.

United States Department of Agriculture. 2015. Packers and Stockyards Annual Report. Grain Inspection Packers and Stockyards Administration. Washington, DC.

United States Standards for Grades of Carcass Beef. 2000. Washington, DC: USDA-AMS.

United States Standards for Grades of Feeder Cattle. 2000. Washington, DC: USDA-AMS.

Wulf, D. M. 1999. Techniques to Identify Palatable Beef Carcasses: MARC Tenderness Classification, SDSU Colorimeter and Near-Infrared Spectrophotometry Systems. Proc. The Range Beef Cow Symposium XVI. Greeley, CO.

10

The Global Beef Industry

Cattle, including domestic water buffalo, contribute food, fiber, fuel, and power to many of the nearly 7.4 billion people throughout the world. Beef producers in the United States are significantly influenced by such global events as international trade, drought, hunger, population changes, political pressures, business opportunities, new sources of breeding stock, and disease problems. It is important for U.S. beef producers to understand the global beef industry not only because they are affected by it in the short term, but also because the opportunity to grow the business will be driven by global beef demand.

Beef and other meats have been adopted into the diet because their assortment of amino acids more closely matches the needs of the human body than does the assortment of amino acids in foods originating from plant sources. Furthermore, vitamin B12, which is required in the human diet, may be obtained from foods of animal origin but not from those of plant origin. Beef and other animal foods also efficiently provide a variety of micronutrients (iron, zinc, etc.), offer a high degree of nutrient density (a high proportion of the recommended daily allowance of several nutrients relative to the number of calories per serving), and provide the unique taste desired by consumers.

Animals are integrated into the agricultural systems of cultures worldwide because much of the earth's landmass is unsuitable for growing cultivated crops. Approximately two-thirds of the world's 323 billion acres are permanent pasture, range, and meadows; of the total land area, only 3.4 billion acres are suitable for producing cultivated plants that can be directly consumed by humans. Most of the untillable acres can produce roughage in the form of grass, forbs, and browse that is digestible by grazing ruminants. Cattle and sheep are the most important domesticated species for this purpose. Appropriately managed, the grazing ruminant provides a highly sustainable and low input source of food in a variety of climatic conditions.

Agriculturalists produce what consumers want to eat, as reflected by the prices consumers can and are willing to pay. In most countries, as per-capita income rises, consumers increase their consumption of meat and animal products, which are generally more expensive on a per-pound basis than products derived from cereal grains. Food producers, processors, and services are responsive to consumer demands. The desire for lower fat content in the diet led to the provision of 1% and skim milk products, the reduction in fat from retail beef, and changes in the USDA grading standards. However, in developed countries, taste is equally as important, if not more so, as nutrition in consumer food selections.

Management practices are also sensitive to consumer demand. Cattle can produce large quantities of meat without grain feeding. Consumers in Argentina, South America, Australia, and New Zealand often prefer beef from grass-fed animals. Consumers in the United States, Canada, and the Pacific Rim typically prefer grain-fed beef. Only 20% of the concentrated feedstuffs (grain and protein meals) fed to livestock

are utilized by beef cattle. Less than 2 lb of concentrate is used per pound of live weight produced, which is lower than for broilers or hogs. The amount of grain feeding in the future will be dictated by cost and availability of feed grains, consumer disposable income and taste preferences, and the cost of non-feed inputs (e.g., fossil fuel, governmental regulation, etc.).

NUMBERS, PRODUCTION, CONSUMPTION, AND PRICES

World Cattle Numbers

There are approximately 1.45 billion cattle in the world. As indicated in Table 10.1, the leading countries in cattle numbers are Brazil (217 mil.), India (214 mil.), China (114 mil.), the United States (89 mil.), Ethiopia (54 mil.), and Argentina (51 mil.).

Table 10.1
Global Cattle Inventory, Beef Production, and Per-Capita Beef Availability

Region/Country	Human Population (million)	% of World Total	Cattle (million)	% of World Total	Beef Production (million lbs)	% of World Total	Per-Capita Production (lb/person/year)
North America							
Canada	35.9	0.5	12	0.8	233	0.2	65
Mexico	127.0	1.7	32	2.2	398	0.3	34
United States	321.8	4.4	89	6.1	25,790	18.3	82
Africa							
Ethiopia	99.4	1.4	54	3.7	751	0.5	10
Kenya	46.0	0.6	19	1.3	937	0.7	21
Nigeria	182.2	2.5	20	1.4	888	0.6	5
South Africa	54.5	0.7	14	1.0	1,876	1.3	35
Tanzania	53.5	0.7	22	1.5	660	0.5	13
Asia							
Bangladesh	162.0	2.2	24	1.7	427	0.3	3
China	1,376	18.7	114	7.9	14,095	10.0	11
India	1.311	17.8	214	14.7	2,131	1.5	3
Indonesia	257.5	3.5	17	1.2	1,203	0.9	6
Pakistan	188.9	2.6	38	2.6	1,792	1.3	19
Europe							
France	64.4	0.9	19	1.3	3,087	2.2	56
Germany	80.7	1.1	13	0.9	2,439	1.7	30
Ireland	4.7	0.1	7	0.5	1,141	0.8	50
Russian Fed.	143.4	2.0	14	1.0	3,601	2.6	36
Turkey	78.7	1.1	14	1.0	1,916	1.4	23
United Kingdom	64.7	0.9	10	0.7	1,867	1.3	41
Oceania							
Australia	23.9	0.3	29	2.0	5,110	3.6	89
New Zealand	4.5	0.1	10	0.7	1,243	0.9	105
South America							
Argentina	43.4	0.6	51	3.5	6,221	4.4	121
Brazil	207.8	2.8	217	14.9	21,329	15.1	87
Columbia	48.2	0.7	23	1.6	1,870	1.3	36
Uruguay	3.4	0.0	12	0.8	1,109	0.8	57
Venezuela	31.1	0.4	15	1.0	1,138	0.8	44
World	7,349.5		1,452		141,058		

Source: FAO, FAS-USDA.

World Meat Production

Although cattle numbers are abundant in India, the available beef meat (3 lb) per capita is very low because cattle are sacred to Hindus and, as such, they do not consume beef (Table 10.1). India is the world's largest exporter of beef and sells approximately one-half of its total production to the international market. Cattle and buffalo are used heavily for draft purposes and meat from these animals is considered a by-product. Buffalo supply 75–90% of the agricultural power in several developing countries. This power indirectly provides food—such as rice and other cereal grains—for these heavily populated areas of the world.

Interestingly, the world's largest producer of beef is the United States that is also the world's largest beef importer. As the world's fourth-largest beef exporter focused on selling high-quality grain-fed beef, the United States offsets its domestic demand for manufacturing beef utilized in ground beef trade by importing lower priced grass-fed beef and selling higher priced whole muscle cuts into the export markets.

World Beef Consumption

Per-capita consumption is not measured directly but rather estimated via the proxy metric of per-capita beef supply. Per-capita beef supplies are highest in Argentina, New Zealand, Australia, Brazil, and the United States (Table 10.1). Table 10.1 identifies leading countries in per-capita beef supplies and other countries whose consumption is highly dependent on imports. Cattle numbers in Australia, Argentina, Brazil, Ireland, New Zealand, and Uruguay exceed their human population, so these countries are large exporters of beef even though their per-capita beef consumption is high. Meat and beef production is relatively high in China; however, per-capita consumption is relatively low because of the extremely large population in that country.

INTERNATIONAL TRADE

Table 10.2 indicates the importance of beef cattle and beef products in international trade. Fresh and chilled beef exports are the highest value generator for the U.S. industry, followed by hides, edible offal, and tallow/fat.

Table 10.2
U.S. Exports and Imports of Primary Beef Products—2015

Commodity	Export Value ($)	Import Value ($)
Total agricultural products	150.5 billion	111.7 billion
Total animal products	17.27 billion	14.18 billion
Meat and meat products (excluding poultry)	12.43 billion	9.10 billion
Beef and veal (fresh/chilled)	5.45 billion	6.7 billion
Edible offal, variety meat	870.9 million	—
Other products		
Tallow and fat	267 million	—
Hides	1.30 billion	53.2 million
Breeding cattle	21.5 million	—
Semen	166 million	37.4 million
Embryos	11.2 million	—

Source: USDA-FAS.

Table 10.3
U.S. Beef and Veal Imports/Exports and Their Relationship to Domestic Production and Disappearance

	Beef and Veal Imports and Exports as a Percent of U.S. Beef Products				
Year	Beef/Veal Export (mil. lb)	Beef/Veal Imports (mil. lb)	Net Beef/Veal Imports	Exports as a % of Production	Imports as a % Disappearance
1965	54	942	888	0.27	4.56
1970	40	1,816	1,776	0.18	7.55
1975	53	1,782	1,729	0.21	6.69
1980	177	2,085	1,908	0.80	8.70
1985	332	2,091	1,759	1.37	8.02
1990	1,006	2,356	1,350	4.36	9.67
1995	1,820	2,104	284	7.13	8.24
2000	2,516	3,032	516	9.32	11.02
2003	2,523	3,006	483	9.51	11.05
2004[1]	461	3,679	3,218	1.86	14.82
2005	697	3,598	2,901	2.79	14.42
2010	2,300	2,298	−1.7	8.66	8.65
2015	2,266	3,370	1,104	9.49	14.11
2017[2]	2,430	2,670	240	9.49	10.42

[1]The low imports in 2004 are due to the single case of BSE identified in the United States on Dec. 23, 2003, which resulted in the closure of the international market to U.S. beef.
[2]Projected.
Source: Adapted from USDA and LMIC.

Various animal disease issues, including foot and mouth disease (FMD), bovine spongiform encephalopathy (BSE), and avian influenza, cloud the international trade data from 2000 to 2005. The most substantial of these factors was the diagnosis of a single U.S. dairy cow (imported from Canada) with BSE on December 23, 2003. This incident resulted in the near-complete cessation of U.S. beef exports and would require nearly a decade to regain the volume and value of beef exports that had been achieved prior to the BSE diagnosis.

Industry participants recognize the huge economic impact resulting from trade barriers resulting from both sanitary issues as well as international policy decisions by the United States and her trading partners. Additionally, shifts in domestic herd size affect beef supplies available for export. For example, as the U.S. cow herd contracted for approximately 15 years following 2000, domestic beef available for grinding to be supplied to the hamburger trade was severely diminished, which resulted in increased imports of lean beef. Table 10.3 describes the relationship of imports and exports to U.S. domestic beef production. While the United States imports significant amounts of manufacturing beef to be combined with the trimmings from fed cattle carcasses, the volume differential is vastly offset by the value of the whole muscle cuts exported by the United States to premium markets overseas.

Imports

Figure 10.1 illustrates the value of cattle, beef, and beef by-products imported into the United States. Beef and live cattle typically dominate the total value of imported beef items. The United States imports beef to meet the deficit in ground and processed beef as well as to maintain trade relations with other countries. The major suppliers of beef to the United States by volume are described in Figure 10.2. Australia is the highest volume beef supplier to the U.S. market with New Zealand and Canada each supplying about half as much as Australia.

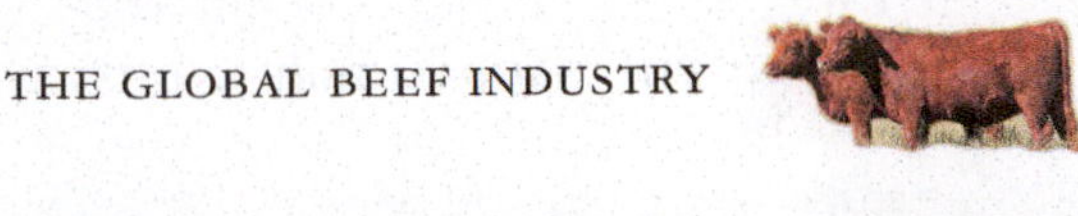

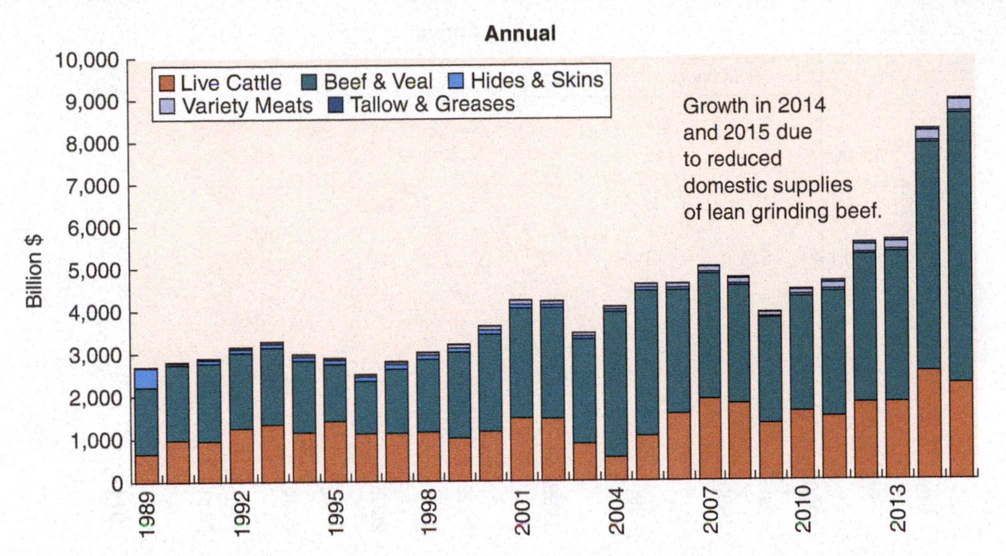

Figure 10.1
U.S. beef and beef by-product values.
Source: Adapted from USDA and LMIC.

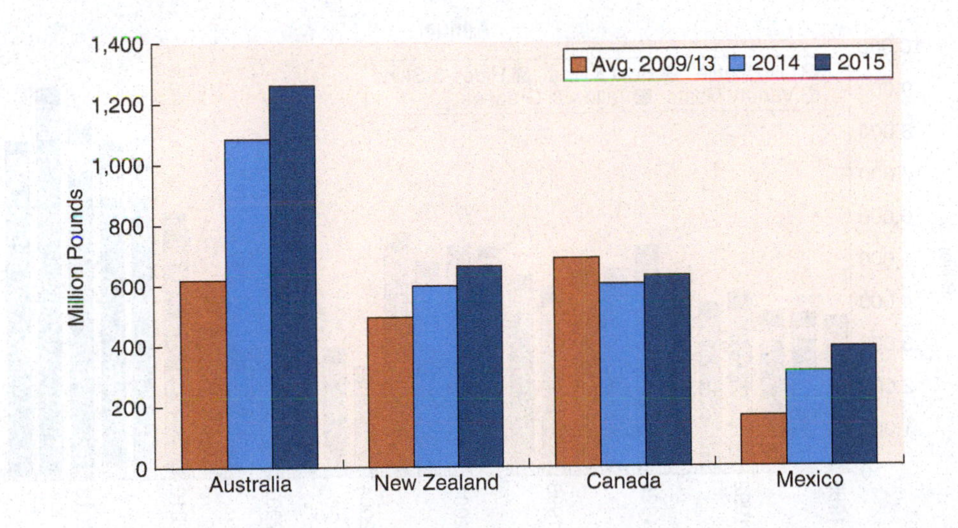

Figure 10.2
Leading sources of beef imported by the United States.
Source: Livestock Marketing Information Center.

The United States also imports live cattle predominately from its two North American Free Trade Agreement (NAFTA) partners—Canada and Mexico. The cattle volume imported from Canada and Mexico has been important over the years to provide sufficient volume to fill the capacity of the U.S. feedlot system and beef supply chain infrastructure (Figure 10.3). With the exception of the suspension of live cattle trade with Canada during the BSE issue in 2004, feeder cattle and fed cattle trade with the two closest U.S. trading partners has vacillated based on available inventories, exchange rates, and comparative prices (Figure 10.3). While the importation of both beef and cattle is of concern to some elements of the industry, they are critical components of the ability of the United States to retain a dominant position as the world's supplier of premium beef.

Exports

The beef export market is important to the U.S. beef industry for several reasons. First, hides, variety meats, and fat have higher values in foreign markets than in the domestic market and second, there is great demand for U.S. fed beef in countries where the standard of living is increasing. Finally, beef and beef by-product exports have a significant influence on U.S. cattle prices.

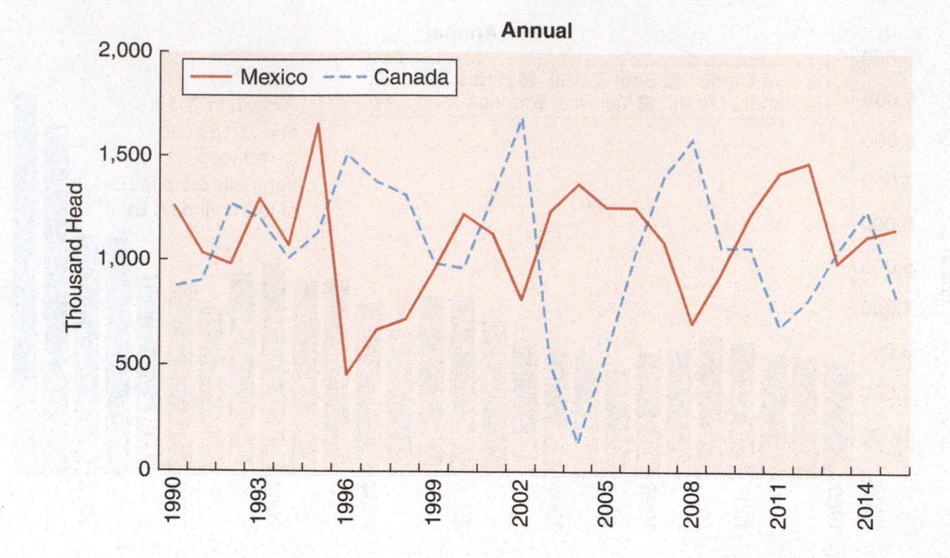

Figure 10.3
Cattle imports from Canada and Mexico.
Source: Livestock Marketing Information Center.

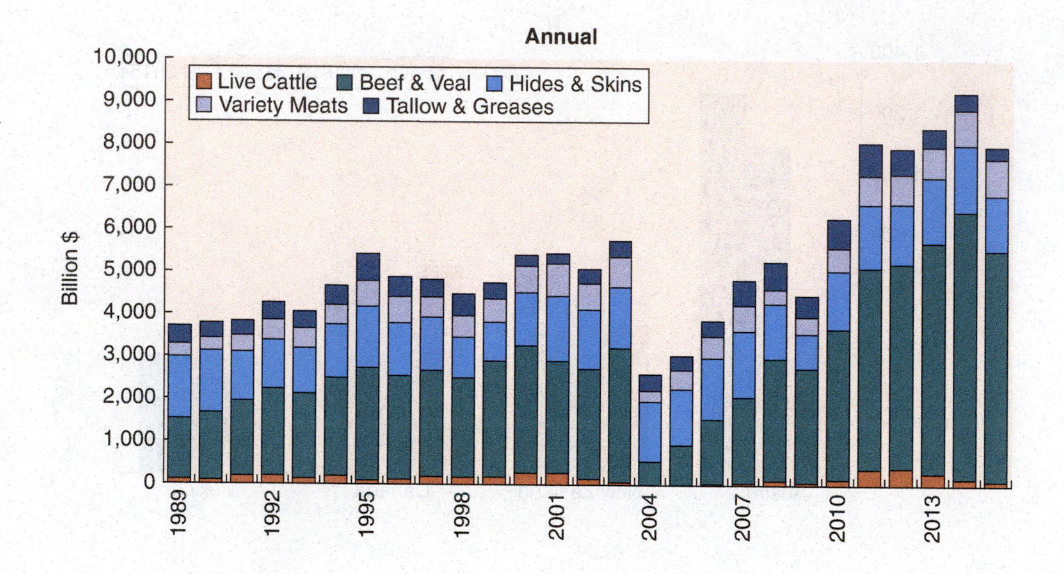

Figure 10.4
U.S. beef and beef by-product export values.
Source: Livestock Marketing Information Center.

Figure 10.4 shows the value of beef products in the export market. Beef and cattle hides comprise the biggest percentage of export revenues. Beef export values grew steadily from 1989 to 2003, prior to the BSE diagnosis of the single dairy cow in a Washington dairy at the end of 2003. Once the United States had established the safety of its beef supply, beef exports began a period of rapid growth that helped fuel the significant levels of profitability experienced by the industry from 1995 to 2016. From 2010 to 2016, the United States exported approximately 2.3 billion pounds of beef annually. The primary reason for a marked increase in exports has been the promotional activities of the U.S. Meat Export Federation in a number of overseas markets. The leading buyers of exported U.S. beef and variety meats values at more than $6 billion are Japan, Mexico, Canada, South Korea, and Hong Kong (Table 10.4).

The impact of biosecurity was never more pronounced for the U.S. industry than during the announcement on December 23, 2003 that an imported dairy cow from Canada had presumably tested positive for BSE. The short-term market effect was the suspension of beef trade from the United States, resulting in a net effect of a $13–15/cwt decline in the price of fed cattle and a $32/cwt decline in the price of 550-pound feeder steers. The loss of the export market in 2004 resulted in a loss of $4.7 billion to the U.S. beef industry, while the Canadian industry lost an estimated $7 billion. Furthermore, more than 4,500 U.S. workers lost jobs in the packing and processing sector as the result of trade suspension.

The impact of a strong international market on the price of individual beef cuts and beef by-product items is illustrated in Table 10.5. The beef export market provides beef producers the opportunity to capture maximum value and to increase the wealth generation capacity of the industry.

The export market is increasingly important to the U.S. cattle industry accounting for nearly 300 dollars value per head of fed steer and heifer. To assure sustained access to international markets, U.S. exporters must continue to strive for perfection in assuring their clients that their beef products are safe and wholesome. Furthermore, customer service and product value must be exceptional to maintain a competitive market position.

The United States has developed better hazard-analysis critical control point protocols and food safety intervention techniques to reduce microbiological

Table 10.4
VALUE OF U.S. BEEF EXPORTS TO PRIMARY CUSTOMERS

Global (bil. $)	Japan (mil. $)	Mexico (mil. $)	Canada (mil. $)	South Korea (mil. $)	Hong Kong (mil. $)	Middle East (mil. $)[1]
2006	66	1,169	433	0.61	15	114
2008	383	1,399	716	294	43	148
2010	639	819	733	518	159	261
2012	1,032	822	1,177	582	343	331
2014	1,579	1.166	1,030	847	1,151	274
2015	1,280	1,092	900	810	800	290

[1]Primarily variety meat trade.
Source: Adapted from U.S. Meat Export Federation and USDA.

Table 10.5
IMPACT OF U.S. BEEF EXPORTS ON THE PRICE OF VARIOUS BEEF CUTS AND BY-PRODUCTS

Item	U.S. Value $/kg	International Value U.S. $/kg	Exports as % of Production	International Clients	Wholesale cut of Origin
Short Ribs	6.50	7–10	>80%	Asia	Rib/Plate
Tongue	2–3	8–12	>90%	Japan	Variety Meat
Outside Skirt	11	13–14	>25%	Japan, Korea, Mexico, Central/South America	Other
Short Plate	2.5	3–4	>80%	Asia	Plate
Chuck eye roll	5	6–7	>25%	Asia	Chuck
Tripe	0.50–0.70	2–2.50	>90%	Mexico, South America	Variety Meat
Chuck flap tail	6	9.50–10.50	>25%	Asia	Chuck
Rib-eye roll	15	17.50–18.50	>6%	Asia, Middle East, Caribbean	Rib

Source: U.S. Meat Export Federation.

contamination of beef. Sustained efforts in these and similar areas yield beneficial results in the world market. To effectively compete in the global beef market, the U.S. industry must develop an export mentality, work more aggressively on quality issues to maintain consumer trust and competitive advantage, and respond proactively to the innovations of competitors.

A discussion of major beef-producing and -consuming nations follows. Nations are categorized as beef suppliers (Argentina, Australia, Brazil, European Union (EU), and New Zealand), NAFTA partners (Canada, Mexico, and the United States), and beef buyers (China, Japan, South Korea, and Hong Kong).

BEEF SUPPLIERS

Argentina

Population—43.4 mil.	Percent population in agriculture—2.4%
Number cattle per capita—0.85	Land area—1.07 mil. sq. mi.
Number cattle per sq. mi.—48	Cattle population—51.1 mil.
Carcass beef produced—6.21 bil. lb	Per-capita beef consumption—121 lb (carcass weight)
Carcass beef exported—51 mil. lb	Exports as a percent of production—9%

Beef is merchandised in Argentina predominately through local butcher shops (70%) with the remainder sold through supermarkets. Consumers prefer the convenience and familiarity of a local butcher and given that they have the highest per-capita beef consumption in the world, they have distinct preferences. Preferred retail beef cuts include traditional grilling cuts, ground beef, and round cuts to prepare milanesas (breaded veal). Butchers buy carcass sides from distributors and packers. Those with sufficient scale purchase live cattle and have them custom harvested. Supermarket chains are vertically integrated with arrangement tying them from the finishing phase to retail. Small and medium supermarkets usually debone half carcasses at the stores and sell cuts either in trays or via in-store full service meat cases.

Argentina's landscape can be divided into four major regions: (1) northern Argentina, which has a hot and humid climate; (2) the Pampa, a fertile plain that spreads west beyond Buenos Aires; (3) the Andine, which is the mountainous western region; and (4) the Patagonia, a dry windswept plateau in southern Argentina.

The Pampa and northern Argentina produce most of the beef and cereal grain production that are the country's leading farm products. The vast majority of Argentine beef was grass fed prior to 1990. However, grass-finishing systems were replaced over the next two decades with grain finishing feedyards that today account for nearly 85% of beef finishing in the country. Feedyard growth was spurred by the profitability challenges faced by the farming sector as well as the high costs associated with commodity transportation from point of production to market terminals. As such, cattle feeding became a more profitable approach to marketing grain. Furthermore, Argentina's infrastructure in highways, rail systems, and port facilities is not sufficiently developed to enhance competitive advantage. Cattle feeders typically purchase feeder cattle weighing 450 to 500 pounds and add 250 to 350 pounds of gain on high-energy rations to attain finished weights of 800 to 850 pounds that is preferred by the domestic market. Cattle produced for the EU markets are finished by 30 feedyards certified by the EU and typically feed to heavier finished weights of 950 to 1,000 pounds.

Historically, FMD has limited potential export markets to neighboring South American trade partners or to canned-cooked products in FMD-free regions of the world. In 1997, the United States opened trade to fresh/frozen beef from FMD-free zones within Argentina. However, FMD outbreaks in 2000–2001 closed the door to fresh/frozen beef from Argentina and Uruguay. The ban on fresh beef from Argentina into the United States was upheld from 2001 to mid-2015. The disease outbreak cost Argentina and Uruguay about one-third of their export volume, valued at $50 to $100 million annually. The losses of international markets bankrupted several meat packers and put increasing pressure on a marginal infrastructure. Coupled with the general collapse of the Argentine economy in 2002, the beef industry lost considerable opportunities to grow market share.

To regain FMD-free status, a national vaccination program was instituted to manage FMD. At a cost of $0.70 per head, the program required an expenditure of $35 million. With FMD-free status, Argentina became particularly competitive in lower-priced beef items with regard to trade in South Korea, Taiwan, Mexico, and Japan. By 2005, Argentina had become the third leading beef exporting nation. However, in response to rising domestic beef prices in 2006, the Argentine government instituted a 6-month ban on beef exports and then followed that policy with the implementation of a 15% tax on all beef exports. These moves created economic disincentives to which beef producers responded by shifting from cattle to soybeans eventually converting millions of acres from permanent pastures to soybean production. These policies were responsible for the industry toppling down the list of leading exporters to 11th by 2014. With 600 packing plants located in the country and one-third of them under federal inspection and only 100 certified for export, these policies forced the shutdown of 50 of the largest volume plants. The United States lifted its ban to Argentinian beef in July 2015, which provided some incentive for rebuilding the industry. Perhaps no other major beef producing nation has felt the sting of poor governmental policy than Argentina. To achieve its beef trade potential, Argentina must attain a number of economic and policy reforms. Monetary policy will be critical in determining how significant a role Argentina will play in the export market.

Australia

Population—23.9 mil.	Percent population in agriculture—2.6%
Number cattle per capita—1.22	Land area—2.97 mil. sq. mi.
Number cattle per sq. mi.—9	Cattle population—29.3 mil.
Carcass beef produced—5.11 bil. lb	Per-capita beef consumption—89 lb (carcass weight)
Carcass beef exported—3.83 bil. lb	Exports as a percent of production—75%

Australia is approximately the same size as the United States, excluding Alaska. However, the U.S. population is nearly 15 times that of Australia. Because of the dry conditions that prevail over the interior, Australians tend to be most concentrated in cities along the southeastern coastal region. In fact, more than 80% of the population resides in the states of Victoria, New South Wales, and Tasmania, plus the cities of Adelaide, South Australia; and Brisbane, Queensland. The remainder of the continent is sparsely inhabited. A vast majority of the cattle, sheep, wheat, cotton, fruit, and vegetable production occurs in the eastern half of the country. The distribution of cattle and feedyards in Australia are provided in Table 10.6.

Settlement of Australia was accomplished via the establishment of very large ranches or stations. Many of these enterprises remain in business. The largest landholders and beef producers are listed in Table 10.7. Some 75,000 individual enterprises raise cattle accounting for almost one-quarter of Australia's farm gate receipts.

Australia is an export-dependent economy. For example, Australia exports approximately 75% of its total beef production. Nearly 30% of Australian feeder

Table 10.6
DISTRIBUTION OF CATTLE AND FEEDYARDS BY AUSTRALIAN STATE

	% Cattle	% Feedyards
Western Australia	7.2	2.4
Northern Territory	7.2	—
South Australia	4.1	3.4
Queensland	44.3	57.8
New South Wales	19.6	30.0
Victoria	14.4	6.4
Tasmania	3.2	—

Table 10.7
LEADING LANDHOLDERS AND BEEF PRODUCERS IN AUSTRALIA

Rank	Company	Landholdings (mil. hectares)	Cattle (1,000 head)	Head Office
1	Nebo Holdings	11.7	500	Brisbane
2	Kidman Holdings	10.8	137	Adelaide
3	AACo, Ltd.	6.5	414	Brisbane
4	North Australia Pastoral Co.	5.7	188	Brisbane
5	Consolidated Pastoral Co.	5.2	242	Sydney
6	Heytesbury Beef	3.3	230	Perth
7	Colonial Agricultural Co.	2.0	128	Sydney

Source: Based on data from Australian Department of Primary Industries.

cattle (2.8 million head) are finished in feedlots. About 40% of beef exports are sold as frozen, with the remainder as chilled product. About 60% of beef exports are from grass-fed cattle and are targeted to the U.S. ground beef trade. The remainder (grain-fed) is directed at the Japanese market. Depending on market conditions, the United States and Japan will switch positions as the leading export market for Australian beef.

The primary markets for Australian beef are listed in Table 10.8. In addition to a thriving beef export business, Australia exports nearly 1.4 million head of live cattle annually (Table 10.8).

Almost all of the beef imported by the United States is utilized as hamburger. Australian exports of beef to the United States are dramatically influenced by the cattle cycle. When U.S. production reaches peak levels, the demand for imported beef declines. Australia also has the world's largest live cattle export market, exporting 1.4 million head to overseas markets (Table 10.8). About two-thirds of the live export trade originates from western Australia and the Northern Territory.

Traditionally, the Australian cattle production system has produced grass-fed beef (Figure 10.5). The Australian climate is drought-prone and the cattle inventory has been very responsive to changes in precipitation patterns. For example, the extended drought of 1895–1903 resulted in a 40% decline of the national cattle herd. The 1982–1983 drought resulted in an inventory loss of 11%, and the drought of 2002–2003 dropped the herd by 5%. However, as a result of increasing demand for higher quality products, its feedlot industry has been developed over the past several decades. Most of the feedlots are located in New South Wales and southern Queensland.

Table 10.8
Leading Destinations for Australian Exports of Beef and Live Cattle

Destination	Beef Exports (%)	Destination	Live Cattle Exports (%)
United States	34.9	Indonesia	54.2
Japan	22.5	Vietnam	22.5
Korea	11.6	China	5.8
China	9.3	Israel	4.8
Middle East	4.1	Malaysia	3.8

Source: USDA-FAS.

Figure 10.5
A group of Shorthorn cattle grazing in Australia. Forage resources are the foundation of the Australian industry.

The Australian industry is characterized as being highly innovative in many areas. From the development of an excellent national sire evaluation program to incorporation of concepts such as national identification and quality assurance to enhance its international competitiveness, the Australian beef industry is a major competitor on the global stage. Its long-range industry plan is focused around a model designed to improve the industry's competitiveness in five core areas—leadership and collaboration, market growth and diversification, improving consumer trust, enhancing the integrity and efficiency of the beef value chain, and increasing profitability and productivity. Without question, the Australian beef industry will be one of the world's leaders.

Brazil

Population—207.8 mil.
Number of cattle per capita—1.04
Number cattle per sq. mi.—66
Carcass beef produced—21.3 bil. lb
Carcass beef exported—3.53 bil. lb

Percent population in agriculture—14.5%
Land area—3.29 mil. sq. mi.
Cattle population—217 mil.
Per-capita beef consumption—87 lb (carcass weight)
Exports as a percent of production—16%

The scope of Brazil's national cattle herd as the world's largest coupled with an aggressive plan to become more export focused provides reason to be wary of Brazil as a significant competitor in international beef trade. In fact, during disrupted trade from North America in 2004, Brazil became the planet's leading supplier of beef and has maintained a relatively dominant position ever since. The Brazilian industry has several distinct advantages: relatively cheap land, significant grassland resources, and availability of low-cost labor. However, Brazil lags in production efficiency but is making substantial strides increasing productivity per hectare by more than 25% over the past decade. Brazil accounts for approximately 15% of the world's cattle and a comparable 15% of total beef production while the United States has only 6% of the global cattle herd but produces 18% of the total beef supply. Historically, expansion occurred in the mostly undeveloped regions of São Paulo, Goiás, and Mato Grosso. The cattle distribution in Brazil can be described regionally. In the west central region are 30% of the cattle, with the remainder located in the southeast (23%), northeast (17%), north (15%), and south (15%). The west central region hosts the largest share of the national cattle herd concentrated between Mato Grosso do Sul and north of Mato Grosso. The northern region is a major supplier to the domestic market.

The southern areas of Brazil are significant beef and dairy production regions and are responsible for exporting beef to Mercosul countries as well as the Middle East, China, and the United States. The Southeast is another important export-driven region. The packing infrastructure is heavily concentrated, with the largest companies controlling a high percentage of the total harvest. These companies have taken on an integrated structure where ownership from conception forward is the norm. It takes Brazil about three times as long to move cattle through the grass-based production and processing chain as it does in the grain finishing system favored in the United States. While the predominate production system is grassland based, approximately 4 million head of cattle are finished in confinement or semi-confinement. The industry is currently in an expansion phase of its feedlot sector, which intensifies production, reduces pressure on sensitive ecosystems, and increases efficiency. The goal is to double the capacity of the feedyard sector in the near term.

Brazil is the largest country in South America and has 5.8 million farms. The Nellore breed genetically influences approximately 85% of Brazil's beef herd. European and British origin breeds are more numerous in the southern states where export markets are the focus, while northern and central Brazil is populated by cattle of Bos indicus influence.

FMD also presents a barrier to the world market for Brazilian beef. The Brazilian government has worked to achieve FMD-free status for the states of Rio Grande do Sul and Santa Catarina as the result of an intensive vaccination program. Approximately 85% of Brazil's cattle are FMD-free as the result of vaccination. An outbreak in late 2005 has prevented Brazil from attaining full status as FMD-free. Part of the strategy has been the incorporation of a national identification and traceback system for the beef industry. Given these efforts, Brazil was able to attain certification and access to many international markets. While U.S. markets have remained closed to fresh beef from Brazil, it is expected that normalization of trade will occur sometime in the second half of 2017 that will set Brazil up to compete with Australia as the primary supplier of lean beef trimmings to the American buyer.

European Union

Population—505.1 mil.	Percent population in agriculture—5.8%
Number of cattle per capita—0.2	Land area—1.6 mil. sq. mi.
Number of cattle per sq. mi.—56	Cattle population—89.4 mil.
Carcass beef produced—16.9 bil. lb	Per-capita beef consumption—34 lb (carcass weight)
Carcass beef exported—2.2 bil. lb	Exports as a percent of production—13%

The EU is comprised of 28 member nations: Austria, Belgium, Bulgaria, Croatia, Czech Republic, Cyprus, Denmark, Estonia, Finland, France, Germany, Greece, Hungary, Ireland, Italy, Latvia, Lithuania, Luxembourg, Malta, the Netherlands, Poland, Portugal, Romania, Slovakia, Slovenia, Spain, Sweden, and the United Kingdom. The countries of the EU share common agricultural policies, including high support price programs, protection against competition from imports, and export restrictions (subsidies).

Beginning in 1989, U.S. beef was banned from the EU market because of the use of growth promotants in U.S. production systems. Trade representatives from the United States argue that the ban is not scientifically based and is a trade barrier disguised as a food safety concern. In July 1997, the World Trade Organization ruled in favor of the United States and ordered the EU to lift its ban or pay damages to the United States. The EU appealed the decision, but the decision in favor of the United States was upheld. The EU refused to submit to WTO authority and moved to permanently ban trade in beef grown with the use of hormonal growth implants.

In response, the United States initiated a system of "carousel" retaliation against a variety of EU products via the imposition of stiff tariff levels. Recognizing the challenges of gaining access to the EU market, U.S. exporters opted to focus on eastern European markets (Russia in particular) outside of the EU. In 1999, USDA introduced an EU access initiative called the Non-Hormone Treated Cattle (NHTC) program, which certified beef as being raised without the use of growth promotants for export to the EU. While many U.S. producers chose not to participate, the NHTC offered a route to access the EU beef market and has allowed the U.S. to become the fourth-largest exporter to the union. U.S. exports specifically target the foodservice and white table cloth trade and deliver grain-fed product. This contrasts with the grass-fed leaner product from Brazil, the largest exporter to the EU. U.S. beef sells at a significant price premium to South American beef. The beef industry in the EU was shaken by two events: BSE and FMD. While both of these events originated in Great Britain, the diseases spread to multiple sites in Continental Europe before full containment. The FMD outbreak in 2001 resulted in the sacrifice of 3.7 million animals, including 579,000 head of cattle in the United Kingdom for a loss valued at $2.85 billion.

The effects of the BSE and FMD situation were devastating for farmers, beef demand, and consumer confidence. The EU nearly lost its export market, prices fell by 15–30%, per-capita consumption of beef dropped by 10% (the equivalent of UK annual beef production), and live cattle trade was severely restricted. The EU beef industry has struggled to regain momentum following the aforementioned setbacks. From 2007 to 2014, the beef herd in the EU declined by approximately 4% in the full 28 member states while the 15 founding states experienced a decline of 6%. The leading beef producers in the EU are France (34%), Spain (15%), the United Kingdom (13%) and Ireland (9%).

The EU will likely rely increasingly on imported beef to meet its full demand needs in the future.

New Zealand

Population—4.5 mil.	Percent population in agriculture—6.4%
Number cattle per capita—2.5	Land area—59 thousand sq. mi.
Number cattle per sq. mi.—172	Cattle population—10.2 mil.
Carcass beef produced—1.45 bil. lb	Per-capita beef consumption—105 lb (carcass weight)
Carcass beef exported—1.27 bil. lb	Exports as a percent of production—85%

New Zealand has been described as the "best farm on earth" (Figure 10.6). Agriculture is very productive but is limited by landmass and proximity to major markets.

Figure 10.6
Upland grazing on the South Island of New Zealand.

To an even greater extent than Australia, New Zealand's beef industry is export driven. The climate on both the north and south islands allows production of vast quantities of forage. However, there is tremendous range in enterprise conditions, from improved pasture under irrigation to dryland range conditions. Herd productivity would be comparable, if not superior, to the average U.S. cow-calf enterprise.

The feedlot industry is very small, with only one feedlot of any consequence (capacity of 10–15,000 head). New Zealand exports are targeted to the United States, China, Japan, and Korea. The five largest markets for New Zealand beef are North America (United States—50% and Canada—6%), China (10%), Japan (6%), South Korea (6%), and Taiwan (5%). Marketing efforts are focused on creating an image of environmental compatibility with an emphasis on grasslands. This "green and clean" message has worked well with modern consumers.

The dairy industry is a major supplier of beef, with milking cows outnumbering beef females by more than double. Of the total export value, dairy and red meat products are the leaders. With only about 10% of the national cow herd in the form of beef cows, more than two-thirds of beef production originates from the dairy sector, mostly in the form of Holstein bulls finished on pasture. The mix of cattle contributing to the export trade is approximately one-quarter steers and heifers finished on grass, another 35% grass-finished bulls, and the remainder cows. As such, a very high percentage of beef imported from New Zealand is used in hamburger trade.

While New Zealand is constrained by land mass and geographic proximity, it is an important case study for those who value innovation, creativity, and adaptability in agricultural operations as well as marketing.

NAFTA PARTNERS

The North American Free Trade Agreement was implemented on January 1, 1994 to eliminate tariffs and other trade barriers between the United States, Canada, and Mexico. The inter-relationship of the beef industries of these nations will have significant impact on world markets.

Canada

Population—35.9 mil.	Percent population in agriculture—1.8%
Number cattle per capita—0.34	Land area—3.85 mil. sq. mi.
Number cattle per sq. mi.—3	Cattle population—12.2 mil.
Carcass beef produced—2.32 bil. lb	Per-capita beef consumption—65 lb (carcass weight)
Carcass beef exported—0.83 bil. lb	Exports as a percent of production—36%

As a North American Free Trade Agreement partner, Canada's beef industry is of significant interest. The fourth-largest nation on earth in terms of land mass, Canada has a relatively small human population. Nearly 80% of Canadians live within 200 miles of the U.S.–Canadian border. Similar to the United States, the average cow herd in Canada is small at approximately 61 head.

The Canadian beef industry is concentrated in the western provinces, while a majority of the population lives in the east. More than 70% of the Canadian beef cow herd is located in the three western most provinces: Saskatchewan (31%), Alberta (39%), and British Columbia (4%). The growth of the feedlot industry has centered in southern Alberta, which produces two-thirds of Canadian fed cattle. The cost of producing fed cattle in this region is considered among the lowest in North America. Almost 60% of Canadian beef harvest occurs in Alberta.

Beef produced in western Canada requires two transport days to reach the population centers of eastern Ontario. The evolving market will likely yield a scenario where eastern Canadian beef buyers will be supplied by the United States, while cattle and beef produced in the western provinces will be marketed into the United States. Beef shipped from Omaha can arrive in Toronto or Montreal in half the time required from a Calgary origination.

Canada is a net exporter of both live cattle and beef. Prior to the BSE case in Canada, beef exports to the United States grew rapidly, with Canada challenging Australia as the leading beef supplier. Of Canada's total beef exports, nearly 80% is destined for the U.S. market. However, Canada also imports live cattle from the United States, mostly from the Pacific Northwest regions (Figure 10.7). The primary suppliers and customers relative to Canada's beef imports and exports are described in Table 10.9.

The diagnosis of a BSE case in a Canadian cow in 2003 created havoc in the industry. The Canadian government estimated the total losses to their domestic beef industry at $7 billion. With the loss of its live cattle market to the United States, Canadian cattle inventories climbed 9% from 2003 to 2004, placing additional

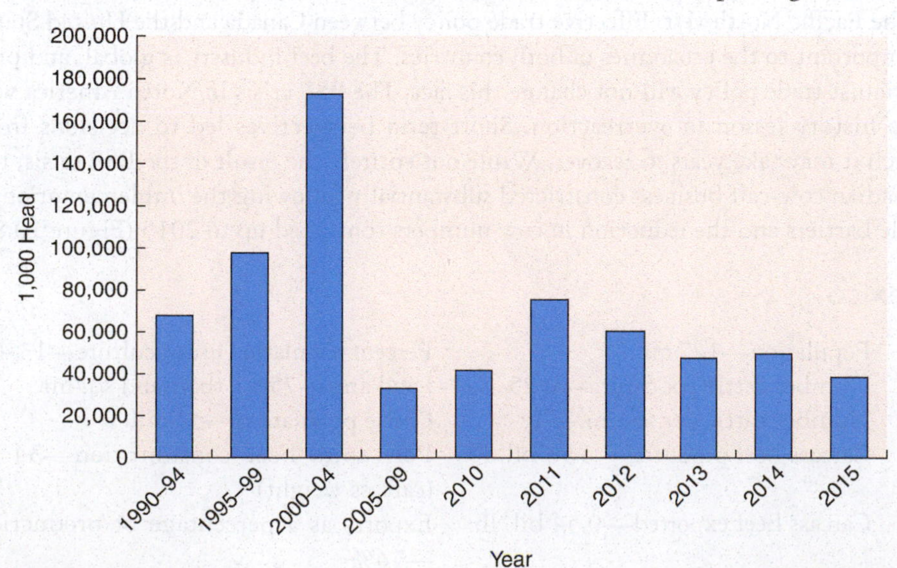

Figure 10.7
Imported cattle from the United States to Canada.
Source: Livestock Marketing Information Center.

Table 10.9
Canadian Beef Exports and Imports

Export Customer	Percent	Import Supplier	Percent
United States	79	United States	57
China	5	Australia	20
Mexico	5	New Zealand	12
Japan	4	Uruguay	9
Hong Kong	4	Brazil	1

Source: Based on multiple USDA databases.

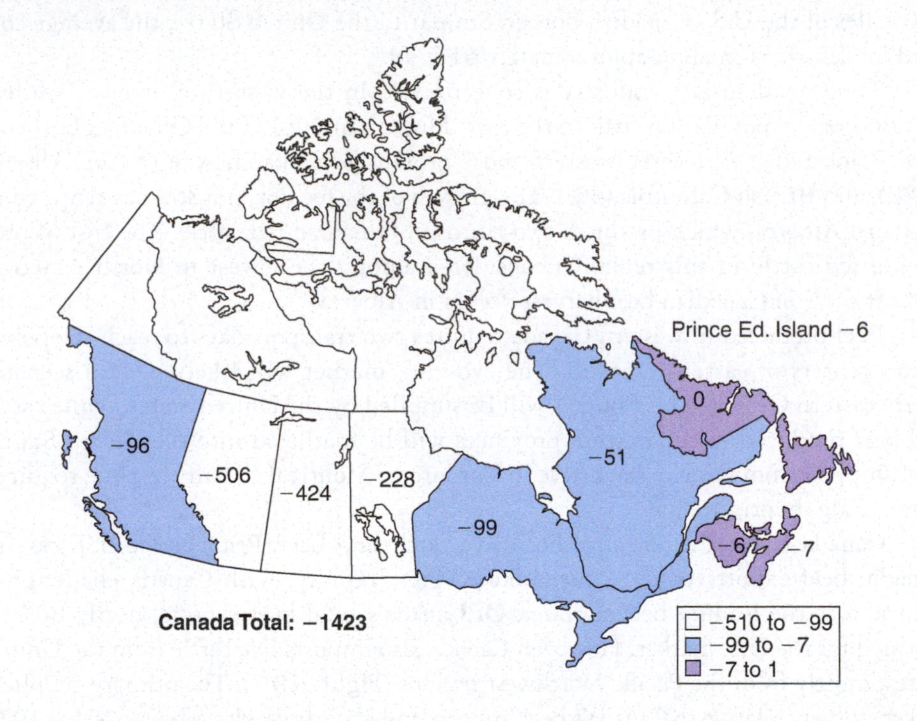

Figure 10.8
Changes in Canadian beef herd inventory.
Source: Livestock Marketing Information Center.

downside pressure on cattle prices. In 2004, fed cattle and feeder cattle prices dropped 30% and 40%, respectively, as a result of the closure of U.S. borders to both cattle and beef. Closed borders not only hurt the Canadian industry, but also put extreme pressure on the feeding and packing sectors of the U.S. industry, particularly those located in the Pacific Northwest. Effective trade policy between Canada and the United States is important to the economies of both countries. The beef industry is global, and protectionist trade policy will not change this fact. The BSE crisis in North America will be a history lesson in overreaction. Short-term perspectives led to decisions from which it may take years to recover. While not entirely the result of the BSE crisis, the Canadian cow-calf business constricted substantially following the implementation of trade barriers and the reduction in cow numbers continued up to 2015 (Figure 10.8).

Mexico

Population—127 mil.	Percent population in agriculture—13.4%
Number cattle per capita—0.25	Land area—756.1 thousand sq. mi.
Number cattle per sq. mi.—42	Cattle population—32 mil.
Carcass beef produced—3.98 bil. lb	Per-capita beef consumption—34 lb (carcass weight)
Carcass beef exported—0.54 bil. lb	Exports as a percentage of production—14%

Mexico can be described as having three geographic regions: the northern arid and semi-arid lands, the central high plateau, and the tropical regions. The northern Mexican states account for a majority of cattle exports to the United States. Seventy percent of USDA-approved packing plants are also located in the northern states. The northern states of Mexico are similar to the southern border states of the United States and account for about 30% of the Mexican herd. Breeding programs in this region are characterized by the use of Hereford, Angus, Charolais, and Simmental cattle, as well as Zebu crosses.

The central region accounts for 26% of the national herd but the genetic base is largely Criollo or Zebu. The tropical region has 44% of the total cattle inventory where Zebra and Continental–Zebu crosses are most typical.

Mexico is an importer of quality beef from the United States accepting $1.1 billion worth of beef and edible by-product in 2015 and an exporter of feeder cattle (Table 10.10). The cattle exported to the United States are typically light feeders with 60% of the volume weighing less than 400 pounds and the remainder weighing in the 400–700-pound range. Cattle exported to the United States must meet the following stipulations:

- Permanently identified;
- Retained in the export region for a minimum of 60 days prior to shipment under the provision that the region is disease free;
- Tested free of tuberculosis and treated for ticks prior to export; and
- Heifers must be spayed.

Mexico's beef industry is capable of meeting only 80% of the domestic beef demand. Mexico has emerged as the second-largest market for U.S. beef and variety meat. A strengthening economy and the fact that Mexico's domestic beef industry is not able to meet domestic demand are the reasons for a strong U.S. export market. As the middle class continues to grow in Mexico, the demand for modern food retailing and family-style foodservice establishments has also grown, accounting for about 30% of imported U.S. beef. The top 10 beef items exported to Mexico are the gooseneck round, clod, inside round, tripe, knuckle, brisket, skirt meat, chuck roll, lips, and liver.

The growth of Mexico's beach resort industry, as well as the Hotel, Restaurant and Institutional trade in the major urban centers, also offers export opportunities focused on the higher-valued middle meats.

Three challenges have been faced by the Mexican beef industry: drought, an unstable valuation of the peso, and a grass-based industry with limited grain feeding

Table 10.10
TRADE IN CATTLE AND BEEF FROM MEXICO TO THE UNITED STATES

Year	Cattle Exported to U.S. (1,000 hd)	Beef Exported to U.S. (million lbs)
1990	1,261.2	0.7
1995	1,653.4	4.6
2000	1,222.6	11.5
2005	1,256.4	22.8
2010	1,222.2	87.5
2012	1,468.2	198.4
2013	989.4	206.9
2014	1,115.8	253.7
2015	1,154.6	321.4

Source: Livestock Marketing Information Center.

potential. The development of the feedyard business in Mexico is relatively recent and is concentrated in four regions—Guadalajara, Jalisco (24%); Monterrey, Nuevo Leon (16%); Mexicali, Baja California (15%); and Sonora (13%).

Beef exports from Mexico to the United States are miniscule in volume. As such, Mexico experiences a very large beef trade deficit with the United States. If the Mexican economy can find a position of positive growth, its importance as a trading partner will increase. If investment in the packing, processing, and value-added capability of the Mexican industry fails to occur, then Mexico could be left in the unenviable position of providing raw good inputs (feeder cattle) while importing higher priced, value-added beef. Because of a strengthening economy, a culture that prefers beef in the diet, and an increasing per-capita demand for red meat and the low transportation margin, Mexico will continue to be a key customer for U.S. beef in the long term.

United States

Population —321.8 mil.	Percent population in agriculture—1.5%
Number cattle per capita—0.27	Land area—3.6 mil. sq. mi.
Number cattle per sq. mi.—25	Cattle population—89.3 mil.
Carcass beef produced—25.8 bil. lb	Per-capita consumption—82 lb (carcass weight)
Carcass beef exported—2.26 bil. lb	Exports as a percent of production—9%

Prior to trade disruptions in 2004, the United States exported nearly 10% of its annual beef production, resulting in net imports accounting for approximately 2% of its beef measured as a percentage of production. The trends in beef imports and exports have been most significantly impacted by trade disruptions and changes in the domestic cow herd that caused the rise in beef imports beginning in 2012 as vendors scrambled to find sufficient grinding beef to meet demand (Figure 10.9).

The international market is critical to the U.S. industry. The total value of the export of cattle, beef, and beef by-products was nearly seven times greater in 2003 than it was in 1975. Following the market disruptions of 2004, U.S. beef and beef by-product exports have rebounded dramatically (Table 10.11). Beef exports were valued at $5.4 billion in 2015 following the record-setting export year of $6.3 billion in 2014 (Table 10.11). The percentage of the total accounted for by each product category was: beef and veal—68%, hides—12%, variety meat—11%, tallow and fat—3%, and live cattle—1%.

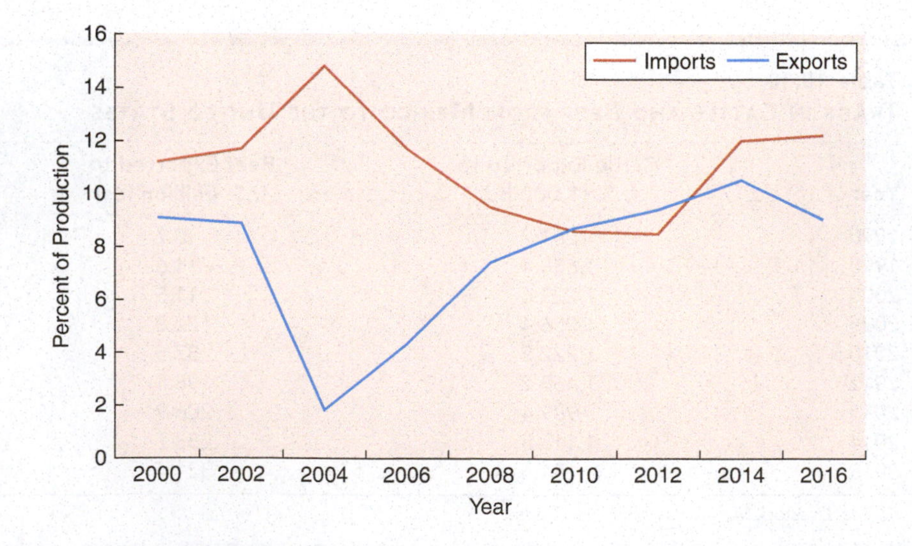

Figure 10.9
U.S. beef and veal imports and exports as a percent of production (carcass weight).
Source: Adapted from USDA.

Table 10.11
VALUE OF BEEF, CATTLE, AND BY-PRODUCT EXPORTS (MIL. $)

Year	Cattle/Calves	Beef/Veal	Hides/Skins	Variety Meat	Tallow, Greases, Lard	Total
1965	17	23	100	56	226	421
1975	77	70	270	110	332	859
1985	122	467	1,036	249	582	2,456
1995	86	2,646	1,444	613	661	5,451
2000	271	3,049	1,310	555	337	5,523
2005	7	930	1,315	447	337	3,037
2010	133	3,537	1,373	541	693	6,278
2013	280	5,427	1,545	730	432	8,414
2014	150	6,297	1,576	838	385	9,246
2015	99	5,431	1,300	871	276	7,978

Source: Adapted from USDA and Livestock Marketing Information Center.

The primary advantages of U.S. beef in the world market include the (1) ability to purchase individual items in volume, (2) tenderness and flavor, (3) high value for the price, (4) overall product quality, (5) confidence in product safety, and (6) positive image of the United States and its grading system.

Problems identified by foreign buyers that lowered their desire to purchase U.S. beef include (1) external fat exceeds purchase criteria, (2) inconsistent and/or heavy box or whole-muscle weight, (3) inadequate customer service, (4) excessive seam fat, (5) inadequate shelf life, (6) excessive packaging purge, (7) poor box condition, (8) leaks in vacuum packaging, (9) poor overall workmanship, and (10) inadequate labels both in terms of language and information.

Australia and New Zealand, in particular, have a stronger level of service orientation in the export market than does the United States. However, the United States has a distinct advantage in the ability to deliver shipments of specific cuts as opposed to delivery of full carcasses in boxed form. As competition increases for gaining share of the world's beef market, attention to customer service, improved quality, and better value delivery will be required. It will become increasingly important to invest in international marketing if the U.S. beef industry wants to ensure profitability and its role as a world leader.

BUYERS

China

Population—1.38 bil.
Number of cattle per capita—0.1
Number of cattle per sq. mi.—3
Carcass beef produced—14.1 bil. lb
Carcass beef exported—13 mil. lb
Percent population in agriculture—31.4%
Land area—33.69 mil. sq. mi.
Cattle population—114 mil.
Per-capita beef consumption—11 lb (carcass weight)
Exports as a percent of production—<1%

With the world's largest human population, China is a tempting target for the world's beef export market. The Chinese population and farmland are both concentrated in the eastern one-third of the country. Only 13% of China's land is arable. However, due to the very long growing season in the major farming region, China's agricultural sector is highly productive. Traditionally, beef was a by-product from the use of cattle as draft animals. The preferred source of animal protein is pork with beef accounting for approximately 15% the consumption of pork.

Throughout the 1990s, China maintained policies supportive to increasing beef production. Cattle and buffalo numbers grew by 21% from 1994 to 1996. From 1996 to 2002, beef production increased by 36%. However, the last decade has not sustained the same level of growth. An annual calf crop of approximately 49 million head is not sufficient to grow the national herd and thus some 200,000 head of live cattle are imported from three primary suppliers—Australia, New Zealand, and Uruguay. In 2014, China and Australia announced a free trade agreement that allows for 1 million head of live cattle imports. Chinese agricultural policy favors the use of forages and pasture to support the growing national cattle herd. Many analysts expect that Chinese beef imports will increase as demand for scarce resources limits growth of domestic beef production and as disposable income rises as a result of economic growth. China imports are expected to reach 600,000 tons of grass-fed beef annually, primarily from Australia with Uruguay, New Zealand, Argentina and Canada rounding out the top five. The United States established normal trade relations with the Chinese in 2000. As part of the agreement, beef tariffs were reduced from 45% to 12% over a five-year period. However, following the BSE incident in 2003, China closed its borders to U.S. beef and beef by-products. Reopening of the Chinese market to U.S. beef was announced in October of 2016. Given the size of the Chinese market with annual retail sales valued at more than $431 billion, it is no surprise that the world's largest retailers compete for increasing share of the market. The rapid growth of the Chinese economy has also been a beacon to world trade with a GDP that increased by double digits from 2005 to 2010. While the subsequent rate of growth has slowed with loss of momentum from 2011 to 2015, the dramatic size of the five major urban markets—Shanghai, Tianjin, Beijing, Guangdong, and Zhejiang—has generated rapid growth and innovation of the food retailing sector. Key retailers include those owned by U.S., French, German, Japanese, Taiwanese, Malaysian, and Thai companies as well as their Chinese owned competitors. Reopening China to U.S. beef would be a substantial advantage and one that would fuel increased profitability for the American cattle industry.

Japan

Population—126.6 mil.
Percent population in agriculture—3.8%
Number of cattle per capita—0.03
Land area—140.7 thousand sq. mil.
Number of cattle per sq. mi.—27
Cattle population—3.9 mil.
Carcass beef produced—1.12 bil. lb
Per-capita beef consumption—20 lb (carcass weight)
Carcass beef exported—1.99 mil. lb
Exports as a percent of beef produced —<1%

The opening of the Japanese beef market in April 1991 was a landmark in the development of the U.S. beef export market to the Pacific Rim. The combination of Japan's economic strength and growing demand among Japanese consumers for beef products has made Japan the focus of many beef exporting nations. Furthermore, Japan's beef and dairy industry is only capable of supplying about one-third of the domestic beef demand. Japan is the most valuable international customer of the U.S. beef industry. Japan is the world's third-biggest beef importer purchasing nearly $3.5 billion annually with 90% of the trade occurring with Australia and the United States. Trade with Japan is an enigma given its historic commitment to protecting its domestic beef industry by imposing significant tariffs, the complexity of its highly segmented trading and distribution system, and the lack of uniformity in policy with various trading partners. Economic growth that once characterized Japan has diminished and Japanese population growth has been negative since 2009, which reduces the potential for long-term beef demand.

The domestic beef supply chain is driven by significant price differentials between types of cattle and their respective carcass grades. The traditional and highly valued product is wagyu that holds a distinct price advantage compared to other domestic sources of beef (Table 10.12) as well as to imported beef. Wagyu is a highly marbled product that is extremely tender and is prepared either with rapid cooking techniques or in raw form. For the average consumer, wagyu is reserved for special celebratory occasions and the highest end foodservice establishments that feature traditional yakiniku (prepared over an open flame) and shabu-shabu dishes (cooked in a boiling broth similar to fondue style).

Japanese consumers are heavily concentrated in urban areas where living space is at a premium. Approximately one-half of Japan's population lives in three urban centers: Tokyo, Osaka, and Nagoya. Because of space limitations, the degree of home perishable food storage is minimal. Consumers traditionally purchase food in one-day increments to assure freshness. Shelf life, color, flavor, and presentation continue to be important influences on the buying decisions of Japanese customers. Because of the important role that food plays in Japanese culture, products with highly perceived quality are required to capture market share.

U.S. exporters also had to learn to effectively do business with Japanese customers. Japanese people are very conscious of preserving harmony in both professional and social encounters. Business negotiations require a great deal of patience, respect for experience, awareness of the focus on group rather than individual needs, and the requirement of establishment of personal trust prior to formal business arrangements.

Imported beef supplies are consumed in highest volume in family-style and fast-food retail outlets as well as at-home consumption. The United States holds a demand advantage compared to Australian product due to the palatability advantages of grain-fed versus grass-fed beef. A favorite fast-food item is a beef bowl dish called gyudon. This and similar products utilize end cuts from the chuck and round, which have added demand to beef cuts from the United States that are lower valued in its domestic markets. Ground beef is growing as a beef item in the Japanese market but is mostly supplied by Australian grass-fed supplies or domestic cull dairy cows.

The top 11 beef items exported to Japan are the short plate, chuck eye roll, chuck short ribs, tongue, outside skirt, rib-eye roll, hanging tenders, boneless short ribs, steak ready strip loin, chuck flap tail, and intestine. About one-half of U.S. beef is purchased at retail with the remainder marketed via foodservice. The foodservice industry in Japan has continued to grow and accounts for half of total beef sales. Chilled and frozen beef each account for about one-half of Japan's beef imports.

Table 10.12
PRICE SEGMENTATION BETWEEN DOMESTIC JAPANESE BEEF CARCASSES

Class	Grade	Yen/kg
Wagyu steer	A-3	2,220
Wagyu steer	A-2	2,056
Wagyu heifer	A-3	2,159
Holstein steer	B-2	1,076
Holstein cow	C-2	769
F1 cross steer	B-3	1,636
F1 cross steer	B-2	1,533
F1 cross heifer	B-3	1,593
F1 cross heifer	B-2	1,466

Source: USDA.

The most significant competition in the Japanese beef market is Australia. The Australian market has been the historic leaders in providing beef to Japan. The United States took over the lead position in 1996 but then relinquished its entire market share due to the 2003 BSE issue. Since then, the United States has steadily recaptured share from Australian exporters. U.S. chilled products sell at a decided advantage to Australian beef. For example, in 2015, U.S. chuck eye roll sold at a 200 yen/kg advantage and strip loins sold at a 1,600 yen/kg premium to Australian-sourced product.

Where Australia has a distinct advantage in terms of its trade agreement with Japan. Australia has an 8-percentage advantage in tariffs on chilled beef and 11-percentage on frozen beef. Despite trade liberalization, Japan imposes a 38.5% tariff on chilled and frozen beef imports. This tariff places significant downside pressure on the price of imported beef. Japanese consumers are hyperconscious of food safety. In fact, beef consumption drops in Japan every time a food safety problem occurs almost anywhere in the world. Marketing efforts that identify the producer, farm, and production practices of the beef purchased are taking hold in Japan. Consumer confidence is the primary goal of every competitor for the Japanese market. As Japanese consumers become more focused on value, price, and food safety, they will demand greater assurances from those who supply beef relative to the "3 Ks"—Kirei (clean), Kenko (healthy), and Kodawari (good value).

Japan is destined to rely on food producers outside its borders. With an average farm size of approximately 3 acres, Japan cannot produce enough food for itself. However, the Japanese market requires suppliers with an excellent food safety track record, the ability to adapt to Japanese culture, and the staying power to evolve into a trusted trade partner.

Republic of Korea

Population—50.3 mil.
Number of cattle per capita—0.04
Number of cattle per sq. mi.—84
Carcass beef produced—740 mil. lb
Carcass beef exported—37 mil. lb
Percent population in agriculture—6.1%
Land area—38.2 thousand sq. mi.
Cattle population—3.2 mil. hd
Per-capita beef consumption—32 lb (carcass weight)
Exports as a percent of production—<5%

South Korea is the fourth-largest international market for U.S. beef. The Korean market is heavily focused on HRI trade, as foodservice accounts for better than 60% of Korean beef consumption. The United States holds 41% of Korea's beef trade value and 36% of the volume. By comparison, Australia accounts for 53% of beef import value and 57% of the total tonnage. The top 10 beef items imported from the United States are short rib, chuck roll, rib finger, chuck short rib, chuck eye roll, shank, outside skirt, intestine, brisket, and back rib.

Liberalization of trade beginning in 2001 caused many Korean beef producers to dramatically decrease their herd inventories. Domestic cattle inventories declined 18.5%. Domestic cattle production in Korea originates almost entirely from small farms producing less than 5 head per farm. The predominant breed of domestic cattle is Hanwoo. In an effort to protect domestic cattle producers during the phase-in of trade liberalization, the Korean government established increased price supports, country-of-origin labeling of beef in restaurants, development of specialized marketing infrastructure to promote Hanwoo as a branded product, and incentives to improve the quality of Korean beef. Nonetheless, domestic cattle stocks have continued to decline that opens additional opportunities for beef exporting countries. A significant policy shift occurred in 2005 that eliminated a separate retail distribution system for imported beef that had limited U.S. beef offerings to only about 10% of retail outlets.

Table 10.13
AVERAGE RETAIL PRICES FOR VARIOUS BEEF SOURCES IN KOREA (WON PER 100 GRAMS)

Source and Description	2013	2015
Hanwoo—Grade 1	3,109	3,739
Hanwoo—Grade 3	2,154	2,522
U.S.—chilled	2,334	2,812
U.S.—frozen	1,487	1,722
Australian—chilled	1,903	2,206
Australian—frozen	1,249	1,452

Source: Adapted from multiple sources.

The change opened an additional 30,000 markets for imported beef. In 2005, Hanwoo beef had a nearly three times price premium compared to U.S. beef. By 2015, that premium had shrunk to a 1.3 times U.S. beef advantage. Price comparisons for Korean Hanwoo, and Australian and U.S. beef are provided in Table 10.13. While domestic beef holds the advantage, it is clear that the flavor profile of U.S. product provides a decided premium to Australian product.

As compared to tariffs in Japan, the United States has a decidedly better trade arrangement in South Korea as compared to Australia. In 2016, the import duty on U.S. beef was 26.6% while Australian exporters were assessed a 32% duty. Current trade agreements assure a continued reduction of the beef tariff such that by 2025, the import duty on U.S. beef will be only 2.7%. While economic and political considerations are always at play, beef demand strength coupled with a limited capacity for beef production growth in the domestic market provides a scenario that warrants continued investment in growing market opportunities in South Korea.

TRADE POLICY AND OTHER FACTORS

Animal products, including beef, are nearly always preferred throughout the world when they are available and consumers have a relatively high level of disposable income. Availability of animal products is influenced by the degree of consumer affluence, the percentage of the population working in agriculture, the cultural norms of a particular society, and the degree of agricultural production, processing, and marketing infrastructure development. However, trade policy and macroeconomic factors play a major role in the movement of beef through international market channels.

Issues beyond the control of beef producers can have significant impacts on trade. For example, the U.S. beef industry lost an entire quarter of chilled beef exports in 2015 as a result of a labor dispute that shut down shipping from West Coast ports. Beef supplies both domestically and from competitors are affected by drought, animal disease, cost of production inputs, and an assortment of mitigating factors. Factors as diverse as currency exchange rates, oil prices, economic trends in both the United States and its trading partners, and shifts in international political winds all have an impact on trade. In essence, the opportunities are too great to not participate in the global beef market but participation does not come with a guarantee of security and stability.

Volatility is a given in international markets. For example, the price of hides and offal is driven by international market factors and is inherently volatile; from 2014 to 2016, the per cwt price dropped by nearly $7 that equates to a nearly $100 per head loss in value.

Trading partners are not all equally reliable. Russia, which only has the production capacity to meet two-thirds of its domestic beef demand, is viewed as a potentially profitable market but it is also an unreliable trading partner. For example, in August 2014, Russia banned beef imports from the United States, EU-28, Canada, Australia, and Norway in a counter-sanctions embargo.

Unfortunately, the response to foreign trade barriers is to initiate retaliatory tariffs. While these arguments sell well in the heat of political grandstanding, they do stand the test of economic evaluation. For example, in 2009, the Obama administration imposed "safeguard" tariffs on tires imported from China to save just over 1,000 jobs. However, the cost per job saved was nearly $900,000. Furthermore, the cost of tires then climbed in the United States costing consumers some $1.1 billion annually and due to reduced spending in other areas of the economy ultimately resulted in the loss of approximately 2,500 jobs. In 2002, the Bush administration implemented a round of tariffs on imported steel to save 1,700 jobs. Unfortunately, the price of steel rose that resulted in an estimated loss of more than 10,000 jobs each in California, Illinois, Michigan, Ohio, and Texas. The unintended consequences of protectionism do not justify the action.

With more than 95% of the world's consumers living outside the United States, the development of bilateral agreements designed to reduce trade barriers offers the greatest opportunity to the beef industry and other segments of the U.S. economy. For example, the proposed Trans-Pacific Partnership (TPP) creates an agreement among 12 nations, including the United States, Japan, Malaysia, Singapore, Vietnam, Mexico, Canada, and New Zealand, to eliminate more than 18,000 tariffs placed on American-made goods. The agreement reduces current Japanese tariffs on U.S. beef from 38.5% to 9%, thus opening the door for more trade with the largest customer of the industry. Furthermore, the agreement assures science-based phytosanitary standards, eliminates export subsidies, and protects trade from reactionary disruptions.

While the U.S. consumer remains the core customer, the beef industry must develop exceptional export capabilities (Figure 10.10) to assure the continuation of farming and ranching for generations to come.

Figure 10.10
An example of a U.S. beef product in a European meat case.

SELECTED REFERENCES

Publications

Action: Expanding foreign markets for U.S. meat products (periodical). Denver, CO: U.S. Meat Export Federation.

Agricultural trade highlights. 2015. Washington, DC: USDA-FAS.

Argentina economic trends report. 2015. U.S. Embassy, Buenos Aires, Argentina.

Argentina livestock and beef market situation. 2015. Secretariat for Agriculture, Buenos Aires, Argentina.

Australian livestock and meat corporation annual report. 2014–15. Sydney, Australia.

Band, S., Goddard, M., and Hygate, L. 1992. *Targeting the Japanese beef market*. Sydney South, Australia: Meat Research Corporation.

Canada's beef industry. 2015. Canfax Research Services. Calgary, Alberta, Canada.

Cattle and beef: Impact of the NAFTA and Uruguay round agreement on U.S. trade. 1997. Washington, DC: International Trade Commission.

FAO statistical databases. (2016). Rome, Italy: FAO.

FAPRI 2016 international agricultural outlook. 2016. Ames, IA: Iowa State University and Columbia, MO: University of Missouri.

Garcia, J.J. 1996. *Argentina's position relative to the global beef market*. Master's thesis. Fort Collins, CO: Colorado State University.

Gleeson, T.D., McDonald, S., Hooper, P., & Martin, P. 2003. *Australian beef industry: Report on the Australian agricultural and grazing industries survey of beef producers*. Canberra, Australia: Australian Bureau of Agricultural and Resource Economics.

Global Agricultural Information Network. 2016. Livestock and Products Reports by Country. United States Department of Agriculture: Foreign Agriculture Service. Washington, D.C.

International agricultural trade reports. 1998–2016. Washington, DC: USDA-FAS.

Japanese business and etiquette and protocol. 1990. Washington, DC: EPISTAT International, Inc.

Livestock, dairy, and poultry situation and outlook. 2016. Washington, DC: USDA-ERS.

Livestock Marketing Information Center. 2016. USDA-ES. Denver, CO.

Morgan, J.B., Smith, G.C., Belk, K.E., & Neel, S.W. 1994. *International beef quality audit*. Fort Collins, Co: Colorado State University and U.S. Meat Export Federation.

New Zealand Meat Producers Board. 2015. Annual Report. Wellington, New Zealand.

Peel, D.S., Mathews, K.H., and Johnson, R.J. 2011. *Trade, the expanding Mexican beef industry, and feedlot and stocker cattle production in Mexico*. United State Department of Agriculture. Economic Research Service. Washington, DC.

Smeaton, D.C. (Ed.). 2003. *Profitable beef production in New Zealand*. Wellington, New Zealand: New Zealand Beef Council.

The value of U.S. beef and beef variety meat exports. 2016. Denver, CO: U.S. Meat Export Federation.

U.S. meat export analysis and trade. 1995–2016. USDA-AMS, Des Moines, IA; Iowa State University, Ames, IA; USMEF, Denver, CO.

United States Trade Representative. 2015. The Trans-Pacific Partnership. Department of Commerce. Washington, DC.

Variety meats from the USA—A buyer's guide. 2003. Denver, CO: U.S. Meat Export Federation.

World agricultural production. 2016. Washington, DC: USDA-FAS.

World livestock situation. 2016. Washington, DC: USDA-FAS.

World population profile. 2016. Washington, DC: U.S. Department of Commerce.

11

Reproduction

Reproductive success sets the stage for the entire beef industry and requires an integrative systems approach focused on selecting and developing breeding stock, assuring successful breeding season outcomes, and then managing gestation, calving, and rebreeding (Figure 11.1). A live calf born and weaned from each breeding female each year is the primary objective for successful cattle reproduction. However, cows are not managed as individuals but as a herd, so the economic evaluation of total herd reproductive performance is most critical. Reproductive efficiency in cattle, as measured by the number of calves born and weaned each year per 100 females in the breeding herd, is considered one of the most important economic factor in cattle production. Reproduction is at least twice as important to profitability as growth or carcass for cow-calf producers who sell their calves at weaning. It is essential for producers to understand reproductive principles, for they are the focal point of overall animal productivity.

Producers can economically manage their herds for optimum reproductive rates. Management decisions that result in breeding animals having early puberty, high conception rates, minimum calving difficulty, and early rebreeding are most critical. Reproductive losses can be minimized by keeping cattle healthy, providing adequate nutrition, maintaining a strong herd health program, selecting genetically superior animals, and providing good management at parturition.

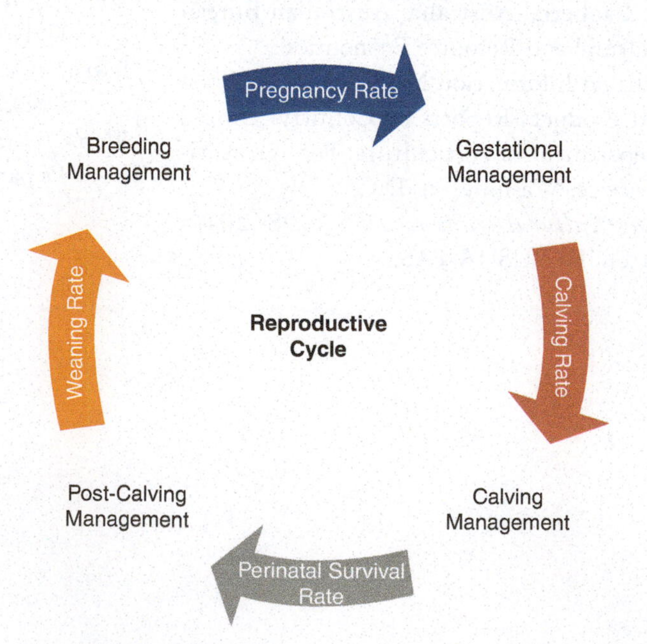

Figure 11.1
Beef cattle reproductive cycle.

STRUCTURE AND FUNCTION OF THE REPRODUCTIVE ORGANS

The Reproductive Organs of the Cow

Figure 11.2 shows the reproductive organs of the cow and their location in the body. The reproductive organs consist of a pair of ovaries (suspended by ligaments just behind the kidneys), and a pair of funnel-shaped tubes (infundibulum), which are part of the oviducts that lead directly into the uterine horns. The two uterine horns merge together to form the uterine body. Collectively, the two uterine horns and the uterine body comprise the uterus (womb). The uterus leads into the cervix, which has a folded surface surrounded by muscles. The cervix opens into the vagina, a relatively large

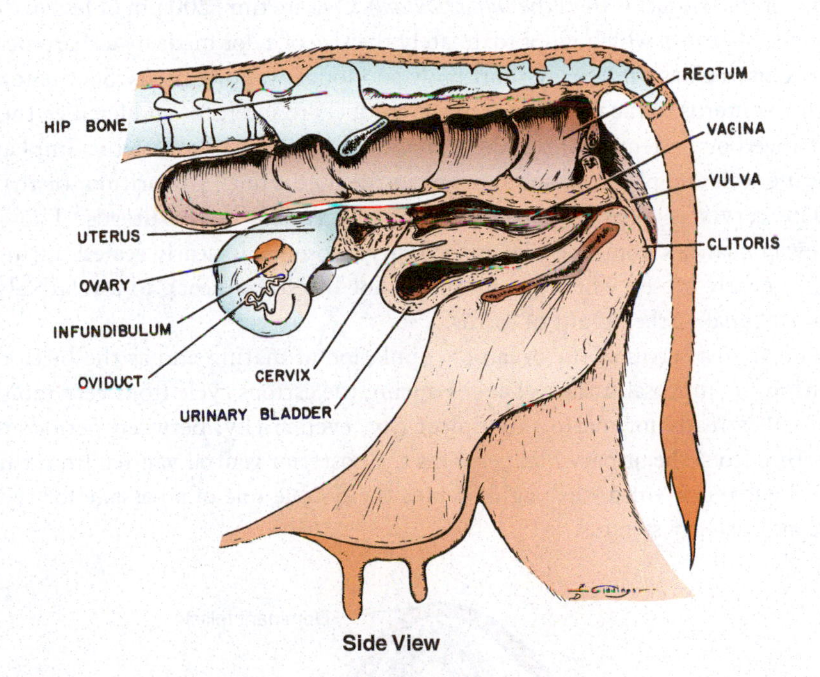

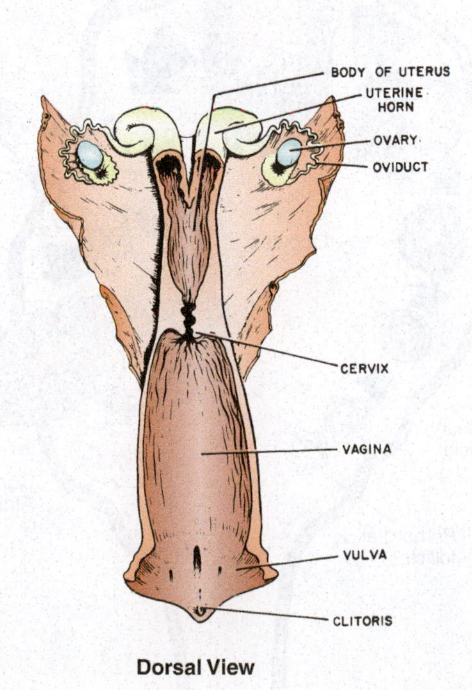

Figure 11.2
Reproductive organs of the bovine female.
Source: Colorado State University.

canal or passageway that leads posteriorly to the external parts—the vulva and the clitoris. The urinary bladder empties into the vagina through the urethral opening.

The ovaries produce eggs (female sex cells, also called *ova*) and hormones. Each egg, or ovum, is individually generated by follicles within the ovary. All of the follicles that a female will ever produce are present at birth as immature follicles. The majority of these follicles (>99%) grow to various stages, cease growth, deteriorate, and are absorbed (Figure 11.3). Less than 1% of follicles will mature (Graafian follicles), reaching a maximum size of about 0.4 in. in diameter, and go on to ovulate. During ovulation, the egg is released from the mature follicle and the follicle develops into a "yellow body," or corpus luteum, which becomes a vitally important structure if pregnancy occurs.

The oviducts receive the ova immediately after the ova leave the ovaries through the open end of the oviduct (called the *infundibulum*). Ova are tiny, 200 μm or less in diameter (200 μm = 1/5 mm), which is approximately the size of a dot made by a sharp pencil.

The uterus has a relatively short body and longer uterine horns. Spermatozoa pass through the uterus to the upper ends of the oviducts to intercept and fertilize the ovum. The fertilized ovum (embryo) travels to the uterus, develops into a fetus, implants into the uterine wall and placenta, and remains in the uterus until parturition (birth).

The cervix is a narrow opening from the vagina to the uterus. The cervical passageway changes from one that is so tightly closed it is nearly sealed during pregnancy, or nearly closed when the animal is not in estrus (heat), to a relatively open, very moist canal at the height of estrus.

The vagina serves as the organ of copulation at mating and as the birth canal at parturition. Its mucosal surface changes during the estrous cycle from very moist when the animal is ready for mating to almost dry, even sticky, between periods of heat. The urethra from the urinary bladder joins the posterior ventral vagina; from this juncture to the exterior vulva, the vagina serves the double role of a passage for the reproductive and urinary systems.

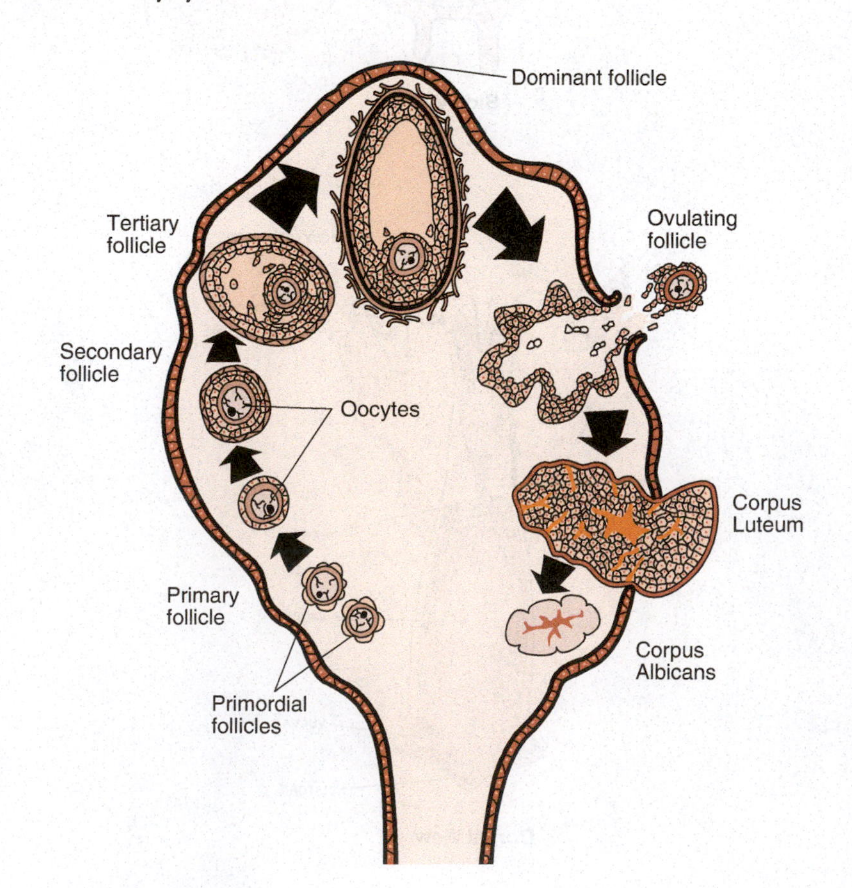

Figure 11.3
A cross section of the bovine ovary showing how a follicle develops to full size and then ruptures, thus allowing the egg to escape (ovulation). The follicle then becomes a "yellow body" (corpus luteum) which is actually orange-colored in cattle. The corpus luteum degenerates in time and disappears. Many follicles cease development, stop growing, and disappear without ever reaching the mature stage.
Source: Sean Field.

A highly sensitive organ, the clitoris is located at the lower tip of the vagina. The clitoris is the homologue of the penis in the male (e.g., it comes from the same embryonic source as the penis). There is some research indicating that clitoral stimulation or massage following artificial insemination will increase conception, but it has not been well verified.

Ovarian hormones (progesterone and estrogen) fluctuate during the estrous cycle in the mature heifer or cow. Estrogen is produced by the follicle in response to follicle-stimulating hormone (FSH) production from the anterior pituitary. More specifically, FSH stimulates follicular development on the ovary and thus production of estrogen. The increased levels of estrogen in the bloodstream stimulate the female to display behavioral estrus. Luteinizing hormone (LH) is produced by the same cells of the anterior pituitary that produce FSH. For the most part, luteinizing hormone and FSH levels parallel one another in the bloodstream throughout the estrous cycle. Luteinizing hormone has three primary functions. One is to stimulate ovulation of a properly developed follicle during estrus; the second is to control the formation of the corpus luteum; and the third is to stimulate progesterone production from the luteal cells of the corpus luteum. The ovarian hormones estrogen and progesterone feed back to control LH and FSH secretion by controlling the secretion of gonadotropin releasing hormone (GnRH) from the hypothalamus.

When estrogen levels increase, just prior to estrus and during estrus, it stimulates a surge of GnRH and thus a significant release of FSH and LH that results in follicle maturation and ovulation. Following ovulation, the cells that once produced estrogen have been destroyed and are luteinizing to form the luteal cells of the corpus luteum (CL). Thus, estrogen production diminishes rapidly. Formation of a mature CL occurs during the next 4–5 days of the estrous cycle. As the corpus luteum matures, progesterone production increases to a peak on about day 10–12 of the estrous cycle and remains elevated until approximately day 17 of the estrous cycle. This elevated level of progesterone feeds back to the hypothalamus to diminish GnRH secretion. Thus, LH and FSH secretions also decrease. If the uterus of the cow has not recognized a pregnancy by day 17, the uterus begins to produce the hormone prostaglandin $F_{2\alpha}$. Prostaglandin ($PGF_{2\alpha}$) acts upon the corpus luteum to cause CL regression. As a result of CL regression, progesterone levels decrease abruptly, removing the inhibition on GnRH, LH, and FSH secretion. As a result, GnRH secretion increases. Thus, LH and FSH secretions increase, and the increased level of FSH in the bloodstream stimulates new follicular development and increased estrogen production, resulting once again in estrus (Figure 11.4).

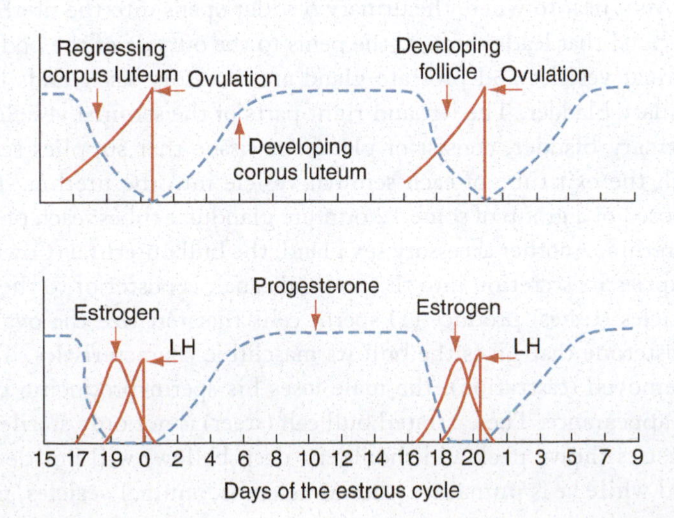

Figure 11.4
The major hormone and ovarian changes in the cow's estrous cycle.

Ovulation occurs in the cow approximately 24–30 hours after the beginning of estrus. During ovulation, the follicle ruptures because of the interaction of hormones rather than bursting as a result of built-up pressure inside. When ovulation occurs, the follicular fluid surrounding the ovum carries the ovum into the infundibulum of the oviduct. After the ovum is released from the follicle, the remaining cells of the follicle luteinize during the next 4–5 days to form the corpus luteum. The corpus luteum is about twice the size and about the same shape as the mature follicle, but no longer contains the egg. The importance of the corpus luteum is to produce high levels of progesterone that maintain the optimum uterine environment for embryo development to allow the embryo to signal its presence. If an embryo does not signal its presence by approximately day 17 of the estrous cycle, the uterus begins to secrete prostaglandin $F_{2\alpha}$ to stimulate regression of the corpus luteum and initiation of a new estrous cycle.

If pregnancy occurs, the corpus luteum remains present on the ovary and continues to produce progesterone at high levels for the duration of pregnancy. These high levels of progesterone suppress GnRH and thus LH and FSH production, resulting in low levels of estrogen production from the ovary. Thus, cows do not display heat during pregnancy.

The process of follicle development, ovulation, and corpus luteum development and regression is illustrated in Figure 11.3. The cow will exhibit estrus approximately 12 hours (a range of 10–27 hours); the average length of the estrous cycle is 21 days (a range of 19–23 days). The major changes in ovarian structure and function in relation to the estrous cycle are shown in Figure 11.4.

The changing length of day also influences the estrous cycle, the onset of pregnancy, and the seasonal fluctuations in male fertility. Changing day length acts both directly and indirectly on the animal. It acts directly on the hypothalamus by influencing the secretion of hormones, and indirectly by affecting plant growth, thus altering the amount and quality of the available nutrition. Increasing length of day is associated with increased reproductive activity in both males and females.

The Reproductive Organs of the Bull

Figure 11.5 shows the reproductive organs of the bull. They consist, in part, of two testes (testicles) that are bean-shaped organs held in the scrotum. Male sex cells (called *sperm* or *spermatozoa*) are formed in tiny seminiferous tubules of the testis. Sperm from each testicle then pass through very small tubes into an epididymis, which is a tube that is attached to the exterior of the testicle. Each epididymis leads to a larger tube, the vas deferens (also called the *deferent duct* or *ductus deferens*). The two vas deferens converge from the left and right sides of the body to connect with the urethral canal at its upper end, very near to where the urinary bladder opens into the urethra. The urethra is a large canal that leads through the penis to the outside of the body.

The seminal vesicles and prostate gland are accessory sex glands found at the base of the urinary bladder. The left and right parts of the seminal vesicles, which lie against the urinary bladder, consist of glandular tissue that supplies secretion that moves through the exit tube of each seminal vesicle into the urethra. The prostate gland is composed of a group of some 12 or more glandular tubes, each of which empties into the urethra. Another accessory sex gland, the bulbourethral (Cowper's) gland, which also empties its secretion into the urethral canal, is posterior to the prostate.

The testicles (testes) produce (1) sperm cells that fertilize the ova and (2) the hormone testosterone that gives the bull its masculine characteristics. Thus, if both testicles are removed (castration), the male loses his sperm-producing capacity and his masculine appearance. The castrated bull calf (steer) is not only sterile but masculine characteristics (heavy neck and shoulders, deep bellow) will not develop. If the calf is castrated while he is immature, the vas deferens, seminal vesicles, prostate, and

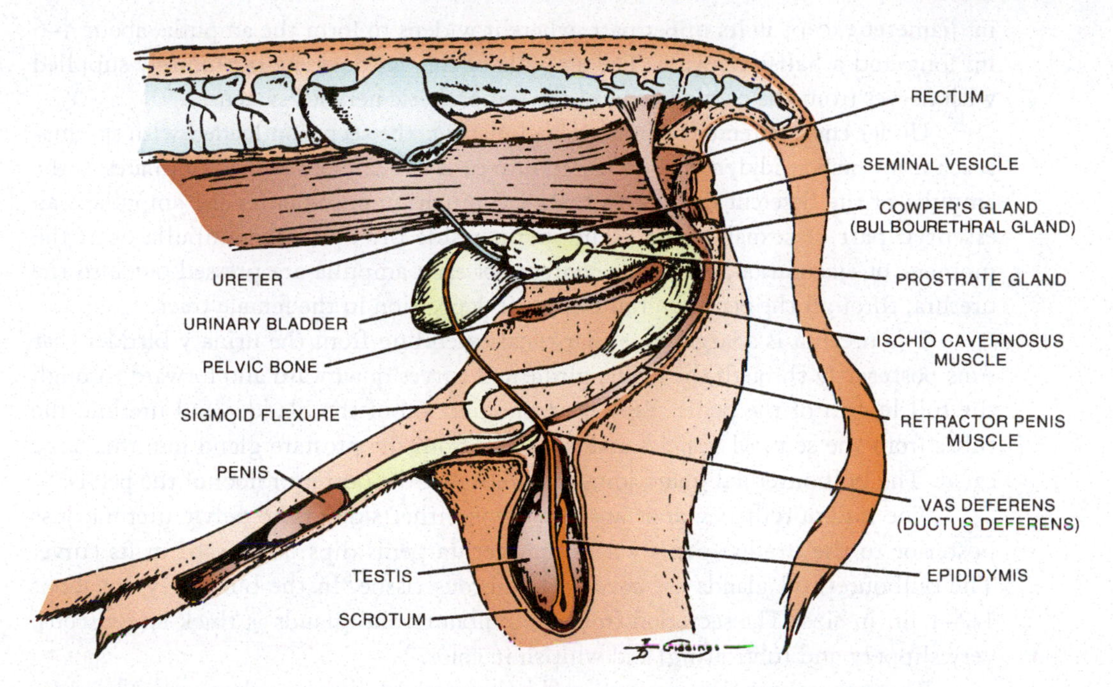

Figure 11.5
Reproductive organs of the bull.
Source: Colorado State University.

bulbourethral glands all but cease further development. If castration is done in the mature animal, the remaining genital organs tend to shrink in size and function.

The scrotum is a two-lobed sac that contains and protects the two testicles. The scrotum also regulates temperature of the testicles, maintaining them at a temperature lower than body temperature (3–7°F lower). When environmental temperature is low, the tunica dartos muscle of the scrotum contracts, drawing the testicles toward the warmer temperature of the body; when the environmental temperature is high, this muscle relaxes, permitting the testicles to drop away from the body and its warmth. This heat-regulating mechanism of the scrotum begins at about the time of puberty. When the environment is so hot that the testicles cannot cool sufficiently, the formation of sperm is impeded and a temporary condition of lowered fertility is produced.

Occasionally during fetal development, one or both of the testicles fail to descend from the body cavity into the scrotum. The animal, known as a cryptorchid, is sterile if both testicles are retained, but is usually fertile if one testicle descends. Masculine characteristics develop because testosterone is produced.

Within each testicle, sperm cells are generated in the seminiferous tubules and testosterone is produced in the cells between the tubules, which are called *interstitial cells* or *cells of Leydig*.

Sperm cells from the testicle undergo maturation in the epididymis. In passing through this long tube, the sperm acquire more capacity to fertilize ova. Sperm taken from the part of the epididymis nearest the testicle are not likely to be able to fertilize ova, whereas those taken from areas farther along this long, winding tube increasingly show the capacity to fertilize.

In the sexually mature bull, sperm reside in the epididymis in vast numbers. Sperm confined to the epididymis for long periods of time degenerate and are absorbed in the part of the epididymis farthest from the testicle.

The vas deferens are transportation tubes that carry sperm-containing fluid from each epididymis to the urethra. The vas deferens join the urethra near its origin as the urethra leaves the urinary bladder. In the mature bull, the vas deferens is about 0.3 in.

in diameter except in its upper part, where it widens to form the ampulla, about 4–6 in. long and a half-inch wide. The ampulla of the vas deferens is profusely supplied with nerves from the pelvic plexus of the sympathetic nervous system.

Under the excitement of anticipated mating, the secretion loaded with spermatozoa from each epididymis is propelled into each vas deferens and accumulates in the ampulla of the deferent duct. This brief accumulation of semen in the ampulla is an essential part of sexual arousal. The sperm reside briefly in the ampulla until the moment of ejaculation, when the contents of each ampulla are pressed out into the urethra, through the penis, en route to their deposition in the female tract.

The urethra is a large, muscular canal extending from the urinary bladder that runs posteriorly through the pelvic girdle and curves downward and forward through the full length of the penis. Very near the junction of the bladder and urethra, the tubes from the seminal vesicles and the tubes from the prostate gland join this large canal. The bulbourethral gland joins the urethra at the posterior floor of the pelvis.

The bulbourethral glands are located on either side of the pelvic urethra, just posterior to the urethra-penis where the urethra-penis dips downward in its curve. The bulbourethral glands are covered by fibrous tissue. In the bull, they are about 1/2–1 in. in size. The secretion from the bulbourethral glands is thick and viscous, very slippery and lubricating, and whitish in color.

The seminal vesicles, prostate, and bulbourethral glands are known as the accessory sex glands. Their primary functions are to add seminal volume and nutrition to the sperm-rich fluid coming from the epididymis. The bull has a semen volume of 3–10 ml per ejaculate with 4–18 billion sperm in each ejaculate.

The penis serves as (1) a passage for urine, (2) a passage for the products originating in other reproductive organs, and (3) an organ of copulation. It is a muscular organ characterized especially by its spongy, erectile tissue, which fills with blood under considerable pressure during periods of sexual arousal, making the penis rigid and erect.

The penis of the bull is about 4–5 ft. in length and 1.5 in. in diameter, tapering to the free end, and it is S-shaped when relaxed. This S-curve, or Sigmoid flexure, is eliminated when the penis is erect. The S-curve is restored after copulation, when the relaxing penis is drawn back into its sheath by a pair of retractor muscles. The free end of the penis is termed the *glans penis*. Only a small portion of the penis of the bull extends beyond its sheath during erection. The full extension awaits the thrust after entry into the vagina has been made.

The testicles produce hormones under stimuli coming from the anterior pituitary (AP) situated at the base of the brain. The AP elaborates two hormones important to male performance—luteinizing hormone (LH) and follicle-stimulating hormone (FSH). These two hormones are known as gonadotropic hormones because they stimulate the gonads (ovary and testicle). LH stimulates testosterone production by the interstitial cells. FSH stimulates cells in the seminiferous tubules to develop into functional spermatozoa. There is some evidence that FSH also influences testosterone levels. It secretes releasing factors through portal vesicles that affect the anterior pituitary and its production of FSH and LH.

Puberty in the Heifer

Age at puberty is a major determinant of lifetime reproductive efficiency of cows. **Puberty** in the heifer is attained when she will stand to be mounted and she ovulates for the first time. This may occur in some heifers when they are 6–8 months of age. Occasionally, these heifers become pregnant prior to weaning if they are not separated from late-weaned bull calves. Usually, the heifers that show early estrous cycles discontinue these cycles until they approach 12–13 months of age. Research studies have shown that this first estrus is not accompanied by an ovulation in nearly 25% of the heifers.

Table 11.1
Breed Comparisons: Testicle Size of Yearling Bulls and Its Relationship to Age of Puberty in Heifers

Breed	No. Heifers	Heifer Age at Puberty (days)	No. Bulls	Scrotal Circumference of Yearling Bulls Average (cm.)	Range (cm.)
Gelbvieh	81	341	22	34.8	30.2–42.2
Brown Swiss	126	347	19	34.3	31.0–29.6
Red Poll	95	352	20	33.5	29.7–37.1
Angus	24	372	79	32.8	26.2–38.4
Simmental	157	372	28	32.8	26.2–39.1
Hereford	27	390	55	30.7	26.2–36.1
Charolais	132	398	31	30.4	25.4–37.6
Limousin	161	398	20	30.2	24.4–34.3
Average		368		32.3	

Source: USDA-MARC Beef Research Program, Progress Report No. 1.

Puberty is influenced primarily by age and weight of heifers in addition to breed, biological type (mature cow weight), and other factors. Weight may be a more important factor than age, so heifers should reach an adequate weight by 13–14 months of age if they are to calve at 2 years of age. The target breeding weight of heifers should be 60–65% of the average mature cow weight in the herd. For example, if cow weight were 1,150 lb (BCS = 5), the desired heifer weight would be approximately 700 to 750 lb.

Puberty in the Bull

Puberty in the bull occurs when viable sperm are first produced. It occurs at approximately 12 months of age, although it can vary in individual bulls several months before or after this age depending on biological type (primarily frame size and potential mature weight), nutrition, and health status. A semen collection can verify puberty; however, scrotal circumference is a good predictor of puberty. Most bulls have reached puberty when scrotal circumference measures approximately 26 cm. However, this does not mean bulls are satisfactory breeders at this scrotal circumference, as sperm evaluation (number, activity, and abnormalities) and other reproductive traits must be evaluated.

There are scrotal circumference differences both between and within breeds as noted in Table 11.1. This table also shows that as breed average scrotal circumference increases, heifers reach puberty earlier. The correlation between heifer age at puberty and yearling bull scrotal circumference is 0.98 among breed averages.

BREEDING

Natural Service

The term natural service implies the bull is responsible for breeding cows as opposed to the producer's using artificial insemination. Approximately 95% of beef cows are bred by natural service.

To help eliminate reproductive losses due to poor fertility, bulls should be evaluated for breeding soundness 30–60 days prior to the breeding season. A breeding soundness evaluation (BSE) consists of a (1) physical examination of the bull with emphasis on

Table 11.2
MALE REPRODUCTIVE TECHNOLOGY ADOPTION BY HERD SIZE

	1–49	50–99	100–199	200+
Semen test	11	33	46	57
Culture sires for Trich.	7	12	15	25

Source: USDA-NAHMS.

the reproductive system, (2) measurement of scrotal circumference, and (3) evaluation of at least one semen sample to evaluate sperm morphology and motility. An experienced veterinarian or a qualified reproductive physiologist should conduct the BSE.

Reproductive efficiency could improve if more producers regularly evaluated the breeding soundness of their bulls. Approximately 30% of all tested bulls are classified as questionable or unsatisfactory breeders. As size of herd increases, the likelihood of semen testing sires prior to breeding season increases significantly while larger herds are more likely to have an annual trichomoniasis evaluation of sires to assure improved reproductive rates (Table 11.2).

Physical Examination

A physical examination should include observation of all conditions that might interfere with the bull's ability to locate cows in heat and breed them. A review of previous disease problems and any recent stressful conditions is useful. For example, high temperatures resulting from infection and extremely low temperatures causing frostbite of the testicles can cause structural defects in sperm cells.

A visual appraisal for unthriftiness, body condition, and structural soundness is also made. Bulls should be in good body condition, neither too thin nor too fat. Feet, legs, and eyes are among the most important structural traits that should be sound.

The reproductive system is examined by rectal palpation. Size, shape, and consistency of the prostate, seminal vesicles, and ampullae are noted, and the internal inguinal rings are examined for size. The external genitalia (scrotum, penis, and prepuce) are examined for any structural abnormalities or adhesions when the semen sample is collected. The pelvic area can also be measured.

Scrotal Circumference

Scrotal circumference is primarily related to sperm production and semen volume. Scrotal circumference appears to be related to age of puberty in heifers (Table 11.1). This relationship also indicates that as scrotal circumference in bulls increases, their daughters reach puberty at earlier ages. Scrotal circumference is measured in centimeters (cm) by using the following simple procedure. The testicles are palpated gently but firmly down into the bottom of the scrotum. A scrotal tape is placed up over the testicles-scrotum and tightened loosely around the scrotal neck, close to the body wall. The tape is then slid slowly and carefully downward. The sliding loop will enlarge as the size of the testicles-scrotum increases until the largest circumference is attained (Figure 11.6). No additional tension should be placed on the free end of the tape. Once the tape has reached the largest circumference, it will drop off and the size can be read directly from the tape.

Table 11.3 provides a guide for determining acceptable scrotal circumference for bulls of different ages as well as standards for seminal quality. A bull should only be purchased if he has passed a breeding soundness exam conducted by an experienced veterinarian.

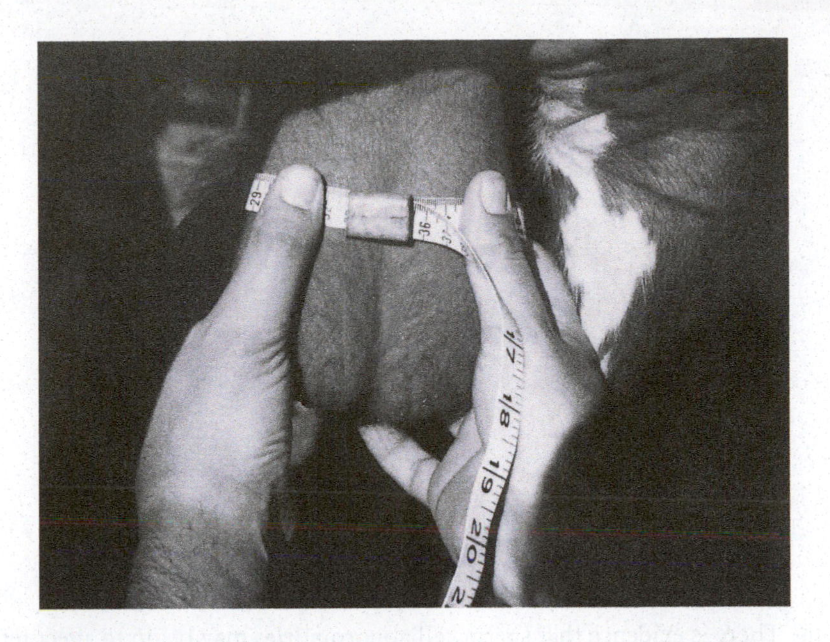

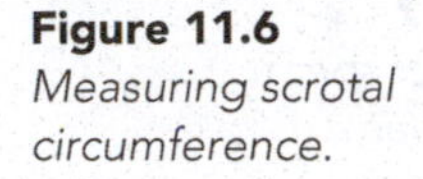

Figure 11.6
Measuring scrotal circumference.

Table 11.3
Evaluation Criteria For Scrotal Circumference, and Semen Morphology and Motility For Breeding Soundness Evaluation in Bulls

Category	Threshold
Scrotal circumference	30 cm at ≤ 15 months of age 31 cm at > 15 ≤ 18 months 32 cm at > 18 ≤ 21 months 33 cm at > 21 ≤ 24 months 34 cm at > 24 months
Sperm morphology	≥ 70% normal sperm
Sperm motility	≥ 30% individual motility and/or "Fair" gross motility

Semen Evaluation

A semen sample is collected through the use of an electroejaculator. Temperature control of the semen sample is necessary to properly evaluate motility. Also, the sample must be kept clean and free of water, urine, or chemical disinfectants. The semen sample is evaluated for motility and morphology. Motility can be estimated by observing the mass movement of sperm under the microscope. Sperm cell motility is more critically evaluated by observing the movement of individual sperm cells.

Morphology, or sperm cell structure, is evaluated by placing a small amount of semen and stain on a glass slide, gently mixing them, and spreading them into a thin film. A special microscope is used to evaluate the sperm cells; at least 100 cells are counted and classified as normal or as one of several abnormal types. Sperm cell abnormalities fall into two classifications: primary abnormalities, which are generally the more severe morphologic defects, and secondary abnormalities. Secondary abnormalities, such as droplets, are usually temporary and are a reflection of sexual immaturity. Normal spermatozoa and several common abnormalities are shown in Figure 11.7.

Although many other factors are related to high conception rates, the number of abnormal sperm cells can be an additive factor in a bull's infertility if they become too

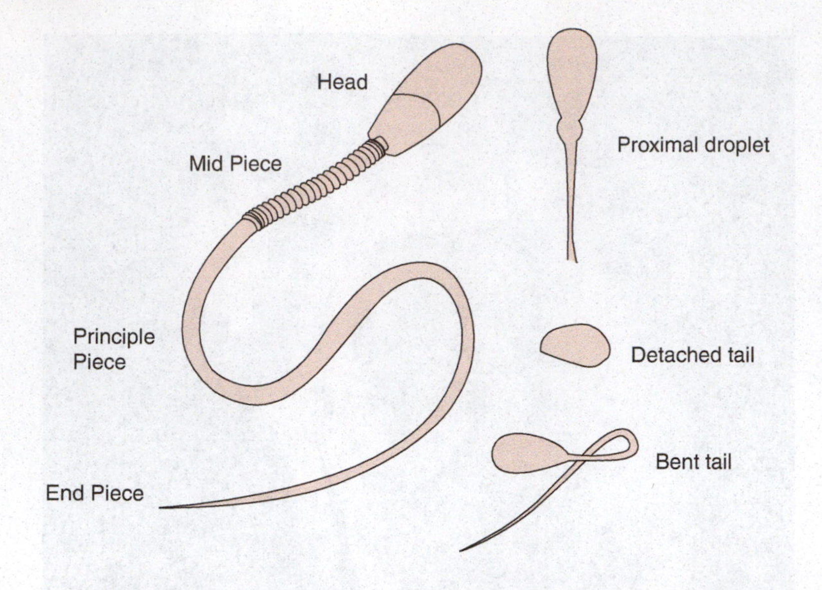

Figure 11.7
Examples of sperm abnormalities.
Source: Sean Field.

numerous. There is evidence that sperm cell abnormalities may begin to affect fertility when 25% or more of the cells are abnormal.

A breeding soundness examination classifies the bull into one of three categories: (1) satisfactory potential breeder, (2) unsatisfactory potential breeder, or (3) classification deferred. Bulls must pass the physical examination and equal or exceed the minimum thresholds in each of the categories identified in Table 11.3 before they can be classified as "satisfactory breeders."

When evaluating scrotal circumference and semen characteristics of young bulls near 1 year of age, it should be kept in mind that bulls of this age will generally improve in scrotal size and semen quality with further maturity. Scrotal circumference at 12 months of age is approximately three-quarters of the potential size at maturity. Development of the epididymis and other sperm-carrying ducts improves in most bulls up to 15 months of age. Due to immaturity, some young bulls receive "classification deferred" on the BSE at 12–14 months of age even though many of these bulls will pass a BSE at a later time (usually a few weeks). This is why a retest is common for a certain percentage of yearling bulls.

Bulls passing a breeding soundness examination may later experience fertility problems. Significant changes in scrotal temperature, caused by disease and extremely high or low ambient temperatures, can cause reduced fertility or temporary or permanent sterility. Scrotal frostbite, scrotal sunburn, severe insect bites on the scrotum, and certain diseases or infections can have a detrimental effect on semen quality. For example, foot rot in the bull can cause two breeding problems: (1) a physical difficulty in breeding females because of lameness, and (2) an elevated body temperature that may reduce semen quality and fertility. After some of environmental insults to testicular function, it may take from 2 to 12 months to restore normal fertility.

Nutrition in young bulls has a significant effect on testicular function. Diets adequate in the basic nutrients are needed to help initiate puberty. However, several studies show that feeding high-energy diets can lower semen production and semen quality (Coulter and Kastelic, 1997). Feeding programs for growing yearling bulls should utilize moderate-energy diets that achieve a target live weight at the beginning of the breeding season (e.g., 1,000–1,200 lb at 14–16 months of age). Bull performance tests or developing show bulls where weaned calves are fed to gain more than 3 lb per day or where body fat is increased beyond minimum levels increase the risks of impaired fertility in bulls.

Table 11.4
Average Number of Females per Bull and Average Bull Inventory (N) by Herd Size

Sire Age	1–49	50–99	100–199	200+
Yearling (<24 mos)	15 (0.1)	16 (0.5)	18 (1.1)	19 (3.6)
Mature (2 yr. and older)	22 (1.3)	27 (2.6)	27 (4.9)	26 (15.6)

Source: USDA-NAHMS.

Libido (sex drive), mating ability, and mating capacity are important factors affecting calf crop percentage; however, there is no simple, successful measurement of these characteristics in individual bulls. It is recommended that virgin, yearling bulls be mated to one or two females prior to the beginning of the breeding season. Observation of their mating behavior may allow producers to avoid serious problems during the subsequent breeding season. The mating experience of the yearling bull may help him be a more successful breeder early in the breeding season.

Cow-to-Bull Ratios

Table 11.4 shows the average number of females per bull as well as herd sire inventory size as herd size changes. Yearling bulls should be expected to have fewer females to service than a mature sire.

Many cow-calf producers could reduce production costs by increasing the cow-to-bull ratios in their herd. Individual bulls can successfully impregnate 50 to 60 cows in a 60-day breeding season if management is superior; however, the typical rule of thumb for cow-to-bull ratio is 30:1. This ratio is used in single sire herds as a base unless the mating capacity of individual bulls is known. Environmental conditions may direct an upward or downward adjustment of the 30:1 ratio. Dominance of individual bulls in multiple sire breeding herds exists, and individual bulls will rank differently over several years as to the proportion of calves they sire.

Bulls can be used naturally in a synchronization program and calf crop percentage can be kept high. In this situation, one bull should be used on one group of 15 to 30 cows kept in a relatively small pen or drylot. It has been well demonstrated that a bull can settle a relatively large number of cows in a 5-day period; however, there is considerable variation among bulls. Giving the bull a 10-day rest between cycles is recommended.

ARTIFICIAL INSEMINATION

Artificial insemination (AI) is a process whereby semen is deposited in the female reproductive tract by artificial techniques rather than by natural mating. Successful AI was first accomplished in cattle in the early 1900s. A successful AI program is the result of the combined effects of semen quality, female fertility, accuracy of heat detection, and skill of the inseminator.

The primary advantage of AI is the extensive use of outstanding bulls to optimize genetic improvement. For example, a bull used in natural service typically sires 10–50 calves per year over a productive lifetime of 3–8 years. In an artificial insemination program, a bull can produce 200–400 units of semen per ejaculate, with four ejaculates typically collected per week. If the semen is frozen and stored for later use, hundreds of calves can be produced by a single sire (one calf per 1.5 units of semen), and many of these offspring can be produced long after the sire is dead. AI also can be used to control reproductive diseases, and sires can be used that have been injured or are dangerous when used naturally.

Artificial insemination and length of breeding season can affect the number of resulting pregnant cows. A long breeding season (120 days or more, under natural service) extends available bull power, whereas a short breeding season (45 days) places heavy demand on available bulls. Cow-to-bull ratios would vary under these two extremes. Length of breeding season can affect AI technicians. Heat checking and technician efficiency can become lax with extended breeding seasons. AI technicians can tire and affect conception rates when large numbers of cows are bred in a few days with or without estrous synchronization.

AI programs can lower calf crop percentage unless high levels of management are given to heat checking, semen quality and handling, and insemination technique. A producer considering implementing an AI program may want to shorten a long breeding season by gradually reducing the number of days of natural service over a period of 2–3 years. For some operations, a 25-day AI plus a 25–50-day natural service breeding season seems to work well. Other operators may use a 4-day AI program plus a 50-or 60-day natural service breeding.

Artificial insemination is typically used in conjunction with synchronization to help shorten the breeding season, to increase the percentage of calves conceived in the first 21 days of the breeding season, and to help make AI more cost effective by limiting the number of days requiring heat detection. It is very important to recognize that the use of AI demands excellent management of cattle before, during, and following breeding. AI increases labor requirements compared to natural service. Furthermore, AI cannot be successful without excellent nutritional management, implementation of effective herd health programs, access to cattle handling facilities, and superior planning on the part of the management team.

Increasingly, producers are implementing timed AI protocols (TAI) coupled with synchronization to improve both cost effectiveness and convenience. Research has shown that TAI compared to natural service increases percent calves weaned by 6%, a tighter calving season distribution, nearly 40 pounds increased weaning weight resulting from the combined impact of better genetics and calves born earlier in the season, and nearly $50 per head improved gross revenue per cow exposed (Lamb, 2015).

Costs and returns should be carefully assessed before implementing an AI program (Table 11.5). In the scenario described in Table 11.5 sufficient economic benefit exists to warrant implementation of an AI program. However, in the same scenario, if the pregnancy rate is reduced by 5%, then the gains are wiped out and replaced with nearly a $20 per head loss. In the U.S. beef industry, AI is not widespread with 5% of cows and 15% of heifers bred artificially.

Semen Collection and Processing

There are several different methods of collecting semen. The most common method is the artificial vagina (Figure 11.8), which is similar to the natural vagina. The semen is collected when a bull mounts an estrous female or by training him to mount another animal or object (Figure 11.9). When the bull mounts, the person collecting the semen directs the bull's penis into the artificial vagina and the semen accumulates in the collection tube.

Semen can also be collected by using an electroejaculator. A probe is inserted into the rectum and an electrical stimulation causes ejaculation. This is used most commonly in bulls that are not easily trained to use the artificial vagina or where semen is collected infrequently.

If semen is collected too frequently, the numbers of sperm per ejaculate decreases. Semen from the bull is typically collected twice a day for 2 days a week. After the semen is collected, it is evaluated for volume, sperm concentration, motility of the sperm, and sperm abnormalities. The semen is usually mixed with an extender that

Table 11.5
Comparison of AI to Natural Service

Artificial Insemination	Synch Protocol	Synch Drug Cost Per Hd.	Pregnancy %	Times Thru Chute	Semen Cost Per Straw	Technician/ Labor Cost Per Head
Cows	CO-Synch	18.50	50	3	$20	$17.50
Heifers	MGA–Lutalyse	16.25	50	3	$20	$17.50

Natural Service	
Pregnancy %–cows	90
Pregnancy %–heifers	85
Average calf weaning weight (lbs)	525
Calf price ($/cwt)	$200
Cow-to-bull ratio	20:1
Bull purchase price ($/hd)	$3,000
Bull salvage value ($/hd)	$1,800
Number of years bull is used	3
Annual cost to maintain bull ($/hd)	$600
Annual bull ownership cost *(purchase–salvage)/3*	$400
Risk of bull loss (death, injury, infertility) (0.20 [1/2 of annual bull purchase price + salvage value])	$480

Increased Expenses	Per Head	Enhanced Revenue	Per Head
Synchronization products	$18.50 cows $16.25 heifers	Additional calf sale weight in 50% of crop	20 lbs
Semen Cost	$20	Value of additional weight	$20
Technician feeds and labor	$17.50	Change in calving percent with 90% weaning rate	+5%
		Value of additional calves	$47.25
Total cost	$56 cows $53.75 heifers	Total revenue	$67.25
Reduced Revenue	**Per Head**	**Reduced Expenses**	**Per Head**
Cull bull sales	$6	Fewer cleanup bulls equivalent to cow to bull ratio of 25:1	$14.80
Net Increase in Profit	**$20.05 cows** **$22.30 heifers**		

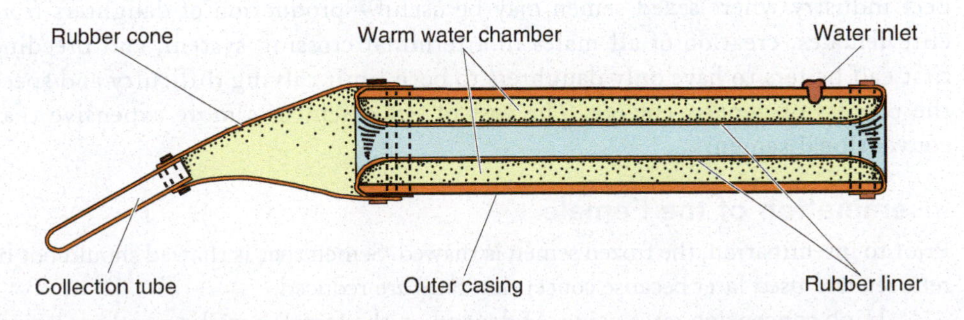

Figure 11.8
An artificial vagina used in collecting semen from the bull.
Source: Colorado State University.

dilutes the ejaculate to a greater volume. This greater volume allows a single ejaculate to be processed into several units of semen, where one unit of semen is used each time a female is inseminated. The extender is usually composed of nutrients such as milk and egg yolk, a citrate buffer, antibiotics, and glycerol. The amount of extender used

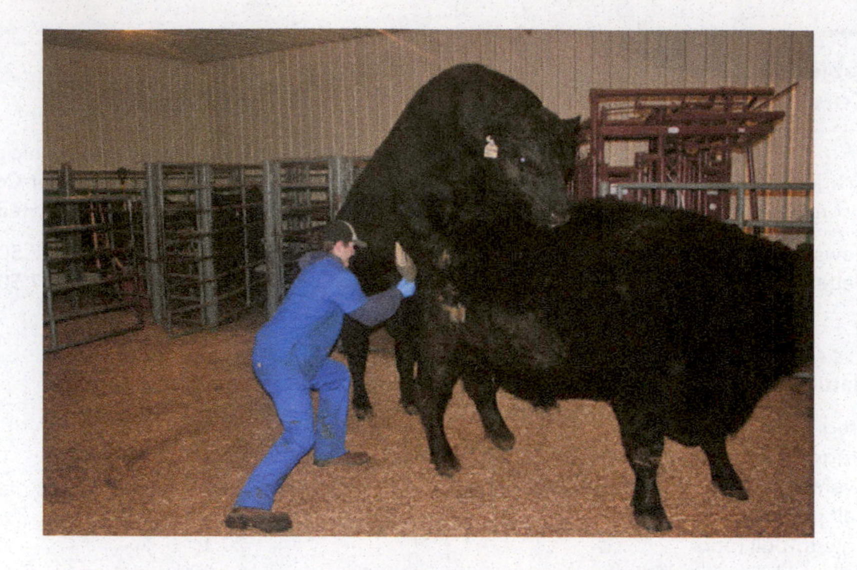

Figure 11.9
Collecting semen from a bull using the artificial vagina.
Source: Genex Cooperative, Inc.

is based on the projected number of viable sperm available in each unit of extended semen. For example, each unit of semen for insemination in cattle should contain at least 10 million motile, normal spermatozoa.

Some semen is used fresh. However, because semen can only be stored fresh for 2–3 days, most of it is frozen in liquid nitrogen and stored in plastic straws. Straws are the most common method of storing and using frozen semen. The semen can be stored in this manner for an indefinite period of time and still retain its fertilization capacity. Most cattle are inseminated with frozen semen rather than fresh semen.

Sexing Semen

USDA research has identified a process where X-chromosome-bearing sperm and Y-chromosome-bearing sperm can be sorted from one another. The process utilizes flow cytometry that sorts at a speed of 10 million cells per hour (about a normal AI dose). While the technique is promising, there are two concerns: (1) sperm viability is slightly compromised, and (2) the number of sperm sorted is lower than the number needed for routine artificial insemination. As a result of these factors, pregnancy rates from the use of sexed semen are about 80% of what is normally expected from an AI breeding program.

This technology is most attractive to dairy producers where production of females is highly preferred to bull calves. However, there are situations in the beef industry where sexed semen may be useful—production of daughters from elite females, creation of all males in a terminal crossing system, and breeding first calf heifers to have only daughters to both limit calving difficulty and speed the process of genetic change. The cost of sexed semen is more expensive than conventional semen.

Insemination of the Female

Prior to insemination, the frozen semen is thawed. Semen that is thawed should not be refrozen and used later because conception rates are reduced.

High conception rates using AI depend on the female's cycling and ovulating; accurately detecting estrus; using semen that has been properly collected, extended, and frozen; thawing and handling the semen satisfactorily at the time of insemination; insemination techniques; and avoiding stress and excitement to the animal being inseminated.

Semen handling technique is a critical component of a successful AI program. The temperature in a liquid nitrogen tank warms as semen is moved toward the top of the tank. For example, at approximately 3 to 4 in. from the top of the tank, the temperature warms above the critical control point and thawing may be initiated.

If semen is transferred between tanks, the exchange should occur quickly and away from direct sunlight. When removing individual straws from the tank at breeding time, minimize the amount of time the semen rack is elevated in the neck of the tank. Furthermore, it is important to frequently monitor the level of nitrogen in the tank and to keep the tank on a pallet to avoid corrosion.

Detecting Estrus

Estrus must be detected accurately because it signals the time of ovulation and determines the proper timing of insemination. The best indication of estrus is the condition called standing heat, in which the female stands still when mounted by a male or another female. However, a number of indicators can be observed as females approach estrus (Table 11.6).

Cows are typically checked for estrus twice daily, in the morning and evening. They are usually observed for at least 30 minutes to detect standing heat. Other observable signs are restlessness, roughed-up tailhead, being followed by a group of bull calves, attempting to mount other cows, pink swollen vulva, and a copious, clear mucous discharge from the vagina. Heifers in heat will often vocalize more than their herd mates.

Some producers use sterilized bulls or hormone-treated cows as heat checkers in the herd. These animals are sometimes equipped with a head harness that greases or paint-marks the cow's back when she is mounted. Other estrous detection methods include chalking or painting tailheads and using color indicative patches (Figure 11.10) or electronic transmitting patches.

Table 11.6
Indicators Prior, During, and Following Standing Estrus (Hours per Phase)

Prior to Standing Heat (6–10 Hours Prior)	During Standing Heat (6–24 Hours)	Following Standing Heat (1–10 Hours Following)	Following Standing Heat (24–72 Hours Later)
Will not stand to be mounted	Stands to be mounted – *the definitive indicator*	Will not stand to be mounted	Will not stand to be mounted
Vocalizes	Congregates with other sexually responsive cows	Clear mucosal discharge	Bloody mucosal discharge
Nervous/restless	Nervous/restless		
Sniffs other cattle	Clear mucosal discharge from vulva		
Attempts to mount other females	Rides other cows in standing heat		
Vulva is moist, red and somewhat swollen	Vulva is moist, red and somewhat swollen		

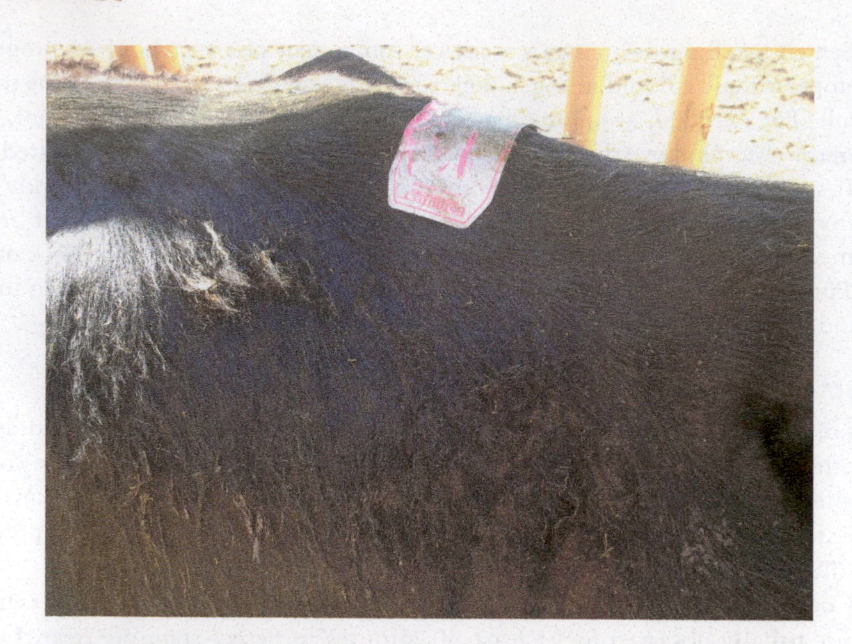

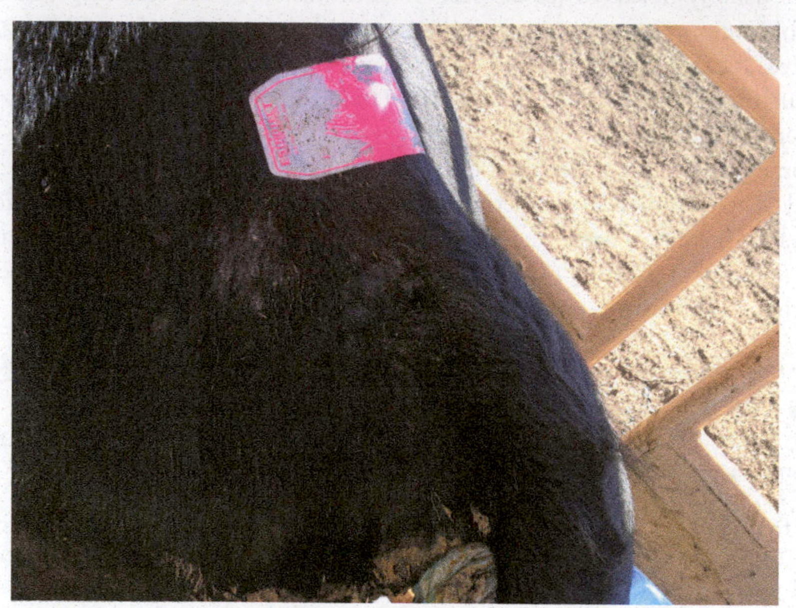

Figure 11.10
Heat detection patches improve the accuracy of estrus detection. When a patch is silver, the female has not been ridden and is assumed to not be in heat (a). However, when the silver overcoat has been rubbed off to reveal mostly pink color as a result of riding behavior, the female is assumed to have been in heat (b).

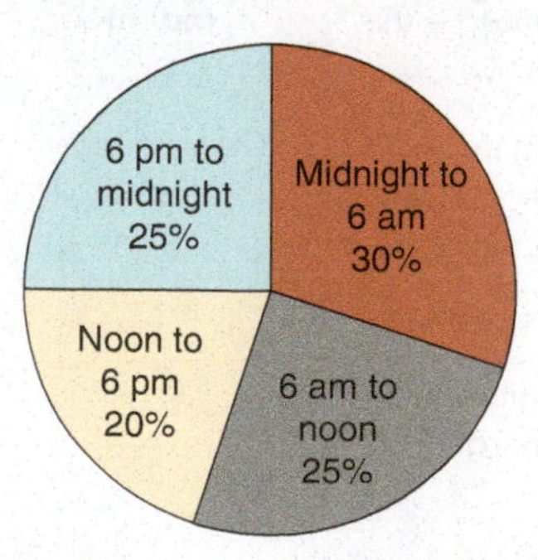

Figure 11.11
Effect of time of day on expression of estrus.

Proper Timing of Insemination

Females should be have a body condition score (BCS) of 5 or 6 to assure a high pregnancy rate. The length of estrus and time to ovulation is quite variable in cows and heifers. This variability poses difficulty in determining the best time for insemination. Females tend to concentrate estrus behavior in the early morning hours (Figure 11.11).

An additional challenge is that the egg and sperm are short-lived when put into the female reproductive tract. It is estimated that the ova is viable for 6–10 hours following ovulation and sperm are viable for 24–30 hours in the female reproductive tract. Also, estrus is sometimes expressed without ovulation occurring; occasionally the reverse of this occurs.

Time of insemination should occur 6–8 hours prior to ovulation because sperm require 2–6 hours in the female tract before they are fully capable of fertilization. Cows found in estrus in the morning are usually inseminated that evening, and cows in heat in the evening are inseminated the following morning. Because ovulation occurs 24–30 hours after the onset of heat, insemination should occur near the end of estrus (Figure 11.12). Superior estrus detection is critical in the absence of the use of exogenous synchronization. (Figure 11.13)

Cows are penned and inseminated in a chute or breeding box that restrains the animal. The advantage of the breeding box is that is provides a complete enclosure which has a calming effect (Figure 11.14). The most common insemination technique

Pre-Heat	Standing Heat	No Egg Until End of Phase	Egg Viable
6–10 hrs	18 hrs	12 hrs	6–10 hrs

Too early to breed	Can breed	Optimal time to breed	Can breed	Too late

Figure 11.12
Optimal time for insemination relative to estrus detection.

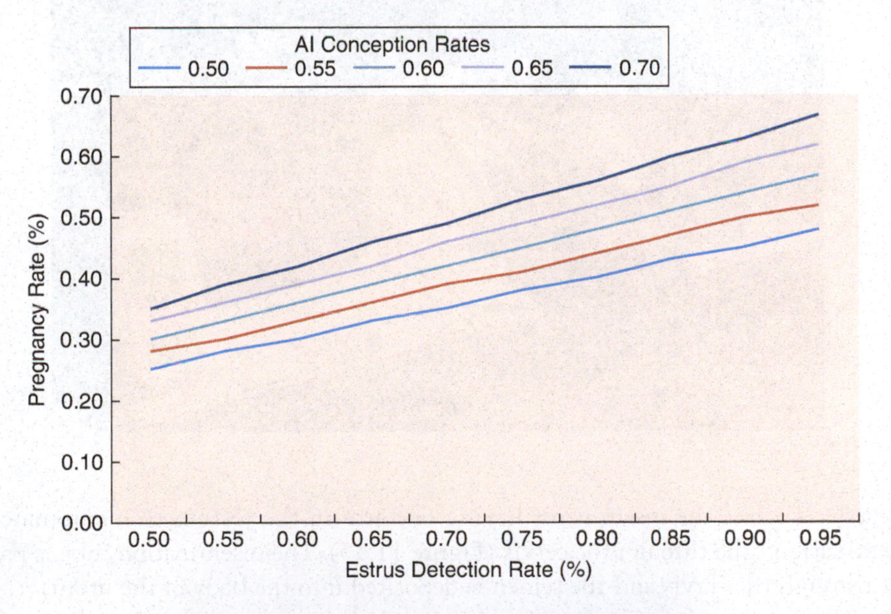

Figure 11.13
Effect of estrous detection success rates on pregnancy rates.

Figure 11.14
A breeding box provides an ideal environment for the successful execution of artificial insemination protocols.

Figure 11.15
Artificial insemination of the cow (a) and equipment needed to artificially inseminate a cow (b).

in cattle involves the inseminator having one arm in the rectum to manipulate the insemination tube through the cervix (Figure 11.15). The insemination tube is passed just through the cervix and the semen is deposited into the body of the uterus.

The amount of semen that is sold domestically, exported, and custom frozen for various breeds is shown in Table 11.7. Semen is available commercially from several artificial insemination companies (see the beef AI organizations listed in the Appendix).

SYNCHRONIZATION OF ESTROUS

Estrous synchronization involves controlling or manipulating the estrous cycle so that females in a herd express estrus at approximately the same time. It is a useful part of an AI program because checking heat and breeding animals, particularly in large pasture areas, is time-consuming and expensive. Estrous synchronization is a tool used in

Table 11.7
Units (1,000) of Beef Semen Sold Domestically, Exported, and Custom Frozen

	Domestic Sales			Export Sales[1]			Custom Frozen Sales		
Breed	2014	2004	1996	2014	2004	1996	2014	2004	1996
Angus	1,518	815	557	2,635	240	104	1,078	751	600
Red Angus	127	87	52	318	123	116	139	104	78
Simmental	208	80	102	75	39	64	178	105	79
Hereford	109	31	68	80	47	45	162	132	47
Charolais	17	12	18	49	31	20	102	117	45
Brahman	6	9	27	108	39	46	65	63	29
Maine Anjou	12	6	7	—	—	—	126	121	98
Gelbvieh	12	6	15	—	4	5	66	49	51
Total	2,208	1,102	903	3,381	581	438	2,704	2,420	1,762

[1]Dollar value for export sales in 2014 was $12.8 mil.
Source: Based on National Association of Animal Breeders.

Table 11.8
Most Recommended Synchronization Systems for Beef Heifers and Cows

Heifers	Cows
Heat detection and AI	*Heat detection and AI*
1 dose prostaglandin (requires early heat detection)	Select Synch
MGA and Prostaglandin	Select Synch and CIDR
7-day CIDR and Prostaglandin	6-day CIDR and Prostaglandin
Heat detection and initial AI and cleanup AI	*Heat detection and initial AI and cleanup AI*
MGA and Prostaglandin	Select Synch and CIDR
14-day CIDR and Prostaglandin	Select Synch
Select Synch + CIDR	6-day CIDR and Prostaglandin
Timed AI	*Timed AI*
CO-Synch and CIDR	7-day CO-Synch and CIDR
MGA and Prostaglandin	5-day CO-Synch and CIDR
14-day CIDR and Prostaglandin	

successful embryo transfer programs and it can be used with natural service where bulls are used intensively in breeding cows for a few days. Synchronization of estrus is used on less than 15% of the total U.S. cow herd but is more likely to be adopted by larger herds.

Success of a synchronization program depends on many factors such as having an excellent management plan, appropriate cattle handling facilities, availability of trained labor, accurate heat detection, and access to skilled AI technicians if that technology is chosen over natural service. Additionally, body condition, health status, time from calving, and cyclicity of females is critical to success. One of the most serious considerations affecting a herd's success with estrous synchronization is selecting the proper method (Table 11.8).

Prostaglandin

In 1979, prostaglandin was cleared for use in cattle. **Prostaglandins** are naturally occurring fatty acids that have important functions in several of the body systems. The prostaglandin that has a marked effect on the reproductive system is prostaglandin F_2 alpha. Lutalyse®, Estrumate®, and Prostamate® are commonly used prostaglandins.

The corpus luteum (CL) controls the estrous cycle in the cow by secreting the hormone progesterone. Progesterone prevents the expression of heat and ovulation. Prostaglandin destroys the CL, thus eliminating the source of progesterone. About 3 days after the injection of prostaglandin, the cow will be in heat. For prostaglandin to be effective, the cow must have a functional CL. It is ineffective in heifers that have not reached puberty or in noncycling mature cows. In addition, prostaglandin is ineffective if the CL is immature or has already started to regress. Prostaglandin is, then, only effective in heifers and cows that are in days 5–18 of their estrous cycle. Because of this relationship, prostaglandin is given in either a one- or two-injection system separately or in combination with other products that influence the estrous cycle.

GnRH controls the follicular phase of the bovine estrous cycle by stimulating a surge of LH to cause ovulation of the dominant follicle even in the presence of progesterone. The ovulation of a mature follicle must occur so that a CL can form at the site of follicular rupture.

Progestins used in synchronization mimic the effect of progesterone to suppress ovulation. Progestins support the luteal phase of the estrous cycle and create an environment for continued follicular growth. Combinations of prostaglandins, GnRH, and progestins allow for the simultaneous control of both the luteal and follicular phases of the estrous cycle in cattle.

Controlled Internal Drug Release (CIDR)

CIDRs (Controlled Internal Drug Release) are an intravaginal progesterone insert approved by the FDA for use in beef cattle in 2002 as a tool in estrous synchronization protocols. Marketed as the CIDR® Cattle Insert, the device is a T-shaped nylon structure coated with a layer of silicone containing 10% progesterone by weight. The device is inserted into the vagina with a lubricated applicator, and grasping and pulling the flexible tail of the CIDR accomplishes removal. Minimal levels of vaginitis may be observed, but do not interfere with fertility. Retention rates typically exceed 95% when correct insertion protocols are followed. CIDR protocols can be incorporated with a number of prostaglandin and or GnRH strategies (Table 11.8).

Heifer Protocols (Figure 11.16)

One dose prostaglandin (heat detection) Cost–low, Labor–high This protocol is useful for heifers that have begun cycling as a functional CL must be present for the system to work. Thus, heat detection ahead of the breeding season to confirm that heifers are cycling is strongly suggested. Normal heat detection coupled with artificial insemination of females demonstrating standing heat occurs from the onset of the program with prostaglandin administration occurring on day 5 followed by another 7 days of heat detection and artificial breeding. Pregnancy rates to AI of 30–40% are typical.

MGA and Prostaglandin (heat detection) Cost–low, Labor–moderate Feeding **MGA (melengestrol acetate)** to suppress heat is approved only for heifers. This protocol requires feeding MGA for 2 weeks followed by a 19-day delay before administering one dose of prostaglandin followed by a standard procedure of heat detection and breeding from day 33 to day 39 of the protocol. This system requires that MGA can be fed to heifers in such a manner to assure uniform consumption. AI pregnancy rates of 55–60% are anticipated from well-managed programs.

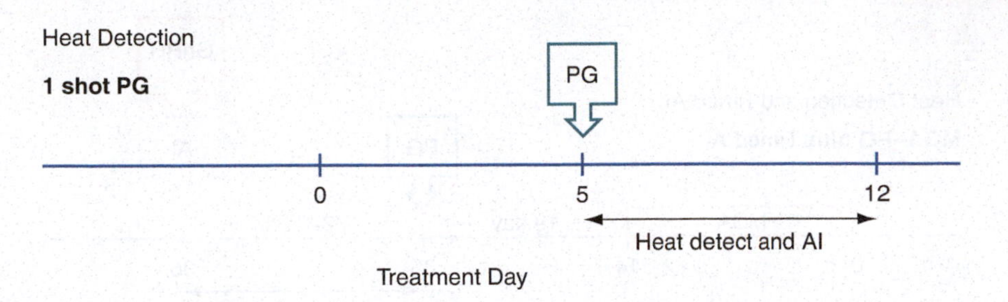

Figure 11.16
Recommended estrous synchronization programs for heifers.

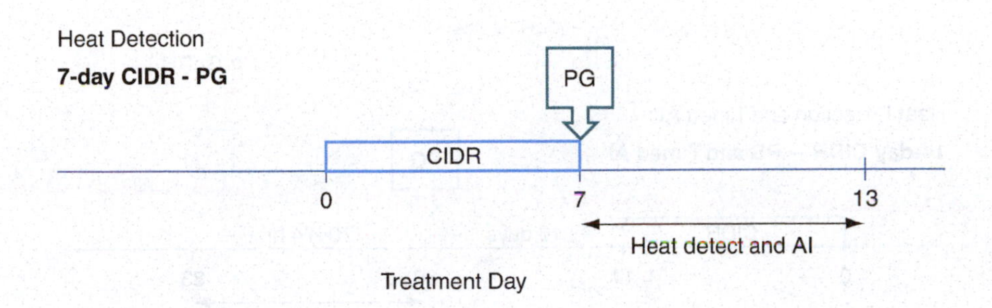

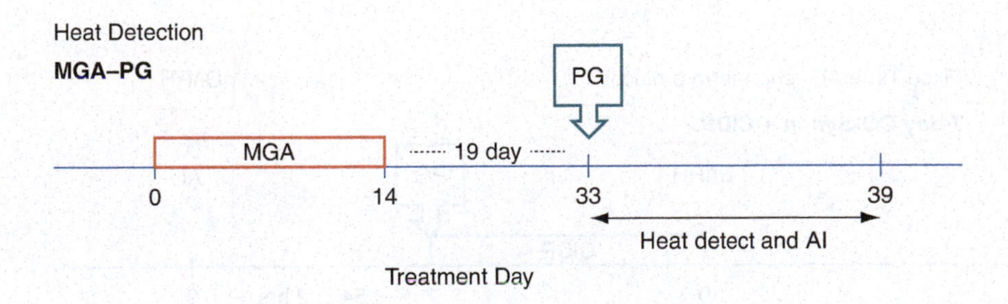

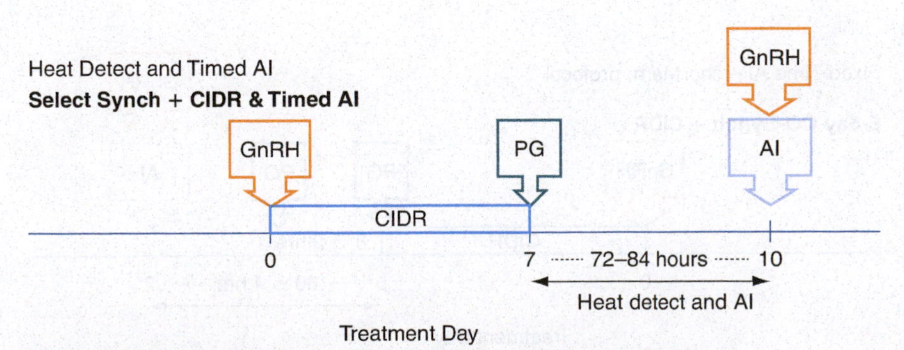

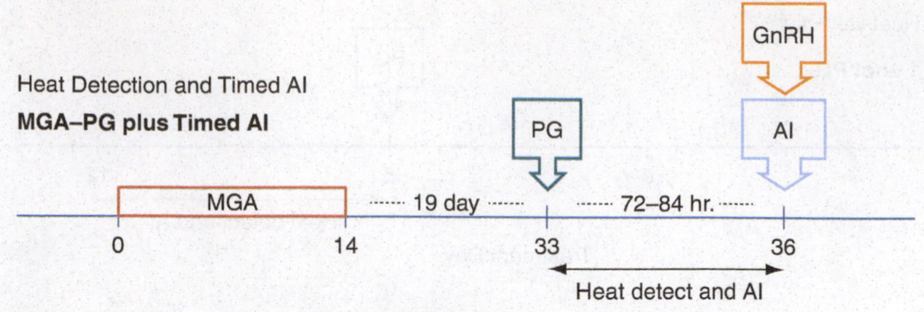

Heat detect and AI d. 33 to 36 and timed AI all non-responders 72–84 after PG with GnRH at timed AI.

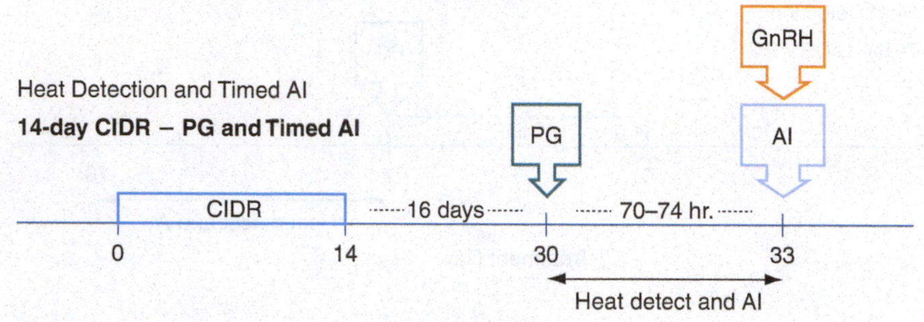

Heat detect and AI d. 30 to 33 and timed AI all non-responders 72 hr. after PG with GnRH at timed AI.

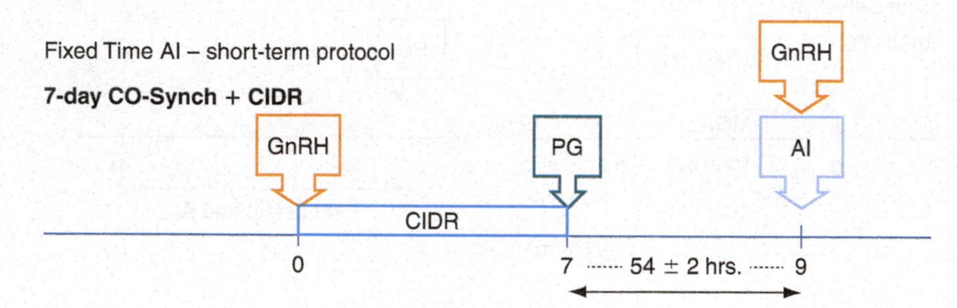

Perform timed AI at 54 ± 2 hr. after PG with GnRH at timed AI.

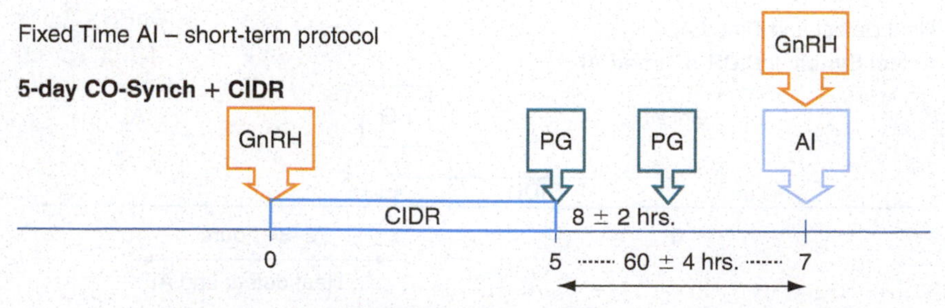

Perform timed AI at 60 ± 4 hr. after CIDR removal with GnRH at timed AI. Two injections of PG 8 ± 2 hr. apart are required.

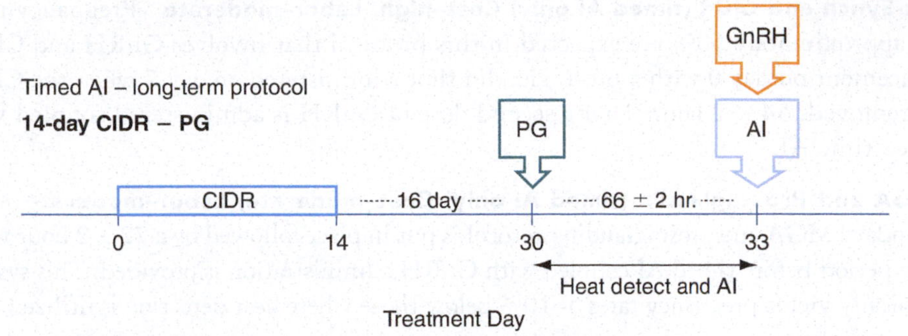

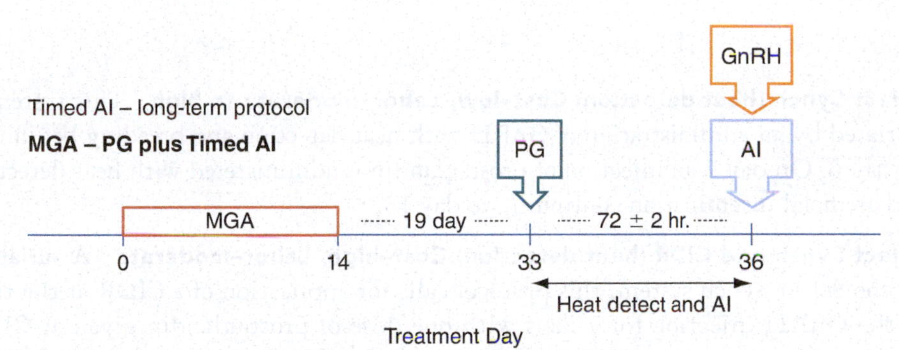

7-day CIDR and Prostaglandin (heat detection) Cost–moderate, Labor–moderate In this protocol, a CIDR is inserted on day 0 and then removed on day 7 at which time a single dose of prostaglandin is administered. Normal heat detection and insemination then takes place out to day 13 of the protocol. Pregnancy rates to AI of approximately 40% are expected.

14-day CIDR and Prostaglandin (heat detection and timed AI) Cost–moderate, Labor–moderate Coupled with both heat detection and timed AI, this system involves 14-day CIDR application followed by a 16-day waiting period with a prostaglandin dose administered on day 30. Heat detection and artificial insemination takes place for the next 70–74 hours. All non-respondent females are then administered an injection of GnRH on day 33 and bred fixed time AI. Pregnancy rates of 55% are anticipated.

Select Synch and CIDR (heat detection and timed AI) Cost–high, Labor–moderate In this system, a dose of GnRH is administered at the time of CIDR insertion. CIDR removal occurs on day 7 at which time a dose of prostaglandin is given followed by a 72–84 hour normal heat detection and AI protocol. On day 10, females that have not been bred are given a second dose of GnRH coupled with fixed time AI. Fifty-five to sixty percent pregnancy rates to AI are achievable.

MGA and Prostaglandin (heat detection and timed AI) Cost–moderate, Labor–moderate This is the same protocol as the standard MGA-P system described above with the addition of an administration of GnRH to all heifers that have not yet responded on day 36 coupled with a fixed time insemination. This system yields pregnancy rates in the 60–65% range when excellent management is in place.

Co-Synch and CIDR (timed AI only) Cost–high, Labor–moderate Pregnancy rates of approximately 50% are expected in this protocol that involves GnRH and CIDR placement on day 0 with a prostaglandin shot administered on day 7 when the CIDR is removed. 54 ± 2 hours later a second dose of GnRH is administered coupled with fixed time AI.

MGA and Prostaglandin (timed AI only) Cost–moderate, Labor–moderate The standard MGA and prostaglandin protocol is put in place followed by a 72 ± 2 hour waiting period before timed AI coupled with GnRH administration is provided. This system typically yields pregnancy rates 5–10% below those where heat detection is utilized.

14-day CIDR and Prostaglandin (timed AI only) Cost–moderate, Labor–moderate This protocol is the same as the system described above with the exception of a 66 ± 2 hour waiting period before timed AI in conjunction with a GnRH injection occur. Pregnancy rates 5–10% lower than systems incorporating heat detection are expected.

Cow Protocols (Figure 11.17)

Select Synch (heat detection) Cost–low, Labor–moderate to high This system is initiated by an administration of GnRH with heat detection and breeding beginning on day 6. On day 7 an injection of prostaglandin is administered with heat detection and artificial insemination continuing to day 13.

Select Synch and CIDR (heat detection) Cost–high, Labor–moderate A variation on the Select Synch system, this protocol calls for application of a CIDR at the time of the GnRH injection for 7 days with one dose of prostaglandin given at CIDR removal. Heat detection and breeding then occurs till day 13. The use of the CIDR will increase response of non-cycling females.

6-day CIDR and Prostaglandin (heat detection) Cost–moderate, Labor–high A second variation on the Select Synch system involves administering an injection of prostaglandin to all females followed by 3 days of heat detection and breeding before insertion of the CIDR. This system has the potential to lower total cost by reducing the number of females that undergo the full protocol if a sufficient number of females are bred in the first 3 days of the system.

Select Synch (heat detection and timed AI) Cost–low, Labor–moderate to high An injection of GnRH is the first step followed by a dose of prostaglandin on day 7. Heat detection and AI breeding begins on day 6 and continues to day 10 at which time a second dose of GnRH is given to non-responding cows coupled with fixed time AI.

Select Synch and CIDR (heat detection and timed AI) Cost–high, Labor–moderate In this variation of the Select Synch system, a CIDR is inserted at the same time as the first administration of GnRH. The CIDR is removed on day 7 and is followed by a 72–84 hour period of heat detection and insemination. Those females not bred by day 10 then received the second dose of GnRH coupled with timed AI.

6-day CIDR and Prostaglandin (heat detection and timed AI) Cost–moderate, Labor–high An initial dose of prostaglandin is administered followed by a 3-day heat detection and insemination period followed by the execution of the Select Synch and CIDR system with timed AI.

7-day CO-Synch and CIDR (timed AI) Cost–moderate, Labor–moderate In this system, GnRH and CIDR insertion occur to start the program with a dose of prostaglandin administered on day 7 when the CIDR is removed. Following a 60–66 hour waiting period, all females are then given a second dose of GnRH and fixed time inseminated.

Figure 11.17 *Recommended estrous synchronization programs for cows.*

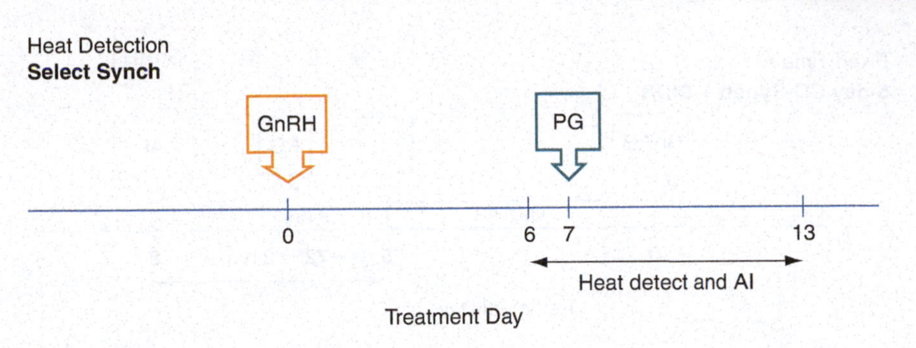

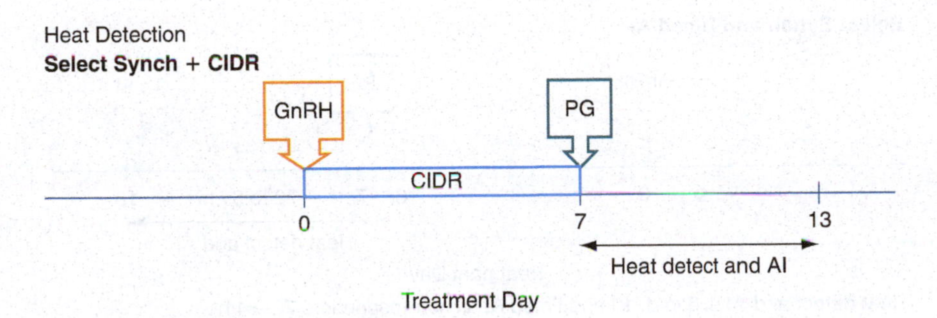

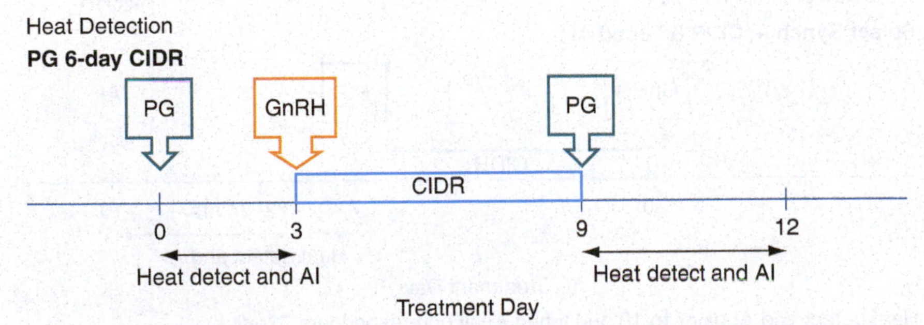

Note: CIDR is applied to non-responders on day 3 with removal on day 9. Heat detection and AI from day 9 to 10

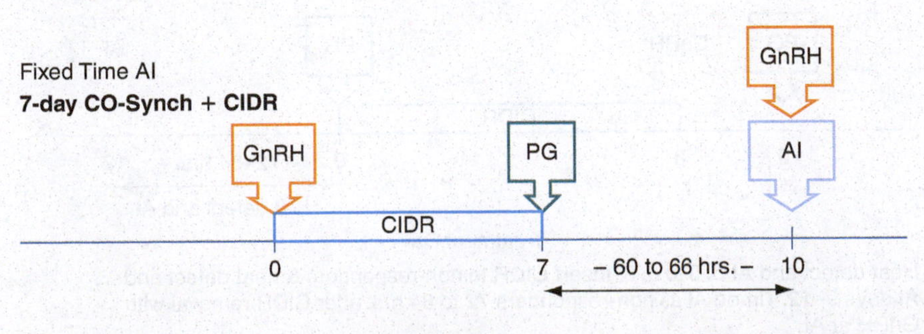

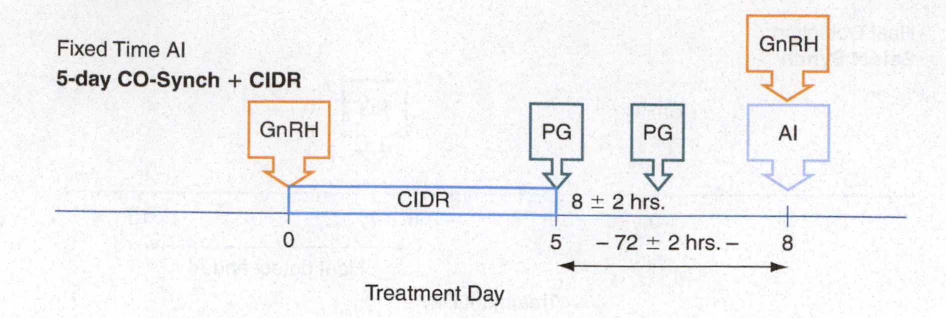

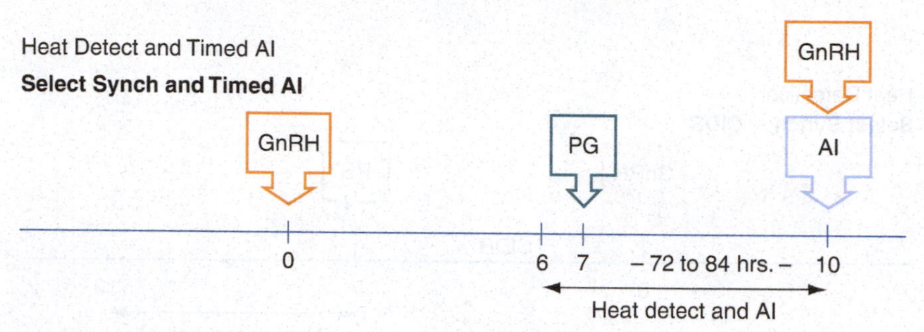

Heat detect and AI d. 6 to d. 10 and timed AI all non-responders 72–84 hr. after PG with GnRH at AI

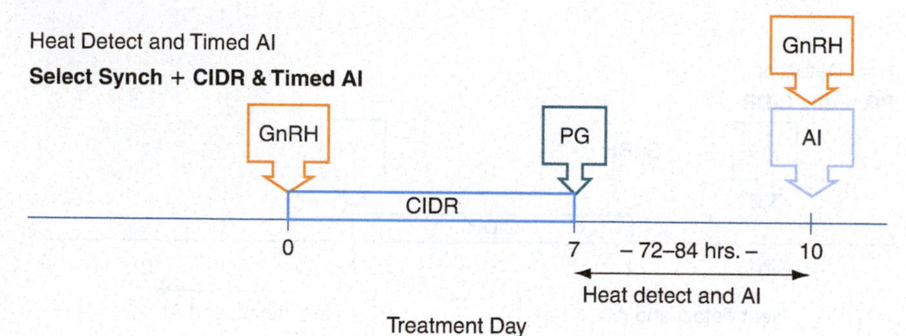

Heat detect and AI day 7 to 10 and timed AI all non-responders 72–84 hr. after PG with GnRH at timed AI

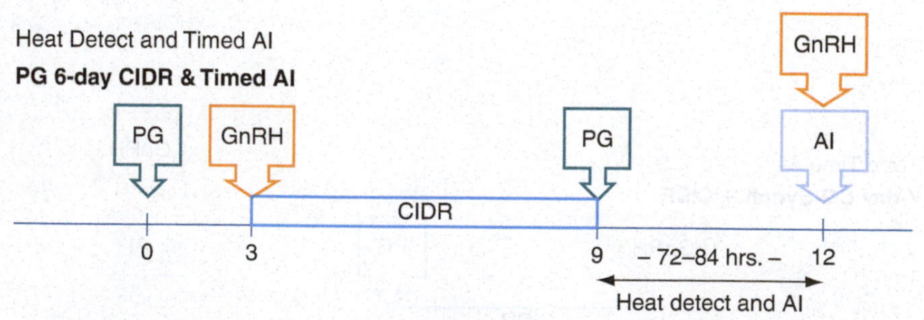

Heat detect and AI d. 0 to d. 3. Insert CIDR to non-responders & heat detect and AI days 9–12. Timed AI all non-responders 72 to 84 hrs. after CIDR removal with GnRH at AI.

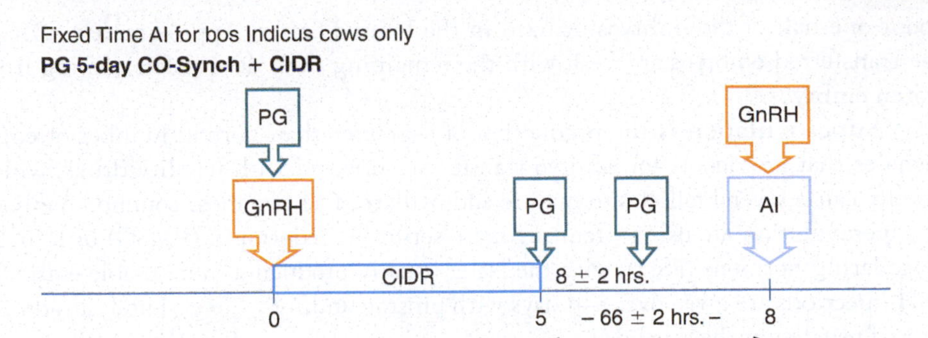

5-day CO-Synch and CIDR (timed AI) Cost–high, Labor–high Under this protocol, GnRH and CIDR insertion initiate the process with CIDR removal occurring on day 5 coupled with an initial dose of prostaglandin. A second dose of prostaglandin is given 8 ± 2 hours later. The second injection of GnRH and timed AI occur 72 ± 2 hours following CIDR removal. This protocol can be administered to heifers with the exception of timed AI occurring 60 ± 4 hours following CIDR removal.

Pregnancy rates to AI should range from 50 to 60% with these cow systems assuming excellent semen quality, technician skill, and superior cow management.

Synchronization with Natural Service

Some producers cannot economically use AI but would like the benefits of estrous synchronization. Research has shown that natural service can be used with estrous synchronization if managed properly. One bull per fifteen to twenty females in a small pasture (or drylot), and rotated every 24 hours with a rested bull, is recommended during the synchronization period of 4–5 days. Observations have shown that one bull may service 5–20 females in a 24-hour period. Pregnancy rates during the synchronized period have ranged from 60–80% and 75–95% during a 30-day breeding season.

Bulls can be used in all synchronization programs. The most popular programs are (1) feed MGA for 14 days, (2) wait 17 days, and (3) then place the bulls with the females. The advantages of this program are low drug costs, no heat detection, and less demand on the bulls in a short time period.

EMBRYO TRANSFER

The primary function of embryo transfer is to increase the reproductive rate of valuable females by tenfold or more in a given year and fivefold or more in a cow's lifetime. Even greater increases in reproductive rates can be expected as new technologies are improved. Other uses of embryo transfer are to obtain offspring from infertile cows and to export or import breeding stock to reduce disease risk.

Embryo transfer is sometimes referred to as ova transplant or embryo transplant. In this procedure, an embryo in its early stage of development is removed from its own mother's (the donor's) reproductive tract and transferred to another female's (the recipient's) reproductive tract. The first successful embryo transfers were accomplished in rabbits in 1890 and in cattle in 1951. Commercial embryo transfer companies have been established in the United States and several foreign countries. More than 576,000 embryos are transferred annually throughout the world with about one-half of the embryos utilized in the United States. Approximately 40% of the transferred embryos are fresh with the remaining 60% of transfers coming from frozen embryos.

Superovulation is the production of a greater-than-normal number of eggs. Females that are donors for embryo transfer are injected with fertility drugs, which usually cause several follicles to mature and ovulate. The two most common methods of superovulation are using pregnant mare serum gonadotropin (PMSG) or follicle-stimulating hormone (FSH), with the latter usually producing more usable embryos. FSH injections are given over 3–4 days with prostaglandin $F_{2\alpha}$((e.g., Lutalyse) administered usually on the third day.

A nonsurgical embryo collection procedure occurs, where a flexible rubber tube (Foley catheter) with three passageways is passed through the cervix and into the uterus. A rubber balloon, which is built into the anterior end of the tube, is then inflated to about half the size of a golf ball so that it expands to fill the uterine lumen and to prevent fluid from escaping around the edges. There are two holes in the tube anterior to the balloon that lead into separate passageways, one for fluid entering the uterus and the other for fluid draining from the uterus. A balanced salt solution (to which antibiotics and heat-treated serum are added) is placed in a container and held about 3 feet above the cow. The container is connected to the Foley catheter by an inflow tube. A second tube is connected to the other passageway of the catheter to drain off the medium that has washed (flushed) the ova and embryos out of the uterus. The solution is then collected in tall cylinders holding about 2 pints of fluid. Filtering the collected fluid through a cup-sized container with a fine stainless-steel filter in the bottom and then searching for them with a microscope isolate the embryos.

At approximately 12 hours and 24 hours after first standing estrus, the donor is bred artificially. Usually frozen-thawed semen is used and 2–4 straws of high-quality semen is recommended for each insemination. The recipients are usually synchronized with a prostaglandin. The recipient cow must be in estrus within 1.5 days of the donor's estrus for best results. Recipients are selected for calving ease, high milk production, and excellent health status.

On an average, seven embryos are collected with two to four calves resulting per superovulation if fertile donors are utilized in a well-managed embryo transfer program. A potential donor cow usually produces more embryos if she is 3–10 years of age, has calved regularly each year of her productive life, usually conceives in two services or less, has exhibited regular estrous cycles, comes from fertile blood lines, did not retain her placenta, and has no history of calving difficulties. The bull, to be bred to the donor cow, should have a history of excellent semen production and a high conception rate determined by his semen having been used in an AI program.

The key to the justification of embryo transfer is identifying genetically superior cows and bulls.Procedures for identifying genetically superior cattle are given in detail in Chapters 4 and 12. Making repeat matings where genetically superior calves have previously been produced can eliminate much of the guesswork. Embryo transfer is usually confined to seedstock herds, where genetically superior females can be more easily identified and high costs can be justified. Seedstock producers must evaluate the marketability of the embryo transfer calves. Some genetically superior embryo transfer calves may not sell for a sufficient amount to cover costs. Some embryo transfer calves that are not genetically superior may be merchandised for high prices because of a demand that has been previously created.

Donor selection that combines expected high levels of financial return, genetic superiority, and high reproductive potential indicates that the numbers of donors worthy of an embryo transfer program are relatively few. This "relatively few" number could be justified into several thousand head, which is less than 1% of the total U.S. cow herd.

Although embryo transfer was done surgically in the past, today new nonsurgical techniques are used. The embryos are transferred by way of an artificial insemination gun shortly after being collected. The recipient females need to be in the same stage (within 36 hours) of the estrous cycle for highly successful transfers to occur. Large numbers of females must be kept for this purpose, or estrous synchronization of a smaller number of females will be necessary. The embryos can be frozen in liquid nitrogen and remain dormant for years or decades.

Although the conception rate is lower for frozen embryos and higher for fresh embryos, ongoing research is narrowing the difference in conception rate. In cattle, approximately 85% of frozen embryos are normal after thawing; however, 40–55% of those normal embryos result in confirmed pregnancies at 60–90 days. In contrast, fresh embryos transferred the same day of collection have a pregnancy rate of 55–65%.

While there is no decline in the embryo recovery rate the first three times a donor is superovulated, additional superovulation treatments result in a reduced number of embryos for some donors. Most donors can be superovulated three or four times a year. Season, breed, and lactating versus dry donors appear to have little effect on the success of embryo transfer. Surveys have shown that 2% of the recipients abort between 3 and 9 months of gestation, 4% of the embryo transfer calves die at birth, and another 4% die between birth and weaning—representing a total loss of 10% between 3 months gestation and weaning. These losses may be similar to those experienced in a typical natural service program.

Recent advances in reproduction technology include the birth of the first test-tube calf (fertilization occurred outside the cow's body) in 1981 and the first identical twin calves resulting from embryo splitting. Splitting one embryo to produce two identical offspring is one of the initial steps in the cloning of cattle and other intriguing aspects of genetic manipulation.

CLONING

Cloning is a technique of molecular biology that produces a duplicate of the DNA or genes of an individual animal. Dolly is the noted sheep produced in 1997 in Scotland, being the first mammal cloned from an adult cell. Using a nuclear transfer procedure, the nucleus from a mammary gland cell was inserted into an empty egg cell. Cell division was initiated and the embryo was transplanted into another recipient ewe. Dolly was born genetically identical to the donor of the adult cell. This genetic and reproductive manipulation opens the door to the numerous possibilities of research in improving the productivity of cattle.

In 1998, a Holstein bull calf, "Mr. Jefferson," was cloned in Virginia. This clone was produced by nuclear transfusion from a fetal cell rather than an adult cell link, as was the case with Dolly. Cloning is a relatively rare event in the industry although companies such as Trans-Ova provide commercial cloning service. However, the cost of cloning is very high, $20,000 or higher. As a result, only extremely valuable breeding stock would be worthy of undergoing the procedure. While the technology is controversial in some circles, FDA has taken the position that there are no consumer safety issues relative to the use of the technology in the livestock industry.

PREGNANCY

When the sperm and egg unite (fertilization), conception occurs and pregnancy (gestation) is initiated. The new organism migrates through the oviduct to the uterus within 2–5 days, at which time it has developed to the 16- or 32-cell stage. The chorionic and amniotic membranes develop around the new embryo and attach it to the uterus. The embryo (later called the fetus) obtains nutrition and discharges wastes through these membranes. This period of attachment (from the thirtieth to the thirty-fifth day of pregnancy) is critical; if the uterine environment is not favorable, embryonic mortality will occur. It is vital that management practices protect the female early in pregnancy by providing feed of sufficient quality and by minimizing stress. The female should enter the breeding season in a thrifty, gaining condition and maintain that condition throughout the first weeks of pregnancy.

The embryonic stage in the life of an individual embryo is defined as that period in which the body parts differentiate to the extent that the essential organs are formed. At approximately 45 days, the embryo becomes designated as a fetus. The fetal period, which lasts until birth, is mainly a time of growth. Gestation length in cattle is approximately 285 days with a range of 275–295 days.The gestation table that appears on the inside front cover of this book makes it easy to calculate projected calving dates when breeding dates are known.

Determination of Pregnancy

In some cow-calf operations, it is economically feasible to determine pregnancy in heifers and cows prior to their anticipated time of calving. This allows nonpregnant females to be evaluated for culling because they will not produce a calf to offset production costs. While pregnancy testing is still poorly adapted in small herds, approximately half of herds with an inventory of 100–199 head utilize the procedure while more than 70% of herds over 200 head have adopted it as standard practice.

Pregnancy in cows is most commonly determined by physical examination. The examiner's arm is inserted into the cow's rectum and the reproductive tract is palpated for pregnancy indications. The palpator wears a protective covering (e.g., a rubber or plastic sleeve) over the arm and hand. A lubricant such as liquid soap is used on the sleeve for ease of entry into the cow's rectum.

Palpation takes only a few seconds when the cow is properly restrained and the palpator is experienced. The restraining chute should have a front wall or gate and a bar just above the hocks. The latter prevents the animal from kicking the palpator. A gate located at the rear of the chute allows the examiner to enter and exit with ease. The gate, which swings across the chute, also provides a front entrance for cows coming into the chute behind the palpator. Facilities should accommodate the safety of the palpator and the proper handling of the animals.

Palpators must have a thorough knowledge of the female reproductive system and the changes that occur during pregnancy. They should be experienced in palpation and should be palpating cows on a regular basis. Technique and practice are important in the accurate diagnosis of pregnancy. Individuals palpating only a few cows once a year are likely to make a large number of errors. The use of ultrasound technology via transrectal screening can greatly enhance the accuracy of pregnancy detection once the technician is very well trained.

A skilled and practiced palpator can determine pregnancy as early as 30 days after breeding. Palpation at this early stage, however, should be accompanied by good breeding records that tell the palpator the approximate breeding date of the animal. Most palpators prefer breeding to end approximately 45 days prior to palpation.

Table 11.9 gives the fetal size and some other identifying characteristics used in pregnancy diagnosis.

Gestation Length and Losses

Historically, gestation length has been reported as 283 days for the European breeds. However, the introduction of the Continental breeds extended gestation length several days, so the average gestation length in the total population of cattle today is 285–286 days. The Brahman breed has a longer gestation length, averaging 290 days.

Table 11.9
Fetal Size and Characteristics Used in Determining Pregnancy

Days of Gestation	Weight	Identifying Characteristics
30	1/100 oz	One uterine horn slightly enlarged and thin; embryonic vesicle size of large marble. Uterus in approximate position of nonpregnant uterus. Fetal membranes of a 30–90-day pregnancy may be slipped between fingers.
45	1/8–1/4 oz	Uterine horn somewhat enlarged, thinner walled, and prominent. Embryonic vesicle size of hen's egg.
60	1/4–1/2 oz	Uterine horn 2 1/2 to 3 1/2 in. in diameter; fluid filled, and pulled over pelvic brim into body cavity. Fetus size of a mouse.
90	3–6 oz	Both uterine horns swollen (4–5 in. in diameter) and pulled deeply into body cavity (difficult to palpate). Fetus is the size of a rat. Uterine artery 1/8–3/16 in. in diameter. Cotyledons 3/4–1 in. across.
120	1–2 lb	Similar to 90-day, but fetus more easily palpated. Fetus is size of small cat with the head the size of a lemon. Uterine artery 1/4 in. in diameter. Cotyledons more noticeable and 1 1/2 in. in length. Horns are 5 to 7 in. in diameter.
150	4–6 lb	Difficult to palpate fetus. Uterine horns are deep in body cavity with fetus size of a large cat. Horns 6–8 in. in diameter. Uterine artery 1/4–3/8 in. in diameter. Cotyledons 2-2 1/2 in. in diameter.
180	10–16 lb	Horns with fetus still out of reach. Fetus size of a small dog. Uterine artery 3/8–1/2 in. in diameter. Cotyledons more enlarged. From sixth month until calving, a movement of fetus may be elicited by grasping the feet, legs, or nose.
210	30–40 lb	From 7 months until parturition, fetus may be felt. Age is largely determined by increase in fetal size.
240	40–60 lb	Uterine artery continues to increase in size—210 days, 1/2 in. in diameter; 240 days, 1/2–5/8 in. in diameter; 270 days, 1/2–3/4 in. in diameter.
270	60–100+ lb	

Source: Adapted from *Determining Pregnancy in Beef Cattle.* John Berelly and L. R. Sprott (2002). Texas Cooperative Extension, College Station, B-1077, 20 pp.

Losses during gestation are usually low, averaging between 2 and 3% reduction in calf crop percentage. The embryo may fail to implant for various reasons or abortion can occur during any stage of pregnancy. Abortion may result from physical trauma or disease problems. Serious diseases such as brucellosis, leptospirosis, IBR (infectious bovine rhinotracheitis), and others can reduce the calf crop percentage by as much as 50%. Herd health programs(discussed in Chapter 16) should be planned on the basis of herd needs and surrounding conditions. If herd health is not given serious attention, gestation losses will be much higher than the 2–3% normally expected.

There are fewer death losses in crossbred calves than in purebred calves, demonstrating the effect of heterosis.

CALVING

Factors Affecting Calf Losses at or Shortly After Birth

Parturition (birth) marks the termination of pregnancy. Next to open cows, calf losses at birth are typically the second most important reason for low percent calf crops. Calving difficulty (dystocia) accounts for most calf deaths within the first 24 hours of calving and most calving difficulty occurs in 2-year-old heifers (Table 11.10).

Table 11.10
CALVING INTERVENTIONS RESULTING FROM DYSTOCIA IN HEIFERS AND CALVES (2007 AND 1997)

Level of intervention	Heifers (%) 2007	Mature cows (%) 2007	Heifers (%) 1997	Mature cows (%) 1997
None	88	96	83	97
Easy pull	8	3	11	2
Hard pull	3	1	5	1
Caesarian section	1	0	1	0

Source: USDA-NAHMS.

However, producers have placed significant selection pressure on calving ease—reducing intervention rates in heifers from 17% to 12% from 1997 to 2007. Mature cow dystocia has remained constant at approximately 4%.

Labor progresses through three stages. Stage 1 is most noticeable in females calving for the first time and is characterized by the heifer/cow isolating herself from the herd, exhibition of restlessness, frequent getting up and lying down, and lasting 2–6 hours in duration. Stage 2 begins with cervical dilation and the water sac extending through the vulva. This stage is considered active labor, and once initiated, cows should make visible progress toward delivery every 30 minutes, while heifers should make significant progress toward delivery every hour. The final labor stage is characterized by the expulsion of the fetal membranes and should be completed within 12 hours following birth of the calf. The fetal membranes should not be pulled from the tract by hand, as this procedure may result in remnant necrotic tissue in the uterus that might lead to serious infection.

Handling Calving Difficulties

The challenges posed by calving difficulty are twofold. The cow-calf producer must know how to handle calving difficulties when they occur as well as how to prevent or minimize them. In most cases, the problem is simply that the calf is too large at birth, or the birth canal (pelvic opening) of the cow is too small, or both. Research conducted by the Meat Animal Research Center (MARC) has shown a fourfold increase in calf death loss when calving difficulty is experienced. Even though it is considered the most important single factor contributing to calving difficulty, calf birth weight has accounted for less than 10% of the variation observed in calving difficulty in cows 4 years old and older. The relationship is much higher in younger cows.

Most calving difficulty losses can be prevented with timely and correct obstetrical assistance. Cows about to calve generally exhibit certain behavior patterns and physical characteristics. For example, they may have a red, swollen vulva, show expulsion of the water sac (water bag), become separated from the herd, turn their heads toward their flank, and have a full, tight udder. Manual assistance should be given if the cow has not calved an hour after she first starts the abdominal presses.

The membranes that form around the embryo in early pregnancy and attach to the uterus are collectively called the *placenta*. This organ of pregnancy produces hormones—estrogens and progesterones. A proper balance of estrogens and progesterones in the initiation of parturition is attained when the former predominate in quantity. The uterine muscles become sensitive to the hormone oxytocin, which is produced in the hypothalamus and released by the posterior pituitary. Under the stimulus of oxytocin, the weak, rhythmic contractions of the uterus that prevail through most of pregnancy become pronounced and cause labor pains, and the parturition process is initiated. Parturition is a synchronized process. The cervix, until now tightly closed, relaxes. The relaxation of the cervix, along with the pressure generated by the uterine muscles on the contents of the uterus, permits the passage of the mature fetus into and through the vagina. Another hormone, relaxin, also aids in parturition. Relaxin, which originates in the corpus luteum or placenta, helps to relax the cartilage and ligaments in the pelvic region and thereby to increase the size of the pelvic opening.

At the beginning of parturition, the calf typically assumes a position that will offer the least resistance as it passes into the pelvic area and through the birth canal. The normal presentation of the fetus is where the front feet are extended with the head between them (Figure 11.18). Occasionally, a calf may present itself in one of several

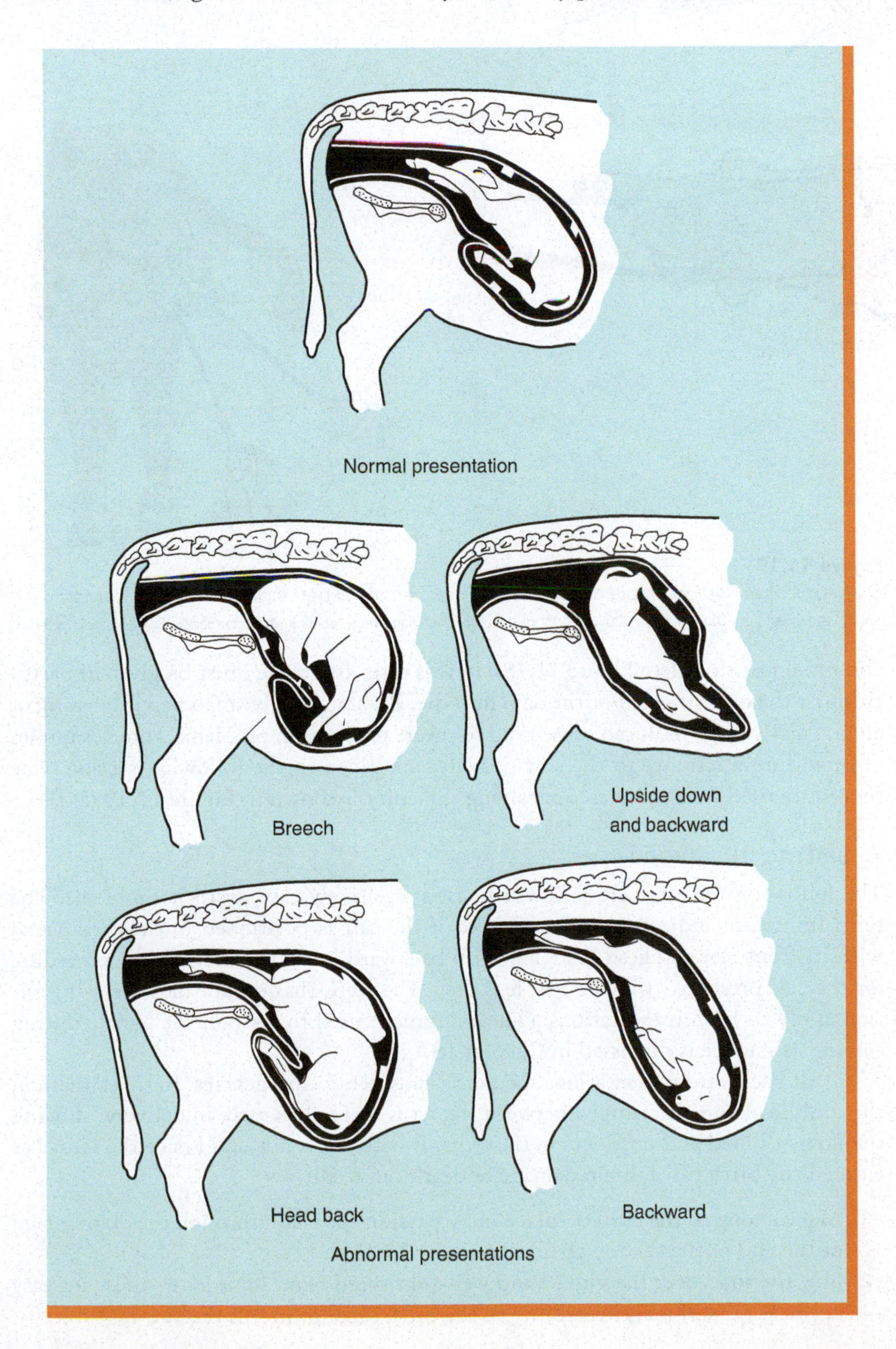

Figure 11.18
Abnormal presentations of the calf at parturition, with guidelines in assisting the difficult deliveries.
Source: Texas A&M University.

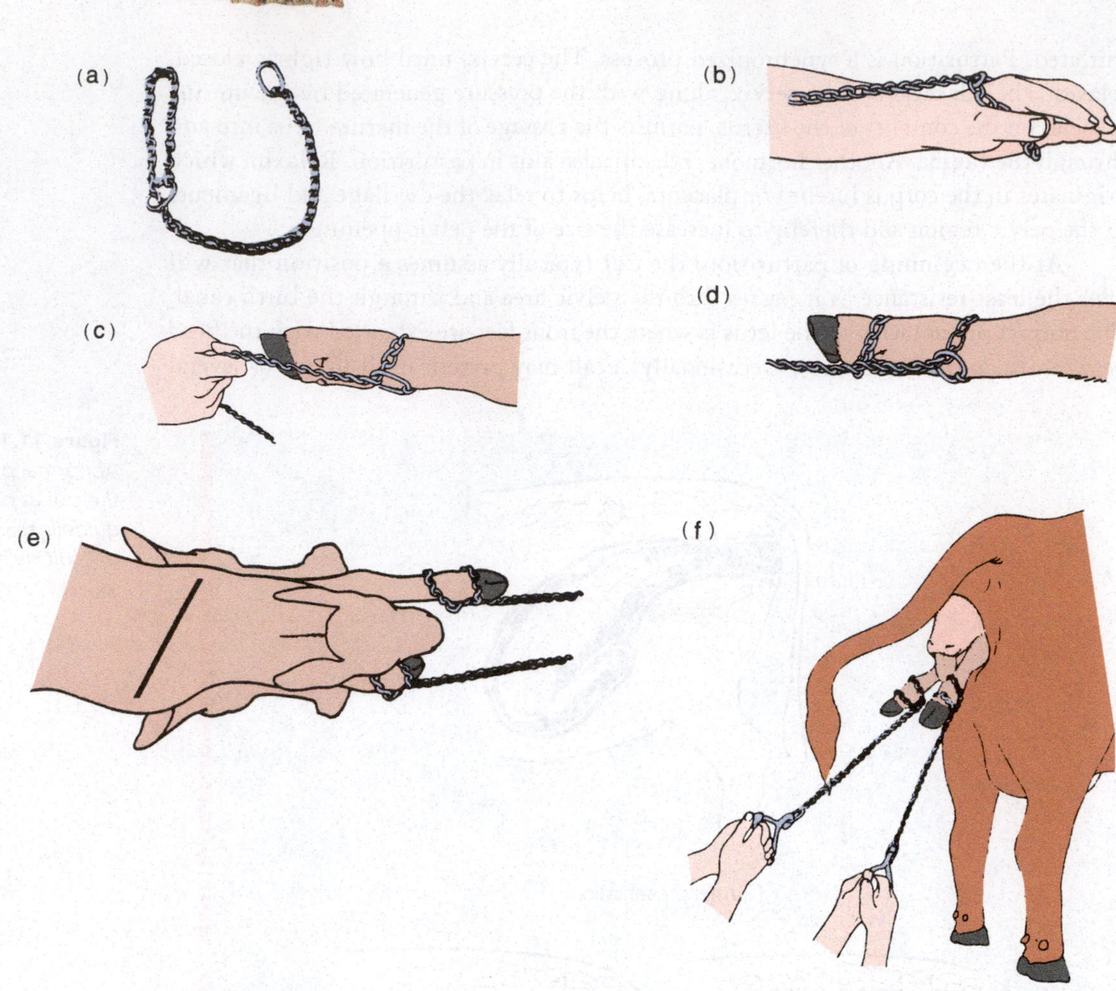

Figure 11.19

Diagrams of providing simple assistance to a beef female experiencing difficulty in calving.

Source: Adapted from Battaglia, R.A. 2001. *Handbook of livestock management techniques.* Upper Saddle River, NJ: Prentice-Hall.

abnormal positions (see Figure 11.18); in this case, assistance must be given at parturition or the offspring and/or the dam may die. An abnormally small pelvic opening or an unusually large fetus can cause mild to severe parturition problems. Some examples of providing assistance to the cow or heifer are given in the following section. The procedure used in simple cases of calving difficulty is shown in Figure 11.19 (a–f).

Assisting the Delivery

The following ten steps are followed to assist relatively easy dystocia problems. The term frontward indicates how to proceed if the calf is positioned in the birth canal with its front feet and head first. The term backward is used to indicate the procedure for the calf presented with its rear legs first. The steps that do not indicate calf position apply to both presentations. (The equipment used to restrain the female during calving assistance is discussed in Chapter 18.)

All backward presentations should be considered emergencies. In this situation, the umbilical cord is cramped between the fetus and pelvis early in delivery, slowing the flow of blood and nutrients to the fetus. If delivery is not rapid once the fetus has entered the birth canal, brain damage or death can result.

1. Make a loop in the obstetrical chain by passing the chain through the oblong ring at the end of the calving chain (Figure 11.19a).
2. Slip the loop over the gloved and well-lubricated hand in order to allow for easy application and maneuverability in the birth canal or uterus (Figure 11.19b).

3. Attach the loop of the chain to one leg of the calf and slide it up on the cannon bone 2–3 in. above the dewclaws. It may be necessary to maintain a slight tension on the chain so that it does not slip off the leg (Figure 11.19c).
4. Half-hitch the chain between the dewclaw and hoofhead. The half-hitch can be made on your hand outside the cow and then applied to the leg. The hitch in a hard pull helps to distribute the stress imposed on the bones over the two locations instead of just one, reducing the possibility of bone fracture (Figure 11.19d). Do not apply a single loop of the calving chain around both legs at the same time. The danger of breaking one or both legs is very high, especially if the birth is difficult and a hard pull is used.
5. Repeat steps 1–4 for the second leg.
6. Before applying the handles to the calving chains, make sure that the chains will pull from the bottom of the leg (dewclaw side). This will ensure that the legs will be pulled straight and not at an angle.
7. Attach two handles to the calving chains and pull gently, making sure that the loop and half hitch of the chain have not slipped from the desired position on the calf's leg.
8. Some calves can be delivered by pulling both legs evenly; however, it is best to pull alternately on one leg and then the other a few inches at a time. When the legs are "walked out" in this manner, the shoulders or hips are allowed to pass through the pelvic girdle one at a time (Figure 11.19e).

   Calf forward: If the shoulders become lodged in the pelvis, apply traction to the calf's head to reduce the compaction of the head against the sacrum (top) of the birth canal and to reduce the dimensions of the shoulder and chest region. The chain can be applied by making a loop as described in step 1 and applying it to the nose and muzzle of the calf (between the muzzle and the eyes). Care should be taken when pulling on the nose to prevent the nose or jaw from being broken.

   Calf backward: In a breech delivery, the extraction of the fetus is against the normal direction of hair growth. The birth canal should be liberally lubricated with an obstetrical or mild household soap dissolved in warm water. It may be necessary to rotate the fetus about one-eighth of a turn to take full advantage of the greatest diameter of the cow's pelvis. If delivery proves extremely difficult, a Caesarean section is probably necessary and should not be delayed.
9. Once the calf's legs are exposed, the calf should be pulled downward (toward the cow's hocks) at a 45-degree angle (Figure 11.19f). As the calf's head and shoulders come through the birth canal, the chance of uterine or cervical lacerations is greatest. This damage may lead to infection and future reproductive problems. Because pressure dilates the cervix and birth canal, traction that is applied gradually can usually prevent such damage. If assistance is given to a cow too early, the slow application of traction would not interfere with normal cervical dilation and would minimize the potential of damaging the birth canal.
10. One or two individuals using chains with manual strength should be able to pull a calf out. If the birth is extremely difficult, however, it may be necessary to use a mechanical calf puller. When this situation arises, it is best to seek experienced help. A calf puller that is used incorrectly can cause permanent damage to both cow and calf. It has been demonstrated that the leverage exerted by calf pullers can pull over a ton of dead weight.

Other important considerations for calving management include the following basic rules.

- Avoid the use of sawdust in calving pens to reduce dust and small particulates that may be inspired by the newborn.

- Avoid moving the calving female multiple times during stage two of labor as a means to reduce duration of labor.
- Rupturing the water bag does not accelerate labor.
- Soap and water is not an effective lubricant and may interfere with the natural lubrication of the birth canal.
- Only apply pulling pressure when the cow is pushing and stop pulling once the last rib has been delivered. This will avoid a premature rupture of the umbilicus and provide the calf time to have fluids clear from the nasal passage.
- Following delivery, do not hang the calf upside down or swing it to clear fluids from the lungs. These practices actually make it more difficult for a calf to draw its first breath.

Preventing Calving Losses

Dystocia should be prevented as cows experiencing dystocia have calf losses 4–5 times greater than calves born without difficulty. Calf survival is less when colder temperatures are combined with precipitation, so calving later in the spring may be part of a good preventative program.

Personal attention to calving (particularly with heifers) should not become lax toward the end of a long calving season, when calving competes with the planting of crops and night-time hours. When management inputs are lowered, greater calf losses occur. The Konefal method of daytime calving claims that feeding cows at 11:00 A.M. to noon and again at 9:30–10:00 P.M. keeps the cows busy feeding and results in most calves arriving in the daytime hours. Research data verify this relationship, though not all cows respond the same. Predicting calving dates is difficult even when the exact day of mating is known. About one-fifth of calves can be expected to arrive on the predicted birthday, with about 80% arriving within +/– 5 days of predicted birth date and 95% within +/– 10 days.

Birth weight, particularly in 2-year-old heifers, is the primary cause of dystocia. Birth weights have been increasing significantly in the U.S. cattle population over the past several years, primarily because producers have been using more bulls from larger breeds and bulls from breeds where more selection pressure for growth and frame size has occurred for several years. Birth weight appears to be more highly influenced by genetics than by environment within the same year. High levels of feed (15–18 lb TDN) during gestation only increase birth weights 3–5 lb over cows fed low levels of feed (7–8 lb TDN). There is also little difference in calving difficulty in cows and heifers on different levels of feed during gestation. There may be different birth-weight levels from cattle of similar genetic background raised in contrasting regions of the United States. These differences have been reported to be as large as 7–20 lb. The reason for the differences is not known, though it has been suggested that nutrition and temperature may affect birth weight. Large yearly differences in birth weights within the same herd, where breeding is essentially the same, suggest that environment plays a significant role in these differences.

There is evidence that the level of crude protein in the gestation of the pregnant heifer can cause differences in the birth weight of calves and in the degree of calving difficulty. Bellows et al. (1982) report birth weights of 73 lb and 84 lb and calving difficulties of 42% and 58% in heifers fed 86% and 145% of the NRC crude protein, respectively. Additional research on protein levels does not show this large a difference in calving difficulty. Recommendations are to feed heifers a balanced ration and to avoid feeding lower protein levels to circumvent calving problems. Producers should be aware that they can overfeed protein not only by feeding excessive amounts of protein supplement but also by feeding large amounts of certain forages such as alfalfa hay.

Selection pressure using calving ease direct and calving ease maternal EPDs is an effective strategy to minimize dystocia. The advent of multiple trait selection tools and selection indices allows producers the ability to avoid calving difficulty without sacrificing profitable levels of growth and muscularity. While the use of terminal breed sires on most British breed first calf heifers should be avoided as a risk mitigation strategy, there is no evidence to suggest that the use of extremely light birth weight breeds is necessary. There is little justification for selecting bulls from breeds known for extremely light birth weights because serious reductions in weaning weight and feedlot gain will result. Furthermore, calves with very low birth weights experience higher rates of morbidity and mortality.

Another factor affecting birth weight is age of cow. Two-year-old heifers have calves weighing an average of 8 lb less than calves from mature cows. This relationship needs to be taken into account when evaluating birth-weight records. Bull calves weigh about 5 lb more than heifer calves, which is why more bull calves require assistance at birth.

In general, the size of the pelvic opening (birth canal) ranks as an important factor affecting calving difficulty in the dam. It is generally true that dams with larger pelvic openings have less trouble at calving. However, within all increments in size of pelvic area, calving difficulty increases with increased birth weight.

It is a questionable practice to include pelvic measurements in a female selection program. The end results apparently do not justify the cost. It has also been determined that cows with larger pelvic areas generally give birth to heavier calves. Pelvic area is largely a function of the size of the female, and larger females have heavier calves. It is not known for certain whether selection for increased pelvic size can offset the increased birth weight.

There is evidence that selection for skeletal size (e.g., hip height) will account for most differences in pelvic area. Thus, cattle of larger skeletal size typically have larger pelvic areas. However, the larger the skeletal size, the greater the mature weight, which results in larger calves at birth, and the larger mature weight may not be profitable to maintain. The conflict between pelvic area and birth weight exists, whether pelvic area is measured directly or indirectly.

Many beef producers believe that adequate exercise in the beef female is needed to prevent calving difficulty. The reasoning is that muscle tone and strength are needed for the female to expel the calf properly. Research at Miles City does not support this concept. Cows kept in feedlots during the last 3 months of gestation had essentially no difference in calving difficulty when compared with cows that exercised daily (Table 11.11). Cows that were forced to exercise by walking to water required approximately 30% more feed to maintain the same body weight compared with the cows in the feedlot. Exercise might benefit beef females kept in confinement facilities for extended periods of time; however, the confinement raising of beef cows is not common.

Table 11.11
EFFECT OF EXERCISE ON CALVING DIFFICULTY

Activity	No.	Gestation Length (days)	Birth Wt. (lb)	Assisted (%)	Calving Difficulty (score)
Forced[1]	30	280	74	24	1.3
Restricted[2]	31	280	71	26	1.5

[1]Walked 2 miles daily during the last 90 days of gestation.
[2]Held in feedlots during the last 90 days of gestation.
Source: USDA: U.S. Livestock and Range Research Station, Miles City, Montana.

Calf Losses After Birth Until Weaning

Calf losses from 24 hours after birth until weaning are usually in the range of 3–6%. Higher losses may occur when severe weather or disease problems exist. In the Great Plains and in some western regions of the United States, for example, spring blizzards have caused calf death losses from 10 to 50%.

However, disease appears to cause the largest number of losses in most herds, particularly infectious calf scours and pneumonia.The prevention and treatment of these and other diseases are discussed in detail in Chapter 16.

Management Programs to Reduce Calf Losses

Management programs have been implemented in several herds of cattle to reduce calf death losses by 16–22%. In one case involving several herds, the calf death loss was significantly reduced by implementing the following management practices and calving techniques: (1) improved calving facilities, (2) improved sanitation, (3) treatment of sick calves with nutrient and electrolyte drench, (4) closer observation during calving, (5) a herd vaccination program, (6) bull selection for lighter birth weights, and (7) improved nutrition of the cow herd. Thus, the producer who gives careful attention to calving losses can reduce them significantly.

REBREEDING

Heifers that calve early in their first and second calving seasons continue to calve early throughout their lifetime. Heifers calving late in the calving season are more likely to be open as 3-year-olds, especially in a short breeding season of 60 days or less. Heifers can be selected for early pregnancy, at pregnancy test time, when more heifers have been saved earlier as potential replacements.

Heifers typically have a longer postpartum interval than cows. This appears to be an age-of-cow effect; however, there is evidence that nutrition plays a role as well (Table 11.12). Heifers fed together with cows typically do not get an equal share of feed. Thus, energy restriction delays the onset of estrus. Feeding 2- and 3-year-old cows separately from mature cows from calving through breeding will usually shorten the postpartum interval in the young cows. The practice of feeding young cows and mature cows separately occurs in approximately one-third of the U.S. cow herds.

A longer postpartum interval in heifers emphasizes the need to have heifers calve at the beginning of the calving season. Breeding heifers 2–3 weeks prior to

Table 11.12
EFFECTS OF GESTATION FEED LEVEL ON REPRODUCTION IN HEIFERS AND COWS

Gestation Feed[1]	Dam	Postpartum Interval (days)	In Heat by Beginning Breeding Season (%)	October Pregnancy[2] (%)
Low	Heifer	100	17	50
	Cow	58	93	83
High	Heifer	77	47	78
	Cow	60	88	81

[1]Low = 8.0-lb TDN; high = 15.0-lb TDN fed during last 90 days of gestation.
[1]After a 45-day AI period.
Source: USDA: U.S. Livestock and Range Research Station, Miles City, Montana.

the cow herd and selection for early pregnancy can accomplish this objective. Breeding heifers earlier than the cow herd lengthens the total breeding and calving seasons. This should be analyzed carefully so that the benefits are greater than the possible additional costs.

The postpartum interval is highly related to the body condition score (BCS) of the cows. Cows that have longer postpartum intervals will also have longer calving intervals. Research has shown that about half the number of cows in "thin" condition will be cycling 60 days after calving as compared with cows in "good" body condition. However, cows in too high or fat body condition reflect an uneconomical feed level that may also negatively affect reproductive efficiency.

The body condition scoring system is described in Table 11.13, where the numerically higher scores identify cows that have more fat on the back, over the ribs, around the tailhead, and in the brisket. Figure 11.20 gives a visual perception of cows representing several body condition scores.

Extreme calving difficulty and delayed obstetrical assistance extend the postpartum interval. Results from Miles City have shown that a 10-minute increase in duration of labor can lengthen the interval from calving to first estrus by 2 days, reduce the percentage of females exhibiting estrus the first 21 days of the breeding season by 7%, and decrease the percentage that become pregnant during a 45-day AI period by 6%. Pulling the calf is recommended if the cow has not had her calf after 1 hour after the second stage of labor began. The second stage starts at the first abdominal press and ends with the birth of the calf.

Table 11.13
SYSTEM OF BODY CONDITION SCORING (BCS) FOR BEEF CATTLE

BCS	Description
1	*Emaciated*—Extremely emaciated, no palpable fat detectable over spinous processes, transverse processes, hip bones, or ribs. Tailhead and ribs project prominently.
2	*Poor*—Somewhat emaciated, but tailhead and ribs are less prominent. Individual spinous processes are rather sharp to the touch, but there is some tissue cover over dorsal portion of ribs.
3	*Thin*—Ribs are individually identifiable, but not quite as sharp to the touch. Obvious palpable fat along spine and over tailhead with some tissue cover over dorsal portion of ribs.
4	*Borderline*—Individual ribs are not visually obvious. The spinous processes can be identified individually via palpation, but feel rounded rather than sharp. Some fat cover over ribs, transverse processes, and hip bones.
5	*Moderate*—Generally good overall appearance. On palpation, fat cover over ribs feel spongy and areas on either side of tailhead have palpable fat cover.
6	*High moderate*—Firm pressure needs to be applied to feel spinous processes. A high degree of fat is palpable over ribs and around tailhead.
7	*Good*—Fleshy and obviously carries considerable fat. Very spongy fat cover over ribs and around tailhead. In fact, "rounds" or "pones" beginning to be obvious. Some fat around vulva and in crotch.
8	*Fat*—Very fleshy and over-conditioned. Spinous processes almost impossible to palpate. Large fat deposits over ribs, around tailhead, and below vulva. "Rounds" or "pones" are obvious.
9	*Extremely fat*—Extremely wasty and patchy and looks blocky. Tailhead and hips buried in fatty tissue and "rounds" or "pones" of fat are protruding. Bone structure no longer visible and barely palpable. Mobility might even be impaired by large fatty deposits.

Source: Adapted from Richards et al., 1986. *Journal of Animal Science.* 62:300.

BCS - 3

BCS - 4

BCS - 5

BCS - 6

BCS - 7

Figure 11.20
Examples of body condition scores.

ADAPTION OF REPRODUCTIVE MANAGEMENT TECHNOLOGIES

The incorporation of reproductive management technologies by the U.S. beef industry is highly variable by both herd size and region of the country (Tables 11.14 and 11.15). Estrus synchronization and artificial insemination are more widely accepted in the western half of the United States and largest herds although the overall acceptance by the industry is relatively low. The use of pregnancy testing and body condition scoring along with semen testing are more widely accepted.

The primary barriers to adoption are related to labor and time constraints with nearly half of producers citing them as the reason for leaving the technology (Table 11.16). Of all the limitations facing a cow-calf producer, time is the greatest constraint given the vast myriad of responsibilities and activities that must be coordinated to assure success. Cost and complexity are also mitigating factors that slow technology adoption rates.

Table 11.14
Reproductive Technology Adoption By Herd Size (% of enterprises)

Technology	1–49	50–99	100–199	200+
Estrus Synchronization	6	10	15	19
AI	6	8	16	20
Pregnancy Check	11	30	48	72
BCS	10	19	27	34
ET	<1	4	3	5

Source: USDA-NAHMS.

Table 11.15
Reproductive Technology Adoption By Region (% of enterprises)

Technology	West	Great Plains and Midwest	Southwest	Southeast
Estrus Synchronization	11	11	7	6
AI	14	11	5	5
Pregnancy Check	42	29	18	11
BCS	20	21	27	34
Semen Test	31	34	15	9

Source: USDA-NAHMS.

Table 11.16
Reasons Technologies Not Adopted (% of enterprises)

	Labor/Time Constraints	Cost Constraints	Lack of Facilities	Too Complicated
Estrus Synchronization	39	17	10	17
AI	38	21	11	16
Pregnancy Check	38	20	11	16
BCS	40	17	8	18
Semen Test	34	25	9	16

Source: USDA-NAHMS.

SELECTED REFERENCES

Publications

American Breeders Service. *A.I. management manual* (3rd ed.). 1991. DeForest, WI: American Breeders Service.

Applied reproductive strategies in beef cattle. 2010. Conference Proc. San Antonio, TX: Beef Reproduction Task Force.

Applied reproductive strategies in beef cattle. 2013. Conference Proc. Staunton, VA: Beef Reproduction Task Force.

Applied reproductive strategies in beef cattle. 2014. Conference Proc. Stillwater, VA: Beef Reproduction Task Force.

Applied reproductive strategies in beef cattle. 2015. Conference Proc. Davis, CA: Beef Reproduction Task Force.

Assam, S.M., et al. 1993. Environmental effects on neonatal mortality of beef calves. *Journal of Animal Science.* 71:282.

Beal, W.E. 2001. Synchronization of estrus in beef heifers. *American Red Angus Magazine*, April. pp. 26–30.

Bellows, R.A., Short, R.E., & Richardson, G.V. 1982. Effects of sire, age of dam, and gestation feed levels on dystocia and postpartum reproduction. *Journal of Animal Science*. 58:18.

Beverly, J.R. 1979. *Recognizing and handling calving problems*. College Station, TX: Texas A&M University.

Coulter, G.H. 1997. *Bull fertility: BSE abnormalities*. Proc. Range Beef Cow Symposium. Rapid City, SD: CO, NE, SD, WY Extension Services.

Coulter, G.H., & Kastelic, J.P. 1997. *The testicular thermoregulation–management interaction in the beef bull*. Proc. BIF Conference. Dickinson, ND.

Chenoweth, P.J., Spitzer, J.C., & Hopkins, F.M. 1992. *A new breeding soundness evaluation form*. Proc. Society for Theriogenology. San Antonio, TX. p. 63–70.

Cupps, P.T. 1991. *Reproduction in domestic animals*. New York: Academic Press.

Geary, T.W., Downing, E.R., Bruemmer, J.E., & Whittier, J.C. 2000. Ovarian and estrous response of suckled beef cows to the Select Synch estrous synchronization protocol. *Professional Animal Scientist*. 16:1.

Hafez, E.S.E. (Ed.). 1993. *Reproduction in farm animals* (6th ed). Philadelphia: Lea & Febiger.

Healy, V.M., et al. 1993. Investigating optimal bull: Heifer ratios required for estrus-synchronized heifers. *Journal of Animal Science*. 71:291.

Johnson, S.K. (Ed.). 2015. 2016 Protocols for the synchronization of estrus and ovulation in beef cows and heifers. Kansas State University Agric. Exp. Sta. and Coop. Ext. Publ. MF2573.

Lamb, G.C. 2015. *Economics of AI versus natural service: Using decision-aid tools*. Proc. Applied Reproductive Strategies in Beef Cattle. Davis, CA.

Martin, L.C., et al. 1992. Genetic effects on beef heifer puberty and subsequent reproduction. *Journal of Animal Science*. 70:4006.

National Association of Animal Breeders. 2015. Semen sales report. Electronic Resource Guide. www.naab-css.org/sales.

Parish, J.A. 2016. *Economic comparisons of artificial insemination vs. Natural mating for beef cattle herds*. Mississippi State University Extension.

Patterson, D.J., et al. 1992. Management considerations in heifer development and puberty. *Journal of Animal Science*. 70:4018.

Ritchie, H.D., & Anderson, P.T. 1992. *Calving difficulty in beef cattle. Part I. Factors affecting dystocia. Part II. Management considerations*. Stillwater, OK: Beef Improvement Association.

Schillo, K.K., Hall, J.B., & Hillman, S.M. 1992. Effects of nutrition and season on the onset of puberty in the beef heifer. *Journal of Animal Science*. 70:3994.

Seidel, G.E. 1997. *Current status of sexing bovine semen*. Proc. BIF Research Symposium. Dickinson, ND.

Seidel, G.E., & Seidel, S.M. 1991. *Training manual for embryo transfer in cattle*. Rome, Italy: Food and Agriculture Organization of the United Nations.

Senger, P.L. 2003. *Pathways to pregnancy and parturition*. Pullman, WA: Current Conceptions, Inc.

Sprott, C.R., Harris, M.D., Richardson, J.W., Gray, A.W., & Forrest, D.W. 1998. Pregnancy to artificial insemination in beef cows as affected by body condition and number of services. *Professional Animal Scientist*. 14:231.

Staigmiller, R.B., Short, R.E., & Bellows, R.A. 1990. *Developing replacement beef heifers to enhance lifetime productivity*. Buzeman, MT. Montana Agricultural Experiment Station and USDA-ARS (Fort Keogh Livestock and Range Research Lab).

Strohbehn, D.R. 2004. *Estrus synchronization: The practical side*. Proc. 20th Annual 4-State Beef Conference. Iowa State University, Ames, IA.

Stevenson, J.S., Thompson, K.E., Forbes, W.L., Lamb, G.C., Greiger, D.M., & Corah, L.R. 2000. Synchronizing estrus and (or) ovulation in beef cows after combinations of GnRH, Norgestomet, and Prostaglandin F_{22} with or without timed insemination. *Journal of Animal Science*. 78:1747.

Summers, A.F., & Funston, R.N. 2015. Fetal programming: Implications for beef cattle production. Proc. Applied Reproductive Strategies in Beef Cattle. Davis, CA.

Thrift, F.A., Franke, D.E., & Thrift, T.A. 2002. The issue of dystocia expressed when sires varying in percent bos indicus inheritance are mated to bos taurus females. *Professional Animal Scientist*. 18:18–25.

Whittier, J. 1990. *Body condition scoring of beef and dairy cows*. Columbia, MO: University of Missouri.

United States Department of Agriculture. 2009. *Part II: Reference of Beef Cow-calf Management Practices in the United States, 2007–08*. APHIS: NAHMS. Washington, D. C.

United States Department of Agriculture. 2009. *Part III: Changes in the U.S. Beef Cow-calf Industry, 1993–2008*. APHIS: NAHMS. Washington, D. C.

United States Department of Agriculture. 2010. *Part V: Reference of Beef Cow-calf Management Practices in the United States, 2007–08*. APHIS: NAHMS. Washington, D. C.

12

Genetics and Breeding

Robert Bakewell was an English farmer born in 1725 who would become the manager of his family farm in 1760. He had traveled throughout Europe studying the best agricultural practices being employed on the continent before spending time at the side of his father learning the intricacies of their farm in Leicestershire. Bakewell was a master innovator who designed vastly improved irrigation systems and methodologies to enhance the productivity of pasturelands. But his greatest achievement was to bring a systematic and disciplined approach to livestock breeding and selection. The traditional randomized breeding methodology had been to leave males and females grouped together throughout the year. He championed the concept of deliberate mating systems where specific males were selected to be mated to specific groups of females judged to be of like kind and type. Since Bakewell's time, knowledge about the functionality of genetic power has grown dramatically with applications from the realms of biology, mathematics, and computing technologies. Effective selection and cattle breeding are founded on knowledge of genetic principles, a disciplined and systematic approach to using information in decision making, and working toward a specific set of breeding objectives.

GENETIC PRINCIPLES

Cells and Chromosomes

A species has a fixed number of chromosomal pairs with cattle possessing 30 pairs. The genetic code is carried within the structure of the chromosome located in the nucleus of the cell. **Chromosomes** are composed of deoxyribonucleic acid (DNA) that is structured as two strands of nucleotides wrapped around each other and connected at the base. The structural configuration is termed "*double-helix.*" Through an intricate process of replication called mitosis, each somatic cell as it divides during growth and repair has the ability to create additional cells with exactly the same genetic structure as the parent cell.

However, in the process of reproduction, gametes (spermatocytes and oocytes) are created via a distinct process called meiosis that allows each parent to contribute a sample half of their genetic code to the next generation.

Fertilization—the union of the sperm and egg—restores the full chromosome complement. This random segregation and recombining of chromosome pairs provides the source of new genetic combinations available for selection.

The two members of each typical pair of chromosomes in a cell are alike in size and shape and carry genes that affect the same hereditary characteristics. Such chromosomes are said to be homologous. The genes are points of activity found in each of the chromosomes that govern the way in which traits develop. The genes form the coding system that directs enzyme and protein production and thus controls the expression of a specific trait or characteristic.

The DNA strands contained in the nucleus are estimated to be approximately 6 feet in length. Segments of DNA are referred to as genes. Each gene is a specific DNA sequence that codes for a particular protein. The location of a gene on the chromosome is called the locus. DNA is composed of deoxyribose sugar, phosphate, and four nitrogenous bases. The combination of deoxyribose, phosphate, and one of the four bases is called a nucleotide; when many nucleotides are chemically bonded to one another, they form a strand that composes one-half of the DNA molecules. (A molecule formed by many repeating sections is called a polymer.)

The four bases of DNA—adenine (A), thymine (T), guanine (G), and cytosine (C)—are the components that hold the key to inheritance. In the two strands of DNA, A is always complementary to (pairs with) T, and G is always complementary to C. During meiosis and mitosis, the unwinding and pulling apart of the DNA strands replicate the chromosomes and a new strand is formed alongside the old. The old strand serves as a template, so wherever an A occurs on the old strand, a T will be directly opposite it on the new, and wherever a C occurs on the old strand, a G will be placed on the new. Complementary bases pair with each other until two entire double-stranded molecules are formed where originally there was one.

The genetic code carries information to trigger formation of specific proteins. For each of the 26 amino acids of which proteins are made, there is at least one "triplet" sequence of three nucleotides. For example, two DNA triplets, TTC and TTT, code for the amino acid lysine; four triplets, CGT, CGA, CGG, and CGC, all code for the amino acid alanine. If we think of a protein molecule as a word, and amino acids as the letters of the word, each triplet sequence of DNA can be said to code for a letter of the word, and the entire encoded message—the series of base triplets—is the gene.

The processes by which the code is "read" and protein is synthesized are called transcription and translation. To understand these processes, another group of molecules, the ribonucleic acids (RNAs), must be introduced. There are three types of RNA: transfer RNA (tRNA), which identifies both an amino acid and a base triplet in mRNA; messenger RNA (mRNA), which carries the information codes for a particular protein, and ribosomal RNA (rRNA), which is essential for ribosome structure and function. The DNA template codes all three RNA forms.

The first step in protein synthesis is transcription. Just as the DNA molecule serves as a template for self-replication using the pairing of specific bases, it can also serve as a template for the mRNA molecule. Messenger RNA is similar to DNA but is single-stranded, coding for only one or a few proteins. The triplet sequence that codes for one amino acid in mRNA is called a codon. There are also three "stop" codons that serve as punctuation in the genetic code and do not carry code that triggers specific protein synthesis (UAA, UAG, and UGA). Through transcription, the encoded message held by the DNA molecule becomes transcribed onto the mRNA molecule. The mRNA then leaves the nucleus and travels to an organelle called a ribosome, where protein synthesis will actually take place. The ribosome is composed of rRNA and protein.

The second step in protein synthesis is the union of amino acids with their respective tRNA molecules. The DNA codes the tRNA molecules. They have a structure and contain an anticodon that is complementary to an mRNA codon. Each RNA unites with one amino acid. This union is very specific such that, for example, lysine never links with the tRNA for alanine but only to the tRNA for lysine.

The mRNA attaches to a ribosome for translation of its message into protein. Each triplet codon on the mRNA (which is complementary to one on DNA) associates with a specific tRNA bearing its amino acid, using a base-pairing mechanism similar to that found in DNA replication and mRNA transcription. This matching of each tRNA with its specific mRNA triplets begins at one end of the mRNA and continues down its length until all the codons for the protein-forming amino acids are aligned in the proper order. The amino acids are chemically bonded to each other by so-called peptide bonding as the mRNA moves through the ribosome, and the fully formed protein disassociates from the tRNA–mRNA complex and is ready to fulfill its role as a part of a cell or as an enzyme to direct metabolic processes.

There are approximately 3 billion pieces of information in the bovine genome, and both the environment and thousands of intracellular proteins influence their expression. The sheer volume of information complicates the process of mapping and understanding the bovine genome. Once the genome is reasonably well understood, it is quite possible that genes and markers will have different effects in different breeds of cattle and in the multitude of environments in which they exist. In combination, these factors complicate the ability of cattle breeders to fully quantify the potential of specific parents.

Sex Determination

The genetic sex of the calf is determined at the time the sperm and egg unite at fertilization. One pair of the 30 chromosomes is known as sex chromosomes. They are designated as the *X* and *Y* chromosomes, where the female has two X chromosomes and the male has an X and a Y. The sex of the calf depends on whether the sperm fertilizing the egg carries an X or a Y chromosome. Because the cow transmits only X chromosomes to each of her offspring, an egg fertilized by an X-bearing sperm always yields a heifer (XX). An egg fertilized by the Y-bearing sperm produces a bull calf (XY). Thus, the sire determines the sex of the calf.

Because the bull can transmit either an X or a Y chromosome to each calf, the probability of his transmitting either one is 50%. The sex ratio of all calves would be expected to be 50% bulls and 50% heifers. There is evidence, however, that shows the ratio of 105 bull calves to every 100 heifer calves at birth and even a higher ratio of bulls at conception. The reason for these differences is not known.

The bull possesses a smaller amount of genetic material than the cow because the Y chromosome is only about 40% as long as the X chromosome. About 6% of the total genetic material is carried on the sex chromosomes in the cow, whereas in the bull, it is 4%. Traits carried on the sex chromosomes are called sex-linked traits. Sex-limited inheritance involves the expression of phenotype in only one gender. For example, expression of milk production phenotype is limited to females. There is also the case of sex-influenced inheritance in which gene expression differs between males and females. A good example in cattle is the inheritance of scurs (small, horn-like growths) where the allele affecting the trait is recessive in females but dominant in males. A male with one copy of the allele will be scurred while the female must have two copies to express the trait.

Genes

Genes are specific units of inheritance that are located on or as a part of the chromosome. This gene location is called a **locus**, Latin for "a place." Because chromosomes occur in pairs, so do genes. Pairs of genes occupying the same loci (position) on the chromosomes are called alleles. These genes control metabolic function and body development, which the cattle producer measures in a variety of ways, such as reproductive performance, body weight gain, and carcass composition. Multiple pairs of genes located on several chromosome pairs control most beef cattle traits.

Inheritance with One Pair of Genes

The fundamental importance of applying genetic principles to the beef industry is the capacity to estimate outcomes from predetermined matings of selected animals. A simple application relates to the determination of coat color. For example, black pigmentation is dominant to red. The allele coding for black color (the dominant form of

Table 12.1
Different Phenotypes and Genotypes with Black and Red Gene Inheritance Resulting from Mating the Following Bulls and Cows

Bull		×	Cow(s)		=	Calves	
Genotype	Phenotype		Genotype	Phenotype		Genotypes	Phenotypes
BB	Black	×	BB	Black	=	all BB	black
BB	Black	×	Bb	Black	=	$\frac{1}{2}$ BB $\frac{1}{2}$ Bb	black black
BB	Black	×	bb	Red	=	all Bb	black
Bb	Black	×	Bb	Black	=	$\frac{1}{4}$ BB $\frac{1}{2}$ Bb $\frac{1}{4}$ bb	black black red
Bb	Black	×	bb	Red	=	$\frac{1}{2}$ Bb $\frac{1}{2}$ bb	black red
bb	Red	×	bb	Red	=	all bb	Red

the gene) is represented as *B* while the recessive form coding for red is represented as *b*. In this case, while neither the dominant nor the recessive condition is inherently better or worse than the other, market signals from some beef brands have favored black-hided cattle as a proxy for more complete knowledge about carcass quality potential. Because the allele (*B*) is completely dominant, any animal carrying at least one copy of B will be black. Thus, the homozygous dominant (*BB*) and the heterozygote (*Bb*) will express black coat color. In the case of the heterozygote, the recessive gene is carried but not expressed. On the other hand, the homozygous recessive (*bb*) will always be red. Various matings of the preceding three types of animals for black and red color will produce different phenotypes and genotypes (see Table 12.1). **Phenotype** is the physical expression of a trait that can be observed or measured and is influenced by both genetic and environmental influence. **Genotype** is the actual genetic makeup of an individual animal. As indicated in the table, cattle with the same phenotype can differ in genotype; thus, breeding performance can differ as well. For example, Bb × Bb (all parents are black) gives a 3:1 phenotypic ratio (3 black and 1 red) and three different genotypes in the calves (BB—homozygous black, Bb—heterozygous black, and bb—homozygous red).

Not all genes have dominant or recessive effects. An excellent example is color in Shorthorns, which is controlled primarily by a single pair of genes where red designated by R is not completely dominant to white designated as r. Homozygous dominant pairings (RR) are red, heterozygotes (Rr) are roan colored, and homozygous recessives (rr) are white.

In the case of the heterozygote, the white gene expresses itself but other genes are involved that determine the amount of red and white in roan animals (Rr). There are numerous coloration and spotting patterns in cattle that involve several pairs of genes.

Table 12.2 identifies several common cattle characteristics that appear to be controlled by a single pair of genes. The genetic defects or abnormalities identified in this table are discussed in more detail later in this chapter.

Inheritance with Two Pairs of Genes

The following example uses two pairs of genes that influence hide color and the presence or absence of horns. In this case, black is dominant to red and the polled condition is dominant to horns; thus, B = black, b = red, P = polled, and p = horned. If a heterozygous bull is mated to a heterozygous female, a number of possible outcomes

Table 12.2
Traits Controlled or Largely Influenced by One Pair of Genes

Trait	Type of Gene Action
Black, red color	Black (B) dominant to red (b)
Color in Shorthorns	Red (R) has no dominance over white (r)
Color dilution[1]	Dilution (D) dominant to non-dilution (d)
Pigmentation, albino	Normal pigmentation (A) dominant to albino (a)
Polled, horned condition	Polled (P) dominant to horned (p) in British breeds
Shorter dwarf, normal size	Normal size (D) dominant to dwarf (d)
Hypotrichosis (short hair or hairlessness), normal	Normal (H) dominant to hypotrichosis (h)
Hydrocephalus, normal	Normal (H) dominant to hydrocephalus (h)
Osteopetrosis (marble bone disease), normal	Normal (O) dominant to osteopetrosis (o)
Syndactyly (mulefoot), normal	Normal (S) dominant to mulefoot(s)
Arthrogryposis (palate-pastern syndrome), normal	Normal (A) dominant to palate-pastern (a)
Double muscling,[2] normal	Normal (D) dominant to double muscling (d)

[1]Black color is diluted to gray when DD or Dd exists with BB or Bb; red color is diluted to yellow when DD or Dd exists with RR or Rr.
[2]Recessive inheritance shown in this table is typical of the British breeds. In other breeds (e.g., Piedmontese), the double muscling gene appears to be dominant. Other pairs of genes also modify the expression of double muscling.

of the phenotype and genotype of progeny are possible. Because one chromosome of each pair goes into the formation of sperm and eggs, there are four genetically different kinds of sperm and four genetically different kinds of eggs that can be formed. When fertilization occurs, each of the genetically different kinds of sperm has an equal chance of uniting with each of the genetically different kinds of eggs. With these various matings, different genotypes and phenotypes can result in the offspring (Table 12.3).

A different type of gene action will give the same genotypic ratio, but the phenotypic ratio will be different. An example is color in Shorthorns with polled, horned condition:

RrPp (roan, polled bull) × RrPp (roan, polled cows)

Table 12.3
Genotypes and Phenotypes in Calves Resulting from Mating Animals with Two Heterozygous Gene Pairs (Bb × Bb)

		Ova			
		BP	**Bp**	**bP**	**bp**
Sperm	**BP**	BBPP[1] black, polled[2]	BBPp black, polled	BbPP black, polled	BbPp black, polled
	Bp	BBPp black, polled	BBpp black, horned	BbPp black, polled	Bbpp black, horned
	bP	BbPP black, polled	BbPp black, polled	bbPP red, polled	bbPp red, polled
	bp	BbPp black, polled	Bbpp black, horned	bbPp red, polled	bbpp red, horned

[1]Genotype:
(Number of different genotypes) 1 (BBPP), 2 (BBPp), 2 (BbPP), 4 (BbPp), 1 (BBpp), 1 (bbPP), 2 (Bbpp), 2 (bbPp), 1 (bbpp)
[2]Phenotype:
(Number of different phenotypes) 9 black, polled; 3 black, horned; 3 red, polled; 1 red, horned

Phenotypic ratio of the calves:
3 red, polled
6 roan, polled
3 white, polled
1 red, horned
2 roan, horned
1 white, horned

Genotypic ratio of the calves:
Same as in previous example with black, red color and polled, horned condition (1:2:2:4:1:1:2:2:1).

Where both pairs of genes show no dominance, the phenotypic and genotypic ratios would be the same (1:2:2:4:1:1:2:2:1).

The inheritance of the horned or polled condition is actually more complex than previously presented for British breeds of cattle. Scurs (horn-like growths) are inherited differently and the African horn gene is involved in the inheritance of horns for Zebu-type cattle such as Brahman. The primary genes involved in these three situations are:

P = polled gene
p = horned gene

S = scur gene
s = absence of scur gene

A = African horn gene
a = absence of African horn gene

Table 12.4 shows the expected phenotypes resulting from the different genetic combinations. Note that the phenotypic expression is different in cows and bulls when the Ss or Aa gene combination occurs. The inheritance of scur may involve more genes than presented here; however, the one pair of genes explains most of the inheritance involved. Horned cattle may carry the gene for scur, but it is visually hidden by the horn gene and horn growth.

Continuous Variation and Many Pairs of Genes

Most economically important traits in beef cattle, such as growth rate and carcass composition, are controlled by many pairs of genes; therefore, it is necessary to expand one's thinking beyond inheritance involving one pair and two pairs of genes. Consider even a simplified example of 20 pairs of heterozygous genes (one gene pair on each pair of 20 chromosomes) affecting yearling weight. The estimated number of genetically

Table 12.4
Inheritance for Scurs[1] and African Horn Gene

Genotype	Phenotype	
	Cows	Bulls
PP (or Pp) SS	Scurred polled	Scurred polled
PP (or Pp) Ss	Smooth polled	Scurred polled
PP (or Pp) ss	Smooth polled	Smooth polled
pp SS (or Ss or ss)	Horned	Horned
PP (or Pp) AA	Horned	Horned
PP (or Pp) Aa	Horned	Polled
PP (or Pp) aa	Polled	Polled
pp AA (or Aa or aa)	Horned	Horned

[1]Scurs are incompletely developed horns that are usually not attached to the skull. They vary in size from scab-like growths to nearly the size of horns.

Table 12.5
NUMBER OF GAMETES AND GENETIC COMBINATIONS WITH VARYING NUMBERS OF HETEROZYGOUS GENE PAIRS

No. Pairs of Heterozygous Genes	No. Genetically Different Sperm or Eggs	No. Different Genetic Combinations (Genotypes)
1	2	3
2	4	9
n	2^n	3^n
20	$2^n = 2^{20}$ = approx. 1 mil.	$3^n = 3^{20}$ = approx. 3.5 bil.

Table 12.6
NUMBER OF GAMETES AND GENETIC COMBINATIONS WITH EIGHT PAIRS OF GENES WITH VARYING AMOUNTS OF HETEROZYGOSITY AND HOMOZYGOSITY

Bull Cow	Aa aa	Bb Bb	Cc CC	Dd Dd	Ee Ee	FF FF	GG gg	Hh Hh	Total
No. different sperm for bull[1]	2	2	2	2	2	1	1	2	64
No. different eggs for cow[2]	1	2	1	2	2	1	1	2	16
No. different genetic combinations possible in calves[3]	2	3	2	3	3	1	1	3	324

[1]Total = 2 × 2 × 2 × 2 × 2 × 1 × 1 × 2 . . . = 64.
[2]Total = 1 × 2 × 1 × 2 × 2 × 1 × 1 × 2 . . . = 16.
[3]Total = 2 × 3 × 2 × 3 × 3 × 1 × 1 × 3 . . . = 324.

different gametes (sperm or eggs) and the genetic combinations are shown in Table 12.5. Remember that for one pair of heterozygous genes, there are three different genetic combinations; for two pairs of heterozygous genes, there are nine different genetic combinations.

Most beef cattle would likely have some gene pairs heterozygous and some homozygous, depending on the mating system being utilized. Table 12.6 shows the number of gametes and genotypes, where eight pairs of genes are either heterozygous or homozygous and each gene pair is located on a different pair of chromosomes.

MATING STRATEGIES

Mating strategies are designed to enhance genetic superiority and/or create hybrid vigor. Genetic improvement is realized in many herds by utilizing a combination of selection and mating strategies. Strategies of mating are identified primarily by animal performance or the genetic relationship of the animals being mated.

Mating strategies can be determined on an individual animal basis where specific sires are chosen to mate with specific dams with the goal of realizing a desired outcome in the progeny. The available approaches to this strategy include assortative mating (positive or negative) and **random mating**.

Assortative mating takes the approach of pairing similar phenotypes/genotypes (positive) or dissimilar types (negative). This approach requires breeders to make

reasoned decisions based on objective performance records. An example might be to mate sires known for high marbling to females that also have high breeding values for intramuscular fat deposition. This approach is typically taken when rapid genetic progress is desired and so attempts are made to produce progeny with extreme performance in one or several traits. It should be noted that positive assortative mating also increases variability.

When breeders mate the highest marbling sires to the lowest marbling females, they would be using negative assortative mating. This approach results in the production of progeny with intermediate levels of performance compared with the parents. This strategy slows the speed of genetic change but reduces variation in the resulting progeny.

Random mating is another option that can be very effective in seedstock herds, particularly when a herd has been highly selected over time. This approach is the most typical strategy in many commercial herds where detailed cow records are unavailable and selection pressure is mostly focused on picking sires. This strategy recognizes and accepts the random nature of genetic inheritance.

The two major systems of mating, based on relationship, are inbreeding and outbreeding.

Inbreeding is the mating of animals more closely related than the average of the breed or population. The two different forms of inbreeding are as follows:

1. Intensive inbreeding—the mating of closely related animals whose ancestors have been inbred for several generations.
2. Linebreeding—a mild form of inbreeding where inbreeding is kept relatively low while a high genetic relationship to an ancestor or a line of ancestors is maintained.

Outbreeding involves the mating of animals less closely related than the average of the breed or population. The five different forms of outbreeding are as follows:

1. Outcrossing—the mating of unrelated animals within the same breed.
2. Grading up—the mating of purebred sires to nondescript or grade females and their female offspring generation after generation.
3. Linecrossing—the crossing of rather distinct lines (that may or may not be inbred) of the same breed.
4. Crossbreeding—the mating of animals of different established breeds. (A detailed discussion is provided in Chapter 13.)
5. Species cross—the crossing of animals of different species (e.g., cattle to bison).

Because mating systems are based on the relationship of animals being mated, it is important to understand more detail about genetic relationship. Proper pedigree evaluation also involves understanding relationships. Relationship is best described by whether the genes of an animal (or animals) exist primarily in a heterozygous or homozygous condition. Figure 12.1 roughly depicts levels of heterozygosity and homozygosity created by various matings systems. For an excellent discussion on computation of relationship, see *Understanding Animal Breeding* by Bourdon (2000).

Inbreeding is the mating of relatives with the goal of increasing homozygosity. Inbreeding is utilized for several reasons. First of all, inbreeding can be utilized to increase uniformity. This is particularly effective for traits controlled by a limited number of genes, such as coat color or the absence of horns. For traits affected by many gene pairs (polygenic), uniformity is enhanced only at very high levels of inbreeding. Crossing highly inbred lines to maximize hybrid vigor has been effectively utilized by the crop industry. Inbreeding is also a useful strategy in the identification of which animals are carriers of genetic defects.

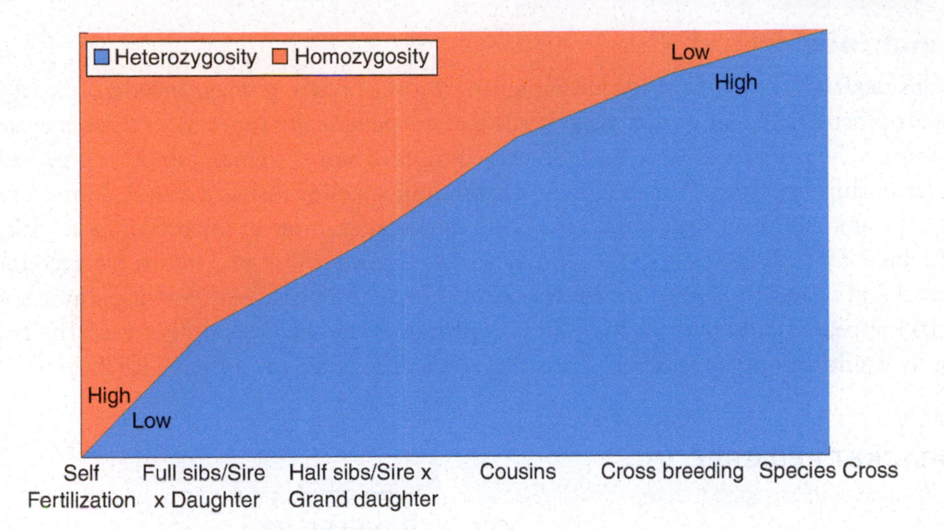

Figure 12.1 *Relationship of mating systems to the amount of heterozygosity or homozygosity.*

The drawbacks to intensive inbreeding include the production of progeny with deleterious recessive alleles (increasing homozygosity not only pairs desirable genes but also those that are undesirable) and reduced general performance as a result of inbreeding depression.

Development of Inbred Lines

There are numerous, genetically different inbred lines that can be produced in a given population such as a breed. The number of different inbred lines is 2^n where n is the number of heterozygous gene pairs. For example, with two pairs of heterozygous gene pairs, there are 2^2 or 4 different inbred lines that are possible. For example, with AaBb genes, the different, completely homozygous lines that can result are AABB, AAbb, aaBB, and aabb.

There have been research projects involving cattle in which highly inbred lines were produced. It was expected that the crossing of these inbred lines would produce results similar to hybrid corn. However, the creation of inbred lines of cattle did not become common practice due to the factors listed below:

1. Inbreeding depression will usually negatively affect the economics of a cattle operation. Increased inbreeding is usually detrimental to reproductive performance and pre-weaning and post-weaning growth. Inbred cattle are more susceptible to environmental stresses. While 60–70% of the inbred lines show the detrimental effects of increased inbreeding, 30–40% of the lines show no detrimental effects with some lines demonstrating improved productivity.
2. In studies conducted at the San Juan Basin Research Center in Colorado, the inbred lines showed a yearly genetic increase of 2.6 lb in weaning weight over a 26-year period, while the line crosses made a 4.6-lb increase over the same time period. Heterosis is demonstrated in the line crosses, and the 4.6-lb increase is typical of what breeders might expect from using intense selection in an outbred herd.
3. Inbreeding quickly identifies some desirable genes as well as undesirable genes, particularly the recessive genes that are hidden in the heterozygous state but become deleterious once they are homozygous.
4. Inbred bulls with superior performance are most likely to have superior breeding values that will result in more uniform progeny with high levels of genetically influenced productivity.
5. Crossing of inbred lines results in heterosis.

Linebreeding

Consideration should be given to implementing linebreeding when breeders have difficulty introducing sires from other herds that are genetically superior to those they are raising. The primary objective of a linebreeding program is to maintain a high genetic relationship of current progeny to an outstanding ancestor (usually a sire). Inbreeding is kept at a relatively low level. Figure 12.2 shows an example of linebreeding in which the bull Rito 2RT2 is only 9% inbred, yet has a relationship of 31% to his ancestor, Band 234 of Ideal 3163, which contributes genes to him by three different pathways. If 3163 appeared only once in Rito 2RT2's pedigree, the relationship between the two bulls would be only 12% if 3163 was three generations removed from 2RT2.

Figure 12.2
Pedigree showing linebreeding.

BRACKET PEDIGREE

BAND 234 OF IDEAL 3163
Q A S TRAVELER 23-4
SIRE
Q A S BLACKBIRD EVE 601 1
R R TRAVELER 5204
SHOSHONE VANTAGE JB23
ERISKAY OF ROLLIN ROCK 3302
ERISKAY OF ROLLIN ROCK 7003
RITO 2RT2
BAND 234 OF IDEAL 3163
TEHAMA BANDO 155
DAM
TEHAMA BLACKCAP G373
RITA 0B5 OF 8E23 BANDO
Q A S TRAVELER 23-4
RITA 8E23 OF 5H28 TRAVELER
RITA 5H28 OF 3A12 RITO 9J9

ARROW PEDIGREE

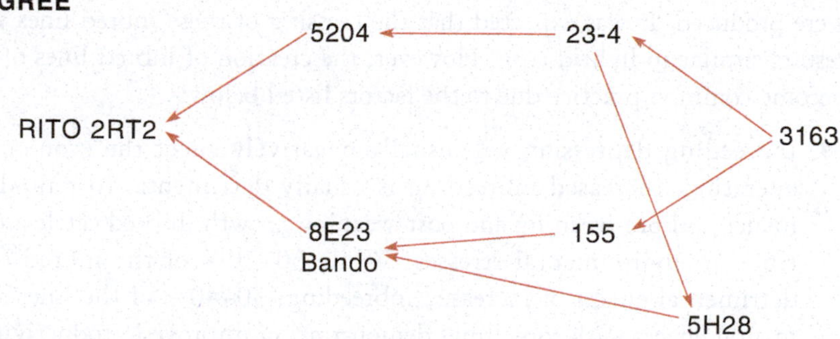

Inbreeding of RITO 2RT2

ancestor	n	n¹	1	(1 + Fa)	
23-4	1	2	1	—	$(1/2)^4$ = 6.2
3163	2	2	1		$(1/2)^5$ = 3.1
					9.3%

Relationship of RITO 2RT2 to 3163

ancestor	n	n¹	#	(1 + Fa)	
3163	3	0		—	$(1/2)^3$ = 12.5
3163	3	0		—	$(1/2)^3$ = 12.5
3163	4	0		—	$(1/3)^4$ = 6.1
					31.1%

Table 12.7
ZOOLOGICAL CLASSIFICATION OF CATTLE

Kingdom *Animalia*: Includes all animals in the animal kingdom
Phylum *Chordata*: Animals that have a backbone (vertebrae) or the rudiment of a backbone (chorda)
Class *Mammalia*: Warm-blooded, hairy animals that suckle their young from a milk-producing mammary gland
Order *Artiodactyla*: Even-toed, hoofed mammals
Family *Bovidae*: Ruminants having hollow, up-branched horns and a placenta type of numerous cotyledons; family also includes wild cattle, bison, and buffaloes[1]
Genus *Bos*: Ruminant quadrupeds, including domestic and wild cattle having a stout body and curved, hollow horns that stand out laterally from the skull
Species *Taurus*: Includes most cattle found in the United States, including their European ancestors
Indicus: These are the Zebu (humped) cattle, including the Brahman breed in the United States

[1]The domesticated species are *Bibos banteng* (the Bali cattle found primarily in Indonesia), *Bibos frontalis* (the Mithan cattle in southeastern Asia), *Poephagus grunniens* (the Yak of Tibet), and *Bubalus bubalis* (all breeds of river and swamp buffalo that are concentrated primarily in Asia). The genus and species of the American Bison is Bison bison.

Outbreeding

Species crossing is occasionally used to introduce new sources of genetic material in developing new breeds or to make cattle more adaptable to certain environmental conditions. Bison crossed with cattle is an example of a species cross. **Bos Indicus × Bos Taurus** are common crosses in the Gulf Coast states, though some question this as a true species cross since they both have the same number of chromosomes and interbreed freely. Table 12.7 shows how cattle species fit into the zoological classification.

Outcrossing involves the mating of unrelated animals and is the most widely used mating system by both commercial and seedstock producers. Outcrossing results in a higher level of heterozygosity. With respect to deleterious recessives, outcrossing does not eliminate these undesirable genes, but rather masks their effects and maintains them in the population disguised in the heterozygote. So long as these recessives occur in low frequencies, they typically have limited impact. The usefulness of outcrossing is primarily dependent on the effectiveness of selection (selection differential × heritability).

Crossbreeding

There are two primary reasons for using crossbreeding in cattle: (1) breed complementarity and (2) heterosis. Breed complementarity implies using breeds in a crossbreeding program so that their strengths and weaknesses complement one another. No one breed is superior in all desired production characteristics; therefore, planned crossbreeding programs, using breed complementarity, can significantly increase herd productivity.

Crossbreeding, when properly managed, allows for the effective use of heterosis (hybrid vigor), which is used extensively in the poultry and swine industries. Heterosis is the change in productivity in the crossbred progeny compared with the average of breeds used in the cross. The calculation of heterosis for weaning weight is shown in Table 12.8, which also indicates the importance of using highly productive cattle from breeds being utilized in the crossbreeding program. Highly productive crossbred cattle result from crossing highly productive purebred cattle, usually through purebred bulls. The use of crossbreeding in commercial breeding programs is discussed in Chapter 13.

Table 12.8
Computation of Heterosis for Weaning Weight

Breed or Category	Low Breed Productivity	High Breed Productivity
Angus	410	510
Hereford	390	490
Average of the two breeds	400	500
Average of crossbreeds	420	525
% heterosis	20/400 = 5%	25/500 = 5%

TRAITS AND THEIR MEASUREMENT

One of the challenges for cattle breeders is to identify those traits that are most important to profitability, and to focus selection pressure on them. More than 70 different traits have been identified that either have or would have associated genetic prediction estimates. This much information is overwhelming. The breeder must identify economically relevant traits—ERT (Table 12.9). These are traits that have a direct impact

Table 12.9
Economically Relevant Traits and Associated Indicator Traits Used to Make Estimates

Economically Relevant Trait	Indicator Traits
Calving ease	Birth weight Calving ease score Gestation length
Sale weight (varies based on goals of enterprise)	Birth weight Weaning weight adjusted to 205 days Yearling weight adjusted to 365 days Live weight at harvest Carcass weight Cull cow weight
Cow herd maintenance requirements	Mature cow weight Milk production Body condition score
Breeding female longevity	Age at puberty Frequency of calving Calving interval Cow mortality Pregnancy data
Marbling score and USDA Quality Grade distribution	Marbling score collected at packing plant Percent intramuscular fat measured ultrasonically in live animals Backfat thickness
Retail yield and USDA Yield Grade distribution	Rib-eye area (carcass) Carcass backfat Carcass weight Rib-eye area (ultrasound) Backfat (ultrasound)
Feedyard feed intake and efficiency	On feed weight Rate of gain Relative feed intake

Source: Adapted from Golden et al. (2000).

on profitability via influence on either cost of production or gross revenues. Many traits are associated with economically important traits, but they are indirect measures. These indirect predictor traits are termed indicator traits (Table 12.9). Data collected on indicator traits provide important inputs in calculating more accurate genetic estimates of the ERT.

Most economically important traits of beef cattle can be classified under (1) reproductive performance, (2) weaning weight, (3) yearling weight, (4) feed efficiency, (5) carcass merit, (6) longevity, (7) conformation, (8) freedom from genetic defects, (9) disposition, and (10) environmental adaptability.

Reproductive Performance

Reproductive performance, typically measured as percent calf crop by commercial cow-calf producers, has the highest economic importance when compared with other beef cattle traits (such as growth or carcass). Most cow-calf producers have a goal for percent calf crop weaned (e.g., the number of calves weaned compared with the number of cows in the breeding herd) of 85% or higher. Beef producers also desire each cow to calve every 365 days or less and to have a calving season for the entire herd of less than 90 days. Some reproductive traits (e.g., calving interval) have low heritabilities, whereas other reproductive traits (e.g., scrotal circumference and birth weight) have relatively high heritabilities. Reproductive traits with low heritabilities can be most effectively improved by changing the environment (e.g., providing adequate nutrition and maintaining good herd health practices) or by incorporating mating systems that increase heterosis. However, the development of genetic predictors of heifer pregnancy rate and lifetime female productivity will aid in the improvement of reproductive performance.

Reproductive performance can be improved through breeding methods by crossbreeding to obtain heterosis for percent calf crop weaned, using bulls with EPDs favorable for direct calving ease and maternal calving ease, avoidance of extremes in birth weight, and by selecting bulls that have a relatively large scrotal circumference. Scrotal circumference has a high heritability (50%), and bulls with a larger scrotal size (over 32 cm for yearling bulls) produce a larger volume of semen and have half-sister heifers that reach puberty at earlier ages than heifers related to bulls with a smaller scrotal size. Scrotal circumference is a threshold trait, as there is no apparent advantage in increasing the size beyond 38–40 cm.

Work by Doyle et al. (2000) and Evans et al. (1999) suggests that heifer pregnancy is approximately 12–14% heritable. Stayability, or the likelihood of cows producing at least five calves, is also sufficiently heritable to warrant selection pressure. Rebreeding rate as a 2-year-old was found to be quite lowly heritable and, as such, is best improved via management.

Reproductive tract scores (1–5) can be taken by rectal palpation of the reproduction tract when heifers are 13–15 months old. Scores of 4 and 5 indicate the heifer is cycling or close to cycling. Heifers with immature ovaries and reproductive tracts receive a score of 1–2. Heritability of reproductive tract score is 30% and thus the trait will respond to selection.

Weaning Weight

Weaning weight, as measured objectively, reflects the milking and mothering abilities of the cow and the pre-weaning growth rate of the calf. Weaning weight is commonly expressed as adjusted 205-day weight, where weaning weight is adjusted for age of calf and age of dam. This adjustment puts all weaning weight records on a comparable basis, since older calves will weigh more than younger calves and mature cows (5–9 years of age) will milk heavier than young cows (2–4 years of age).

The weaning weight of a calf is usually computed by dividing the calf's adjusted weight by the average weight of other calves in the herd and then expressed as a ratio. For example, a calf with a weaning weight of 550 lb in a herd averaging 500 lb has a ratio of 110. This calf's weaning weight ratio is 10% above the herd average. Ratios can be used primarily for selecting cattle within the same herd and contemporary group. Comparing ratios between herds is misleading from a genetic standpoint because most of the differences are caused by differences in environment. Bulls raised in different herds can be compared for genetic differences in weaning weights by evaluating their weaning weight EPDs. Weaning weight is 30% heritable and will respond to selection.

Yearling Weight

Yearling weight measures weaning weight and post-weaning gain to a weight that approaches slaughter weight. Post-weaning growth may take place on a pasture or in a feedlot. Usually animals with relatively high post-weaning gains make efficient gains at a relatively low cost to the producer.

Post-weaning gain in cattle is usually measured in pounds gained per day after a calf has been on a feed test for 100–120 days. Weaning weight and post-weaning gain are usually combined into yearling weight, that is, the adjusted 365-day weight.

Average daily gains for 140 days and adjusted 365-day weight both have high heritabilities (40%), so genetic improvement can be quite rapid when selection is based on post-weaning growth or yearling weight.

Feed Efficiency

Feed efficiency is usually measured by pounds of feed required per pound of live weight gain. Specific records for feed efficiency can only be obtained by keeping records on the amount of feed consumed by individual animals. With the possible exception of some bull-testing programs, determining feed efficiency on an individual animal basis is challenging. However, systems have been developed that automatically measure feed intake and associated data in real time (Figure 12.3) usually not economically feasible.

The interpretation of feed efficiency records can be rather confusing depending on the endpoint to which animals are fed. The feeding endpoint can be a certain number of days on feed (e.g., 140 days) to a specified slaughter weight (e.g., 1,200 lb), or to a carcass compositional endpoint (e.g., low Choice quality grade or 0.40 in. of fat over the rib eye). Most differences shown by individual animals in feed efficiency are related to pounds of body weight maintained through feeding periods and daily rate of gain or feed intake

Figure 12.3
Feed intake and feed efficiency data has been difficult and costly to collect. New technologies are becoming available that provide real-time automated data collection for feed intake.

of each animal. Cattle fed from a similar initial feedlot weight (e.g., 600 lb) to a similar slaughter weight (e.g., 1,200 lb) will demonstrate a high correlation between rate of gain and efficiency of gain. In this situation, cattle that gain faster will require fewer pounds of feed per pound of gain. Thus, a breeder can select for rate of gain and thereby make genetic improvement in feed efficiency. However, when cattle are fed to the same compositional endpoint (approximately the same carcass fat), the difference in amount of feed required per pound of gain is relatively small.

The heritability of feed efficiency is high (45%), so selection for more efficient cattle can be effective. It is logical to use the genetic correlation between gain and efficiency when possible because of the expense of obtaining individual feed efficiency records.

Carcass Merit

Carcass merit is presently measured primarily by carcass weight, tenderness, quality grade, and yield grade (the latter two are described in detail in Chapter 17). Many cattle breeding programs seek to produce cattle that will have quality grade Choice and have yield grades from 1.5 to 3.4. Visual or objective measurements of backfat thickness on the live animal and measurements of hip height can assist in predicting the yield grade at certain slaughter weights. Visual estimates can be relatively accurate in identifying actual yield grades if cattle differ by as much as one yield grade.

The most accurate measure of quality grade results from carcass evaluation. Traditionally, steer and heifer progeny of different bulls are fed and harvested to best identify genetic superiority or inferiority of bulls for both quality grade and yield grade. Heritabilities of most beef carcass traits are high (over 40%), so selection can result in marked genetic improvement for these traits.

An innovation in the assessment of carcass traits is the use of ultrasound technology (Figure 12.4). The correlations between ultrasonic measures of carcass traits on an individual animal and that animal's own carcass performance measured in the cooler are reasonably high. Correlations between ultrasonic measures of marbling, fat thickness, and rib eye on yearling Angus bulls and cooler measures on steer progeny for the same

Figure 12.4
The use of chute-side ultrasonic measurement of body composition is a means to enhance genetic selection for carcass traits.

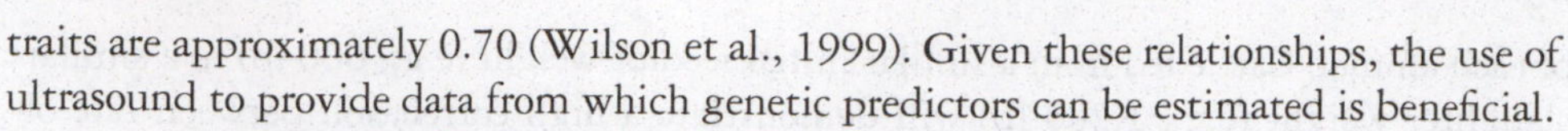

traits are approximately 0.70 (Wilson et al., 1999). Given these relationships, the use of ultrasound to provide data from which genetic predictors can be estimated is beneficial.

Tenderness can be measured by the Warner–Bratzler Shear test, which measures the pounds of force needed to cut through cores of meat, or by a trained panel of people who evaluate the palatability characteristics (juiciness, flavor, and tenderness) of meat. However, these methods of tenderness evaluation are slow and expensive. Furthermore, there are no direct economic incentives for producers to focus their attention on tenderness even though variation in tenderness affects the eating experience of beef consumers.

More objective measurements for carcass tenderness and muscling are needed. Tenderness is not highly correlated to marbling nor is muscling highly correlated to actual rib-eye area or rib-eye area per cwt of carcass. Visually appraising muscling by evaluating stifle thickness is more highly correlated to trimmed box beef yield than is rib-eye area.

Longevity/Stayability

Longevity, which measures the length of productive life, is an especially important trait for cows. Bulls are usually kept in a herd for only a few years, or inbreeding may occur. Some highly productive cows remain in the herd until 15 years of age or older, while other highly productive cows have been culled from the herd prior to reaching 3 or 4 years of age. Little direct selection for longevity in cows exists because few cows remain highly productive past the age of 10 years. Some cows are culled because of problems such as skeletal unsoundness, poor udders, or unhealthy eyes or teeth. Most cows that leave the herd early have poor reproductive performance (open or late pregnancy).

Some producers need to improve their average herd weaning weights as rapidly as possible rather than improve longevity. In this situation, a relatively rapid turnover of cows is needed.

Some selection for longevity occurs because highly productive cows that stay in the herd for a long period of time leave more numbers of potential replacement heifers. Some beef producers attempt to identify bulls that have highly productive, relatively old dams. Certain conformation traits such as skeletal soundness and udder soundness may be evaluated to extend the longevity of production. Crossbreeding will increase longevity as crossbred cows usually retain their teeth longer and in better condition than straightbred cows.

A few breeds have included stayability EPD in their sire summary. This EPD predicts genetic differences in the likelihood or probability that daughters of bulls will remain in production until they are at least 6 years of age. Cows culled from the herd prior to 6 years of age are costly because in most herds, it takes two to four calves produced per cow to cover the cost of the replacement heifer minus the salvage value of the cow. Longevity/stayability has a heritability of approximately 20%, so genetic change can be made in this trait by including it in a selection program.

Conformation

Conformation is the form, shape, and visual appearance of an animal. How much emphasis to put on conformation in a beef cattle selection program has been and continues to be controversial. Some producers feel that putting a productive animal into an attractive package contributes to additional economic returns. It is more logical, however, to place more selection emphasis on traits that will produce additional numbers of calves and pounds of lean growth for a given number of cows. Placing some emphasis on conformation traits such as skeletal (particularly feet and legs), udder, and eye and teeth soundness is justified. Conformation traits tend to be assessed subjectively that complicates development of meaningful genetic estimates that are typically objectively measured. Conformation differences such as in fat accumulation or predisposition to fat can be used effectively to make meaningful genetic improvement in carcass composition.

Traits such as stature, length, body capacity, and muscling score have heritabilities of 0.40 or higher. Udder scores that assess attachment, depth, and teat size have heritability estimates between 0.35 and 0.40. Leg structure traits such as pastern and hock angulation have heritability estimates approximating 0.15.

Genetic Defects

Genetic defects, including those identified previously in the sections on longevity and conformation, need to be considered in breeding productive beef cattle. Cattle have numerous known hereditary defects, as more than 200 different defects have been identified. Most of them, however, occur infrequently and are of minor concern. Some defects increase in their frequency, so selection needs to be directed against them. Most genetic defects are determined by a single pair of recessive genes. When one of these hereditary defects occurs, it is a logical practice to cull both the cow and the bull.

Some genetic defects that are observed in cattle today include the following.

Achondroplasia (bulldog dwarfism)—The homozygote may be aborted at 6–8 month's gestation and has a compressed skull, nose divided by furrows, and a shortened upper jaw, giving the bulldog facial appearance. This defect is inherited as an incomplete dominant. The most common type of dwarfism is snorter dwarfism, in which the skeleton is quite small and the forehead has a slight bulge. Some snorter dwarfs exhibit a heavy, labored breathing sound. This defect was common in cattle in the 1950s, but has decreased significantly since that time. Also occurring is the long-head dwarf (simple recessive) and the compress (comprest) dwarf inherited as incomplete dominance.

Alopecia—A lethal abnormality very similar to hypotrichosis. It takes a laboratory analysis to distinguish between the two. It has only been observed in Polled Herefords since 1988. Calves have kinky, curly hair that is soon lost in patches around the head, neck, and shoulders. Skin changes and anemia occur in all cases, with death occurring before 7 months of age as a result of anemia. This defect is believed to be a simple recessive.

Ankylosis—This defect results from an abnormal union of any of the joints in the calf's body. Cleft palate frequently occurs in this recessively inherited defect.

Arthrogryposis multiplex (palate-pastern syndrome)—Defect in which the pastern tendons are contracted and the upper part of the mouth is not properly fused together (cleft palate). Calves are typically born dead or perish soon after birth, legs are crooked, and minimal muscle development is observed.

Brachygnathia inferior (parrot mouth)—Cattle have a short lower jaw (simple recessive inheritance).

Contractual arachnodactyly—Calves are viable at birth, can walk, nurse, and survive. However, contraction of upper limb joints cause the affected animal to assume an abnormal crouched posture. Ambulation will improve with age but these cattle will retain a slender and angular appearance without normal expression of muscularity, often display poor foot structure, and exhibit poor levels of productivity.

Cryptorchidism—defect is the retention of one or both testicles in the body cavity.

Dermoid (feather eyes)—Skin-like masses of tissue occur on the eye or eyelid. Animals may become partially or completely blind. The mode of inheritance is polygenic.

Double muscling—defect is evidenced by an enlargement of the muscles, with large grooves between the muscle systems. It is especially noticeable in the hind leg. Double-muscled cattle usually grow slowly, and their fat deposition in and on the carcass is much less than that of normal beef animals.

Hydrocephalus—A bulging forehead due to fluid that has accumulated in the brain area is the typical symptom of this defect. Calves with arthrogryposis, hydrocephalus, or osteopetrosis usually die shortly after birth.

Hypotrichosis (hairlessness)—a low-frequency, non-lethal genetic disorder characterized by partial to almost complete lack of hair. It is inherited as a simple recessive, and was first identified in Hereford cattle in 1934.

Hypotrichosis ("rat-tail")—form of congenital hypotrichosis, this defect is characterized by the colored hair anywhere on the body being short, curly, malformed, and sometimes sparse. An abnormal tail switch occurs. Calves have slower rates of gain between weaning and yearling ages compared with normal calves from similar mating. This defect has occurred from mating of Simmental with Angus. This abnormality is controlled by interaction between two loci where at least one gene is for black color and must be heterozygous at the other locus involved.

Neuraxial edema—Calves are normal size at birth, but may not be able to get up or lift their head. Muscle spasms of neck and legs may last for 1–2 minutes (simple recessive).

Osteopetrosis (marble bone disease)—characterized by the marrow cavity of the long bones being filled with bone tissue. All calves with osteopetrosis have short lower jaws, a protruding tongue, and impacted molar teeth.

Progressive bovine myeloencephaly (weaver calf)—Calves start developing a weaving gait at 6–8 months of age and get progressively worse until death at 12–20 months (simple recessive).

Protoporphyria (photosensitivity)—Cattle are sensitive to sunlight and develop scabs and open sores when exposed to sunlight. This defect is inherited as a simple recessive.

Syndactyly (mulefoot)—one or more of the hooves are solid in structure rather than cloven. Mortality rate in calves is high.

When an abnormal calf is born, the producer should first try to determine whether the abnormality is genetic. Some abnormalities are environmentally caused, such as crooked calf disease. The crooked-leg condition makes one suspicious of a genetic abnormality; however, the cause is the cow eating the lupine plant during a certain stage of pregnancy (Chapter 15).

A producer should not panic if a genetically abnormal calf is born in the herd. A commercial producer might consider selling the cow and replacing the bull. A seedstock producer may need to take a more serious look at the problem and take additional steps to eliminate or reduce the incidence of the genetic problem.

The genetic abnormalities that have caused the greatest concern in recent years were identified earlier in the chapter (Table 12.2). Several of these abnormalities reached unexpected levels of occurrence due to a selection preference for the carrier animal or a noted bull being used extensively through AI. In the latter situation, a carrier bull can sire several thousand calves before the abnormality expresses itself. Obviously, one-half of the bull's calves would carry the undesirable recessive. However, this would not be known until one of his carrier daughters was mated to another animal carrying a similar undesirable gene. Even then, the chance of an abnormal calf occurring is one chance in four for each carrier × carrier mating.

Occasionally, it is deemed desirable and necessary to progeny test a bull from unknown or known carrier parentage. It is generally accepted that the level of testing should be at the 1% level of probability; that is, there is only one chance in a hundred that a bull could be tested this extensively and still be carrying the undesirable recessive gene.

Using the inheritance of the mulefoot abnormality (Table 12.2), a bull (S?) could be mated to eight females showing the mulefoot condition (ss) or 16 known carriers (Ss) or 32 of his own daughters (1/2 SS and 1/2 Ss). If no mulefoot calves were identified, the bull would be tested to the 1% level of probability (Table 12.10).

Disposition

Disposition or **docility** refers to the level of calmness or excitability of cattle. This trait is best evaluated when producers are near cows during calving or when working cattle through handling and restraining facilities. Obvious differences in the behavior of cattle exist that affect their disposition (Chapter 18).

Evidence suggests that both the genetic makeup of cattle and the environmental conditions to which they are exposed determine disposition. There are breed differences, line of breeding differences within a breed, and individual differences in disposition. Cattle with very poor dispositions are sometimes culled to prevent human injury or to decrease repair costs of fences and working facilities. How producers handle their cattle also has a significant impact on their disposition (Figure 12.5).

Adaptability

The Brahman breed and Brahman crossbreds are used widely throughout the southeast and Gulf Coast areas because of the breed's adaptability to high environmental temperatures and high humidity. Most of the Bos taurus breeds are not adapted to these subtropical conditions; thus, their performance is lower than the Zebu cattle.

Table 12.10
PROBABILITY OF DETECTING A BULL AS A CARRIER OF AN UNDESIRABLE RECESSIVE GENE, USING THREE KINDS OF TEST MATINGS

	Mating Bull to . . .		
No. Matings	Double Recessive	Known Heterozygote	Own Daughter
1	0.5	0.25	0.125
2	0.75	0.44	0.235
3	0.87	0.58	0.330
4	0.94	0.68	0.414
5	0.97	0.76	0.487
6	0.98	0.82	0.551
7	0.99	0.87	0.607
8	0.996	0.90	0.656
9	0.998	0.925	0.699
10	0.999	0.944	0.737
15	—	0.987	0.865
20	—	0.997	0.931
25	—	0.9993	0.965
30	—	—	0.982
35	—	—	0.991
40	—	—	0.995
50	—	—	0.999

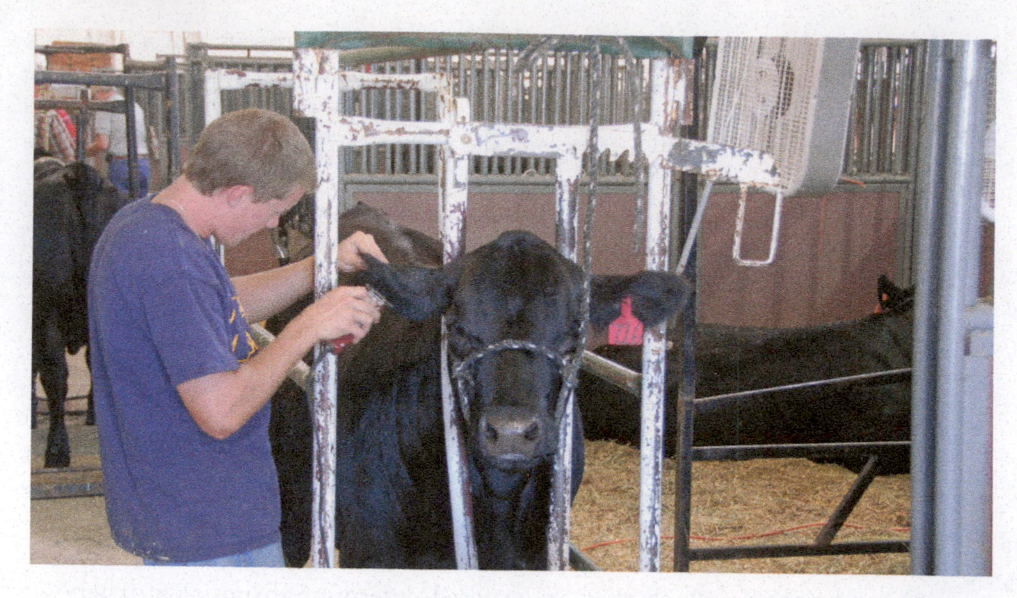

Figure 12.5
Disposition in cattle has a genetic basis; however, the major influence on disposition is how the cattle are handled.

Bos taurus cattle can function much better in colder environments than the Bos indicus breeds. Newborn Brahman calves are unable to maintain thermostability in cold (5°C) environmental temperatures.

Adaptability of different biological types will vary depending on the environment where they are expected to perform. For example, cows of large mature weight and high milk production will likely have below-average reproductive performance in a low-cost grazed forage environment. In this situation, there would be high breakeven prices of weaned calves. The cows are adapted to an environment where large amounts of harvested and (or) supplemental feed must be provided to achieve satisfactory levels of reproductive performance.

IMPROVING BEEF CATTLE THROUGH BREEDING METHODS

Genetic improvement in beef cattle occurs by selection and the choice of mating system. Significant improvement by selection results when the selected animals are superior to the herd average and when the heritabilities of traits are medium to high (above 30%). Factors affecting the rate of genetic improvement from selection include genetic variation, heritability, selection differential, and generation interval.

Genetic Variation

Many pairs of genes influence most economically important traits in beef cattle. This genetic variation, along with different environmental effects, causes considerable variation in cattle traits. Reports of yearling weights of bulls in excess of 1,700 lb, rib-eye areas larger than 20 sq. in., and 205-day weaning weights over 800 lb demonstrate that high levels of productivity are possible with exceptional genetic and environmental conditions. These extremely high levels of productivity are usually not economically feasible because maximum profitability is reached before maximum levels of productivity (Chapter 3).

Differences in productivity are noticeable even under similar environmental conditions. It is not uncommon for a commercial cow-calf producer to find weaning weight differences of over 100 lb in calves within the same herd.

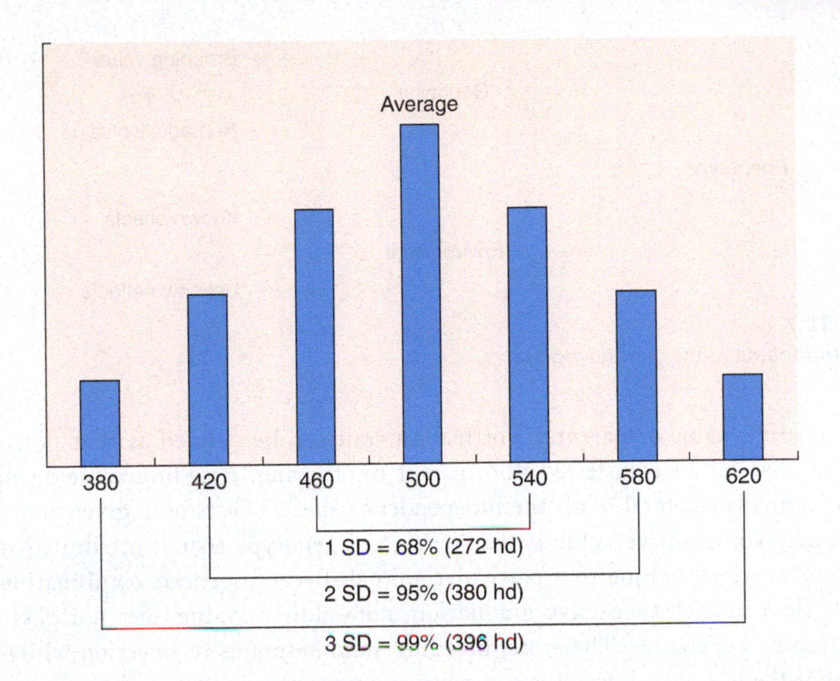

Figure 12.6
A normal bell-shaped curve for weaning weight showing the number of calves under each standard deviation area of the distribution (400 calves in the herd and a S.D. of 40 pounds).

As noted earlier, many economically important traits in beef cattle show continuous variation primarily because many pairs of genes control them. As these genes are expressed and as the environment influences these traits, producers usually observe and measure large differences in the performance of animals for any one trait. For example, if many calves are weighed at weaning (approximately 205 days of age) in a single herd, there would be considerable variation in the calves' weights. Distribution of weaning weights of the calves would be similar to the example shown in Figure 12.6. The bell-shaped curve distribution demonstrates that most of the calves are near the herd average (440 lb), with relatively few calves having extremely high or low weaning weights when compared at the same age.

Figure 12.6 shows the use of the statistical measurement of **standard deviation (SD)**, which describes the variation or differences in a herd (in this case, the average weaning weight is 500 lb and the calculated standard deviation is 40 lb). Using herd average and standard deviation, the variation in weaning weight shown in Figure 12.6 can be described as follows:

$$500 \text{ lb} \pm 1 \text{ SD}(40 \text{ lb}) = 460 - 540 \text{ lb}$$
(68% of the calves are in this range)

$$500 \text{ lb} \pm 2 \text{ SD}(80 \text{ lb}) = 420 - 580 \text{ lb}$$
(95% of the calves are in this range)

$$500 \text{ lb} \pm 3 \text{ SD}(120 \text{ lb}) = 380 - 620 \text{ lb}$$
(99% of the calves are in this range)

One percent of the calves (4 calves in a herd of 400) would be on either side of the 380–620-lb range. Most likely, two calves would be below 380 lb and two calves would weigh more than 620 lb.

Genetic improvement ultimately requires that producers understand the basic model listed in Figure 12.7. Phenotype is the observed measurement or characterization of a specific trait and its expression is influenced by genotype that results from both the cumulative effects of the animal's individual genes for the trait and the effect of the gene combinations. Environmental effects can be thought of as the summation of all non-genetic influences that affect the phenotype for a trait.

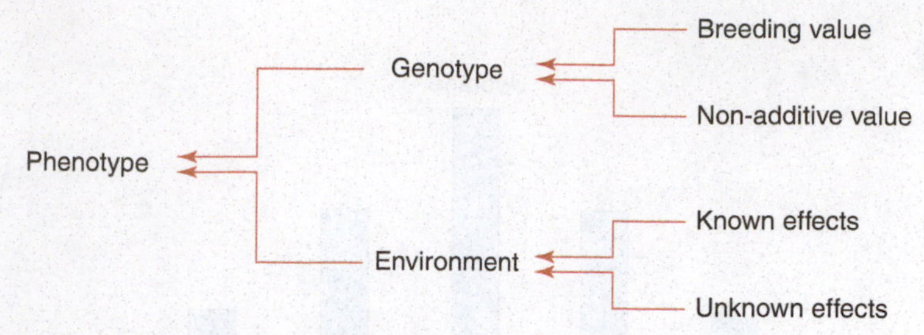

Figure 12.7
The components of the genetic model.

Breeding value or parental worth for a trait can be defined as that portion of genotype that can be transferred from parent to offspring. Breeding value in mathematical terms is the total of all the independent genetic effects on a given trait of an individual. Non-additive value is that portion of genotype that is attributed to the gene combinations unique to a particular animal. Because genetic combinations are reestablished in each successive graduation, non-additive value does not pass from generation to generation. Therefore, breeding value responds to selection while non-additive value is accessed via choice of mating system (i.e., crossbreeding).

There are two basic types of environmental effects—known and unknown. Known effects have an average effect on individuals in a specific category. Examples include age, age of dam, and gender. Calves born earlier in the calving season and thus older at time of weaning typically weigh more than their younger counterparts. Unknown effects are random in nature and are specific to an individual phenotype. Known effects can be quantified and used to adjust phenotypic measures to allow more accurate selection. Unknown effects are more difficult to account for, but breeders can use management to minimize their impacts.

Considerable variation exists in most traits of importance. Phenotypic standard deviations for selected beef cattle traits are shown in Table 12.11.

Table 12.11
Averages and Phenotypic Standard Deviations for Selected Beef Cattle Traits

Trait	Average	Standard Deviation
Birth weight	90 lb	10 lb
Body condition score (BCS)[1]	5 units	0.7 units[1]
Scrotal circumference	33 cm	2.5 cm
Weaning weight	525 lb	45 lb
Daily gain (feedlot)	3.0 lb/day	0.3 lb
Yearling weight	985 lb	70 lb
Yearling hip height	49 in	1.4 in.
Mature cow weight	1250 lb	125 lb
Slaughter weight	1210 lb	95 lb
Dressing percentage	60.6%	1.9%
Carcass weight	735 lb	63 lb
Marbling score[2]	5.0 units	0.59 units
Retail product (0 fat trim)	65.8%	3.2%
Fat thickness (12th rib)	0.26 in.	0.12 in.
Rib-eye area	12.5 sq. in.	1.3 sq. in.
Tenderness (WBS)	11.2 lb	2.4 lb

[1]Based on a visual score (range of 1–9).
[2]4.0 to 4.9 = slight; 5.0 to 5.9 = small.
Source: Adapted from USDA: ARS: MARC (data from 12 breed groups and three composite populations).

Heritability

A heritability estimate is a percentage figure that indicates what proportion of variation in a trait is due to heredity. Heritability identifies that portion of variation that is passed on from parent to offspring. Table 12.12 lists heritability estimates for some economically important traits for beef cattle.

Table 12.12
HERITABILITY ESTIMATES FOR IMPORTANT BEEF CATTLE TRAITS

Trait(s)	Heritability[1] (%)
Reproductive	
Age at puberty	40
Weight at puberty	50
Scrotal circumference	50
Breeding soundness examination (BSE)	10
Primary sperm abnormalities	30
Reproductive tract score	30
First service conception rate	25
Stayability	20
Calving ease (direct and maternal)	15
Weight at puberty	40
Gestation length	40
Birth weight	40
Pelvic area	50
Body condition score	40
Calving interval	10
Multiple births	05
Heifer pregnancy	14
Growth	
Weaning weight	30
Milk production	20
Post-weaning ADG (feedlot)	45
Post-weaning ADG (pasture)	30
Efficiency of feedlot gain	45
Maintenance (MEm)	50
Yearling weight	40
Yearling hip height	40
Mature weight	50
Carcass	
Carcass weight (at similar age)	40
Marbling	35
Fat thickness	45
Percent retail product	25
Rib-eye area	25
Tenderness	
Shear force (WBS)	40
Sensory panel	10
Other Traits (functional, convenience, longevity)	
Cancer eye susceptibility	30
Horn fly resistance	60
Pink eye susceptibility	25
Pulmonary arterial pressure	40
Disposition (docility)	40
Longevity/stayability	20
Sheath area (Zebu)	45
Udder attachment	20
Teat size	50

[1]Heritability estimates below 20% are considered low; those 20–39% are considered medium; and those 40% and higher reflect highly heritable traits.

Selection Differential

Selection differential is sometimes referred to as **reach** because it shows how superior (or inferior) the average of selected individuals is to the average of the group from which they were selected.

Predicting Genetic Change

An example of a selection differential is shown in the following calculation of genetic change.

Genetic change per year = *(heritability × selection differential)/generation interval*

Bulls

Selected bulls average	545 lb
Of bulls in herd	400 lb
Selection differential =	145 lb
Heritability	0.30
Total genetic superiority =	43.5 lb
Only half passed on (43.5 ÷ 2) =	21.7 lb

Heifers

Selected heifers	460 lb
Average of heifers in herd	400 lb
Selection differential =	60 lb
Heritability	0.30
Total genetic superiority =	18.0 lb
Only half passed on (18 ÷ 2) =	9.0 lb

Mating bulls to heifers

Fertilization combines the genetic superiority of both parents
21.7 lbs + 9.0 lbs = approximately 30 lb.

Generation Interval

Generation interval is the average age of the parents when the calves are born. The selected heifers in the previous example represent only about 15% of the total cow herd, so the 30-lb estimate is for one generation. This much selection would have to be practiced over approximately 6 years to replace the entire cow herd. Thus, 30 lb ÷ 6 years = 5 lb per year. This example demonstrates how generation interval affects the yearly genetic change.

Multiple Trait Selection

The example with weaning weight represents genetic change if selection is for only one trait. If selection is practiced for more than one trait, genetic change is $1/\sqrt{n}$, where n is the number of traits in the selection program. If four traits are in the selection program, the genetic change per trait would be $1/\sqrt{4} = {}^{1}/_{2}$. This means that only one-half of the progress would be made for any one trait, compared to giving all the selection to one trait. This reduction in genetic change per trait should not discourage producers from multiple trait selection. Maximizing genetic progress in a single trait selection usually is associated with lowering productivity in some other economically important traits.

Table 12.13
Examples of Favorable and Unfavorable Genetic Correlations in Beef Cattle

Favorable	Unfavorable
Weaning weight and post-weaning gain (+.44)	Birth weight and calving ease (−.74)
Yearling weight and post-weaning gain (+.81)	Post-weaning gain and calving ease (−.54)
Post-weaning gain and feed per lb. of gain (−.64)	Yearling weight and mature cow weight (+.72)
	Backfat and marbling (+.35)

Genetic and Phenotypic Correlations

The strength of relationship between specific traits can be measured in three ways—phenotypic, genetic, and environmental correlations. Phenotypic correlations measure how performance in two traits changes in the same or opposite directions because of genetic and environmental influences on the traits. A genetic correlation measures the strength of relationship between the breeding values of two or more traits and results when genes affect more than one trait. An environmental correlation measures the relationship between environmental effects on traits. Correlations range from +1.0 to −1.0 and these relationships may help or hinder genetic improvement. Correlations in the range of +0.20 to −0.20 are considered low and relatively unimportant, whereas correlations greater than +0.50 and lesser than −0.50 are considered important. The direction of a correlation should not be viewed as indicating favorability as there are favorable genetic correlations that are both positive and negative. On the other hand, there are unfavorable relationships with both positive and negative sign (Table 12.13).

The genetic correlation between rate of gain and feed per pound of gain (from similar beginning weights to similar end weights) is negative. This is desirable from a genetic improvement standpoint because animals that gain faster require less feed per pound of gain. This relationship is also desirable because rate of gain is easily measured whereas feed efficiency is an expensive trait to measure. Yearling weight and mature weight are positively correlated with birth weight; that is, as yearling weight increases, birth weight and mature weight also increase. This may pose a potential problem in that excessive birth weight may result in increased calving difficulty and feed requirements increase as mature weights increase. The positive genetic relationship between backfat thickness and marbling challenges the ability of the industry to meet the dual market demand for beef that is both flavorful and relatively lean. Research studies and observations in progressive cow herds show that improvement can be made in selecting for these economically important traits that have challenging correlations. For example, it is possible to combine moderate birth weights (65–80 lb) with rapid feedlot gains (≥3.0 lb/day). Another example shows that it is feasible to produce carcasses that have Choice quality grades and also have Yield Grade 2 carcasses. Optimum combinations of these traits where extremes are avoided should be the selection goal.

SELECTION PROGRAMS

Types of Selection

There are three fundamental approaches to selection: (1) tandem, (2) independent culling level, and (3) selection index.

Tandem is selection for one trait at a time. When the desired level is achieved in that one trait, then selection is practiced for a second trait. Tandem is the least effective of the three selection methods. Because of unfavorable genetic correlations, the prolonged selection for one trait may yield deficiencies in another. While progress

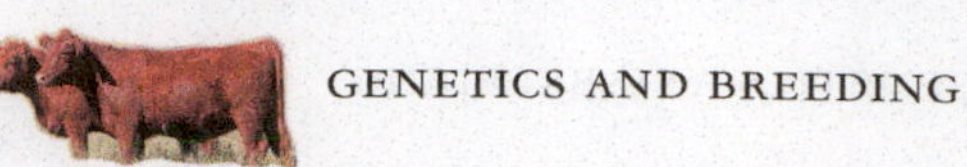

Table 12.14
Independent Culling Level Selection in Yearling Bulls

		Bull				
Trait	Culling Level	A	B	C	D	E
Birth weight (lb)	85 (max)	[105]	82	85	83	80
Weaning weight (lb)	500 (min)	550	[495]	505	570	600
Post-weaning gain (lb/day)	3.0 (min)	3.7	3.6	3.1	3.3	3.5
Scrotal circumference (cm)	30 (min)	34	37	31	[29]	35

Boxed values indicate performance that necessitates culling under the selection criteria.

in the selected trait may be significant in the short run, it is very difficult to hold a trait at an optimal level over time.

Independent culling level establishes minimum culling levels for each trait in the selection program. Even though it is not as effective as the selection index, independent culling level is the most common type of selection in use today. Its major disadvantage is that an animal slightly below the minimum culling level in one trait but highly superior in other traits would be culled (Table 12.14). For example, in this case, bull B is culled for being 5 lb under the minimum weaning weight despite being superior in the other three traits.

The **selection index** is the most effective selection type when selecting for two or more traits. As noted earlier, caution should be exercised about genetic change when many traits are included in the selection program. A selection index is not easily constructed; it requires highly involved statistical methods in order to put the heritabilities, economic values, genetic correlations, and variabilities of several traits into a single formula. An example of a selection index is: Index = yearling weight – 3.2 (birth weight), as mentioned in Chapter 11. This index is effective in selecting for an optimum combination of higher yearling weights and lower birth weights.

Perhaps the longest application of a selection index to a cattle population is that utilized by Landcorp Farming, Ltd. in New Zealand. The breeding objective, defined in 1976, and applied continuously to the herd was:

$$H = 0.53 \times L \times D_p (4.8F - 1) + 0.06 \times M \times D_m$$

Where: H = net lifetime income per cow
0.53, 0.06 = the net income (1976 NZ$/kg carcass) from the slaughter of young stock and cull cows, respectively
L = market weight (kg) of progeny at 30 months of age
D_p, D_m = dressing percentage of young stock and cows, respectively
F = net fertility
M = sale weight (kg) of cows

In the calculation, 4.8F – 1 is the total number of saleable calves per the lifetime of a female minus her own replacement. To account for nutrient intake, the gross income from young stock and cows was adjusted 11% and 32%, respectively. Following the development of the breeding objective, selection indexes were formulated to predict overall breeding values for each animal. Selection of replacement bulls and heifers took place at 1 year of age and were based on the following:

$$I = 40.4F_D + 0.0398\ MWW + (-0.2274WW_I) + 6191\ YW_I$$

Where: I = income
F_D = fertility of the dam

WW_I = weaning weight of the individual
YW_I = yearling weight of the individual

The results of this breeding approach were genetic trend increases of direct and maternal weaning weight of 0.72 and 0.33 lb per year. Yearling weight breeding values rose 1.67 lb annually, and fertility—defined as the number of calves weaned per cow exposed to mating—increased the equivalent of one additional calf per 100 head of females bred. The overall index breeding value increased $4.32 (1976 NZ) per year. Thus, as a result of selecting for the defined breeding objective, each new heifer entering the herd was worth $4.32 more than the heifers entering the herd in the preceding year (Enns and Nicoll, 1997).

NATIONAL SIRE EVALUATION

At one time, genetic comparisons were limited to within-herd comparisons, central bull test stations, and highly structured progeny tests. These efforts, while useful, did not yield the desired accuracy and breadth of prediction across populations. With the advent of improved statistical programs such as Best Linear Unbiased Prediction (BLUP), genetic evaluation systems on a national level could be conducted. These contemporary genetic prediction efforts have been embraced by nearly all of the beef breed associations.

A more detailed discussion of using EPDs in beef cattle selection is presented in Chapter 4. National cattle evaluation programs (NCE) are based on a set of guidelines and protocols established by the Beef Improvement Federation. The key to NCE development is the widespread use of AI and the use of common sires across herds to allow for the creation of linkages between a large number of herds. Furthermore, it is absolutely critical that breeders collect and report accurate data on as many animals as possible for the traits in question. Finally, it is critical that breeders report accurate contemporary groupings to assist in the calculation of the most reliable genetic predictions possible.

To calculate an EPD for a particular trait, performance information can be utilized from the individual, its siblings, its ancestors, and best of all, from its progeny. Because selection of replacement animals occurs prior to their actually becoming parents, the use of several different EPD formats is made. These include parent EPDs for those animals with progeny data and non-parent EPDs for those animals without corresponding progeny data. Non-parent EPDs can be calculated not only for those animals that have individual data on a trait, but also for those animals where the data are not yet available. For example, a yearling weight EPD can be computed for a weanling calf by utilizing only pedigree data. These are referred to as pedigree estimates.

The incorporation of molecular breeding values into EPD estimates yields a more accurate value referred to as a genomically enhanced EPD.

An emerging trend in sire evaluation is to create and report economic indexes. The dairy industry has successfully utilized this approach for a number of years and its application in the beef industry can help to aggregate a number of EPDs into one value reported as differences in economic performance. Examples in the beef industry include the Weaning Calf Index (Angus) that accounts for differences in pre-weaning performance based on revenue and cost adjustments related to birth weight, weaning weight growth, milk production, and mature weight; the All-Purpose Index (Simmental) that evaluates sires for use in herds where they are bred to both heifers and mature cows, replacements are retained, and market progeny is retained through the feedyard phase of production; the Mainstream Terminal Index (Limousin) that measures economic differences when progeny is sold on a typical industry grid pricing structure; and the Efficiency Profit Index (Gelbvieh) that allows economic comparisons for differences in feed efficiency.

Breeders are cautioned to avoid intensive single-trait selection even on an index basis. Keeping focus on the whole system continues to be an important component of a profitable breeding plan.

Table 12.15
POSSIBLE CHANGE VALUES FOR EPDS OF VARIOUS TRAITS AND ASSOCIATED ACCURACIES

BIF Accuracy	Birth Weight	Weaning Weight	Yearling Weight	Maternal Milk	Fat Thickness	Rib-eye Area	Marbling Score
0.0	3.0	12	15	9	0.03	0.31	0.24
0.2	2.4	9	12	7	0.02	0.25	0.19
0.4	1.8	7	9	5	0.02	0.19	0.14
0.6	1.2	5	6	4	0.01	0.12	0.10
0.8	0.6	2	3	2	0.01	0.06	0.05
0.9	0.3	1	2	1	0.00	0.03	0.02
1.0	0.0	0	0.00	0	0.00	0.00	0.00

Source: Reprinted with permission from Red Angus Sire Summary, 2003.

Table 12.16
PROPORTION OF PROGENY EXPECTED TO CARRY A GENE(S) GIVEN THE FREQUENCY IN THE PARENTAL GENOME

		Proportion (%) of Progeny Expected:		
Sire	Dam	0	1	2
0	0	100		
0	1	50	50	
0	2		100	
1	0	50	50	
1	1	25	50	25
1	2		50	50
2	0		100	
2	1		50	50
2	2			100

Associated with each EPD is a measure of accuracy. The range of values for accuracy is 0–1, with values closer to 1 indicative of estimates in which a high level of confidence can be given. In essence, accuracy increases as more data are used in making the estimate. High accuracies indicate that the estimate is more reliable and that it is less likely to change as more information becomes available.

Another way to report accuracy is with possible change values (Table 12.15). Assume that an animal has an EPD for yearling weight of +60 and an associated accuracy of 0.80. Given the possible change value of 3 from Table 12.16, we can be 68% confident that the true genetic value of the animal for yearling weight lies in the range of +60 ± 3.

Understanding and utilizing accuracy information allows breeders to make appropriate selection decisions based on their own tolerance for risk.

GENETIC TESTING AND MARKER-ASSISTED SELECTION

DNA information in the beef cattle industry can be applied to a variety of problems—parentage testing, identification of genetic defect carriers, and improving selection. Furthermore, genetic information at the molecular level provides an opportunity to gain information on hard-to-measure traits early in an animal's life before having the opportunity to express the trait, to better understand carcass trait differences in live animals, and to estimate variation in sex-limited traits.

Parentage determination utilizes specific variation in the DNA sequence tested for affiliation with a specific phenotype to establish the sire and/or dam of a calf. Accurate parentage verification is important in assuring effectiveness of selection based on EPDs that depend on **pedigree** information, determination of the correct sire of an offspring when multiple sire mating systems are utilized or to determine whether the offspring originated from the AI or natural service sire, and to create opportunities for retroactive selection. For example, if a herd is using multiple sire pastures and retaining ownership of progeny to the packing plant, parentage technology allows a producer to sort out above- and below-average performers in feedlot and carcass traits and to then use parentage identification via DNA analysis to determine the sires of those animals, ultimately leading to a decision as to which sires should be retained in the bull battery.

For simply inherited traits such as color or horned status, genetic tests can determine whether or not an animal is a heterozygote (e.g., black hided sire that carries the red gene). Genetic testing can also identify carriers of deleterious genes. The emergence of Arthrogryposis Multiplex (AM) and Hydrocephalus (NH) in Angus cattle as announced by the American Angus Association in 2008 and 2009 threatened to create market disruptions to the seedstock sector. However, given the availability of genetic testing technologies, genetic markers were identified to allow clear separation between carriers of the defect (heterozygotes) and normal genotypes. Thus, affected animals could be efficiently culled from the breeding population without condemning entire genetic lines. By comparison, when dwarfism impacted the Hereford breed in the 1940s and early 1950s, genetic tests were not available and breeders had to rely on imperfect pedigree analysis and the use of time-consuming designed matings to determine carriers. Without the ability to efficiently and accurately identify individual carriers, entire genetic lines fell into disfavor.

Relative to most traits of interest, breeders rarely have the luxury of selecting directly a specific gene given that most economically important traits are affected by a multitude of genes. **Marker-assisted selection** (MAS) is based on the identification of specific regions of chromosomes where genes affecting economically important traits are located, and then identifying MAS individuals with favorable allelic combinations.

DNA testing can be utilized to help establish differences in genetic potential for specific quantitative traits by computing a molecular breeding value (MBV). While these techniques do not yield estimates that are superior to EPDs, when combined with the performance data utilized in a traditional EPD estimate computation, the resulting genomically enhanced EPDs have improved accuracy. Thus, the use of genetic markers provides significant value to national cattle evaluation efforts.

According to the National Beef Cattle Evaluation Consortium, the available approaches to incorporating MBV information with EPDs include the following:

- Compute EPD and MBV separately and then combine the results into an index.
- Determine specific single nucleotide polymorphisms (SNP) for use in combination with data tied to pedigrees.
- Include MBV in a national cattle evaluation as a correlated trait. In this approach, as the genetic correlation of the MBV and the trait of interest increase, the accuracy of the estimate improves. This is of particular value in selecting young animals and especially for traits not expressed until relatively late in life (marbling, tenderness, reproductive fitness, etc.).
- Utilize MBV as external EPDs (outside the population), so the accuracy of an estimate is individualized to a specific animal based on the relationship between the individual and the training population.

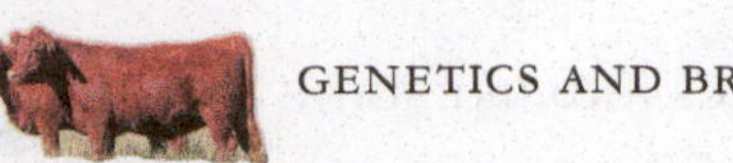

DNA sequencing by itself will not provide a simplistic approach to selection. The use of MAS will make cattle breeding more effective by providing more information for inclusion in National Cattle Evaluation systems, but MBV will not be a standalone solution.

The usefulness of any MBV ultimately depends on the proportion of genetic variation accounted for by the marker panel. However, genetic markers are only useful within the specific population where it was developed. Thus, MBVs are quite breed specific and thus must be developed separately for breeds. However, by knowing the frequency of a gene in both parents, the likelihood of the gene being carried by the progeny can be estimated (Table 12.16). If the economic benefits of producing cattle carrying particular genes are sufficiently high, then the investment in DNA testing will be warranted.

Genomics technologies are expensive and thus wide-scale incorporation beyond elite seedstock production is unlikely until scalability decreases the unit cost substantially. However, multiple opportunities exist that may drive incorporation beyond the seedstock producer. These include opportunities to utilize genetic testing to identify lines of cattle that have inherent disease resistance capability, possess unique attributes that contribute to improved nutritional value of beef products, or offer unique attributes of value for two or more segments of the beef supply chain. Furthermore, genomics offers potential as a means to sort out genetic differences for traits that are difficult or expensive to measure or are sex-linked. Specifically, feed efficiency, reproductive merit, and longevity would be enhanced if molecular influences can be illuminated.

SELECTED REFERENCES

Publications

American Angus Association. 2015. *Sire summary report*. St. Joseph, American Angus Association located in St. Joseph, MO.

American Gelbvieh Association. 2015. *Sire summary report*. Westminster, CO.

American Simmental Association. 2015. *Sire summary report*. Bozeman, MT: ASA.

Beef Improvement Federation. 2010. *Guidelines for uniform beef improvement programs*. Manhattan, KS.

Bourdon, R.M. 2000. *Understanding animal breeding*. Upper Saddle River, NJ: Prentice Hall.

Brethour, J.R. 2000. Using serial ultrasound measures to generate models of marbling and backfat thickness changes in feedlot cattle. *Journal of Animal Science.* 78:2055.

Brinks, J.S., & Knapp, B.W. 1975. *Effects of inbreeding on performance traits of beef cattle in the western region*. Fort Collins, CO: Colorado Agricultural Experiment Station Tech. Bull. 123.

Davis, G.P., & DeNise, S.K. 1998. The impact of genetic markers on selection. *Journal of Animal Science.* 76:2331.

Dekkers, J.C.M. 2004. Commercial application of marker- and gene-assisted selection in livestock: Strategies and lessons. *Journal of Animal Science*. 82(E. Suppl.):E313–E328.

Doyle, S.P., Golden, B.L., Green, R.D., & Brinks, J.S. 2000. Additive genetic parameter estimates for heifer pregnancy and subsequent reproduction in Angus females. *Journal of Animal Science.* 78:2091.

Enns, R.M., & Nichol, G.B. 1997. *Index selection in practice—A New Zealand case study*. Proc. BIF Research Symposium. Dickinson, ND.

Evans, J.L., Golden, B.L., Bourdon, R.M., & Long, K.L. 1999. Additive genetic relationships between heifer pregnancy and scrotal circumference in Hereford cattle. *Journal of Animal Science.* 77:2621.

Golden, B.L., Garrick, D.J., Newman, S., & Enns, R.M. 2000. *Economically relevant traits: A framework for the next generation of EPDs*. Proc. BIF Research Symposium. Wichita, KS.

Green, R.D. 1996. *How will DNA technology impact beef cattle selection?* Proc. 2nd Congress on Genetics and Reproduction of the Brazilian Zebu Breeders Assoc. Uberaba, Brazil.

Gregory, K.E., & Cundiff, L.V. 1980. Crossbreeding in beef cattle: Evaluation of systems. *Journal of Animal Science.* 51:1224.

Gregory, K.E., Cundiff, L.V., & Koch, R.M. 1995. *Composite breeds to use heterosis and breed differences to improve efficiency of beef production.* Clay Center, NE: MARC.

Griffin, D.B., Savell, J.W., Recco, H.A., Garrett, R.P., & Gross, H.R. 1999. Predicting carcass composition of beef cattle using ultrasound technology. *Journal of Animal Science.* 77:889.

Hickman, C.G. (Ed.). 1991. *Cattle genetic resources: World animal science, B7.* New York: Elsevier Science Publishers.

Legates, J.E. 1990. *Breeding and improvement of farm animals.* New York: McGraw-Hill.

Long, C.R. 1980. Crossbreeding for beef production: Experimental results. *Journal of Animal Science.* 51:1197.

National Beef Cattle Evaluation Consortium. 2013. *Delivering genomics technology to the beef industry.* www.nbcec.org.

National Beef Cattle Evaluation Consortium. 2012. Beef Sire Selection Manual. www.nbcec.org.

North American Limousin Foundation. 2015. *Sire evaluation report.* Denver, CO.

Olson, T. (1994, October). The genetics of coat color inheritance in cattle. *American Hereford Journal.* 32.

Red Angus Association of America. 2015. *Sire evaluation report.* Denton, TX.

Reverter, A., Johnston, D.J., Graser, H.U., Wolcott, M.L., & Upton, W.H. (2000). Genetic analysis of live-animal ultrasound and abattoir carcass traits in Australian Angus and Hereford cattle. *Journal of Animal Science.* 78:1786.

Schalles, R.R. 1986. *The inheritance of color and polledness in cattle.* Bozeman, MT: American Simmental Association.

Spangler, M.L. & L. Schiermiester. 2014. DNA tests for genetic improvement of beef cattle. NebGuide, University of Nebraska Extension, G1856.

Tess, M.W., & Thrift, F.A. 1992. Genetic aspects of beef production in the southern region. Auburn, AL: Auburn University.

Willham, R.L. 1982. Genetic improvement of beef cattle in the United States: Cattle, people, and their interaction. *Journal of Animal Science.* 45:659.

Wilson, D., Rouse, G., Hays, C., & Hassen A. 1999. *Genetic evaluation of Angus ultrasound measure.* Proc. BIF Research Symposium. Roanoke, VA.

Wyse, R. 2001. *A livestock genomics institute.* Proc. Intl. Livestock Congress. Houston, TX.

13 Cattle Breeds

Much of the history and lore of the cattle business is tied to the development, propagation, and promotion of specific beef breeds. A cattle breed is defined as a race or variety related by descent and similar in certain distinguishable characteristics. More than 250 breeds of cattle are recognized throughout the world, and several hundred other varieties and types have not been given breed names.

Some of the oldest breeds introduced in the United States were officially recognized as breeds during the mid-to-late 1800s. Most of these breeds originated from crossing and combining existing strains of cattle. When a breeder or group of breeders decided to establish a breed, creating clear differentiation from other breeds was of paramount importance. Major emphasis was placed on establishing readily distinguishable visual characteristics, such as specific colors, color patterns, the presence or absence of horns, and identifiable differences in size, form, and shape.

Breeds in the United States are categorized as either **Bos Taurus** or **Bos Indicus**. *Taurus* and *Indicus* reflect different species and even though species crosses such as cattle × bison typically have depressed reproductive rates, this is not the case for matings involving Bos taurus and Bos indicus. Typical Bos taurus breeds include the British breeds (e.g., Angus, Hereford, and Shorthorn) and Continental breeds (e.g., Simmental, Limousin, and Charolais). The most common Bos indicus breeds in the United States are the Gray and Red Brahman, with the Nellore, Gyr, and Indu–Brazil being less numerous.

Relatively new cattle breeds, such as Brangus, Santa Gertrudis, and Beefmaster, were formed in the United States during the 1950s by crossing Brahman with one or more of the British breeds. Other breeds and composites of breeds have and are being developed to take advantage of the desirable characteristics of multiple gene pools. Soon after some of the first breeds were developed, the word **purebred** was attached to them. Herd books and registry associations were established to assure the "purity" of each breed and to promote and improve each breed. *Purebred* refers to purity of ancestry as established by the pedigree, which shows that only animals recorded in that particular breed have been mated to produce the animal in question. Purebreds, therefore, are cattle from various breeds that have individual pedigrees recorded in their respective breed registry association.

When viewing a herd of purebred Angus, Charolais, or another breed, the uniformity (particularly that of color or color pattern) is noted. Because of this uniformity of one or two characteristics, the word *purebred* has come to imply genetic uniformity (homozygosity) of all characteristics. A breed may be homozygous for a few qualitative traits (e.g., color and horns or polled), but they are highly heterozygous for quantitative traits (e.g., birth weight, weaning weight, and others). High levels of homozygosity occur only after several generations of close inbreeding (e.g., father–daughter and brother–sister). Intensive inbreeding has not occurred in cattle breeds due to the deleterious effects as described in Chapter 12. Additionally, if breeds were to become highly homozygous, genetic change and improvement would be very difficult due to the lack of variation from which to select.

BREED VARIATION

The genetic basis of cattle breeds is not well understood by many beef producers. Often the statement is made, "There is more variation within a breed than there is between breeds." The validity of this statement needs to be carefully examined. Considerable variation exists within a breed for most of the economically important traits. Table 13.1 illustrates the average, minimum, and maximum EPDs of four traits from three popular breeds of cattle. The full distributions of each of these traits would be arrayed as a bell-shaped curve and the ranges in Table 13.1 clearly illustrate the diversity of performance within specific breeds.

The statement, "There is more variation within a breed than between breeds," is not necessarily true. Differences between breed averages as well as variation in performance between individuals within breed-specific populations provide opportunities for commercial producers to make effective genetic decisions. Figure 13.1 shows breed comparisons for retail product produced from cattle harvested at the same age. Note in this comparison that there is more variation between the extremes in breed averages (Jersey and Charolais) compared with the variation within any one breed. Thus, if a cow-calf manager wants to make substantial progress in retail yield from a Hereford–Angus cross cow herd then utilization of Continental breed genetics will be the most effective path to attain the goal.

Major U.S. Beef Breeds

Shorthorn, Hereford, and Angus were the major beef breeds in the United States during the early 1900s. During the 1960s and 1970s, the number of cattle breeds in the United States remained relatively stable at 15 to 20. Today, more than 80 breeds of cattle are available to U.S. beef producers. However, only 10 to 15 make a significant contribution to the total number of cattle in the United States. Why is there large importation of the different breeds from several countries? Following are several possible reasons.

Table 13.1
Distribution of Performance in Traits for Three Beef Breeds

	CED			YW			Milk			Marb.		
	Ave.	Min.	Max.	Ave.	Min.	Max.	Ave.	Min.	Max.	Ave.	Min.	Max.
Red Angus	+5	−17	+23	+91	−18	+170	+20	−13	+45	+0.47	−0.30	+1.75
Hereford	+1	−18	+13	+79	−15	+150	+20	−15	+54	+0.08	−0.50	+1.05
Brangus	+4	−4	+15	+39	−37	+104	+10	−10	+25	+0.01	−0.30	+0.70

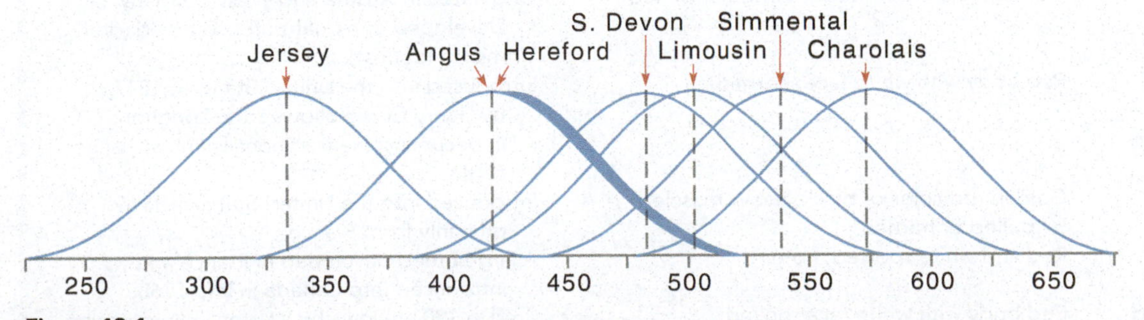

Figure 13.1
Retail product (pounds) at 457 days of age for cattle slaughtered from several breeds. Retail product is boneless, closely trimmed (to 0.3 in. fat) cuts from the carcass.
Source: USDA-ARS-MARC.

1. Feeding higher energy diets, a practice started in the 1940s, resulted in many overfat cattle—cattle that had been previously selected to finish on grazed forage. Therefore, a need was established to identify cattle that could produce a higher percentage of lean to fat at desired market weights under higher concentrate feeding regimes.
2. Economic pressures to produce more weight in a shorter period of time created a need for cattle with more milk and more growth.
3. An opportunity was available for some breed promoters to capitalize on merchandising a certain breed as being the ultimate in all production traits. This opportunity could be merchandised easily because there was little comparative information on breeds.

Table 13.2 gives some distinguishing characteristics and brief background information for the beef cattle breeds in the United States while Figure 13.2 provides

Table 13.2
BACKGROUND AND DISTINGUISHING CHARACTERISTICS OF BEEF BREEDS, COMPOSITES, AND HYBRIDS CONTRIBUTING TO THE U.S. BEEF SUPPLY

Breed	Distinguishing Characteristics	Background
	Annual registrations > 9,000	
Angus	Black color; polled	Originated in Aberdeenshire and Angushire of Scotland. Imported into the United States in 1873.
Beefmaster	Various colors; horned	Developed in the United States from Brahman, Hereford, and Shorthorn breeds. Selected for its ability to reproduce, produce milk, and grow under range conditions.
Brahman	Various colors, with gray predominant; horned. They are one of the Zebu breeds that have the hump over the top of the shoulder. Most Zebu breeds also have large, drooping ears and loose skin in the throat and dewlap.	The Brahman is a Zebu breed developed in the United States from cattle imported from India and Brazil in the early 1900s. These cattle are heat-tolerant and well adapted to the harsh conditions of the Gulf Coast region.
Brangus	Black; polled	U.S. breed developed around 1912—3/8 Brahman and 5/8 Angus.
Charolais	White color; heavy muscled; horned or polled	One of the oldest breeds in France. Brought into the United States soon after World War I, but its most rapid expansion occurred in the 1960s.
Gelbvieh	Golden colored or black; horned or polled	Originated in Austria and West Germany. Developed as a dual-purpose breed (draft, milk, and meat).
Hereford	Red body with white face; horned	Introduced into the United States in 1817 by Henry Clay. Followed the Longhorn in becoming the traditionally known range cattle.
Limousin	Golden red color or black; heavy muscled; polled or horned	Introduced into the United States in 1969, primarily from France.
Maine–Anjou	Red and white spotted; horned	A large breed developed in France and introduced into Canada in late 1968.
Polled Hereford	Red body with white face; polled	Bred in 1901 in Iowa by Warren Gammon who accumulated several naturally polled cattle from horned Hereford herds.

Table 13.2
(Continued) Background and Distinguishing Characteristics of Beef Breeds, Composites, and Hybrids Contributing to the U.S. Beef Supply

Breed	Distinguishing Characteristics	Background
Red Angus	Red color; polled	Founded as a performance breed in 1954 by selecting the genetic recessives from black Angus herds.
Salers	Uniform mahogany red; medium to long hair; horned	Raised in the mountainous area of France, where they were selected for milk, meat, and draft power.
Shorthorn (includes polled and milking)	Red, white, or roan in color; horned or polled	Introduced into the United States in 1783 under the name "Durham." Most prominent in the United States around 1920. Illawarra is an Australian breed of dairy cattle similar to the Milking Shorthorn. Illawarra and Milking Shorthorn have been crossed with the U.S. Shorthorn in recent years.
Simmental	Yellow to red and white color pattern; polled or horned	A prominent breed in Switzerland and France. First bull arrived in Canada in 1967. Originally selected as a dual-purpose breed for milk and meat.
	Annual registrations of 2,000–9,000	
Balancer	Black or red	25–75% Gelbvieh with the remainder Angus or Red Angus
Braford	Usually red with white markings; Brockled face and pigmented eyes are common; Brahman-type ears are apparent; polled or horned	U.S. composite; composition of Brahman, Hereford, Polled Hereford, and Zebu breeds varies
Braunvieh	Various shades of brown, lighter-colored around muzzle and inside legs; horned	Imported from Switzerland in 1983; Similar to Brown Swiss, however they have been selected to be more dual purpose (meat and milk)
Chianina	White or black with black eyes and nose; horned or polled; extremely tall	An old breed originating in Italy; acknowledged as the largest breed, with mature bulls weighing more than 3,000 lb
Dairy Breeds	Holstein (black and white), Jersey (fawn with black switch), Guernsey (light red and white), Brown Swiss (solid blackish or brown), Ayrshire (mahogany and white spotted), and Red and White (red and white)	Have been selected for milk production with meat production considered a by-product of the dairy industry; of the approximate 9 million dairy cows, the Holstein is the most numerous
LimFlex	Black or red	25–75% Limousin and 25–75% Angus or Red Angus with a maximum of other breed contribution of 12.5%
SantaGertrudis	Red; horned	First U.S. breed of cattle; developed on the King Ranch in Texas—5/8 Shorthorn and 3/8 Brahman
SimAngus™	Red or Black	At least 1/4 Simmental and 1/4 Angus or Red Angus and no more than 3/4 Simmental or no more than 3/4 Angus or Red Angus and the sum of Simmental and Angus or Red Angus blood in the animal is at least 3/4.
Tarentaise	Solid wheat-colored hair ranging from light cherry to dark blond; horned	Mountain cattle derived from an ancient Alpine strain in France; originally a dairy breed where maternal traits were emphasized

(continued)

Table 13.2
(Continued) Background and Distinguishing Characteristics of Beef Breeds, Composites, and Hybrids Contributing to the U.S. Beef Supply

Breed	Distinguishing Characteristics	Background
	Annual registrations < 2,000	
Amerifax	Solid red or black; polled	Originated in the United States in 1977; purebreds are 5/8 Angus and 3/8 Beef Friesian
Ankina	Black; polled	Originated from crossing Angus and Chianina breeds
Ankole Watusi	Claimed as the world's longest horned cattle	Originated in Africa; approximately 120 purebreds in the United States in 1990; registry formed in 1983
Barzona	Dark reddish-brown; polled or horned	Developed by the Bard Family in 1942 in the intermountain desert area of Arizona; selected for hardiness traits under rigorous range conditions; foundation stock came from Angus, Africander, Hereford, and Santa Gertrudis breeds
Beefalo (also referred to as Cattalo)	Various colors; polled or horned	Originated in California where the bison was crossed with domestic cattle; term *cattalo* or *catalo* originated in Canada, where similar crosses were made earlier
Beef Friesian	Black and white markings similar to Holstein, although more black color	Imported to the United States from Ireland in 1971; selected for more muscling than U.S. Holstein
Belgian Blue	White, blue, or black and white; horned and polled; heavily muscled	Imported from Belgium through Canada in 1986
Belted Galloway	Black with white belt; polled	Imported from Scotland in 1950
Blonde d'Aquitaine	Fawn colored, sometimes with a reddish tinge; heavily muscled; horned	Originated in France and live cattle imported into the United States in 1973; semen imported prior to that time
Bonsmara	Dark red, horned, and polled	South African breed developed by Jan Bonsma and imported to the United States in 1996; composite of 5/8 Africander, 3/16 Hereford, and 3/16 Shorthorn; tropically adapted
Brah–Maine	Dark red to black and white; horned	U.S. composite; 5/8 Maine–Anjou and 3/8 Brahman
Brahmental	Various colors	Originated in United States by crossing Brahman and Simmental; must have a minimum of 1/4 Brahman and a minimum of 3/8 Simmental
Brahmousin	Red with black pigmentation; horned	U.S. composite; 5/8 Limousin and 3/8 Brahman
Bralers	Red; horned	U.S. composite; 5/8 Salers and 3/8 Brahman
British White	White with black markings (ears, eyes, nose, and feet); polled	Imported from England; claimed to have the longest documented history of any specific breed of cattle
Buffalo (Bison)	Not typically considered a breed; however, registry associations have been formed	Bison are wild oxen belonging to the cattle family; native to North America; numbered more than 20 million head in the 1850s
CASH	Various colors	Composite breed combining Charolais, Angus, Swiss, and Hereford
Charbray	Various colors; horned	Cross between Charolais and Brahman
Chargrey	Various colors	Developed in Australia by crossing Charolais and Murray Grey breeds

Table 13.2
(Continued) Background and Distinguishing Characteristics of Beef Breeds, Composites, and Hybrids Contributing to the U.S. Beef Supply

Breed	Distinguishing Characteristics	Background
Charswiss	Various colors	Cross between Brahman and Brown Swiss
ChiAngus	Black; polled	U.S. composite; maximum of 3/4 and 1/4 of Chianina and Angus, respectively
Chiford	Marked like Hereford with pale fawn, cream, to dark red color; polled or horned	Out of registered Chianina bull or cow and Hereford or Hereford cross cow or bull; no more than 75% Chianina
Chimaine	Various colors; horned or polled	U.S. composite; any percentage up to and including 3/4 Chianina
Criollo/Corriente	Multiple color patterns; horned	Descendants of cattle brought to the New World by the Spanish explorers, likely related to Florida Cracker cattle; mostly used for roping and rodeo; demand for ~40,000 per year
Cracker Cattle	Various colors	Not designated as a breed; however, a breeder's association was formed to preserve descendants of the Spanish cattle brought to Florida; also called *Florida Scrub* or *Piney Woods* cattle
Devon	Dark red (North Devon) to light red or brown (South Devon); horned	North Devon is an old breed originating in England; South Devon is more of a dual-purpose breed; the Devon first came to America with the Pilgrims in 1623
Dexter	Black; claimed as smallest breed in the United States (cows, 650 lb; bulls, 850 lb)	Originated in Ireland; introduced into the United States in 1912
Galloway	Polled, with the majority solid black in color; some dun-colored, and others white with black noses, ears, and feet; Belted Galloway is black with a distinctive white belt; polled	An old breed from the Scottish province of Galloway; long, burly hair has made it adaptable to the harsh climates of the North; the Belted Galloway has a separate breed association; it has the same origin as the Galloway, but had an infusion of the belted cattle in the seventeenth or eighteenth century
Gelbray	Various colors; polled or horned	Originated in the United States by crossing the Gelbvieh and Brahman breeds; Gelbvieh (maximum 3/4; minimum 5/8) and Brahman (maximum 3/8; minimum 1/4); Registered in American Gelbvieh Association
Hays Converter	Various colors	Developed in the 1950s by Harry Hays, former Canadian Minister of Agriculture, by crossing Hereford, Holstein, and Brown Swiss breeds
Herens	Rusty brown to black with some white on the udder; horned	Mountain cattle from Switzerland
Irish Blacks	Black; polled	Originated in Ireland
Longhorn	Multicolored; characteristically long horns	Came to West Indies with Columbus; brought to the United States through Mexico by the Spanish explorers; were the noted trail-drive cattle from Texas into the Plains states
Mandalong Special	Primarily a solid, reddish-tan color; polled Large but refined in bone	Developed in Australia by combining Brahman, Shorthorn, Charolais, Chianina, and British White breeds

(continued)

Table 13.2
(Continued) Background and Distinguishing Characteristics of Beef Breeds, Composites, and Hybrids Contributing to the U.S. Beef Supply

Breed	Distinguishing Characteristics	Background
Marchigiana	Light gray to almost white; tail switch is dark colored; horned	Originated in the Marche and surrounding areas of Italy; resembles Chianina in color and conformation, which implies possible intermixing with them
MARC I		A composite developed at the Meat Animal Research Center; combines Swiss, Limousin, Charolais, Hereford, and Angus breeds
MARC II		A composite developed at the Meat Animal Research Center; combines Hereford, Angus, Simmental, and Gelbvieh breeds
MARC III		A composite developed at the Meat Animal Research Center; combines Red Poll, Hereford, Angus, and Pinzgauer breeds
Murray Grey	Silver gray; polled	Developed in Australia from Shorthorns crossed on Angus
Normande	Coat colors may be yellowish or blond, very dark brown, or almost black, and white; colors occur in spots and speckles	Originated in France; first 7/8 Normande calf was born in the United States in 1976
Piedmontese	Extremely heavily muscled; color varies; horned	Imported from Italy
Pinzgauer	Reddish chestnut with white markings on rump, back, and belly; horned	A hardy breed developed in the Pinza Valley of Austria and in areas of Germany and Italy; introduced into North America in 1972
Ranger	Various colors	Developed in the 1940s–1960s in several western U.S. cattle herds; crosses of Hereford, Red Angus, Shorthorn, Scotch Highland, and Brahman are primary breeds represented
Red Brangus	Red; polled	No specific percentage of Brahman and Angus, though generally considered about 50% of each
Red Poll	Red; polled	Introduced into the United States in the late 1800s; originally a dual-purpose breed but now considered a beef breed
Romagnola	Gray with black muzzle; horned; heavily muscled	Originated in Italy; imported to the United States in 1974
Romosinuano	Red (light and dark); polled	Bos taurus breed that is adapted to tropical conditions; developed in Costa Rica and Columbia; introduced into the United States as embryos in 1992 from Costa Rica and Venezuela
RX_3	Red; polled	A composite breed combining 1/4 Hereford, 1/2 Red Angus' and 1/4 Red and White Holstein
Salorn	Various colors	U.S. composite; 5/8 Salers and 3/8 Longhorn
Santa Cruz		U.S. composite developed by the King Ranch in Texas; involves crosses of Santa Gertrudis, Red Angus, and Gelbvieh;
Scotch Highland	Golden with long, shaggy hair; horned	Bred in the highlands of Scotland; imported to the United States in 1894

Table 13.2
(Continued) Background and Distinguishing Characteristics of Beef Breeds, Composites, and Hybrids Contributing to the U.S. Beef Supply

Breed	Distinguishing Characteristics	Background
Senepol	Mostly polled; color varies from light tan to dark red	Developed in the Virgin Islands in the 1900s by crossing the Red Poll and N'Dama cattle
Simbrah	Various colors; polled or horned	Purebreds are 5/8 Simmental and 3/8 Brahman; registered in the American Simmental Association
Tuli	Color ranges from silver through golden brown to rich red (red predominates); polled or horned	An early maturing, medium-sized pure African Sanga (*Bos taurus*) breed; claimed to have heat and tick resistance similar to *Bos indicus* breeds
Wagyu	Solid, dull black to solid light brown with reddish tinge; horned or polled	Native cattle of Japan; three breeds are recognized: Japanese Black (predominate), Japanese Brown, and Japanese Polled; some U.S. producers are using Wagyu to access the Japanese market where highly marbled beef is economically rewarded
Welsh Black	Black; horned	Imported from Wales in 1963
White Park	White with characteristic black ears and muzzle; polled or horned	Imported from England in 1940; U.S. breed association formed in 1975
Zebu	Varied colors; horned; characteristic hump over shoulders	Originated in India; imported from Mexico to the United States in the 1800s and in 1926; later importations came from South America; approximately fifty strains of Zebu (*Bos indicus*); six strains (Gyr, Gray Zebu, Nellore, Guzerat, Indu–Brazil, and Red Zebu) are most numerous in the United States

There are other breeds or types such as Aubrac, Buelingo, Charmaine, Corriente, Dutch Belted, El Monterey, Gascone, Geltex, Kerry, Lincoln Red, Lineback, Loala, Luing, MRI, Norwegian Red, Parthenais, Simbrangerford, Texon, Water Buffalo, and other composites with specific names that are not identified in the table, but little is currently known about them.

A - Angus

B - Brangus

C - Balancer

D - Beefmaster

Figure 13.2
Representative images of the leading U.S. beef breeds in the United States.
Source: American Angus Association, International Brangus Breeders Association, American Gelbvieh Association, Beefmaster Breeders United, JD Hudgins, Inc., American International Charolais Association, American Gelbvieh Association, American Hereford Association, North American Limousin Foundation, Red Angus Association of America, Santa Gertrudis Breeders International, Altenburg Super Baldy Ranch

E - Brahman

F - Charolais

G - Gelbvieh

H - Hereford

I - Limousin

J - Red Angus

K - Santa Gertrudis

L - Simmental

Figure 13.2 *(continued)*

representative images. For excellent photos and additional information, visit the breed Web site built at Oklahoma State University (www.ansi.okstate.edu/breeds/).

The relative importance of the various breeds' contributions to the total beef industry is best estimated by the registration numbers of the breeds (Table 13.3). These 12 breeds would be considered the major breeds of beef cattle in the United States, as they contribute most of the genetics to the commercial cow-calf segment.

Although registration numbers are for purebred animals, they reflect commercial cow-calf producers' demand for different breeds. Trends in registration numbers over the past several decades illustrate the shifts in demand for specific breed types. For example, the combined registrations of Continental breed cattle peaked in 1995 and then declined into 2000. A period of stabilization followed, but then registration numbers declined again over the past decade. The combined

Table 13.3
Major U.S. Beef Breeds Ranked by Annual Registration Numbers (in thousands)

Breed[1]	Year 2014	2003	2000	1990	1980	Date Association Formed
	thousand head					
Angus	282.9	281.7	260.9	159.0	257.6	1883
Hereford/Polled Hereford[2]	71.4	71.2	84.9	170.5	353.2	1881
Simmental	49.0	47.0	43.1	79.3	66.1	1969
Red Angus	48.2	42.2	39.6	15.4	12.5	1954
Charolais	33.5	47.1	42.7	44.8	23.0	1957
Gelbvieh	33.7	31.7	26.3	22.8	NA	1971
Limousin	20.0	45.0	48.8	71.6	13.8	1968
Brangus	22.9	22.3	26.9	32.1	24.5	1949
Beefmaster	16.0	21.3	32.3	38.4	30.0	1961
Shorthorn	13.5	22.4	18.6	18.0	19.4	1872
British Breeds	416.0	417.5	404.0	362.9	642.7	
Continental Breeds	136.2	170.8	160.9	218.5	102.9	

[1]Only breeds with more than 15,000 annual registrations for 2014 are listed.
[2]Hereford and Polled Hereford are combined. Prior to 1996, they were separate breed associations and some animals were registered in both associations.
Source: Based on breed association data and the National Pedigreed Livestock Association.

registrations of British breed cattle have been increasing since 1990s, fueled by significant growth in Angus and Red Angus registrations. The success of Angus cattle in the marketplace was founded on several innovations including the creation of Certified Angus Beef™ (CAB), Angus Herd Improvement Records (AHIR), the world's largest performance database and national cattle evaluation, and a set of marketing, promotion, and education strategies that helped to drive commercial demand. The success of CAB discussed in Chapter 9 was followed by a host of other quality-focused brands that also required predominately black-hided cattle as the beef source. As a result, premiums and demand for black cattle at the feeder and fed cattle phases of production grew. The demand for Angus-influenced and black-hided genetics led nearly all of the Continental breed associations to create appendix registries featuring composites containing both Continental and Angus (black or red) influence.

It is reasonable to expect that, in future years, some breeds will become more numerous while others will decrease significantly in number. These changes will be influenced by economic conditions as well as by genetic improvements in the economically important traits to meet industry demand. Seedstock breeders focused on meeting the needs of the commercial sector will be the winners in the long run, regardless of the breed or breeds they utilize.

BREED EVALUATION FOR COMMERCIAL PRODUCERS

Breeds exist primarily to be used as genetic resources by commercial producers. Most commercial beef producers use crossbreeding because they can take advantage of the heterosis in addition to genetic improvement from selecting within two or more breeds. A crossbreeding system should be determined by availability of breeds (and biological types within breeds) and how well adapted they are to the commercial producers' feed supply, market demands, and other environmental conditions. A good example of adaptability is the Brahman breed, which is more heat- and humidity-resistant than most other cattle breeds. Because of its higher resistance, the level of productivity in the southern and Gulf regions of the United States is higher for the Brahman, Brahman crosses, and other breeds that include Zebu breeding.

Most commercial producers travel less than 150 miles to purchase their bulls for natural service. Therefore, producers should critically assess the breeders, breeds, and biological types available within a logical distance of their individual operations. This assessment, in most cases, will establish the range of options available upon which to build a breeding program.

Breed Differences

Breeds should be chosen for a crossbreeding system based on how well the breeds and their respective biological types complement each other along with achieving cost-effective levels of heterosis. Breed differences have been evaluated most extensively in the Germplasm Evaluation (GPE) program at the U.S. Meat Animal Research Center (MARC) located at Clay Center, NE. Top cross performance of more than 25 different sire breeds have been evaluated in five cycles of the GPE program (Table 13.4).

Hereford–Angus reciprocal F1 crosses (Hereford sires mated to Angus dams and vice versa) were produced using semen from the same sires. Data were pooled

Table 13.4
SIRE BREEDS IN THE MARC EVALUATION PROGRAM[1]

Breed Cross Groups	Cycle I (1970–1972)	Cycle II (1973–1974)	Cycle III (1975–1976)	Cycle IV (1986–1990)	Cycle V (1992–1996)	Cycle VI (1997–1998)	Cycle VII (1999–2000)	Cycle VIII (2000–2002)
F1 crosses from Hereford or Angus dams[2]	Hereford Angus Jersey South Devon Limousin Simmental Charolais	Hereford Angus Red Poll Brown Swiss Gelbvieh Maine–Anjou Chianina	Hereford Angus Brahman Sahiwal Pinzgauer Tarentaise	Hereford[3] Angus[3] Longhorn Salers Galloway Nellore Shorthorn Piedmontese Charolais Gelbvieh Pinzgauer	Hereford Angus Tuli Boran Brahman Belgian Blue Piedmontese	Hereford Angus Wagyu Norwegian Red Swedish Red & White Friesian	Hereford Angus Red Angus Limousin Charolais Simmental Gelbvieh	Hereford Angus Brangus Beefmaster Bonsmara Romosinuano
Three-way crosses out of F1 dams[4]	Hereford Angus Brahman Devon Holstein	Hereford Angus Brangus Santa Gertrudis						

[1]The Germplasm Evaluation program at the U.S. Meat Animal Research Center, Clay Center, NE.
[2]For example, Jersey sire × Hereford dam.
[3]Hereford and Angus sires, originally sampled in 1969, 1970, and 1971, have been used throughout the program. In Cycle IV, a new sample of Hereford and Angus sires produced after 1982 was used and compared with the original Hereford and Angus sires.
[4]For example, Brahman sire × (Jersey × Hereford) dam.
Source: USDA-ARS-MARC.

over cycles by adding the average differences between Hereford–Angus reciprocal crosses and other breed groups (two-way and three-way F1 crosses) within each cycle to the average of Hereford–Angus reciprocal crosses over the three cycles. Females produced by these matings were all retained to evaluate age and weight at puberty and reproduction and maternal performance through 7 or 8 years of age. Data are presented for breed crosses (two-way and three-way) grouped into several biological types based on relative differences (X = lowest; XXXXXX = highest) in growth rate and mature size, lean-to-fat ratio, age at puberty, and milk production (Table 13.5).

Although the information in Table 13.5 is useful, it should not be considered the final answer for decisions on breeds to use. First, a producer needs to recognize that this information reflects breed averages; therefore, there are individual animals and herds of the same breed that are much higher or lower than the average ranking given. This demonstrates that there are several different biological types within the same breed. Also, the level of nutrition under these studies was relatively high, so breeders under different environmental conditions would need to account for the impact of those

Table 13.5
Breed Crosses Grouped into Six Biological Types on the Basis of Four Major Criteria

	Traits Used to Identify Biological Types[1]			
Breed Crosses Grouped into Biological Types	Growth Rate and Mature Size	Lean-to-Fat Ratio[2]	Age at Puberty	Milk Production
Jersey (J)	X	X	X	XXXXX
Longhorn	X	XX	XXX	XX
Hereford–Angus (HA)	XXX	XX	XXX	XX
Red Poll (RP)	XX	XX	XX	XXX
Devon (D)	XX	XX	XXX	XX
Shorthorn	XXX	XX	XXX	XXX
Galloway	XX	XXX	XXX	XX
South Devon (Sd)	XXX	XXX	XX	XXX
Tarentaise (T)	XXX	XXX	XX	XXX
Pinzgauer (P)	XXX	XXX	XX	XXX
Brangus (Bn)	XXX	XX	XXXX	XX
Santa Gertrudis (Sg)	XXX	XX	XXXX	XX
Beefmaster	XXX	XX	XXXX	XX
Sahiwal (Sw)	XX	XXX	XXXXX	XXX
Brahman (Bm)	XXXX	XXX	XXXXX	XXX
Nellore	XXXX	XXX	XXXXX	XXX
Branvieh (B)	XXXX	XXXX	XX	XXXX
Gelbvieh (G)	XXXX	XXXX	XX	XXXX
Holstein (Ho)	XXXX	XXXX	XX	XXXXXX
Simmental (S)	XXXXX	XXXX	XXX	XXXX
Maine–Anjou (M)	XXXXX	XXXX	XXX	XXX
Salers	XXXXX	XXXX	XXX	XXX
Piedmontese	XXX	XXXXXX	XX	XX
Limousin (L)	XXX	XXXXX	XXXX	X
Charolais (C)	XXXXX	XXXXX	XXXX	X
Chianina (Ci)	XXXXX	XXXXX	XXXX	X

[1]Breed crosses grouped into several biological types based on relative differences (X = lowest, XXXXXX = highest).
[2]Steers were slaughtered at 15 months of age.
Source: Adapted from USDA-ARS-MARC Research Progress Reports.

differences, particularly as it relates to reproductive performance. Figures 13.3 to 13.15 show considerable genetic variation between and within breeds for several economically important traits. In these figures, the averages for the various breed crosses are shown on the lower horizontal axis. The spacing on the vertical axis is arbitrary but the ranking of the biological types (separate bars) from the top to bottom reflects increasing increments. Breed rankings within each biological type are noted within each bar.

These frequency curves are shown for selected breeds to compare to the average of Hereford and Angus (the base comparison). The frequency curves reflect the expected distribution of breeding values for the breeds assuming a normal distribution

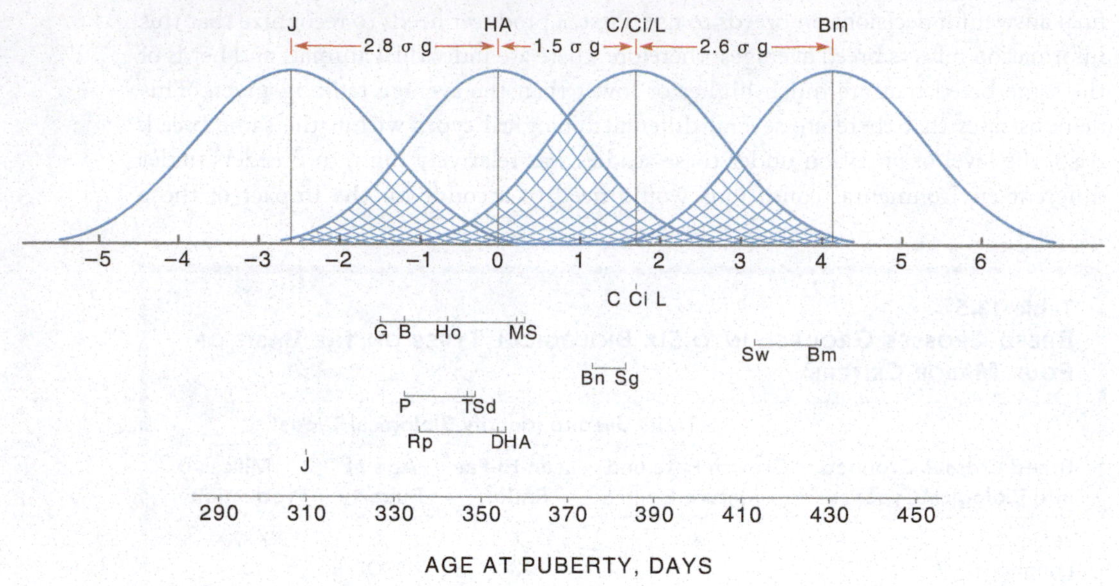

Figure 13.3
Variation between and within breeds for age at puberty. See Table 13.5 for breed abbreviations.
Source: USDA-ARS-MARC.

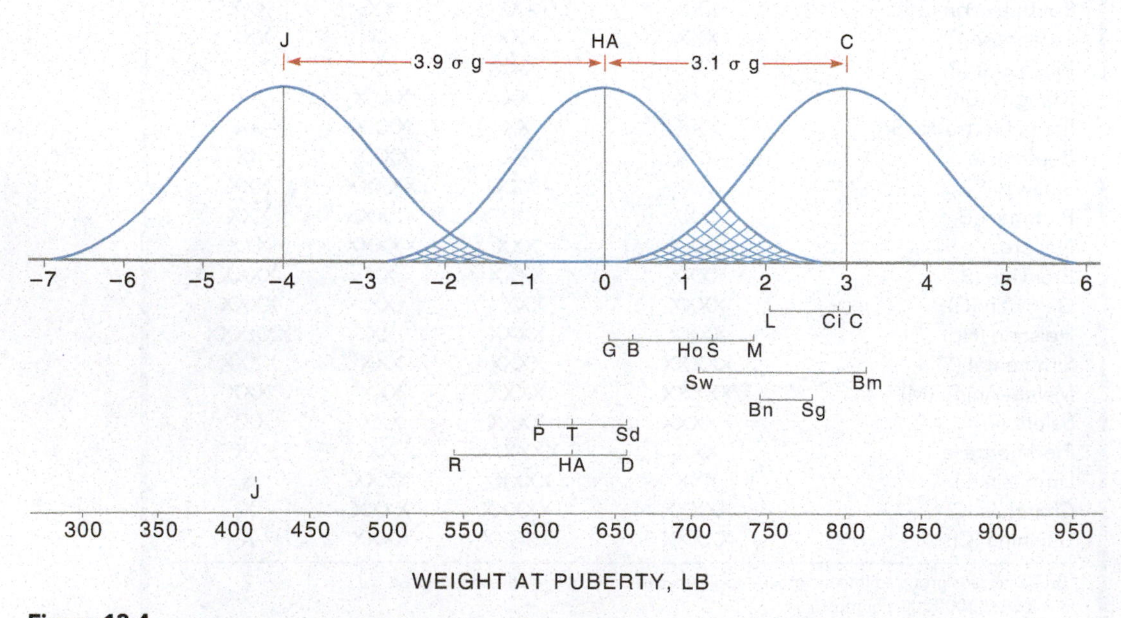

Figure 13.4
Variation between and within breeds for weight at puberty. See Table 13.5 for breed abbreviations.
Source: USDA-ARS-MARC.

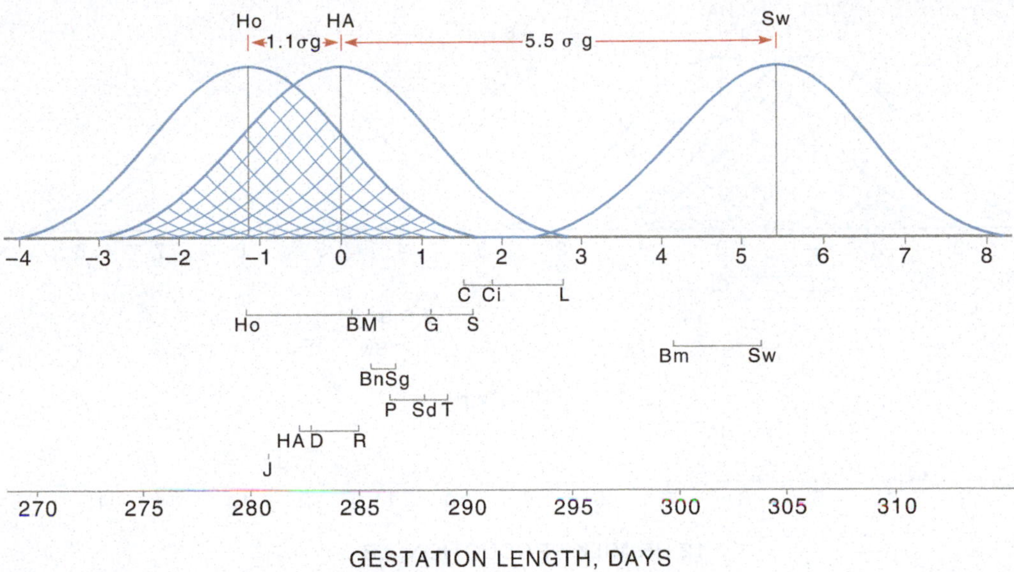

Figure 13.5
Variation between and within breeds for gestation length. See Table 13.5 for breed abbreviations.
Source: USDA-ARS-MARC.

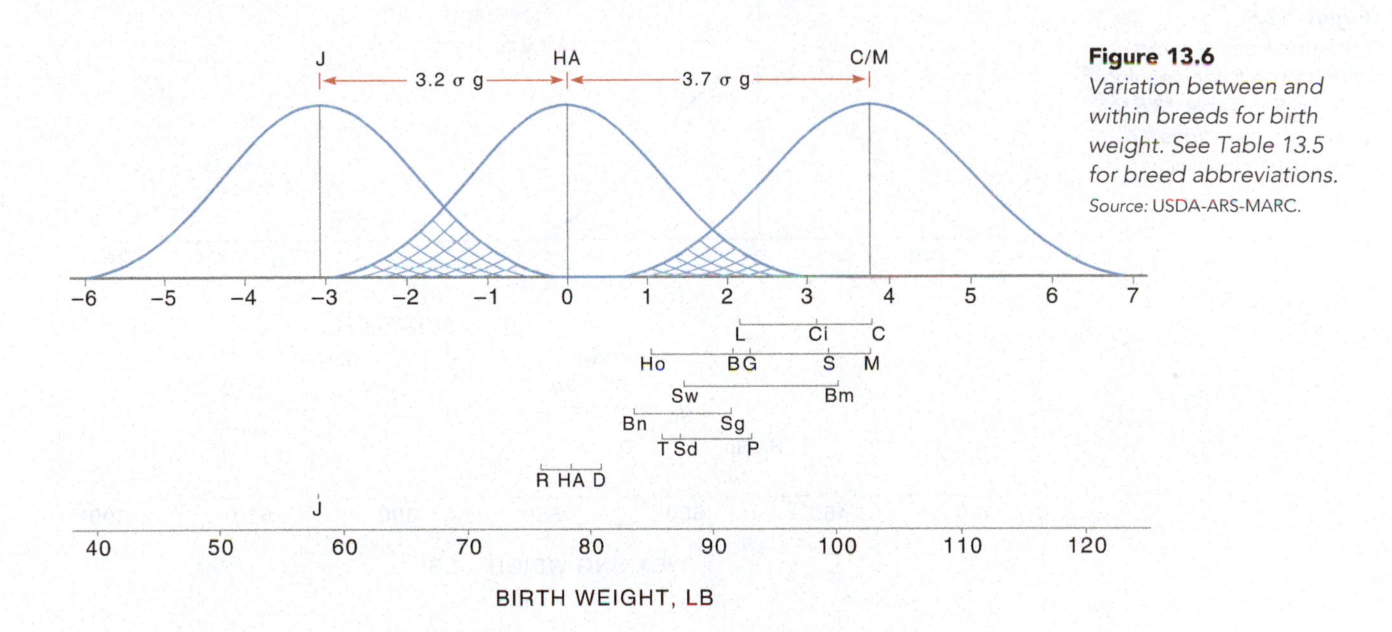

Figure 13.6
Variation between and within breeds for birth weight. See Table 13.5 for breed abbreviations.
Source: USDA-ARS-MARC.

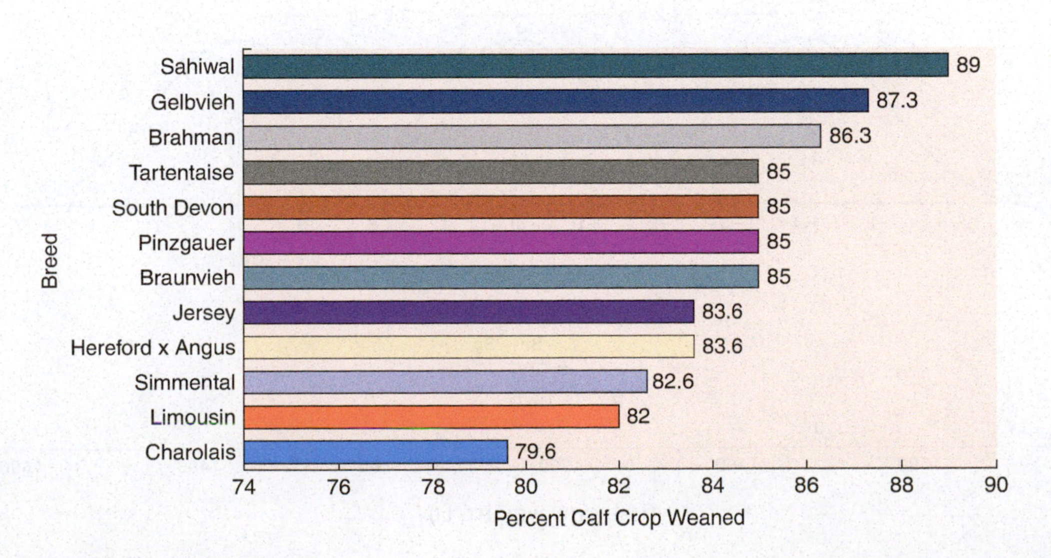

Figure 13.7
Breed of sire of dam means for percentage calf crop weaned per cow exposed to breed. See Table 13.5 for breed abbreviations.
Source: USDA-ARS-MARC.

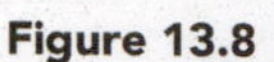

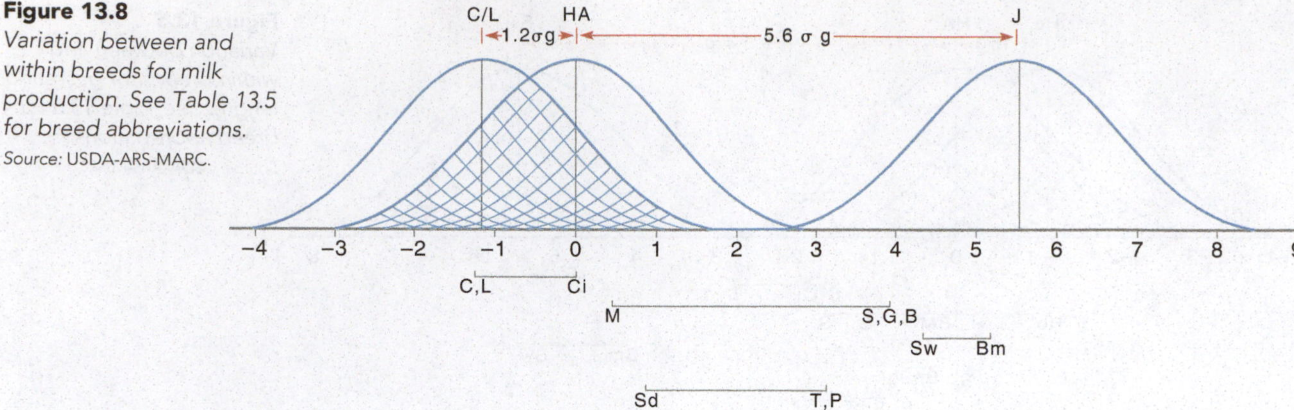

Figure 13.8
Variation between and within breeds for milk production. See Table 13.5 for breed abbreviations.
Source: USDA-ARS-MARC.

Figure 13.9
Variation between and within breeds for weaning weight. See Table 13.5 for breed abbreviations.
Source: USDA-ARS-MARC.

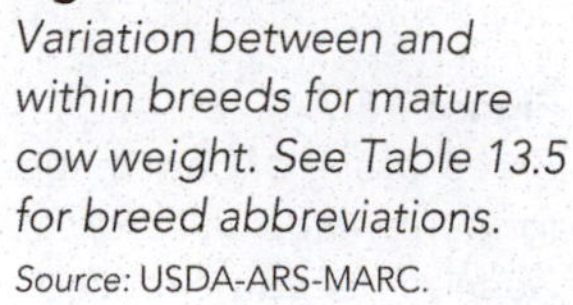

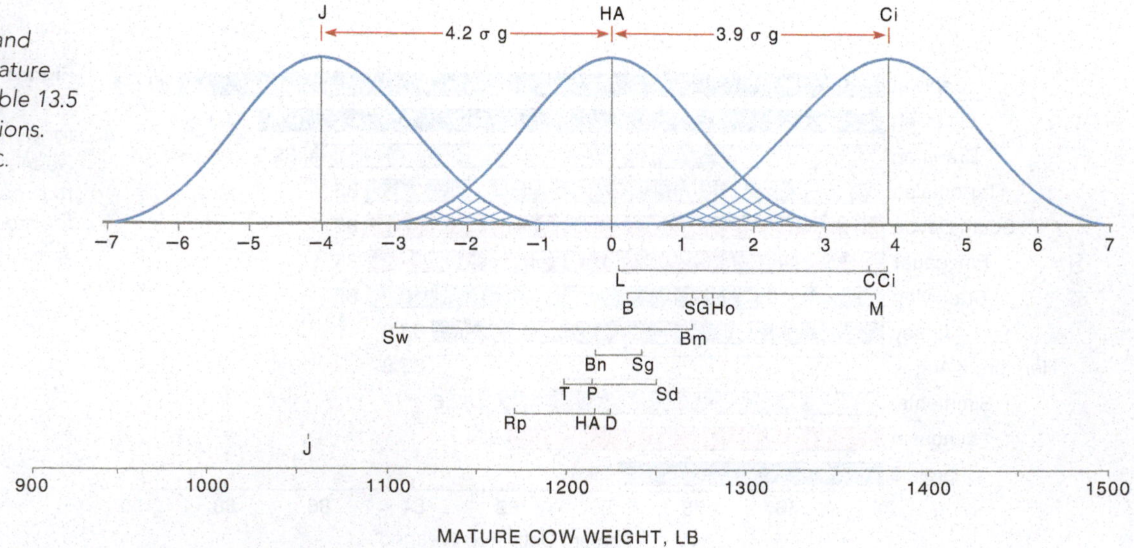

Figure 13.10
Variation between and within breeds for mature cow weight. See Table 13.5 for breed abbreviations.
Source: USDA-ARS-MARC.

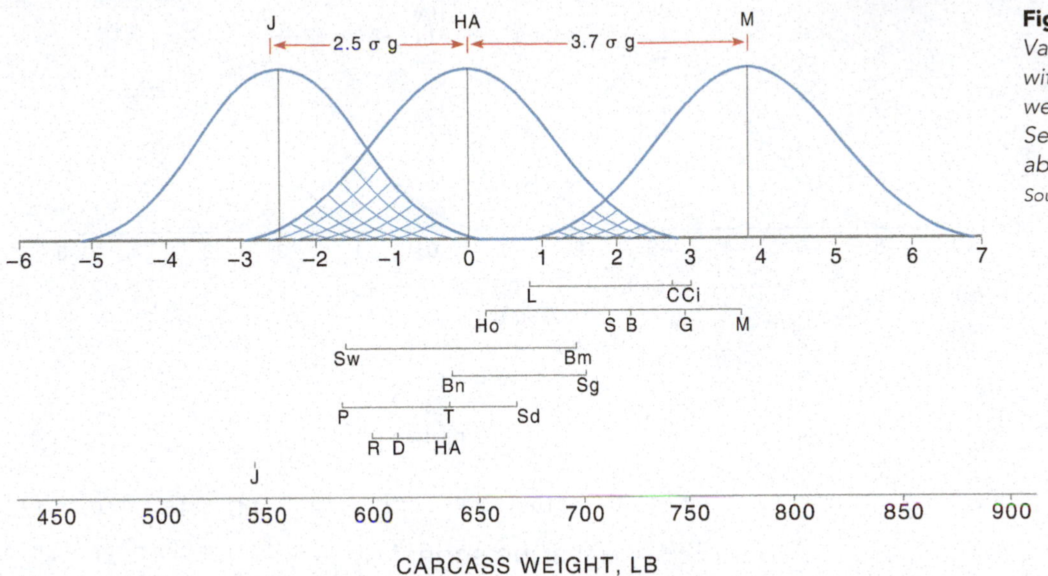

Figure 13.11
Variation between and within breeds for carcass weight at 15 months of age. See Table 13.5 for breed abbreviations.
Source: USDA-ARS-MARC.

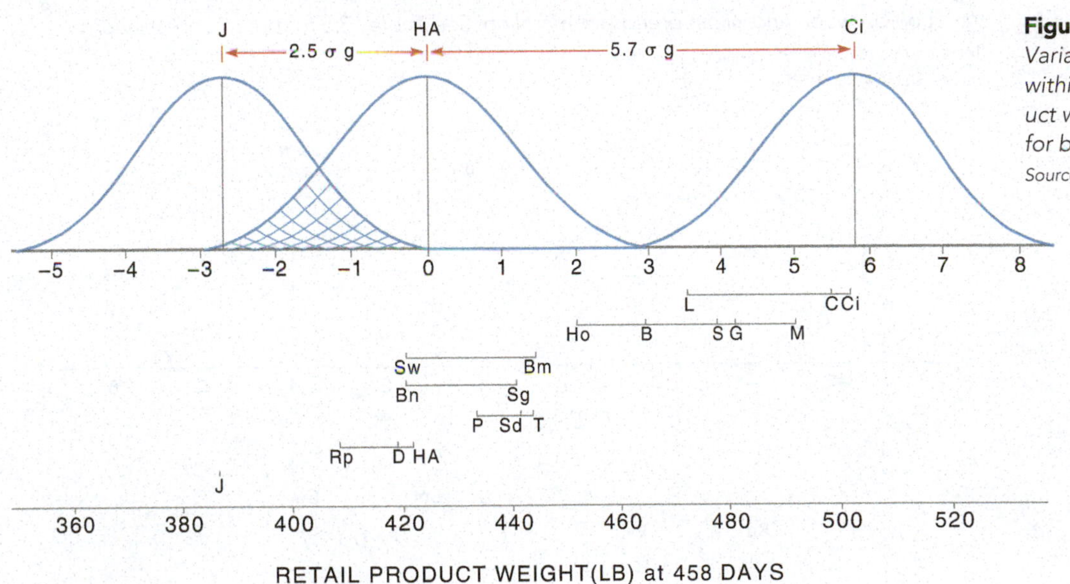

Figure 13.12
Variation between and within breeds for retail product weight. See Table 13.5 for breed abbreviations.
Source: USDA-ARS-MARC.

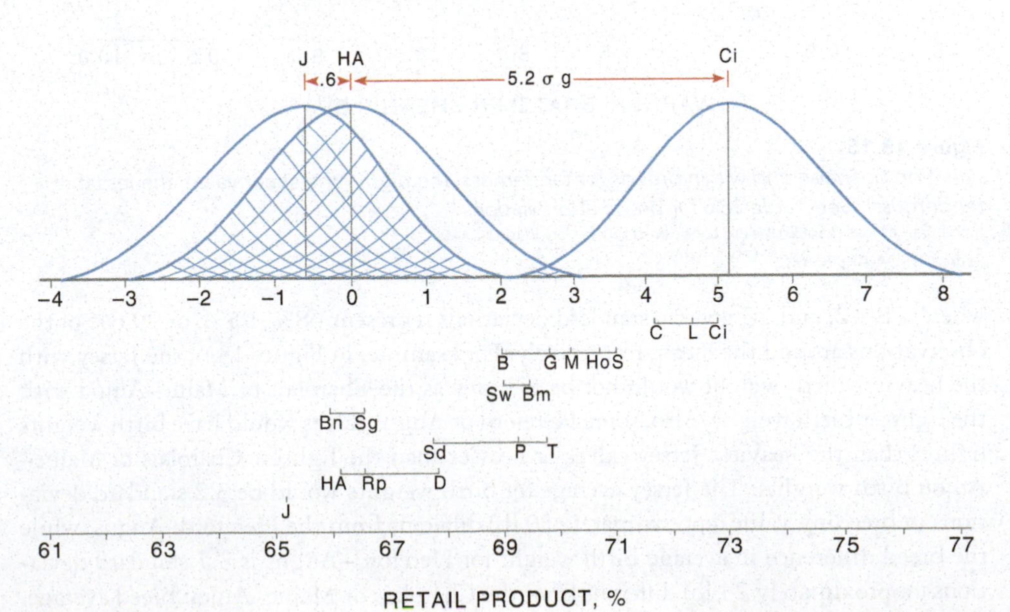

Figure 13.13
Variation between and within breeds for percent retail product. See Table 13.5 for breed abbreviations.
Source: USDA-ARS-MARC.

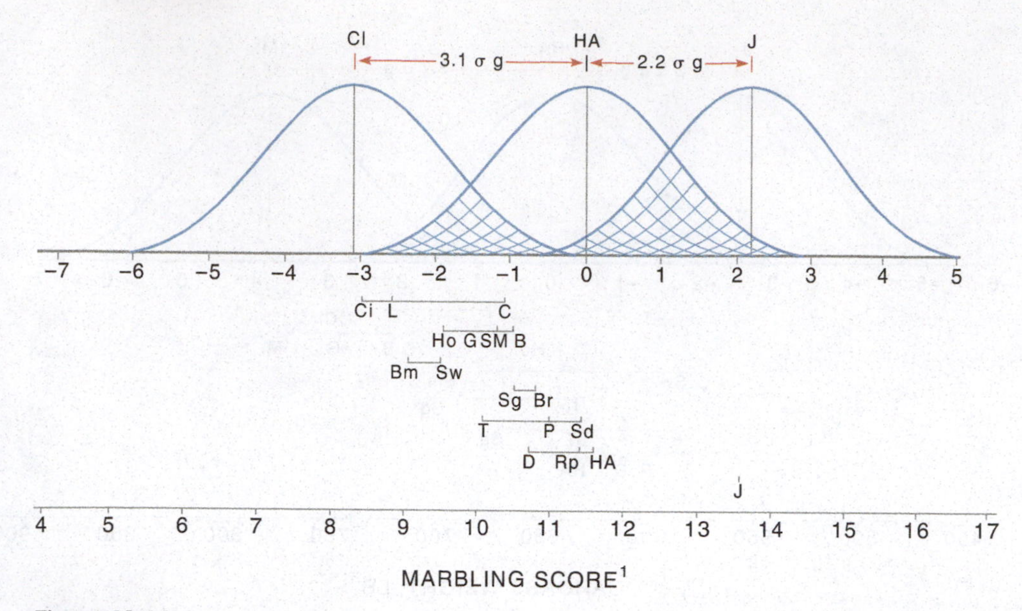

Figure 13.14

Variation between and within breeds for marbling. See Table 13.5 for breed abbreviations.

[1]Minimum scores: Select = 8; Choice = 11.

Source: USDA-ARS-MARC.

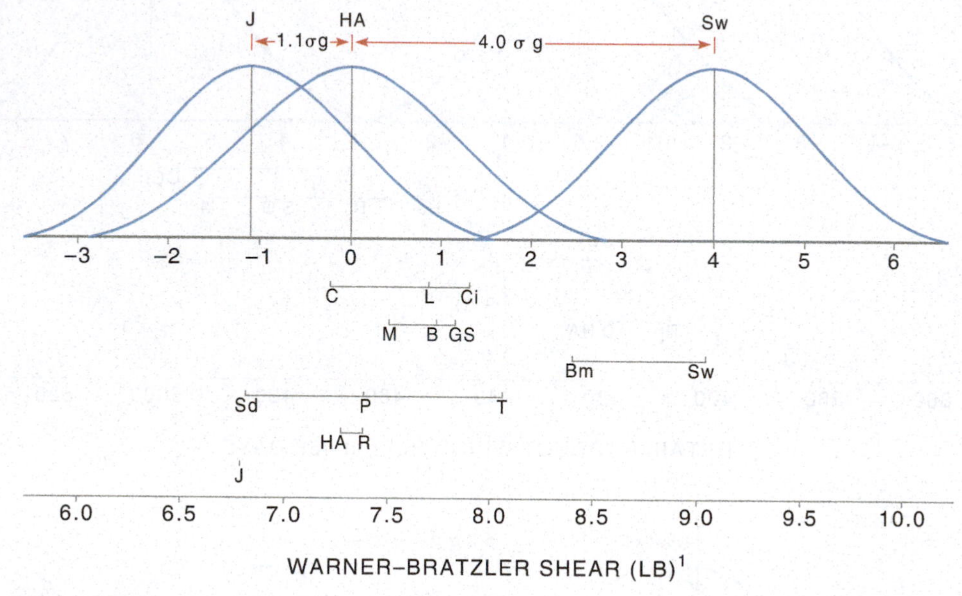

Figure 13.15

Variation between and within breeds for tenderness (the higher the shear value, the less the tenderness). See Table 13.5 for breed abbreviations.

[1]Shear values less than 8.0 lbs are considered satisfactory in tenderness.

Source: USDA-ARS-MARC.

where ±1, ±2, and ±3 genetic standard deviations represent 68%, 95%, or 99.6% of the observations around the mean, respectively. For example, in Figure 13.6, the Jersey with the heaviest birth weight would not be as heavy as the Charolais or Maine–Anjou with the lightest birth weight. Also, some Hereford or Angus calves would have birth weights lighter than the heaviest Jersey calves or heavier than the lightest Charolais or Maine–Anjou birth weights. The Jersey average for birth weights would be 3.2 standard deviations in breeding value (approximately 20 lb) different from the Hereford–Angus, while the breed difference in average birth weight for Hereford–Angus is 3.7 standard deviations (approximately 24 lb) different from the Charolais or Maine–Anjou breed average.

A comparison of Figures 13.3 to 13.15 shows that there are genetic antagonisms between traits when comparing breed averages. Breeds that excel in retail product from birth to market age (Figure 13.12) sire progeny with heavier birth weights (Figure 13.6), greater calving difficulty, reduced calf survival, and reduced rebreeding in dams; produce carcasses with lower marbling (Figure 13.14) but very acceptable meat tenderness (Figure 13.15); tend to reach puberty at an older age (Figure 13.3); and generally have heavier mature weights (Figure 13.10). Heavier mature weight and high milk production increase output per cow (e.g., calf weight and slaughter weight of cull cows) but they also increase nutrient requirements for maintenance and lactation so that differences in life cycle biological efficiency are generally small. An economic evaluation in most cow-calf environments usually favors an optimum or balanced combination of these traits where extremes are avoided.

A careful analysis of the information in Table 13.5 and Figures 13.3 to 13.15 shows that no one breed is superior in all these economically important traits. This gives an advantage to commercial producers using a crossbreeding or a composite breeding program if they select breeds whose superior traits complement each other. The Angus and Charolais breeds provide an excellent example of breed complementation; when crossed, they complement each other in terms of both quality grade and yield grade. The superior marbling (quality grade) of the Angus is combined with the high level of retail yield of the Charolais.

Some caution should be exercised in the breed comparisons noted in Figures 13.3 to 13.15, as these breeds were given an excellent forage and nutrition program. The breeds could rank differently in an environment characterized by a less plentiful, but lower-cost supply of forage and feed.

Producers need to use some of the previously described methods to identify superior animals within the breed (Chapter 4). It should also be recognized that average breed performance in various traits will change with time, depending on the improvement programs, breeding objectives, and market targets used by leading breeders. For example, breed comparisons in the MARC germplasm study that were conducted in the 1970s and then reevaluated in the late 1990s illustrate the impact of selection over time on breed performance. Figures 13.16 and 13.17 demonstrate breed average comparisons for birth weight and finished weight. Over the 20-year period of the comparison, economics favored larger finished weights across the industry; at the same time, bull buyers were pressuring seedstock producers of

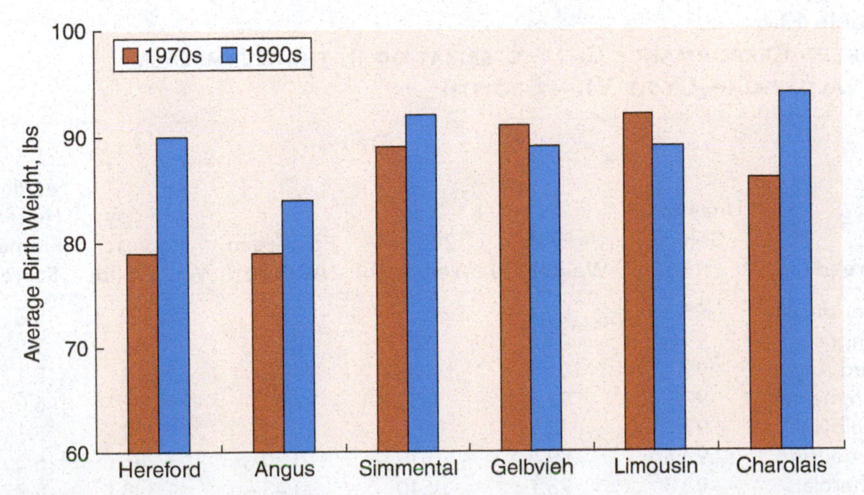

Figure 13.16
Comparison of breed means for birth weight over time.
Source: Adapted from USDA-ARS-MARC.

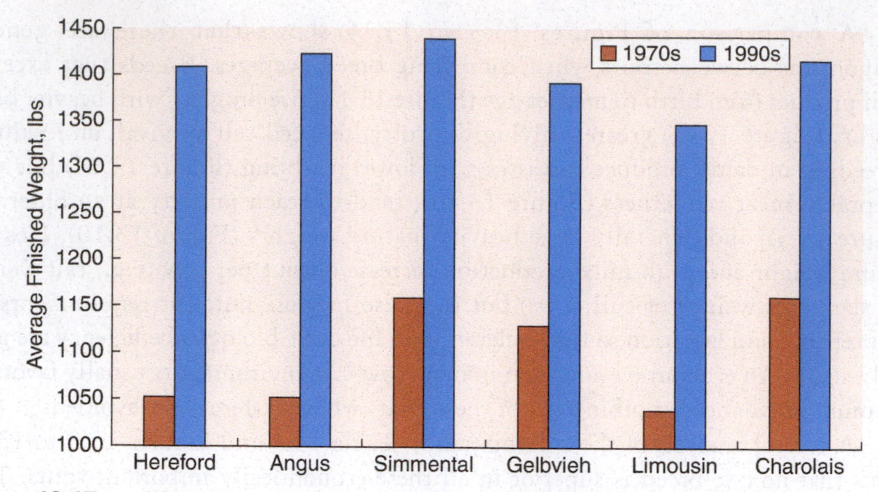

Figure 13.17

Comparisons of breed means for finished weight over time.

Source: Adapted from USDA-ARS-MARC.

maternal and dual-purpose breeds to avoid calving difficulty. The result was an increase in all breeds for market weight (Figure 13.17) while limiting increases in birth weight (Figure 13.16). Breeds differences narrowed and rankings reordered as a result. For example, over the time period between the two studies, Hereford cattle experienced a dramatic increase in market weight but lost their comparative advantage in birth weight as compared to the Continental breeds. On the other hand, breeds that had excelled in market weight in the 1970s lost ground to the British breeds as the result of breeders of Continental breeds (with a focus on either maternal or dual-purpose traits) putting downside pressure on mature weight and birth weight. Tables 13.6 and 13.7 highlight some of the breed comparisons of calf crops in 1999 and 2000 at MARC. The differences between breed means for birth weight, growth performance, and frame score have narrowed over time. However, the compositional and carcass trait differences between breeds continue to be relatively wide. Nonetheless, the advantages of heterosis and breed complementarity are significant and warrant the use of effective combinations of breeds by the industry.

Table 13.6

BREED PERFORMANCE CHARACTERIZATION IN THE GERMPLASM EVALUATION—CYCLE VII—GROWTH

	Trait					
Breed	Unassisted Calving (%)	Birth Weight (lb)	200-day Weight (lb)	Post-wean ADG (lb)	445-day Harvest Weight (lb)	Yearling Heifer Frame Score
Hereford	95.6	90.4	524	3.46	1,322	5.6
Angus	99.6	84.0	533	3.40	1,365	5.3
Red Angus	99.1	84.5	526	3.40	1,333	5.1
Simmental	97.7	92.2	553	3.47	1,362	6.0
Gelbvieh	97.8	88.7	534	3.33	1,312	5.7
Limousin	97.6	89.5	519	3.30	1,285	6.0
Charolais	92.8	93.7	540	3.43	1,348	5.9

Source: USDA-ARS-MARC, 2004.

Table 13.7
Breed Performance Characterization in the Germplasm Evaluation—Cycle VII—Carcass

	Trait					
Breed	Carcass Weight (lb)	USDA Choice (%)	USDA Yield Grade	Fat Thickness (in.)	Rib-eye Area (in.^{2})	Warner–Bratzler Shear
Hereford	803	65.4	3.19	0.50	12.3	9.1
Angus	836	87.6	3.44	0.58	12.8	8.9
Red Angus	811	89.9	3.44	0.53	12.1	9.1
Simmental	829	65.7	2.72	0.37	13.6	9.5
Gelbvieh	800	57.7	2.60	0.35	13.4	10.0
Limousin	795	56.9	2.43	0.37	13.9	9.5
Charolais	826	61.9	2.53	0.34	13.7	9.6

Source: USDA-ARS-MARC, 2004.

Breed differences must be accounted for as commercial breeders utilize expected progeny differences from various national cattle evaluation systems. Thus, each year a table of breed adjustments is provided by USDA to allow across-breed EPD comparisons (Table 13.8). Across-breed EPD adjustments provide a tool that better allows commercial bull buyers the opportunity to make selection decisions by establishing a means to put EPDs from different breeds on a common scale. Angus is established at par to provide a base point for comparisons. Assume that a cow-calf producer is considering a Charolais bull and a Simmental bull each with a yearling

Table 13.8
Additive Adjustment Factors to Estimate Across-Breed EPDs

Breed	Birth Wt. (lbs)	Weaning Wt. (lbs)	Yearling Wt. (lbs)	Maternal Milk (lbs)	Marbling Score (5 = small 00)	Rib-eye Area (in.^{2})	Fat Thick. (in)	Carcass Wt. (lbs)
Angus	par	par	par	par	par	par	par	par
Hereford	2.7	−4.4	−26.6	−17.8	−0.32	−0.10	−0.053	NA
Red Angus	3.4	−25.7	−30.9	2.4	−0.32	0.03	−0.023	−6.2
Shorthorn	5.1	−30.7	−12.3	4.6	−0.24	0.31	−0.107	−11.6
South Devon	3.6	−8.0	−25.9	2.4	−0.09	0.21	−0.129	−22.3
Beefmaster	5.7	36.1	32.3	11.9	NA	NA	NA	NA
Brahman	10.9	47.5	9.2	23.6	−0.83	−0.11	−0.146	−28.5
Brangus	3.9	13.9	5.1	4.6	NA	NA	NA	−12.5
Santa Gertrudis	6.9	41.4	42.2	14.2	−0.62	−0.06	−0.097	−5.4
Braunvieh	2.5	−22.1	−49.3	−0.4	NA	NA	NA	NA
Charolais	8.6	39.6	40.8	7.3	−0.39	0.98	−0.207	5.4
Chiangus	3.5	−26.9	−38.8	0.2	−0.40	0.34	−0.114	−20.9
Gelbvieh	2.7	−21.5	−30.4	1.6	−0.33	0.65	−0.117	−22.6
Limousin	3.0	−17.0	−42.0	−8.8	−0.60	0.8	NA	−13.4
Maine-Anjou	5.0	−24.5	−35.0	−3.6	−0.60	0.78	−0.192	−23.6
Salers	2.2	−4.1	−26.3	4.9	−0.14	0.85	−0.203	−29.7
Simmental	3.6	−4.8	−9.5	3.6	−0.38	0.43	−0.137	3.8
Tarentaise	3.1	28.3	9.6	23.4	NA	NA	NA	NA

Source: USDA-ARS-MARC.

weight EPD of +80. To determine the anticipated difference in yearling performance between the progeny of these two bulls, the yearling EPD must be adjusted for breed difference. From Table 13.8, the yearling adjustment for Charolais is +40.8 and for Simmental, the adjustment factor is –9.5. Thus, the adjusted yearling EPD for each is as follows:

Charolais: $80 + 40.8 = 120.8$

Simmental: $80 - 9.5 = 70.5$

Anticipated difference in progeny performance when both are mating to cows of a different breed: 120.8 – 70.5 = 50.3 lbs more for progeny of the Charolais sire.

In a second example, consider that a breeder is choosing between a Shorthorn bull with a weaning EPD of +40 and a South Devon bull with a weaning EPD of +18. The anticipated difference in weaning weight from progeny resulting from both bulls being mated to cows of a different breed would be as follows using the adjustment factors for weaning weight in Table 13.8:

Shorthorn: $40 - 30.7 = 9.3$

South Devon: $18 - 8 = 10.0$

Anticipated difference in progeny performance for weaning weight is 10 – 9.3 = 0.7 lb advantage to the South Devon sire. In practice, the two sires will be expected to produce progeny of nearly identical weaning weight.

Using Breeds in Crossbreeding Systems

Producers have a multitude of breeding system options from which to choose. The simplest approach is straight breeding throughout the herd. The primary advantage is convenience. However, when this choice is made, a producer gives up two very powerful genetic tools: heterosis and breed complementarity. Effective crossbreeding systems take advantage of both. Before a producer implements a crossbreeding system, it is important to address the following questions.

1. What is the most appropriate grazing system for the enterprise? What are the issues associated with duration, frequency, and timing of the grazing system that may limit or be in conflict with the mating system alternatives?
2. What proportion of total income is derived from the cow-calf enterprise and what times of the year do other enterprises compete for a producer's time?
3. What is the cost and availability of labor?
4. How much complexity can management effectively handle?
5. How important is simplicity and convenience to implementing the breeding program?
6. What is the cost and availability of bulls from desired breeds or composites?
7. What is the feasibility and cost effectiveness of utilizing artificial insemination?
8. What is the marketing plan for the enterprise?

Once the aforementioned questions have been addressed then a crossbreeding system can be designed that optimizes the use of heterosis and breed complementarity in harmony with the total management system. Heterosis is the performance advantage of crossbred progeny compared with the average of the parental breeds involved in the cross for a particular trait (Figure 13.18). Heterosis is expressed in individuals (crossbred progeny), via maternal influence (crossbred or composite cows), and in a smaller number of traits via paternal effect (crossbred or composite sires). Heterosis is highest for the traits related to reproductive rate, cow lifetime productivity, and calf survival rate. Moderate positive effects are also obtained in the growth traits (Table 13.9).

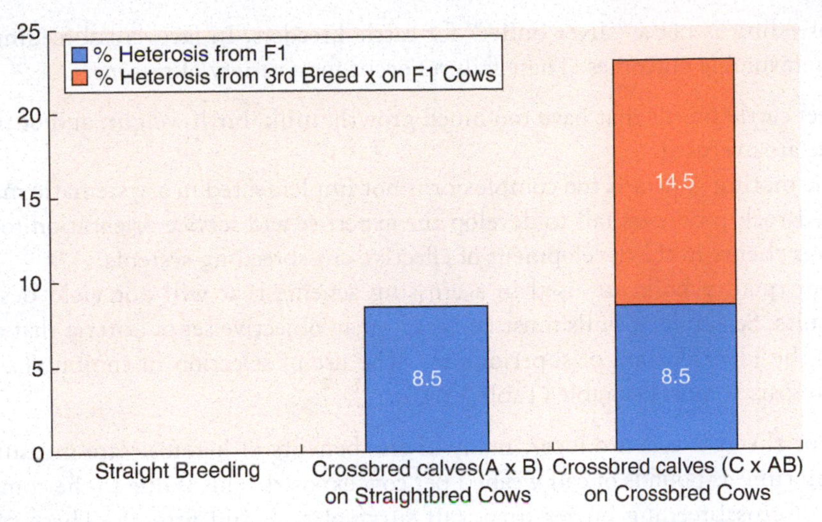

Figure 13.18
Heterosis for pounds of calf per cow exposed to mating.
Source: USDA.

Table 13.9
Average Levels of Heterosis for Traits of Beef Cattle

Trait	Individual Heterosis (%)	Maternal Heterosis (%)	Total Heterosis (%)
Calving	0	6	6
Weaning rate	0	8	8
% reaching puberty by 15 mo.	15	–	15
Survival to weaning	3	1	4
Calving difficulty	2	0	2
Birth weight	4	2	6
Weaning weight	5	6	11
Yearling weight	4	–	4
Feed conversion	−2	–	−2
Carcass weight	3	–	3
USDA quality grade	2	–	2
USDA yield grade	5	–	5
Trimmed retail cuts	3	–	3
Milk production	–	–	0
Calf weaned wt/cow exposed	–	–	18
Cow forage intake	–	–	2
Cow longevity	–	–	38
Cow lifetime productivity	–	–	23

Source: Based on Kress and MacNeill (1999).

The concept of breed complementarity is founded on the premise that no one breed excels in all economically important traits. As such, the systematic matching of breed strengths and weaknesses can yield progeny with optimal and highly desirable trait combinations.

However, implementation of an effective and profitable crossbreeding system requires thorough planning. It should also be recognized that a crossbreeding system usually increases the level of management and must be implemented under the resource and forage constraints of a particular farm or ranch.

The goal of a commercial breeding program is to assure that the cow herd is well matched to the environment with particular attention to the forage resource.

Crossbreeding is not a "silver bullet" for cattle breeders. In fact, crossbreeding may yield unfavorable outcomes. These failures occur for a variety of reasons.

1. Beef cattle breeds that have too much growth, milk, birth weight, and/or mature size are overused.
2. The mating system is too complex or is not implemented in a systematic manner.
3. Seedstock producers fail to develop the expertise and service orientation to assist their clients in the development of effective crossbreeding systems.
4. Poor quality bulls are used in a crossing scheme that will not yield desirable results. Selection of bulls must be based on an objective set of criteria that allows for the identification of superior sires. The use of selection in conjunction with heterosis is most desirable (Table 13.10).

For the cow-calf producer, many of the benefits of heterosis are measured at weaning time as pounds of calf weaned per cow exposed. This is due to the combined effects of crossbreeding on fertility, calf survival rate, and growth. These cumulative benefits from several crossing systems are listed in column two of Table 13.9. Note that the greatest advantage is derived from crossbred calves weaned from crossbred cows.

When choosing a crossbreeding system, producers are advised to focus on two general targets: producing profitable cows well suited to the forage production of a particular ranch and that can be executed in harmony with the grazing plan, and producing profitable progeny that fits the demands of a particular consumer market. Most crossbreeding studies have been conducted at university or USDA research farms. Thus, some have questioned the applicability of the results in real-world conditions. However, a 3-year study conducted by California State University, Chico, at a well-known ranch involving 400 breeding females in the first year and 600 cows in years 2 and 3 provides an integrated systems evaluation of straightbred versus F1 cross progeny. The study was designed to compare straightbred Angus progeny to F1 progeny resulting from crossing Hereford bulls on Angus cows under commercial production conditions extending from the ranch through the carcass phase of production. Results are summarized below:

- F1s had a 10-pound and $12 per head advantage at weaning.
- During backgrounding, the F1s had an additional advantage of 10 pounds and $12 per head.

Table 13.10
Biological Traits of Importance in Cattle Production

	Heterosis	Heritability
Traits at the ranch		
Percent calf crop weaned	high	low
Cow longevity	high	low
Sale weight	moderate	moderate
Cow lifetime productivity	high	low
Traits at the feedyard		
Growth rate	moderate	moderate
Disease resistance	moderate/high	low
Traits at the product level		
Palatability/tenderness	low[1]	moderate/high
Cutability	low[1]	moderate/high

[1]Breed complementarity is very useful in balancing these two traits.
Source: Based on Field and Cundiff, 2000.

- F1s had a cost of gain advantage of nearly $4 per head in the feedyard.
- There were no meaningful differences in carcass weight or yield grade but straightbred Angus had a decided advantage in marbling performance with a total carcass advantage of $15.60 per head.
- Overall, the advantage of the F1s compared to the straightbreds in a vertically coordinated marketing program was $30 per head (weaning through carcass). Note that this study did not evaluate the maternal benefits of heterosis although heifer pregnancy rates for the F1s had a 7% advantage.

Replacement heifer production considerations and the simplicity of the crossing system need to be evaluated (Table 13.11). Furthermore, one of the major challenges facing commercial cow-calf producers is that of overcoming the genetic antagonisms between maternal, growth, and carcass traits. The utilization of breed complementarity is very important in resolving these antagonisms (Table 13.11).

Figure 13.19 illustrates the capacity of various forage environments to sustain differing levels of mature size and milk production. In limiting environments, profitable levels of reproduction can best be sustained by cows with relatively small mature size and moderately low to low levels of milk production. As precipitation and forage production increase, there is less risk associated with larger mature sizes and higher levels of milk production. However, the risk is not zero. In any environment when periods of drought or extreme cold are experienced, those females with the highest nutritional requirements (large size and heavy milk production) are the most likely to suffer reduced reproductive rates due to the increased stress. Consistent high levels of heterosis can be maintained generation after generation if crossbreeding systems such as those shown in Figures 13.20, 13.21, 13.22, and 13.23 are utilized. The attributes of the breeding systems diagrammed in Figures 13.20 to 13.23 are listed in Table 13.12.

Table 13.11
Comparison of Various Crossbreeding Systems for Heterosis, Breed Complementarity, Replacement Production, and Simplicity

System	Heterosis	Breed Complementarity	Production of Replacement	Simplicity
Rotation				
2-breed	16	—	+	+
3-breed	20	—	+	—
2-breed w/F1 sires	22	+	+	+
Static terminal sire				
3-breed	20	+	+	—
Buy purebred females	24	++	—	++
Buy crossbred F1 females	28	++	—	++
Rotate sire breed				
3-breed	16	—	+	+
Rotational terminal sire				
2-breed	21	+	+	+
Composite				
2-breed	14	+	+	++
3-breed	17	+	+	++
4-breed	20	+	+	++
Multiple sire breed				
2-breed with crossbred females	10	—	+	+

Source: Based on Bourdon (2000).

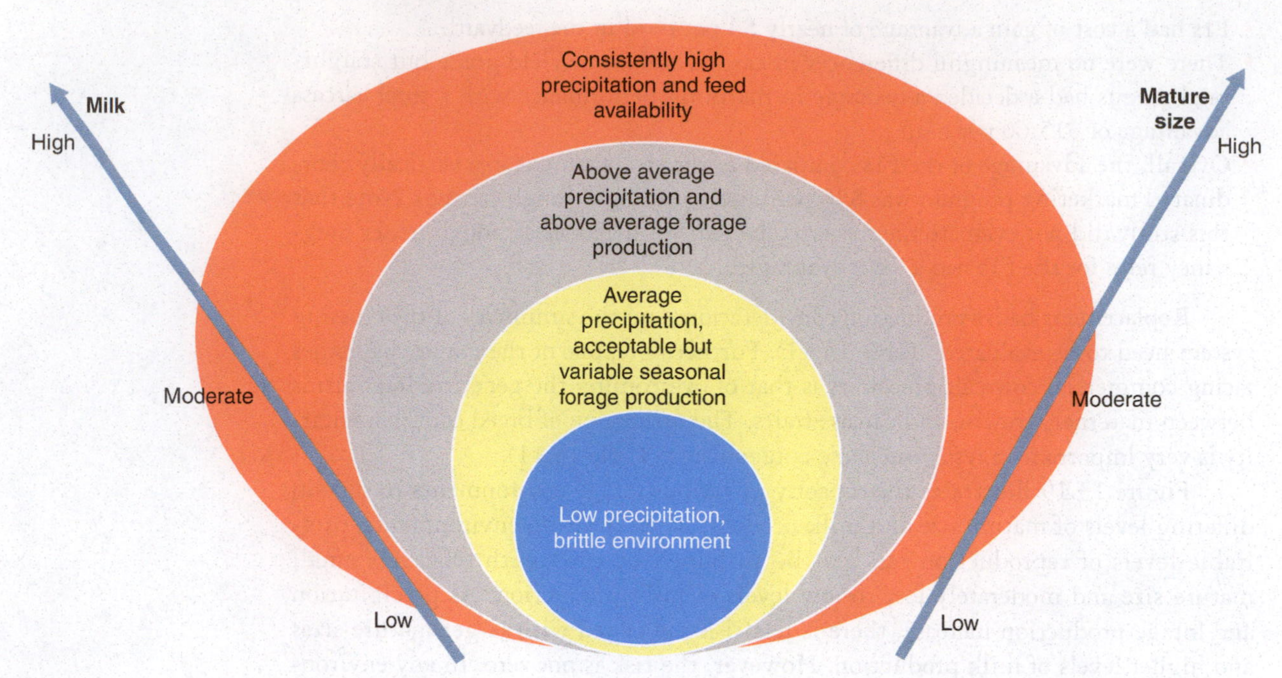

Figure 13.19
Factors associated with matching biological type to diverse environments.

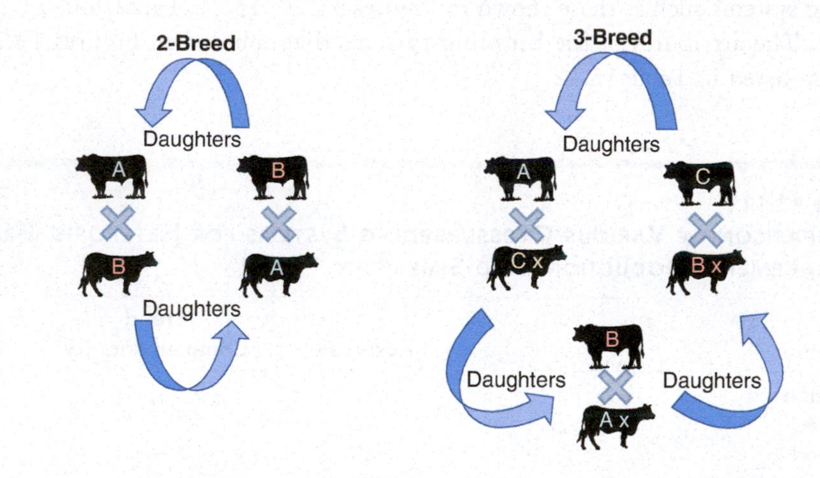

Figure 13.20
Two-breed and three-breed rotational crossbreeding systems.

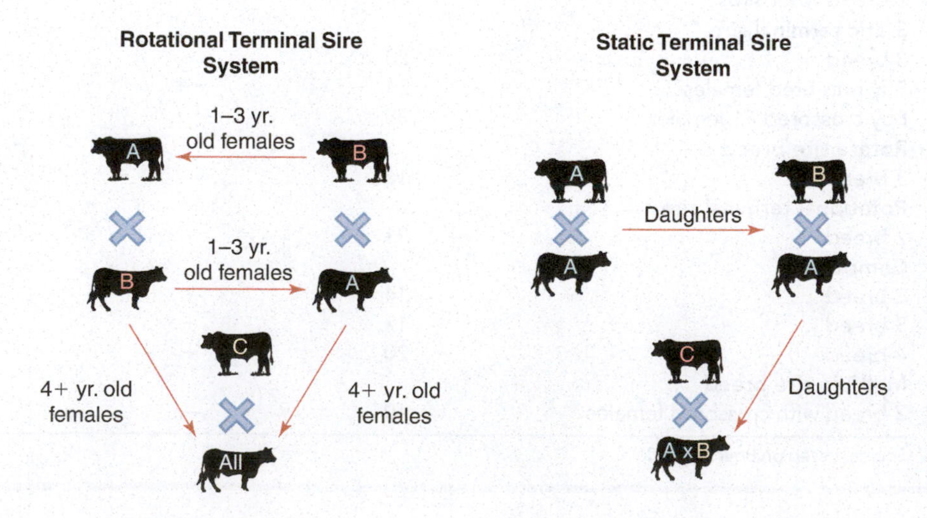

Figure 13.21
Terminal crossbreeding systems.

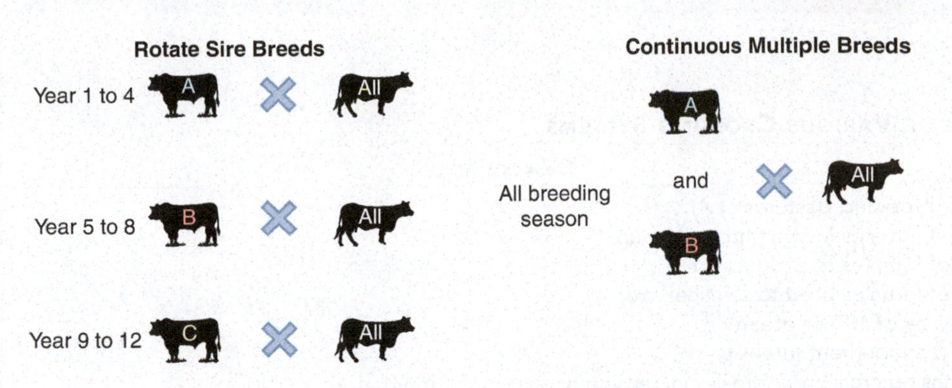

Figure 13.22
Low management crossbreeding systems.

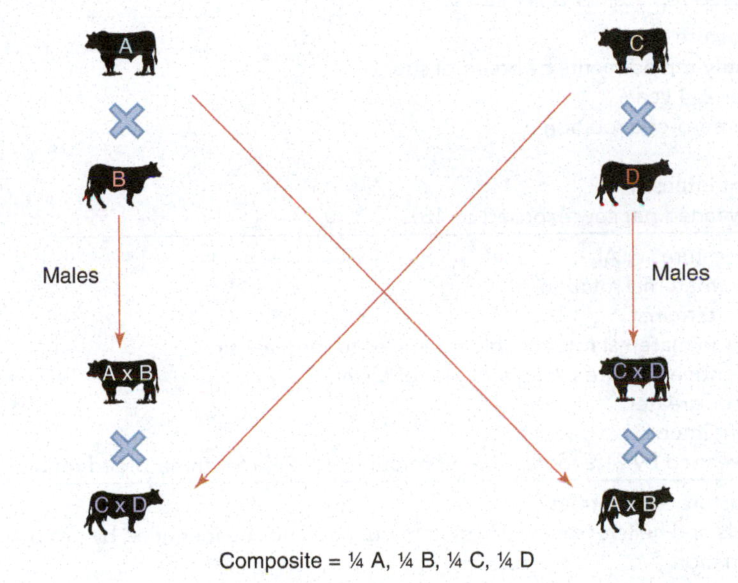

Figure 13.23
Composite breeding system.

Table 13.12
DESCRIPTION OF VARIOUS CROSSING SYSTEMS

System	Description
Two-Breed Rotation	Two breeding pastures or AI Replacement female identified by breed of sire Breeds should be similar for size and milk production Herd size of 50 or greater Generation-to-generation variation may be large, use of F1 sires would diminish variation Moderate management intensity Increase lb calf weaned per cow exposed by 16%, 220 lb with F1 sires
Three-Breed Rotation	Three breeding pastures or AI Replacement females identified by breed of sire Breeds should be similar for size and milk production Females should be bred to the breed of sire to which they are least related Herd size of 75 or greater Generation-to-generation variation may be large High management intensity Increase lb calf weaned per cow exposed by 20%

(continued)

Table 13.12
(Continued) Description of Various Crossing Systems

System	Description
Three-Breed Static Terminal Sire	Three breeding pastures or AI 25% of females in straightbred group 30% of females to produce F1 45% of females bred to terminal sire Herd size of 100 or greater High management intensity Purchasing crossbred females increasing breed complementarity Increase lb of calf weaned per cow exposed by 20%, 24% if purebred females are purchased, 28% if purchased replacements are crossbred
Rotate Sire Breed	One breeding pasture No need to identify replacements by breed of sire Each sire used for 2–4 years Approximates three-breed rotation Any herd size Low management intensity Increase lb calf weaned per cow exposed by 16%
Rotational Terminal Sire	Three breeding pastures or AI 45% of females in maternal rotation 55% of females in terminal Youngest females in maternal rotation, older females to terminal sire Replacements identified by sire breed and year of birth Herd size of 100 or greater High management intensity Increase lb calf weaned by 21% for two-breed rotation and 24% for three-breed rotation
Composite	One breeding pasture after formation Replacement bulls and heifers from within system but need not be identified by sire breed following formation Any herd size can use the composite after formation Availability may be limited Genetic info (EPD) may be limited Generational consistency comparable to purebreds Low management intensity following formation Increase lb of calf weaned per cow exposed by 14, 17, and 20% for two- three-, and four-breed composites
Multiple Sire Breed with Crossbred Females	One breeding pasture Produce own replacements, sire breed identification not required Any herd size Between and within generation variation may be high Low management intensity Increase lb calf weaned per cow exposed by 10%

Source: Based on Bourdon (2000); Field and Cundiff (2000); Sire Selection Manual (2010).

In many environments, the use of heat-adapted breeds is required. However, feedlot managers typically perceive southern-origin feeder cattle to be less desirable than northern-origin calves for health, feeding performance, and carcass merit. In these cases, it would be possible that a British breed bull would be used as a terminal sire. The same situation might occur where high percentages of Continental breed herds were attempting to hit a quality-/marbling-based target. In this case, a British purebred or British cross F1 might be deemed appropriate. For example, one survey of feedlot managers who fed a significant number of southern-origin feeder calves found that they desired either the same amount of heat-adapted breed influence (24%) or less

Table 13.13
Crossbreeding Systems for Single-Sire Herds

Crossbreeding System	Year	1	2	3	4	5	6	7	8	9	10	11	12
Two-breed rotation	Breed of sire	Angus	Angus	Angus	Angus	Hereford	Hereford	Hereford	Hereford	Angus	Angus	Angus	Angus
Three-breed rotation	Breed of sire	Angus	Angus	Angus	Angus	Hereford	Hereford	Hereford	Hereford	Gelbvieh	Gelbvieh	Gelbvieh	Gelbvieh
Two-breed composite	Breed of sire	Brangus	Brangus	Brangus	Brangus	Brangus	Brangus	Brangus	Brangus	Brangus	Brangus	Brangus	Brangus

To enhance heterozygosity, the sire of any one breed should be changed every two to three breeding seasons.
Source: Adapted from Lamberson, B. (1990). *Crossbreeding systems for small herds of beef cattle.* Extension Guide G2040. Columbia, MO: University of Missouri.

heat-adapted breed influence (76%). In the same survey, 85% of respondents desired more Angus influence and 33% desired more Continental breed influence.

There are numerous herds in the United States having 20–40 cows per herd involving only one breeding pasture. Table 13.13 shows several simplified crossbreeding systems that can be used. First system involves the rotation of two breeds, second system rotates three breeds, while the third system is simplified by using a composite breed. In each system, a new bull is introduced after every 2 years to avoid mating heifers back to their sire.

The single-sire rotation is expected with yield 59% of maximum individual heterosis and 47% of maximum maternal heterosis. This compares with 72% of maximum individual heterosis and 56% of maximum maternal heterosis that can be obtained from a two-breed rotation in a large herd or through artificial insemination. The single-sire rotation in a three-breed rotation should produce 27% of maximum individual and 60% of maternal heterosis.

Single-sire rotations can increase productivity and profitability of small beef herds. Choice of biological type is important in addition to obtaining the heterosis by crossing the breeds. Biological type and performance should be similar for all the breeds that are used. This is important, as bulls will be bred to both heifers and cows for calving ease and the biological type must match the most economical feed and production environment.

While commercial producers can use crossbreeding, purebred breeders cannot do so without affecting breed purity. Therefore, commercial producers have an advantage over purebred breeders in being able to utilize more genetic resources to make genetic improvements. Selection and crossbreeding must be used effectively to optimize productivity, otherwise market targets and high levels of profitability will not be achieved (Figure 13.24).

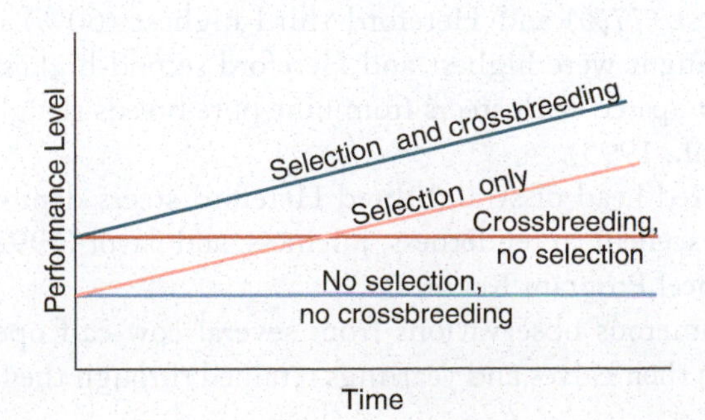

Figure 13.24
Improvement in performance with various combinations of selection and crossbreeding.

Although traits with a low heritability respond very little to genetic selection, they show a marked improvement in a sound crossbreeding program. Carcass traits associated with growth rate (carcass weight per day of age, fat thickness, and retail product weight at the same age) will show some heterosis. Commercial producers need to (1) select sires carefully in order to improve traits with a high heritability, and (2) use a well-planned crossbreeding program in order to use heterosis effectively while matching biological types to their most cost-effective environments.

BREED EVALUATION FOR LOW-COST PRODUCTION

Comparing breeds is a sensitive issue among cattle breeders and cow-calf producers. These comparisons are usually avoided unless the discussion is with people having similar viewpoints. However, comparing breed averages is needed to initially assess the genetic input into total beef management systems. Equally important, if not more important, is selecting the biological types within the breeds that match low-cost production, contribute to consistently higher net profit, and produce highly palatable consumer products.

There are beef industry leaders who propose reducing the number of breeds to four or five to improve the uniformity and consistency of costs, profits, and beef products. Also, a reduction in the number of biological types within breed would be needed to improve uniformity and consistency.

A review of the research on evaluating low-cost commercial cow-calf operations that also have excellent feedlot and cattle performance of their cattle strongly suggests that (1) Angus, Red Angus, and Hereford should be numbered among the breeds to be utilized in an integrated management system, and (2) the identification of moderate or optimum biological types within the breeds is equally as important as the breeds selected. The following information gives evidence to support these statements.

Angus × Hereford was the leading breed cross in U.S. cow herds among more than 200 different breed combinations. Red Angus showed the largest increase in cow-calf herds when compared with nine other leading beef breeds. Angus and Hereford breeds ranked No. 1 or No. 2 among the most popular breeds in all five regions in the United States (1994 and 1995 Cattle-Fax Cow-Calf Producer Surveys).

Calving Difficulty (Dystocia)

Angus had the least (32%) and Herefords the third-least (49%) calving difficulty in 2-year-old heifers of nine different breeds (Gregory et al., 1995).

Feed Efficiency

Hereford and Angus breeds ranked first and second, respectively, when comparing nine breeds for gain efficiency (live weight gain/Mcal of ME, g) at endpoints of "small" marbling score or 4% fat in the longissimus dorsi muscle (Gregory et al., 1995).

Carcass Traits

Angus were highest (77%) and Hereford third-highest (60%) in percent USDA Choice carcasses; Angus were highest and Hereford second-highest in sensory panel tenderness when compared with steers from nine pure breeds slaughtered at 438 days of age (Gregory et al., 1995).

Several hundred head of straightbred Hereford steers demonstrated that this breed sample was excellent in tenderness, juiciness, and flavor (1992, 1993, and 1995 Colo. State Univ. Beef Program Reports).

There are numerous observations from several cow-calf operations with low production costs on their calves and yearlings retained through the feedlot and sold on the grid that support the concept of optimum production. Many of these cows are Angus × Hereford crosses, and have a moderate biological type (mature weight of 1,000–1,150 lb), milk production that produces 500–550 lb weaning weights, and a weaned calf breakeven price of <$0.65/lb. These same feeder cattle gain rapidly in the feedlot, and when harvested at 0.4 in. of fat, they produce optimum carcass weights of 700–800 lb.

Longevity

Angus–Hereford cross females (compared with several breed crosses) excelled in lifetime productivity based on pounds of calf weaned per cow exposed (MacNeil, 1993).

However, the British breeds are less competitive with other breeds in terms of percent retail yield, muscularity, and growth rate.

BREED EVALUATIONS TO IMPROVE CONSUMER MARKET SHARE

The primary components of beef palatability are tenderness, flavor, and juiciness. While most consumers rate tenderness as the most important attribute, the combination of all three attributes is necessary to improve or stabilize consumer market share.

Marbling has a positive relationship to juiciness, flavor, and tenderness; therefore, increased levels of marbling will improve the palatability of beef. Differences in marbling account for approximately 10–15% of the differences in tenderness. While this relationship between marbling and tenderness is not very high, it is sufficient in magnitude to continue its inclusion in the quality grades. Additional measures of tenderness are needed. Currently, the Warner–Bratzler Shear (WBS) measures tenderness even though it is a rather slow, expensive process. There are large breed differences in WBS values (Figure 13.15). Even more important are the sire differences for WBS within a breed. It has been shown that more uniform and consistently tender beef can be produced by using bulls whose progeny had low (highly tender) WBS values.

Breeds vary in muscle fiber color, which is related to the ability to deposit marbling. Heavily muscled breeds, such as Belgian Blue and Piedmontese, have primarily white muscle fibers. The muscle metabolism of these breeds uses glucose and they deposit very little intramuscular fat. Breeds such as Wagyu and Angus have primarily red muscle fibers. Their muscle metabolism involves most fatty acids, which results in the ability to deposit relatively large amounts of marbling.

A literature review of the breed differences for genetic merit relative to carcass traits (Marshall, 1994) allows a ranking of breeds for marbling score and pounds of retail product yield (Table 13.14). If the goal is to meet market specifications by optimizing marbling and retail yield, it becomes obvious that the use of multiple breeds is important. As Table 13.14 illustrates, those breeds excelling in marbling are less proficient at retail yield and vice versa.

A comparison of various breed crosses and their ability to hit USDA Yield and Quality Grade targets is provided in Table 13.15. These comparisons illustrate the need for a 25–50% Continental influence in combination with 75–50% British input to minimize non-conformance (MARC II and MARC III). This same performance could be achieved in 75% of the progeny produced by terminal sires bred to rotational-cross or composite females. In an effort to capture the market potential of a 50% Continental, 50% British breed combination, the American Simmental Association, the North American Limousin Foundation (LimFlex), and the American Gelbvieh Association (Balancer) have developed programs that promote the production of F1 bulls.

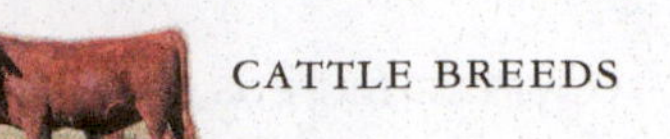

Table 13.14
TOP 10 BREEDS FOR MARBLING SCORE AND RETAIL PRODUCT

Rank	Breed	Marbling Score	Retail Product (lb)	Breed	Retail Product (lb)	Marbling Score
1	Jersey	Modest^{14}	363	Charolais[3]	482	Small^{19}
2	Red Angus	Small^{74}	389	Charolais[4]	460	Small^{24}
3	Angus	Small^{64}	389	Piedmontese	460	Small^{06}
4	Shorthorn	Small^{62}	431	Salers	453	Small^{11}
5	South Devon	Small^{50}	416	Chianina	453	Slight^{44}
6	Hereford × Angus[1]	Small^{47}	396	Holstein	453	Small^{21}
7	Hereford × Angus[2]	Small^{39}	436	Maine–Anjou	451	Slight^{96}
8	Santa Gertrudis	Small^{34}	418	Simmental	444	Small^{06}
9	Pinzgauer	Small^{30}	411	Nellore	440	Small^{01}
10	Galloway	Small^{25}	400	Gelbvieh	438	Small^{03}

[1]Original H × A in MARC germplasm evaluation.
[2]Current H × A in MARC germplasm evaluation.
[3]Current Charolais in MARC germplasm evaluation.
[4]Original Charolais in MARC germplasm evaluation.
Source: Adapted from Marshall (1994).

Table 13.15
CONFORMANCE OF VARIOUS BREED CROSSES AND COMPOSITES TO YIELD AND QUALITY GRADE TARGETS IN STEERS PRODUCED AT THE U.S. MEAT ANIMAL RESEARCH CENTER

	MARC I[1]	MARC II[2]	British	Continental	MARC III[3]
≥70% Yield Grade (YG) 1 & 2	83	56	38	89	52
≥70% Quality Grade (QG) Ch & up	43	55	70	30	66
% Non-conform YG	17	34	62	11	47
% Non-conform QG	57	45	30	70	34
Deviation from acceptance non-conform (30%)					
YG	0	4	32	0	17
QG	27	15	0.4	40	5
Total	27	19	33	40	22

[1]MARC I =1/4 Charolais, 1/4 Limousin, 1/4 Braunvieh, 1/8 Angus, 1/8 Hereford.
[2]MARC II = 1/4 Gelbvieh, 1/4 Simmental, 1/4 Hereford, 1/4 Angus.
[3]MARC III = 1/4 Pinzgauer, 1/4 Red Poll, 1/4 Hereford, 1/4 Angus.
Source: Based on MARC data.

INTEGRATED PRODUCTION SYSTEMS

When grading economic efficiency in integrated production systems (cow-calf through carcass) involving five breeds (Angus, Hereford, Limousin, Simmental, and Charolais) and their two- and three-breed rotational crosses, Angus and Herefords were the most economically efficient purebreds when valued by carcass weight. When steers were slaughtered at a constant fat finish (low choice), purebred Angus and Herefords and their crosses spent the fewest days on feed and had the lowest costs. Angus and Herefords graded choice at lighter weights (Tess et al., 1993).

Angus × Hereford cows gave the highest economic performance of five biological types of cows under either (1) no resource restraints or (2) fixed forage resource base.

In the latter situation, net profit varied by $22,000 under 2,700 AUM of range forage (Davis et al., 1994).

Researchers at MARC have indicated that if cows of large size and high milk production are not given high levels of feed, postpartum interval will increase and conception rates will decline. The F1 cows of each breed cross in each cycle of the MARC program were given the same feeding regime. Breeds having large mature weight and high milk production were provided a nutritional regime for high reproductive rates in addition to meeting requirements for growth, maintenance, and lactation (Cundiff et al., 1991).

There are real challenges in combining breeds and biological types to obtain an optimum combination of maternal (primarily reproduction), growth, and carcass traits. Cundiff (MARC) noted the following: "Unfortunately, breeds (and sires within breeds) that excel in growth and retail product also: (1) sire progeny with higher birth weights, increased calving difficulty, reduced calf survival, and reduced breeding of dams, (2) tend to be older at puberty, and (3) have heavier mature weight that increases maintenance feed requirements."

These antagonisms between maternal traits and growth and (or) carcass traits can be resolved by either (1) selecting for an optimum combination of these traits while avoiding the extremes, or (2) using terminal-cross bulls on cows that are superior for maternal traits and above average for feedlot and carcass traits. There are advantages and disadvantages for each of these two systems that can be evaluated by cow-calf producers by determining how they affect long-term profitability.

While breed average evaluations are important criteria to evaluate, it is just the beginning of evaluating and using genetic resources. Selecting the biological types within the various breeds that best match genetics to low-cost production environments while producing retail products that are high in palatability is the highest priority.

Optimum levels of selection and heterosis are challenging to determine because there are few research studies that focus on the evaluation of genetic optimums. For example, how does a producer determine if a two-breed composite cow with approximately 12% heterosis is more profitable than a three-to-four-breed crossbred cow that achieves 18–20% heterosis? There is evidence that a two-breed should utilize similar biological types; however, it is not clear if three-to-four-breed crossbred cows should be composed of similar widely divergent biological types.

Melton and Colette (1993) raise questions about what economic evaluations best rank and evaluate breeds. They state that simple measures of efficiency (e.g., output–input ratios) or annual net returns may not measure relative values and breed rankings. They suggest that the computation of multigenerational net present values, under alternative price scenarios and prevailing production conditions, be used to rank breeds.

Commercial cow-calf producers can best evaluate breeds of cattle and biological types within breeds by measuring their contribution to low breakeven prices of calves, yearlings, and carcasses. The combination of breakeven price evaluations with high continuing net profits gives assurance in making correct management decisions.

SELECTED REFERENCES

Publications

American Gelbvieh Association. 2001. *The Smart-Cross™ handbook*. Westminster, CO: AGA.

Beef research progress reports No. 1 (Apr. 1982); No. 2 (Dec. 1985); No. 3 (June 1988); No. 12 (May 1990); and No. 4 (May 1993). Clay Center, NE: MARC.

Beef sire selection manual. 2010. National Beef Cattle Evaluation Consortium.

Bourdon, R.M. 2000. *Understanding animal breeding*. Upper Saddle River, NJ: Prentice-Hall.

Briggs, H.M., & Briggs, D.M. 1980. *Modern breeds of livestock*. New York, NY: Macmillan.

Cundiff, L.V., Gregory, K.E., Koch, R.M., & Dickerson, G.E. 1986. *Genetic diversity among cattle breeds and its use to increase beef production efficiency in a temperate environment*. Proc. Third World Congress on Genetic Application to Livestock Production. Lincoln, NE.

Cundiff, L.V., Gregory, K.E., & Koch, R.M. 1991. Reproduction and maternal characteristics of diverse breeds of cattle used for beef production. *The American Beef Cattlemen*.

Cundiff, L.V., Wheeler, T.L., Gregory, K.E., Shackelford, S.D., Koohmaraie, M., Thallman, R.M., Snowder, G.D., & Van Vleck, L.D. 2004. *Preliminary results from cycle VII of the cattle germplasm evaluation program*. Progress Report No. 22. Clay Center, NE: MARC.

Daley, D., Field, T., & Taylor, R. 1995. *New cattle breeds and lines: Composites, synthetics, and hybrids*. Fort Collins, CO: Colorado State University.

Davis, K.C., Tess, M.W., Kress, D.D., Doornbos, D.E., & Anderson, D.C. 1994. Lifecycle evaluation of five biological types of beef cattle in a cow-calf range production system. II. Biological and economic performance. *Journal of Animal Science*. 72:2591.

Fennewald, D. 2003 (February). Increasing consistency a must for all segments. *Gelbvieh World*. 32–33.

Field, T.G., & Cundiff, L.V. 2000. Designing breeding systems that work. *Beef: Cow-calf*. Minneapolis, MN.

Franke, D.E. 1997. Postweaning performance and carcass merit of F1 steers sired by Brahman and alternative subtropically adapted breeds. *Journal of Animal Science*. 75:2604.

Green, R.D., Field, T.G., Hammett, N.S., Ripley, B.M., & Doyle, S.P. 1999. *Can cow adaptability and carcass acceptability both be achieved?* Proc. Western Section of American Society of Animal Science. Provo, UT.

Gregory, K.E., Cundiff, L.V., & Koch, R.M. 1995. *Composite breeds to use heterosis and breed differences to improve efficiency of beef production*. Clay Center, NE: MARC.

Hickman, C.G. 1991. *Cattle genetic resources: World animal science, B7*. New York, NY: Elsevier Science Publishers.

Kress, P.D., & MacNeill, M.D. 1999. *Crossbreeding beef cattle for western range environments* (2nd ed.). Ardmore, OK: Samuel Roberts Noble Foundation.

Lamberson, B. 1990. *Crossbreeding systems for small herds of beef cattle*. Extension Guide G2040. Columbia, MO: University of Missouri.

MacNeil, M. 1993. *Comparison of breed-type for lifetime productivity*. Miles City, MT: Fort Keogh Research Report.

Marshall, D.M. 1994. Breed differences and genetic parameters for body composition in beef cattle. *Journal of Animal Science*. 72:2745.

Melton, B.E., & Colette, W.A. 1993. Potential shortcomings of output: Input ratios as indications of economic efficiency in commercial beef breed evaluations. *Journal of Animal Science*. 71:579.

Mason, I.L. 1988. *A world dictionary of livestock breeds, types, and varieties* (3rd ed.). Wallingford, Oxon, UK: CAB International.

O'Connor, S.F., et al. 1997. Genetic effects on beef tenderness in Bos indicus composite and Bos taurus cattle. *Journal of Animal Science*. 75:1822.

Payne, W.J. & Hodges, J.A. 1997. *Tropical cattle: Origins, breeds and breeding policies*. Ames, IA: Iowa State University Press.

Randel, R.D. 1994. *Factors affecting calf crop: Unique characteristics of Brahman and Brahman based cows*. Boca Raton, FL: CRC Press.

Rouse, J.E. 1970. *World cattle I: Cattle of Europe, South America, Australia, and New Zealand*. Norman: University of Oklahoma Press.

Rouse, J.E. 1970. *World Cattle II: Cattle of Europe and Asia*. Norman: University of Oklahoma Press.

Rouse, J.E. 1973. *World cattle III: Cattle of North America*. Norman: University of Oklahoma Press.

Slaven & Associates. 2003 (January). Feedyards seek improvements in southern feeder cattle. *Gelbvieh World*. 40–44.

Smith, G.C. 1993. *Assuring the consistency and competitiveness of correct biological types of cattle*. Proc. 42nd Annual Florida Beef Cattle Short Course, University of Florida. Gainesville, FL.

Tess, M., Lamb, M., & Robison, O. 1993. Comparison of breeds and mating systems for economic efficiency in cow-calf production. *Montana Ag Research*. 10:22.

Van Vleck, L.D., Splan, R.K., & Cundiff, L.V. 1999. *Genetic correlations between carcass traits and heifer productivity traits*. Proc. BIF Annual Research Symposium. Roanoke, VA.

Walker, H. 1989. *Blue Book of beef breeds*. Allen, KS: PAW Publishing.

14
Nutrition

Beef cattle nutrition involves the chemical and physiological processes by which cattle utilize available feeds to maintain body functions, reproduce, lactate, grow, and produce a desirable end product. Cattle have inherited certain genetic potentials; however, the development of these genetic potentials into economically important traits is highly dependent on the nutritional environment to which cattle are exposed.

NUTRIENTS

A **nutrient** is any feed constituent that functions in the support of life. Beef cattle need nutrients to live and the feeds they consume provide these nutrients. Nutrients are composed of at least 20 of the more than 100 known chemical elements. These 20 elements and their chemical symbols are calcium (Ca), carbon (C), chlorine (Cl), cobalt (Co), copper (Cu), fluorine (F), hydrogen (H), iodine (I), iron (Fe), magnesium (Mg), manganese (Mn), molybdenum (Mo), nitrogen (N), oxygen (O), phosphorus (P), potassium (K), selenium (Se), sodium (Na), sulfur (S), and zinc (Zn).

The six basic classes of nutrients—water, carbohydrates, fats, proteins, minerals, and vitamins—are found in varying amounts in animal feeds.

Water

Water contains hydrogen and oxygen. The terms water and moisture are used interchangeably. Typically, water refers to drinking water, while moisture is used in reference to the amount of water in a given feed or ration. The remainder of the feed, after accounting for moisture, is referred to as **dry matter**. Moisture is found in all feeds, ranging from 10% in air-dry feeds to over 80% in fresh green forage. Cattle consume several times more water than dry matter each day and will die from lack of water more quickly than from lack of any other nutrient. Water in feed is no more valuable than water from any other source. This is important to understand to properly assess feeds that vary in their moisture content.

Water has important body functions because it enters into most metabolic reactions, assists in transporting other nutrients, helps maintain normal body temperature, and gives the body its physical shape (it is the major component of cells).

Water quality is important to cattle performance and water sources should be critically assessed. Taste, odor, degree of particulate matter, and degree of contamination affect the quality of water.

Carbohydrates

Carbohydrates contain carbon, hydrogen, and oxygen in either simple or complex forms. The more simple carbohydrates, such as starch, supply the major energy source for cattle rations, particularly feedlot rations. The more complex carbohydrates, such as cellulose, are the major components of the cell walls of plants. These complex carbohydrates are not as easily digested as simple carbohydrates.

Fats

Fats and oils, also referred to as lipids, contain carbon, hydrogen, and oxygen, though there is more carbon and hydrogen in proportion to oxygen than with carbohydrates. Fats are solid and oils are liquid at room temperature. Fats contain 2.25 times as much energy per pound as do carbohydrates.

Fats comprise fatty acids and glycerol:

Example:

$$3C_{17}H_{35}COOH + C_3H_5(OH_3) = C_{57}H_{110}O_6 + 3H_2O$$

Stearic acid (a fatty acid) + glycerol = stearin (a fat) + water

There are saturated and unsaturated fats, depending on their particular chemical composition. Saturated fatty acids have single bonds tying the carbon atoms together (e.g., -C-C-C-C-), whereas unsaturated fatty acids have one or more double bonds (e.g., -C=C=C-C-). The term polyunsaturated fatty acid is applied to those having more than one double bond. While more than 100 fatty acids have been identified, only three or four have been determined as dietary essentials. The two apparent functions of the essential fatty acids are as (1) precursors of prostaglandins and (2) structural components of cells.

Proteins

Proteins always contain carbon, hydrogen, oxygen, and nitrogen and sometimes iron, phosphorus, and/or sulfur. Protein is the only nutrient class that contains nitrogen. Proteins in feeds, on an average, contain 16% nitrogen, which is why feeds are analyzed for the percent of nitrogen in the feed (the percentage is multiplied by 6.25 to convert nitrogen to percent protein).

Proteins are composed of various combinations of some 25 amino acids. Amino acids are called the "building blocks" of the animal's body because they make up the muscle mass plus other parts. Amino acids carry the amino group (NH_2) in each of their chemical structures. There are many different combinations of amino acids that can be structured together.

Minerals

Chemical elements other than carbon, hydrogen, oxygen, and nitrogen are called minerals. They are inorganic because they contain no carbon, whereas organic nutrients contain carbon. Some minerals are referred to as macro (required in larger amounts) and others are called micro or trace (required in smaller amounts).

Minerals have many body functions, including bone structure, body water balance, oxygen transport and transfer, and metabolic reactions.

Vitamins

Vitamins are organic substances containing carbon, hydrogen, and oxygen. There are 16 known vitamins that are important in animal nutrition. Vitamins are required by beef cattle in very small amounts. They function in regulating certain body processes toward normal health, growth, and reproduction. Vitamin A is typically the only vitamin of concern in most beef cattle rations.

PROXIMATE ANALYSIS OF FEEDS

The nutrient composition of a feed cannot be determined accurately by visual inspection. The value of a feed can be measured by a proximate analysis, which separates feed components into groups according to their feeding value. The analysis is based on an

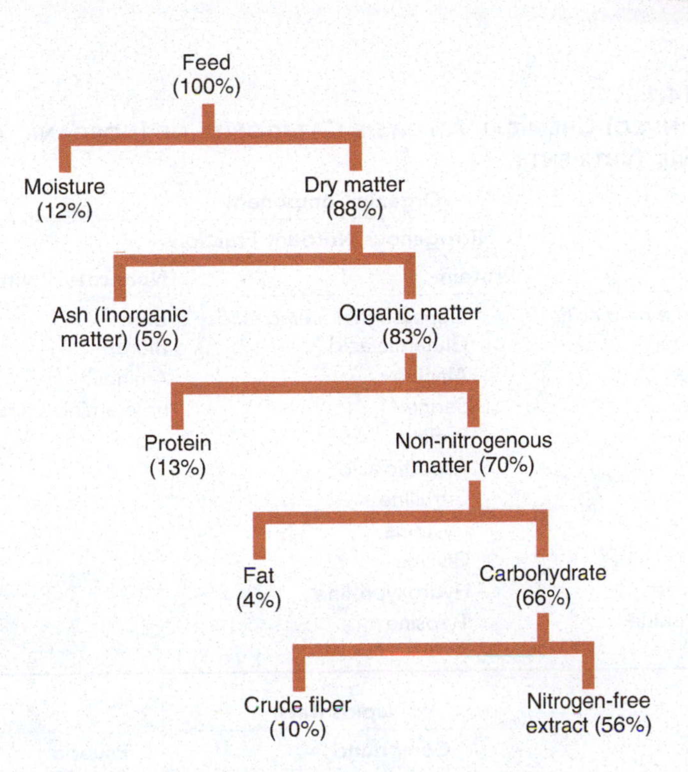

Figure 14.1
Proximate analysis showing the inorganic and organic components of a feed on a natural or air-dry basis.

analysis of a feed sample, so it is no more accurate than how representative the sample is of the entire feed source.

The groups measured in a proximate analysis are water, crude protein, crude fat (sometimes referred to as ether extract), crude fiber, nitrogen-free extract, and ash (minerals). Figure 14.1 shows these components resulting from a feed that had a laboratory analysis of 88% dry matter, 13% protein, 4% fat, 10% crude fiber, and 56% nitrogen-free extract (NFE) on a natural or air-dry basis. The analysis might be reported on a dry-matter basis of 12% moisture, 14.8% protein, 4.5% fat, 11.4% crude fiber, and 63.6% nitrogen-free extract. Therefore, caution needs to be exercised in interpreting proximate analysis results because laboratories report their analytical values in different ways.

The proximate analysis for the six basic nutrients does not distinguish the various components of a nutrient. For example, the ash content of a feed does not tell the amount of calcium, phosphorus, or other specific minerals. Table 14.1 lists the organic and inorganic nutrients that can be characterized by chemical analysis.

Table 14.1
Chemical Analysis Categories of Inorganic and Organic Nutrients

Inorganic Components	
Macro Minerals	**Micro Minerals**
Calcium (Ca)	Cobalt (Co)
Phosphorus (P)	Iron (Fe)
Sodium (Na)	Zinc (Zn)
Chlorine (Cl)	Iodine (I)
Potassium (K)	Manganese (Mn)
Magnesium (Mg)	Sulfur (S)
	Copper (Cu)
	Molybdenum (Mo)
	Selenium (Se)

(continued)

Table 14.1
(Continued) Chemical Analysis Categories of Inorganic and Organic Nutrients

Organic Components		
Nitrogenous Nutrient Fractions		
Protein		**Nonprotein Nitrogen**
Essential amino acids	*Nonessential amino acids*	Urea
Tryptophan	Glutamic acid	Biuret
Histidine	Alanine	Amines
Arginine	Serine	Free amino acids
Threonine	Proline	
Lysine	Aspartic acid	
Leucine	Citrulline	
Isoleucine	Cysteine	
Valine	Glycine	
Methionine	Hydroxyproline	
Phenylalanine	Tyrosine	
Taurine		

Lipids (fats)		
Simple	**Compound**	**Psuedo**
Fatty acids	Neutral fats	Vitamin A, D, E, K
	Sterols	Carotene

Carbohydrates				
Crude Fiber		**Nitrogen-Free Extract**		
Polysaccharides		*Monosaccharides*	*Polysaccharides*	*Water-soluble vitamins*
Cellulose	Hemicellulose	Simple sugars	Starches	Vitamin C
				B-vitamins

THE RUMINANT DIGESTIVE SYSTEM

Beef cattle are **ruminants**, which means they ruminate or chew a bolus of feed called a **cud**. Cattle eat the feed and form it into boluses that pass to the stomach. After eating for a while, cattle usually lie down, regurgitate the boluses, and proceed to chew the feed, breaking it into smaller pieces. A beef cow normally ruminates approximately 8 hours each day. As the animal ruminates, large amounts of saliva (35–40 gallons per day) are secreted from the salivary glands in the mouth. This amount varies with the diet being fed. For example, forage diets stimulate increased saliva production. Saliva is highly alkaline and it neutralizes (buffers) the large amount of acid produced in the ruminant stomach during digestion.

The digestive tract, sometimes referred to as the alimentary canal or tract, includes those organs from the mouth to the anus through which feed passes (see Figure 14.2). Table 14.2 identifies the primary parts of the digestive tract and some of their capacities.

The process of digestion is twofold: (1) the physical breaking down of the feed into smaller pieces and (2) the chemical breaking down of more complex nutrients into simpler forms that can be absorbed by the digestive tract and utilized on the cellular level by the animal. Simply stated, proteins are converted to amino acids; starch

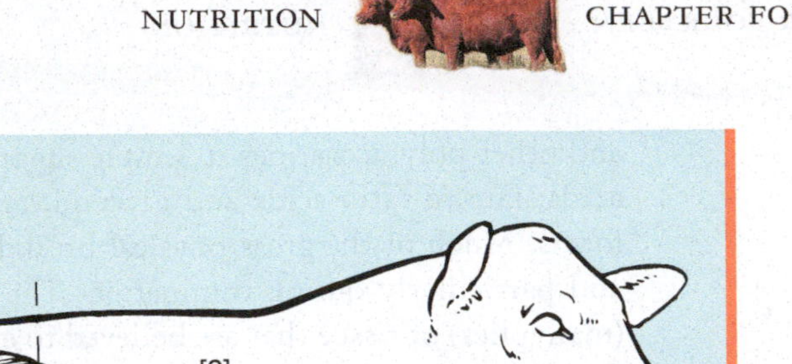

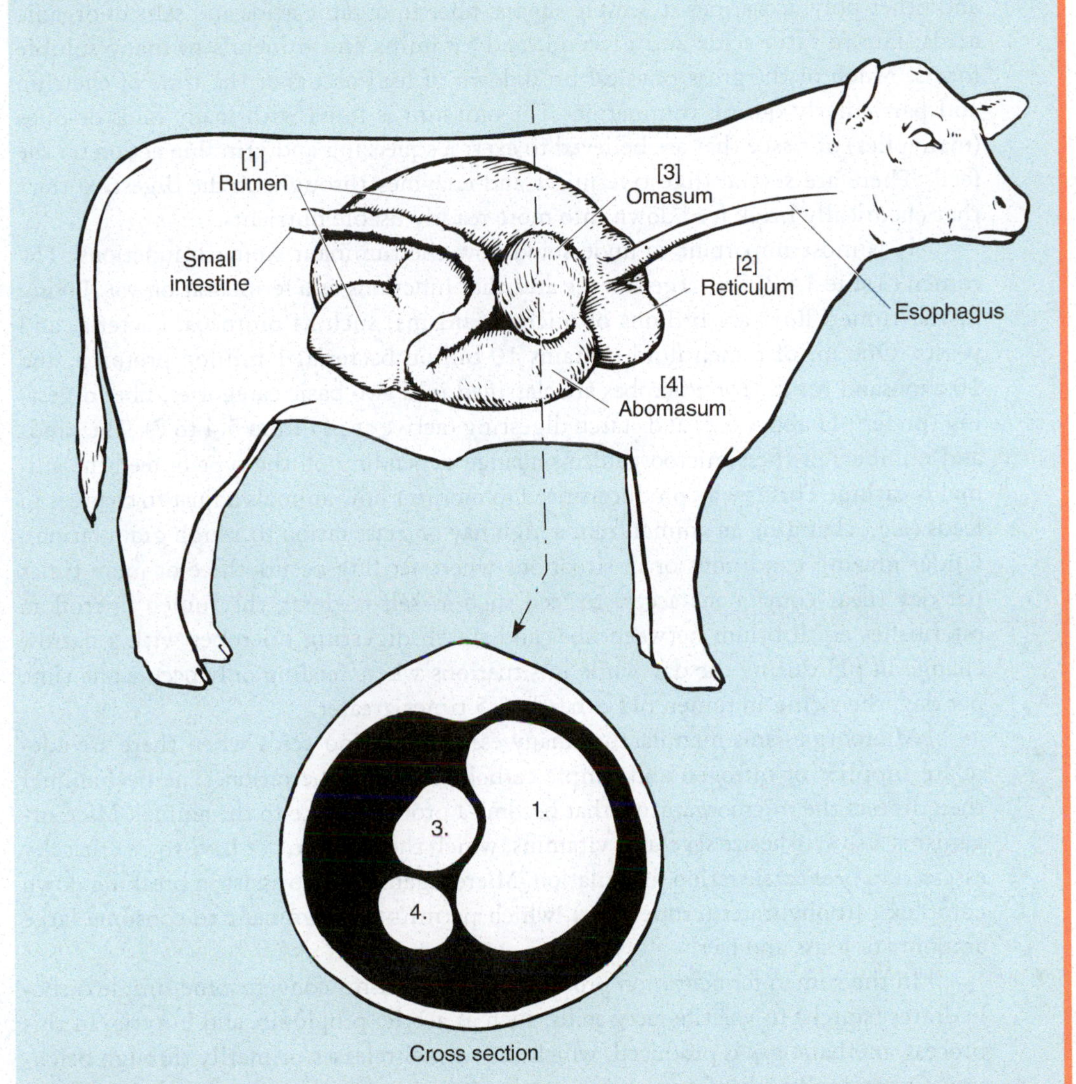

Figure 14.2
Beef cattle digestive tract.

Table 14.2
Parts and Capacity of the Digestive Tract of a Cow

Part	Capacity (gal)
Mouth	—
Esophagus	—
Stomach (four compartments)	—
Rumen (paunch)	(50)
Reticulum (honeycomb)	(2)
Omasum (manyplies)	(4)
Abomasum (true stomach)	(4)
Small intestine	15
Large intestine	10
Anus	—

and other polysaccharides to simple sugars; fiber to organic acids and salts of organic acids; fats to fatty acids and glycerol; and vitamins and minerals to many soluble forms. Much of the gross physical breakdown of feed occurs at the time of chewing and particularly during rumination. The omasum is lined with many folds or plies (manyplies) of tissue that are believed to exert a squeezing and grinding action on the feed. There are several digestive juices and enzymes throughout the digestive tract that chemically break feed down into more readily usable nutrients.

It is most important to understand how the ruminant stomach functions. The rumen (Table 14.2) is a large storage area that functions as a fermentation vat. Living in the rumen fluid are billions of microorganisms, such as protozoa, bacteria, and yeasts. One ml of rumen fluid contains 10 billion bacteria, 1 million protozoa, and 10 thousand fungi. The microbes are classified into two basic categories: fiber digesting (prefer pH above 6.2) and starch digesting (active at pH from 5.4 to 7). The kinds and numbers of these microorganisms change depending on the type of feed the animal is eating. This is why producers need to monitor how animals adjust to changes in feeds (e.g., changing an animal from a high hay or grass ration to a high grain ration). Under grazing conditions or in situations where feeding occurs three or more times per day (near-continuous access to feed such as self-feeders), the rumen microflora establishes equilibrium between fiber and starch digesting microbes with a narrow change in pH during the day while in situations where feeding only occurs one time per day, the swing in rumen pH is nearly 3.5 times greater.

Microorganisms manufacture many essential amino acids when there are adequate supplies of nitrogen and simple carbohydrates in the ration. The beef animal then digests the microorganisms that become a protein source to the animal. Microorganisms also synthesize several B vitamins, which therefore do not have to be critically assessed in beef cattle ration formulation. Microorganisms also assist in breaking down complex carbohydrates (crude fiber), which permits the ruminant to consume large amounts of grass and hay.

In the rumen fermentation process, microorganisms convert some simple carbohydrates (starch) to volatile fatty acids, such as acetic, propionic, and butyric. In this process, methane gas is produced, which the animal releases primarily through belching. Occasionally, the gas-releasing mechanism does not function properly and gas accumulates in the rumen, causing a condition called *bloat*. Death will result if gas pressure builds to a high level and interferes with adequate respiration.

Rumen microflora may also be disrupted by environmental stress or transportation stress. These events require that the post-stressor feeding regimen be structured to allow rebuilding of a normal microbial population within the rumen.

DIGESTIBILITY OF FEEDS

Digestibility refers to the percentage of a nutrient in a feed that is absorbed from the digestive tract. Different feeds and nutrients vary greatly in their digestibility. Many feeds have been subjected to digestion trials, where feeds of known nutrient composition have been fed to cattle. Feces have been collected and the nutrients in the feces analyzed. The difference between nutrients fed and nutrients excreted in the feces is considered the digestibility of the feed.

ENERGY EVALUATIONS OF FEEDS

Energy needs of cattle account for the largest portion of feed consumed. Several systems have been devised to evaluate feedstuffs for their energy content. The most common energy system is **total digestible nutrients** (TDN), typically expressed in pounds, kilograms, or percentages. TDN is calculated after obtaining the proximate analysis and digestibility figures for a feed. The formula for calculating TDN is as follows:

$$\text{TDN} = (\text{digestible crude protein}) + (\text{digestible crude fiber}) + (\text{digestible nitrogen-free extract}) + (\text{digestible crude fat} \times 2.25)$$

TDN is roughly comparable to digestible energy (DE), but it is expressed in different units. TDN and DE both tend to overvalue roughages compared to the NE system.

Even though there are some apparent shortcomings in using TDN as an energy measurement of feeds, it works well in balancing rations for cows. There is more precision in the energy measurement of feeds in using the net energy (NE) system, so it should be used for growing cattle and in feedlot rations. The TDN system overestimates the energy value of roughages for weight gain compared to NE. NE measures energy values in megacalories per pound or kilograms of feed. The calorie basis, which measures the heat content of feed, is as follows:

Calorie (Cal)—amount of energy or heat required to raise 1 g of water 1°C.

Kilocalorie (Kcal)—amount of energy or heat required to raise 1 kg of water 1°C.

Megacalorie (Mcal)—equal to 1,000 kilocalories or 1,000,000 calories.

Figure 14.3 shows various ways that cattle utilize energy in feeds. Also shown are various energy measurements of feeds that reflect the types of energy utilization by the beef animal. Gross energy (GE) is the quantity of heat (calories) released from the complete burning of the feed sample in an apparatus called a *bomb calorimeter*. GE has little practical value in evaluating feeds for cattle because the animal does not metabolize feeds in the same manner as a bomb calorimeter. For example, oat straw has the same GE value as corn grain. DE is GE of feed minus fecal energy. Metabolizable energy (ME) is GE of feed minus energy in the feces, urine, and gaseous products of digestion.

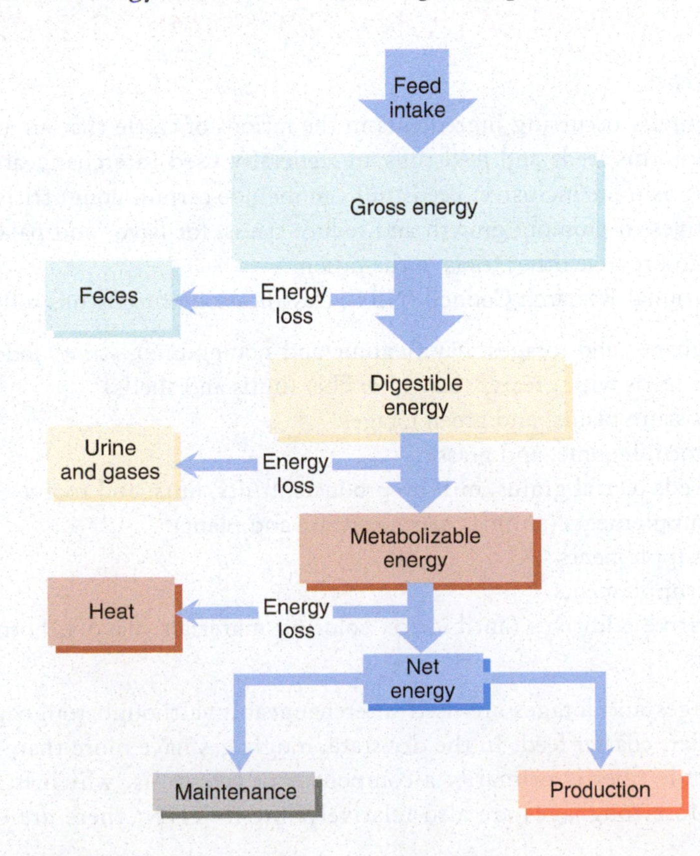

Figure 14.3
Measures of energy and energy utilization by cattle.

Table 14.3
Comparison of the TDN and NE Energy Systems for a Yearling Steer with Varying Rates of Gain

	Rate of Gain (lb)		
Energy System	0	2	2.9
TDN (lb)	6.4	12.8	15.0
NEm (Mcal)	6.2	6.2	6.2
NEg (Mcal)	—	4.3	6.5
Total lb feed			
TDN basis	8.2	16.5	19.3
NE basis	7.2	15.0	19.0
Feed per lb gain (NE)	—	7.5	6.6
Feed per lb gain (TDN)	—	8.2	6.6

NEm (net energy for maintenance) and **NEg (net energy for gain)** are more commonly used for formulating rations for feedlot cattle than any other energy system. NEm in feedlot cattle rations is the amount of energy needed to maintain a constant body weight. Animals of known weight, fed for zero energy gain, have a constant level of heat production. The NEg measures the increased energy content of the carcass after feeding a known quantity of feed energy. All feed fed above maintenance is not utilized at a constant level of efficiency. Higher rates of gain require more feed per unit of gain since composition of gain varies with rate of gain.

The TDN and NE systems are compared in Table 14.3. Information in the table is based on feeding a simple ration of ground ear corn (90%) and supplement (10%) to a yearling steer (772 lb) for different rates of gain.

FEEDS: CLASSIFICATION AND COMPOSITION

Classification

Feeds are naturally occurring ingredients in the rations of cattle that are used to sustain life. The terms feeds and feedstuffs are generally used interchangeably, though the latter term is more inclusive. Feedstuffs can include certain nonnutritive products such as additives to promote growth and reduce stress, for flavor and palatability, to add bulk, or to preserve other feeds in the ration.

The National Research Council (NRC) classification of feedstuffs follows.

1. Dry roughages and forages: hay (legume and nonlegume), straw, fodder, stover, and other feeds with greater than 18% fiber (hulls and shells)
2. Range, pasture plants, and green forages
3. Silages (corn, legume, and grass)
4. Energy feeds (cereal grains, mill by-products, fruits, nuts, and roots)
5. Protein supplements (animal, marine, avian, and plant)
6. Mineral supplements
7. Vitamin supplements
8. Nonnutritive additives (antibiotics, coloring materials, flavors, hormones, and medicants)

Roughages and forages are used interchangeably, although roughages usually imply a bulkier, coarser feed. In the dry state, roughages have more than 18% crude fiber. The crude fiber is primarily a component of cell walls, which is not highly digestible. Most roughages are also relatively low in TDN; there are exceptions, though, such as corn silage, which has over 18% crude fiber and approximately 70% TDN.

Feedstuffs that contain 20% or more protein are classified as protein supplements (e.g., soybean meal and cottonseed meal). Feedstuffs having less than 18% crude fiber and less than 20% protein are classified as energy feeds. Concentrates, such as the cereal grains, are typical energy feeds, which is reflected by their high TDN values.

Nutrient Composition

Feeds are analyzed for their nutrient composition. The ultimate goal of nutrient analysis of feeds is to predict the productive response of cattle when they are fed rations of a given nutrient composition.

The nutrient compositions of some of the more common feeds utilized for cattle are shown in Table 14.4. The information in the table represents averages of numerous feed samples. Feeds are not constant in composition, so analyses should be obtained whenever economically feasible. Because an analysis is not always feasible because of lack of available laboratories and insufficient time, feed analysis tables are the next best source of reliable information on nutrient composition of feeds. It is not uncommon to expect the following deviations of actual feed analysis from the table values for several feed constituents: crude protein (±15%), energy values (±10%), and minerals (±30%).

Table 14.4
Nutrient Composition of Selected Feeds Commonly Used in Beef Cattle Rations

Feed	Dry Matter (%)	TDN (%)	NEm[1] (Mcal/cwt)	NEg[2] (Mcal/cwt)	NEl[3] (Mcal/cwt)	Crude Protein (%)	By-pass (%)	ADF[4] (%)	NDF[5] (%)	Ca (%)	P (%)
Alfalfa hay (early bloom)	90	59	59	28	59	19	20	36	46	1.41	0.26
Alfalfa (dehydrated)	92	61	62	31	61	19	60	34	45	1.42	0.25
Alfalfa (fresh)	24	61	62	31	61	19	18	34	46	1.35	0.27
Barley grain	89	84	92	61	87	12	28	7	20	0.06	0.38
Beet pulp (dried)	91	75	79	50	77	11	44	26	46	0.65	0.08
Bermuda grass (hay, coastal)	89	56	56	23	56	10	20	36	73	0.47	0.21
Birdsfoot trefoil (grazed)	22	66	68	38	67	21	20	31	47	1.78	0.25
Bluegrass (grazed, Kentucky)	36	69	71	43	70	15	20	32	60	0.37	0.30
Bromegrass (grazed, early vegetation	30	64	65	36	65	15	22	33	54	0.45	0.34
Bromegrass (hay)	89	55	55	21	55	10	33	41	66	0.40	0.23
Clover, red (hay)	88	55	55	21	55	15	28	39	51	1.50	0.25
Corn (whole grain)	88	87	96	64	90	9	58	3	10	0.02	0.29
Corn (grain rolled)	88	89	99	67	93	9	52	3	10	0.02	0.30
Corn silage (mature)	34	72	75	47	74	8	28	32	54	0.28	0.23
Corn stover (no ears or husks)	80	59	59	28	59	5	30	43	70	0.35	0.19
Cottonseed meal	92	80	86	56	83	46	50	20	28	0.21	1.19
Dicalcium phosphate	96	—	—	—	—	—	—	0	0	22.00	18.65
Distillers grain, corn, dry	91	95	107	73	99	30	58	16	44	0.09	0.35
Distillers grain, corn, wet	36	96	109	74	100	30	47	16	44	0.09	0.35
Distillers grain, corn, with solubles	91	96	109	74	100	31	55	14	30	0.21	0.82

(continued)

Table 14.4
(Continued) Nutrient Composition of Selected Feeds Commonly Used in Beef Cattle Rations

Feed	Dry Matter (%)	TDN (%)	NEm[1] (Mcal/ cwt)	NEg[2] (Mcal/ cwt)	NEl[3] (Mcal/ cwt)	Crude Protein (%)	By-pass (%)	ADF[4] (%)	NDF[5] (%)	Ca (%)	P (%)
Distillers dried solubles	93	87	96	64	91	32	40	7	22	0.35	1.20
Fescue KY31 (fresh)	29	64	65	36	65	15	20	32	64	0.48	0.37
Fescue KY31 hay (mature)	88	52	52	16	51	11	30	42	73	0.45	0.26
Limestone (ground)	98	0	0	0	0	0	0	0	0	34.00	0.02
Meadow hay (native, intermount)	90	50	50	12	49	7	23	44	70	0.61	0.18
Molasses (cane)	76	75	79	50	77	5	0	0	0	1.00	0.10
Oats (grain)	89	76	81	52	78	13	18	15	28	0.05	0.41
Oats (silage)	35	60	60	30	60	12	21	39	59	0.45	0.31
Prairie hay (mid bloom)	91	50	50	12	49	7	37	47	67	0.40	0.15
Sagebrush (grazed)[2]	50	50	50	12	49	13	—	30	38	1.00	0.25
Sorghum stover (no heads)	87	55	55	21	55	5	—	41	65	0.49	0.12
Sorghum silage	32	59	59	28	59	9	30	38	59	0.48	0.21
Soybean meal (solvent)	91	84	92	61	87	49	35	10	15	0.38	0.71
Soybean (straw)	88	42	43	0	40	5	—	54	70	1.59	0.06
Sudangrass (hay)	88	57	57	25	57	9	30	43	67	0.50	0.22
Timothy (hay, full-bloom)	88	57	57	25	57	8	30	40	65	0.43	0.20
Wheat (hard grain)	89	88	98	65	91	14	28	6	14	0.05	0.43
Wheat (grazed early)	21	71	74	46	73	20	16	30	50	0.35	0.36
Wheat (straw)	91	42	43	0	40	3	60	57	81	0.16	0.05
Wheatgrass, crested (early)	37	60	60	30	60	11	25	28	50	0.46	0.32
Wheatgrass, crested (full-bloom)	50	55	55	21	55	10	33	36	65	0.39	0.28
Wheatgrass, crested (hay)	92	54	54	20	54	10	33	36	65	0.33	0.20

[1]Net energy maintenance.
[2]Net energy growth.
[3]Net energy lactation.
[4]Acid detergent fiber (feed digestibility).
[5]Neutral detergent fiber (voluntary intake and energy availability).
Source: Adapted from multiple sources.

Digestible protein is included in many feed composition tables, but because of the large contribution of body protein to the apparent protein in the feces, digestible protein is more misleading than crude protein. For this reason, crude protein is more commonly used in formulating rations.

Digestible protein (DP) can be calculated from crude protein (CP) content by using the following equation (%DP and %CP are on a dry-matter basis):

$$\%DP = 0.9\ (\%CP) - 3$$

Four measures of energy value (TDN, NEm, NEg, and NEl) are shown in Table 14.4. TDN has been the industry standard in the past and many cow-calf producers are familiar with it as a measure of energy value. However, the most recent NRC standards utilize the NE system as a more precise tool for ration formulation. Historically, crude fiber was used to describe the poorly digested carbohydrate levels of feed. Improved laboratory techniques have produced better measures—acid detergent fiber

(ADF) that is related to feed digestibility and neutral detergent fiber (NDF) that is related to feed intake and NE availability.

Management for High Nutrient Content of Feeds

How feeds are grown, harvested, and preserved can have a significant effect on total nutrients produced per unit of land, nutrient content of feeds, and safety of the feeds at the time they are being fed to cattle. Some major management practices utilized to enhance nutrient content and utilization of feeds are presented here. Hays and crop residues are covered in Chapter 15.

Silages

Silages (Figure 14.4) are produced from green forage crops that are compressed and stored under anaerobic (oxygen-free) conditions in upright or horizontal silos. The feed is preserved through an acid fermentation process that prevents mold and other spoilage from occurring.

The most common crops used for silage are corn, sorghums, grasses, legumes, and some small grains. These must contain enough fermentable carbohydrate to allow sufficient lactic acid production to register a pH of 4. With some forages, grain may be added to achieve this effect. Haylage is forage that is intermediate between hay and silage in terms of moisture content (40–60%), whereas the moisture content of silage is usually 60–75%.

Silage comprises a large part of the feeding program of many beef cattle operations, primarily because TDN and beef produced per acre are greatest with silage. Caution should be exercised, however, because silage is one of the most variable feedstuffs in nutrient content and storage loss. The nine factors that interact to affect silage quality are type of silage crop, weather, stage of maturity, moisture, additives, fineness of chopping, packing and filling the silo, silo structure, and feed-out of the silage.

Moisture content of silage is the single most important factor affecting silage quality. Ensiling above 72–74% moisture can produce seepage, undesirable butyric acid, high fermentation losses, and reduced intake of silage by cattle. Ensiling below 50–55% moisture creates problems in eliminating air (poor compaction) and achieving sufficient fermentation and low pH.

An optimum stage of maturity should be selected that will make the best compromise between increased dry-matter yield and decreased digestibility. Table 14.5 gives the recommended maturity stages for several silage crops. A fine chop—one-fourth to one-half inch—is recommended for most crops. Hollow-strawed cereals need to be chopped finer than solid-stemmed corn and sorghum. Crops that are chopped too coarse will have more air trapped when ensiled and cattle will refuse more silage when it is fed.

Figure 14.4
Corn silage stored in a concrete bunker-style pit.

Table 14.5
STAGE OF MATURITY RECOMMENDATIONS FOR SEVERAL SILAGES

Crop	Recommended Stage of Maturity
Alfalfa	Late bud to one-tenth bloom
Cereal grains (barley, oats, and wheat)	Soft-dough kernels
Corn	Kernels fully dented
Grasses (summer and perennial)	Prior to head emergence
Sorghum	Soft-dough kernels

A recent innovation in silage harvesting is the use of mechanical processing during which freshly cut silage is crushed and sheared by being put through a set of differential speed rollers. This process increases the surface area of the corn plant as a means of enhancing fiber digestion while the crushing of the kernels increases starch digestibility.

Type of silo can affect the quality of silage as losses can be 30% or higher (Table 14.6). Silo construction cost and silage losses need to be compared.

Packing and filling the silo should be done to eliminate air pockets while also exposing a small surface area. After filling a trench or bunker silo, the surface should be sealed to keep out air, rain, and snow and thereby reduce spoilage loss. Producers should grow those silage crops that are best suited agronomically for their area, selecting crops that can economically produce the greatest amount of TDN per acre. Feed-out should allow the silage to be fresh and not exposed to air for extended periods of time. This is especially critical when the weather is warm, when the growth of mold and other spoilage organisms is stimulated. Best practices for silage bunker management are provided in Table 14.7.

There are four general categories of silage additives: (1) acids for direct acidification, (2) preservatives (sterilants and fatty acids), (3) feedstuffs (molasses, NPN, grain), and (4) fermentation aids (enzymes, inoculants, and antioxidants). In excess of 200 different silage additives/preservatives are commercially available. Before using an additive, however, the costs and returns should be carefully evaluated. Claims made for additives should be based on well-documented research. Additives should not be used as a substitute for poor management. Quality silage should have an aroma that smells sweet coupled with a distinct vinegar odor. A *Corn Silage Processing Score* is a test that estimates the percentage of starch in silage that has been processed for optimal digestion. A score of 70% or better is considered ideal, while scores of less than 50% are unacceptable. This approach allows a quick determination early in the harvest season as to the quality of silage processing so that corrective action can be taken if needed.

The amounts of grain and moisture in silage have major influences on its feed value, and both should be used to determine the dollar value of silage. The value of corn

Table 14.6
SILO TYPE AND SILAGE LOSSES

Type of Silo	Dry Matter Loss (%)
Oxygen-limiting	3–8
Concrete upright	5–15
Trench or bunker	12–25
Open stack	20–40

Table 14.7
BEST PRACTICES FOR MANAGING SILAGE STORAGE IN BUNKERS

Train employees and depend on experienced people to build silage mounds.
Monitor silage density to assure optimal pack of the mound.
Moderate fresh plant delivery rates from the field so that correct packing can be attained.
Spread forage consistently while optimizing forage push-up and packing rates.
Increase the weight of vehicles used to pack forage and increase the number of passes during packing.
Shape the surface crown so that water drains away from the mound.
Use two sheets of barrier material to seal the mound and overlap sheets by 5–6 feet to assure effective oxygen barrier.
Overlap barrier sheets so that water runs across the material without contacting plant material.
Barrier material should extend 6 feet off the silage pile.
Barrier materials should be held down by uniform placement of weights—sandbags or large tire sidewall discs are excellent weights.

Source: Adapted from multiple sources.

silage as a standing crop can be determined in the following three ways: (1) leave a few rows for later grain harvest to determine grain yield per acre, (2) harvest by hand a small plot (1/100 acre) and determine grain yield, or (3) use relationship of 1 ton of 30% dry-matter silage for each harvested foot of plant, not counting the tassel. The last option assumes an average stand, so adjustments would have to be made for light or heavy stands.

Corn grain yield estimates in silage can be used to estimate the value of the silage. For example, if it is estimated that there are 6 bushels of number 2 corn in a ton of wet silage (30–35% dry matter), then the cash corn price times the 6 bushel plus harvesting costs would give a reasonable estimate on the value of the silage (e.g., corn selling for \$3 per bu × 6 bu corn per ton = \$18 + \$4 harvest cost = \$22 for 35% dry-matter silage).

If the amount of grain in the silage is not known, a rough estimate of silage value is eight times the market value of the corn grain. If the silage is already in storage, it is worth about 10 times the market price of corn.

Sorghum silage should be priced similar to corn of the same moisture content. Forage sorghums with fairly high grain yield usually have 80–90% the value of corn silage per unit of dry matter. Sudan and Sudan–sorghum crosses or varieties, with low grain yields, may have only 65–80% the value of corn silage per unit of dry matter.

Grains and Concentrates

Concentrates are high-energy feeds (mostly feed grains and their by-products) that contain less than 20% protein and less than 18% fiber. Corn and sorghum (milo) are the most common feed grains in cattle rations, with barley, oats, wheat, and other concentrates being of lesser importance.

Table 14.8 shows the relative feeding values of several grains. The relative value, using the information in the table, is calculated in the following manner:

$$\text{Relative value of grain} \times \text{corn cost per lb} = \text{relative cost of grain per lb}$$

For example, to determine the price to pay for barley relative to corn if corn is \$3.10 per bu or 5.5¢ per lb (price of corn), then 5.5¢ per lb × 90% = 4.95¢ per lb or 48 lb per bu × 4.95¢ per lb = \$2.38 per bu for barley. Barley would be purchased if the price is less than \$2.38 per bu compared with corn at \$3.10 per bu.

The maximum level recommendations for the grains listed in Table 14.8 are for high-concentrate rations. They would be higher for oats, rye, and wheat if grain were being limit-fed in the ration.

Table 14.8
Relative Feeding Values of Various Grains for Cattle

Grain	Relative Value (% lb-for-lb basis)	Maximum Level (%)	Bushel Weight (lb)
Barley	88–90	100	48
Corn	100	100	56
Corn and cob meal	85–95	100	70
Milo	85–95	100	56
Oats	70–90	25	32
Rye	80–85	20	56
Wheat	100–105	50	60

Feedlot cattle consume most of the feed grains in the beef production system, although concentrates can be fed to cows during 1–3 months of the calving–breeding seasons, when energy requirements are highest and grains are a cost-effective source of energy.

Feed grains are harvested and stored in a relatively dry condition (less than 20% moisture); otherwise, the grain will spoil and lose its nutrients and feeding acceptability to cattle. Another factor associated with spoilage of grains is the production of mycotoxin from molds growing in feedstuffs. The major mycotoxin of concern in regard to corn is aflatoxin. These pathogens can be produced on growing plants or in storage conditions. Aflatoxin may cause reduced growth rate, feed consumption, and liver damage in cattle consuming contaminated feeds. The maximum allowable level of aflatoxin is 20 ppb for immature cattle, 100 ppb for breeding cattle, and 300 ppb for finishing cattle. Utilization of an on-farm quality assurance program that establishes a feed sampling and testing protocol is highly recommended.

Most grains are processed in one of the following ways: (1) dry process (ground or rolled), (2) steam flaking, (3) oxygen-limited reconstitution (adding water, then storing in an airtight area), (4) early harvested (ground and ensiled in trench or airtight silos), (5) early harvested with acid treatment of grain, and (6) pelleted. Corn is the only grain that can be fed satisfactorily in whole form to feedlot cattle. Whole corn is usually fed to feedlot cattle on high (80–90%) concentrate rations. It is essential that milo be processed to improve its digestibility. Other advantages in processing grains are improved gain, efficiency of gain, and increased feed intake by cattle. Using high-moisture grains reduces field losses during the time of harvesting.

Increased animal performance must be tempered with energy costs in processing. Currently, steam flaking is the most intensive grain-processing alternative; however, it is feasible only in large-sized feedlots (20,000 head and above). The primary benefit of steam flaking is that it increases the feed value of corn by 18%. The critical factors of the process that impact the quality of steam-flaked corn include steam chest temperature, duration of steaming, the corrugation, gap, and tension of the rollers. Steam-flaked corn should be fed relatively soon after leaving the mill (Figure 14.5) or allowed to dry before being placed into storage.

Caution should be exercised if wheat is fed by keeping it below 40% of the total ration. It can cause compaction if ground too fine and acidosis if cattle are not accustomed to it. However, much higher levels of steam-flaked wheat have been successfully fed by increasing the roughage and ionophore levels in the finishing ration.

Protein Supplements

Protein supplements are dry or liquid feedstuffs that contain 20% or more protein. They may be of plant origin (soybean meal, cottonseed meal, linseed meal, corn gluten

Figure 14.5
High technology feed milling allows more efficient and cost-effective rations to be designed and delivered.

meal, etc.) or animal origin (blood meal, dried skim milk, etc.). Because of concerns about BSE, the feeding of ruminant by-products to cattle is prohibited.

Molasses is commonly used as the base for liquid supplements. Soybean meal is the most widely used protein supplement in the United States. Cottonseed meal, however, is commonly used in the South.

Urea (Figure 14.6) is not a protein supplement but it is a source of nitrogen (42–45%) for protein synthesis by rumen bacteria. Urea has a protein equivalent of

Figure 14.6
Urea is used as a supplemental protein source.

262–281%, since 1 lb of urea contains as much nitrogen as 2.62–2.81 lb of protein. Urea works well with other plant proteins to lower the cost of protein in the ration. It works best in feedlot rations that are relatively high in energy because rumen organisms need a readily available energy source to convert the urea nitrogen into microbial protein. Urea should not be utilized as a supplement for low-quality forages, as assuring a consistent intake of urea is difficult to achieve.

General guidelines for using urea or other sources of nonprotein nitrogen (NPN) are as follows: (1) a maximum of one-third of total nitrogen in the ration should come from urea; (2) urea should not represent more than 10–15% of a typical protein supplement; (3) it should not be more than 1% of a diet or 3% of a concentrate mixture; and (4) it should not be more than 5% of a supplement when used with low-grade roughages. Urea poisoning can occur when its level in the ration is too high or when it is not mixed properly.

Vitamin Supplements

The only vitamin supplement of general practical importance to beef cattle is vitamin A (precursor is carotene). Vitamin A or carotene content in feeds depends largely on maturity, harvest conditions, and length and conditions of storage. Where roughages being fed contain a bright green color or are being fed as immature fresh forage (e.g., pasture), there will probably be sufficient vitamin A value in them to meet the animal's requirement. If there is a question about vitamin A availability, a feed analysis may be needed or vitamin A supplementation is usually economical. Vitamin A is generally included in all commercial protein supplements for growing–finishing cattle.

Mineral Supplements

Mineral supplements can be simple or complex, the latter occurring when several trace minerals are included. In most areas of the United States, mineral supplementation of salt, phosphorus, and calcium is adequate. Their level in the soil on which feeds are grown or to other environmental factors largely determines trace minerals needed in feeds. However, there are areas in the United States where mineral deficiencies or toxicities exist (see Figure 14.7). In these areas, additional mineral supplementation should be undertaken and methods to reduce toxicity effects should be employed. Trace minerals, if necessary, are usually supplied in a trace mineralized salt.

Determining the appropriate mineral supplementation program requires a critical analysis of both the nutritional needs of cattle and the cost of delivering appropriate mineral supplements. Water and forage samples should be analyzed on at

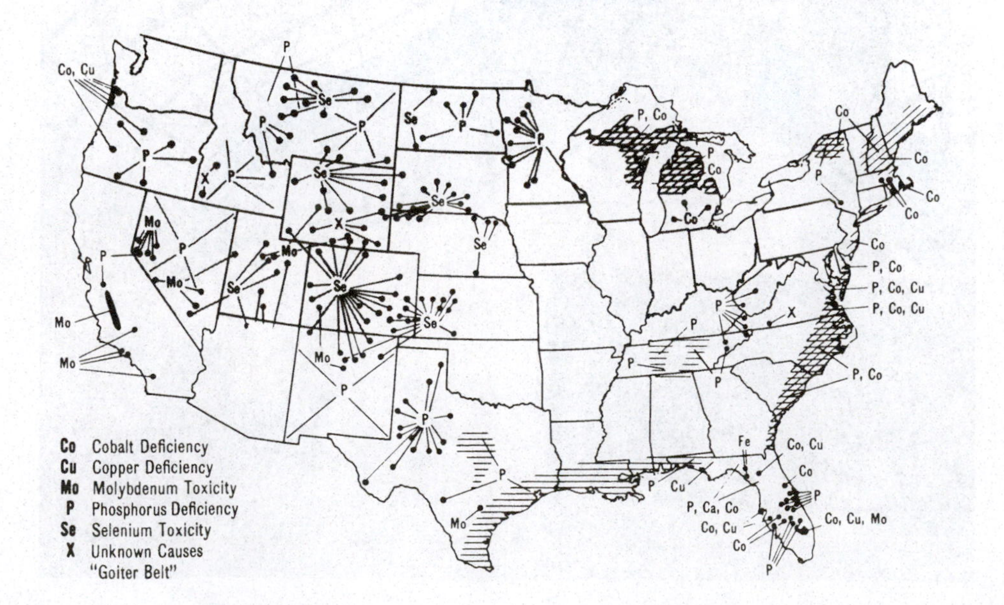

Figure 14.7
Areas of mineral deficiency or toxicities.
Source: USDA.

Table 14.9
Disorders Associated with Minerals in Beef Cattle

Mineral	Disorder
Calcium	Milk fever (hypocalcemia)—paralysis, circulatory collapse, coma and death; most cases within 24 hr. following calving
Magnesium	Grass tetany (hypomagnesemia)—staggering, incoordination, exitability, profuse salivation; typically occurs in lactating cattle grazing lush pasture, especially small grains (oats, barley)
Manganese	Deficiencies associated with suppressed estrus expression, reduced fertility, and poor skeletal growth
Copper	Deficiencies associated with increased retention of placentas, retarded growth, poor hair coat, diminished fertility
Zinc	Deficiencies associated with reduced immune response, poor fertility, reduced feed efficiency and growth; increased incidence of foot rot
Selenium	Toxicity—alkali disease, blind staggers, death
Cobalt	Deficiency prohibits formation of vitamin B_{12} and leads to emaciation and anemia

Source: Adapted from multiple sources.

least an annual basis to provide trend lines and to provide a starting place to identify potential problems. Late fall is a good time to collect the samples for spring calving herds to allow sufficient time to make adjustments, if needed.

Soil tests are also useful in helping to monitor mineral issues. Minerals may be antagonistic to one another. For example, high levels of molybdenum may interfere with copper absorption.

The mineral requirements for cattle vary with their stage of production. The two months leading up to calving as well as the same time period following parturition is a critical time for mineral supplementation. The method of mineral feeding is also critical to assure that each animal in the herd has access to and consumes appropriate supplement levels.

Bioavailability, or ease of absorption of mineral sources, is also an important consideration. For example, the bioavailability of calcium from dicalcium phosphate is high, but it is low from hay. The supplementation of copper as cupric chloride or cupric sulfate is highly available, while cupric oxide is quite low in bioavailability.

The failure to provide appropriate (neither too high nor too low) mineral levels can lead to a number of disorders (Table 14.9).

NUTRIENT REQUIREMENTS OF BEEF CATTLE

In formulating rations, producers must know the nutrient requirements of beef cattle in addition to knowing the nutrient composition of feedstuffs. The National Research Council (NRC) publications are the generally accepted standards that identify the nutrient requirements of beef cattle. Table 14.10 illustrates the nutrient requirements of growing and finishing cattle within a few selected weights and levels of productivity. Table 14.11 provides an overview of the requirements of breeding cattle. More extensive tables of nutrient requirements and feed compositions can be found in the National Research Council's *Nutrient Requirements of Beef Cattle.*

RATION FORMULATION

Rations should be formulated or balanced to meet the nutrient requirements of animals in different stages of production. Another important consideration is the cost of the ration with an assurance that it is palatable and will not cause any serious digestive disturbances or toxic effects.

Table 14.10
Nutrient Requirements for Growing and Finishing Cattle

Live Weight (lb)	TDN (% DM)	Daily Gain (lb)	Dry Matter Intake (lb/day)	Protein Intake (lb)	Crude Protein (%)	NEm (Mcal/lb)	NEg (Mcal/lb)	Ca (%)	P (%)
1,100 lb at finishing (28% body fat—feedlot steers and heifers) or maturity (replacement females)									
605	50	0.7	16.3	1.2	7.2	0.45	0.20	0.22	0.13
	70	2.9	16.9	2.1	12.7	0.76	0.48	0.49	0.24
	90	4.0	14.7	2.6	17.8	1.04	0.72	0.72	0.34
660	50	0.7	17.5	1.2	7.1	0.45	0.20	0.21	0.13
	70	2.9	18.0	2.2	12.3	0.76	0.48	0.45	0.23
	90	4.0	15.7	2.7	17.1	1.04	0.72	0.66	0.32
715	50	0.7	18.5	1.3	6.9	0.45	0.20	0.20	0.13
	70	2.9	19.1	2.2	11.5	0.76	0.48	0.42	0.21
	90	4.0	16.7	2.6	15.9	1.04	0.72	0.61	0.30
770	50	0.7	19.6	1.3	6.8	0.45	0.20	0.19	0.12
	70	2.9	20.2	2.2	10.9	0.76	0.48	0.39	0.20
	90	4.0	17.6	2.6	14.8	1.04	0.72	0.56	0.28
825	50	0.7	20.6	1.3	6.6	0.45	0.20	0.19	0.12
	70	2.9	21.3	2.2	10.3	0.76	0.48	0.37	0.19
	90	4.0	18.6	2.6	13.9	1.04	0.72	0.52	0.26
880	50	0.7	21.7	1.4	6.5	0.45	0.20	0.19	0.12
	70	2.9	22.4	2.2	9.8	0.76	0.48	0.34	0.18
	90	4.0	19.5	2.5	13.1	1.04	0.72	0.48	0.25
1,300 lb at finishing (28% body fat—feedlot steers and heifers) or maturity (replacement heifers)									
715	50	0.8	18.5	1.3	7.3	0.45	0.20	0.22	0.13
	70	3.2	19.1	2.5	13.0	0.76	0.48	0.49	0.24
	90	4.5	16.7	3.0	18.3	1.04	0.72	0.72	0.34
780	50	0.8	19.8	1.4	7.1	0.45	0.20	0.21	0.13
	70	3.2	20.4	2.5	12.1	0.76	0.48	0.45	0.23
	90	4.5	17.8	3.0	16.9	1.04	0.72	0.66	0.32
845	50	0.8	21.0	1.4	6.9	0.45	0.20	0.21	0.13
	70	3.2	21.7	2.5	11.4	0.76	0.48	0.42	0.22
	90	4.5	18.9	3.0	15.7	1.04	0.72	0.60	0.30
910	50	0.8	22.2	1.5	6.7	0.45	0.20	0.20	0.13
	70	3.2	22.9	2.4	10.7	0.76	0.48	0.39	0.20
	90	4.5	20.0	2.9	14.6	1.04	0.72	0.56	0.28
975	50	0.8	23.4	1.5	6.6	0.45	0.20	0.20	0.13
	70	3.2	24.1	2.5	10.2	0.76	0.48	0.37	0.19
	90	4.5	21.0	2.9	13.7	1.04	0.72	0.52	0.26
1040	50	0.8	24.5	1.6	6.5	0.45	0.20	0.19	0.13
	70	3.2	25.3	2.4	9.6	0.76	0.48	0.34	0.19
	90	4.5	22.1	2.8	12.9	1.04	0.72	0.48	0.25

Source: Adapted from *Nutrient Requirements of Beef Cattle.* (1996). Washington, DC: National Research Council and Feedstuffs, 2016.

Rations should be balanced for energy (using TDN, ME or NEm, and NEg), crude protein, vitamin A, calcium, and phosphorus. Salt is usually provided on a free-choice basis or mixed into the ration. An assessment should be made to determine whether trace minerals are needed.

PRICING FEEDSTUFFS

Given the significant influence of feed cost on the overall profitability of beef production, developing balanced rations at the least cost is an important management strategy. Commodity feeds may be priced on "as is" basis (water is included in the weight)

Table 14.11
Daily Nutrient Requirements for Breeding Cattle (Heifers, Cows, and Bulls)

Weight (lb)	Daily Gain (lb)	Dry Matter Consumption (lb)	Total Crude Protein (lb)	TDN (lb)	ME (Mcal)	Ca (gm)	P (gm)	Vitamin A (1,000 IU)
Growing Heifer Calves and Yearlings (see Table 14.10)								
Two-Year-Old Heifers—Last Third of Pregnancy								
1,000[1]	0.73	20.7	1.8	11.7	20.9	29	22	31
1,100[1]	0.80	22.3	1.9	12.6	22.5	31	22	33
1,200[1]	0.88	23.7	2.0	13.3	24.2	33	24	34
1,300[1]	0.95	25.2	2.1	14.1	25.6	34	25	35
1,400[1]	1.02	26.6	2.2	14.8	26.9	37	27	37
Cows Nursing Calves—Average Milking Ability[2]—First 3–4 Months Postpartum								
1,000	0	23	1.9	12.5	20.9	24	17	36
1,200	0	26	2.1	13.9	23.4	27	19	41
1,400	0	29	2.3	15.4	25.7	30	21	45
Cows Nursing Calves—Superior Milking Ability[3]—First 3–4 Months Postpartum								
1,000	0	25.4	2.6	14.9	24.9	35	24	37
1,200	0	28.4	2.8	16.4	27.3	37	25	42
1,400	0	31.3	3.0	17.8	29.7	40	27	47
Dry Pregnant Mature Cows—Middle Third of Pregnancy								
1,000	0	16.7	1.2	8.2	13.4	14	14	23
1,200	0	19.5	1.4	9.5	15.6	17	17	26
1,400	0	23.3	1.6	11.4	18.7	21	21	30
Dry Pregnant Cows—Last Third of Pregnancy								
1,000	0.9	21.0	1.6	10.9	18.3	23	14	25
1,200	0.9	24.1	1.8	12.6	20.9	27	17	28
1,400	0.9	27.0	2.1	14.2	23.9	32	21	32
Bulls—Growth and Maintenance (Moderate Activity)								
1,400	1.0	26.8	2.0	15.0	24.6	26	23	48
1,600	1.0	29.7	2.2	16.6	27.2	29	26	53
1,800	0	28.9	2.0	14.0	23.0	27	27	51
2,000	0	31.3	2.1	15.2	24.5	30	30	55
2,200	0	33.6	2.3	16.3	26.7	33	33	60

[1]Mature weight potential.
[2]10 lb of milk per day.
[3]20 lb of milk per day.

or on dry-matter (DM) basis (weight of water is removed). Pricing feeds on a DM basis is a better approach to understanding the real volume of the feedstuffs being purchased. A ration that is priced on an "as is" basis can be converted to DM price by dividing the "as is" price by the percentage of dry matter in the feed. For example, if the "as is" price is $5.80 per cwt with a DM of 72%, then the price on a DM basis is equal to $8.05 per cwt ($5.80/0.72). In other words, high-moisture feeds priced on an "as is" basis tend to be overpriced. Feed pricing should be conducted on an equal DM basis when comparing various feeds for inclusion in a ration.

Conducting appropriate feed testing is also critical to avoiding pricing problems when a feed differs from the promised DM. For example, assume that 100 tons of hay

(advertised as 990% DM) can be purchased for $80 per ton. However, the feed is actually 85% DM. The following formula can be used to make the necessary price adjustment:

$$\frac{\textit{Price—Nutrient A}}{\text{Dry Matter A}} = \frac{\textit{Price—Nutrient B}}{\text{Dry Matter B}}$$

Step 1: $\frac{\$80/\text{ton}}{90\%\ \text{DM}} = \frac{x}{85\%\ \text{DM}}$

Step 2: $x = \frac{\$80 \times 85\%}{90\%\ \text{DM}}$

Step 3: $x = \frac{\$68}{90\%}$

Step 4: $x = \$75.56/\text{ton}$

Given the discrepancy between the advertised DM and the actual DM, the feed is only worth $75.56 per ton due to the variation in moisture content. If the sale occurs on the 5% DM discrepancy, the buyer is actually purchasing 5 ton of water at the $80/ton price.

Balancing Cow Rations Using TDN

TDN is a satisfactory energy basis for balancing rations for cows. Table 14.12 shows an example of balancing a cow's ration where silage is the primary source of feed. The calculations used in the table are as follows:

Line a: TDN requirement (Table 14.11).

Line b: Calculate lb of corn silage dry matter (DM) needed by dividing lb of TDN required by % TDN in corn silage (13.9 ÷ 0.72 = 19.3).

Line c: Calculate lb of "wet" corn silage needed by dividing lb of corn silage DM by % DM in corn silage (19.3 ÷ 0.34 = 56.8 lb).

Line d: Crude protein (CP) requirement from Table 14.11.

Line e: Calculate lb of CP supplied by corn silage by multiplying lb of corn silage DM fed (line b) by % CP in corn silage (19.3 × 0.08 = 1.54).

Line f: Calculate CP deficiency by subtracting lb of CP supplied on *line e* from the CP requirement on *line d* (2.10 − 1.54 = 0.56).

Line g: Calculate lb of CP supplement needed by dividing lb of CP deficiency by % CP in the supplement (0.56 ÷ 0.35 = 1.60).

Table 14.12
Beef Cow Ration with Corn Silage and Protein Supplement

1,200-lb Cow (average milk, early lactation)	Amount
a. TDN required (lb/day)	13.9
Corn silage needed:[1]	
b. lb of silage (dry matter [DM]/day)	19.3
c. lb of wet silage/day	56.8
Crude protein (CP) supplement needed:[2]	
d. CP required (lb/day)	2.10
e. CP supplied by corn silage (lb/day)	1.54
f. CP deficiency (lb/day)	0.56
g. lb of protein supplement/day	1.60

[1]Composition of corn silage taken from Table 14.4: dry matter, 34%; TDN, 72%; and crude protein, 8%.
[2]Assumes a protein supplement of 35% CP.

This ration is now balanced for TDN and CP; however, it does not meet the minimum DM requirement of 26 lb (Table 14.11). To meet the DM requirement, one would increase the silage fed to approximately 77 lb. Fulfilling the DM requirement is not a major concern in feeding cows as long as energy, protein, mineral, and vitamin requirements are met. Salt, phosphorus, and possibly vitamin A would be supplied on a free-choice basis to meet the cow's nutrient requirements.

It is not uncommon to have DM vary in the same silage crop. Amount of wet silage fed should be adjusted as moisture content changes. If moisture changes are not considered, then critical nutrients will be oversupplied or undersupplied.

Using the same cow requirements as shown in Table 14.12, the CP deficiency may be filled using a legume such as alfalfa hay. The Pearson Square method can be used to determine the amount of alfalfa hay needed, along with corn silage, to meet the DM and CP requirements. An example using Pearson's Square follows:

On the left-hand side of the square are the CP percentages of corn silage and alfalfa hay taken from Table 14.4. The 8.1% in the middle of the square is the CP requirement (expressed in %) of the 1,200-lb cow. This is calculated as follows:

$$\frac{2.10 \text{ lb CP requirement}}{26 \text{ lb dry} - \text{matter requirement}} = 8.1\%$$

On the right-hand side of the square are parts of corn silage and parts of alfalfa hay of the total ration. These parts are obtained by subtracting, diagonally, the smallest % from the largest %. Table 14.13 permits a check of the calculations.

Least-cost rations can be calculated using a computer program that can analyze many feed sources and their nutrient composition and cost. Some rough approximations of least-cost rations can be hand-calculated by determining energy and protein costs on several readily available feeds. Because energy and protein costs make up the largest part of the total nutrient costs, using the information in Table 14.14 and Table 14.15 can be helpful. The information in these tables is for example purposes only. Individual producers should substitute feeds common to their area, the energy and protein compositions of the feeds, and the current feed cost.

Formulating Feedlot Rations

The NE system is typically used when formulating feedlot rations (Figure 14.8). In example 1, a ration is evaluated in terms of adequacy in NE and crude protein for a 700-lb medium frame steer expected to gain approximately 3 lb per day during a 140-day finishing period. The requirements for this steer per NRC

Table 14.13
Checking the Pearson Square Method Calculations

	Part of Ration		Protein	
Feed	%	Lb	%	Lb
Corn silage	99	25.7	8.0	2.05
Alfalfa hay	1	0.3	19.0	0.05
Total or average	100	26.0		2.10

Table 14.14
Comparative Energy Costs Using TDN and NEg

Feedstuff	Cost/Ton ($)	% TDN	Cost/lb TDN ($)	NEg (Mcal/lb)	Cost/Mcal Neg ($)
Alfalfa hay[1]	110	59	0.093	0.28	0.196
Barley (48 lb/bu)	90	84	0.054	0.61	0.074
Corn (56 lb/bu)	90	87	0.055	0.64	0.070
Corn silage (28% dry matter)	25	72	0.017	0.47	0.026
Cottonseed meal	165	78	0.106	0.54	0.153
Dehydrated alfalfa	140	61	0.115	0.31	0.226
Soybean meal	205	84	0.122	0.61	0.168
Wheat (60 lb/bu)	117	88	0.066	0.65	0.090

[1]Example of calculation: Alfalfa hay at $\frac{\$110/\text{ton}}{2{,}000} = \$0.055/\text{lb}$; TDN $= \frac{0.055}{0.59} = \$0.093/\text{lb}$; NEg $= \frac{0.055}{0.28} = \$0.14/\text{Mcal}$.

Table 14.15
Calculation of Value of Protein Supplements on the Basis of Crude Protein

Supplement	% Crude Protein	Cost/Ton ($)	Cost/Lb Crude Protein ($)
Alfalfa hay	19	110	0.289
Cottonseed meal	45	165	0.183
Dehydrated alfalfa	19	140	0.368
Soybean meal[1]	50	205	0.205

[1]Example of calculation: Soybean meal at $\frac{\$205/\text{ton}}{2{,}000} = \$0.1025/\text{lb}$; CP cost $= \frac{0.1025}{0.59} = 0.205/\text{lb}$.

Figure 14.8
A mixed feedyard ration delivered to the bunk.

requirements are 19.1 lb DM, 2.2 lb CP, 5.85 Mcal, NEm, and 6.35 Mcal, NEg. Table 14.16 shows pounds of feeds being fed and their conversion from an "as-fed" to a "dry-matter" basis. The conversion is calculated by multiplying the pounds "as fed" by its dry-matter content (Table 14.4).

Table 14.16
FEED FED TO A 700-LB STEER

Feed	Lb Fed	Dry Matter %	Dry Matter lb
Corn silage (mature)	9.0	34.0	3.06
Corn	10.0	88.0	8.80
Chopped alfalfa hay	2.0	90.0	1.80
Beet pulp	1.5	91.0	1.36
32% protein supplement	1.0	90.0	0.90
Total			15.92

Table 14.17 shows calculations for determining the megacalories for maintenance and gain and amount of protein in the ration (e.g., for corn silage, NEm = 3.06 × 0.75 = 2.29; NEg = 3.06 × 0.47 = 1.13; and CP = 3.06 × 0.08 = 0.24). Values are obtained from Table 14.4.

The questions posed in example 1 are answered in the following steps:

Step 1: Determine NEm and NEg per pound of feed:

$$\text{From Table 14.17, the NEm per lb ration} = \frac{13.55}{15.92} = 0.85$$

$$\text{From Table 14.17, the NEg per lb ration} = \frac{8.76}{15.92} = 0.55$$

Step 2: Pounds of feed needed to meet the maintenance requirement of the 700-lb steer:

$$5.85 \text{ Mcal} \div 0.85 = 6.88 \text{ lb}$$

Step 3: Pounds of feed needed to meet the requirement for 3.0 lb daily gain:

$$6.35 \text{ Mcal} \div 0.55 = 11.54 \text{ lb}$$

Step 4: Total lb of feed the steer must eat to gain 3.0 lb per day:

$$6.88 \text{ lbs} + 11.54 \text{ lbs} = 18.4 \text{ lb}$$

Conclusion: If the steer has a capacity to eat 15.9 lb of dry matter of the ration fed, then a 3.0 lb per day gain will not be realized. The energy density in the ration could be increased by substituting corn grain for corn silage on a lb-for-lb basis until the desired gain is realized. The 1.73 lb of protein supplied in the 15.9 lb of dry matter fails to supply the 2.2-lb protein requirement.

Table 14.17
CALCULATING THE NEm AND NEg IN THE RATION

Feed	Lb Dry Matter	NEm (Mcal)	NEg (Mcal)	Lb Protein
Corn silage	3.06	2.29	1.44	0.24
Corn	8.80	8.45	5.63	0.79
Chopped alfalfa hay	1.80	1.06	0.56	0.34
Beet pulp	1.36	1.07	0.68	0.12
32% protein supplement[1]	0.90	0.68	0.45	0.29
Total	15.92	13.55	8.76	1.78

[1]32% protein supplement contains 0.75 Mcal/lb (NEm) and 0.5 Mcal/lb (NEg).

Example 2 asks how the NE system can be used to predict the amount of daily gain if the feed consumption is known. Using the content of example 1, what would be the expected gain if the 700-lb steer consumed 15.9 lb of the ration in Table 14.16?

Step 1: Pounds of feed needed to meet daily maintenance requirement = 6.88 lb (Step 2 in example 1).

Step 2: Pounds of feed remaining for gain:

15.9 lb − 6.88 = 9.02 lb

Step 3: Mcal of NEg supplied from the remaining amount of feed:

9.02 lb × 0.55(NEg per lb in ration) = 4.96 Mcal

Step 4: Daily gain expected from 4.96 Mcal of NEg for 700-lb steer (NRC requirements):

4.96 Mcal produces approximately 2.4 lb of gain

In example 3, suppose that the genetic ability of the 700-lb steer only allowed the daily consumption of 12.5 lb of the ration. What would be the expected gain? The same steps as in example 2 would be followed:

Mcal of NEg for gain = 3.09 Mcal. Expected gain from 3.42 Mcal (NRC requirements) = approximately 1.55 lb per day.

Most feedlot rations are balanced for vitamin A, calcium, and phosphorus. Potassium becomes important as the level of concentrate increases or when NPN is substituted for intact protein. Sulfur also becomes important when the level of NPN increases in the ration. Most purchased protein supplements have trace minerals added. It is important to balance trace minerals for optimum performance.

Adjusting Rations for Weather Changes

The published nutrient requirements and predicted gains of cattle are the results of research studies in which the animals were protected from environmental extremes. However, environmental factors, especially temperature, can alter both performance and nutrient requirements. Thus, cattle producers should be aware of critical temperatures that affect cattle performance, and then consider changing their feeding program if economics so dictate.

The thermoneutral zone (TNZ) is the range in effective temperature where the rate and efficiency of performance is maximized. Critical temperature is the lower limit of the TNZ and is typified by the ambient temperature below which the performance of cattle begins to decline as temperatures become colder. Figure 14.9 illustrates that

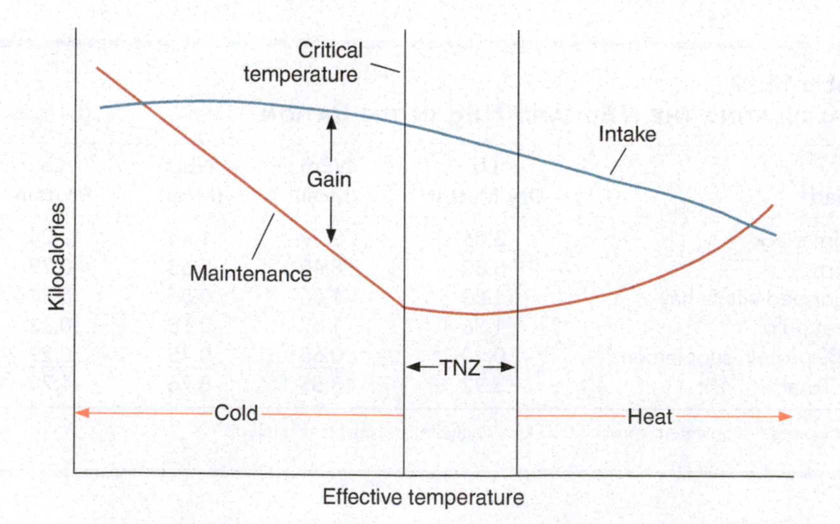

Figure 14.9
Effect of temperature on rate of feed intake, maintenance energy requirement, and gain.
Source: Adapted from Ames (1980).

Table 14.18
Wind Chill Factors for Cattle with a Winter Coat

Wind Speed (mph)	Temperature (°F)												
	−10	−5	0	5	10	15	20	25	30	35	40	45	50
Calm	−10	−5	0	5	10	15	20	25	30	35	40	45	50
5	−16	−11	−6	−1	3	8	13	18	23	28	33	38	43
10	−21	−16	−11	−6	−1	3	8	13	18	23	28	33	38
15	−25	−20	−15	−10	−5	0	4	9	14	19	24	29	34
20	−30	−25	−20	−15	−10	−5	0	4	9	14	19	24	29
25	−37	−32	−27	−22	−17	−12	−7	−2	2	7	12	17	22
30	−46	−41	−36	−31	−26	−21	−16	−11	−6	−1	3	8	13
35	−60	−55	−50	−45	−40	−35	−30	−25	−20	−15	−10	−5	0
40	−78	−73	−68	−63	−58	−53	−48	−43	−38	−33	−28	−23	−18

the maintenance energy requirement increases more rapidly during cold weather than does rate of feed intake. This results in a reduction of gain and more feed required per pound of gain, which typically causes cost per pound of gain to be higher.

Effective ambient temperatures below the lower critical temperature (below the TNZ) constitute cold stress and those above the TNZ constitute heat stress. The term effective ambient temperature is an index of the heating or cooling power of the environment in terms of dry bulb temperature. It includes any environmental factor—such as radiation, wind, humidity, or precipitation—that alters environmental heat demand. Not all effective ambient temperatures have been calculated for cattle, though some combined environmental factors are available. The wind chill index, for example, is shown in Table 14.18. An example in the table shows that with a temperature of 20°F and a wind speed of 30 mph, the effective ambient temperature is –16°F.

The low critical temperature for cattle depends on how much insulation is provided by the hair coat, whether the animal is wet or dry, and how much feed the cow consumes. Table 14.19 shows lower critical temperatures for beef cattle. For example, a cow being fed a maintenance ration may have a low critical temperature of 32°F when dry, but a low critical temperature of 60°F when wet.

The coldness of a specific environment is the value that must be considered when adjusting rations for cows. Coldness is simply the difference between effective temperature (wind chill) and low critical temperature. Using this definition for coldness instead of using the temperature reading on an ordinary thermometer helps explain why the wet, windy days in March may be colder for a cow than the extremely cold but dry, calm days of January.

The major effect of cold on the nutrient requirements of cows is increased need for energy, which usually means the total amount of daily feed must be increased.

Table 14.19
Estimated Low Critical Temperatures for Beef Cattle

Coat Description	Critical Temperature (°F)
Summer coat or wet	59
Fall coat	45
Winter coat	32
Heavy winter coat	18

Feeding tables recommend that a 1,200-lb cow receive 16.5 lb of good mixed hay to supply energy needs during the last one-third of pregnancy. How much feed should the cow receive if she is dry and has a winter hair coat but the temperature is 20°F with a 15 mph wind? The coldness is calculated by subtracting the wind chill or effective temperature (4°F) from the cow's low critical temperature (32°F). Thus, the magnitude of coldness is 28°F. A rule of thumb (more detailed tables are available) is to increase the amount of feed 1% for each degree of coldness or to increase TDN by 1 pound for every 5 degrees below 0. A 28% increase of the 16.5 lb (original) requirement would mean that 21.1 lb of feed must be fed to compensate for the coldness. This example is typical of many feeding situations; however, if the cow were wet, the same increase in feed would be required at 31°F wind chill (28°F of coldness). Similar relationships exist for growing–finishing cattle.

During cold stress, managers can alter schedules to reduce the impact of extreme temperatures by avoiding processing or handling cattle during the coldest time of day (5 A.M. to 10 A.M.) and by feeding late in the day to assist in heat production due to digestive activity during the night.

Management of heat stress is discussed in Chapter 7.

COW-CALF NUTRITION

Development of a nutritional management plan for a cow-calf herd should be based on conditions specific to a particular enterprise. The following information provides baseline information and a summary of the key principles involved in effective nutritional management of beef cows.

Nutritional requirements for TDN, CP, and NE change—as a cow transitions through the process of gestation and as mature cow weight increases, there is a concurrent increase in feed requirements (Figures 14.10 and 14.11). Failure to meet the nutritional requirements of the cow during gestation not only affects her but also her fetus. The concept of fetal programming is based on research that shows that calves born to dams with poor nutrition have poorer productivity and growth throughout their lifetimes and suffer from higher rates of disease. These impacts affect feedlot performance, marbling levels, and lifetime reproductive rates.

Table 14.20 illustrates the changes in nutritional requirements as a 1200-pound cow (approximates the industry average) moves through an annual production cycle. Note the variation that occurs resulting from productive status of the cow—lactation,

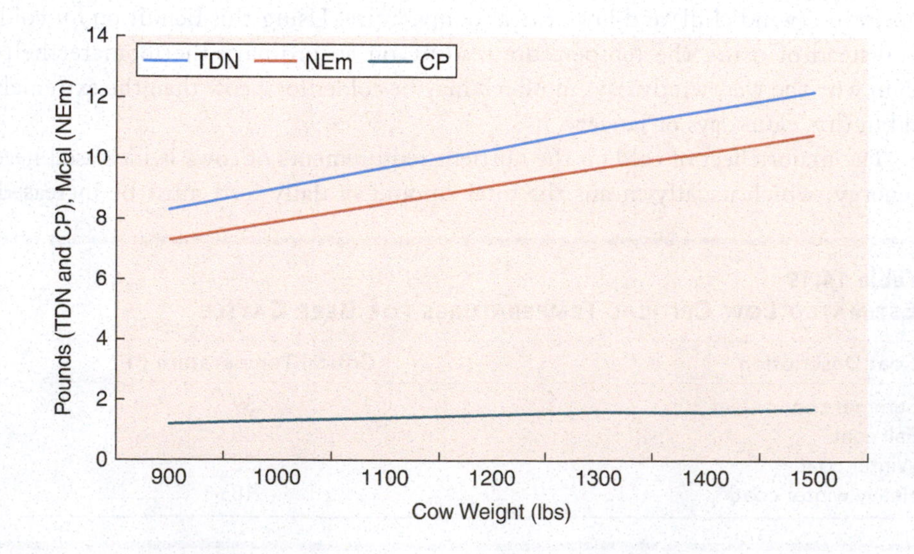

Figure 14.10
Nutritional requirements of gestating cows in the middle third of pregnancy.

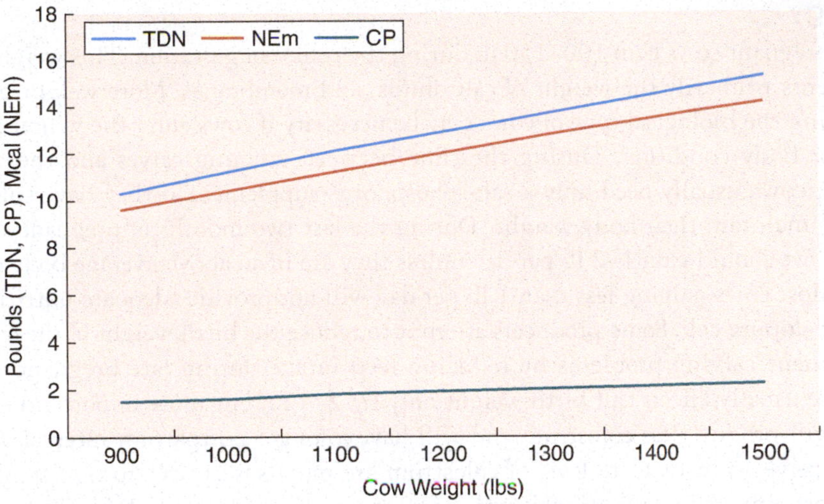

Figure 14.11
Nutritional requirements of gestating cows in the last third of pregnancy.

Table 14.20
Nutritional Requirements of 1,200-lb Mature Cows at Various Months Post Calving

	Months Since Calving				
Requirement	**1–3**	**4–6**	**7–8**	**9–10**	**11–12**
Net Energy (Mcal/day)					
Maintenance	10.3	10.3	8.5	8.5	8.5
Lactation	5.2	3.1	0	0	0
Pregnancy	0	0.09	0.3–0.6	1.2–2.1	3.4–5.4
Total	15.3	13.4	9.0	10.2	13.0
Metabolizable Protein (lb/day)	1.77	1.45	0.97	1.08	1.36
Calcium (g/day)	34	37	26	24	28
Phosphorus (g/day)	24	19	13	15	18

Source: Adapted from multiple sources.

mid-gestation, and late gestation. For example, mineral requirements for calcium and phosphorus are highest during lactation while energy requirements are highest at peak lactation and then rise again as gestation advances to the last trimester.

Cows that are grazing or are being fed harvested forages will have varying levels of intake depending on the type and quality of forage available to them. Forages with higher levels of digestibility have higher levels of consumption (Table 14.21). When forages are of moderate to low quality, meeting nutritional requirements is impacted not only by the quality of feed but also the level of intake.

Table 14.21
Estimated Intake and Digestibility of Various Forage Classes

Forage Class	Intake (% of body weight)	Digestibility (%)
Pasture (lush)	2.75–3.5	65 and up
Pasture (moderate quality)	2.5–3.2	60
Grass Hay (good quality)	2.0–2.5	55
Grass Hay (moderate quality)	1.5–2.0	50
Grass Hay (low quality)	1.0–1.5	40
Straw	<1.0	35

Energy

Most pregnant cows gain 100–150 lb during the course of gestation. This weight gain represents primarily the weight of calf, fluids, and membranes. More weight gain or adjusting the biological type of cows may be necessary if cows enter the winter period in poor body condition. During the time between weaning calves and the winter period, cows usually need only a salt–phosphorus supplement if they have sufficient feed to maintain their body weight. During the last two months of pregnancy, however, cows should gain 1–2 lb per day unless they are in an above-average body condition. Most cows gaining less than 1 lb per day will not provide adequate nutrition for the developing calf. Some producers attempt to reduce the birth weight of the calf and subsequent calving problems by reducing feed intake during late pregnancy. This practice usually affects calf birth weight only by 2–4 lb, but more importantly, these cows will produce less colostrum and will have a longer postpartum interval. Calves that receive an inadequate level of colostrum are more susceptible to disease. One of the most important nutritional management practices is to see that calves receive adequate quantities of colostrum within a few hours after birth. Cows in poor condition (<BCS 5) are less able to produce an adequate level of immunoglobulins in their colostrum and thus their calves are more susceptible to disease.

The first 90 days after calving is the most important nutritional period for the brood cow as she experiences peak lactation while preparing for subsequent rebreeding (Figure 14.12). Note that differences in both mature weight and lactation levels affect nutritional requirements. Cows that are highly variable in mature size and milk production should be sorted into management groups to assure that nutritional needs are met while avoiding both overfeeding and underfeeding part of the herd.

Most calves are born during the late winter through the spring months, when most pastures are at their lowest point nutritionally. Supplemental feed is provided at this time so that cows can produce an adequate supply of milk and still gain body weight. Cows that do not gain weight prior to breeding and during the breeding season are most likely to have poor reproductive performance. Producers may want to weigh a sample of their cows during this period to assess that weight gains are occurring and adjust the amount of feed being provided if necessary.

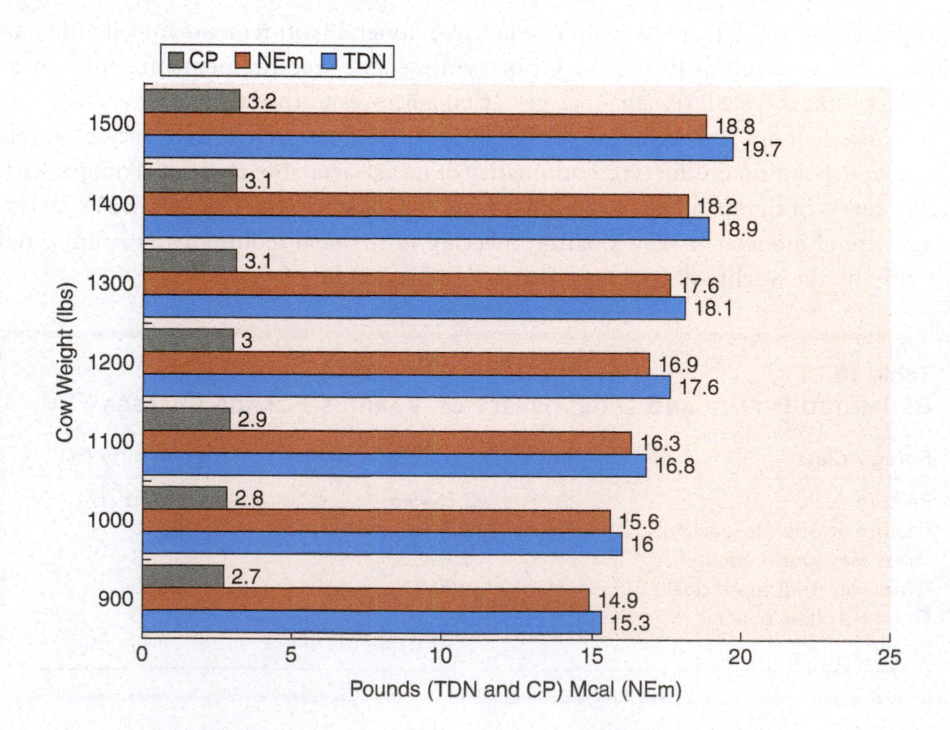

Figure 14.12
Nutritional requirements of lactating cows 90 days postpartum.

A logical management practice involves sorting cows into groups prior to calving through part of the breeding season and providing levels of feed consistent with cow age and body condition. First- and second-calf females and thin cows need to be separated from other cows to assure the feeding levels needed for high reproductive performance. Feed that has been evaluated to be of higher quality should be targeted to younger females and thin cows. Lower quality feeds are then fed to non-lactating, mid-gestation older females in adequate to good condition.

Cow size should be adjusted to fit existing environmental conditions. Moderate-sized and moderate-milking cows are highly profitable in most cow-calf operations. Cow size and milk production may be less in arid and semiarid regions.

Summer forage, water, and salt will provide all of the cow's nutrient requirements if the nutrient sources are available in ample supply. Winter pasture will provide the cow's energy needs if it is present in adequate amounts and is not covered with snow. When adequate amounts of feed are not present, the most cost-effective source of supplemental energy should be provided.

Protein

Figure 14.13 illustrates that during spring and early summer, protein is adequate in most range and pasture grasses. The pregnant cow needs 6.5% protein in her ration; 9% is needed for a lactating cow, and 11% for a growing calf or yearling. During the winter period, protein in mature grass frequently decreases to 4–5%.

Protein is usually more than adequate when feeding alfalfa hay. When grass hay is fed and questions arise about meeting the cow's requirements, a feed analysis should be made. Corn stalks, straw, and other aftermath feeds provide a cheap source of energy, but as a rule they need to be supplemented with protein, especially during the latter part of the grazing period. Underfeeding protein is common and results in decreased performance. Overfeeding protein will not harm animals, but the practice is extremely uneconomical. When a protein supplement is necessary, the cheapest source available should be fed. Often, cottonseed cake, alfalfa hay, or a feed grain is a more economical choice than a commercially prepared supplement. Supplements are available, however, in loose, block, liquid, and range cube forms. Table 14.22 gives a general idea of how much protein supplement to feed. Research at Kansas State University has demonstrated that supplementation to beef cows in adequate-to-good body condition can be delivered every 5–7 days without affecting weight and condition.

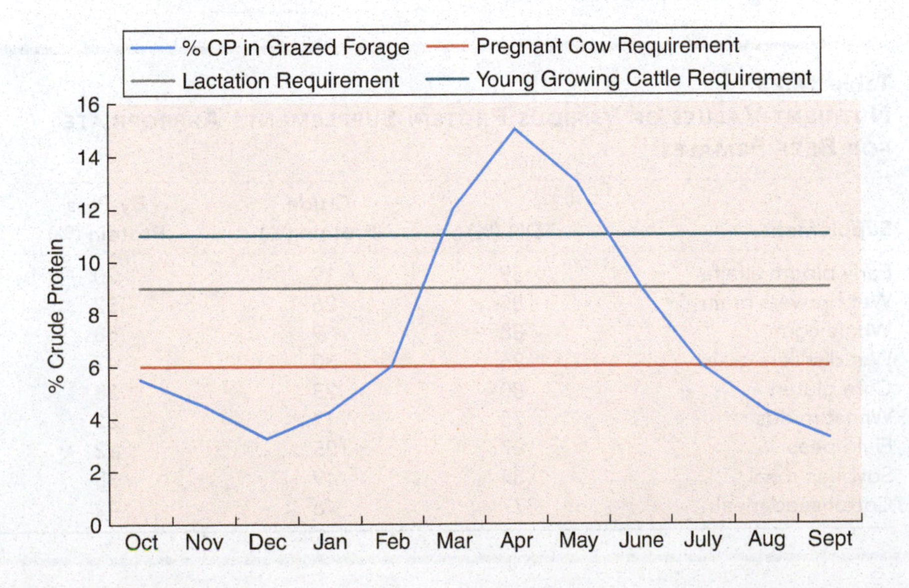

Figure 14.13
Crude protein content in a grazing cow's diet.

Table 14.22
Recommended Levels of Crude Protein for Pregnant and Lactating Cows and Heifers

Item	Crude Protein in Supplement					
	50%	40%	30%	20%	15%	10%
Supplement fed to cows during last 2 months of pregnancy (lbs/cow/day)	0.6	1.0	1.5	2	3.5	7
Supplement fed during lactation and to pregnant heifers (lbs/cow/day)	1.3	1.8	2.3	4	11	20

Many cow-calf producers and yearling operators make a mistake in feeding NPN as the supplemental source of protein when cattle are on an all-forage ration. NPN in the form of urea, nucleic acids, or biuret must be fed along with a readily available source of carbohydrates to be adequately utilized. This readily available carbohydrate source is not provided in adequate amounts with range forage alone. If adequate energy is not available, cattle show little benefit from feeding NPN and the money spent on NPN is not cost effective. Regardless of the source of NPN, it is of questionable benefit in meeting the beef cow's protein needs while grazing dry forage.

Selecting the appropriate protein supplement depends on price, availability, and nutritional value. A list of common sources of supplemental protein is provided in Table 14.23 along with the accompanying levels of TDN, CP, and by-pass protein.

Minerals

Salt, in either loose or block form, should be made available free choice to cattle at all times. The cheapest source should be utilized. Calcium, as shown in Figure 14.14, is present in adequate amounts throughout the year in range and pasture forages. Supplementation is unnecessary.

Figure 14.14 also illustrates that phosphorus is adequate during summer months, but deficient for more than half the year. Inadequate phosphorus can cause reduced growth rates, decreased milk production, and delayed conception. Research shows varied responses to phosphorus supplementation; however, it is recommended as a risk-management practice.

Table 14.23
Nutrient Values of Various Protein Supplements Appropriate for Beef Females

Supplement	TDN (%)	Crude Protein (%)	By-Pass Protein (%)
Early bloom alfalfa	59	19	20
Wet brewers grain	85	26	52
Whole corn	88	9	58
Wet distillers grain	96	30	47
Corn gluten	80	23	28
Wheat midds	75	17	22
Field peas	87	25	22
Soybean meal	84	49	35
Cottonseed meal	77	46	42

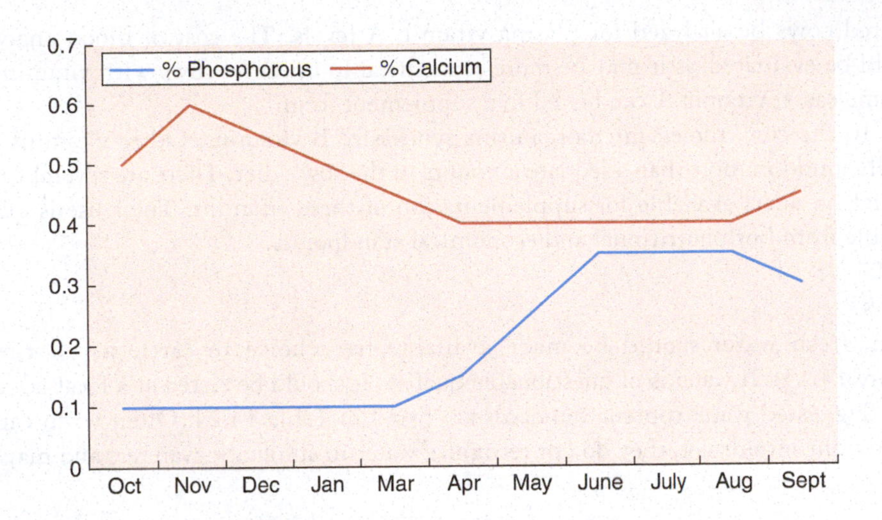

Figure 14.14
Calcium and phosphorous in a grazing cow's diet.

Phosphorus can be supplemented by mixing a phosphorus source with free-choice salt. The most common supplement is one or two parts salt to one part phosphorus source. Some sources of phosphorus include dicalcium phosphate and steamed bone meal. Producers usually find that mixing their own salt–phosphorus supplement and providing it free choice is significantly cheaper than using a commercially prepared supplement.

On some pastures, grass tetany is a problem. This condition is most commonly observed early in the spring on fresh, green, rapidly growing grasses. Affected animals are often seen staggering. The best prevention is to mix magnesium oxide (MgO) in the salt supplement. This should be started several weeks ahead of the potential problem time and continued throughout the grazing period. If palatability is a problem, MgO can be mixed with soybean meal or grain. Usually, no benefit is seen from supplementation of Mg when tetany is not a problem.

Except for a few isolated local deficiencies, all other essential minerals in the range cow diet are present in adequate amounts in the forage. Individual herd and area problems should be considered. As a general rule, cafeteria-style mineral feeding or the feeding of complex mineral mixtures is not necessary or beneficial.

Vitamins

Generally, if the forage being consumed contains green color, it may supply sufficient carotene to meet vitamin A requirements. In forages, beta carotene, which is the precursor of vitamin A, is present. Vitamin A is stored in the liver primarily during the spring and summer months, and this storage usually supplies the animals' needs during the winter. Vitamin A deficiency in cows occurs infrequently. Some pastures during periods of extended drought, however, may not supply adequate amounts of vitamin A. In these cases, 3–6 months hepatic supply is sufficient to meet the cow's requirements until the last few months before calving, at which time the cow may become deficient. She needs it primarily at this time to produce colostrum for the calf and for her own needs. Often a cow has ample vitamin A available, but the calf is born deficient. Colostrum must provide the vitamin A to make up the deficiency. Deficiencies make calves more susceptible to disease, and the deficient cow is more susceptible to retained placentas and infections. The calf's deficiency is supplied only if the colostrum contains adequate vitamin A and the calf receives an adequate amount of colostrum. Most veterinarians and nutritionists recommend an intramuscular injection of vitamin A for the calf at birth, considering this an inexpensive insurance policy. If producers are concerned that their brood cows are deficient, they may have them tested. Current recommendations suggest that blood from five to seven randomly selected cows be analyzed for plasma vitamin A levels. The cost of blood analysis should be evaluated, as it may be more economical to inject all cows with vitamin A. In some cases, vitamin A can be fed in a supplement form.

In the cow, rumen microorganisms synthesize B vitamins. Other vitamins are usually found in more than adequate amounts in the cow's diet. There are several commercial products available for supplementation of these vitamins. Their use is questionable from both nutritional and economical standpoints.

Water

Clean, fresh water should be made available free choice to cattle at all times (Figure 14.15). If water is of questionable quality, it should be tested at a local laboratory. Suggested water content standards are listed in Table 14.24. Often when cattle are brought into drylot, they do not recognize water in automatic waterers and may be

Figure 14.15
Water tank provides continuous supply of fresh water but must be cleaned periodically.

Table 14.24
Maximum Allowable Contents of Minerals and Solids in Water

Substance	Upper Allowable Levels
Aluminum, ppm	5.0
Arsenic, ppm	0.2
Barium, ppm	1.0
Bicarbonate, ppm	1,000
Boron, ppm	5.0
Cadmium, ppm	0.01
Calcium, ppm	100.0
Chloride, ppm	100.0
Chromium, ppm	0.1
Copper, ppm	0.2
Fluoride, ppm	2.0
Iron, ppm	0.2
Lead, ppm	0.05
Magnesium, ppm	50.0
Manganese, ppm	0.05
Mercury, ppm	0.01
Molybdenum, ppm	0.03

Table 14.24
(Continued) Maximum Allowable Contents of Minerals and Solids in Water

Substance	Upper Allowable Levels
Nickel, ppm	0.25
Nitrate-Nitrogen, ppm	20.0
pH	6.0–8.5
Phosphorus, ppm	0.7
Potassium, ppm	20.0
Selenium, ppm	0.05
Sodium, ppm	50.0
Sulfates, ppm	50.0
Total dissolved solids, ppm	10,000
Vanadium, ppm	0.1
Zinc, ppm	5.0
Coliform, #/100 mL	0.5
Fecal coliform, #/100 mL	0.1
Total bacteria, #/100 mL	1,000.0

Source: USDA.

slow to drink from a tank. By running the water fast enough to make noise the first few days, cattle become more easily adapted to water consumption under drylot conditions.

The approximate daily water intake for various classes of beef cattle is shown in Figure 14.16. Water intake of a given class of cattle, under a specific management program, is primarily a function of dry matter intake and ambient temperature. Cattle prefer to drink water that is near the normal temperature of the rumen (100–102°F). It is not necessary to cool water; in fact, doing so may reduce consumption. However, during cold weather, warming water will increase intake. Generally, in cold weather

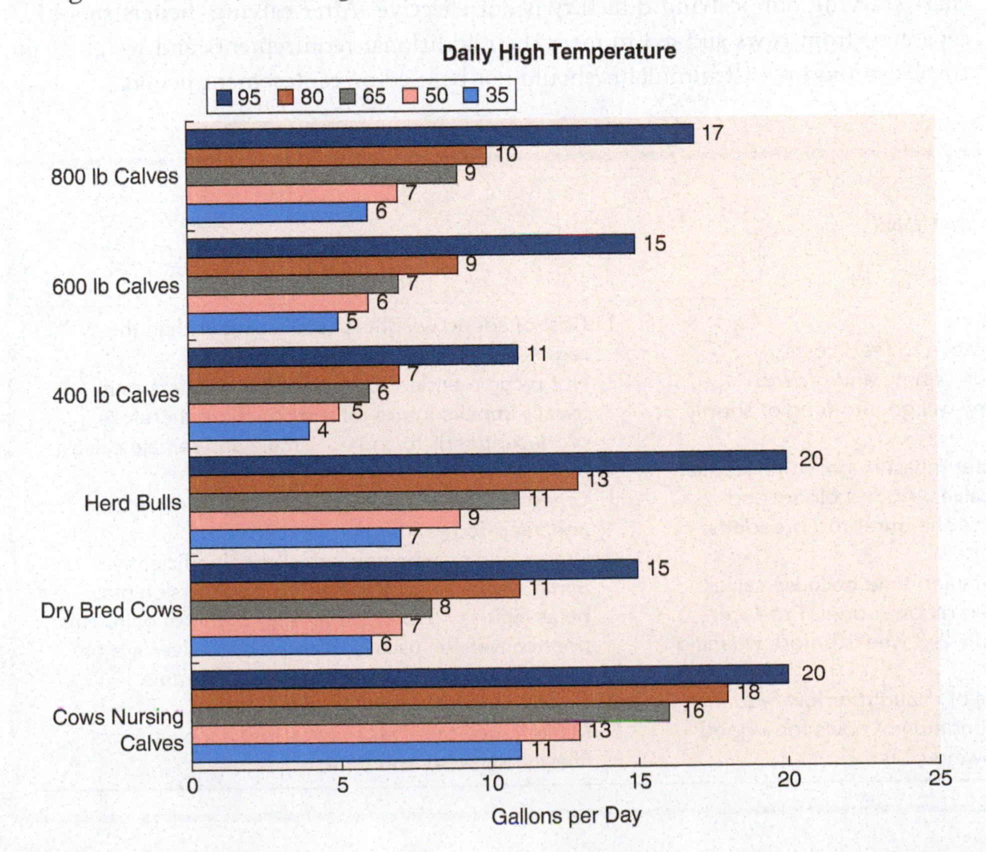

Figure 14.16
Water intake requirements for various classes of cattle.

cattle will consume one gallon of water per hundred pounds of body weight while during the hottest weather consumption will nearly double. Lactating cows require double the level of water consumption as compared to dry females. Feedlot cattle consume approximately 9 gallons per day in the summer and 4.5 gallons daily in the winter.

Creep Feeding Calves

Creep feeding allows calves to have supplemental feed (usually concentrates). Feed is provided in a creep feeder, or some other physical barrier is provided to prevent cows from having access to the supplemental feed. Creep rations can be very simple and use of the cheapest concentrate is usually advisable. Addition of bulky feeds such as oats, beet pulp, and dehydrated alfalfa pellets will prevent digestive problems. A ration such as 50% corn and 50% oats or 50% barley and 50% oats is satisfactory; there is no advantage to feeding a complex creep ration. The creep feeder should be located near the area where cows and calves spend a considerable amount of time, such water or shade. The advantages and disadvantages of creep feeding are pointed out in Table 14.25.

Developing Replacement Females

Replacement heifers require a different level of nutritional management as compared to mature cows. The replacement female has a complex set of nutritional needs driven by the fact that she is still growing while developing the capacity to cycle and then sustain her first pregnancy. During gestation, the replacement heifer has rising requirements for NE and metabolizable protein in each month leading to her first parturition (Table 14.26).

Heifers of 500–700 lb need a 4% higher level of protein in their diet than do mature animals. The concentrate fed should contain enough protein, phosphorus, additional energy, and vitamin A to meet the requirements of the heifer. Heifers should reach 85% of their mature weight by calving time. Again, it should be stressed that "starving out" calving difficulty is not effective. After calving, heifers should be separated from cows and fed to meet their additional requirements and weight gain. Implants and growth stimulants should not be used on replacement heifers.

Table 14.25
Creep Feeding Considerations

Advantages	Disadvantages
1. Weaning weight is increased.	1. Cost of added weight is usually higher than the return.
2. Price of calves is higher relative to feed costs.	2. Not recommended for replacement heifers as it usually impairs future milk production; therefore, producer needs to separate male and female calves if creep feeding is practiced.
3. Useful for calves with skeletal frame and growth potential, particularly if they will go into feedlot shortly after weaning.	3. Creep-fed yearlings weigh only slightly more than noncreep-fed yearlings.
4. Purebred breeder can better measure growth potential (yearling weights) of bull calves. Added bloom and weight give an advantage to the purebred breeder's merchandising program.	4. Creep feed may be converted very inefficiently. Some studies show that the feed conversion may be as high as 15 lb of creep feed per lb of gain. This poor conversion usually occurs when calves are on high milking cows and have access to ample supplies of high-quality forage.
5. Less stress is present at weaning time because calves adjust to their post-weaning ration sooner. Producer may want to creep feed only 2–3 weeks before weaning the calves.	5. Market price of calves may be less if they are too "fleshy" or fat at sale time.
6. Most useful during periods of drought or low feed supply to maintain body condition of cows for a good reproductive rate the following year.	

Table 14.26
Nutritional Requirements of Replacement Heifers (1,200 lb Mature Weight Potential) at Various Months Postconception

	Months Since Conception				
Requirement	1–3	4–5	6–7	8	9
Net Energy (Mcal/day)					
Maintenance	6.15	6.53	6.80	7.10	7.20
Growth	2.36	2.52	2.62	2.70	2.80
Pregnancy	0.14	0.50	1.75	3.45	5.38
Total	8.65	8.55	11.17	13.25	15.38
Metabolizable Protein (lb/day)	0.94	1.00	1.16	1.35	1.58
Calcium (g/day)	20	19.5	26	33	33
Phosphorus (g/day)	12	11.5	16	20	20

Source: Adapted from multiple sources.

The following calculations should be applied to the target weights and days that fit each producer's replacement heifer program:

Example:

Expected mature cow weight	1,000 lb
	× 0.65
Weight desired at breeding	650 lb
Present weaning weight	− 450 lb
Gain needed	200 lb
Days until breeding season	180
Gain needed	1.1 lb per day

These 450-lb medium-framed heifers must gain 1.1 lb per day for most of them to reach puberty prior to the breeding season. To accomplish this, they must consume about 11 lb of dry matter a day that is composed of 80–90% roughage and about 7.7 lb of TDN. A ration with this much TDN will require some concentrate to be supplemented in addition to the forage. Some protein, phosphorus, and vitamin A may be necessary as well, depending on the forage being consumed. Additional considerations related to replacement heifer development can be found in Chapter 5.

YEARLING-STOCKER CATTLE NUTRITION

Yearling-stocker cattle are typically grown on high-roughage rations. Most of these cattle are between 5 and 25 months of age and weigh from 300 lb to 1,000 lb. Yearling-stocker cattle can be raised on grazed forage or in confinement (e.g., in a feedlot); the latter situation is referred to as backgrounded cattle.

Rations for yearling-stocker cattle are formulated using the NE system. See Chapter 6 for a more detailed discussion of nutritional programs.

FEEDING FEEDLOT CATTLE

There are numerous feeding programs utilized by cattle feeders. Individual feeding programs depend on (1) the age, weight, and sex of the cattle; (2) whether feeds are raised or purchased; (3) the relative costs of available feedstuffs; (4) whether one or more groups of cattle will be fed per year using the same facilities; and (5) the length of feeding time.

Simple feeding programs involve feeding silage and grain with an added protein–mineral–vitamin supplement or self-feeding a ration where both rations can remain relatively constant for the entire feeding period. Complex feeding programs combine the altering of concentrate-to-roughage ratios several times (three to five) with formulating least-cost (both ingredient and gain) rations. In the more complex feeding programs, feeding a ration several times a day while carefully monitoring feed bunks and the digestive patterns of cattle maximizes feed intake. An example diet is illustrated in Figure 14.17. The green feed is an example of a starter ration fed to cattle on arrival at the feedyard to transition them to higher energy rations.

Calves are usually grown on higher roughage rations before being fed high-grain finishing rations; otherwise, cattle finish at lighter weights than the preferred slaughter weights. Other large-framed calves, weighing 500–700 lbs at weaning and possessing a high genetic potential for gain, are usually phased into high-energy rations as quickly as possible. A similar feeding program would be used for 700–900-lb, thin to moderately fleshed yearling cattle.

Concentrate-to-roughage ratios can vary from 100% roughage to 100% concentrate to any combination of those two extremes. More conventional rations range from 40% roughage–60% concentrate to 10% roughage–90% concentrate. The most common sources of roughage are silage (essentially 50% grain and 50% forage, on a dry-matter basis), hay, and corn cobs, while the most common concentrates or grains are corn (in the Midwest and northern Great Plains), milo (in the southern Great Plains and Southwest), and barley (in the northwestern United States). The supplement mixture of protein, minerals, vitamins, and possibly additives, usually supplied at 1–2 lb per day, is also considered part of the concentrates.

Properly mixing rations and delivering correct amounts of feed are among the most important concerns in feeding cattle. Where high daily rates of gain are economically feasible, a high level of feed intake in a well-balanced, energy-dense ration is essential. The appetite of the animal largely determines whether large amounts of feed are consumed on a daily basis by each animal in the feedlot. Appetite is influenced by genetic and environmental factors. Genetic selection for gain appears to be primarily a selection preference for cattle that have large appetites. Previous level of

Figure 14.17
Example of the components of a feedyard ration.

nutrition that accounts for compensatory gain, health of cattle, weather, kinds of feed, and condition of feed in the feed bunk all influence the appetite of cattle.

Individuals monitoring feed bunks and observing recent fecal material assess the appetite of cattle and the condition of feed presented to cattle. This type of assessment is critical to proper management of the daily feeding program in the feedlot.

One of the greatest challenges for a feedlot manager is to develop an effective receiving regime for newly arrived feeder cattle. These animals have frequently experienced the stresses of transportation and weaning. As such, they are susceptible to disease and poor performance if not handled appropriately. A general set of guidelines for a receiving ration would be 14–16% crude protein, no more than 50% concentrates (more for yearlings), plus appropriate vitamin and mineral supplementation. Adopting feeder cattle to waterers and feed bunks is also important. In the first 3 days on feed, newly arrived calves eat for an average of 115 min (range of 15 to 225 min) spread over eating frequencies of 11.5 feedings per day (range of 5 to 21). Daily drinking time per day was 7.7 min (range of 1–31) spread over six daily trips to the water source (range of 2 to 18) (Buhman et al., 2000). The high degree of variability in consumption behaviors points to the need for close supervision of newly arrived feeders.

Hot and humid weather conditions also create challenges for feedlot operators. Heat stress not only affects feedlot and carcass performance but, under severe conditions, a high level of mortality may also occur. The key to managing heat stress is to assure highly available sources of fresh water. Sprinklers may be used to wet cattle at a rate of a 1- to 2-minute shower every half hour. Sprinklers should have a nozzle for every 8 to 10 head of cattle, and a flow rate of 2.5 gal/min/nozzle. It is also useful to shift the feeding schedule to coincide with a period of 2 to 4 hours following daily peak temperature. Cattle should not be handled or mixed under these weather conditions. If cattle must be handled, the best time would be from midnight to approximately 8:00 A.M. Shade is also a factor, but the cost of these structures must be balanced with the risk of prolonged, extreme heat conditions. However, in environments where high temperatures are an issue, research has clearly shown that feedlot cattle provided shade have improved rates of gain, have lower incidence of dark cutting carcasses, spend more time lying down, have lowered incidence of bullying behavior, and have lower respiration rates.

SELECTED REFERENCES

Publications

Albin, R.C., & Thompson, G.B. 2010. *Cattle feeding: A guide to management*. Amarillo, TX: Trafton Printing.

Ames, D.R. 1980 (October). Livestock nutrition in cold weather. *Animal Nutrition and Health*. pp. 6–39.

Anderson, D.C. 1978. Use of cereal residues in beef cattle production systems. *Journal of Animal Science*. 46:849.

Bopp, S.G. 2001 (May/June). Feeling the heat. *Bovine Veterinarian*. May/June. pp. 4–8.

Buhman, M.J., Perino, L.J., Galyean, M.L., & Swingle, R.S. 2000. Eating and drinking behaviors of newly received feedlot calves. *Professional Animal Scientist*. 16:241.

Cheeke, P.R. 2005. *Applied animal nutrition*. Upper Saddle River, NJ: Pearson: Prentice Hall.

Church, D.C. 1991. *Livestock feeds and feeding* (3rd ed.). Englewood Cliffs, NJ: Prentice Hall.

Cow-calf management guide and cattle producers' library. Cooperative Extension Service: WA, OR, ID, MT, WY, UT, NV, and AZ.

Ensminger, M.E., Oldfield, J.E., & Heinemann, W.W. 1990. *Feeds and nutrition*. Clovis, CA: Ensminger Publishing Company.

Farmer, S.G., Cochran, R.C., Simms, D.D., Klevesahl, E.A., & Wickersham, T.A. 2000. *Effects of frequency of supplementation on performance of beef cows grazing winter pasture*. Proc. Kansas State University Cattlemen's Day Conference. Manhattan, KS.

Feedstuffs: Reference Issue. 2015. Ingredient Analysis. Bloomington, MN.

Great Plains beef cattle feeding handbook. Cooperative Extension Service: CO, KS, MT, NE, NM, ND, OK, SD, TX, and WY.

Guyer, P., & Henderson, P. Estimating corn and sorghum silage value. *Great Plains Beef Cattle Handbook*, GPE-2401.

Jackson, J.J., Greer, W.J., & Baker, J.K. 2000. *Animal health*. Danville, IL: Interstate Publishers.

Jurgens, M.H. 1993. *Animal feeding and nutrition* (7th ed.). Dubuque, IA: Kendall-Hunt.

Light cattle management seminar. 1982. Fort Collins, CO: Colorado State University Veterinary Medicine and Animal Science Extension Service.

Mitlohner, F.M., Galyean, M.L., & McGlone, J.J. 2002. Shade effects on performance, carcass traits, physiology, and behavior of heat-stressed feedlot heifers. *Journal of Animal Science*. 80:2043–2050.

National Research Council. 1981. *Effect of environment on nutrient requirements of domestic animals*. Washington, DC: National Academy Press.

National Research Council. 2000. *Nutrient requirements of beef cattle*. Washington, DC: National Academy Press.

National Research Council. 1996. *Nutrient requirements of beef cattle* (7th ed.). Washington, DC: National Academy Press.

Preston, R.L. 2016 (March). Typical composition of feeds for cattle and sheep. *Beef*. pp. 48–60.

United States Department of Agriculture. 2008. Part I: Reference of beef cow-calf management practices in the United States, 2007–08. APHIS: NAHMS. Washington, D.C.

United States Department of Agriculture. 2010. Part V: Reference of beef cow-calf management practices in the United States, 2007–08. APHIS: NAHMS. Washington, D.C.

Zinn, R.A., Owens, F.N., & Ware, R.A. 2002. Flaking corn: Processing mechanics, quality standards, and impacts on energy availability and performance of feedlot cattle. *Journal of Animal Science*. 80:1145–1156.

15

Managing Forage Resources

As James Ingalls once wrote, "grass is the forgiveness of nature – her constant benediction. . . Its tenacious fibers hold the earth in its place, and prevent its soluble components from washing into the wasting sea." Grazing is not a recent phenomenon; for millions of years large grazing beasts have utilized grasslands for sustenance. Grazing, like fire, is a natural process characterized by the mutually beneficial relationship between grazers and grasslands. Research has clearly demonstrated that grasslands can be both overgrazed and over-rested with a resulting loss of both diversity and productivity in each case. Grazing animals both domestic and wild are selective in their consumption patterns, alter what they eat by both season and stage of their own production (e.g., lactating versus dry), and display both instinctive and learned behavior as they forage.

Forage from range, pasture, hay, and crop residues are the primary resources upon which the beef cattle industry is based. More than 80% of the feed consumed by cattle is forage, and most of it is consumed in the form of grazing. Forage, nearly undigestible by humans and monogastric animals, can be converted by cattle to a highly preferred consumer product.

Both range and pasture areas in the United States are primary sources of forage but are described somewhat differently. Typically, forage resources considered as rangelands are located in the arid and semiarid lands of the 17 western states (further broken down into the 11 western states and the major six states in the Great Plains area). Pasture lands can be considered as the irrigated pastures of the West in addition to the grazed areas of the east of the 100th meridian in the United States where rainfall is typically sufficient to provide for forage growth.

Rangeland has been defined by the Society for Range Management as land on which the native vegetation (climax or natural potential) is predominantly grasses, grass-like plants, forbs, or shrubs suitable for grazing or browsing and present in sufficient quantity to justify grazing or browsing use. Rangelands include natural grasslands, savannahs, shrub lands, most deserts, tundra, alpine communities, coastal marshes, and wet meadows. While this definition is inclusive for all grazed forage, the discussion in this chapter uses the former definition that separates range and pasture areas.

GRAZED FORAGE RESOURCES

The total land area in the United States is approximately 2.2 billion acres. Approximately, 786 million acres are grazed by livestock in the United States (Table 15.1). An additional 63 million acres is used for hay production. Table 15.2 outlines the changes in lands available to grazing in the United States. The available range, pasture, and forage resources in the United States and their utilization are very diverse, primarily because of the different climatic and soil conditions. Figure 15.1 illustrates the classification of agricultural lands by region of the United States. Note that the three regions with the highest percent of lands classified as grazing lands are the

Table 15.1
AGRICULTURAL AND NONAGRICULTURAL USES OF LAND IN THE UNITED STATES (LOWER 48 STATES)

Major Land Uses	Acres (mil.)	% of Total
Agricultural		
Cropland	408	21.5
Cultivated cropland	335	17.7
Cropland used for pasture	36	1.6
Cropland idled	37	1.6
Pastureland and range	612	32.3
Forestland grazed	127	6.7
Total land grazed	775	40.6
Total agricultural land	1,159	61.2
Nonagricultural		
Forestland not grazed	449	23.7
Urban, transportation, and other built-up areas	107	5.7
Other land (includes 9 mil acres of small water areas)	68	3.6
Recreation and wildlife areas	110	5.8
Total nonagricultural land	734	38.8
Total land area	1,894	100.0

Source: USDA.

Table 15.2
GRAZING LANDS TRENDS IN THE UNITED STATES (1945–2002)

Year	Cropland Pasture	Grassland Pasture and Range	Forestland Grazed
		Million acres	
1945	47	659	345
1954	66	634	301
1964	57	640	225
1974	83	598	179
1982	65	597	158
1992	67	591	145
2002	62	587	134

Source: USDA.

southern plains, the mountain west, and the northern plains. The southeast, Delta region, and the Appalachians have significant forest lands that are utilized, at least in part, for grazing.

Beef cattle derive their nutrients from a ration composed of 83% forage and 17% concentrate. Stocker cattle and cows exist on a ration that is nearly 96% forage. During the relatively short feedlot phase (100–145 days), the ration is approximately 5 to 15% forage.

Annual precipitation, sometimes associated with other climatic influences, can have marked effects on forage supply (Figure 15.2). In the West, high precipitation occurs in the mountains, where much of it accumulates as snowpack in the winter. The snow provides needed irrigation water during the spring and summer runoff. In many regions where precipitation is less reliable and abundant, irrigation water is

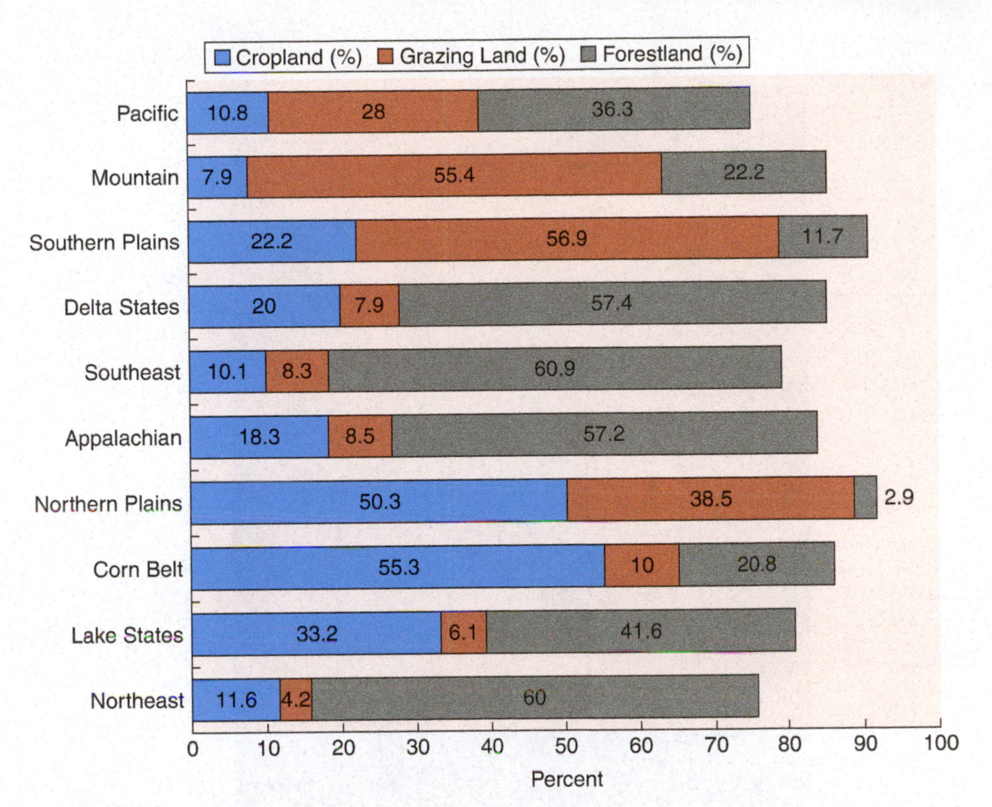

Figure 15.1

Percent of U.S. land in various uses by region of the country.

Source: USDA.

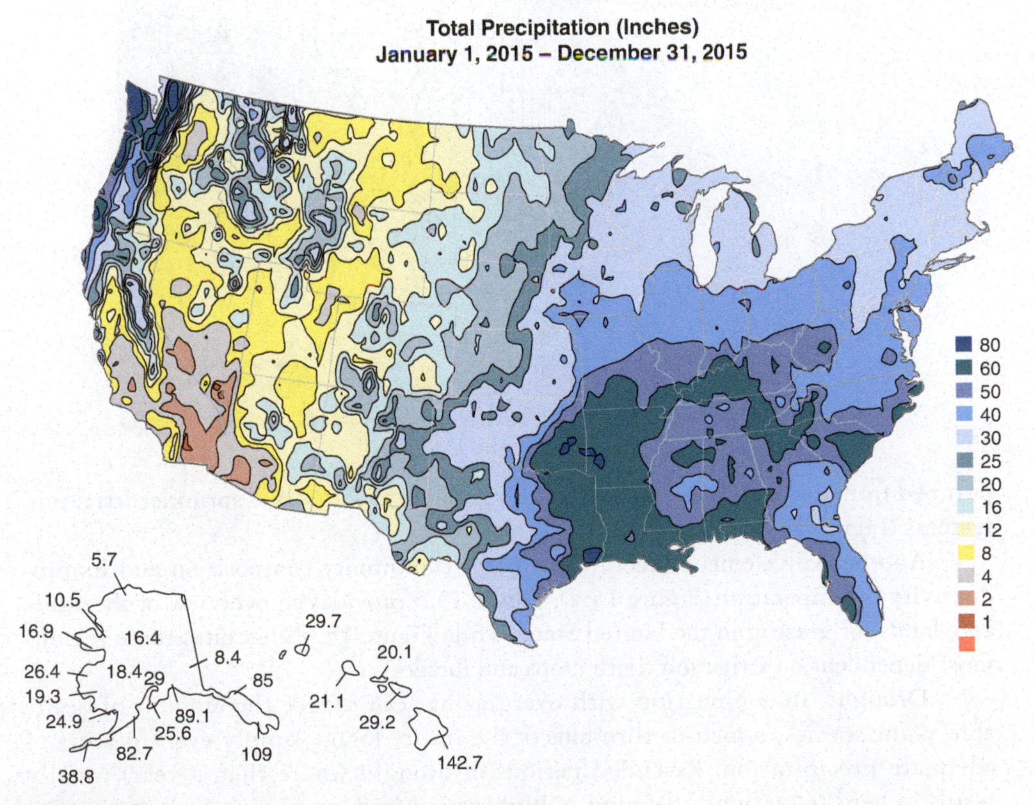

Figure 15.2

Various levels of precipitation by region of the United States.

Source: National Weather Service.

Figure 15.3
Irrigation systems—flood systems that depend on snowpack runoff and center pivot water delivery from pumped wells.

pumped from deep wells to be distributed via center pivot or line sprinkler irrigation systems (Figure 15.3).

Another key element affecting the plant community composition and its productivity is temperature (Figure 15.4). Figure 15.5 provides an overview of the available lands for grazing in the United States while Figure 15.6 illustrates those regions most dependent on irrigation (both crops and forages).

Drought, in conjunction with overgrazing, can reduce the amount of desirable plant species, which in turn affects the future forage supply even in years of adequate precipitation. Extended periods of drought (more than a year) usually result in herd reductions, shipping in high-priced feed, or moving cattle to an area with a more abundant feed supply. Figure 15.7 illustrates the variation in acres

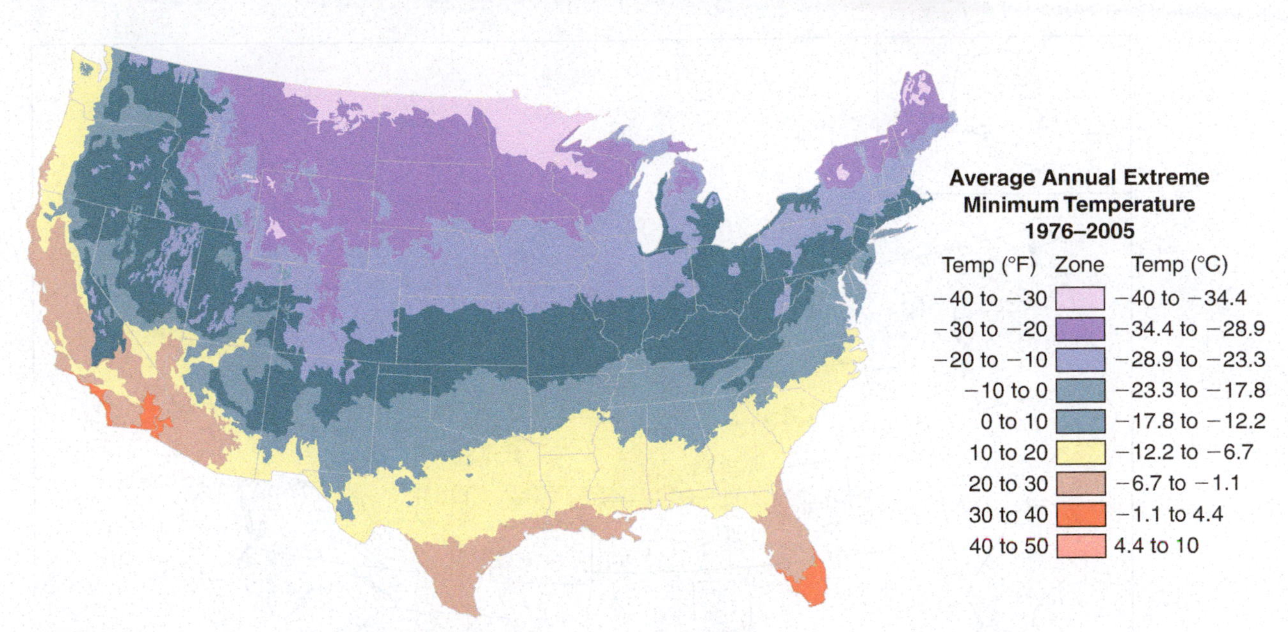

Figure 15.4
USDA plant hardiness zones of the United States.
Source: USDA.

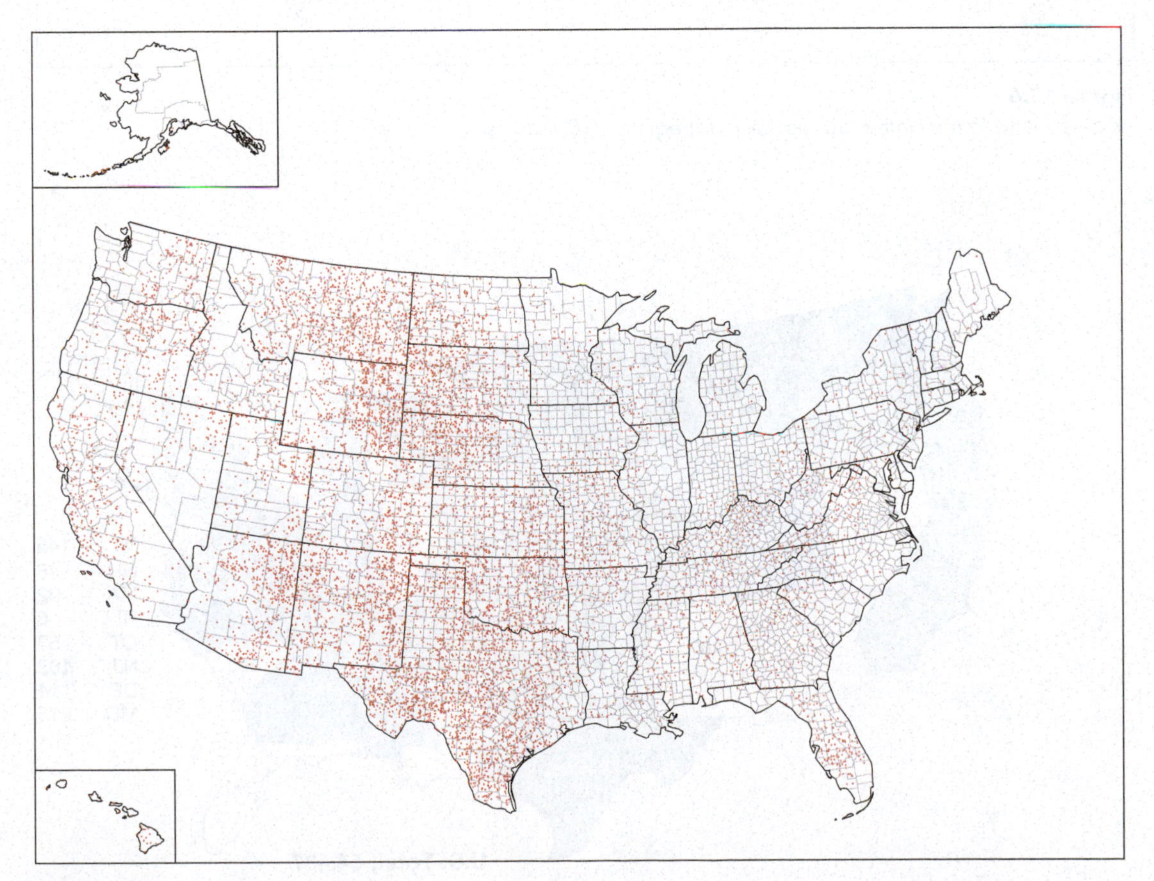

Figure 15.5
Acres of pastureland in the U.S. One dot equals 50,000 acres.
Source: USDA.

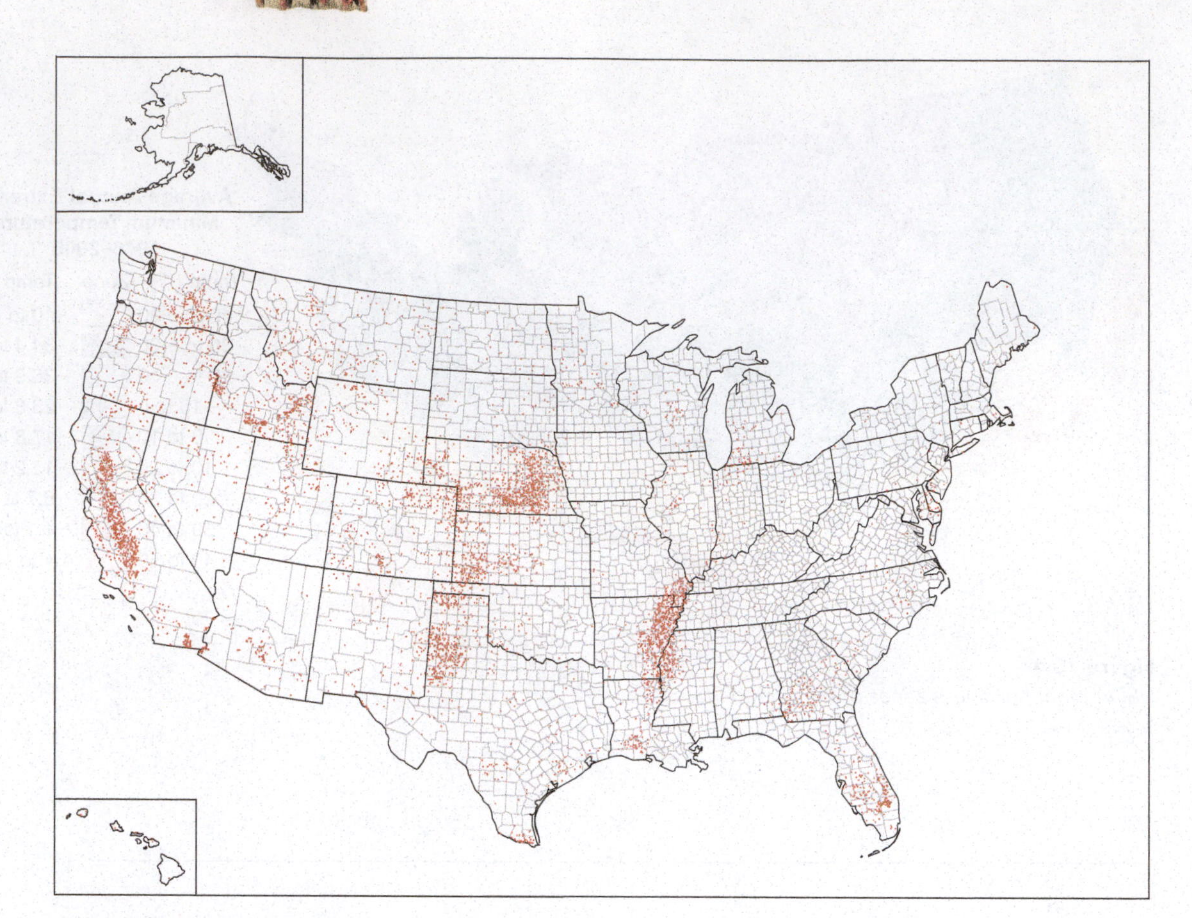

Figure 15.6
Irrigated land in the United States. Each dot equals 10,000 acres.
Source: USDA.

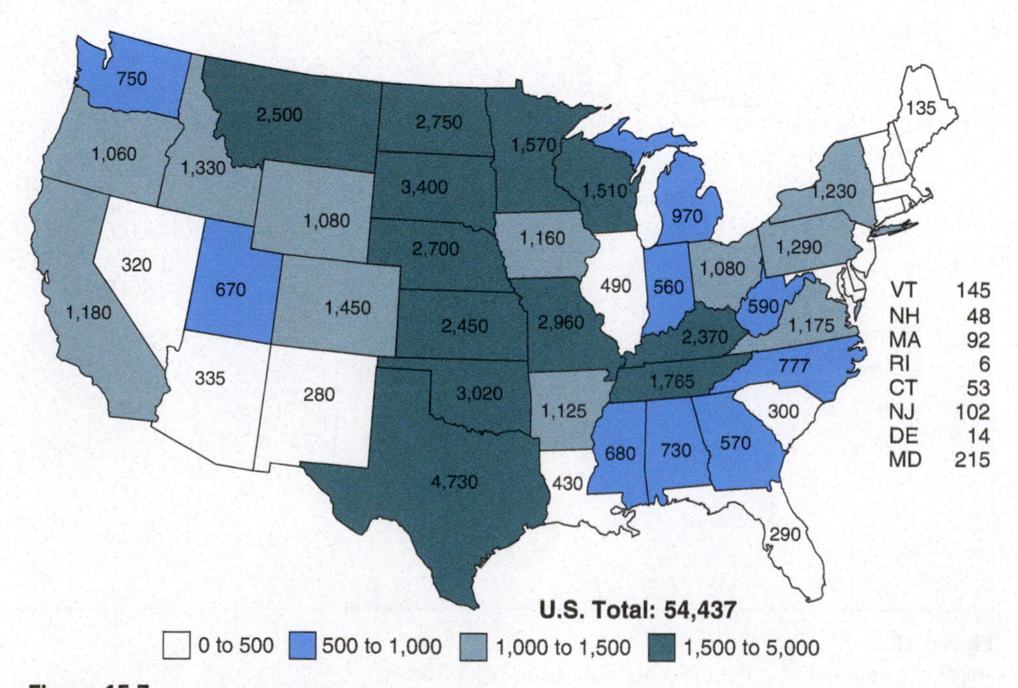

Figure 15.7
Hay production acres by state.
Data source: USDA National Agricultural Statistics Service; compiled by Livestock Marketing Information Center.

dedicated to hay production across the United States. Harvested forage supplies and prices are affected by variation in weather conditions. The major drawback of harvesting forage via mechanical means is cost. It has been estimated that between 50 and 70% of the annual cow cost for intermountain ranchers is tied to the haying enterprise.

PLANT TYPES AND THEIR DISTRIBUTION

The general classifications of plants used primarily for grazed forage are grasses, **forbs**, and **browse**. Some grazing areas contain all of these types of plants; more commonly, however, mixed species of one or two classes are present. Because hundreds of different species for each plant class exist, only some of the more important species in the cow-calf and yearling production areas are discussed here.

Forages are also classified as annuals or perennials and as warm-season or cool-season plants. Cool-season plants make most of their growth during late fall, winter, and early spring, while warm-season plants have their major growth during late spring to early fall. Table 15.3 gives an overview of the significant grasses, forbs, and browse that contribute to beef production in the United States.

Grasses

Table 15.4 gives a more detailed description of major grasses. The most important native grasses are found in the western two-thirds of the United States, while tame pasture predominates from the eastern Great Plains toward the Atlantic seaboard.

The wheatgrasses, bluestems, grama grasses, buffalograss, and switchgrasses are most common on western ranges. Orchardgrass, smooth bromegrass, reed canary grass, the fescues, and bermudagrass (primarily coastal) account for most tame grass production in the United States. Tall fescue is the most widely grown cultivated pasture grass in the country, while smooth bromegrass is the most widely distributed cool-season grass in use.

Forbs

Forbs are herbaceous annual and perennial dicots and monocots other than grasses. Legumes are the most important forbs; cattle producers call numerous other forbs *weeds*. Many forbs are fair-to-good range forage plants. Some highly palatable range forbs are sensitive to overgrazing and so are used as indicators of range condition. Common perennial range forbs include halfshrub sundrop, pitchers sage, compass plant, and slender lespedeza.

Legumes, identified by their butterfly-like flowers, are usually higher in protein than most other grazed plants and are important to the nitrogen-fixation process that enhances soil fertility. The most important forage legumes are alfalfa, the trefoils, the lupines, sweet clover, kudzu, and true clovers. Alfalfa (Figure 15.8) is widely distributed because it is adapted to environmental conditions in nearly all states.

Browse

Browse consists of perennial woody shrubs and vines used as forage by cattle. Some examples of range browse are sagebrush, shadscale, winterfat, mountain mahogany, and four-winged saltbush.

Cattle grazing on range that includes grass, forbs, and browse typically change their consumption pattern of each plant class as the grazing season progresses. Grass comprises a higher percentage of the diet in early summer. In late summer, there is an increase in the consumption of forbs and browse.

Table 15.3
A General Classification of the Major Forages in the United States

	Perennials		Annuals	
Category	**Warm Season**	**Cool Season**	**Warm Season**	**Cool Season**
Grasses	Bahiagrass Bermudagrass Big bluestem Blue grama grass Buffalograss Carpetgrass Caucasian bluestem Dallisgrass Eastern grama grass Indiangrass Johnsongrass Little bluestem Switchgrass	Crested wheatgrass Hardinggrass Intermediate wheatgrass Kentucky bluegrass Orchardgrass Reed canary grass Russian wildrye Smooth bromegrass Tall fescue Tall wheatgrass Timothy Western wheatgrass	Browntop millet Corn Crabgrass Forage sorghum Foxtail millet Grain sorghum Pearl millet Sorghum–Sudan hybrids Sudangrass	Barley Oats Rescue grass Rye Ryegrass Triticale Wheat
Forbs, Legumes	Kudzu Perennial peanut Sericea lespedeza	Alfalfa Alsike clover Birdsfoot trefoil Red clover White clover	Alyceclover Cowpea Korean lespedeza Soybean Striate lespedeza Velvetbean	Arrowleaf clover Bull clover Berseem clover Bigflower vetch Black medic Burclover Button clover Caleypea Common vetch Crimson clover Hairy vetch Hop clover Lappa clover Persian clover Rose clover Subterranean Clover Sweetclover Winter pea
Other Forbs	Compass plant Halfshrub sundrop Pitchens sage Roundhead Slender lespedeza Weeds			
Browse		Four-wing Saltbush Mountain mahogany Sagebrush Shadscale Winterfat Other brushes and Shrubs		

Table 15.4
An Overview of the Production Requirements, Expected Yields, and Nutritional Value of Selected Grasses and Legumes

	Characteristic									
							Nutritional Values (%)			
Grass Legume	Winter Tolerance	Drought Tolerance	Soil	Annual Nitrogen	Harvest Maturity	Average Yield (tons/acre)	NDF	CP	ADF	Uses
Alfalfa	Very good	Excellent	Fertile, well-drained	0[1]	Late, bud early bloom	4–8+	17–28	28–35	38–46	Grazing, hay, silage, green chop
Red Clover	Very good	Good	Fertile, well-drained	0[1]	Early mid-bloom	3.5–5+	17–28	28–35	38–46	Grazing, hay
Birdsfoot Trefoil	Excellent	Very good	Not on droughty soils	0[1]	Early flower	4–6	17–24	28–35	38–46	Grazing, hay
Crownvetch	Very good	Very good	Fertile, well-drained	0[1]	Early to mid-bloom	2.5–3	17–24	28–35	38–46	Grazing, hay
Switchgrass	Very good	Excellent	Widely adapted	0[2]	Early boot	—	7–12	35–40	55–60	Grazing, hay
Big Bluestem	Very good	Excellent	Well-drained to somewhat poorly drained	0[2]	Early boot	—	7–12	35–40	55–60	Grazing, hay
Buffalograss	Very good	Excellent	Not sandy soils	—	Early boot	—	7–12	35–40	55–60	Grazing, hay
Coastal Bermudagrass	Poor	Very good	Heavy soils	300–400	10″ tall	15–20	11–14	35–40	55–60	Grazing, hay green chop
Bahiagrass	Poor	Excellent	Sandy	150–300	—	4–6	11–14	35–40	55–60	Grazing, hay
Tall Fescue	Good	Fair	Tolerates most soils	75–240	Early boot	3–5	11–14	35–40	55–60	Grazing, hay
Orchardgrass	Very good	Very good	Most soils, avoid poorly drained	75–240	Boot	3.5–5	11–14	35–40	55–60	Grazing, hay, silage
Smooth Bromegrass	Very good	Good	Well-drained, silt or clay loam	75–240	Late boot	3.5–5	11–14	35–40	55–60	Grazing, hay, silage
Timothy	Very good	Fair-good	Well-drained, productive silt or clay loam	75–240	Early head	2–3.5	11–24	35–40	55–60	Grazing, hay
Reed Canary grass	Good	Excellent	Widely adapted, good in poorly drained soils	75–240	Early heading	3.5–6	11–14	35–40	55–60	Hay, grazing, silage
Kentucky Bluegrass	Very hardy	Poor	Heavy soil	150+	Boot	—	11–14	35–40	55–60	Grazing
Corn Silage	N/A	Good	Well-drained	100–200	$^1/_2$–$^3/_4$–milk	25	6–10	24–33	43–52	Silage
Sorghum	N/A	Very good	Well-drained	40–60	Soft- to Mid-dough	7–9	9–14	30–38	40–50	Silage
Sorghum–Sudan	N/A	Very good	Well-drained	50–100	24–30″ or boot	10–12	9–12	26–35	45–50	Silage, hay, grazing

[1]Nitrogen application not needed if stand is greater than 50% legume.
[2]None during stand establishment.
Source: Based on Pioneer Forage Manual (1995).

GRAZING MANAGEMENT

Grazing systems are extremely diverse—there are hundreds of combinations of climatic differences, plant species, terrain, types of cattle, and preferences of cattle producers. There is no best grazing system for all producers, so the management team

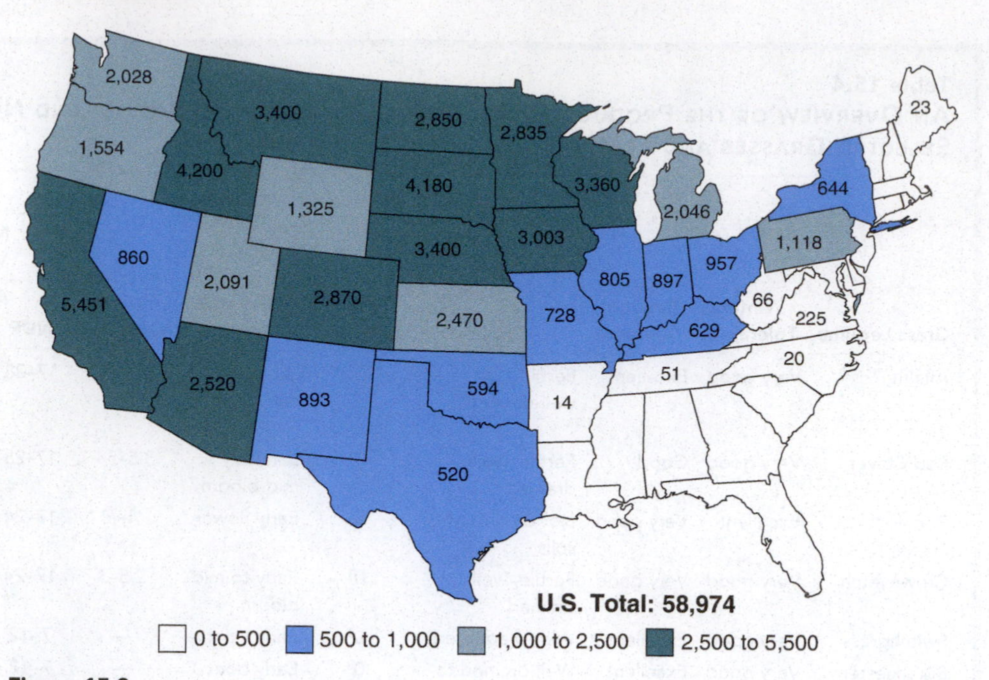

Figure 15.8
Alfalfa hay production by state.
Data source: USDA National Agricultural Statistics Service; compiled by Livestock Marketing Information Center.

should identify the most economical system that best fits each producer's needs and goals. However, management should always focus on the health and care of natural resources.

Plants that are grazed experience disturbance but have the resilience to recover from grazing so long as three factors are managed correctly—timing of grazing, amount of the plant grazed, and the length of time between grazing events. Plants have a productive cycle driven by availability of sunlight, water, and nutrients. A perennial grass, for example, is dormant in the absence of heat, moisture, and key nutrients and shuts down its activities above the ground, maintaining only the root system. Grazing dormant plants has minimal impact on the plant. Once conditions are ripe for growth, the plant uses photosynthesis to form new leaves and pull nutrients and minerals from the soil as it turns from brown to green. In the early stages of growth, the plant is most vulnerable to disturbance. As the growing season continues, it produces the energy required for root growth, leaf development, tiller formation, and seed development. As the seasons and/or conditions change, the plant will return to dormancy.

Several factors allow sustainability of grasslands in the presence of grazing pressure from cattle—bovines are not able to graze plants closer than a couple of inches from the ground, the vast majority of the plant is below ground in the root system and unavailable to grazing, grasses have growth points close to the ground and tend to produce more leaf area than is required for optimal photosynthesis.

A manager's perspective of a landscape impacts decisions and choices amongst alternatives. Moving from the most micro view to a systems viewpoint, scale (Figure 15.9) ranges from an individual plant where an animal grazes a plant to where the impact is immediate but then lessens as the plant recovers. A pasture scale takes into account a community of plants within a pasture and the impact of a herd(s) impact over a grazing period while a ranch scale includes the full set of pastures and the yearlong management of that particular landscape. Finally, a watershed scale incorporates multiple ranches, herds (domestic and wild), and other

Figure 15.9
Landscape and grazing perspectives can be viewed on four scales—individual plant and animal interaction, specific pastures or grazing sites, specific ranches, and watersheds.

human impacts (energy, recreation, etc.). As the scope widens, the impacts develop more gradually but also require more time to take corrective action when mistakes are made.

From an ecosystems perspective, the health of grazing lands is tied to three primary criteria—watershed functionality resulting from a healthy water cycle and soil stability (Figure 15.10), the integrity of nutrient cycles (Figure 15.11) and energy flows, and functional recovery mechanisms that allow rangeland to resist disturbance and remain resilient in the face of changing conditions. Range and pasture grazing systems are primarily a function of three factors: (1) amount of forage produced, (2) forage quality, and (3) the efficiency by which the forage is harvested (Heitschmidt and Walker, 1983). All systems of grazing management are centered on one basic principle: controlling the frequency and severity of defoliation of individual plants. This principle needs to be carefully assessed over several grazing seasons as well as during a single grazing season to allow managers to build in as much freedom from risk as is possible. Effective grazing systems are determined by stocking rates, intensity and frequency of grazing, grazing of single or mixed stands of forage, and other factors that can optimize efficient beef production. Grazing systems, to be effective over the long term, need to be economically and environmentally sound. Furthermore, they should be as easy to implement as possible.

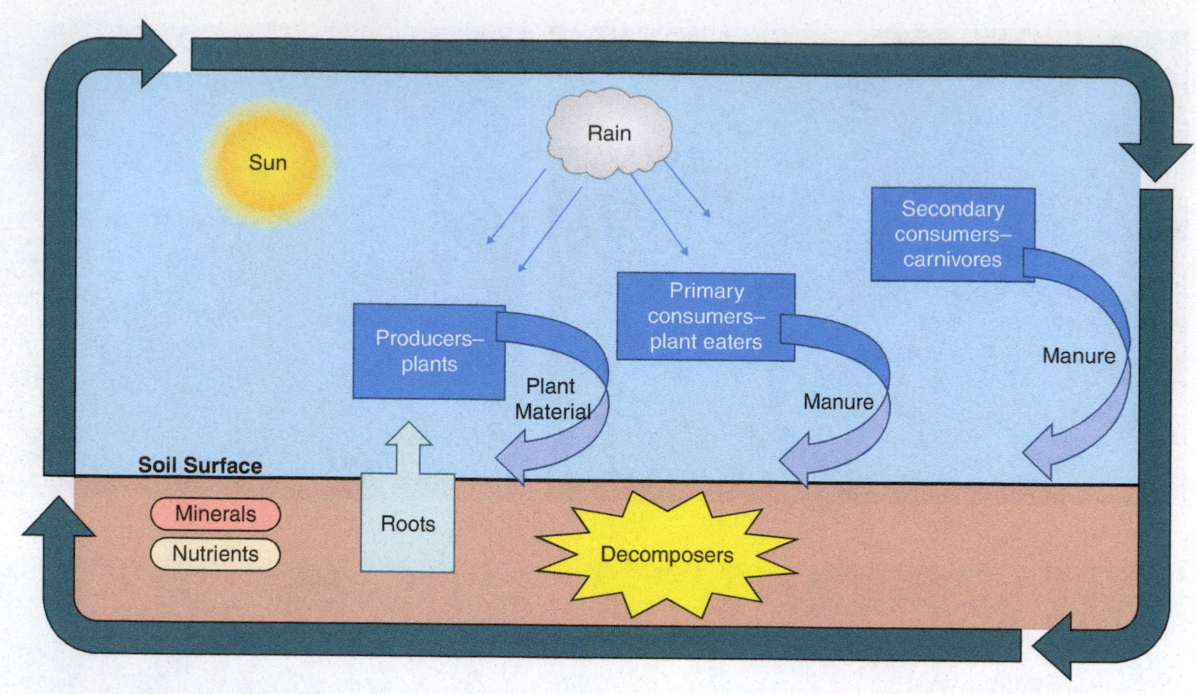

Figure 15.10
The water cycle upon which healthy watersheds are based.

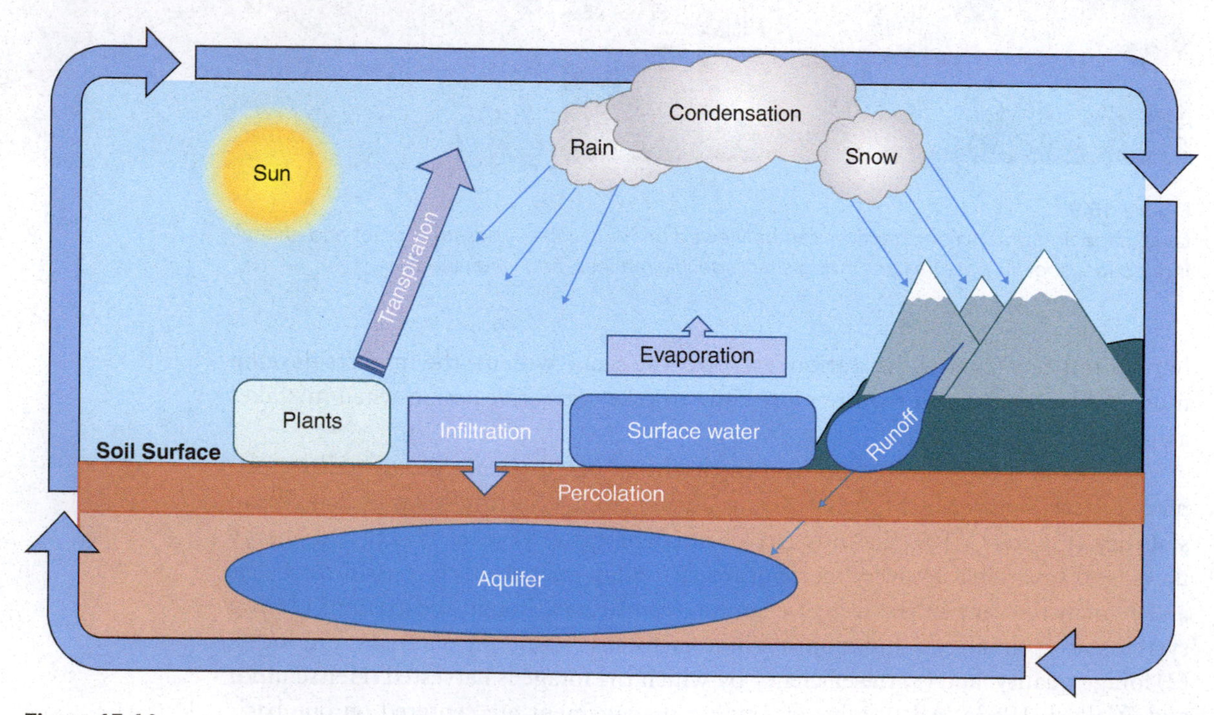

Figure 15.11
The nutrient cycle that drives the grazing ecosystem.

Some grazing intensities and frequencies are reflected in the following commonly used grazing terms. Several of these systems will be discussed in more detail later in the chapter.

Continuous—a system that allows grazing on a specific area throughout the entire grazing season (does not refer to level of use, and does not imply yearlong grazing).

Cell system—the management of a larger acreage in a multi-pasture grazing system typically utilizing high-intensity, short-duration grazing. Such a system often incorporates a common water source and cattle handling facility central to all pastures.

Rotational—cattle are moved from pasture to pasture on a schedule (based on calendar, level of use, or stage of growth). These systems typically rest pastures after they have been grazed.

Deferred—not grazed until the primary forage plants have matured, usually the setting of seed.

Rested—no grazing for at least a 1-year period.

Rested-rotational—incorporates at least one rested pasture on a rotational basis.

Seasonal—grazed for the duration of a particular season of the year.

Short duration—heavy stocking rates for short periods; each unit may be grazed several times during a grazing season.

The objective of most grazing systems is to optimize animal performance and maintain or improve the production potential of the vegetation. The kinds of available animals and the type of vegetation, along with the seasons of grazing, help determine the stocking rate as part of the overall grazing plan. Matching dietary preferences and nutrient requirements of cattle to kinds and quality of vegetation is the first important step in determining the proper grazing management system.

The season of grazing is determined primarily by recognizing two important relationships. First, animal performance is greatest when the forage is grazed when it is growing most rapidly (high in quantity and high in nutrient quality) and before it matures. Second, to obtain grassland improvement, the desirable forage species must be allowed to reproduce.

Several concepts are important to the grassland manager—carrying capacity, utilization, and condition. Carrying capacity is used at the pasture and ranch scale and is defined as the maximum number of individuals of a given species that a grazing site can sustain over a fixed period of time without deteriorating the resource. This is a particular challenge in that differing levels of production will occur not only between sites but also between plots within a site. Also productivity is not fixed but varies based on season, precipitation, weather events such as early frosts or extreme hail, composition of the plant community, and soil type.

Carrying capacity is determined by assessing several sites or a single site and then extrapolating to a larger area (assumptions can become problematic if the sites are not representative of the whole and implies that productivity is stable when in fact the degree of productivity is variable season to season and year to year in most cases). To be effective, carrying capacity must be reassessed on at least a seasonal basis with adjustments made in accordance with changing conditions or to stock at 90% of carrying capacity to build in a buffer against worsening conditions.

Range utilization measures or estimates the percentage of biomass harvested from a site. Utilization on arid and semiarid lands of 30–35% will assist in improving the resource while utilization rates of greater than 40–45% may result in reduced resource health. In higher precipitation areas, utilization of up to 55–60% may be

appropriate. One of the limitations of this metric is that animals rarely graze a pasture uniformly, thus plant selection has an impact and given the variation within a plant community relative to regrowth and recovery, season of use and timing of measurement may skew results.

Range or pasture condition compares the vegetation on a site to the area's potential. The benchmark or potential is often defined as that which existed prior to grazing or prior to a fire event. One of the limitations of condition is that it is based on an assumption that only grazing affects condition; this of course is a poor assumption in light of the multivariate impacts on landscapes.

Pasture and range condition can be evaluated via a 10-factor determination that examines species composition, plant diversity, density and vigor, percent legumes within the stand, plant residue, uniformity and intensity of use, woody canopy, and soil erosion. An example form to record a grazing site estimate is provided in Table 15.5.

While each of these metrics offers information and insight, there is no single measure that provides a complete assessment, so the landscape manager must develop an integrated and responsive perspective that allows as Wyoming rancher Bob Budd puts it "to think at the pace of rocks and mountains while learning to act within our own lifetimes."

The optimum stocking rate (the animal demand per unit area over a period of time) is not a constant; it will vary from season to season and year to year. Grazing pressures that are too high for long periods of time will reduce the cattle's nutrient intake and thus reduce their performance as well as reducing the diversity of plant species on the site. The correct stocking rate for a particular pasture depends on the desired level of animal performance, the ability of the vegetation to withstand grazing, and the range or pasture improvement goals. Carrying capacity is the stocking rate applied to a site with a specific animal density for a specific time without damaging the resource. To determine carrying capacity for a site, producers need to know the following:

1. Annual pounds of forage produced per acre (Table 15.6)
2. The percent of annual production consumed by cattle (continuous grazing—33% utilization; weekly controlled rotation—50% utilization; daily controlled rotation—70% utilization)
3. Daily animal consumption as determined by livestock performance goals

Table 15.5
PASTURE CONDITION SCORESHEET

	Descriptor						Descriptor
Factor	Low range			Score			Upper range
Species composition	Undesirable	0	1	2	3	4	Desirable
Plant diversity	Narrow	0	1	2	3	4	Broad
Plant density	Sparse	0	1	2	3	4	Dense
Plant vigor	Weak	0	1	2	3	4	Strong
% legumes in stand	<10%	0	1	2	3	4	>40%
Plant residue	Deficient	0	1	2	3	4	Excessive
Uniformity of use	Spotty	0	1	2	3	4	Uniform
Intensity of use	Heavy	0	1	2	3	4	Light
Woody canopy	>40%	0	1	2	3	4	<10%
Soil erosion	Severe	0	1	2	3	4	Minimal

Source: Adapted from NRCS.

Table 15.6
Dry-Matter Yield per Acre for Various Forages and Stand Densities in the Northern Plains

Forage Composition	Fair (>25% bare ground)	Good (10–15% bare)	Excellent (<10% bare)
Yield (lbs per acre)			
Bluegrass—white clover	150–250	300–400	500–600
Tall fescue—legume	100–200	200–300	300–400
Smooth brome—legume	150–250	250–350	350–450
Orchardgrass—legume	100–200	200–300	300–400
Mixed stand	150–250	250–350	350–450
Red clover or alfalfa	150–250	200–250	250–300
Native tall warm season	50–100	100–200	200–300

Source: Adapted from NRCS.

4. Grazing season duration as determined by forage availability, rest period, etc. (Morrow and Hermal, 1996, Univ. of Missouri).

Carrying capacity can be roughly calculated by the following formula:

$$\text{Carrying capacity} = \frac{\text{annual forage production} \times \text{seasonal utilization rate}}{\text{average daily intake} \times \text{length of grazing season}}$$

Many range managers estimate stocking rate using the animal unit month (AUM) concept. This approach provides an inexact estimate, but it is the standard in many regions of the world. An AUM is a 1,000-pound mature cow of above-average milking ability, with a calf at side less than 4 months of age with a 400-pound weaning weight that consumes 800 pounds of dry matter in 1-month time. Other classes of cattle have AUM estimates that differ based on weight, age, and productive status (Table 15.7).

If cows are heavier than 1,000 lb, or wean calves heavier than 400 lb, the standard AUM must be modified. A rule of thumb adjustment is to increase the AUM one-tenth for every 100 pounds of cow weight over 1,000 lb and every 100 lb of weaning weight beyond 400 lb. For example, a 1,200-lb spring calving cow weaning a 600-lb calf would require:

$$\text{Cow} = \frac{1{,}200}{1{,}000} = 1.2 \text{ AU}$$

$$\text{Calf} = 0.3 \text{ AU} + \frac{600 - 400}{1{,}000} + 0.5 \text{ AU}$$

$$\text{Pair} = 1.2 + 0.5 = 1.7 \text{ AU}$$

Making stocking rate decisions on the most accurate AUM is critical in assuring both animal response and landscape health. For example, not many cows are 1,000 pounds at maturity—most are between 1,200 and 1,400 pounds, so adjustments have to be made in the form of an animal unit equivalent (AUE). Assume that the average cow weight is 1,250 pounds and thus will consume 1.25 as much as an AUM; put another way 100 cows weighing 1,250 lbs will consume as much forage as 125 cows that weigh the standard 1,000 pounds. A rule of thumb is that a mature cow eats about 80% of her body weight per month, thus the 1,250-pound cow consumes 1,000 pounds per month—200 pounds more than the 1,000-pound cow. Extended out to a 100 head herd equates to 20,000 pounds of dry matter more feed for the larger cows.

Table 15.7
Animal Unit Values (AU) for Various Classes of Beef Cattle

Class of Cattle	Animal Unit
Cow and calf (1,000 lb, above-average milking ability, spring calving)	1.00
Calf (spring born, 3–4 months)	0.30
Replacement heifer (24–36 months)	1.00
Cow (1,000 lb, non-lactating)	0.90
Yearling cattle (12–17 months)	0.70
Weaned calves (<12 months)	0.50
Yearling bulls	1.20
Mature bulls	1.50

Source: Adapted from Blasi and Corah (1993).

Stocking rate can be defined as the number of animals per unit area for a fixed time period. Stocking rate is typically calculated by either AUMs per acre or acres per AUM. Using the AUE above and a grazing season of 2.5 months would yield 125 × 2.5 = 312.5 AUMs. Assuming a range site of 500 acres and a productive capacity of 1,500 pounds of dry matter per acre:

$$312.5 \text{ AUMs}/500 \text{ acres} = .625 \text{ AUMs per acre or } 500 \text{ acres}/312.5 \text{ AUMs} = 1.6 \text{ acres per AUM}$$

The sustainable stocking rate based on the current inventory is calculated as follows:

$$\frac{\text{lbs of DM per acre} \times \text{N acres}}{\text{lbs consumed per month}}$$

500 × 1,500 = 750,000 lbs of total production of which some is palatable and some is not and that if the rule of take half and leave half is applied then only 375,000 pounds is available before applying utilization rate (assume 50%). Accounting for utilization rate leaves 187,500 pounds of available forage. The 100 head herd weighing an average of 1,250 pounds consumes 100,000 pounds of feed per month. Thus, the sustainable stocking rate is equal to 187,000/100,000 = 1.87 months. These adjustments are important to avoid overstocking a site.

Estimates of carrying capacity for various plant communities of differing conditions on several soil types are illustrated in Table 15.8. Table 15.9 shows examples of stocking rate for 500-pound stockers with varying levels of forage production and utilization rates.

Grasses such as bermudagrass, bahiagrass, K. R. bluestem, buffalograss, Kentucky bluegrass, and common curly mesquite are well suited for intensive grazing. Most taller grass species and bunch grasses do not withstand intensive grazing. Where mixed stands of these grasses exist, the species that eventually predominate will depend on the intensity and length of grazing.

Cattle are selective in their grazing for two reasons. First, they respond to topography, distance to water, and other factors. Salt distribution, water distribution, fencing, and forced movement of cattle are four practices utilized to prevent overgrazing by assuring more uniform animal distribution. Second, some species of plants in mixed stands are more palatable than others. Some grazing systems will prevent overgrazing of the more palatable species.

The high level of diversity in weather, soil, topography, plant community composition, and management skill makes it impossible to establish a "one size fits all" approach to grazing management. The following protocols are some of the most frequently utilized

Table 15.8
Average Carrying Capacity for Various Range Sites and Conditions

Range Site	Range Condition: Excellent	Good	Fair	Poor
	AUM/Acre			
Tallgrass prairie				
Clay upland	1.30	0.95	0.65	0.45
Limy upland	1.10	0.90	0.65	0.45
Loamy lowland	2.30	1.70	1.15	0.65
Loamy upland	1.50	1.10	0.75	0.50
Mixed grass prairie				
Clay upland	0.90	0.70	0.55	0.25
Limy upland	0.90	0.70	0.45	0.35
Loamy lowland	1.35	1.05	0.75	0.50
Loamy upland	1.00	0.70	0.55	0.35
Sand prairie				
Subirrigated	2.25	1.75	1.25	0.85
Sandy	0.90	0.70	0.50	0.35
Choppy sands	0.60	0.50	0.35	0.25
Shortgrass prairie				
Loamy upland	0.60	0.50	0.40	0.30
Loamy lowland	0.80	0.70	0.60	0.40
Loamy upland	0.80	0.50	0.40	0.30

Source: Adapted from Blasi and Corah (1993).

Table 15.9
Stocking Rates for Grazing Stocker Cattle on Differing Levels of Pasture Productivity and Utilization Rates

Annual Forage Yield (lbs/year/acre)	Utilization Rate: 20%	40%	60%
	Number of 500-pound stockers per acre		
4,000	0.17	0.34	0.52
8,000	0.35	0.70	1.05
12,000	0.52	1.04	1.56
16,000	0.70	1.40	2.10

strategies but all depend on appropriate stocking rate, responsiveness to changing conditions, and agile, flexible decision-making processes. In addition to the grazing program, mangers must also use animal distribution strategies such as herding, placement of salt and mineral blocks as well as improvements such as burning, brush and noxious weed controls, and controlling access to water and waterways. Furthermore, the successful manager is able to integrate effective monitoring with a robust planning process.

Continuous, Season-Long

No attempt to avoid grazing on any portion of the range or pasture is made. With appropriate stocking rates, typically light, these systems allow maximum dietary

choice during a year or throughout the season of use without damaging the health of the plant community. This system helps assure that both forbs and grasses will be utilized in most cases. Best sites for this approach are even, flat terrain with excellent water availability. However, inappropriate stocking rates can yield significant negative outcomes.

Deferred-Rotation

It involves pasture rotation using at least two pastures with each pasture receiving deferment or freedom from grazing during the growing season every 2–4 years. Under this protocol, plant response is typically better for many species as compared to season-long, continuous systems while cattle performance is either minimal or unaffected. This system is particularly useful in protecting riparian habitat when implemented correctly.

Rest-Rotation

In this approach, multiple pastures are required and grazing is scheduled so that individual pastures are completely rested within a management cycle. For example, in a six-pasture, 3-year system, each pasture would receive 1 year of rest, be continuously grazed for a year or growing season, and be deferred grazed in the other year. These systems are especially useful when protecting riparian areas, stockpiling forage for low precipitation years, and implementing other improvements such as reseeding, burning, or implementing brush controls.

A modification of this system is known as the Santa Rita and for each pasture involves a 12-month rest followed by a 4-month season of grazing, and another year-long rest followed by an 8-month grazing period. This variation is of particular value for creating improvements on sites in poor condition.

Short Duration

In this protocol, the grazing area is divided into multiple smaller units or paddocks that receive more than one period of non-use and grazing during a growing season. Typical systems involve 5–12 cells or subdivisions with grazing periods lasting 3 days to 2 weeks in duration. The speed of rotation is aligned with precipitation and season such that rotations occur more frequently during rapid plant growth and less frequently when growth is slower. Advantages of this system when it is implemented correctly include improved water infiltration, enhanced mineral cycling, better animal distribution, and more even utilization of both palatable and less palatable plants. Stocking rates and stock density are usually higher than in other systems. This system is best suited to areas with extended growing seasons (90 days plus), average rainfall in excess of 20 inches, and forage production approaching a ton per acre.

Paddock number and size as well as determination of forage requirements are important in these systems. Calculations are provided below:

$$\text{Paddock number} = (\text{length of rest period (days)}/\text{length of grazing activity (days)}) + 1$$

$$\text{Paddock size} = \frac{(\text{daily herd forage requirement}) \times (\text{days of grazing period})}{\text{pounds of forage available per acre}}$$

$$\text{Daily forage requirement} = (\text{number of animals}) \times (\text{average weight}) \times (\text{daily utilization rate})$$

Rule of thumb recommendation for utilization rate is 4% of live weight
(2.5% intake + 0.5% trampling rate + 1% buffer)

$$\text{Seasonal forage requirement} = \text{daily forage requirement} \times \text{number of days in grazing season}$$

However, these systems may be expensive to implement because additional labor and fencing are required. Range improvement may occur under short-duration grazing systems if moderate stocking rates, which will give acceptable levels of cattle performance, are utilized. Strong proponents of certain short-duration systems have generated considerable controversy. Much of the attention has focused on the Savory grazing method. Excellent discussions of the Savory grazing method and the basic principles of grazing systems can be found in Heitschmidt and Walker (1983) and Savory (1999).

Range and pasture areas that permit year-round grazing with satisfactory animal performance are usually the most economical for cattle producers. This is true if input costs can be moderated. The primary objective is to have green forage (which is typically higher in nutrient content than dry, mature forage) available to cattle as long as possible during the year. The 300-day grazing system, developed at the University of Arkansas, is designed to manage a cow herd with a minimal level of purchased or mechanically harvested supplemental feeds. While local conditions will determine the level of implementation, the concept is based on the following goals and objectives:

- Establish inventory of forage base
- Determine management practices to increase seasonal grazing from forage base
- Add complementary forages to fill in seasonal gaps
- Plan forage and grazing practices ahead for the year
- Monitor and adjust forages and livestock as needed

To assure the success of the 300-day grazing program, the following parameters need to be in place:

- Good ratio of warm and cool season forage
- Optimum (+) soil fertility
- Electric fence and water placement
- Stocking rate is not excessive (2.7 a/AU)
- Targeted fertilizer application
- Optimum cow herd/calving season management implemented
- Planning one or more seasons ahead to ensure forage is as good as possible in subsequent seasons

Prior research (Hart et al., 1993) emphasized the importance of two requirements of sound forage management: (1) proper stocking rate, and (2) even livestock distribution. This research demonstrated that these two requirements can be achieved independent of grazing system. Also, these researchers noted that reduced pasture size and distance to water may be responsible for the claimed benefits of short-duration and controlled rotation grazing systems. A careful economic evaluation must assess the short-term and long-term benefits of fencing for reduced pasture size, water development, and/or rotation grazing (intensive, less intensive basis, or not at all).

Effective grazing management requires the careful use of range management tools, timing, and knowledge about a particular geographic area. Some of the common misconceptions relative to grazing as noted by Reece (1991) are listed here.

1. Rangeland plants must be grazed to assure plant health and vigor.
2. Late season grazing is more favorable to plant health than grazing early in the growing season.
3. Conventional deferred rotational grazing systems provide all pastures with equal recovery opportunity.
4. Stocking rates can always be increased once grazing systems are implemented.
5. Overgrazing cannot damage dormant vegetation.
6. Increased stocking rates are required to increase total animal production per acre.

7. Continuous grazing is "bad."
8. There is one grazing strategy best suited to all range environments.
9. Production per animal is less critical than production per acre in generating maximum profits.

A review of the extensive literature on grazing management fails to yield a consensus opinion as to the most desirable grazing system. However, the following summary points from Vallentine (1990) are useful.

1. No grazing system eliminates the need for appropriate stocking rates and the application of sound management principles.
2. Grazing systems must be developed on a site-specific basis.
3. A specialized grazing system is only as good as the ability of the range manager.
4. Because of changing environmental conditions, flexibility is key to the success of a grazing system.
5. If a rotational grazing system is utilized, managers should be aware of the following.
    a. The need and cost of fencing
    b. The cost of assuring adequate water for livestock
    c. Variation in grazing capacity of different pastures
    d. The potential effects of drought on the system
    e. Potential effects on wildlife

In the end, excellent grazing management is driven by a set of clear objectives that may include improving profitability by lowering feed costs, by limiting reliance on expensive inputs such as fossil fuels, improving resource quality/health, and optimizing animal performance. The intentional application of knowledge and innovation can yield both desirable animal performance and grazing resource health. The following multi-step stewardship manifesto was developed by the NCBA to help guide the environmental stewardship efforts of the industry:

- Recognize the environment for its varying and distinct properties.
- Manage for the whole resource, including water, soil, topography, plant, and animal communities.
- Realize that natural resources are ever changing and management must adapt.
- Recognize and appreciate the interdependence of ecosystems.
- Recognize that management practices should be site- and situation-specific and must be locally designed and applied.
- Recognize that successful management is an ongoing, long-term process, and commit to stewardship, economic success, and business continuity.
- Strive to develop a management framework that involves family, employees, and business associates so that the entire team is committed to common goals.
- Monitor and document for effective practices.
- Never knowingly cause or permit abuses that result in permanent damage to public or private land.
- Develop ways to communicate and share the vast practical experience of other resource stewards.
- Become involved in organizations that provide an effective way to educate and support individuals.
- Solicit input from a variety of sources on a regular basis as a means to improve the art and science of resource management.
- Help develop public and private research projects to enhance the current body of knowledge.
- Recognize that individual improvement is the basis for any change.
- Communicate with diverse interests to resolve resource management issues.

Range Monitoring

Monitoring of rangeland and pastures requires a fundamental first step—defining the goals (vision of the long-term outcome) and objectives (steps that make progress toward the goals). Developing specific and measurable outcomes that are aligned with the realities of a particular landscape are fundamental to success. Long-term monitoring is sometimes placed on the back burner due to the pressing needs of short-term needs and conditions. However, using repeated ground photography, plant density assessment, measuring streambank stability, and benchmarking the species comprising a plant community are important to building a more vibrant and productive landscape. Short-term measures such as maintaining a management calendar/diary, logging weather data, capturing grazing utilization information, assessing inventories of domestic livestock and wildlife using a resource, and assessing recreational impacts are important to the overall management plan.

Monitoring provides data and information so that trends can be more clearly recognized to evaluate the effectiveness of management decisions, to make timely changes in management, and to assess performance of vegetation, soil, and livestock, as well as providing guidelines to maintain focus on the goals and objectives. Data to be collected, recorded, and analyzed can be categorized into four basic areas—basic information, short-term, long-term, and interpretation. More specifically the data needs within these four categories are as follows:

Basic

- Site description including unit name, pasture identifier, study site identifier, name of data collector and date of collection, and monitoring method(s) used.
- Site location using legal description of GPS, when possible, describes the landform, elevation, percent slope, and average yearly precipitation. Growing conditions, exposure of the slope, soil type, and other climate information is also useful.
- Photo catalogs—to assure usefulness, utilize consistent image capture technique and equipment, identify location and date of each image, capture images at the same site and stage of plant growth each year, maintain consistency of skyline in the background, and capture images at approximately the same time of day. Range specialists should be contacted for assistance in establishing the system.

Short-term

- Landscape appearance description to make an assessment of general forage utilization (high, moderate, low).
- Identify key species to be measured (3–5 of the most prolific species as well as 3–5 indicator species).
- Grazing map development to depict pastures, paddocks, or grazing units with utilization classes and timing recorded.
- Stubble height measurement to determine utilization.
- Pasture use data should include the kind and class of livestock, season of use complete with on and off dates, number of livestock, and where possible, numbers of other grazing wild ungulates, grazing system employed, and rotation pattern.

Long-term

- Plant density count of key species.
- Riparian area stability determination by monitoring streamside communities along a green line.

- Permanent photo plot or photo point transect to establish means to record changing conditions over time.
- Canopy measurement.

Data Interpretation

The grazing response index (GRI) calculation to describe annual use is a critical tool that helps managers quantify the impacts of a grazing system by establishing intensity of utilization, frequency of use during the growing season, and growth–regrowth patterns before and after grazing.

The GRI provides an evaluation of the current grazing season to help aid in planning for subsequent seasons. The GRI combines observations of three factors—frequency, intensity, and recovery. Frequency is the number of times plants are grazed during a season. The maximum number of times that a plant should be grazed is three. Divide the number of days in the grazing period by 7–10 (7 to be most conservative, 7 days for regrowth, 10 for slower growth rate). For example, the frequency scores for 10, 21, and 30 days grazing periods on a pasture would yield scores of 1.4, 3, and 4.3, respectively. Scoring would then occur as follows: 1 to 7 frequency would receive a +1 value, frequency scores of 7.1 to 14 would be assigned a null score, and those over 14 would receive a −1 valuation.

Intensity is the level of leaf removal and can be roughly described as light, moderate, and heavy. Plants with more leaf area are able to regrow more quickly. A rule of thumb is to leave 50% leaf area following grazing. Light grazing that takes 40% or less leaf area receives a +1, moderate leaf removal (41–55%) is assigned a 0, and heavy use (>55% removal) yields a score of −1. Utilizing grazing cages or enclosures spread across the site provides a means to get a representative assessment. Cages should be moved following each season.

Finally, opportunity for plant recovery is assessed. Plant regrowth is the most critical measure for long-term health of the plant community, so scores are doubled based on the following scale:

Full recovery = +2
Partial recovery = +1
Some recovery = 0
Little recovery = −1
No recovery = −2

Recovery depends on providing rest periods to allow for plant regrowth (Table 15.10).

The total GRI is computed by adding the values of frequency, intensity, and recovery. Each of these factors can be managed—shortening the season, reducing

Table 15.10
REST PERIOD RULE OF THUMB RECOMMENDATIONS—NORTHERN GREAT PLAINS

Plant Type	Cool Weather Conditions	Warm Weather Conditions
Grasses—cool season	14	35–50
Grasses—warm season	35–40	21
Legumes	21–28	21–28

Source: Adapted from NRCS.

Figure 15.12
A controlled burn site where old growth and mature forage were sacrificed to provide for new growth.

stocking rate, and shifting season of use either alone or in concert may be required if the overall GRI or any of the components are negative to assure future productivity.

Range Improvements

Range improvements are based on the ecological principle of providing a competitive advantage for desirable plant species by increasing their access to water, nutrients, and sunlight. Simultaneously, undesirable plant types are controlled to limit their access to limited resources. Range improvement programs are typically implemented to increase quantity and (or) quality of forage, enhance livestock/wildlife production, enhance the ability to handle and care for livestock, improve water resources, and minimize erosion.

Range improvement techniques typically can be characterized as biological control (grazing, insects), mechanical control (cutting, bulldozing, chaining, mowing, disking), herbicidal control, prescribed burning (Figure 15.12), seeding, and fertilizing. The range improvement techniques best suited to a particular site are determined by a combination of geographical, ecological, economic, regulatory, and management expertise considerations.

Soil classification occurs using a system of orders, suborders, great groups, subgroups, families, and series. For the purposes of this chapter, only the orders will be mentioned. There are 11 orders of soils: five exist in diverse climates, while the other six are more regionalized. A description of soil types, distribution, and productivity is provided in Table 15.11.

MAJOR GRAZING REGIONS OF THE UNITED STATES

The West

The western half of the United States is highly dependent on snow accumulation in the mountains to supply needed water for irrigation and to enhance spring plant growth. Thus, many cattle producers in the West watch for mountain snowpack reports as they prepare forage management plans for the upcoming spring and summer (Figure 15.13).

Most cattle grazing operations in the Pacific Northwest are cow-calf operations. Nearly three-fourths of the brood cow's diet is from on-site grazed forage with supplemental

Table 15.11

Description of Major Soil Orders of Importance to Grass and Forage Production

Soil Order	% of U.S. Total Soils	Distribution	Productivity	General Description
Entisols	8	River floodplains, rocky soils in mountain areas, beach sands, rangeland	Variable, from highly productive to infertile	Slight soil development, range from sand to river-deposited clay to volcanic ash deposits
Inceptisols	18.2	Middle Atlantic and Pacific states	Highly variable	Weakly developed, quick horizon formation
Andisols	1.9	Hawaii and Pacific Northwest	Highly productive if well managed	Weakly to moderately developed, rapidly weathered, mostly formed from volcanic activity
Aridisols	11.6	Western mountain and Pacific states	Very productive when irrigated	Dry soils, prone to accumulation of soluble and fertilized Salts
Mollisols	25.1	Great Plains	High fertility due to higher humus and nutrient content	Dark soils of grasslands and some hardwood forests, very diverse
Vertisols	1.0	Southeastern and central Texas, western Gulf Coast states	Not easily cultivated, generally fertile, but restriction of root penetration	Originate from limestone, marl or basalt, high swelling-clay content
Alfisols	13.5	North central and mountain states	Productive crop land, irrigation is less critical	Moist mineral soils, often developed under deciduous forests
Spodisols	4.8	New England, mid-Atlantic, and northern Great Lakes states	Requires significant fertilization	High sand content, well leached, strongly acidic
Ultisols	12.8	Southern Atlantic, eastern south central, and Pacific states	Very productive with high management and fertilization	Humid-area soils, clay layers

Source: Based on *Miller and Donahue (1995)*, and *Brady (1990)*.

Figure 15.13
In the Intermountain West, snowpack provides a substantial percentage of water used for irrigation.

hay fed in the wintertime. Northern California, with its mixture of valleys and timbered mountain ranges, has the typical cow-calf operations of the Pacific Northwest states. In the California annual grass type, there is no one predominant type of cattle operation. The mountain grasslands that lie between the valleys and timbered mountains typically are grazed for 2 months each in the spring and fall. Productivity of mountain bunchgrass is high, yet nutritional value drops below adequate levels in late summer and fall as forage matures. During about one-third of the fall period, seasonal rains followed by several days of adequate growing temperatures permit fall regrowth, which has good protein content.

The California annual grasslands, prominent in much of the California foothills, are used during fall, winter, and spring periods. They are a good source of forage in connection with yearling or feeder operations. Coupled with the use of annual grasslands and supplemented rations for feeders or stockers are improved pastures and grazing of crop aftermath. Some cow-calf operators make use of the annual grass type until it matures, then move their herds to higher elevation timber types or to renovated or suitable mountain shrub areas for summer grazing.

Ranchers living within a reasonable distance of mountain-timbered summer range often combine grazing of sagebrush range with mountain summer grazing to achieve forage supplies for all but the winter feeding period. Ranch operations in the heart of sagebrush country try to make this range serve their needs for spring, summer, and fall. Ideally, however, sagebrush range should be used as spring and late fall forage with summer grazing occurring in the mountains. Many producers have small tracts of irrigated meadows interspersed with sagebrush, or they provide supplemental feeds in addition to their sagebrush range.

Throughout the Pacific West, forests provide considerable livestock grazing on a range basis. The most valuable of these for range grazing is ponderosa pine. In prime condition, ponderosa pine is open forest, with an abundance of green feed of good protein content during the hot season. This characteristic gives the ponderosa pine type a premium or extra incremental value over a lower elevation range.

Coastal Douglas fir and interior mixed conifer forests occupy western Oregon, western Washington, and the northerly slopes of all interior mountains. The grazing productivity of the forest floor is poor (0–50 lb of air-dry herbage per acre) prior to logging but may increase greatly following logging. The openness of a particular logging or harvest system regulates the amount of usable ground cover produced. Because these tree species are so shade tolerant, they tend to naturally provide a closed canopy in about 20 years. After tree seedlings are reestablished, the forage produced during this 20-year period is often abundant, but the nutritive quality is variable and uncertain. With better utilization of forest residues, treatment of slash, and reseeding of improved ground cover, possibilities exist for summer season herbage production of 1,200–1,800 lb per acre.

Livestock operators in the Intermountain Region, which consists of Idaho, Nevada, Utah, western Colorado, and Wyoming, make use of seasonally productive rangelands by moving animals from one geographical range to another. The desert ranges are used during the winter (November to April) (Figure 15.14); the foothill or intermediate elevation ranges are used during the spring (April to July), and some are used in the fall (October to November). Mountain ranges are used during the summer and fall (July until about mid-October) (Figure 15.15).

Of great importance is the comparative nutrient value of different forage plants during various seasons and the ability or inability of these forage species to meet the requirements for optimum livestock production. Animals in the Intermountain Area do not need a supplement during the spring and summer range grazing seasons if the plants are growing. During fall and winter, supplements often are necessary because the forage nutrient content is marginal for animal needs, and inclement weather may seriously reduce daily intake.

Figure 15.14
High desert of eastern Oregon used for early spring grazing.

Figure 15.15
Summer grazing in the high country of Colorado.

A scarcity of suitable spring range in the Intermountain Area generally is a limiting factor for successful cattle production. During recent years, it has become common practice to seed depleted foothill range with cool-season species to provide more suitable forage for spring grazing. Grass species such as crested wheatgrass, intermediate wheatgrass, or Russian wild rye often are planted to help provide needed nutrients when native foothill ranges are mature and deficient in nutrients. An alternative, when foothill ranges become dry and dormant, is to move animals to higher elevations where feed is still green and growing. This is sometimes difficult because it puts an additional grazing load on ranges that must also be grazed in the summer.

The vegetation of desert ranges in the Intermountain–Great Basin region is composed primarily of browse species with various quantities of grass. Generally, desert browse plants meet the protein requirements for livestock during gestation and are exceptionally high in carotene. They may be slightly deficient in phosphorus, however, and decidedly low in energy-furnishing constituents. Grasses, during the winter, are markedly deficient in protein, phosphorus, and carotene, but are good sources of energy. Forbs are generally sparse on desert ranges and are unimportant in the diet during winter grazing. If the diet is largely grass, then phosphorus and digestible protein may be markedly deficient; if the diet is largely browse, energy may be deficient.

Animals on many winter ranges normally require a supplement to meet their requirements when properly grazed; however, with increased grazing intensity, the quantity and even the type of supplement needed may change. Overgrazing may create a need for a greater quantity or even a more expensive supplement over a longer period of time.

In the desert areas of the Southwest, small herds are common, with steers purchased during wet periods to use abundantly produced annual grasses and forbs.

Desert ranges can be grazed yearlong. Grasses like black grama cure well and maintain some nutritive value yearlong. Desert grasslands (Figure 15.16) are quite valuable for grazing despite their aridity because most of the grasses are highly preferred. During the growing season, grasses have a crude protein content of 8% or higher, generally at or above minimum levels for livestock. Protein drops off to 4–5% in winter, requiring some supplementation. Desert shrub areas have a more erratic type of forage production. Most of the forage is produced following rains (usually in late winter and spring) and consists of annuals such as alfilaria, six weeks grama, and other annual forbs and grasses. Some shrubs, principally four-wing saltbush, provide nutritious, palatable feed. Four-wing saltbush, for example, retains a yearlong protein content in excess of 11%.

The foothills-type rangeland, which includes pinyon-juniper and chaparral-mountain shrub ecosystems, provides mainly spring–fall grazing for cowherds. The pinyon-juniper is an abundant ecosystem but relatively unproductive of forage. It has historically been heavily grazed because of its proximity to ranch headquarters and the better parts have been plowed for crop production. The chaparral-mountain shrub ecosystems generally provide limited grazing because the dense brush and tree cover make grazing difficult. Their primary use is spring–fall, or spring, summer, and fall.

The mountains of the Southwest commonly support a forest cover of ponderosa pine with occasional open, mountain grassland areas. Part of the precipitation falls in the winter and spring from Pacific storms, and part during the summer from storms originating in the Gulf of Mexico. Grazing by cowherds occurs during the summer and fall.

Throughout the Southwest, brush has increased since domestic livestock grazing began there several hundred years ago. Palatable, perennial grasses have decreased, and

Figure 15.16
Desert range in central New Mexico.

in some areas, like the Desert Shrub, annual grasses and forbs have largely replaced the perennial grasses.

The Great Plains

The Great Plains, which occupies approximately one-third of all U.S. land, includes parts of 10 states (Montana, North Dakota, South Dakota, Wyoming, Nebraska, Kansas, Colorado, Oklahoma, Texas, and New Mexico). This vast area of semiarid, high-plateau grasslands once known as the Great American Desert varies in topography from extremely flat to quite rolling. In general, it is covered with short grasses and mid-grasses, but in the Black Hills of South Dakota, it is covered with trees. On the eastern edge and scattered throughout on deep, sandy soils are large areas of highly productive tallgrass prairie.

The northern Great Plains covers an estimated 300,000 square miles, which is approximately one-tenth of the total land area of the United States. The range in the northern Great Plains occurs in the western three-fourths of North Dakota, South Dakota, and Nebraska, in the eastern two-thirds of Montana, and in the eastern one-half and eastern one-fourth, respectively, of Wyoming and Colorado. Approximately one-fourth of the northern Great Plains area is under cultivation. Production of both cultivated crops and range forage varies greatly from year to year and from one area to another. Droughts are frequent; records indicate that precipitation in certain parts of this area has dropped below 75% of the normal on the average of once in every 5–8 years. Native vegetation can best withstand these successive drought periods.

While cultivated crops may be of importance, especially in local areas, grasslands are the principal resource bases for the agricultural economy prevalent in the northern Great Plains.

Livestock in the northern Great Plains graze the native range species for about 8–9 months of the year, from April to December. During cold and stormy winter months, cattle are generally fed in holding pastures near farmsteads or on range areas where shelter is provided by breaks in the terrain or creek bottoms where trees may be present. Range areas in the northern Great Plains furnish one-half of the nutrients needed by cattle. An additional 30% comes from grassland hay, which is useful during the long winter periods. The rest of the annual feed requirements come from supplemental feeds and cultivated cropland, including crop aftermath.

Some livestock operations in the northern Great Plains are cow-calf enterprises or yearling steer operations, where most producers sell weaned calves and yearlings. In cases where topography is highly variable and vegetation composition is complex, both cattle and sheep can be managed to utilize more fully the range forage resource.

The major range types in the northern Great Plains consist of short grass and mixed grass prairies. The Sandhill ranges are also important vegetation types and occur mainly in west-central Nebraska (Figure 15.17), with smaller areas in adjacent states. Precipitation varies from 15 in. to as high as 20–22 in. annually. Because moisture is rapidly absorbed into the soil, there is little runoff; therefore, effective precipitation is greater in the sand areas.

The sagebrush–saltbush grassland of the northern Great Plains is called the Red Desert and covers some 43,000 square miles. It is the most arid part of the area. Mainly sheep graze the Red Desert in the fall and winter when snow is present.

In the Black Hills of South Dakota and the Bighorn Mountains of Wyoming and other local areas, grasses occur in an open forest type and include ponderosa pine, Douglas fir, and some spruce at higher elevations. This covers almost 20,000 square miles in the northern Great Plains and produces considerable amounts of forage used primarily by cattle during the summer.

The southern Great Plains is a leading cattle producing and livestock farming area that includes over 130 million acres of southeastern Colorado, western Kansas, western

Figure 15.17
Sandhills range in Nebraska.

Oklahoma and Texas, and parts of northeastern New Mexico. Like the Plains area to the north, the climate of the southern Great Plains is extremely variable from year to year. Rainfall is generally limited, humidity is low, and high winds cause high evaporation rates.

Yearlong grazing of range is a common practice in the southern Great Plains, especially with breeding herds. Cultivated crops are often used to supplement winter range, and wheat pasture is an important winter and early spring forage. In the extreme northern part of this area, ranges may be used from April or May to October or November, as is the case in the northern Great Plains.

The vegetation and topography of the southern Great Plains are excellent for cattle. The environment is generally well suited to the production of young animals. Calves are sold primarily as stockers and feeders.

Some supplemental feeding is practiced in winter for increasing protein intake. The use of crop residues in parts of the southern Great Plains and of wheat pasture in fall and spring is common. Using sorghum pastures for late summer grazing is rapidly increasing throughout most of the southern Great Plains.

The tallgrass prairie in the eastern portion of the area is among the world's most productive grassland areas. The vegetation comprises primarily tall grasses, some of which exceed 6 ft in height. The forage of this area can support many cattle. Compared to drier areas, the vegetation does not cure well. Protein supplements are needed throughout the winter months. Areas that once largely grazed steers in the spring and summer are now being converted to cow-calf operations.

The principal range forage types in the western portion of the southern Great Plains consist of short- and mid-grass vegetation that occurs on the heavier textured soils. On more sandy soils along the streams in this area, mid-grass and tall grass vegetation occurs. The short grass vegetation type is dominated by blue grama and buffalograss.

North-Central Area

A large concentration of highly fertile soil and favorable growing conditions is located in the Corn Belt states (Iowa, Missouri, Illinois, Indiana, Ohio, and parts of some adjoining states). In most cases, pasture production cannot compete economically on the most productive land where corn and soybeans can be grown. There are thousands of acres of productive land in the Midwest, where row-crop farming is not feasible on all these acres because of the slope of the land. This land has major erosion problems if it is plowed annually. Conservation compliance may result in more of this land being converted into permanent pasture. Pasture production on this type of land is excellent, and beef cow numbers are high in several of the states in the Midwest.

Smooth bromegrass, orchardgrass, fescue, and Kentucky bluegrass are among the most common grasses in the Midwest. They are grown in pure stand pastures or interseeded with common legumes, such as alfalfa, clovers, birdsfoot trefoil, and lespedeza. These cool-season pastures are highly productive during spring and early summer. Their productivity, however, is low in July and August. Thus, cattle performance declines during the midsummer months. If adequate amounts of rainfall are received during late summer and early fall, there will be a regrowth of highly nutritious cool-season forage that can be utilized (Figure 15.18). Freezing temperatures occurring in the fall terminate new growth, and beef producers rely on the grazing of matured forages and crop aftermath. Harvested forage is then provided when available forage for grazing is depleted, or when snow prevents an adequate amount of dry matter to be grazed.

There is increased utilization of warm-season grasses to complement cool-season forage. For example, in past years the tallgrass prairie extended into the eastern Corn Belt. Switchgrass, Indiangrass, and big bluestem are being utilized in Iowa and Missouri to provide available green forage during July and August. Preference is given to seeding these in pure stands because they are easier to manage. Also, warm-season grasses do not perform well when mixed with cool-season grasses or legumes because of differences in growth patterns and competition.

Warm-season grasses have to be managed differently than the same grasses grown farther west because higher rainfall in the Midwest and competition from weeds is a greater problem. Switchgrass is often the first choice among farmers trying a warm-season grass for the first time because it is easier to establish. Switchgrass, however, is considered a lower-quality grass compared with big bluestem or Indiangrass.

Grazing management of cool-season species is similar to warm-season grasses with one noticeable exception: warm-season grasses store their energy reserve for regrowth above ground in the lower 8–10 inches of the plant, whereas cool-season grasses and legumes place their food reserves in the root systems and crowns. This is the main reason warm-season grasses do not tolerate close grazing.

Figure 15.18
Hay is produced on these Missouri pastures prior to having regrowth grazed by freshly weaned calves and then by dry, pregnant cows.

The South

Until the 1940s, cattle production in the South was incidental to the principal uses of the cleared eastern forestland for raising cotton and tobacco. Prior to this time, the relatively infertile soils were becoming depleted and cash returns were dropping. Ways were sought to rehabilitate seriously eroded cotton and tobacco lands and to provide alternate sources of cash income.

Many eroded croplands in the South were planted in trees; however, this did not solve the cash income problem since 15–20 years were required for the planted trees to reach pulpwood size. Interest in cattle raising as an income source increased, particularly when it was realized that forest soils could produce high yields of forage when soil fertility needs were met. In addition, high-yielding species or strains of grasses and legumes were developed, among them the pioneer Coastal bermudagrass.

As a result of improved pastures in the Southeast, beef cattle numbers increased dramatically. In 12 southeastern states (Florida, Mississippi, Louisiana, Alabama, Georgia, Arkansas, Kentucky, Tennessee, North Carolina, South Carolina, Virginia, and West Virginia), beef cow numbers increased more than fourfold since 1950. There is great diversity in the pasture operations of the Southeast, including many highly developed and intensively managed pastures for purebred, commercial cow-calf, or yearling operations. Kentucky bluegrass is used extensively in the northeastern part of the region. In the southern part of the region, bermudagrass, bahiagrass, and other warm-season grasses are common.

In some areas of the Southeast, cattle may graze entirely on farm woodland areas and other pastures. Many depend largely on industrial forestlands and national forests. It is not unusual for cattle to graze yearlong in the forest with no provision for rotation other than that associated with range burning. Systems for grazing pine forest ecosystems have been developed that are compatible with tree growing. Cattle also graze hardwood forests, but this is often highly damaging to tree growth.

Cattle that can be grazed on green forage most of the year will have excellent performance. Small grains pasture or small grains and ryegrass overseeded into bermudagrass can provide excellent winter forage. Grazing systems in the south are dynamic and vary dramatically due to topography and soil type variability.

Intensive grazing systems can be utilized in this region. These systems can improve cattle productivity and quality of the pasture. Establishing the infrastructure in terms of watering systems that are flexible with pasture movement, the use of high-tensile and electric fencing, and using plantings and cover crops to enhance the forage stand are typically required. In the Gulf Coast region, year-round grazing is an attainable goal with thoughtful planning and well-executed grazing strategies (Figure 15.19).

The Northeast

A high proportion of the farms in the Northeast are dairy farms situated on a combination of hill and valley land, which requires the use of forage grown in rotation to maintain soil productivity. Nearly half of the region's farmland is occupied by hay and pasture crops because most of the land is poorly suited to intensive cultivation. The Northeast has traditionally imported much of its feed grains from the Midwest.

Perennial grasses dominate most haylands and pastures in the Northeast. Timothy, sorghum–sudangrass, smooth bromegrass, and orchardgrass are the most commonly used forage grasses. Except for alfalfa, most perennial seedings include both grasses and legumes. Alfalfa is the most productive legume where it is adapted. It is used as hay, silage, or green chop, or in rotation pastures. Red clover, ladino clover, and birdsfoot trefoil are other legumes successfully grown in the area.

Figure 15.19

Intensive grazing system in the Gulf Coast.

(a) The implementation of an intensive grazing system can yield significant improvements in productivity and health of the pasture system. (b) This system involves the use of high intensive grazing with periods of rest and regrowth – the green paddock to the left has been most recently grazed. (c) Assuring a consistent supply of quality water is critical to success of the system and with the use of pipelines and portable tanks, this objective may be attained.

Source: Photos by John Locke and Corrine Rockemer.

In the southern area of the northeastern United States, cool-season annual grasses are an important source of forage. Winter cereals can be grazed by beef cattle during late fall and early spring. Winter rye can be sod-seeded where bermudagrass is grown to extend the grazing season into the cooler months. Warm-season annual grasses are used throughout the region as a source of supplemental summer feed.

HAYS

Hays are harvested from legumes, grasses, and cereal crops. Alfalfa hay comprises nearly 60% of the more than 60 million acres of hay land in the United States.

The primary objective in haymaking is to reduce the moisture content of green forage, which typically ranges from 65–85% but can be as low as 20%. Hay that is stored with excessive moisture can be destroyed by fire caused by spontaneous combustion, or mold can develop from heating and thus reduce the feeding value. Hay that is harvested too dry is likely to lose many leaves, which are much higher than the stems in protein, carotene, and other nutrients. In addition, hay that is bleached by the sun has a reduced carotene content. Bleaching of hay can be extensive when it is left too long in the windrow or when it is not properly stored outside.

Maturity at harvest has one of the most pronounced effects on the nutrient composition of hay, accounting for approximately 70% of quality differences. The crude protein content of alfalfa, for example, shows the following values on a dry-matter basis: immature, 21.5%; pre-bloom, 19.4%; early bloom, 18.4%; mid-bloom, 17.1%; full bloom, 15.9%; and mature, 13.6%.

Hay exposed to moisture during the haymaking process is subject to extensive leaching of nutrients from the plant material. Studies have shown that alfalfa hay in the swath when exposed to 0.8 in. of rainfall can have its dry-matter yield reduced by 10% and its cellular proteins and carbohydrates reduced by 10–20%. The leaching losses from this amount of moisture are not as large as those that occur when the maturity of plants is allowed to advance 7 days. Rainfall during the curing of hay can maintain moisture levels favorable to microbial decomposition, which causes the shattering of fragile leaves.

It is not uncommon for hay to lose 30–50% of its dry matter from harvesting, storing, and feeding. Small bales of hay can be stored in well-constructed stacks in order to reduce losses in nutrient storage (see the Appendix). Losses from large, round hay bales can be much greater than losses from smaller, square bales unless managed properly. A study conducted in a 19-in. rainfall area showed the following losses in storing large round bales: pyramid, 10.3%; individually separated, 4%; and end-to-end arrangement, 0.8%. If baled hay is to be stored outside, it is important to select a well-drained, gently sloping site with southern exposure. A barrier between the ground and the bale can significantly reduce spoilage and loss. The use of individual bale wrappers or a tarp covering for the stack is recommended especially when bales are in a multilayer formation.

If bale spoilage occurs, losses can be significant. For example, a round bale that is 4 ft wide and 5 ft in diameter will have 13% of the forage in the bale affected by spoilage if the weather penetration is 2 in. By contrast, if the weather penetration is 6 in. deep, 36% of the forage is affected. It is extremely important to use feeding systems that reduce the amount of hay that is trampled and wasted.

Because forage quality is maximized prior to attainment of maximum yields, forage to be utilized as feed for young, growing stock should be harvested early (Figure 15.20) to take advantage of feed value while forage destined for use by mature cattle should be harvested later to gain increased tonnage. Furthermore, stockpiling forage, cool-season species in particular, may be an appropriate option. Stockpiled pastures are saved for late season grazing. Stockpiled legume pastures should not be grazed until after a hard frost occurs.

Measuring Forage Quality

Measuring forage quality is important to the development of accurate rations. A number of measurements contribute to understanding nutritional value including crude protein, insoluble crude protein, neutral detergent fiber, acid detergent fiber, digestible dry matter, net energy, dry matter intake, and relative feed value.

Figure 15.20
Cutting hay from a mixed grass and legume pasture.

Crude protein (CP) is limited, as it is a composite measure of both true protein and nonprotein nitrogen. Insoluble crude protein (ICP), acid detergent insoluble nitrogen, unavailable nitrogen, and heat-damaged protein are indicators of the amount of nitrogen that have linked to carbohydrates as a result of heat damage and become indigestible. This heat damage is most likely to occur in stacks of hay bales of >20% moisture or in silage stored at less than 65% moisture content. Heat-damaged forage becomes discolored and typically has a dark brown or caramelized appearance. Heat damage is most accurately indicated by a ratio of ICP: CP of greater than 0.1.

Traditionally, adjusted CP was used to measure fiber content, but more accurate estimates are obtained via the use of neutral detergent fiber (NDF) and acid detergent fiber (ADF) percentages. NDF measures the cell wall content of the plant and lower values are preferred. Plants harvested at more extended stages of maturity will have increased levels of NDF. ADF measures cellulose, lignin, silica, ash, and ICP. Lower ADF values equate to higher levels of net energy.

Digestible dry matter (DDM) estimates the percentage of forage that is digestible by the ruminant animal. DDM percent = 88.9 – (%ADF × 0.779).

Net energy has replaced total digestible nutrients (TDN) as the best indicator of the energy available to the animal after accounting for losses in waste, gas, and heat production. However, for beef cows on forage diets, TDN is still an appropriate measure. Dry matter intake (DMI) is an estimate of the maximum consumption of forage dry matter and is calculated by:

$$\text{DMI (5\% of body weight)} = 120/\%\text{NDF}$$

Relative feed value (RFV) is a combination of consumption plus digestibility and is used as a reliable method to determine value of forage, particularly legumes or legume–grass mixes. RFV is a measure of energy calculated by the following formula:

$$\text{RFV\%} = \frac{\text{DDM(\%)} \times \text{DMI (\% of body weight)}}{1.29}$$

Because RFV is limited to estimating the energy value of a feedstuff, measures of protein and mineral are also required to create a balanced ration. The guidelines for best use of forages with varying test results are provided in Table 15.12.

Table 15.12
Best Use of Forages of Varying Quality

Best Use	Range/Most Common				
	RFV	ADF	NDF	DDM	DMI
Dairy—high producers	>151	<31	<40	>65	>3.0
Beef—heifers, backgrounding	125–151	31–35	40–46	62–65	2.6–3.0
Beef—lactating cows	102–124	36–40	47–53	58–61	2.3–3.5
Beef—maintenance/non-lactating	87–102	41–42	54–60	56–57	2.0–2.2
Poor—requires supplementation	75–86	43–45	61–65	53–55	1.8–1.9

Source: Adapted from Grant et al. (1997).

CROP RESIDUES

Several crop residues provide a substantial amount of roughage and forage for beef cattle. The most common crop aftermath feeds that can be harvested or grazed are corn stalks (Figure 15.21), straw, or sorghum stubble and soybean residue. Most of these crop residues will contain what is left behind after machine harvesting. It is estimated that the 75 million acres of harvested corn and 35 million acres of harvested grain sorghum can produce approximately 200 million tons of crop residue each year, supplying the winter energy needs for 35–40 million cows. An additional 75 million tons of cereal grain residue (primarily straw) is available in the United States.

Crop residues, because of their relatively low availability of metabolizable energy, are most effectively utilized in maintenance rations for gestating cows when the energy requirement is less than during lactation. Typically, crop residues are low in protein and phosphorus, marginal in calcium, and high in fiber and lignin.

Grazing crop aftermath is usually the most economical method for utilizing feed; however, harvested material can be used when supplemented properly. Snow and mud interfere with the effective utilization of grazed crop residues. During the past several years, more crop residues have been used for cattle feed due to the combined efforts of producers, researchers, and equipment manufacturers.

Figure 15.21
Cows grazing corn stalks during the winter months.

In semiarid regions, residues from feed grains and cereal crops function in water harvest (via snow collection) and erosion protection. Thus, their value as feed is less important than their value as ground cover in minimum tillage systems.

DROUGHT MANAGEMENT

Drought is defined as a prolonged period of several months or more of below normal soil moisture. A severe drought occurs when precipitation is 25% or more below normal. While drought is usually identified as years of low rainfall, drought can occur when the seasonal distribution of rainfall is not favorable for plant growth or when temperatures are high.

Below-normal rainfall throughout the United States and reduced snowpack in the mountains of the arid West can result in extremely low forage production. These drought conditions can occur during one year or can extend through several years. If stocking rate is not reduced during drought, there is a reduction in the pounds of beef produced and usually costs increase. Drought can lead to lowered pregnancy rates, reduced rangeland productivity, and changes in plant community composition.

Drought management cannot always be precisely implemented. However, a critical analysis of weather trends and the development of a plan based on "what if" questioning can help producers mitigate the severity of impact from drought conditions. The first step is to assure that the land resource is continually managed with the goals of establishing organic matter and plant cover appropriate to a healthy rangeland. If soils and plants have already been stressed by poor management practices in the average to above average rainfall years, then a drought year leads to catastrophic conditions for the resource.

Managers should have sufficient monitoring data to be able to establish a general forage grow curve for the ranch or farm. Based on these trends, producers should be able to determine a "date of no return" at which time forage growth is likely to be insufficient even if rainfall occurs.

The adjustment of stocking rate is critical in drought conditions, and changes must be made in a timely fashion to avoid losing flexibility altogether. Drought slows the ability of plants to grow and, therefore, resting pastures become of increasing importance. Movement of cattle more frequently is a likely strategy under these conditions.

Producers should have a destocking strategy in place based on their "what if" planning process. Reacting under stressful conditions is not desirable. An analysis of market conditions and the financial position of the enterprise can help determine whether to buy feed, transport cattle to non-drought areas, or to sell cattle. If selling cattle is the best option, marketing the least productive animals is a good place to start.

In some drought areas, producers will wean calves early so that cows can maintain or increase body condition, thus preventing low reproductive performance the following year. Creep feeding calves prior to early weaning can be done to reduce stress following weaning. The early-weaned calves make an easier transition to the feed after weaning.

Income taxes should be evaluated, as forced liquidation may dramatically increase the tax liability for that year. However, if a federal drought disaster is declared, special tax considerations may change what are the best management options.

MARKET CATTLE PRODUCTION ON GRAZED FORAGE

The vast majority of the beef produced outside of North America is almost exclusively grass fed although the growth of the feedyard business in Australia, Argentina, and Brazil has been substantial. Even in the United States that has a well-developed feedyard business, the industry is largely dependent on forages to sustain the seedstock, and cow-calf and stocker segments of the business. During the 1970s, the production of market cattle with forage-based systems was a topic of much discussion and research. Interest has been reinvigorated as the segment of consumers who identify products based on perceived friendliness to the environment and animal welfare grows. Consumers who make food choices based on a heightened sensitivity to diet–health perceptions have also fueled renewed interest in grass-fed beef production.

Some cattle are fed grain while grazing forage from pastures. Market cattle production from forage occurs when feed grain prices are very high and large losses are experienced by commercial feedlots or when the price of cattle is poor. Feed grain prices are variable depending on yearly production, exports, and government programs. Currently, most market cattle are "fed" cattle, which means they have been fed concentrates in a feedlot. A major challenge in producing forage-fed cattle in extensive forage areas such as the Southeast is the lack of accessible packing plants.

The large renewable forage base for cattle grazing assures people in the United States of a supply of beef even under adverse economic conditions. Most of the nation's beef supply comes from the commercial cattle feeding industry. Without a feeding industry, cows and yearlings would compete for the forage supply. Total annual tonnage of beef would likely be reduced and the supply would be more seasonal. In addition, the palatability characteristics of grass-finished beef would be less uniform than those experienced with grain-fed beef.

Nonetheless, there appears to be a small but growing segment of consumers who desire to purchase beef that has been produced under forage-finishing systems. For those producers who desire to meet this demand, the following questions must be addressed in the formation of a management plan:

Is the seasonal availability of forage sufficient to allow for a grass-based finishing system?

Do the benefits outweigh the costs and will cash flow or debt service be adversely affected?

Can packing, distribution, and marketing services be secured to meet the projected demand?

How strong is consumer demand for the product?

Will the available cattle perform at or above expectation under the grass-based system?

Can the expertise required to make the venture successful be accessed?

Once a producer has made the decision to participate in a forage-finishing beef system, it is of utmost importance that others who have experience in the approach are consulted for advice and assistance. A detailed marketing plan coupled with a complete business plan should be completed before allocating resources to the venture.

Economics Associated with Grazing

The most expensive asset purchased by an agricultural enterprise is land. Real estate market trends have a significant impact on the ability of existing enterprises to expand or for new producers to initiate a farm or ranch. The value of pasture land has increased dramatically since 1997 (Figure 15.22), which has affected the ability of the industry to expand due to a rising cost structure and investment values too high to allow entry of many new cattle producers into the business. The value of pasture land is variable across the United States (Figure 15.23) due to differences in productivity, competition for land, and variation in topography, climate and access to infrastructure.

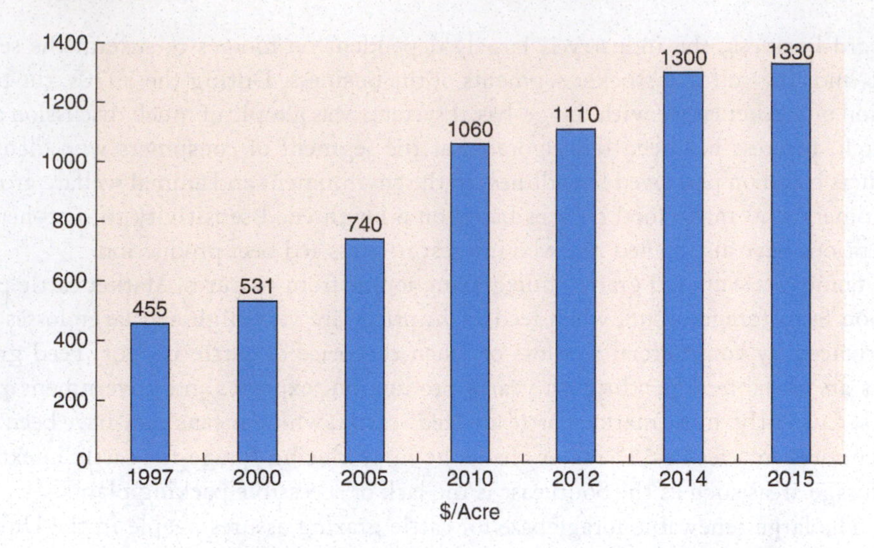

Figure 15.22
Average pasture land prices ($ per acre) in the United States.
Source: USDA.

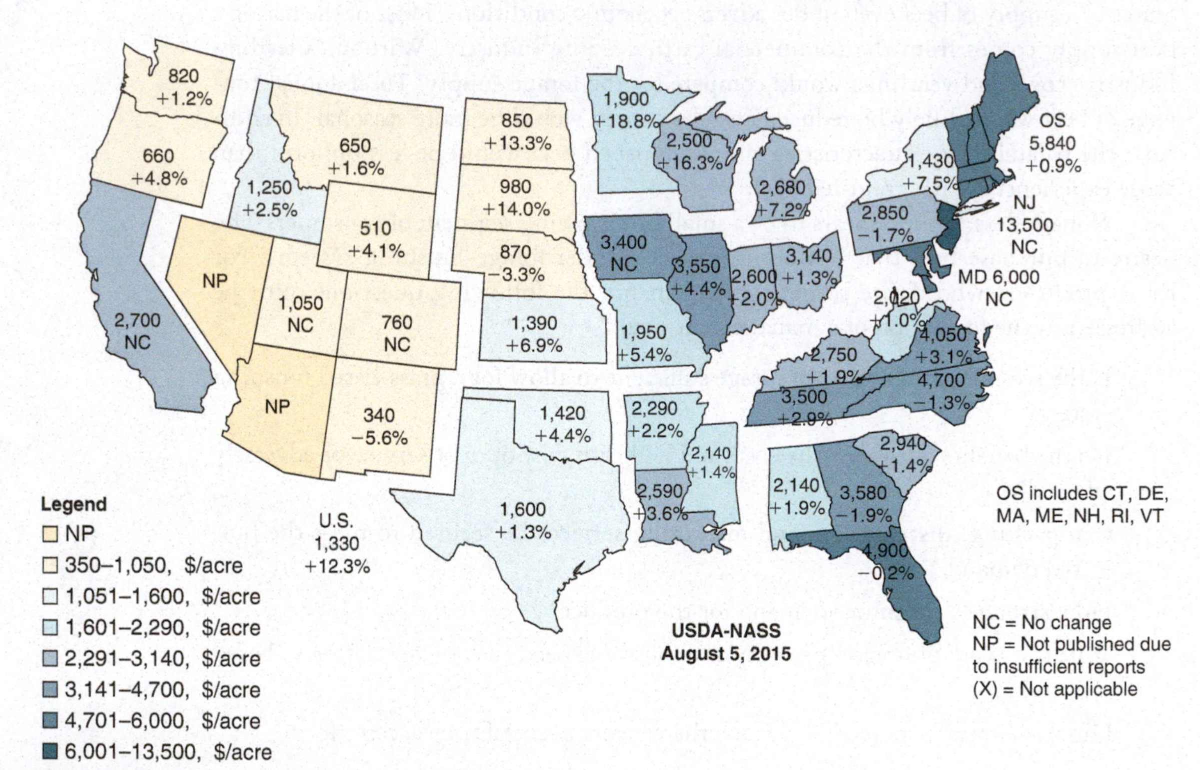

Figure 15.23
Pasture rental prices by state.
Source: USDA.

Factors influencing farm real estate values include pressure from alternative uses, proximity to population centers, availability of water, aesthetic value of the property, governmental regulations, and availability of credit. Given the rather dramatic increased farm real estate valuations in the 1990s and beyond, agricultural producers must carefully assess the costs and benefits associated with land purchases. Because of the generally high cost of purchasing land, many cow-calf producers and stocker cattle operators prefer to lease grazing. Grazing leases on private property are typically evaluated on a monthly cost for an animal unit or for a cow-calf pair. Stocker rates are also tabulated as a monthly fee per animal unit or per head. Changes in average pasture cash rents in the United States are shown in Table 15.13.

Table 15.13
AVERAGE CASH RENT FOR PASTURE PER ACRE (1998–2015)

	($ per acre)
1998	8.80
2000	8.50
2002	9.20
2004	9.60
2006	10.80
2008	10.50
2010	11.00
2012	11.50
2014	12.00
2015	14.00

Source: USDA-NASS.

Grazing Fees

Rangelands in the 17 western states (including the Great Plains) have both private and public ownership. Many ranches have a combination of privately deeded land and grazing permits for use of public lands. Grazing associations also exist where several ranchers cooperatively graze cattle on land privately owned by a grazing association, public grazing land, or a combination of the two. Grazing on federal lands has declined dramatically since the end of WW II—18.2 million AUMs were allowed on the Bureau of Land Management (BLM) permits in the mid-1950s, levels that declined by more than 50% by 2014 when only 8.3 AUMs were allowed.

Public lands are both state and federally owned (Figure 15.24). State lands primarily involve millions of acres. These state lands are usually referred to as "school sections," as revenues from two sections from each township were designated to help support public schools. Cattle producers often have lease agreements for this land with extended leases given to them if certain terms and conditions are met. Lessees own all improvements on the land. If the land is sold or leased to another individual, the original lessee is reimbursed for an appraised value of the improvements.

The U.S. Forest Service (USFS) or the BLM typically manages federally owned grazing lands. Permits to graze Forest Service land are issued to qualifying permittees who have a privately owned livestock operation. Only cattle owned by permittees are allowed to graze the lands managed by the Forest Service. Forest supervisors control the pattern of range use. They may adjust the number of cattle permitted to graze a certain area and the length of time the cattle can be on the range. Grazing fees, however, are established by federal legislation. Therefore, grazing fees and stocking rates are established independently of each other.

Permits can be terminated if the Forest Service decides to use the land for other purposes. In recent years, there has been considerable discussion as to whether federally owned lands should have only a single use (recreation for the general public) or multiple uses, including cattle grazing.

The BLM manages the other public land in a manner similar to that of the Forest Service. Grazing permits are issued, and BLM personnel determine the manner in which the land will be grazed.

Annual grazing fees are assessed to beef producers who have grazing permits. Fees are determined on an AUM basis. Total forage costs on national forests (BLM forage costs are similar) may include one or more of the following components: (1) the annual grazing fee, (2) non-fee costs (investment and maintenance of improvements, herding,

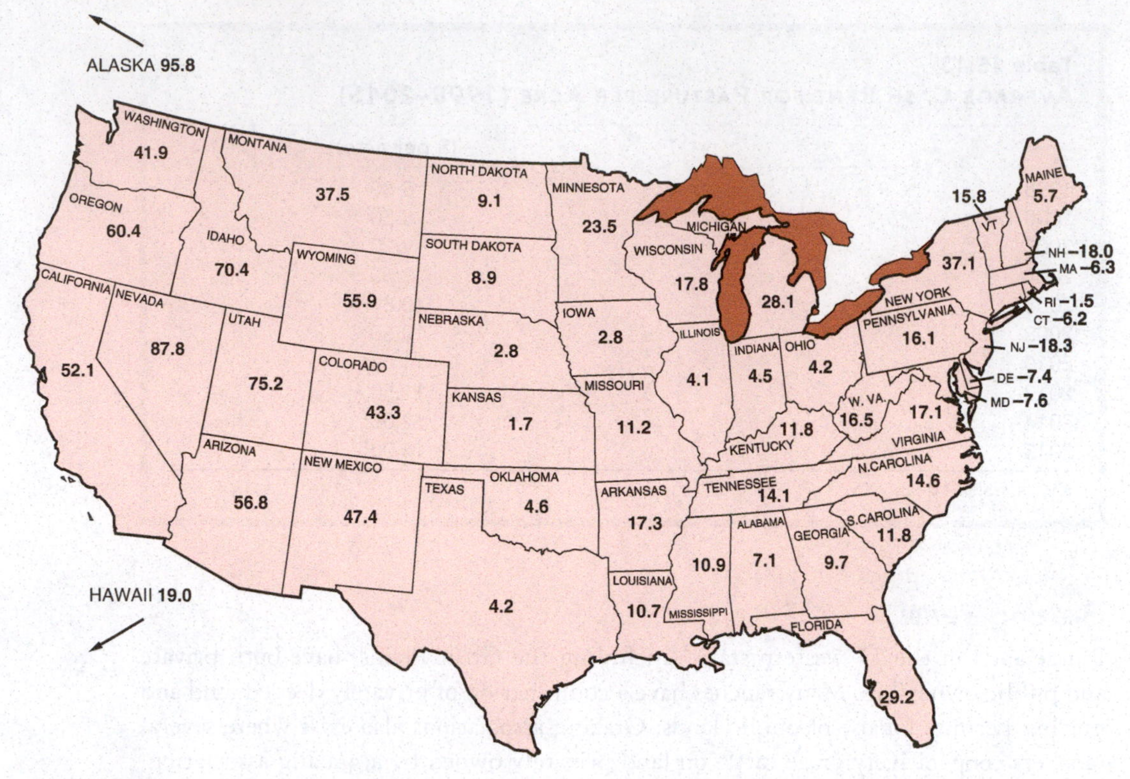

Figure 15.24
Percentage of federal and state government ownership of land.
Source: Adapted from USDA and the Department of Interior.

transportation, and others), and (3) investment in the grazing permit. Grazing fees, which are governed by Congress, are often a source of controversy (Table 15.14).

One of the challenges in determining comparable value between private and public land grazing leases is accounting for non-fee costs and permit value. While the courts have ruled that permit values (economic value associated with holding a federal grazing permit) do not have to be accounted for in determination of federal grazing fees, they still have market value and, as such, remain a point of contention.

Table 15.14
AVERAGE ANNUAL GRAZING FEES FOR FEDERAL RANGE, 1970–2016

Year	Forest Service/BLM Cost per AUM ($)
1970	0.60[1]
1975	1.11[2]
1980	2.36
1985	1.35
1990	1.81
1995	1.61
2000	1.35
2005	1.79
2010	1.35
2015	1.69
2016	2.11

[1]Denotes USFS fee; BLM fee was $0.44 per AUM.
[2]Denotes USFS fee; BLM fee was $1.00 per AUM.

The 1995 report by the Grazing Fee Task Group found that on private leases, landlords provided all or part of property maintenance (51%), daily care of cattle (17%), water supply (53%), liability insurance (41%), utilities (34%), and provision for death losses (6%). On federal lands, neither the BLM nor USFS provides these services. Figure 15.25 illustrates the impact of non-fee costs on the full cost of a grazing lease. Therefore, the non-fee costs must be accounted for when comparing private and public lease rates. The total cost analysis found that 34% of cattle producers on BLM lands and 62% of cattle producers on USFS lands paid more for grazing public lands as compared with private lands.

The Grazing Fee Task Group (1995) made several key recommendations in regard to the grazing fee issue:

1. The grazing fee should be established within the range of $3 to $5 per AUM. This recommendation assumes no allowance for the investments graziers have in grazing permits. Recognition of grazing permit value remains a key issue. The current grazing fee, or an even lower fee, would be appropriate if grazing permit investment were considered.
2. Any base grazing fee should be applied throughout the West.
3. Any base grazing value should be updated annually with the forage value index from the previous year.
4. The BLM and USFS should investigate the potential of implementing a competitive bid system to create a market for public land grazing.

Wheat Pasture

Cattle graze several types of small grain pastures throughout the United States. The most noted of these are the winter wheat pasture areas in several southern Great Plains states where millions of stocker cattle are grazed each year. Wheat pasture is also important in some adjoining states but to a much lesser extent than in Kansas, Oklahoma, and Texas.

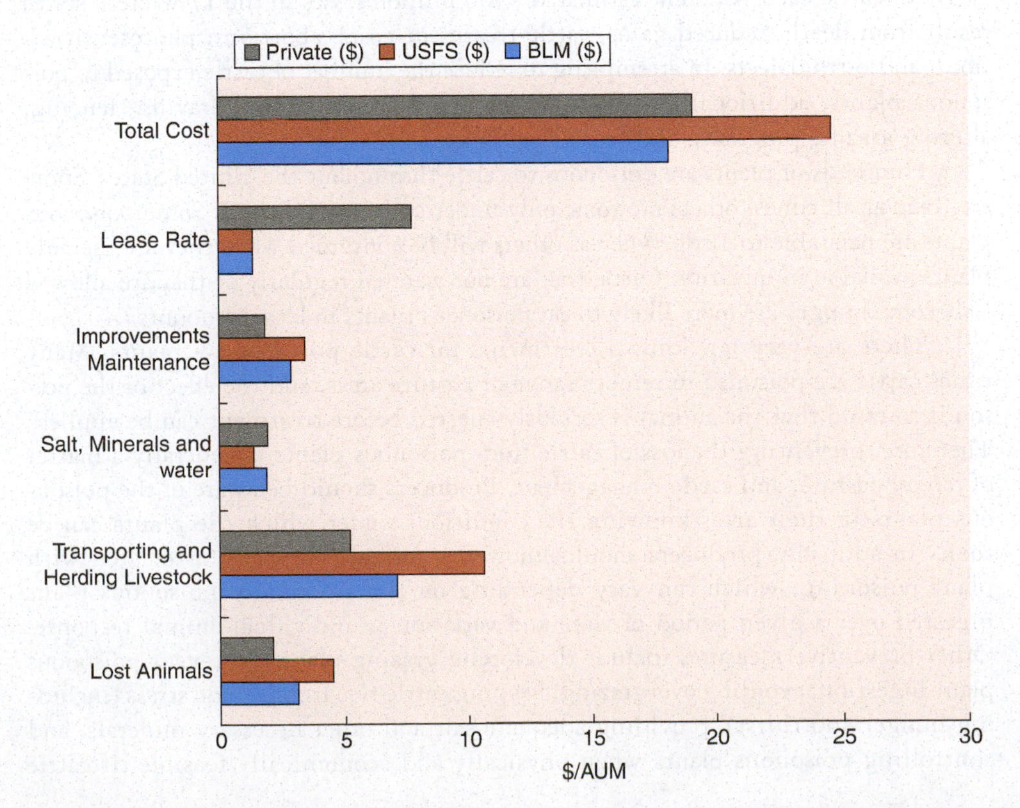

Figure 15.25
Comparison of fee and non-fee costs associated with private and public grazing leases.
Source: Adapted from multiple sources.

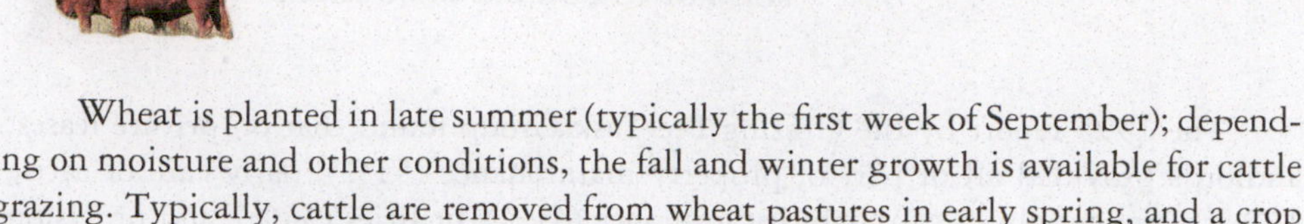

Wheat is planted in late summer (typically the first week of September); depending on moisture and other conditions, the fall and winter growth is available for cattle grazing. Typically, cattle are removed from wheat pastures in early spring, and a crop of wheat for grain is then produced. The usual grazing period is from November to the middle of March. Grazing can continue to the end of May if a decision is made not to harvest a crop of grain.

Daily gains on wheat pasture are usually 1.5–2.5 lb per day. There can be serious problems with bloat; average annual death losses of less than 1% can be expected, although losses may reach 3%. Bloat is most frequent during a 2–4-week period in the early spring when the wheat breaks dormancy. As the wheat matures, bloat is reduced. Some producers feed low-quality roughages (wheat straw or sorghum–sudan hay) in an attempt to improve the utilization of wheat pasture and reduce bloat. Research shows that *ad libitum* consumption of these roughages is low, with little effect on the incidence of bloat.

Stocking rates vary with the amount of top growth available, and more numbers can typically be put on irrigated wheat than dryland wheat. Stocking rates are more critical after the jointing of the wheat as it starts to mature and accelerate growth. If stocking rates are not high enough at this time, the wheat will mature ahead of the cattle's ability to graze it. Thus, the cattle will spot graze and avoid the more mature wheat plants.

Cattle may have to be moved from wheat pasture during periods of heavy rainfall or irrigation to prevent trampling damage and a significant reduction to the wheat crop.

HEALTH PROBLEMS ASSOCIATED WITH GRAZING PLANTS

Poisonous Plants

Poisonous plants represent a major economic loss to the cattle industry, affecting 3–5% of cattle each year. The estimated $300 million losses in the 17 western states result from death, reduced gains, cattle treatment costs, abortions, photosensitizations, and birth defects. In attempting to reduce the number of cattle exposed to poisonous plants, additional economic losses include the costs of spraying, fencing, altering grazing programs, and preventing forage loss.

Hundreds of plants are poisonous to cattle throughout the United States. Some are toxic at all times; others are toxic only under certain conditions. Some poisonous plants are palatable to cattle, whereas others will be consumed when they are the only plants available in quantity. Cattle that are not watered regularly or that are allowed to become hungry are more likely to eat poisonous plants in lethal amounts.

There are very few known treatments for cattle poisoned by plants. Many times cattle are poisoned in remote range or pasture areas, and the effect of the poison is so rapid that the animal is seriously affected before treatment can be applied. Therefore, preventing the loss of cattle from poisonous plants is generally a matter of range, pasture, and cattle management. Producers should be aware of the poisonous plants in their areas knowing the conditions under which the plants can be toxic. In addition, producers should know the animal symptoms associated with plant poisoning, which can vary depending on the amount of poisonous plant ingested over a given period of time and variation in individual animal response. Other preventive measures include developing grazing plans to prevent poisonous plant ingestion, avoiding overgrazing, keeping cattle free from undue stress (including hunger and thirst), providing adequate salt and other necessary minerals, and controlling poisonous plants when physically and economically feasible. If cattle show signs of poisoning, a local veterinarian should be consulted for proper diagnosis, treatment, and prevention.

Table 15.15 lists some common poisonous plants and provides information on their distribution and control as well as the symptoms and treatment of poisoned cattle. However, producers will need more extensive information when plant-poisoning problems are encountered. A good informational source is the USDA Poison Plant Research Laboratory (see the Appendix).

Fescue Toxicity

Fescue toxicity is a serious problem with cattle grazing tall fescue, especially in the tall fescue-growing areas of the eastern United States, where the tall fescue is present on nearly 35 million acres of pasturelands. Fescue toxicity is associated with a fungus called endophyte. The fungus lives within the leaves, stems, and seed of the tall fescue plant and is not visible externally. Tall fescue pasture, hay, and seed contain the fungus, although the fungus dies within about a year in stored seed. It is likely that the fungus causes the grass to produce a toxic compound that is harmful to animals eating the forage. Surveys conducted in several southeastern states show that over 95% of tall fescue pastures are heavily infected with the fungus and it has been estimated to have a negative economic impact in excess of $600 million annually.

Fescue toxicity impacts the productivity and health status of cattle that consume endophyte-infected forage stand. Specific impacts include reduced milk production, diminished feed intake, poor growth rates, and subpar reproductive performance. Cattle consuming endophyte-infected fescue often have increased respiration rates, elevated body temperature, and excessive saliva production. Cattle become less tolerant of heat and will seek shade, stand in water, and increase water intake in an attempt to mitigate heat. Affected animals will have an unhealthy hair coat appearance, may demonstrate nervousness, display lameness, and in extreme cases may slough hoof tissue.

Average daily gain of steers on tall fescue pastures has been poor, usually about 1 lb per day or less for the grazing season. The introduction of endophyte-free fescue has increased gains dramatically. A 4-year grazing study showed significantly higher gains (1.82 lb per day versus 1.09 lb per day) for steers grazed on endophyte-free fescue. Another study with cows and calves showed average daily calf gains of 2.65 lb on endophyte-free fescue and 1.92 lb gains for calves on endophyte-infected grass. The average daily weight changes of cows were 1 lb on endophyte-free fescue and –0.40 lb on endophyte-infected fescue. The average daily gain of steers fed endophyte-free fescue hay was 1.45 lb, but only 0.63 lb for steers fed endophyte-infected hay.

One or more of the following methods can reduce the effects of fescue toxicity:

1. Introduce legumes to dilute tall fescue (e.g., planting ladino or red clover in fescue sod is often practical).
2. Feed hay other than tall fescue in winter, when cattle are grazing toxic fescue pasture. Bermudagrass and sericea lespedeza hay are good choices.
3. Terminate the existing stand and replant infected fescue pastures with warm- or cool-season grasses or stress-tolerant tall fescue with nontoxic endophytes. Old seed, where the fungus has died, can be planted, but the seed rate must be increased to make up for reduced germination. An infected fescue pasture should not be allowed to make seed during the year it is to be destroyed and replanted. The endophyte-free cultivars are less tolerant of drought and overgrazing, so excellent management is required.
4. Introduce the winter-productive Triumph variety of fescue.

Table 15.15
Selected Poisonous Plants[1] That Cause the Largest Economic Loss in Cattle

Plant	Distribution	Cause of Poisoning	Prevention or Control[2]	Sign of Poisoning	Effective Treatment
Arrowgrass	Widely distributed in marshy, alkaline pastures and native hay areas throughout the United States	Hydrocyansic acid	Prevent consumption, especially of plants retarded by drought or frost	Distress, cyanosis, rapid respiration, salivation, convulsions, rapid death	Limited; immediate treatment of solution (10 cc of 10% sodium thiosulfate and 10 cc of 10% sodium nitrate)
Astragulus (nitro-containing)	Western Canada, western United States, and northern Mexico	3-nitropropionic acid or 3-nitro-1-propanol	Use grazing programs that favor more desirable forages; apply herbicides; prevent grazing	*Acute*: Respiratory distress and muscular weakness, primarily in the pelvic limbs *Chronic*: Labored and rapid respiration, wheezing or roaring sound, weakness in pelvic limbs, knuckling at fetlocks, goose stepping, and knocking together of hind feet when walking	None
Bracken fern	Throughout the United States		Prevent overgrazing	High fever; loss of appetite, depression, labored breathing, salivation, nasal and rectal bleeding	None
Grass tetany	(See section in latter part of the chapter)				
Fescue toxicity	(See section in latter part of the chapter)				
Larkspur (two groups—tall and low—based on height at maturity)	Primarily in the 11 western states	Alkaloids	Grazing of infested areas later in grazing season; apply herbicides if economical	Nervousness, staggering gait, salivation, bloat, sudden death	Limited; immediate injections of physostigmine salicylate can be helpful
Locoweed (several species)	Foothills, plains, and semiarid desert areas of the 17 western states		Keep off if infested areas until good feed is available; apply herbicides if economical	Abortion, birth defects, depression, nervousness, loss of sense of direction, poor feed consumption, and poor gains	None
Lupines	Foothills and mountain ranges in sagebrush and aspen areas (most western states)	Alkaloids	Birth defects prevented if consumption not permitted between 40 and 70 days of gestation; apply herbicides if economical	Crooked-legged calves and other birth defects (Figure 15.26); nervousness, loss of muscular control, convulsions, coma, frothing at mouth	None
Milkvetch (several species)	Throughout most of the United States	Miserotoxin or nitro compounds	Prevent grazing as plants are palatable; use herbicides on some species if economical	*Acute*: Tendency to stand on toes, weakness, irregular gait, posterior paralysis *Chronic*: Nervousness, muscular weakness, respiratory distress	None

Table 15.15
(Continued) Selected Poisonous Plants[1] That Cause the Largest Economic Loss in Cattle

Plant	Distribution	Cause of Poisoning	Prevention or Control[2]	Sign of Poisoning	Effective Treatment
Milkweed (several species)	Most of the 17 western states	Glucosidic substances (called cardenolides)	Adequate forage, as plant is unpalatable; herbicides available, but usually not economical	Difficult breathing, (grunting), loss of muscular control, bloat, spasms	None
Nitrate-poisoning	(See section in latter part of the chapter)				
Ponderosa pine needles	Most 11 western states		Needles are usually grazed only when cattle are stressed (due to feed changes, weather, and hunger)	Abortion during last trimester, retained placenta, calves may be weak if near term	None
Selenium-accumulating plants (sometimes called *alkali disease*)	Primarily in North and South Dakota, Montana, Wyoming, Colorado, and Utah	More than 5 parts per million of selenium in plants *Acute*: Selenium accumulation greater than about 100 ppm *Chronic*: Grasses and grains containing 5–40 ppm selenium	Prevent consumption	*Acute*: Loss of appetite, depression, coma, death *Chronic*: Dullness, rough hair coat, emaciation, lameness, possible hoof deformities	Remove from range containing plants with high selenium concentrations
Sweet clover poisoning	Throughout most of the United States	Coumarin in clover is converted to dicoumarin, which prevents the synthesis and metabolism of vitamin K	Avoid feeding moldy sweet clover	Stiffness, lameness, and swellings (blood clots) beneath the skin	Blood transfusions; intramuscular injections of vitamin K
Threadleaf snakeweed (broomweed)	Drier range areas, primarily in the southwestern United States		Avoid grazing; plants susceptible to 2, 4-D	Abortion, sometimes death; weight loss and diarrhea	None
Water hemlock	Throughout most of the United States	Cicutoxin (a highly poisonous unsaturated alcohol that has a strong carrot-like odor)	Provide more palatable forage; prevent cattle from consuming roots brought to the surface by plowing or cleaning ditches; keep animals away (3 weeks) from plants sprayed with herbicide as palatability is increased; apply herbicide if economical	Cattle die suddenly with few observed signs of poisoning	None

[1]Other plants occasionally poisonous to cattle include bitter rubberweed, chokecherry, cocklebur, copperweed, desert baileya, death camas, drymary (inkweed), greasewood, groundsel (riddel and threadleaf), halogeton, hemp dogbane, horsetails, oak, pinque, poison hemlock, rayless goldenrod, St. John's wort, sneezeweed, spring parsley, tansy mustard, tansy ragwort, and western false hellebore.
[2]The use of some herbicides listed may be banned in certain states. Directions should be followed carefully.

Figure 15.26
Brood cows grazing toxic species of lupine plants during day 40 and 100 of pregnancy may result in offspring with congenital defects such as crooked legs and cleft palate.
Source: USDA:ARS.

5. Do not stockpile spring fescue for use in the fall months but fall fescue growth may be stockpiled for winter grazing.
6. Increase grazing intensity to decrease stem and seedhead formation. Use mechanical pasture clipping if necessary.
7. Hay the stand during spring growth prior to seedhead formation and then graze until summer dormancy occurs.
8. Prevent cattle from grazing fescue in the summer months or supplement cattle grazing infected stands with 2 to 5 pounds per head per day.
9. Shift cow herd production cycle so that calving occurs in the fall or winter months only.

Grass Tetany

Grass tetany, also known as winter tetany, grass staggers, wheat poisoning, magnesium tetany, and hypomagnesemia, occurs throughout the United States. Although it is usually observed in the spring, it can occur in the fall and winter. Most often it is observed when cattle are grazing cool-season forages that have a very rapid, lush growth. Cows nursing calves under 2 months of age are most frequently affected by grass tetany.

Many times, the clinical symptoms of grass tetany are not observed, and dead animals are the only history of the problem. Affected animals may become excitable, expressing a wild stare with erect ears, and appear to be blind. They may appear uncoordinated, tending to lean backward or to stumble. An affected animal often has trembling muscles and grinding teeth followed by violent convulsions, deep coma, and death.

Positive diagnosis of grass tetany is difficult because its symptoms are shared with other diseases. A blood test can be helpful, as the serum of magnesium is typically low (the normal level is 2.25 mg per 100 ml of serum, whereas affected animals are usually below 1 mg).

Prevention of grass tetany may involve a combination of several factors. Magnesium mineral (magnesium oxide) should be provided. Legume or legume–grass pasture should be utilized, as tetany seldom occurs where legumes are grazed. Less susceptible animals (heifers, stockers, dry cows, or cows with calves over 4 months) can be grazed on high-risk pastures. Some producers feed a corn silage supplement while cows and calves are on wheat pasture, a practice that greatly reduces the incidence of grass tetany. If soil magnesium is low, a program can be implemented to increase its level in the soil. However, fertilization with magnesium is less effective than supplementing cattle with magnesium.

Early treatment of animals affected by grass tetany is important. Cattle affected by grass tetany that collapse and have been down more than 12–24 hours seldom recover. Animals should be handled gently to prevent excitement. Two hundred cubic centimeters (cc) of a sterile solution of magnesium sulfate (Epsom salts) injected under the skin in at least four different sites will give a high level of magnesium in the blood within 15 minutes.

Nitrate Poisoning

Nitrate poisoning occurs throughout the United States because any forage plant can accumulate nitrate (NO_3) under certain conditions such as drought. Nitrate is lethal to cattle when plants contain more than 0.9% nitrate (NO_3) and sublethal when plants contain 0.5–1.5% nitrate. The nitrate level of plants should be checked if nitrate poisoning is suspected or anticipated.

Nitrates are converted to nitrites, which produce methemoglobin, a type of hemoglobin that cannot carry oxygen. The acute signs of nitrate poisoning in cattle are a blue coloration of mucous membranes, a staggering gait, shortness of breath, and then death. Some chronic signs of poisoning are watering eyes, unthriftiness, reduced milk flow, and reduced gains.

Administration of methylene blue to affected animals will convert methemoglobin back to hemoglobin. In chronic cases, the feed should be changed or mixed to reduce the nitrate intake of cattle. The use of careful testing of feeds for nitrate levels, the frequent observation of cattle, and utilization of appropriate harvesting techniques can minimize the potential problem.

Prussic Acid Poisoning

Prussic acid or hydrocyanic acid (cyanide) can be a problem when feeding sudangrass, forage sorghum, or sorghum–sudangrass hybrids. Prussic acid is not available from normal, healthy plants. The toxic compound is created by enzymatic activity under conditions when plants are stressed by drought or freezing. Symptoms of prussic acid poisoning include excitability, rapid pulse, and muscular tremors followed by staggering and collapse. Treatment must be administered quickly to avoid death.

Prussic acid poisoning can be avoided by not grazing sudangrass or sorghum–sudangrass hybrids until plants are 18–24 inches tall. Heavy stocking rates coupled with rotational grazing are advised to prevent animals from selectively consuming the leaves and growth shoots that have the highest concentration of the compound. The best approach to avoid prussic acid poisoning is to conduct feed analysis on both growing and harvested forms of problematic plants. Hydrocyanic acid concentrations of less than 50 mg% on a dry basis are considered safe while more than 75 mg% is deemed very dangerous.

SELECTED REFERENCES

Publications

Agricultural land values and cash rents. 2016. Washington, DC: USDA-NASS.

Bailey, R.G. 1995. *Description of the ecoregions of the United States.* USDA Forest Service Publication 1391.

Ball, D.M., Hoveland, C.S., & Lacefield, G.D. 1991. *Southern forages.* Norcross, GA: Potash and Phosphate Institute.

Barnes, R.F., Miller, D.A., & Nelson, C.J. (Eds.). 1995. *Forages: An introduction to grassland agriculture* (5th ed.). Ames, IA: Iowa State University Press.

Barnes, R.F., Miller, D.A., & Nelson, C.J. (Eds.). 1995. *Forages: The science of grassland agriculture.* Ames, IA: Iowa State University Press.

Blasi, D.A., & Corah, L.R. 1993. *Assessment of forage resources for determining the ideal cow.* Proc. Cow-Calf Conference III. Salina, KS.

Brady, N.C. 1990. *The nature and properties of soils* (10th ed.). New York: MacMillan.

Colorado Rangeland Working Group. 2009. *Colorado rangeland monitoring guide.* Denver, CO.

Cook, C.W., et al. 1983. *Alternate grass and grain feeding systems for beef production*. Fort Collins, CO: Colorado State University Experiment Station Bull, 579S.

Dagget, D., & Dusard, J. 1995. *Beyond the rangeland conflict: Toward a West that works*. The Flagstaff, AZ: The Grand Canyon Trust.

Forage facts notebook. 1998. Manhattan, KS: Kansas State University Agricultural Experiment Station and Cooperative Extension Service.

Grant, R., Anderson, B., Rasby, R., & Mader, T. 1997. *Testing livestock feeds for beef cattle, dairy cattle, sheep, and horses*. Cooperative Extension Publication G89-915-A. Lincoln, NE: University of Nebraska.

Grazing Fee Task Group. 1995. *The value of public land forage and the implications for grazing fee policy*. New Mexico State University, AES Bulletin 767.

Hancock, A. 2006. *Doing the math: Calculating a sustainable stocking rate*. Central Grassland Range Extension Center. NDSU. Fargo, ND.

Hart, R.H. 1987. *Economic analysis of stocking rates and grazing systems*. Proc. Range Beef Cow Symposium X. Cheyenne, WY.

Hart, R.H., et al. 1993. Grazing systems, pasture size, and cattle grazing behavior, distribution, and gains. *Journal of Range Management*. 46:81.

Heitschmidt, R., & Walker, J. 1983. Short duration grazing and the Savory grazing method in perspective. *Rangelands*. 5:147.

Hermel, S. 1996. Calculating grass needs. *Beef*. Spring Cow-Calf Special.

Holechek, J.L., Pieper, R.D., & Herbel, C.H. 1989. *Range management: Principles and practices*. Englewood Cliffs, NJ: Prentice Hall.

Holland, C., & Kezar, W. (Eds.). 1995. *Pioneer forage manual: A nutritional guide*. Pioneer Hi-Bred International. Des Moines, IA.

Hoveland, C.S. 2000. Achievements in management and utilization of southern grasslands. *Journal of Range Management*. 53:17.

Intermountain planning guide. 2003. Logan, UT: ARS, USDA-NRCS, and Utah State University.

James, L.F., Keeler, R.F., Johnson, A.E., Williams, M.C., Cronin, E.H., & Olsen, J.D. 1980. *Plants poisonous to livestock in the western states*. Washington, DC: USDA, Agric. Info. Bull. 415.

Jennings, J., Troxel, T., Gadberry, S., Simon, K., & Jones, S. 2010. *300 Days grazing program*. University of Arkansas Cooperative Extension Service. Fayetteville, AR.

Know your grasses: A field guide. 2004. Publication B-182. College Station, TX: Texas Cooperative Extension Service.

Launchbaugh, J.L., & Owensby, C.E. 1978. *Kansas rangelands: Their management based on a half-century of research*. Manhattan, KS: Kansas Agricultural Experiment Station Bull. 622.

Martin, J.M., & Rogers, R.W. 2004. Forage-produced beef: Challenges and potential. *Professional Animal Scientist*. 20:3.

Miller, R.W., & Donahue, R.C. 1995. *Soils in our environment* (7th ed.). Englewood Cliffs, NJ: Prentice Hall.

Morrow, R., & Hermel, S. 1996. Calculating grass needs. *Beef*. Spring Cow-Calf Edition.

Neilsen, N.B., & James, L.F. 1992. The economic impact of livestock poisoning by plants. In *Poisonous plants*. Proc. Third International Symposium. Ames, IA: Iowa State Univ. Press.

Pratt, D.W. 2013. *Healthy land, happy families and profitable businesses*. Ranch Management Consultants. Fairfield, CA.

Pastures for profit. 1999. Minnesota Extension Bulletin A3529. University of Minnesota.

Reece, P.E. 1991. *Evaluation and practical use of research results for developing grazing strategies*. Proc. Range Beef Cow Symposium XII. Fort Collins, CO.

Reid, R.L., & Klopfenstein, T.J. 1983. Forages and crop residues: Quality evaluation and systems of utilization. *Journal of Animal Science*. 57:534 (Suppl. 2).

Roath, R. 1997. *Applications of monitoring for producers*. Proc. Range Beef Cow Symposium XV. Rapid City, SD.

Savory, A. 1999. *Holistic management* (2nd ed.). Washington, DC: Island Press.

Sayre, N. 2001. *The new ranch handbook: A guide to restoring western rangelands*. The Quivera Coalition. Santa Fe, NM.

Stevens, R., Aljoe, H., Forst, T.S., Motal, F., & Shankles, K. 1997. How much does it cost to burn? *Rangelands*. 19:2.

Smith, B., Leung, P., & Lore, G. 1986. *Intensive grazing management: Forage, animals, men, profits*. Kamuela, HI: The Graziers Hui.

Tranel, J.E., Sharp, R.C., & Kaan, D.A. 2003. *Custom rates for Colorado farms and ranches in 2000*. Fort Collins, CO: Colorado State University.

United State Department of Agriculture: National Agricultural Statistics Service. 2016. *Farms and land in farms: 2015 Summary*. Washington, DC.

Vallentine, J.F. 1989. *Range development and improvement*. San Diego, CA: Academic Press.

Vallentine, J.F. 1990. *Grazing management*. San Diego, CA: Academic Press.

White, R.S., & Short, R.E. (Eds.). 1988. *Achieving efficient use of rangeland resources*. Miles City, MT: Montana State Agricultural Experiment Station.

16

Herd Health

Profitability of the cattle enterprise is influenced by a diverse set of influences of which none is more important than the execution of an effective health program founded on disease prevention and backed up by the expertise to deal with disease. Preventative measures are typically more cost-effective to implement than attempting to deal with a disease outbreak. This chapter focuses on helping producers understand the major health problems affecting beef cattle and the importance of planning economically effective prevention and treatment programs.

Productive cattle are typically in excellent health. Death loss (**mortality rate**) is the most dramatic sign of health problems; however, lower production levels and higher costs of production due to sickness (**morbidity**) are economically more serious. For example, feedlot cattle that are healthy may return as much as $100 per head more profit as compared to cattle that are sick. The $100 profit is realized from reduced death loss, improved cost of gain, decreased medicine costs, and better carcass value resulting from better health.

Disease is any deviation from normal health in which there are marked physiological, anatomical, or chemical changes in the animal's body. There are two major disease types: noninfectious and infectious. Noninfectious diseases result from injury, genetic abnormalities, ingestion of toxic materials, and poor nutrition. Microorganisms are not involved in noninfectious diseases. Examples of noninfectious diseases are plant poisoning, bloat, and mineral deficiencies. Microorganisms such as bacteria, viruses, and protozoa cause infectious diseases. A contagious disease is an infectious disease that spreads rapidly from one animal to another.

Temperature, pulse, and respiration rate are important vital body signs that can be monitored to assess health. An abnormal temperature is one of the first objective signs of a health problem. When an animal's temperature is above normal, it is considered a fever; when it is below normal, it is called **hypothermia**. Fever is more common than hypothermia. The absence of fever does not mean the animal is not sick in all instances. Infectious diseases may cause fever. Metabolic diseases are usually not associated with fever.

The normal temperature range for cattle is from 100.4°F to 103.1°F; factors such as time of day, amount of physical activity, and others cause normal fluctuations in temperature. When body temperature goes one degree above the normal upper limit, the animal is considered to have a fever. Even with an extremely high fever, the temperature seldom goes above 107°F unless the animal experiences heat stroke, at which time the temperature may exceed 110°F.

Pulse is the rhythmic periodic thrust felt over an artery in time that is associated with the heartbeat. The rate of pulse can be obtained by putting the fingers over the superficial arteries and pressing the arteries against a hard or bony structure (the common location in cattle is the lower part of the jaw or under the tail 2–6 inches from its base). The pulse rate can vary depending on age, size, sex, atmospheric condition, time of day, excitement, and other factors. The normal range for pulse rate (heartbeats per minute) is 40 to 70 in mature cattle and higher in calves.

Respiration is the act of breathing—taking in oxygen and expiring carbon dioxide. Many primary and secondary diseases affect the respiratory system. Respiration rate is the number of inspirations (expansion of chest or thorax area) per minute. The normal range is 10–30 per minute. Some of the same factors identified for temperature and pulse can cause variations in the rate of respiration.

Several other aspects of disease need to be recognized. Some diseases in beef cattle are acute in nature, as they are typically severe but short term. These diseases, such as bloat or enterotoxemia, must be recognized early or prevented altogether to avoid death losses. **Chronic** diseases are lingering dysfunctions that lead to the continual reduction of productivity and health status. Johne's disease is an excellent example of this disease classification.

Some diseases are overt and thus can be clinically diagnosed, as the symptoms are relatively apparent. Subclinical conditions are harder to detect as they may involve slow losses in animal productivity and generalized health status. These diseases require a greater attention to monitoring if they are to be detected early and prior to their having significant negative consequences.

DISEASES AND HEALTH PROBLEMS

Table 16.1 identifies some common disease and health problems encountered by beef cattle. Accurate disease diagnosis is an essential element in a health management program. A thorough examination by a veterinarian plus diagnostic laboratory verification may be necessary to establish a specific disease diagnosis. Once a diagnosis has been made, the veterinarian should prescribe treatment. Specific products used in prevention and treatment should be cleared through a veterinarian so that producers can meet FDA regulations in product use and withdrawal times. Prevention and treatment guidelines are given in Table 16.1 for general information only.

In any system, death losses will occur due to advance age, random events such as blizzards, and normal variation in health status within a population. However, without effective management morbidity rates and death losses can become an

Table 16.1
Common Diseases and Other Health Problems in Cattle

Disease and Cause(s)	Clinical Signs	Prevention	Treatment
Abomasitis Ulceration, sand colic, abomasal bloat in calves, *clostridium perfringens A*	Adults: anorexia, distended right flank, weak, dehydrated. Calves: unthrifty, distended abdomen, soft-discolored feces	Avoid trash (rags, tarps, etc.) in pastures, supplement protein when ration is high in poor quality roughage	Prognosis generally considered unfavorable; abomasotomy possible in calves
Acidosis, Rumenitis, Liver Abscess Complex Overconsuming concentrate after a period of reduced feed consumption; increasing concentrate in diet too rapidly *Fusobacterium necrophorus; Actinomyces pyogenes; Bacteroides*	Initially, animals off feed, clinical signs of "feed intoxication;" acidosis observed following slaughter	Reduction of concentrates (less than 75%), more roughage; Chlortetracycline, 70 mg/head/day; Tylosin, 60–90 mg/head/day; use of buffers; use of ionophores	Antacids; antifermentatives; gastric lavage; fluid therapy; reduce concentrate levels in feed

Table 16.1
(Continued) Common Diseases and Other Health Problems in Cattle

Disease and Cause(s)	Clinical Signs	Prevention	Treatment
Anaplasmosis *Anaplasma marginale*	Anemia, fever, icterus, weakness, and emaciation; use only laboratory diagnosis	Control of insects; care in spreading disease by veterinary instruments; vaccine is available. Chlortetracycline: Cattle up to 700 lb—350 mg/head/day; cattle 700–1,000 lb—500 mg/head/day; cattle 1,500 lb and over—0.5 mg/head/day. Chlortetracycline (for carrier stage of anaplasmosis): 5.0 mg/lb body weight/day for 60 days in the feed	For acute cases, blood transfusions, chlortetracycline, oxytetracycline
Anthrax *Anthrax bacillus*	Sudden death (1–2 hours after infection); "sawhorse on its side" appearance; failure of blood to clot; delayed rigor mortis; can cause disease in humans	Vaccination (recommended only in areas where disease occurs)	Antibiotics and antiserums (success relatively poor); contact state veterinarian; do not move or transport carcasses
Asthma (cow) Occurs in the fall when cows are moved from dry, sparse forage to lush, green pasture; may be caused by an improper amino acid (tryptophan) metabolism	Clinical signs vary from slight to severe; respiratory distress, with abnormal increase in depth and rate of respiration; a grunt usually accompanies respiration	Moderate the feed change; have cows filled with dry feed before moving to lush, green pasture; Ionophores for 5d before change in feed to 7d after change in feed type	None very effective
Blackleg *Clostridium chauvoei* (bacteria)	Muscular depression; gaseous swelling in muscles; lameness	Vaccination of calves at branding and weaning and/or dam precalving	Penicillin
Bloat Variety of causes	Excessive accumulation of gas in the rumen and reticulum causes distension of left side; right side is also distended in more severe cases; labored breathing may occur	Poloxalene drench; if possible, feedlot rations with >10–15% cut or chopped roughage; antifoaming agents; avoid feeding alfalfa hay with high barley rations	Poloxalene oral drench for pasture bloat; antifoaming agents such as vegetable oil or mineral oil; stomach tube; emergency rumenotomy
Bluetongue Virus spread by blood-sucking insects (primarily the genus *Culicoides*)	Varies—high fever; depression; profuse slobbering; crusty muzzle; ulcers of dental pad and lips; poor reproduction; coronary band lesions	Maintenance of closed herd and control of *Culicoides*; vaccine available, but recommended only in infected herd; vaccination of pregnant animals may cause brain damage to fetus	None—disease runs its course in several weeks; death may occur due to secondary infection of pneumonia

(continued)

Table 16.1
(Continued) Common Diseases and Other Health Problems in Cattle

Disease and Cause(s)	Clinical Signs	Prevention	Treatment
BVD (bovine viral diarrhea) Virus; laboratory diagnosis imperative for accuracy	Feeder cattle: ulcerations throughout digestive tract; diarrhea (often contains mucus or blood) Breeding cattle: abortions, repeat breeding	Vaccination prior to exposure; avoid contact with infected animals Annually vaccinate cows 30 days prior to breeding	Limited to supportive therapy
Brisket Disease Congestive right heart failure due to stress of high altitude; greater predisposition if cattle graze on locoweed at high altitudes	Edema in brisket ; jugular vein distension; high neonatal calf mortality	Keep cattle below 5,000-ft altitude; practice genetic selection (primarily bulls) using pulmonary arterial pressure test	Take affected cattle to elevations below 5,000 ft. However, cases have been identified at elevations of less than 5,000 ft.
BRSV (bovine respiratory syncytial virus)	Labored breathing; Pneumonia	Vaccination	Antihistamines Corticosteroids
Brucellosis *Brucella abortus* (bacteria)	Abortions	Calfhood vaccination (in some states) at the age of 4–12 months	Test and slaughter; report reactors to state veterinarian
Campylobacteriosis (formerly vibriosis) *Campylobacter fetus*	Repeat breeding; abortions (1–2%)	Annual vaccination of females and bulls prior to breeding; use of artificial insemination; virgin bulls on virgin heifers; avoid sexual contact with infected animals; cull open cows in infected herds	None (consult herd veterinarian)
Cancer Eye Genetic predisposition; environmental factors (dust, wind, and ultraviolet light) enhance development	Cancerous cell growth on the eye, eyelid, third eyelid, or conjunctivitis	Genetic selection; reduction of other predisposing factors	Surgery; immunotherapy; electrothermal; cryotherapy
Coccidiosis *Eimeria zurnii; Eimeria Bovis* (Protozoa)	Fluid feces, bloody feces, and straining; occasionally central nervous system signs	Avoid crowding, wet pens, wet pastures, filth; remove affected animals from lots in early stages; Ionophores	sulfaquinoxaline (6 mg/lb/day for 3–5 days) and amprolium (10 mg/kg/day for 5 days)
Diphtheria *Fusobacterium necrophorus* (bacteria); tissue damage to laryngeal area	Difficult breathing; painful coughing; hoarseness; rattling noise when breathing	Control other respiratory diseases	Oxytetracycline or procaine penicillin
Enteric colibacillosis or Salmonellosis	Severe gastrointestinal inflammation; diarrhea; dehydration; high death loss of affected animals	Improve absorption of colostral immunoglobulins	Aggressive antibiotic and anti-inflammatory treatment plus supportive fluid therapy

Table 16.1
(Continued) Common Diseases and Other Health Problems in Cattle

Disease and Cause(s)	Clinical Signs	Prevention	Treatment
Enterotoxemia "Overeating." Toxins of *Clostridium perfringens* Type D (bacteria)	Sudden death "downers;" animals usually fed on high-concentrate diets; diarrhea, though many animals die before clinical signs appear	Increase concentrate in diet at slow rate; vaccination	Reduce concentrate; sudden death usually precludes treatment; antiserum
Fescue Toxicity (Chapter 15)			
Foot rot (infectious pododermatitis) *Fusobacterium necrophorus; Bacteroides sp.* (bacteria)	Lameness; foot swelling	Pens and lots free of objects that can injure feet	Topical bacteriostatic agent Sulfonamides Oxytetracycline
Founder (laminitis) Similar to an allergic reaction associated with acidosis complex	Lameness; inflammation between bony part of the foot and the hoof wall; toes may grow long and turn up; painful, straight-legged gait in extreme cases	Keep cattle from over-consuming grain; high level of management of high-concentrate rations; gradual adaptation to high energy feeds	Trim long toes; sell animal for slaughter; antihistamines; steroid therapy may be useful, but not economical
Grass Tetany (Chapter 15)			
Haemophilus Somnus (Infectious thromboembolic meningoencephalitis) *Haemophilus somnus* infections; difficult to diagnose; needs laboratory confirmation	Fever; incoordination; head-pressing; central nervous system signs	Vaccine (two injections)	Sulfonamides Oxytetracycline
Hardware Disease Ingestion of sharp objects that perforate the reticulum and cause severe damage to the abdominal cavity, heart sac, or lungs	Loss of appetite; reduced milk production; abdominal pain; sometimes labored breathing; bloat and diarrhea in chronic cases	Keep wire, nails, and other sharp objects from being eaten by cattle; magnets on feed equipment; intraruminal magnets	Mild cases may heal if animal is stalled and feed intake is reduced; antibiotics; surgery may be performed in severe cases
IBR (Infectious Bovine Rhinotracheitis) Virus. Laboratory diagnosis imperative for accuracy	Pneumonia; fever; vaginitis; infertility and abortion in females; preputial infections in males	Vaccinate cows 40 days prior to breeding; vaccinate feeder cattle prior to exposure; semen from reputable bulls	Oxytetracycline; penicillin to minimize bacterial infections
Johne's Disease, Paratuberculosis *Mycobacterium paratuberculosis* Laboratory confirmation necessary	Chronic diarrhea	Vaccine requires approval of state veterinarian; does not prevent infection; test: remove reactors; hygienic program for calf raising; separate from adult herd	Consult herd veterinarian

(continued)

Table 16.1
(Continued) Common Diseases and Other Health Problems in Cattle

Disease and Cause(s)	Clinical Signs	Prevention	Treatment
Leptospirosis *Leptospira* spp.	Fever; off feed; abortions; icterus; discolored urine	Vaccination at least annually (in high-risk areas, more frequently); proper water management; control rodents; avoid contact with wildlife and other infected animals	Dihydrostreptomycin;
Leukosis (Bovine; BLV) Virus; Laboratory diagnosis necessary	Weight loss; lymph node Enlargement	Prohibit sale of animals for breeding purposes; prevention of infected animals going into non-infected herds	None; cull affected animals
Listeriosis *Listeria monocytogenes*	Fever; circling; one eye or one ear paralyzed; sudden death	No vaccine; organism frequently found growing in moldy silage	Sulfamethazine Oxytetracycline
Lump Jaw (Actinomycosis, Actinobacillosis) *Actinomyces bovis.*	Lumps on bony tissues of head	Removal of objects causing mouth punctures (nails, splinters, grass awns, stemmy feed)	Ethylenediamine Dihydroiodide, 10 mg/head/day; sodium iodide
Malignant Catarrhal Fever Virus; frequently confused with other viral infections	High fever; inflammation of respiratory, digestive, and urinary systems	No vaccine; requires positive diagnosis to differentiate from other diseases; prevent contact with sheep that harbor organism	Treatment generally ineffective
Malignant Edema *Clostridium septicum* (bacteria)	History of wounds; fever and swelling around wound	Vaccination	Penicillin
Mastitis *Staphylococci* spp. *Streptococcus* spp. Others	Inflammation of mammary gland; udder is hard, hot, painful to the touch, and commonly discolored; milk is yellow, thick, and stringy	Reduce injury from rough handling and other types of trauma	Systemic antibiotics; frequent milking, anti-inflammatories
Navel Ill Neonatal septicemia and bacteremia; poor sanitation at calving	Umbilical abscesses and hernias; swollen joints in young calves	Iodine on or within umbilical cord immediately after birth	Penicillin–dihydro-streptomycin combination (1–2 gm streptomycin every 12 hours for 3 days)
Neonatal septicemia or calf septicemia *Septicemic colibacillosis, septicemia salmonellosis, Pasteurellosis*	Occurs when infecting organism spreads via blood stream to multiple organs; very high mortality rates; diarrhea plus increased heart and respiratory rates; weakness and recumbency	Improve colostral absorption of antibodies	Early and aggressive therapy is required; treatment includes fluid therapy, antibiotics, IV plasma, intranasal O_2

Table 16.1
(Continued) Common Diseases and Other Health Problems in Cattle

Disease and Cause(s)	Clinical Signs	Prevention	Treatment
Pine Needle Abortion Cows eat needles or buds from ponderosa pine	Abortion	Prevent consumption of needles	None
Pinkeye *Moraxella bovis* (bacteria, spread by insects, primarily the face fly)	Watery eyes; swelling, corneal opacity and ulceration	Control of flies; isolate infected animals; select breeding cattle with eyelid pigmentation; vaccination as an adjunct to sound management	Oxytetracycline; patch over eye
Pneumonia *Pasteurella* spp. *Haemophilus somnus* (bacteria)	Pneumonia		Broad spectrum antibiotics
Poisoning (plant) (see Chapter 15)			
Polioencephalomalacia H_2S toxicity	Sudden death; blindness; incoordination; "downers"	Gradual adaptation to high energy feeds; thiamine supplementation; decrease sulfur intake in water or feed	Vitamin B complex (thiamine) may be helpful
Prolapse (See uterine and vaginal prolapse.)			
Pulmonary Emphysema (See asthma.)			
Rabies Virus; usually bites from infected mammals	Clinical signs vary from furious to dumb type; can be transmitted to humans; never insert hand in suspect animal's mouth	Vaccination of susceptible animals, especially dogs and cats that act as intermediary between humans and infected mammals when necessary (epidemic)	None; human health hazard
Red Water Disease, Bacillary Hemoglobinuria *Clostridium novyi Type D*	Sudden deaths; red urine; bloody diarrhea; not to be confused with leptospirosis	Vaccination: *Clostridium hemolyticium*	Penicillin or tetracyclines
Ringworm Fungus	Round, crusty, and thickened circles, with hair denuded inside circle; more common in winter months	Avoid contact with infected animals and areas	Local application of fungicide (equal parts of tincture of iodine and glycerin); avoid contact of lesions with hands, as the disease is contagious to humans

(continued)

Table 16.1
(Continued) Common Diseases and Other Health Problems in Cattle

Disease and Cause(s)	Clinical Signs	Prevention	Treatment
Salmonellosis *Salmonella typhimurium* and *S. dublin* (bacteria)	Diarrhea (bloody); high temperature; often confused with coccidia infections; highly fatal	Contaminated lots and feed must be eliminated; vaccinate in heavily infected feedlots	Neomycin, 140 mg per ton of complete feed for calves; Chlortetracycline 70 mg/head/day
Scours *E. coli K99* (bacteria), corona virus, rotavirus (viral), and cryptosporidia (protozoa)	Diarrhea; weakness; dehydration; severe acidosis	Covered later in this chapter	Covered later in this chapter
Tetanus *Clostridium tetani* (bacteria)	Spasms; contractions of voluntary muscles; high mortality rate	Avoid contamination of open wounds; vaccination in high risk areas	Three phases: (1) antibiotics; (2) tranquilizers or chloral hydrate to relax muscle; (3) high doses of tetanus antitoxin (up to 300,000 units every 12 hours); keep cattle in quiet, dark area
Trichomoniasis *Trichomonas fetus* (protozoa) Bulls are asymptomatic carriers	Infertility and abortion (2–4 months) pyometra	Maintain closed herd or introduce only virgin replacement heifers and bulls; cull open cows in infected herds	Cull carrier animals; report to state veterinarian
Tuberculosis	Usually none	Periodic testing; slaughter reactors	Test and slaughter reactors; report to state veterinarian
Ulcers, Gastric Sudden changes of feed ration too high in concentrates; some association with trace mineral deficiency (see acidosis complex)	"Tarry" stools; bloat; sudden death	Change roughage concentrate ratio; reduce stresses; proper balance of trace minerals	None
Urinary calculi, "Water Belly" Change in ration; mineral imbalance	Straining to urinate; dribbling of urine	Salt (NaCl) up to 4–5% of ration; ammonium chloride 0.74–1.25 oz per head per day or at 0.5% of ration; keep calcium levels higher than phosphorus in diet	Surgery
Uterine and Vaginal Prolapse Excessive straining at calving or from coughing or diarrhea	Vagina, cervix, uterus, or all three protrude through vulva	Not known	Disinfect prolapsed organ; reposition organs and suture around vulva

Table 16.1
(Continued) Common Diseases and Other Health Problems in Cattle

Disease and Cause(s)	Clinical Signs	Prevention	Treatment
Vesicular stomatitis (virus)	Inflammation or blistering of tongue, lips, dental pad, nose, teats, and feet	Avoid contact with infected animals	None; report to state veterinarian
Warts, Viral Papillomatosis (virus)	Warts on all parts of body	Sanitation in vaccination; tattooing to prevent transmission of virus from animal to animal; vaccination against warts may be considered	Surgical removal
White Muscle Disease Vitamin E and selenium deficiency	Calves are stiff with arched back; diarrhea can occur; death from starvation	Selenium/vitamin E supplementation in some areas	Sodium selenite-vitamin E in aqueous solution (0.03 mg selenium per lb of body weight)

economic burden. The causes of death losses for calves and breeding stock are provided in Figures 16.1 and 16.2, respectively. In the case of pre-weaned calves in the first 21 days of life, the leading causes of mortality are related to dystocia, weather, and digestive disorders (most typically scours). The primary concern following the first 3 weeks up to weaning time is respiratory disease. Breeding cattle losses are frequently associated with calving difficulty.

The role of the veterinarian and the establishment of a preventative health care plan cannot be overstated for all phases of beef production. However, both utilization of professional veterinary service and implementation of a preventative vaccination system are not uniform in the cow-calf sector. In addition to variation in these practices by region, small herds are the least likely to vaccinate and consult with a veterinarian (Table 16.2).

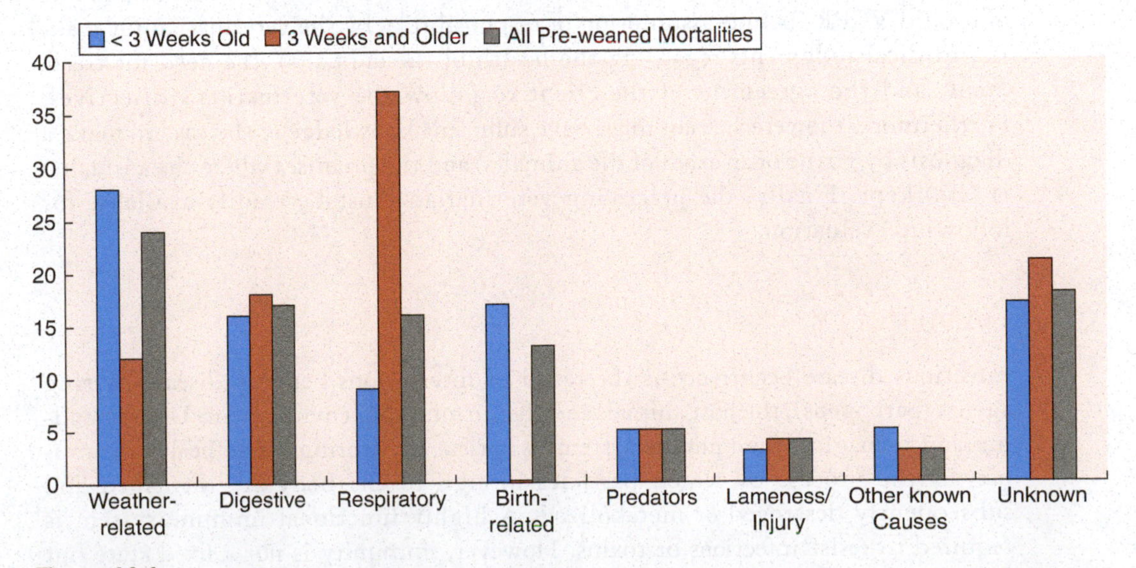

Figure 16.1
Primary causes of death in pre-weaned calves.
Source: NAHMS.

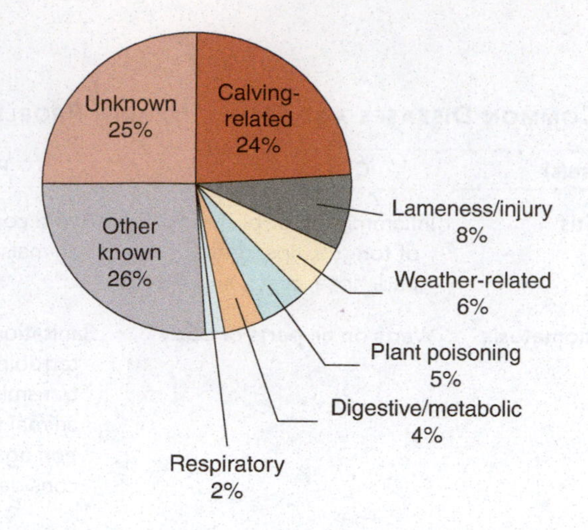

Figure 16.2
Primary causes of breeding cattle death losses.
Source: NAHMS.

Table 16.2
PERCENT OF COW-CALF ENTERPRISES UTILIZING PREVENTATIVE HEALTH PRACTICES ANNUALLY

Size of Herd (N of cows)	*1–49*	*50–99*	*100–199*	*200+*
Administered vaccines	59	87	96	92
Consulted veterinarian	43	63	76	82
Location of herd	*West*	*Central*	*Southeast*	
	76	91	60	

Source: USDA-NAHMS.

Specific treatments and preventive measures should be determined by establishing a valid veterinarian–client–patient relationship (VCPR). The components of a valid VCPR include assumption of responsibility by the veterinarian for making medical judgments regarding the health of the animal(s), the need for treatment, and the agreement of the client to follow the veterinarian's directives. Furthermore, the veterinarian must have sufficient knowledge of the case to make a diagnosis by virtue of an exam of the animal(s) and the premises where the animal(s) is (are) kept. Finally, the practicing veterinarian must be readily available for follow-up evaluation.

IMMUNITY

Infectious disease occurrence is the result of interactions between disease-causing agents (pathogens), the host animal, and the surrounding environment. Understanding and managing the immune system is critical in assuring cattle health. Immunity is the process by which particles foreign to the body are identified and subsequently destroyed or metabolized. A highly functional immune system is required to resist infections or toxins. However, immunity is not a fixed state but varies based on age of the animal, nutritional status, degree of exposure to organisms capable of initiating disease (pathogens), and a host of other stressors. Disease occurs when the pathogens overwhelm the animal's ability to eliminate them. Most frequently, the numbers of pathogens are low enough that the host can neutralize their effects. Environmental factors such as temperature and precipitation conditions can dramatically affect disease occurrence. For this reason, disease incidence may often follow a seasonal pattern.

Immunity takes two forms—the first is natural or native immunity that is present at birth under normal circumstances. For example, skin secretions that coat the respiratory and intestinal tracts and the acidic environment of the stomach are protections against disease native to the animal. The second form is acquired resistance provided by the actions of specialized white blood cells called lymphocytes. These cells are produced by the spleen, lymph nodes, intestine, mammary glands, and respiratory and reproductive tracts. Acquired immunity is activated when the body encounters foreign substances or antigens.

Lymphocytes take two forms—B cells and T cells. B cells secrete antibodies in response to specific antigens that are transferred via body fluids to provide humoral immunity such that free pathogens are recognized by the antibodies and neutralized. T cells provide intracellular protection by stimulating production of substances that directly attack an infected cell. Cell-mediated immunity is especially effective relative to virus infected cells, intracellular bacteria, and cancer. An example of a cell-mediated response is the production of macrophages that engulf foreign bodies (bacteria and viruses) as well as dead or damaged cells and then destroying them by enzymatic action.

Immune function develops in two phases—passive and active. Because newborns are not immediately capable of protecting themselves with their own antibody production, they receive passive immunity in the form of antibody-rich colostrum also known as first milk. Antibodies can only be absorbed intact from the dam's milk by the newborn's digestive system for a period of 12–36 hours post-birth. It should also be noted that a cow needs time to build colostral antibodies to assure passive immunity. Active immunity is attained when the individual can initiate its own antibody production against specific invasive antigens.

Even though immunity is an enhanced defense against disease, it is not absolute in effectiveness. Immunity is acquired via an active response to an antigen (pathogen or invasive foreign body). Memory cells are produced after the first encounter with an antigen that boosts the immune response in subsequent encounters. This "memory system" explains why some vaccines are most effective when administered as an initial dose followed by a booster(s) to achieve the desired level of immunization (Figure 16.3).

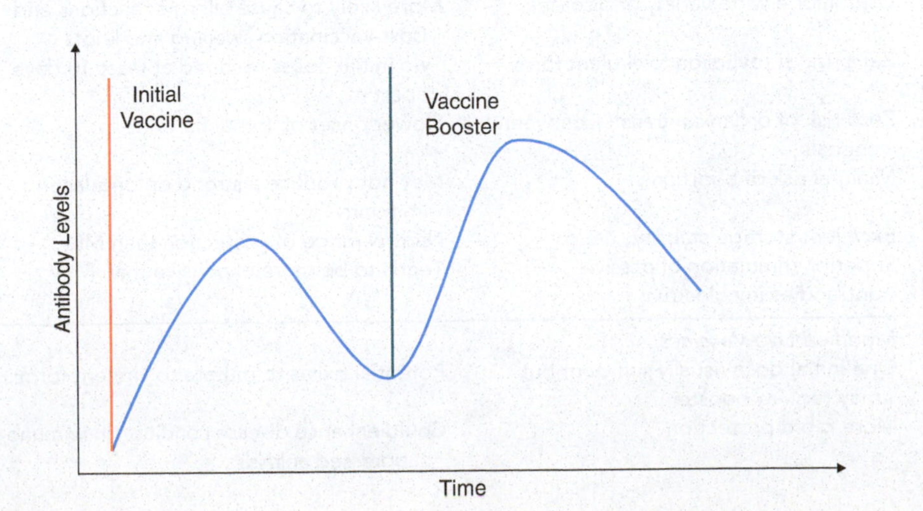

Figure 16.3
Immune response to initial and secondary vaccination.

Vaccines are most effective when administered at a time when the animal's stress is low and before the specific disease challenge has occurred. A veterinarian should be consulted to make recommendations as to appropriate timing of vaccination.

Types of Vaccines

Biologicals are derived from living pathogenic organisms or their metabolic products and are used to enhance an animal's disease resistance. Vaccines, bacterins, and antitoxins are all examples of biologicals. The three common forms of vaccine utilized in the cattle industry are:

1. Killed organism vaccines. They utilize dead organisms or their parts. The chemical antigen remains intact to stimulate an immune response. Effectiveness depends on using large numbers of organisms as well as adjuvants such as aluminum hydroxide or oil to stimulate an immune response.
2. Chemically altered vaccines. They utilize modified live organisms grown in a chemical media that alters the organisms' metabolism to reduce disease causing capacity. Toxicity of a pathogen is neutralized while the antigen's structure remains in place.
3. Modified-live organism vaccines. They contain small quantities of virus or bacteria that have been altered to prevent clinical disease while still allowing infection and multiplication sufficient to elicit the immune response.

Each type of vaccine has its own set of advantages and disadvantages (Table 16.3).

A sound vaccination program is designed to help keep an animal's disease resistance well above the level of disease challenge. Disease occurs when challenge exceeds resistance. Resistance fluctuates over time as a result of environmental stress, poor nutrition, and a variety of other factors that can lower an animal's immunity. Disease challenge is also variable. For example, the disease challenge on the ranch can be considerably different than at a commercial feedlot where cattle from a variety of sources are commingled. In some cases, vaccines need to be administered as a series, and as such revaccination must be timely to insure adequate protection.

Table 16.3
Advantages and Disadvantages of Various Vaccine Types

Advantages	Disadvantages
Killed Vaccines and Toxoids	
Useful for a wide variety of diseases	More likely to cause allergic reactions and post-vaccination swelling and knots
Zero risk of reversion to virulent form	Two initial doses required at least 10 days apart
Zero risk of organism transfer between animals	Slower onset of immunity
Minimal risk of abortion	May not produce a strong or long-lasting immunity
Excellent storage stability	Narrow range of protection than MLV
Superior stimulation of passive antibodies in colostrum	Tends to be more expensive than MLV
Modified Live Vaccines	
One initial dose usually sufficient but may require booster	Potential exists to mutate to virulent form
More rapid protection	Could enhance disease condition in immuno-suppressed animals

Table 16.3
(Continued) Advantages and Disadvantages of Various Vaccine Types

Advantages	Disadvantages
Produces wider spectrum of protection than killed products	Potential for excessive immune response
Less likely to create allergic reactions plus swelling/knots	Some risk of abortion
Less susceptible to passive antibody vaccine block than killed products	Must be handled and mixed with care
Typically less costly	
Chemically Altered Vaccines	
Shares many characteristics of MLV	Protection not as rapid as MLV
Safety similar to killed products	Two initial doses required
More rapid protection than killed products	May not produce a strong or long-lasting response
No risk of reversion to virulent form	Stimulates minimal mucosal immunity unless directly applied to the membrane
Minimal risk of abortion	Must be handled and mixed with care
	Tend to be more costly than MLV

One of the critical control points for vaccine efficacy is temperature both during storage and use. Producers should be disciplined in following storage temperature instructions while verifying that refrigeration units are functioning correctly and maintaining consistent appropriate temperatures. During cattle processing, it is critical to avoid exposure to direct sunlight and to avoid excessive changes in product temperature. Modified live vaccines should be administered within 2 hours of mixing to assure effectiveness.

Establishing a herd health plan should be accomplished with a specific focus on the risks, conditions, and goals of specific enterprises. While larger herds that control the vast majority of the U.S. cow herd have a high likelihood of implementing a preventative vaccination program, managers of small herds are less likely to take advantage of the physiological and economic benefits provided by vaccines. Given the preponderance of herds with fewer than 50 head, Table 16.4 illustrates the opportunity for improved preventative health care in the industry.

Antimicrobials

The judicious use of antimicrobial agents in beef production is a critical tool in assuring the health and well-being of cattle as well as the safety and wholesomeness of the final product. These products are used to inhibit (bacteriostats) or kill (bacteriocides) pathogenic microbes. While the foundation of herd health is placed squarely on disease prevention, it is unrealistic to expect that no animal will become ill. In situations where sickness occurs, it is vital that cattle managers have effective treatment plans in place.

Guidelines for the correct use of antimicrobials are as follows:

- Select the appropriate product in consultation with the herd veterinarian.
- Avoid selection of medically important antibiotics for humans.
- When possible, use laboratory analysis to determine the appropriate treatment agent to assure best response to the specific pathogen in question.
- Whenever possible avoid combining antibiotics in the course of therapy.
- Treat the fewest number of animals possible.

Table 16.4
Percent of Cow-Calf Herds Administering Vaccines by Cattle Class

Vaccine Type	Calves through Weaning	Replacements—Weaning through 1st Calving	Cows	Bulls
General (respiratory and/or reproductive)				
IBR (red nose)	31.7	31.3	24.6	18.2
BVD	36.1	38.8	28.1	24.3
Histophilus somni	17.0	14.6	7.9	5.5
Respiratory				
PI3V	29.6	30.4	22.6	17.6
BRSV	27.4	27.8	21.1	16.2
Pasteurella/ Mannheimia	13.8	8.9	4.5	3.1
Reproductive				
Brucella abortus (bangs)	6.4	17.6	1.0	NA
Leptospira (lepto)	10.5	35.0	31.7	21.2
Campylobacter (vibrio)	NA	22.6	19.0	13.3
Trichomonas (trich)	NA	1.6	1.0	0.7
Clostridial				
Blackleg—Malignant edema—*Cl. novyi—Cl. sordellii* (2 or 4 way)	65.8	32.9	14.5	10.1
Enterotoxemia (*Cl. perfringens C and D*)	40.1	18.9	11.6	8.2
Tetanus (*Cl. tetani*)	19.4	6.8	5.7	3.6

IBR—Infectious bovine rhinotracheitis.
BVD—Bovine viral diarrhea.
PI3V—Parainfluenza 3 virus.
BRSV—Bovine respiratory syncytial virus.
Source: USDA-NAHMS.

- Follow the course of therapy to its conclusion. Stopping therapy earlier is a risk factor for bacteria to develop resistance.
- Follow the label recommendations and maintain records documenting use.
- Extra-label use can only occur under the direction of a licensed veterinarian and under the umbrella of a valid veterinary–client–patient relationship. Such use must conform to the Animal Medicinal Drug Use Clarification Act. Extra-label use of medicated feed additives is strictly prohibited and violates federal law.
- Antibiotics should be limited to prevention or control of disease.

Additionally, it is important that treated animals do not enter the market chain until they have conformed to the appropriate withdrawal time as indicated on the product label. Several considerations are important relative to adhering to withdrawal times:

- All withdrawal times must be calculated from the last day of treatment and for the longest time period of the list of products administered.
- Avoid intermuscular injection of greater than 10 cc per site.
- In those rare cases where a veterinarian has prescribed extra-label usage, the withdrawal time must be extended to assure no violative residues occur—typically this is a period of 60 days beyond the label requirement.

If multiple doses of a single product are given, then the withdrawal time should be sum of the withdrawal period for each administration. For example, assume a 21-day withdrawal period where the protocol requires an initial treatment followed by a second dose 5 days later. In this case, the withdrawal period would be (21–5) + 21 equal to 37 days from the last injection administered. The Food and Drug Administration is initiating regulatory action that will shift the way that antibiotics are utilized in food animal production. The stated goals of the regulatory changes are to eliminate the use of medically important (for humans) antibiotics for growth promotion and to bring the therapeutic (treat, control, and prevent specific diseases) use of antibiotics under the supervision of a veterinarian. Development of valid VCPRs will be fundamental to management success. Furthermore, managers will implement management practices that reduce the need for medically important antibiotics, that develop protocols that only use antibiotics when they offer clear effective advantages, and by assuring adherence to judicious use guidelines.

Veterinary Feed Directive

In 2015, the Food and Drug Administration published the Veterinary Feed Directive (VFD) in the Federal Register that created the following changes to the manner in which antimicrobials can be utilized by the livestock industry by restricting the use of medically important antibiotics to cases that meet the following criteria:

- Necessary for assuring animal health
- Veterinary oversight must be in place

This new policy will require some level of adaptation by the industry but in the long term it will protect access to antibiotics for producers. There are four uses of antibiotics in beef production:

1. Treat animals who have been diagnosed with a specific disease
2. Control the spread of bacterial diseases within a herd
3. Prevent disease in healthy animals
4. Improve productivity and growth/feed efficiency

Under the new regulations, the first three uses are retained under the supervision of a veterinarian and a valid client–patient relationship. Feed grade antibiotics deemed important to human medicine are no longer allowed under any circumstance for the purpose of increasing growth and feed efficiency. Additionally, all extra-label usage is prohibited.

The following medically important products will be affected by the VFD and will require new approvals from FDA to be used as feed grade antibiotics to treat, control, or prevent disease—chlortetracycline, sulfamethazine, neomycin, oxytetracycline, tilmicosin (previously on VFD), tylosin, and virginiamycin.

The following products typically used to control parasites, bloat, and reproductive cyclicity will not be affected—amprolium, bacitracin, bambermycin, decoquinate, fenbendazole, laidlomycin, lasalocid, melengestrol acetate, methoprene, monensin, morantel, poloxalene, ractopamine, and tretrclovinphos.

The keys to complying with the regulation will be record keeping, maintaining a strong relationship with a veterinarian, early diagnosis, and very thorough pre-planning of events such as weaning calves and receiving calves considered to be high risk. In developing plans and executing them under the supervision of the veterinarian, the following questions provide a good basis for discussion:

- Is there a genuine need for the use of antibiotics?
- Are there alternatives to antibiotics?
- Can the withdrawal time be met prior to harvesting the cattle?
- Do product label specifications match the proposed use?

Administering Health Products

Administering health products is a vital part of disease prevention and treatment. Method of administration is important because it affects the speed with which the product enters the animal's system. Health products should always be measured accurately and administered according to the manufacturer's recommendations. Aseptic conditions should be utilized during vaccination because other diseases and infections can be easily transmitted.

Some vaccinations do not result in the animal becoming immunized. This occasionally happens because vaccines are not appropriately selected or handled properly, improper vaccination methods are used, or the individual animals have a unique physiological response. Killed organism and toxoid products require two doses. Failure to administer the second dose would result in immunization failure.

The proper use of animal health products is a requirement to assure superior efficacy and assurance of food safety to consumers. Federally licensed products should always be selected and used in accordance with the label instructions. Specific label information to be considered includes dosage, timing, and route of administration; withdrawal periods and other warnings; storage conditions; and expiration dates. The use of any product in a manner not specified by the label requires a veterinarian's prescription.

Vaccines should never be combined into a single dose, as this will likely dramatically reduce product effectiveness. For reconstituted products, a sanitary transfer needle should be utilized. Reconstituted vaccines tend to lose their effectiveness within an hour after mixing. Direct sunlight and improper storage temperatures during processing often result in poor product performance.

Always mark and separate different syringes used with different products to prevent bacterins or killed-type products from damaging modified live vaccines. Syringes used to administer modified live vaccines should be cleaned only with hot water. Cleaning with disinfectants may destroy the efficacy of vaccines. Select the correct needle size (for subcutaneous, use 16- or 18-gauge needle, $^1/_2$–$^3/_4$ in. length; for intramuscular, use 16- or 18-gauge needle, 1–1$^1/_2$ in. length), assure that the air is out of the syringe prior to injection, and use appropriate injection techniques, route, and site. If using an implant, assure proper placement and good sanitation.

Following are several methods of administering vaccines and therapeutic agents.

Intramuscular (IM)—The preferred injection site is a well-muscled area of the neck. Absorption is rapid as a result of a good blood supply. A 1$^1/_2$ in. needle is commonly used, allowing complete penetration of the skin and partial penetration of the muscle. The volume of the injection in any one site should be moderate (<10 cc in adult cattle) to prevent an abscess from forming. Injection into the hip, lower leg, and shoulder is to be avoided so as not to create an injection blemish.

Intranasal—This method is used to create local resistance to disease in the respiratory tract. The vaccine is administrated via inspired air.

Figure 16.4
Subcutaneous injection using the tenting technique.

Intravenous (IV)—This method is commonly used to administer therapeutic agents to young calves. The drug is rapidly available to the animal's system in larger volume and tissue irritation is avoided. The best site is the jugular vein, located between the neck muscles and throat on the side of the neck (the ideal site is approximately one-third of the distance between the jaw and chest). The site should be cleansed and wetted with alcohol, and a 1.5–2-in. needle should be inserted at a 30-degree angle pointing toward the body. Slight suction on the syringe should allow aspiration of blood into the syringe to verify accurate placement. The injection should be slow and steady.

Oral—Common in lightweight calves, this method is used to administer therapeutic drugs to cattle. Administration is simple in smaller calves. The greatest danger is failure to get the drug far enough into the throat to ensure swallowing; aspiration of the drug can lead to pneumonia or, in large enough quantities, to drowning.

Subcutaneous (SQ)—Injection under the skin results in slow but sustained rate of absorption due to a relatively small blood supply in this part of the anatomy. The injection site is where the skin is loose, usually in the neck. Utilize the "tented" technique whereby loose skin is gathered and pulled away from the body with one hand and the needle is inserted into the skin fold with the other (Figure 16.4). This assures that the product is administered under the skin but not into the muscle. The base of the ear is an alternative subcutaneous injection site.

ESTABLISHING A HERD HEALTH PROGRAM

Beef producers should be knowledgeable of disease agents, contributory risk factors, and other health-related conditions that may be unique to their region. This knowledge should be used in developing a specific health program for each specific enterprise. The importance of effective sanitation and biosecurity practices cannot be overemphasized. The most successful herd health management programs involve collaborative planning by producers and their veterinarians. Veterinarians need to understand the management programs of individual beef cattle operations, especially the major factors affecting profitability. Producers should be aware of their limitations in evaluating, treating, and caring for sick animals and of the proper timing for seeking a veterinarian's assistance. Veterinarians can also serve as liaisons in utilizing the services of other specialists and well-equipped diagnostic laboratories.

It has been said that a "good herd health program does not come in a bottle." A sound herd health program is based on the following components:

1. Sound nutritional regime
2. Continuous personnel training
3. Acquiring cattle only from known sources
4. Sound sanitation management and biosecurity practices
5. Excellent record-keeping system accompanied by a sound monitoring and evaluation system
6. Functional, well-maintained facilities
7. Strong relationship with a professional herd veterinarian
8. Sound preventative vaccination system

Implementation of an immunization protocol requires that producers and their herd health consultants identify and understand the following:

1. Herd health management history (identifying diseases of greatest risk)
2. Marketing goals for the herd and expectations of downstream customers in the supply chain
3. Timing of vaccination to maximize immune response
4. The impact of stress on animal health and immune response
5. Correct vaccine protocols in terms of storage temperature, date of expiration, and route and site of administration
6. Maintenance of biosecurity via strict control over the source of new cattle and good sanitation procedures in facilities, feeding equipment, transportation equipment, and market sites

As prey animals, cattle will attempt to disguise early signs of illness as a mechanism to avoid predation. This behavior makes it difficult for the novice herdsman to easily identify those individuals that may require attention and treatment. However, early recognition of and response to emerging health problems are keys to avoiding the financial losses associated with both isolated and widespread disease in a herd. Some of the early "signals" that an animal may be diseased are rough and/or dull hair coat, loss of appetite, general depression, low head carriage, droopy ears, weight and condition loss, lameness, stiff movement, nasal or ocular discharge, and irregular respiration. However, the need to pursue a specific diagnosis cannot be over emphasized. Cost-effective health programs recommendations require an accurate diagnosis, an excellent record-keeping system, and superior communication between producer and veterinarian. In regard to the responsible use of therapeutic and prophylactic drugs, it is important for beef producers to identify accurate and professional sources of information.

STRESS AND HEALTH

Cattle have passive and active defense mechanisms that can counteract most disease organisms. Exposure to severe stress conditions, however, limits the ability of these defense mechanisms to overcome disease and can lead to sickness or death in animals.

Stress factors include fatigue, hunger, thirst, dust, weaning, castration, dehorning, shipping, mixing with other cattle, unnecessary or abusive handling, adverse weather, parasites, poor sanitation, ammonia buildup, and anxiety. Excessive stress can be avoided by vaccinating and processing calves 30 days prior to weaning, by avoiding inclement weather during cattle handling/processing/transport, and by maintaining a functional working facility where cattle can be handled quietly and gently.

Preconditioning is a complete health management program for reducing sickness and death rate and improving weight gains. Preconditioning can prepare calves to better withstand the stress of movement from their pre-weaning production site into and through the various production and marketing channels. A complete preconditioning program for calves is described here.

Weaning—All calves should typically be weaned at least 30 days prior to their entry into the next production phase and (or) marketing channels.

Nutrition—Calves should be adjusted to trough and bunk and accustomed to a beginning feedlot ration. They should also be trained to drink from a non-stream water source.

Vaccination—Calves should be vaccinated against the Clostridial group, infectious bovine rhinotracheitis, parainfluenza 3, Pasteurella, and bovine respiratory syncytial virus at least 2 weeks prior to weaning.

Deworming—All calves should be dewormed with an injectable, drench, or feed additive anthelmintic when internal parasites are a problem in the area.

External parasite control—All calves should be treated with a recommended, appropriate, and approved external parasiticide at least 2 weeks prior to weaning.

Castration and dehorning—Calves should be castrated and dehorned at least 30 days prior to movement, preferably at 1–2 months of age.

Identification and certification—All preconditioned calves should be identified with an ear tag and should be accompanied by a properly signed certificate that states the specifics relative to the preconditioning process (timing and type of vaccine, etc.).

Producers, buyers, and veterinarians often poorly interpret the concept of preconditioning. Many buyers prefer to buy replacement cattle at the cheapest price and in thin condition. They frequently overlook the immediate health status and prior immunization of the animals they are purchasing, hoping to compensate for health losses through compensatory gains. As a result, the economic success of preconditioning varies considerably. However, preconditioning offers significant economic merit when it is implemented correctly.

Preconditioning programs should be assessed from a cost-to-benefit and risk-analysis standpoint. The specific preconditioning program that is best for a particular enterprise will depend on age of the cattle at time of marketing or movement into the next production phase, marketing channel utilized, the degree of discount/premium associated with preconditioning, and the costs of labor, vaccines, and feed. The effects of sick versus healthy calves in feedlot performance and profitability are illustrated in Table 16.5. The impact of sick cattle on financial performance should not be overlooked. The impact on carcass performance of calves that require multiple treatments is substantial (Figure 16.5).

Quality Assurance Programs

Beef Quality Assurance (BQA) is a nationally coordinated, state-implemented program that provides information and training to U.S. beef producers plus assurance to beef consumers of how animal husbandry techniques borne of a commitment to stewardship can be coupled with scientific knowledge to develop best practice standards and protocols for cattle production. BQA provides the basis for consumers to trust the beef they purchase. The impacts of implementing BQA include, but are not limited to, residue prevention, avoidance of pathogen contamination, and elimination

Table 16.5
COMPARISON OF PERFORMANCE AND PROFITABILITY OF SICK VERSUS HEALTHY CALVES

	Sick	Healthy
Number of head	3,203	9,393
Death loss (%)	3.4	0.5
Average daily gain (lb/day)	2.8	3.0
Total cost of gain ($/cwt)	65.96	56.68
Medicine cost ($/hd)	31.33	0.00
Percent grading		
Choice (%)	29	39
Select (%)	63	56
Standard (%)	8	5
Net return ($/hd)	−31.96	+61.23
Healthy cattle advantage ($/hd)		+93.20

Source: Adapted from Texas A&M Ranch to Rail data.

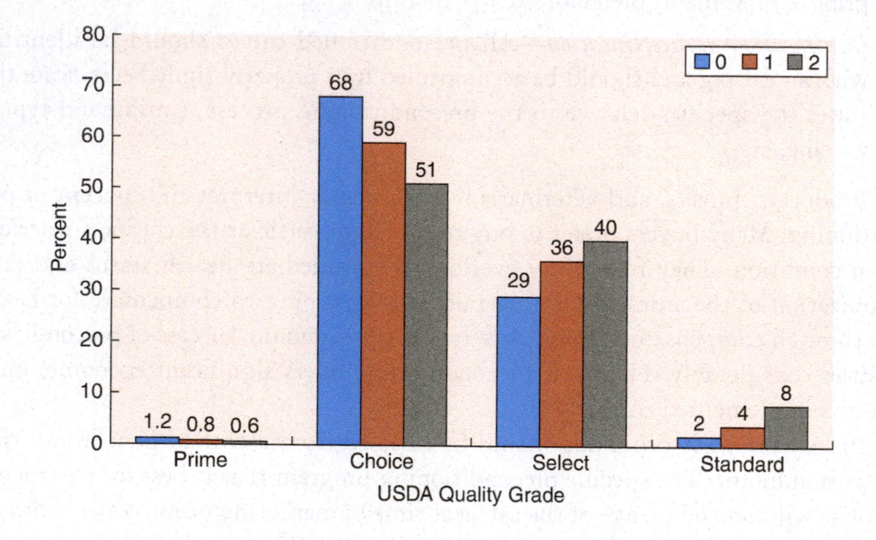

Figure 16.5
Impact of the number of treatments in the feedyard on carcass performance.

of injection site blemishes. BQA efforts incorporate Deming's Total Quality Management (TQM) principles, Hazard Analysis Critical Control Point (HACCP) protocols, and biosecurity principles. There are four basic pillars that underpin BQA programs—food safety, animal care, environmental stewardship, and staff training (Figure 16.6).

BQA programs focus attention on the following components—best practice animal care and handling protocols, herd health planning, management of feedstuffs, control of animal health care products, management systems to assure residue avoidance, effective record keeping, employee training, emergency action planning, and provision of ongoing assessment efforts.

A multitude of organizations offer BQA programs on the state level and many individual cattle enterprises have developed on-farm quality plans as a result. While BQA efforts are voluntary in nature, a number of supply chain programs are emerging that require periodic third-party audits to assure conformance to quality standards

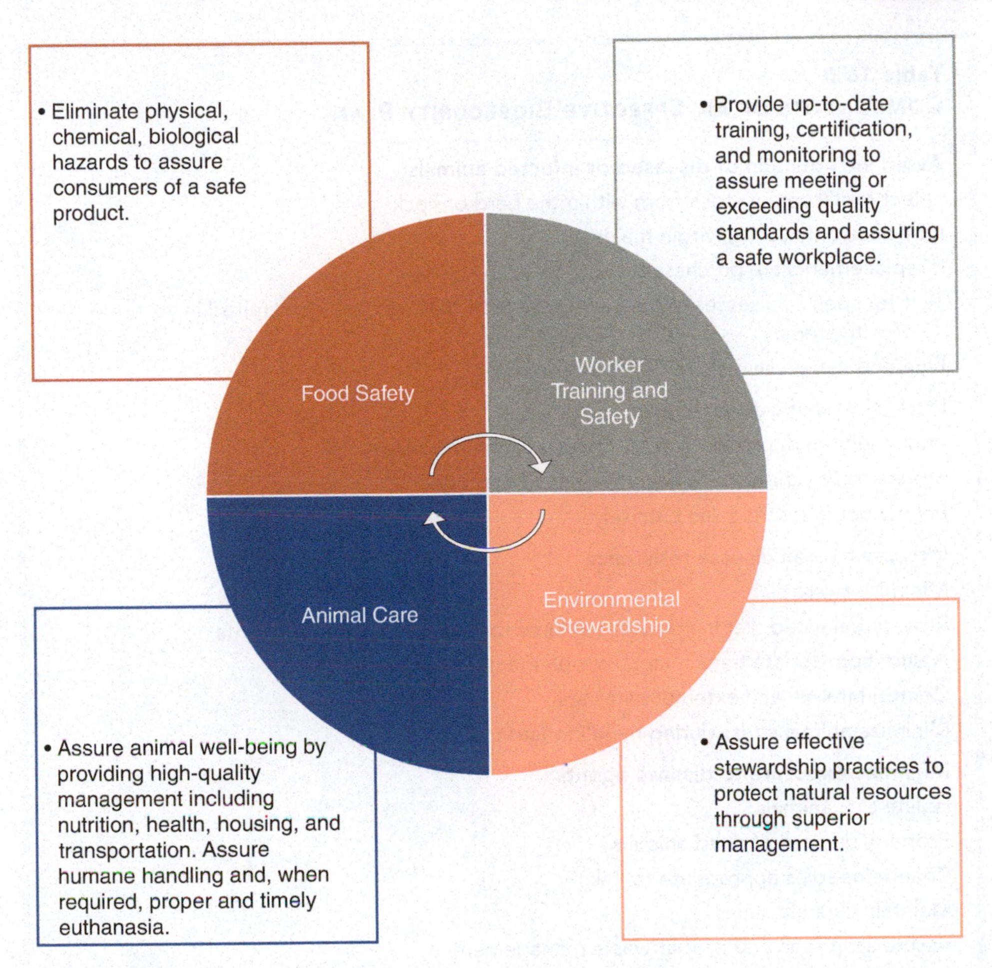

Figure 16.6
Quality assurance systems are founded on the four pillars of food safety, animal care, environmental stewardship, and employee safety.

and protocols. Biosecurity measures are those procedures that can be implemented to minimize the risk of introducing and/or spreading disease (Table 16.6). Biosecurity measures go far beyond vaccination programs to include development of select supplier arrangements, improved sampling and monitoring procedures, and enhanced employee training.

In the case of those situations where highly contagious diseases are involved, additional precautions may be required. These include restricting access to the farm premises, having employees change clothes and disinfect both when entering and leaving the farm, and disinfecting equipment, tack, and vehicles.

The implementation of BQA or other quality programs is becoming recognized as an advantage in the marketplace. Increasingly, packers are requiring that their suppliers document animal care, conformance to pharmaceutical product withdrawal times, rations that conform to ingredient standards, and other quality protocols. These losses to the industry could be considered small compared to the loss of reduced beef consumption if consumers lose confidence in the safety and wholesomeness of beef.

An example of a quality-related defect is related to IM injections. The use of IM injections can lead to associated lesions that result from tissue irritation by the injected substance. In 1990, product audits revealed that nearly 22% of beef top butts contained a blemish resulting from an IM injection. In response, the beef industry mounted an intensive educational effort to reduce IM injections and to move the site of administration

Table 16.6
COMPONENTS OF AN EFFECTIVE BIOSECURITY PLAN

Avoid introduction of diseased or infected animals
Select replacement stock from within the herd or flock
Introduce only young, virgin males
If replacements are purchased, buy only from known sources
Test for specific diseases of concern (recognize that tests may be limited in effectiveness)
Quarantine new animals for 30 days or more
Increase specific disease resistance
Implement an appropriate and effective vaccination program
Store and use the vaccine according to label guidelines
Follow booster schedules precisely
Increase overall disease resistance
Minimize dystocia
Provide balanced diet in adequate volume to meet animal requirements
Assure appropriate trace mineral supplementation
Control internal and external parasites
Minimize animal stress during handling and transport
Minimize exposure to disease agents
Isolate sick animals
Properly dispose of dead animals
Change needles appropriate to risk
Maintain clean facilities
Assure sanitation and proper waste management
Properly control rodents, birds, and insects
Assure fresh water supply
Store feed properly
Test and monitor feedstuffs
Maintain effective record-keeping systems
Avoid outside vehicle traffic through livestock areas
Provide protective clothing and foot covering for visitors

Source: Adapted from Smith (2002).

to an area in front of the shoulder (Table 16.7). As a result of these efforts, by 2000, the incidence had been reduced to 2.5%.

Better than 90% of the lesions identified in the various audits conducted by Colorado State University researchers were classified as "clear scars" or "woody calluses" that would have originated from injections given by cow-calf producers, stockers, or very early in the feeding period. While injection-site lesions are not a food safety concern, they do impact palatability and consumer perception. In fact, injections given by cow-calf producers to calves (Table 16.7) can have negative consequences on the palatability of steaks/roasts from those same animals. The shear force required to tear a steak core taken as far as 3 inches from the center of an injection-site lesion was 5.80 kg (acceptable shear forces are ≤3.86 kg for restaurant quality and ≤4.45 kg for retail trade), according to George et al. (1997). Any IM injection results in tissue irritation and scaring—these scars should be considered a life-long effect (Table 16.8).

Table 16.7
CHANGES IN INJECTION ROUTE AND SITE OF ADMINISTRATION

	NAHMS Beef 1997	NAHMS Beef 2007–08
Route of Administration		
Intramuscular (IM)	71	51
Subcutaneous (SQ)	68	76
Site of Administration—IM		
Neck	35	65
Shoulder	17	13
Upper rear leg/hip	43	20
Lower rear leg	5	2
Site of Administration—SQ		
Neck	78	84
Shoulder	13	11
Upper rear leg/hip	5	3
Lower rear leg	2	0.2

Source: USDA-NAHMS (2009a).

Table 16.8
INCIDENCE OF INJECTION-SITE LESIONS ASSOCIATED WITH PRODUCT ADMINISTRATION AT BRANDING AND WEANING

Intramuscular Injection Type of Product	Incidence of Associated Lesions at Slaughter[1] (%)
Branding (376 days prior to slaughter)	
Clostridial bacterin (2 ml)	72.5
Clostridial bacterin (5 ml)	92.7
Vitamins A and D_3	5.3
Long-lasting oxytetracycline	51.2
Weaning (225 days prior to slaughter)	
Clostridial bacterin (2 ml)	46.3
Clostridial bacterin (5 ml)	79.5
Vitamins A and D_3	10.0
Long-lasting oxytetracycline	92.3

[1]Ages at slaughter from 12 to 24 months.
Source: Adapted from Colorado State University, 1997.

Minimizing injection-site lesions can be accomplished by adopting the following management strategies.

1. Administer clostridial bacterins subcutaneously in the neck using the tented techniques (Figure 16.4) or at the base of the ear.
2. Avoid unnecessary repeat or multiple injections of clostridial bacterins.
3. Avoid IM injections whenever other routes of administration are listed in the label recommendations.
4. Inject no more than 10 ml per injection site (less in light calves).
5. Change needles at least every 15 injections (more often if cattle are dirty).
6. Appropriately discard bent and/or used needles. Never straighten and reuse.
7. Choose products that have low-volume doses whenever possible.

PARASITES

Parasites cause millions of dollars of loss to the cattle industry annually. These losses take the forms of reduced weight gains, increased amount of feed per pound of gain, lower milk production, reduced hide value, additional trim on carcasses, and death. The losses are not always apparent and may be excessive before cattle producers recognize there is a serious problem. There are two basic types of parasites: external and internal.

External Parasites

The two primary groups of external parasites are (1) insects (flies, lice, and mosquitos) and (2) arachnids (ticks and mites). External parasites live off the flesh and/or blood of cattle. They can mechanically transmit the organisms that cause pinkeye, mastitis, anaplasmosis, blue tongue, and other infectious diseases to cattle. Buzzing and biting of flies annoy cattle, decreasing gains and increasing feed costs. Lice and mites cause cattle to rub, which occasionally results in large areas of bare skin. Fence maintenance costs increase when posts and wire are used to satisfy the itching caused by certain lice and mites.

Lice are flat, wingless insects with three pairs of legs. The body of the louse is divided into three segments, with the body of most lice being less than 3/16-inch long. Lice are classified as biting lice or sucking lice. The blood-sucking lice are bluish-slate in color. Lice spend their entire lives on the host and are especially abundant during the winter when the host's hair is longer and thicker.

Mites have four pairs of legs and undivided bodies. They are microscopic in size and are whitish in color. They live off the skin tissue of the host and cause cattle diseases such as scab, mange, itch, or scabies. Mites are spread by contact with infected animals or from objects that have been in contact with infected animals. Increasing levels of confinement heighten this contagious aspect. However, seasonal environmental effects are also realized. The psoroptic mite causes common scab. If cattle scab is diagnosed, a report should be made to the state veterinarian. The mites feed on the skin of the host, which becomes covered with scabs that form over the mites, and the hair comes out in patches. Scabs usually appear first on the neck and around the base of the horns.

Grubs infest cattle via the attachment of heel fly eggs to the hair of the animal (Figure 16.7). Larvae migrate down the hair and penetrate the skin. The larvae pass through the warble stage beneath the skin of the back. They make breathing holes

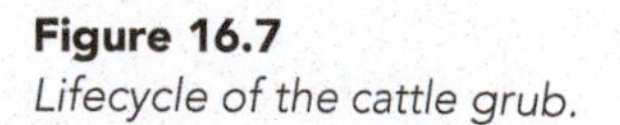

Figure 16.7
Lifecycle of the cattle grub.

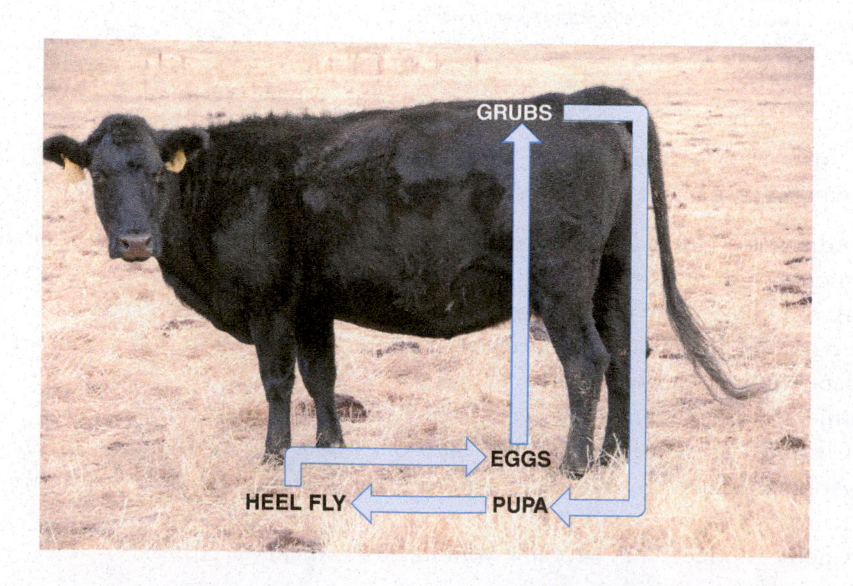

through the hide that provide a migratory route to the outside several weeks later. Larvae first appear on cattle in the southern regions in September and on cattle in the northern regions in January or later. Grubs emerge from the animal two months later.

Controlling External Parasites

Table 16.9 lists several commonly used insecticides for controlling external parasites. The most popular method of application is the use of systemic insecticides. In this situation, the material is applied to only one area of the body (usually the back) and the insecticide is absorbed, spreading throughout the entire system (Figure 16.8). Spraying, fogging, or using back rubbers, ear tags, injections, or dust bags can also apply insecticides. They can also be applied by preparing a large quantity of solution and placing it in a below ground vat. The cattle are then either herded into the vat one by one or lowered into the vat using a hydraulic cage. This is the most thorough method of applying insecticides; however, it is usually the most expensive and least practical method for most cattle operations. The most commonly reported fly control methods are topical products (dust bags, dips, sprays, and backrubs) and treated ear tags, with 60.6 and 30.8% of operations utilizing these control methods, respectively, according to NAHMS data.

Insecticides are used to break the life cycle of the parasite. Figure 16.7 shows the life cycle of the cattle grub and Figure 16.8 demonstrates a topical treatment given to break

Table 16.9
PREVENTION AND CONTROL OF EXTERNAL PARASITES IN CATTLE

Parasite	Prevention	Treatment	Comments
Grubs	Learn life cycle of grubs and treat animals accordingly	Co-Ral, Ivomec, Spotton, Tiguron, Warbex, Dectomax	Follow labels carefully; some products have side reactions; should be used only on certain ages or types of cattle
Lice	Sanitation; use of back rubbers or periodic spraying with insecticide	Pour-ons, Ivomec, Lysoff, Neguvo, Permectin, Tiguron, Warbex, Ivomec Eprinex Injection: Ivomec Sprays: Taktic, Prozap	Watch withdrawal periods for all products used to control parasites
Mange	Sanitation; avoid contact with infected animals	Check with local veterinarians	Mange is a reportable disease for which animals are quarantined
Flies (house, stable)	Sanitation, primarily by removing manure and cleaning other areas where flies breed and multiply	Supplemental use of insecti-cides or biological controls can be effective	Do not contaminate feed or water
Flies (face, horn)	Apply ear tags or tape for pastured cattle	Cyano benzeneacetate (Ectrin), Cyfluthrin (Cutter Gold), Cyhalothrin (Saber Extra), Cypermethrin + Chlor-Pyrifos (Max-Con), Diazinon (Terminator), Fenthion (Cutter Blue), Permethrin, Atroban Extra, Deckem, Ectiban, Expar Extra	In some areas flies have developed resistance to certain products; producers should alternate products and control measures; the latter could include sprays, dust bags, back rubbers, pour-ons, or feed additives; remove ear tags before slaughter

Figure 16.8
Topical application of a parasiticide.

the life cycle. Several methods may have to be implemented to control external parasites in any one operation. Continuous inspection for parasite infestation is the best indicator of effective control. It is important to remember that insecticides are toxins and, as such, should be handled only by trained individuals following proper procedures and utilizing the correct equipment. Failure to do so may result in serious human health consequences.

Internal Parasites

While the effects of external parasites are often obvious to cattle producers, the impact of internal parasites can easily escape notice. Internal parasites are present inside the animal, but the parasites and their eggs are microscopic in size. The economic loss to cattle producers resulting from internal parasites is often a slow, draining, continuous process that goes unnoticed because of the difficulty in measuring it. The impacts of internal parasite infestation may include loss of body condition, reduced pregnancy rates, and poor growth performance.

The most common class of internal parasites is the roundworm (Nematodes) with the other two basic classes being tapeworms (Cestodes) and flukes (Trematodes). Typical internal parasites are identified in Table 16.10.

Controlling Internal Parasites

In most areas, internal parasites cannot be totally eliminated. Cattle producers, however, should work to keep the level of infection below the level of economic loss. Table 16.11 illustrates the deworming practices adopted on cow-calf enterprises in the United States.

Control of internal parasites involves interrupting their life cycle. This can be accomplished in several ways: (1) the presence of unfavorable climatic conditions; (2) the development of resistance in cattle; (3) the management of cattle to prevent their ingestion of infective organisms; (4) the destruction of intermediate hosts and environmental features hospitable to the parasite outside the cattle's bodies; and (5) the therapeutic chemical treatment of cattle.

Table 16.10
COMMON INTERNAL PARASITES

Common Name	Scientific Name	Length and Shape	Location	Effect
Large stomach worm (barber pole or wire worm)	*Haemonchus placei*	15–40 mm	Abomasum	Sucks blood; anemia, diarrhea
Medium stomach worm (brown stomach worm)	*Ostertagia ostertagi*	6–9 mm	Abomasum	Reduction of nutrient absorption; profuse, watery diarrhea
Small stomach worm (bankrupt worm)	*Trichostrongylus axei, Trichostrongylus colubriformis*	5 mm	Abomasum	Reduction of nutrient absorption; profuse, watery diarrhea
Cooperids	*Cooperia* (various species)	5–8 mm	Small intestine (first 10–20 feet)	Profuse diarrhea; weight loss
Hookworm	*Bunostomum phlebotomum*	9–18 mm	Small intestine (first few feet)	Sucks blood; anemia; weight loss
Intestinal threadworm	*Strongyloides papillosus*	3.5–6 mm	Small intestine	Intestinal hemorrhages
Thread-necked intestinal worm	*Nematodirus (helvetianus)*	12–25 mm	Small intestine	Diarrhea; loss of appetite; weakness
Nodular worm	*Oesophagostonum radiatum*	12–15 mm	Colon	Loss of appetite and weight; diarrhea; weakness
Lungworm	*Dictyocaulus viviparous*	80 mm	Trachea and bronchioles	Coughing; rapid, shallow breathing; secondary bacterial infections
Tapeworm	Most common is *Moniezia expansa* and *Moniezia benedeni*	Several feet	Small intestine	Reduced nutrient absorption; occasional diarrhea
Common liver fluke	*Fasciola hepatica*	30 mm (leaflike)	Liver (bile ducts)	Hemorrhage of liver; hyperplastic and enlarged bile ducts
Giant liver fluke	*Fascioloides magna*	75 mm (oval)	Liver	Similar to common liver fluke
Lancet fluke (lesser liver fluke)	*Dicrocoelium dendriticum*	12 mm (long and slender)	Liver	Similar to common liver fluke
Stomach flukes (conical flukes or amphistomes)	*Paramphistomes* (several species)	15 mm	Rumen and reticulum	Diarrhea; loss of appetite and weight

Wet and warm weather favor proliferation of certain internal parasites. Areas that are colder and more arid (such as the western states) and northern Great Plains do not have the same internal parasite problems as those found in the southern states. While cattle producers cannot control the climate, they can use weather to assist in controlling parasites. In areas where internal parasite control may not be economically feasible, producers should conduct fecal egg counts to determine if a treatment program is needed.

Table 16.11
FREQUENCY OF DEWORMING VARIOUS CATTLE CLASSES BY COW-CALF ENTERPRISES

Cattle Class	Never	Occasionally (<1 × per yr.)	Annually	>1 × per yr.
Unweaned calves	38	8	31	23
Replacement heifers weaned but not yet calved for the first time	25	6	29	40
Weaned stocker calves	41	5	29	25
Cows	13	5	38	44

Source: USDA-NAHMS.

Some cattle are more resistant to internal parasites than are other cattle. This resistance may be limited to an immunity developed due to earlier infections. Younger cattle and cattle stressed with poor nutrition and diseased conditions are less resistant to internal parasites. Close confinement and overgrazing are both conducive to spreading parasitic infections. Under these conditions, cattle are more likely to eat infected manure or graze the lower parts of plants where parasites are more numerous. Pasture rotation can help reduce the number of parasites by causing infective organisms to die before they are ingested.

In some cases, the intermediate host can be destroyed, thereby reducing the number of parasites. An example is the snail, which is the intermediate host for the fluke. Scattering manure, renovating pastures, plowing, and draining marshes and swamps all upset the hatching process and reduce the concentration of infective organisms. When controlling parasites, it is important to consider the impacts on other organisms in the ecosystem. Some insects may provide useful functions including burying nutrient-rich manure into the ground and inhibiting pest-type insects.

Anthelmintic drugs are used to kill parasites inside cattle. However, these chemical dewormers have varying degrees of effectiveness on different parasites and different forms of the same parasite (Table 16.12). Dewormers come in drench, feed additive, injection, or bolus form.

Table 16.12
SELECTED ANTHELMINTIC DRUGS USED TO CONTROL INTERNAL PARASITES

Chemical	Trade Name	Form	Effective Against
Thiabendazole	TBZ E-Z-Ex	Paste Pellet/Block	*Haemonchus, Ostertagia, Trichostrongylus* (adult and immature forms), *Cooperia, Nematodirus, Oesophagostomum radiatum*
Levamisole	Levasole Tramisol	Injectable Bolus Drench Soluble powder Feed additive Oral gel	*Haemonchus, Ostertagia Trichostrongylus* (adult and immature forms), *Cooperia, Nematodirus, Oesophagostomum, Bunostomum, Chabertia, Dictyocaulus*
Morantel tartrate	Rumatel	Feed additive	*Haemonchus, Ostertagia, Trichostrongylus, Cooperia, Nematodirus, Desophayostomum*
Doramectin	Dectomax	Injectable	*Ostertagia* + 35 stages of internal and external parasites
Ivermectin	Ivomec-F	Injectable	*Haemonchus, Ostertagia, Trichostrongylus, Cooperia, Oesophagostomum, Bunostromum, Nematodirus* (adults only), *Dictyocaulus, Fascuola hepatica*

The best time to deworm depends on the type of operation and its location. In general, the best time to deworm a cow herd is just before calving. Calves may be dewormed when they are worked prior to weaning. In areas where internal parasites have not been a serious problem, a periodic fecal check for parasite eggs and a comparison of the performance of some treated and non-treated animals are useful in determining whether a total treatment program is necessary. Because feedlot cattle originate from many sources, a routine deworming of incoming cattle from geographic problem areas should be considered.

Avermectins are a class of parasiticides effective against both internal and external parasites. However, this class of products and all other products should be evaluated in terms of their biological and economic effectiveness.

COW-CALF HEALTH MANAGEMENT PROGRAMS

Respiratory, reproductive, and digestive diseases continue to be the major cause of disease loss in beef cattle. The cost of respiratory diseases alone to cattle producers, in terms of treatment, weight loss, death loss, and culling in weaning calves, is at a third of a billion dollars annually.

Bovine respiratory disease (BRD) is seldom the result of a single factor. BRD usually is caused by a combination of stress, viral infection, and invasion of the lungs by pathogenic bacteria such as *Pasteurella* and *Haemophilus*. Stress undermines the natural defenses built into the lining of the trachea and bronchi, and respiratory viruses (such as IBR, PI3, BVD, and BRSV) further damage these natural defenses. Ultimately, pathogenic bacteria find a wide-open road into the lungs where they localize, multiply, and cause BRD, pneumonia, or shipping fever.

Immunity against IBR, BVD, BRSV, and PI3 can be enhanced by the administration of MLV vaccines to cattle. Modified live and inactivated virus vaccines are available in single and combination forms. The routes of administration of these vaccines are IM (IBR, PI3, BRSV, BVD) or intranasal (IBR, PI3 only). Both IM and intranasal vaccines provide adequate immunity. Best protection is obtained by administering vaccines to healthy, non-stressed animals.

Controlling Major Cattle Diseases

An example health program focused on calves from birth through the weaning stage of production is provided in Figure 16.9. It should be recognized that this system may not meet the needs of specific herds. Regional and operational differences must be considered and producers should always consult a veterinarian in the creation of a workable herd health plan.

In addition to the guidelines provided in Figure 16.9, when designing and implementing a calf program producers should take the following factors into consideration:

- Some cattle operators and veterinarians prefer not to administer BVD simultaneously with IBR-PI3 modified live vaccine IM because of the added stress on the animal. This is true in very young or heavily stressed animals, which can be vaccinated against BVD 2–3 weeks later; in this case, at weaning time. It must be emphasized that calves stressed in any form (e.g., castrated, dehorned, branded, shipped, and so forth) should not be vaccinated against BVD; otherwise, adverse reactions could occur.
- *Pasteurella* and *Haemophilus somnus*: Two injections, 2–4 weeks apart, are needed whenever *Pasteurella* and *Haemophilus* bacterins are used. If the first injection is given at this time, follow it with a booster injection at weaning time.

CALF AT BIRTH

Iodine navel, Apply ID tag, Dehorn if needed [1], Castrate [1], Provide colostrum to stressed calves

CALF FROM BIRTH TO START OF DAM'S BREEDING SEASON

Castrate and dehorn if not done earlier, implant non-replacement breeding stock if marketing plan warrants, vaccinate all calves in accordance with herd health plan (consider 4-way MLV: IBR, BVD, PI3, BRSV + 2-way latest generation Mannheimia-Pasturella + 4-or 7-way tissue friendly Clostridial)

BVD-PI test calves – sort positives, isolate both calves and dams and cull from herds

CALF 14–28 DAYS PRIOR TO WEANING

Re-vaccinate according to herd health plan (consider 4-way MLV: IBR, BVD, PI3, BRSV + 2-way latest generation Mannheimia-Pasturella + 4-or 7-way tissue friendly Clostridial, 5-way Lepto administered to replacement females), implant/re-implant according to marketing plan, check for presence of horn buds and testicles that may have been missed by earlier protocols and take corrective action coupled with administration of tetanus antitoxin

CALF AT WEANING (background on home ranch)

Consider revaccinated with MLV: 4-way IBR, BVD, PI3, BRSV, no others required if two pre-weaning vaccines were administered and calves will not be commingled with un-vaccinated cattle, treat for external, internal parasites if needed, Bang's vaccinate replacement females 60 days post-weaning

OR

CALF AT WEANING (feedlot)

Administer MLV: 4-way IBR, BVD, PI3, BRSV, apply highly efficacious dewormer, implant if one was not administered pre-weaning otherwise delay for 1.5–2 months

[1]Complete procedure within 30–45 days of birth.

Figure 16.9

Herd health protocol for calves from birth through weaning.

- If the immunization program is started early enough, both the initial and booster shots may be administered prior to weaning. This will reduce stress at weaning and reduce shrink and amount of time required to regain weaning weight. This regimen is subject to local practices, individual herd health needs, and various combinations of vaccines. The above recommendations, early immunization, processing of cattle, and other management practices are designed to reduce stress at weaning.

Increasingly the market is rewarding cattle that have been managed under a verified herd health program that has been designed to assure that calves leaving the ranch are best prepared for transition to either the stocker or feedyard phase of production. These programs may specify both the diseases to be vaccinated against as well as the particular brand of product to be administered (Table 16.13).

Table 16.13
Specified Calfhood Vaccination Programs

Program Name	Timing of Administration	IBR-PI3-BVD-BRSV	Clostridial/ Blackleg	*Mannheimia* (Pasteurella) *haemolytica* bacterin/ leukotoxoid	Parasiticide	Implant Protocol
VAC 24[1]	2–4 months	IBR-PI3-BVD-BRSV	Clostridial 7-way	Mannheimia haemolytdica and/or P. multicoida	optional	not required
PrimeVAC™24	1×, early in life, @ herd of origin	Bovi-Shield GOLD, Cattlemaster GOLD, INFORCE 3 or Resvac/ Somubac	ULTRABAC, ULTRACHOICE OR ONE SHOT ULTRA	Recommended: ONE SHOT or ONE SHOT ULTRA	recom-mended	Recom-mended—Synovex C
VAC 34[1]	2–6 weeks prior to shipping	IBR-PI3-BVD-BRSV	Clostridial 7-way	Mannheimia haemolytdica and/or P. multicoida	optional	not required
PreVAC™34	1× @ herd of origin 2–6 weeks prior to shipping	INFORCE 3 and Bovi-Shield GOLD BVD or Bovi-Shield GOLD 5 or Cattlemaster GOLD or Resvac/Somubac	ULTRABAC, ULTRACHOICE OR ONE SHOT ULTRA	ONE SHOT or ONE SHOT ULTRA	recom-mended	Recom-mended—Synovex C
VAC 34+[1]	VAC 34 plus initial IBR-P1_3-BVD-BRSV at branding	IBR-PI3-BVD-BRSV		Mannheimia haemolytdica and/or P. multicoida	optional	not required
PreVAC+™34	Processed 2× at herd of origin with last complete processing 2–6 weeks prior to shipping	Two doses: 1. INFORCE 3 and Bovi-Shield GOLD BVD or Bovi-Shield GOLD 5 or CattleMaster GOLD 2. Followed by Bovi-Shield GOLD 5 or CattleMaster GOLD or Resvac 4/Somubac	ULTRABAC, ULTRACHOICE OR ONE SHOT ULTRA	ONE SHOT or ONE SHOT ULTRA	recom-mended	Recom-mended—Synovex C
VAC 45	*Option 1*: 2–6 weeks prior to weaning and at weaning w/ 2–6 weeks between vaccinations *Option 2*[1]: @ weaning and 2–6 weeks post-wean with >14 days prior to shipping *Retain*: 45 days post-weaning	Two doses—IBR-PI3-BVD-BRSV	Two doses—Clostridial 7-way	Mannheimia haemolytdica and/or P. multicoida	optional	not required

(Continued)

Table 16.13
(Continued) Specified Calfhood Vaccination Programs

Program Name	Timing of Administration	IBR-PI3-BVD-BRSV	Clostridial/ Blackleg	*Mannheimia* (Pasteurella) *haemolytica* bacterin/ leukotoxoid	Parasiticide	Implant Protocol
WeanVAC[cr]45	Processed 2× at herd of origin at or near weaning *Retain*: 45 days post-weaning	Two doses: 1. INFORCE 3 and Bovi-Shield GOLD BVD or Bovi-Shield GOLD 5 or CattleMaster GOLD 2. Followed by Bovi-Shield GOLD 5 or CattleMaster GOLD or Resvac 4/Somubac	ULTRABAC, ULTRACHOICE OR ONE SHOT ULTRA	ONE SHOT or ONE SHOT ULTRA	One dose – Dectomax or Valbazen	Recom-mended—Synovex C, Synovex S, or Synovex H
VAC PRECON	Vaccinated 2×, 2–6 weeks apart, last vaccination > 14 days prior to shipping *Retain*: 60 days post weaning prior to shipping	Two doses—IBR-PI3-BVD-BRSV	Two doses—Clostridial 7-way	Mannheimia haemolytdica and/or P. multicoida	optional	not required
StockerVAC[cr] Precon	Purchased calves processed 2× with last processing >14 days prior to shipment *Retain*: 60 days after weaning prior to shipping	Two doses: 1. INFORCE 3 and Bovi-Shield GOLD BVD or Bovi-Shield GOLD 5 or CattleMaster GOLD 2. Followed by Bovi-Shield GOLD 5 or CattleMaster GOLD or Resvac 4/Somubac	ULTRABAC, ULTRACHOICE OR ONE SHOT ULTRA	ONE SHOT or ONE SHOT ULTRA	One dose—Dectomax or Valbazen	Recom-mended—Synovex C, Synovex S, or Synovex H

[1]Must be home-raised.

Determining whether or not to participate in these programs requires that managers weigh the costs and benefits as well as the fit with their overall marketing and management plans.

Additional Health Management Tips

Management of Calves at Weaning

1. Calves should be eating some dry feed 2–4 weeks prior to weaning.
2. Vaccination procedures should be reviewed and changed as necessary, depending on the health conditions of specific lots of cattle, environmental conditions, and the prevalence of various diseases in the immediate area.
3. An adequate fresh water supply is essential, preferably from a source that cattle can see or hear running.
4. Vitamin A prior to or at weaning is generally recommended.
5. Check feed and water consumption—both should increase during the weaning period.

6. Provide good, high-quality hay. Calves should be consuming 2–3% of body weight of feed before either feed or water is medicated.
7. Check calves two or three times daily.
8. Seek professional help from a veterinarian when needed and before a major problem arises.

Transportation and Transition to Feedyard

1. Minimize stress factors as follows:
   a. It is crucial to get calves moving through marketing channels quickly.
   b. Avoid crowding and bruising.
   c. Avoid conditions of extreme temperature variations, dust, or wetness.
   d. Feed and water calves before shipping.
2. Other factors to improve health:
   a. Upon arrival at the feedlot, cattle should first be fed hay prior to having access to water. Be sure to have adequate water facilities available. Begin a limited feeding of grain and protein supplement.
   b. Segregate sick animals.
   c. Tractor exhaust stacks must be tall enough for gases to clear the trailer well.
   d. Avoid ammonia buildup in trucks, yards, barns, and sheds from excess urine, manure, and moisture. Ammonia contributes to respiratory disease.
   e. Start adequate treatment promptly. Identify sick cattle and treat as recommended by a veterinarian. Maintain and utilize accurate records.

Cow Herd, Replacement Heifers, and Bulls

It must be emphasized that vaccination and adequate handling of calves are part of, but not a substitute for, a total herd health management program. It is essential that an adequate breeding herd vaccination program be implemented for maximum benefits to be expected from vaccinating or preconditioning calves or both.

A sound long-term herd health program depends on the systematic monitoring of all aspects of enterprise management that have an impact on animal well-being. This monitoring process includes observation of BCS changes, avoidance of nutrient or environmental conditions that create toxic or deficient states, awareness of actual growth rates relative to projected outcomes, and coordination of nutritional, genetic, reproductive, and biosecurity management. Suggested herd health protocols relative to breeding cattle are provided in Figure 16.10.

Frequently, veterinarians recommend that IBR, BVD, and *Clostridial* immunizations be administered during the last trimester of pregnancy because it conveys a greater passive immunity to the calf. One should be cautioned, however, that if IBR is administered during the last trimester of pregnancy, that it be a killed product, chemically attenuated, or an intranasal vaccine.

BVD vaccines are modified live virus vaccines and ordinarily should not be administered to pregnant animals. There are experimental studies and reports of practitioners having administered BVD vaccines during the last trimester of pregnancy to increase passive immunity in the calf without adverse effects. Killed BVD vaccines are also available and have different recommendations. Caution is emphasized that all biological products should be administered in accordance with the recommendations of the manufacturers and in consultation with a veterinarian. Annual boosters are frequently recommended. It is important, however, that cattle operators consult with a veterinarian as to the appropriate schedule of immunization for their herds.

BREEDING FEMALES

Prior to Entering the Herd: Replacements or New Females First Breeding

BVD-PI test; MLV: IBR and BVD, Vibrio, Lepto and either 4/7/8-way clostridial, 21–28 days pre-breeding

Pre-breeding/Post-calving: Resident females

21–28 days pre-breeding administer MLV: IBR and BVD (every 3–5 years), Vibrio, Lepto, Deworm (fall calvers)

Post-breeding

Pregnancy diagnosis, scour vaccine as needed (heifers require initial dose), Deworm as needed (spring calvers), Grub and Lice control as needed

Pre-calving

3–7 weeks ahead of calving administer scour vaccine booster, Lice control as needed

BREEDING MALES

End of Growing Phase

BVD-PI test, MLV: IBR, BVD plus 4/7/8-way clostridial

Pre-breeding

21–28 days pre-breeding, MLV: IBR, BVD (every 3–5 years), Vibrio, 4/7/8-way Clostridial, Deworm as needed, Lice control, Breeding Soundness Exam

Figure 16.10
Preventative health care protocols for breeding stock (heifers, cows, and bulls).

Brucellosis Status

Cow-calf producers should pay special attention to *brucellosis* because the infection still exists in several herds in some states. Regulations regarding vaccination and testing are expected to change. If new cattle are to be brought into a herd, they should originate from a negative herd, be isolated from other animals in the herd of destination, and be retested before being commingled with the new herd. Vaccination should be used as recommended by a veterinarian. Vaccinate all eligible heifers for *brucellosis* and purchase only vaccinated heifers.

Calf Scours

Causes Since diarrhea in the calf is second only to reproductive diseases as a cause of losses to the cattle industry, it is discussed here in more detail than some of the other diseases. The four major causes of diarrhea are colibacillosis, enterotoxemia, salmonellosis, and coccidiosis. A summary of each of the latter three diseases is presented in Table 16.1.

Colibacillosis is by far the most important cause of diarrhea in calves. Although this type of scours can affect calves up to a month of age, it is most commonly observed in calves from birth up to 2 weeks of age. Scours, in general, may be caused by bacteria, viruses, or parasites (Table 16.14). Vaccination or immunity against one type has little effect against the others, and many different stress factors may predispose the calf to these organisms (Table 16.15). The variety of causative organisms and mitigating stress factors make the calf scours disease complex difficult to prevent or treat.

There are two clinical forms of calf scours. One is sudden death, often without signs of sickness and before diarrhea has had a chance to develop. It occurs because the

Table 16.14
Causes of Calf Scours

Pathogen	% of Cases
Bacteria[1]	22
Viruses[2]	35
Parasites[3]	24
Other	19

[1]Primarily *E. coli K99* and *clostridial* strains.
[2]Rotavirus and coronavirus.
[3]Cryptosporidia and coccidia.

Table 16.15
Potential Stress Factors that can Contribute to the Incidence of Calf Scours

Management Area	Specific Issues
Sanitation	Overcrowding, poor ventilation, poor drainage
Climate	Wet weather, extreme temperatures
Nutrition	Poor quality colostrum, insufficient colostrum supply/intake, parasite infestation, poor general nutrition, changes in ration, overeating, poorly balanced diet, appropriate pre- and post-calving dam nutrition
Dystocia	Stress on cow and calf

organisms, or toxins produced by them, gain access to the blood, and death occurs within 4–24 hours. The other and more common clinical form of scours is diarrhea. Stools vary in consistency from watery to semisolid and are lighter in color than normal. This is in contrast to coccidiosis and enterotoxemia, in which the feces are darker than normal and may contain visible blood. In the early stages of the chronic form, the calf continues to eat but later becomes dehydrated and weak and then stops eating. The course of the disease generally is 2–3 days, but may vary from 1–7 days.

Prevention Scours occurs when a calf's exposure to the disease exceeds the calf's resistance to the disease. Prevention of calf scours occurs when exposure is decreased and resistance is increased.

Calving and raising calves in dry, clean areas can decrease exposure to calf scours organisms. If cows have to calve in a confined area, provide a dry, clean individual stall for the cow. The cow and calf should be moved to clean pasture as soon as possible after calving. Calving in corrals or keeping cows and calves in corrals increases the risk of exposing the animals to scour-causing organisms.

Bringing new cows or calves into the herd just prior to calving or during calving increases the risk of exposure to calf scours. Calves needed for grafting would best be selected within the herd rather than purchasing calves from outside the herd.

The major factor in increasing a calf's resistance to scours is to give the calf the first-milking colostrum, preferably within 6 hours after the calf is born. Colostrum contains immunoglobulins (antibodies) needed by the calf, as the calf is born with little natural immunity against disease organisms. Because of rapid changes in the digestive system, the calf's ability to absorb antibodies diminishes rapidly. Immunoglobulin absorption at calving is 100%; absorption rates decline to 50, 25, and 5% at 6, 12, and

Table 16.16
Effect of Dystocia on Calf Vigor and Immunoglobulin Concentration

	Calving Difficulty		
	Unassisted	Easy Pull	Hard Pull
Time from delivery to standing (min.)[1]	40	51	84
Calf serum IgG (mg/ml)[2]	2,401	2,191	1,918
Calf serum IgM (mg/ml)[2]	195	173	136

[1]A measure of calf vigor.
[2]Assisted females were milked out immediately after calving and calves were fed the colostrum.
Source: Adapted from Odde (1989).

24 hours post calving, respectively. Furthermore, the acidosis experienced by calves that require assistance at birth inhibits antibody absorption (Table 16.16).

Cows can be vaccinated 30–40 days prior to calving, with a repeat injection 2 weeks following the first injection. The antibodies produced are for specific vaccine organisms only, and scours could still occur if caused by other organisms. Immunity, even for the specific organisms, is effective only when the calf consumes the colostrum and the antibodies are absorbed from the digestive tract.

A genetically engineered oral vaccine to prevent *E. coli K99-specific* infections can be given to calves shortly after they are born.

Treatment Diarrhea occurs because of alterations in intestinal function resulting from increased secretion from the body or decreased absorption of water and electrolytes (sodium, chloride, potassium, and bicarbonate) from the intestinal lumen. These important electrolytes are lost in the feces.

Loss of bicarbonate is a primary cause of acidosis. Potassium loss causes lethargy and muscle weakness in the affected animal. Loss of water and sodium chloride results in dehydration. Treatment should be directed toward correction of the dehydration, acidosis, and electrolyte loss.

During diarrhea, cells that line the intestine are sloughed off, and it takes 2–3 days for new cells to develop. The major aspects of treatment, therefore, are to keep the calf alive while the affected intestinal cells are replaced and to prevent secondary diseases such as pneumonia.

Most dehydrated calves suffer from hypothermia (i.e., the body temperature is lower than normal). Therefore, the calf should be provided with supplemental heat in a dry location.

Clinical signs of dehydration occur when fluid loss reaches approximately 6% of body weight. Fluid losses of 8% result in depression, sunken eyes, dry skin, and difficulty standing upright. With 10% fluid loss, the legs will be colder than the rest of the body. A 12% fluid loss usually results in death.

There are several commercial electrolyte powders for administering fluids to the calf. If they are not available, producers can economically prepare satisfactory mixtures as follows:

Formula 1

1 tablespoon baking soda
1 teaspoon salt

Formula 2

1 package (1 oz) fruit pectin
1 teaspoon of Lite® salt

8 oz (250 cc) dextrose
(do not use table sugar)
Add warm water to make a gallon
Administer up to 1 quart orally
every 4–6 hours

2 teaspoons baking soda
1 can beef consomme soup
Add warm water to make 2 quarts
Give 1 quart orally every 4–6 hours

Either of these homemade products administered at appropriate intervals with recommended volumes will provide an adequate nutrient supply over a period of 24–48 hours. Do not give milk or milk replacers to the calf at the same time as administering electrolytes (however, electrolytes do not replace the need for nutrition). Electrolytes interfere with the normal digestion of milk. Also, whole milk and milk replacers should not be altered as fluid sources because of altering digestive enzymes needs. Whole milk or milk replacers can be alternated at appropriate intervals with oral electrolytes as long as fluid needs are being met. In general, a calf requires 10–12% of body weight in oral fluids every 24 hours. Return the calf to the cow as soon as it is able to follow its mother.

Antibiotics are given systemically during fluid therapy to prevent secondary infections. They should be given under the direction of a veterinarian at proper treatment levels each day for at least 3 days or until the calf has recovered.

Economic Impact Research has shown that calves with inadequate colostrum intake have five times greater risk of pre-weaning death, six times greater risk of morbidity in the first month of life, and a three times greater risk of sickness prior to weaning. Calves that attained sufficient colostrum intake weighed nearly 30 pounds more at weaning and only 5% will require any kind of pre-weaning treatment.

The prevention of disease requires a two-pronged approach—increase resistance to pathogens and reduce disease challenge through systematic management. Effective health management is fundamental to assure enhanced animal well-being, improved cattle performance, higher value products, and sustained levels of profitability.

SELECTED REFERENCES

Publications

A new benchmark for the U.S. beef industry. 2005. Final Report of the National Beef Quality Audit. National Cattlemen's Beef Association. Centennial, CO.

Battaglia, R.A., & Mayrose, V.B. 2002. *Handbook of livestock management techniques.*
Upper Saddle River, NJ: Prentice Hall.

Beef quality assurance assessors guide to a cow-calf assessment. 2011. National Cattlemen's Beef Association. Centennial, CO.

Beef quality assurance assessors guide to a feedyard assessment. 2011. National Cattlemen's Beef Association. Centennial, CO.

Busby, W.D., Strohbehn, D.R., Beedle, P., & Corah, L.R. 2004. *Effect of postweaning health on feedlot performance and quality grade.* White Paper. Iowa State University. Ames, IA.

Byford, R.L., Craig, M.E., & Crosby, B.L. 1992. A review of ectoparasites and their effect on cattle production. *Journal of Animal Science.* 70:597.

Gaafar, S.M., Howard, W.E., & Marsh, R.E. (eds.). 1985. *Parasites, pests and predators.* New York: Elsevier Science.

Garry, F.B. 2000. *Newborn calf infectious diseases: The importance of distinguishing etiologies.* White Paper. Department of Clinical Sciences, Colorado State University. Fort Collins, CO.

Garry, F.B., & Odde, K.G. 1991. *Calf scours: Causes, treatment, and prevention.* Proc. Integrated Resource Management Program, Colorado State Univ. Ext. Service.

George, M.H., Tatum, J.D., Smith, G.C., & Cowman, G.L. 1997. Injection-site lesions in beef subprimals: Incidence, palatability, consequences, and economic impact. *Compendium's Food Animal Medicine and Management.* 19:2.

Griffin, D., & Mayer, J. 2002. *Feedlot animal health*. Proc. Cattlemen's College Workshop. Denver, CO.

Improving the quality, consistency, competitiveness and market share of fed beef. 2000. Final report of the Third National Quality Fed-Beef Audit, National Cattlemen's Beef Association. Englewood, CO.

Jackson, N.S., Greer, W.J., & Baker, K. 2000. *Animal health*. Danville, IL: Interstate.

Kahn, C.M., & Line, S. 2010. *The Merck veterinary manual*. NJ: Merck and Co. Inc.

Keeler, R.F., Van Kampen, K.R., & James, L.F. (eds.). 1978. *Effects of poisonous plants on livestock*. New York: Academic Press.

Kimberling, C.V. 1981. *Brucellosis*. Proc. Range Beef Cow Symposium VII, Dec. 7–9.

McCurnin, D.M. 1990. *Clinical textbook for veterinary technicians*. Philadelphia: W.B. Saunders.

McNeill, J. 2000. *Ranch to rail program summary*. College Station, TX: Texas A&M University.

Odde, K.G. 1989. *Survival of the newborn calf: Colostrum and other factors*. Range Beef Cow Symposium XI: 76–81.

Pierson, R.E., Kainer, R.A., & Teegarden, R.M. 1981. *Classifications of pneumonias in cattle*. Proc. Range Beef Cow Symposium VII, Dec. 7–9.

Roeber, D.L., Speer, N.C., Gentry, J.G., Tatum, J.D., Smith, C.D., Whittier, J.C., Jones, G.F., Belk, K.E., & Smith, G.C. 2000. Feeder cattle health management: Effects on morbidity rates, feedlot performance, carcass characteristics, and beef palatability. *Professional Animal Scientist*. 17:39.

Smith, D. 2002. *Biosecurity principles for livestock producers*. Publication G1442. University of Nebraska-Lincoln, NC.

Thomas, H.S. 2009. *The cattle health handbook*. Adams, MA: Storey Publishing.

USDA. 2008. *Beef 2007–08 Part I: Reference of Beef Cow-calf Management Practices in the United States*. NAHMS:APHIS:VS.

USDA. 2009. *Beef 2007–08 Part II: Reference of Beef Cow-calf Management Practices in the United States*. NAHMS:APHIS:VS.

USDA. 2009a. *Beef 2007–08 Part III: Changes in the U.S. Beef Cow-calf Industry*. NAHMS:APHIS:VS.

USDA. 2010. *Beef 2007–08 Part IV: Reference of Beef Cow-calf Management Practices in the United States*. NAHMS:APHIS:VS.

USDA. 2010. *Beef 2007–08 Part V: Reference of Beef Cow-calf Management Practices in the United States*. NAHMS:APHID:VS

Ward, J.K., & Nielson, M.K. 1979. Pinkeye (Bovine infectious keratoconjunctivitis) in beef cattle. *Journal of Animal Science*. 38:1179.

17

Growth, Development, and Beef Cattle Type

Growth and development of beef cattle is affected by nutrition, genetics, health status, and environmental influence. Additionally, bovine growth and development affects carcass merit, marketing decisions, and profitability. Given the interconnectivity of this topic with many aspects of beef cattle management, some aspects of growth and development are also covered in previous chapters. While beef cattle type receives less attention in the contemporary industry than it has received historically, it is still important to understand how perceptions about desired type changed the beef cattle business over time.

GROWTH AND DEVELOPMENT

Beef type is defined as the desired or ideal shape and form of the beef animal. Historically, the industry has vacillated as to preferences for large versus small, lean versus fat, horned versus polled, as well as for differences in hide color, color markings, and a host of other visual attributes. Beef conformation (structure) of the live animal or carcass results from numerous factors including growth and development. During the progression of life span, cattle undergo changes in form and composition. **Growth** is an increase in weight and dimension until mature size is reached. Growth results from several physiological processes—increases in cell number (hyperplasia), changes in cell size (hypertrophy), and protein deposition. Finally, development is defined as the directive coordination of all diverse processes until maturity is reached. It involves growth, cellular differentiation, and changes in body shape and form. For the purpose of this discussion, growth and development are integrated into a single process.

Prenatal Growth

Prenatal growth, or embryological development, is initiated as a spherical mass of cells organizes itself via morphogenesis into functional tissues and organs (Figure 17.1). The endoderm transforms into the digestive tract, lungs, and bladder; the mesoderm gives rise to the skeleton, skeletal muscle, and connective tissue; and the ectoderm yields the skin, hair, brain, and spinal cord. The genetic code directs growth, development, and differentiation, primarily via protein synthesis.

The three phases of prenatal life—the zygote, the embryo, and the fetus—are discussed in Chapter 11. The relative size of the fetus changes during gestation, with the largest increase in weight occurring during the last 3 months of pregnancy (Figure 17.2). The fetus undergoes marked changes in shape and form during prenatal growth and development. Early in the prenatal period, the head is much larger than the body. Later, the body and limbs grow more rapidly than the other body parts. The order of tissue growth follows a sequential trend determined by physiological importance, starting with the central nervous system and progressing to skeletal tissue, tendons, muscles, intermuscular fat, and subcutaneous fat.

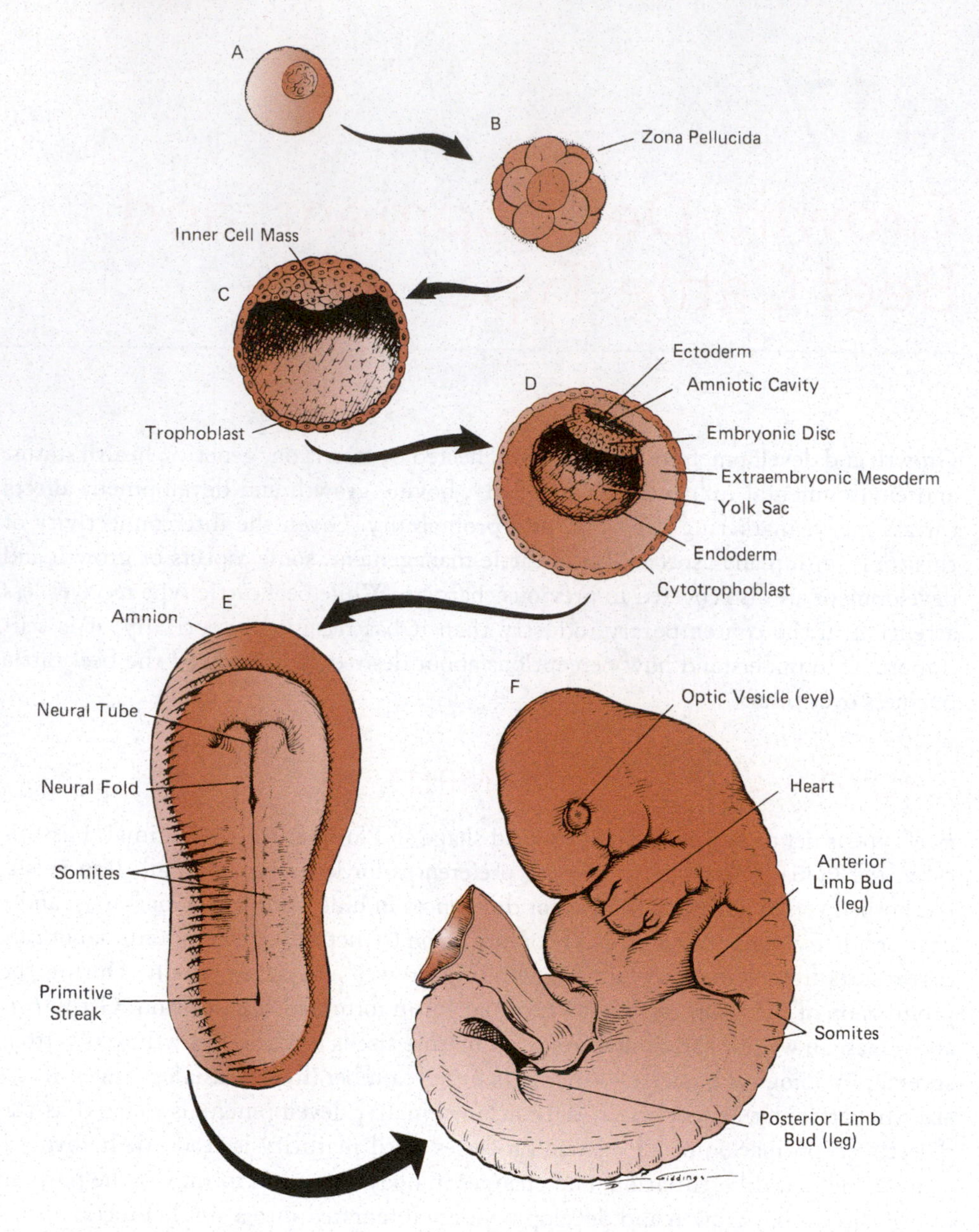

Figure 17.1

The morphogenesis of a single egg cell (A) into a morula; (B) then to a blastocyst; (C) and (D) show the stage at which the two cavities have formed in the inner cell mass: an upper (amniotic) cavity and a lower cavity yolk sac. The embryonic disc containing the ectoderm and endoderm germ layers is located between cavities. (E) is a cattle embryo showing the neural tube and the somites. (F) illustrates the development of a 14-day cattle embryo.

Source: Colorado State University.

During the first two-thirds of the prenatal period, most of the increase in muscle weight is the result of hyperplasia (an increase in the number of muscle fibers). During the last 3 months of pregnancy, hypertrophy (an increase in the size of muscle fibers) represents most of the muscle growth. Individual muscles differ in their rate of growth; the larger muscles of the legs and back have the greatest rate of postnatal growth.

Birth

After birth, the number of muscle fibers does not increase significantly; therefore, postnatal muscle growth is primarily by hypertrophy. All muscle fibers appear to be

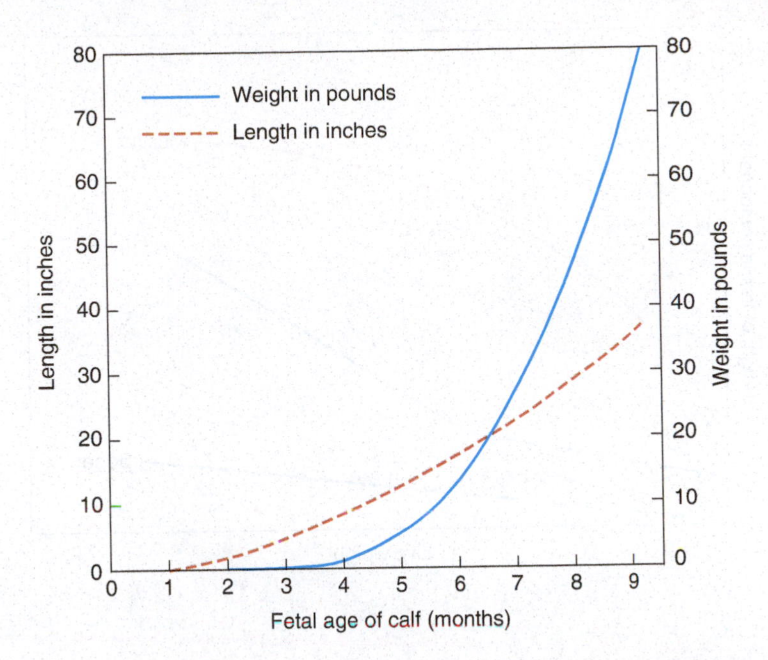

Figure 17.2
Growth of the fetus.
Source: Colorado State University.

red or fast-twitch type at birth, but shortly thereafter some of them differentiate into white and intermediate muscle types.

At birth, the various body parts have considerably different proportions when compared to mature body size and shape: the head is relatively large, the legs are long, and the body is small. In the mature animal, in contrast, the head is relatively small, the legs are relatively short, and the body is relatively large. Birth weight represents approximately 6–8% of the mature weight, while leg length at birth is approximately 60% and height at withers is approximately 50% of those same measurements at maturity. Hip width and chest width at birth are approximately one-third of the same measurements at maturity. This shows that the distal parts (leg length and wither height) are developed earlier than the proximal parts (hips and chest).

Live Weight Gain

Beef cattle growth is typically observed or measured by evaluating body weight changes per unit of time or by plotting body weight against age. Growth, expressed as average daily gain, weight per day of age, or pounds of lean (muscle) per day of age, is particularly useful as a management tool. It is also useful to describe growth curves, which are similar in shape for various breeds and types of cattle but contain noticeable differences relative to mature size and rate of maturity. Growth curve analysis receives increased emphasis as altering the shape of curves or achieving the best combination of rapid early growth in slaughter cattle and reducing mature size in breeding cattle become more important.

Carcass Composition

A low proportion of bone, a high proportion of muscle, and an optimum amount of fat characterize a superior carcass. During the past two to three decades, carcass composition has been improved in beef cattle primarily by reducing the amount of fat in the carcass.

Figure 17.3 shows expected changes in fat, muscle, and bone as cattle increase in live weight during the linear phase of growth. Fat growth begins rather slowly, and then increases exponentially as the animal enters the fattening phase. Bone has a smaller relative growth rate (2 lb of muscle to 1 lb of bone) than either fat or muscle. The muscle-to-bone ratio in the young calf is approximately 2:1; the ratio at a slaughter weight of 1,100–1,200 lb for the typical slaughter steer is 3.5 to 4:1.

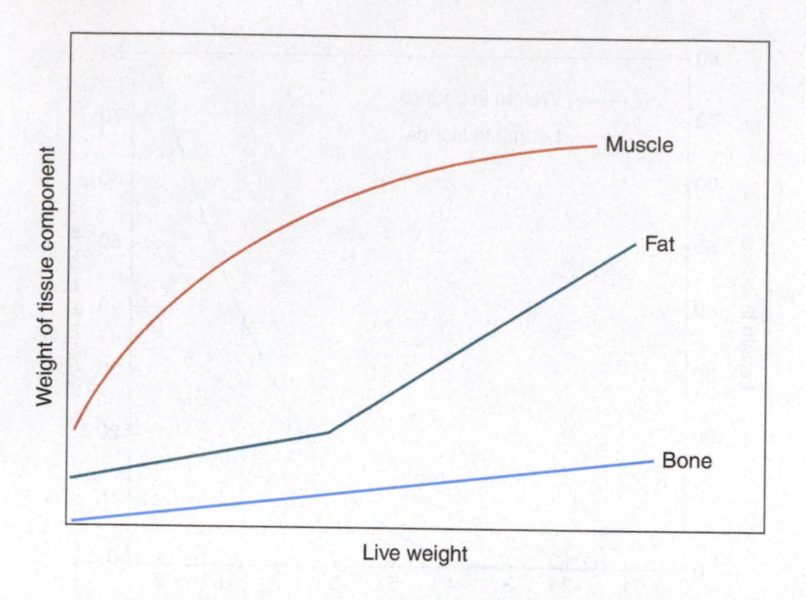

Figure 17.3
Tissue growth relative to live weight.

Sex affects tissue growth primarily in terms of the fat component. Heifers fatten at an earlier weight than steers, which is why they are slaughtered at a weight 100–200 lb less than steers of a similar frame size. Bulls have less fat than both steers and heifers at the same slaughter weight while also being superior in muscle mass.

Fat Growth and Development

Plane of nutrition or level of energy in the ration changes the relative amount of fat that is deposited. Cattle fed high-energy diets deposit fat earlier than cattle on lower-energy diets, even when the cattle are evaluated at the same live weights.

The site of fat deposition as live weight increases is of high economic importance to the beef industry. Current market preferences favor an optimum combination of intramuscular fat (marbling) and other fat depositions (subcutaneous, cavity, and seam) that allow a carcass to attain a USDA Quality Grade of at least low Choice while having an assigned USDA Yield Grade of 3 or better. The use of both within-breed variation and between-breed differences can be employed to enable the consistent production of carcasses that meet both quality and cutability standards.

Fat partitioning among the various fat depots results from both genetic and environmental influences. Both genetic and environmental tendencies to earlier fattening result in increased proportion of subcutaneous fat to intermuscular fat, while delayed or decreased fattening has the opposite effect. Kidney, pelvic, and heart fat (KPH) generally grows at the same relative rate as total fat, although there are some breed, sex, and plane of nutrition effects. Some evidence indicates that during the fattening phase, the amount of KPH fat increases at a faster rate than intermuscular fat. Dairy breeds deposit a higher proportion of internal fat and a lower proportion subcutaneously than traditional beef breeds. Limited data indicate that at equal fatness the fat partitioning in steers, heifers, and bulls is quite similar.

Cattle deposit fat at relatively high levels in the flank and brisket as well as along the loin and rib from the tailhead if the diet is of sufficient energy level. Fat depositions are minimal at the forearm and stifle. Thus, these are excellent anatomical reference points to evaluate muscle, independent of fat differences.

Muscle Growth and Development

The proportion of muscle in the carcass varies indirectly with fat, and a higher proportion of fat is associated with a lower proportion of muscle and vice versa.

Muscle has a much faster relative growth rate than bone (Figure 17.3). Muscle weight relative to live weight or muscle-to-bone ratio can be used as valuable measurement of muscle yield.

Relatively large genetic differences exist for muscling in cattle. The current population of cattle in the United States has sufficient muscling variation to allow changes in muscle via selection or cross-breeding. Muscle yield relative to live weight of more than 33–50% (muscle-to-bone ratio ranges of 3.5:1 to 5.0:1) has been observed in steers and heifers of slaughter weight without experiencing the extremes of double-muscling. These differences are economically important, particularly if more boneless cuts are merchandised.

Extreme muscling, as observed in double-muscling types, is associated with several problems of fitness, such as increased susceptibility to stress and reduced reproductive efficiency. There may be some latitude in increasing the muscle-to-bone ratio in slaughter cattle without encountering negative side effects.

Individual muscles have different relative growth rates. The newborn calf has relatively well-developed muscles in the hind legs. As maturity continues, there is a progressive rise in relative growth rate from the rump to the neck, with the shoulder girdle and neck muscles having the highest growth rates of all muscles. Also, the front limb muscles develop more rapidly than the hind limb muscles as the weight of the animal shifts progressively forward. Muscle growth rate is relatively high in the abdominal and thoracic regions.

Muscle Distribution

Individual muscle dissection has shown a similarity of proportion of muscles to the total body muscle weight. Thus, increasing muscle weight in areas of high-priced cuts at the expense of muscle weight in lower-priced cuts is not commercially feasible. Some small yet statistically significant differences in muscle distribution have been shown, but the differences do not seem to be economically significant for the industry. Double-muscled animals show the greatest evidence of a higher proportion of muscle weight in higher-priced cuts.

Bone Growth and Distribution

Bone growth has the slowest growth rate when compared to muscle and fat (Figure 17.3). There are only minor differences in the relative growth rates of different bones, although the growth rate for limb bones is somewhat slower. Although breed differences in bone distribution have been reported, they have little commercial value.

Age and Teeth Relationship

Evaluating teeth can be used to estimate the age of cattle as well as to determine their ability to graze effectively with advancing age. This process is often referred to as "mouthing" and has been used in the industry to cull breeding cattle or to classify stock of unknown birthdate into age groups. Cattle are sometimes "mouthed" to classify them into appropriate age groups for specific show ring classes or to determine whether breeding animals of unknown age have reached a stage where culling is appropriate.

Table 17.1 provides guidelines used in estimating the age of cattle by evaluating condition of their incisors (front cutting teeth). There are no upper incisors and the molars (back teeth) are not commonly used to determine age. Figure 17.4 shows the incisor teeth of cattle of different ages. When the grazing area is sandy and the grass is short, the wearing of the teeth progresses faster than that described in Table 17.1. Also, some cows may be "broken-mouthed," which means that they have lost some of their permanent incisors.

Table 17.1
Estimating the Age of Cattle by Observing Their Teeth

Age	Description of Teeth
Birth	Usually only one pair of middle incisors
1 month	All eight temporary incisors
$1^1/_2$–2 years	First pair of permanent (middle) incisors
$2^1/_2$–3 years	Second pair of permanent incisors
$3^1/_4$–4 years	Third pair of permanent incisors
4–$4^1/_2$ years	Fourth (corner) pair of permanent incisors
5–6 years	Middle pair of incisors begins to level off from wear; corner teeth may also show some wear
7–8 years	Both middle and second pairs of incisors show wear
8–9 years	Middle, second, and third pairs of incisors show wear
10+ years	All eight incisors show wear

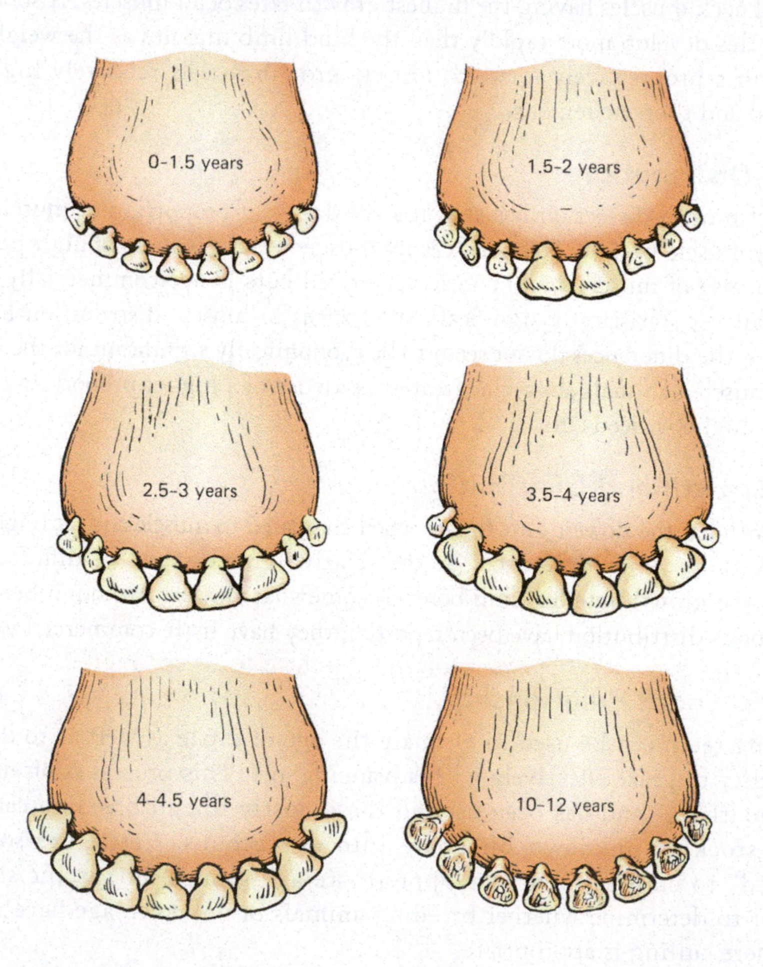

Figure 17.4
Incisor teeth of cattle of different ages.

BEEF TYPE

The term beef type is defined as the ideal or standard of perfection that combines all characteristics contributing to an animal's usefulness for a specific purpose. **Conformation** broadly refers to the form and skeletal structure of the animal. Both **type** and conformation

involve external characteristics of the animal that can be evaluated visually or measured objectively. Most general descriptive terms involve length, height, thickness, and general symmetry.

One of the first-known writings on livestock judging is Bourgelat of Lyon's *Traite de la Conformation Exterieure du Cheval,* written in 1768. Since that time, there has been much interest in the visual appraisal of cattle. Conformation (or form and shape) has been widely used in evaluating beef cattle and carcasses.

However, there are wide differences of opinion about the relationship of form and function in beef cattle. The roles of visual appraisal and measurements of body form continue to be debated as to their value in breeding programs, market grades, and defining "ideal" types of cattle. Obvious differences in cattle form and function are to be expected given the wide diversity of uses for which cattle have been developed over the ages. Where cattle have been used as draft animals, the form generally has been one of large, massive size. Cattle selected primarily for milk production have developed into more angular animals. Cattle developed for maximum meat production express more thickness and total volume of muscle. Dual-purpose cattle, where the emphasis has been placed on meat, milk, and possibly draft, typically reflect some combinations of body form unique to each of those functions being combined. Human beings have directed the change in cattle form under certain environmental conditions and are thus responsible for most differences in cattle types observed today (Figure 17.5). One study evaluated the type changes in the Hereford breed from the 1950s to the 1990s. Hereford sires from the 1950s, 1970s, and 1990s were randomly mated to commercial Hereford cows similar in age, pedigree, and frame size from a single Nebraska ranch. A sample of the progeny is shown in Figure 17.6. Their feedlot and carcass performance records are highlighted in Table 17.2.

While some of these relationships have been observed and validated, the issue is far from resolved. For example, some cattle producers argued strongly that certain animal characteristics were more attractive or "more pleasing to the eye," even though they had little or no relationship to productivity. Those who could verbalize with great conviction, and particularly those who became recognized authorities in beef evaluation, greatly influenced the opinions of many others in the industry. As a result, the perpetuation of both valid and invalid opinions about the value of beef type and conformation continues.

Part of the beef production goal should be to put a highly productive animal into a merchandisable package that will command the highest market price. Using meaningful performance records and effective visual appraisal best identifies productivity of breeding and slaughter cattle. It is well demonstrated that performance records are much more effective than visual appraisal in improving beef cattle productivity. In fact, objective measures of genetic merit in the form of expected progeny differences accurately predicted the differences between the three generations of Hereford cattle (Table 17.3). These results demonstrate the effectiveness of expected progeny differences. The actual yearling performance was likely depressed by the limited energy backgrounding ration provided to the cattle from weaning to one year of age.

Visual appraisal, however, is important in evaluating functionality, primarily in identifying skeletal soundness and health status. Visual appraisal is also useful in evaluating differences in composition as it relates to muscularity and fat deposition. Both objective records (such as ultrasound readings for fatness) and visual appraisal are used in evaluating finished animals. Finished cattle that are purchased live have been visually assessed by an experienced buyer who estimates carcass merit based on visual indicators.

(a)

(b)

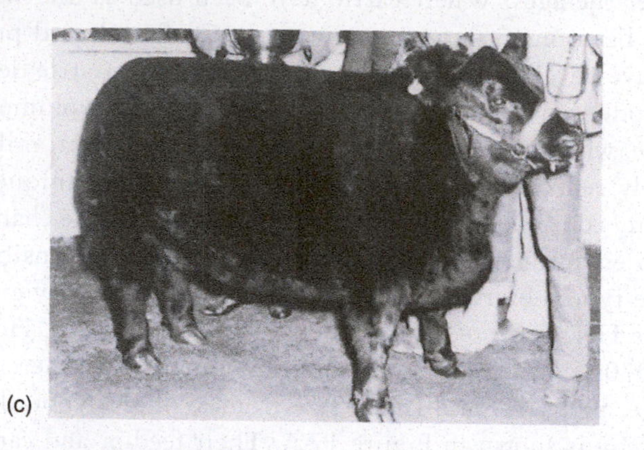

(c)

(d)

Figure 17.5

Body type changes, reflecting breeders' preferences, have been dramatic over the past 100 years—even within the same breed. (a) Angus steer Black Prince, imported from Scotland to United States was the champion steer, weighing approximately 2,400 lbs at the 1883 Chicago Fat Stock Show. (b) Champion steer at 1912 Chicago International (1,630 lbs). Also champion at the 1913 Chicago International weighing approximately 150 lbs less than in 1912. (c) Champion steer at 1958 Chicago International, weighing less than 900 lbs. (d) Preferred type of Angus steer for the 1980s and 1990s, weighing approximately 1,200 lbs. Note the type resemblance of steers (a) and (c); also steers (b) and (d).

Source: American Angus Association.

Figure 17.6
Left to right: 1970s sired steer, 1950s sired steer, and a 1990s sired steer.

Table 17.2
Average Growth and Carcass Performance of Three Generations of Hereford Cattle[1]

	1950s	1970s	1990s
Birth weight (lb)	82.5	85.9	91.4
Live weight @ .45 in. external fat (days)			
Steers (lb)	1,083	1,214	1,275
Heifers (lb)	970	1,013	1,138
Frame size	3.7	4.9	5.5
Days on feed @ .45 in. external fat (days)	109	111	111
Average daily gain (lb)	3.81	4.01	4.25
Carcass weight			
<600 lb (%)	50	50	0
600–800 lb (%)	15	85	0
>800 lb (%)	0	74	26
Rib-eye area (in.)	11.5	11.8	12.3
Quality Grade Choice (%)	33	50	52
Select (%)	67	47	48
Standard (%)	0	3	0

[1]Reflects only sire-effect differences.
Source: Based on Tatum and Field (1996).

Table 17.3
Expected Versus Actual Differences in Growth Traits as Projected by EPDs

	1950–1970		1970–1990		1950–1990	
Trait	Actual	Projected	Actual	Projected	Actual	Projected
Birth weight (lb)	+4	+6	+5	+4	+9	+10
Weaning weight (lb)	+35	+32	+26	+25	+61	+57
Yearling weight (lb)	+30	+54	+46	+50	+76	+104

Source: Based on Tatum and Field (1996).

PARTS OF THE BEEF ANIMAL

Figure 17.7 identifies the parts of the beef animal. Figure 17.8 provides additional descriptors of conformation often used to describe cattle.

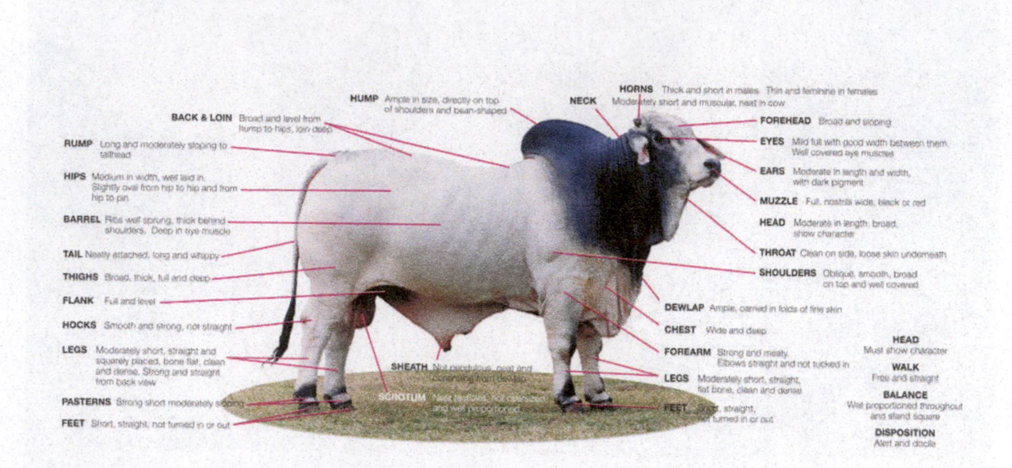

Figure 17.7
Parts of the beef animal.
Source: American Brahman Breeders Association.

Figure 17.8
Parts of the beef animal used to identify body dimensions (side view).

CARCASS CONFORMATION

When the USDA beef carcass quality grades were established in 1927, conformation was included as a factor in grade standards. In 1975, however, conformation was eliminated from quality grade standards because research studies had shown that it has no positive effect on beef palatability.

Fat thickness, carcass weight, and area of the rib eye are the carcass characteristics related to carcass conformation that are presently included in yield grade standards. Research reports disagree about the importance of carcass conformation, as described in the USDA quality grade standards prior to 1975, and yield of retail cuts. Reports range from a slightly positive relationship between conformation and retail cut yield to no relationship.

Fat greatly influences the form and shape of the carcass. Subcutaneous fat affects conformation the most; however, intermuscular fat also causes muscles to have an appearance of greater fullness and depth. There are large differences in the muscle-to-bone ratios in the carcass. These widely different ratios can be observed visually by noting the muscling areas of the carcass that are not greatly influenced by subcutaneous and intermuscular fat. Some packers and retailers still consider carcass conformation important from a marketing standpoint. For example, several branded beef programs discriminate against dairy carcass types. Preference is given to a thicker carcass with a thick, plump round.

CONFORMATION OF MARKET CATTLE

Visual appraisal appears to be more accurate for assessing differences in carcass composition than for other economically important traits. The accuracy of individual animal estimates for quality grade is low; however, live animal estimates of carcass yield grades are more accurate. This accuracy is dependent on the experience of the appraiser and the variability in yield grades of cattle.

After becoming familiar with the external parts of the animal, effective visual appraisal involves understanding anatomical reference points that delineate major meat cuts as well as the physiology of muscle growth and fat deposition. A carcass is evaluated after an animal has been harvested, eviscerated, and split into two halves. When the carcass is evaluated, it is hanging by the hind leg from the rail in the packing plant. This makes it difficult to perceive how the carcass would appear as part of the live animal standing on all four legs. Figure 17.9, which shows the location of wholesale cuts on the slaughter steer, helps to understand the major carcass component parts of the live animal.

The carcass is composed of fat, lean (red meat), and bone. The beef industry goal is to produce large amounts of highly palatable lean and optimal amounts of fat and bone. Effective visual appraisal of carcass composition requires knowing the body areas of the live animal where fat deposits and muscle growth occur. Compositional differences are contrasted in Figures 17.10 and 17.11.

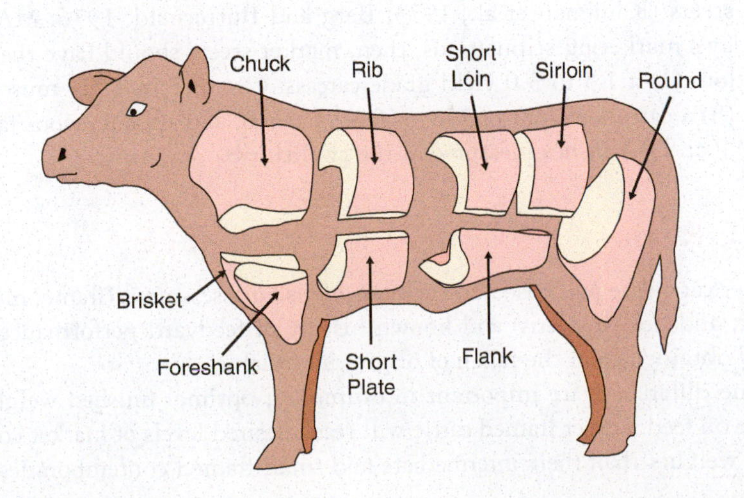

Figure 17.9
Location of the wholesale cuts on the live steer.

Figure 17.10
Yield Grade 4 or 5 live steer depiction.

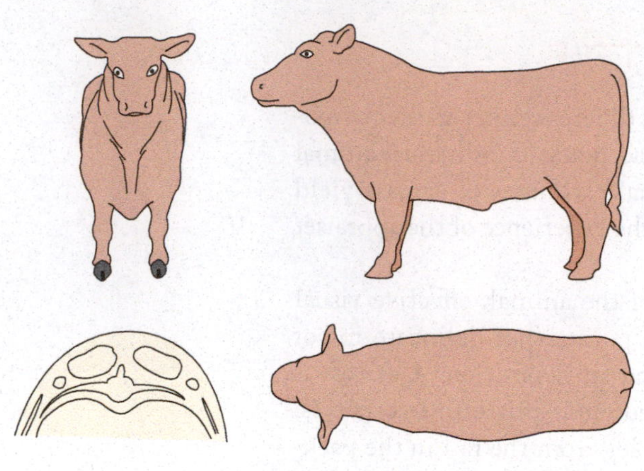

Figure 17.11
Yield Grade 2 live steer depiction.

Muscling is best described by muscle-to-bone ratios, where observed differences typically range from 2.5:1 to 4.5:1 (and even higher for double-muscled steers). Muscling differences of 2.5 to 3.0:1 versus 4.0 to 4.5:1 can be visually appraised in slaughter cattle with a reasonable degree of accuracy. Although it is commonly assumed that large differences exist in muscle distribution in cattle (e.g., that a higher percentage of carcass weight is in the hind limb muscle than in the forelimb muscle or vice versa), there is no evidence to indicate any significant differences in muscle distribution in slaughter steers (Kauffman et al., 1973; Berg and Butterfield, 1976; MARC data). Under today's marketing stipulations, then, market steers should have the following conformation: (1) a 1.5 to 3.0 yield grade carcass; (2) a 3.5 to 4.5:1 muscle-to-bone ratio; and (3) a carcass weight of 750–950 lbs (steers), and subcutaneous fat depth of between 0.2 and 0.5 inches measured at the twelfth rib.

CONFORMATION OF FEEDER CATTLE

Most feeder cattle are purchased on the basis of visual assessment (frame, muscularity, hide color, and health status) and known history of feedyard performance on cattle previously obtained from the ranch of origin, if available.

Frame differences are important in estimating optimal finished weights and to assess time on feed. Larger framed cattle will reach desired levels of market composition at heavier weights than their intermediate and small framed contemporaries and their feedlot gains are usually higher when fed to a constant time on feed or weight-constant basis. When cattle are fed to a similar compositional endpoint, efficiency differences between frame sizes are minimized. However, large-framed cattle are typically faster growing than small cattle. This implies rather large differences in final carcass weights, which could have economic significance depending on carcass weight specifications.

Large differences in rate and efficiency of gain exist in cattle of similar frame size and environmental background. Visual appraisal is not able to detect these differences with any high degree of accuracy.

Frame size, whether appraised visually or measured as hip height, can be used as a management tool in determining the logical slaughter weight of cattle in the feedlot. Figure 17.12 shows compositional differences of different frame sizes of feeder cattle fed to different slaughter weights. No specific frame size is superior to the others. Feed resources, market weight preferences, and other production costs determine the frame size preference of a specific cattle feeder. A visual evaluation of frame size and condition can be used to sort

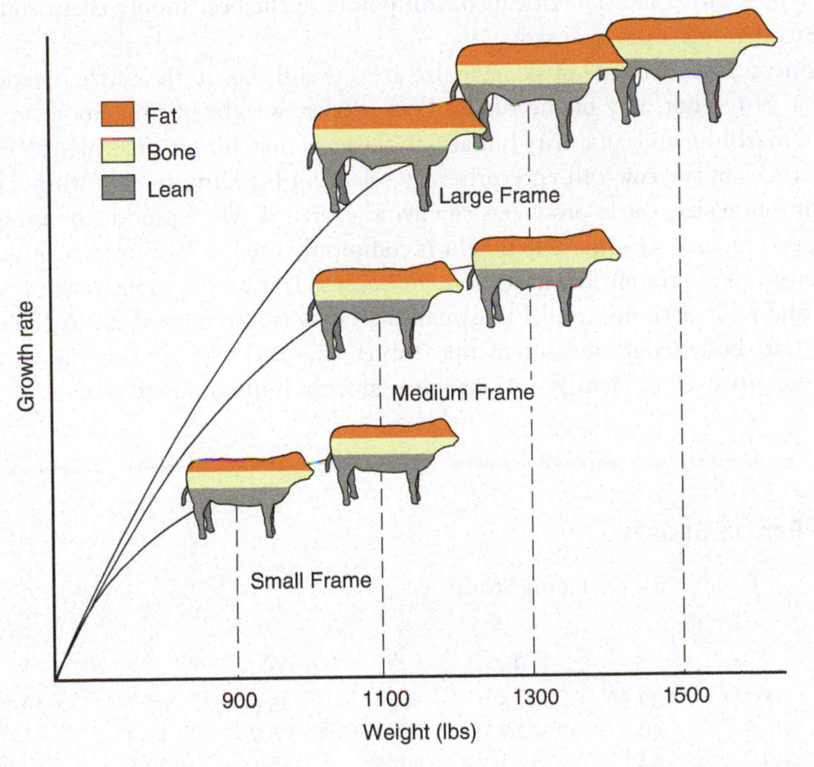

Figure 17.12
The relationship of frame size and weight to carcass composition in beef steers.
Source: Colorado State University.

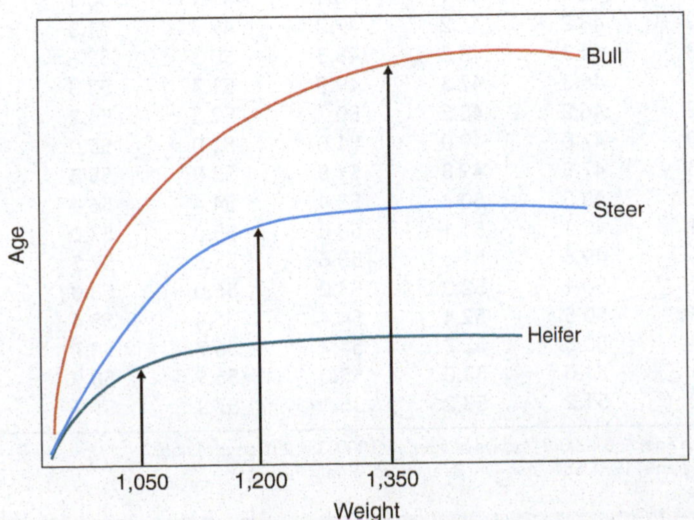

Figure 17.13
The influence of sex on body (carcass) composition at various weights.

feeder cattle into separate feeding groups. Days on feed can be projected for each group to reach similar yield grade and quality grade endpoints (Daley et al., 1983b).

Cattle producers should recognize how sex differences affect conformation, particularly body composition. Figure 17.13 shows the compositional differences of heifers, steers, and bulls at different weights and explains why heifers are typically marketed 100–150 lb lighter than their counterpart steers.

CONFORMATION OF BREEDING CATTLE

Cattle breeders should study carefully the value of conformation in carcasses, slaughter, and feeder cattle before making decisions on how conformation characteristics can be used in a breeding program. Decisions made at the breeding level are manifested long into the future as they impact not only replacement breeding stock but also every step in the supply chain. Seedstock and cow-calf producers who understand the physiology of beef cattle and the critical control points in the beef supply chain contribute the greatest value to the industry.

Linear measurements of skeletal size are a useful tool to the cattle breeder, providing a gauge not only of the likely ideal market weight of progeny at optimum levels of marbling and cutability but also in terms of matching replacement females to the resources of the cow-calf enterprise. By selecting breeding stock within a desired range of frame size, cattle producers can avoid extremes while producing appropriate market and mature weights. Hip height is commonly used to measure size of skeleton. Body length is a questionable measurement because it is quite proportional to skeletal height and adds little additional information. There is also more difficulty in obtaining accurate body length measurements. Tables 17.4 and 17.5 provide the hip height measurements used to identify various frame sizes in bulls and heifers.

Table 17.4
Bull Hip Height (Inches) and Frame Scores

Age in Months	Frame Score[1]								
	1	2	3	4	5	6	7	8	9
5	33.5	35.5	37.5	39.5	41.6	43.6	45.6	47.7	49.7
6	34.8	36.8	38.8	40.8	42.9	44.9	46.9	48.9	51.0
7	36.0	38.0	40.0	42.1	44.1	46.1	48.1	50.1	52.2
8	37.2	39.2	41.2	43.2	45.2	47.2	49.3	51.3	53.3
9	38.2	40.2	42.3	44.3	46.3	48.3	50.3	52.3	54.3
10	39.2	41.2	43.3	45.3	47.3	49.3	51.3	53.3	55.3
11	40.2	42.2	44.2	46.2	48.2	50.2	52.2	54.2	56.2
12	41.0	43.0	45.0	47.0	49.0	51.0	53.0	55.0	57.0
13	41.8	43.8	45.8	47.8	49.8	51.8	53.8	55.8	57.7
14	42.5	44.5	46.5	48.5	50.4	52.4	54.4	56.4	58.4
15	43.1	45.1	47.1	49.1	51.1	53.0	55.0	57.0	59.0
16	43.6	45.6	47.6	49.6	51.6	53.6	55.6	57.5	59.5
17	44.1	46.1	48.1	50.1	52.0	54.0	56.0	58.0	60.0
18	44.5	46.5	48.5	50.5	52.4	54.4	56.4	58.4	60.3
19	44.9	46.8	48.8	50.8	52.7	54.7	56.7	58.7	60.6
20	45.1	47.1	49.1	51.0	53.0	55.0	56.9	58.9	60.9
21	45.3	47.3	49.2	51.2	53.2	55.1	57.1	59.1	61.0

[1]Frame Score = −11.548 + 0.4878 (ht) − 0.0289 (days of age) + 0.00001947 (days of age)^2 + 0.0000334 (ht) (days of age).
Source: Guidelines for Uniform Beef Improvement Programs, 1990. Stillwater, OK: Beef Improvement Federation.

Table 17.5
HEIFER HIP HEIGHT (INCHES) AND FRAME SCORES

Age in Months	Frame Score[1]								
	1	2	3	4	5	6	7	8	9
5	33.1	35.1	37.2	39.3	41.3	43.4	45.5	47.5	49.6
6	34.1	36.2	38.2	40.3	42.3	44.4	46.5	48.5	50.6
7	35.1	37.1	39.2	41.2	43.3	45.3	47.4	49.4	51.5
8	36.0	38.0	40.1	42.1	44.1	46.2	48.2	50.2	52.3
9	36.8	38.9	40.9	42.9	44.9	47.0	49.0	51.0	53.0
10	37.6	39.6	41.6	43.7	45.7	47.7	49.7	51.7	53.8
11	38.3	40.3	42.3	44.3	46.4	48.4	50.4	52.4	54.4
12	39.0	41.0	43.0	45.0	47.0	49.0	51.0	53.0	55.0
13	39.6	41.6	43.6	45.5	47.5	49.5	51.5	53.5	55.5
14	40.1	42.1	44.1	46.1	48.0	50.0	52.0	54.0	56.0
15	40.6	42.6	44.5	46.5	48.5	50.5	52.4	54.4	56.4
16	41.0	43.0	44.9	46.9	48.9	50.8	52.8	54.8	56.7
17	41.4	43.3	45.3	47.2	49.2	51.1	53.1	55.1	57.0
18	41.7	43.6	45.6	47.5	49.5	51.4	53.4	55.3	57.3
19	41.9	43.9	45.8	47.7	49.7	51.6	53.6	55.5	57.4
20	42.1	44.1	46.0	47.9	49.8	51.8	53.7	55.6	57.6
21	42.3	44.2	46.1	48.0	50.0	51.9	53.8	55.7	57.7

[1]Frame Score = −11.7086 + 0.4723 (ht) − 0.0239 (days of age) + 0.0000146 (days of age)^2 + 0.0000759 (ht) (days of age).
Source: Guidelines for Uniform Beef Improvement Programs, 1990. Stillwater, OK: Beef Improvement Federation.

Table 17.6
AGE-OF-DAM ADJUSTMENT FACTORS FOR HEIGHT AT WEANING

Age of Dam (years)	Bulls (weaning ht)	Heifers (weaning ht)
2 and 13 or older	1.02	1.02
3 and 12	1.015	1.015
4 and 11	1.01	1.01
5 through 10	(no adjustment)	

Source: Guidelines for Uniform Beef Improvement Programs, 1990. Stillwater, OK: Beef Improvement Federation.

Weaning height for bulls and heifers can be adjusted for age of calf and age of dam. To adjust heights to 205 days for sex, (1) multiply the number of days *under 205* by 0.033 for bulls or 0.025 for heifers and add that figure to the actual height or (2) multiply the number of days *over 205* by 0.033 for bulls or 0.025 for heifers and subtract the result from the actual height. To adjust for age of dam, multiply the adjusted height for sex by the age-of-dam factor (Table 17.6).

Yearling height measurement can be adjusted for age of bull or heifer by using the same age-adjustment factors as for 205-day adjustments (e.g., for bulls, 0.033 in./day; for heifers, 0.025 in./day).

Comparing Table 17.4 with Table 17.5 notes sex differences affecting hip height. Steers are intermediate between bulls and heifers when compared in the 5–21 months age range. Steers have 1 in. more than heifers and 1 in. less than bulls, when compared at the same age.

Hip height (frame score) is related to sexual maturity (puberty), carcass composition, and mature weight (Figure 17.14). As Table 17.7 demonstrates, when frame score increases, the live weight and associated carcass weight at which the animal is likely to reach a desirable composition also increases progressively.

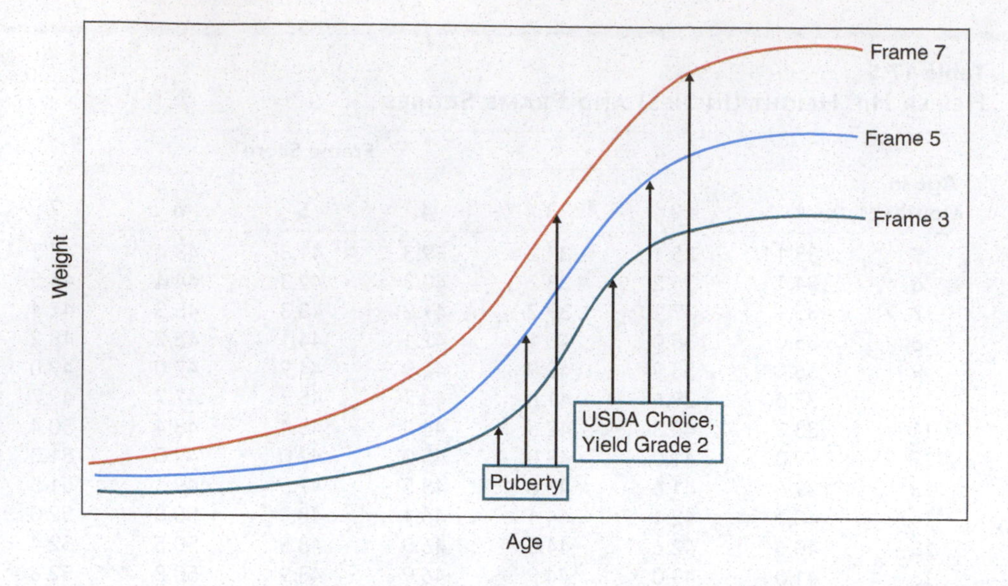

Figure 17.14
Frame size, weight, and age influence on maturity and carcass traits.

Table 17.7

Relationship of Frame Score to Slaughter and Carcass Weights of Steers Grading USDA Choice with Backfat Equivalent To USDA Yield Grade of 3.0

Frame Score	Slaughter Weight (lb)	Carcass Weight (lb)
2	800–900	500–575
3	950–1,050	600–675
4	1,100–1,200	700–775
5	1,250–1,350	800–875
6	1,350–1,450	875–925
7	>1,450	>925

Using Standards to Meet Market Targets

A case in point for the development of both live animal and carcass specifications to meet a specific market demand is Certified Angus Beef™. The live specifications include at least 51% black-hided and having Angus-type characteristics. The carcass specifications are: modest or higher degree of marbling, medium or fine marbling texture, Yield Grade 3.9 or better, A maturity, at least moderately thick muscling, no evidence of Brahman influence (minimal hump <2 in.), no blood spots, and no dark cutters.

Approximately 8% of all U.S. cattle qualify for the Certified Angus Beef™ label (CAB®). Less than 20% of cattle meeting the live specification also qualify based on carcass performance. The primary reason that more than 75,000 cattle studied by Iowa State University scientists failed to meet the specifications was insufficient marbling (84%). Failure to meet Yield Grade standards accounted for an additional 14% of the failure rate. While the genetic correlation between subcutaneous fat and marbling deposition is nearly zero, there is a tendency for CAB® acceptance rates to increase with declining cutability (i.e., 11% acceptance for Y.G. 2.5 and 33% for Y.G. 3.5).

This analysis points to the need for increased selection to hit these carcass specifications as well as the need to better manage individual cattle to meet market targets. For example, it is more profitable to harvest cattle that meet a marbling target at desirable levels of composition rather than allowing them to become excessively fat prior to harvest. The dilemma is finding ways to accurately assess both marbling and compositional differences in live cattle. In an attempt to overcome this challenge, management systems have been developed that focus on more precise management of specific cattle types that attempts to narrow the variation in performance within a pen or group of cattle. These approaches enable cattle producers to meet demand for improved red meat yields as a result of an increase in case-ready processing. The net effect of case-ready beef production is that trim losses will exclusively be borne at the packing level. As such, those animals with less trim and superior red meat yield are likely to receive significant premiums in the marketplace.

Precision management involves the application of a variety of technologies to diverse populations of cattle with a goal of sorting these animals into outcome groups. These groups can be appropriately managed by matching nutrition, implant protocols, and harvest timing to the growth potential of the animal. These systems utilize ultrasound measures of fat and muscle, video image analysis of body dimension, weight measures, and hip heights to sort cattle. These systems are typically tied to electronic identification approaches and to integrated information management databases. There is evidence that such approaches can add to profitability by sorting cattle into more uniform outcome groups.

Skeletal Soundness

Visual appraisal is used extensively by beef producers to evaluate skeletal soundness as it relates to longevity and productivity. Table 17.8 provides feet and leg descriptions commonly used to define skeletal soundness. Preferred structure of feet and legs as evaluated from the side profile is defined by imagining a line that extends from the top of the shoulder to the middle of the front foot and a line extending from the midpoint between hooks and pins that extends through the middle of the hind foot. When viewing cattle from the front or the rear view, it is generally accepted that desired structure is such that the toes of the feet are pointing forward with minimal deviation from center point (a line drawn from point of shoulder extending through midpoint of the knee and center of the forefoot or a line drawn from point of hip that extends through the center of the hock and the midpoint of the hindfoot).

However, exactly what constitutes skeletal soundness is not well defined. The following points about skeletal soundness should be considered.

1. Cattle with hock angles greater than 150 degrees appear to be more predisposed to becoming "stifled." Cattle that are post-legged have more serious problems than those with sickle-hocked condition (Woodward, 1968).
2. Excessive hoof growth can result in lameness and economic concern if foot trimming is necessary. Hoof growth appears to be highly heritable (Brinks et al., 1979).
3. "Founder," where lameness is typically involved, can be a relatively serious condition under feedlot conditions. This structural soundness condition, usually observed when cattle are on high-energy rations, appears to have a genetic basis.

Table 17.8
DESCRIPTORS OF DEVIATIONS IN FEET AND LEG STRUCTURE

	Set Excessively Forward of Ideal	*Set Excessively Backward of Ideal*
Front leg	Buck-kneed	Back at the knees
Rear leg	Sickle-hocked	Post-legged
	Midpoint of foot outside ideal	*Midpoint of foot inside ideal*
Front feet	Toed-out	Toed-in
Rear feet	Cow-hocked	Wide at the hocks

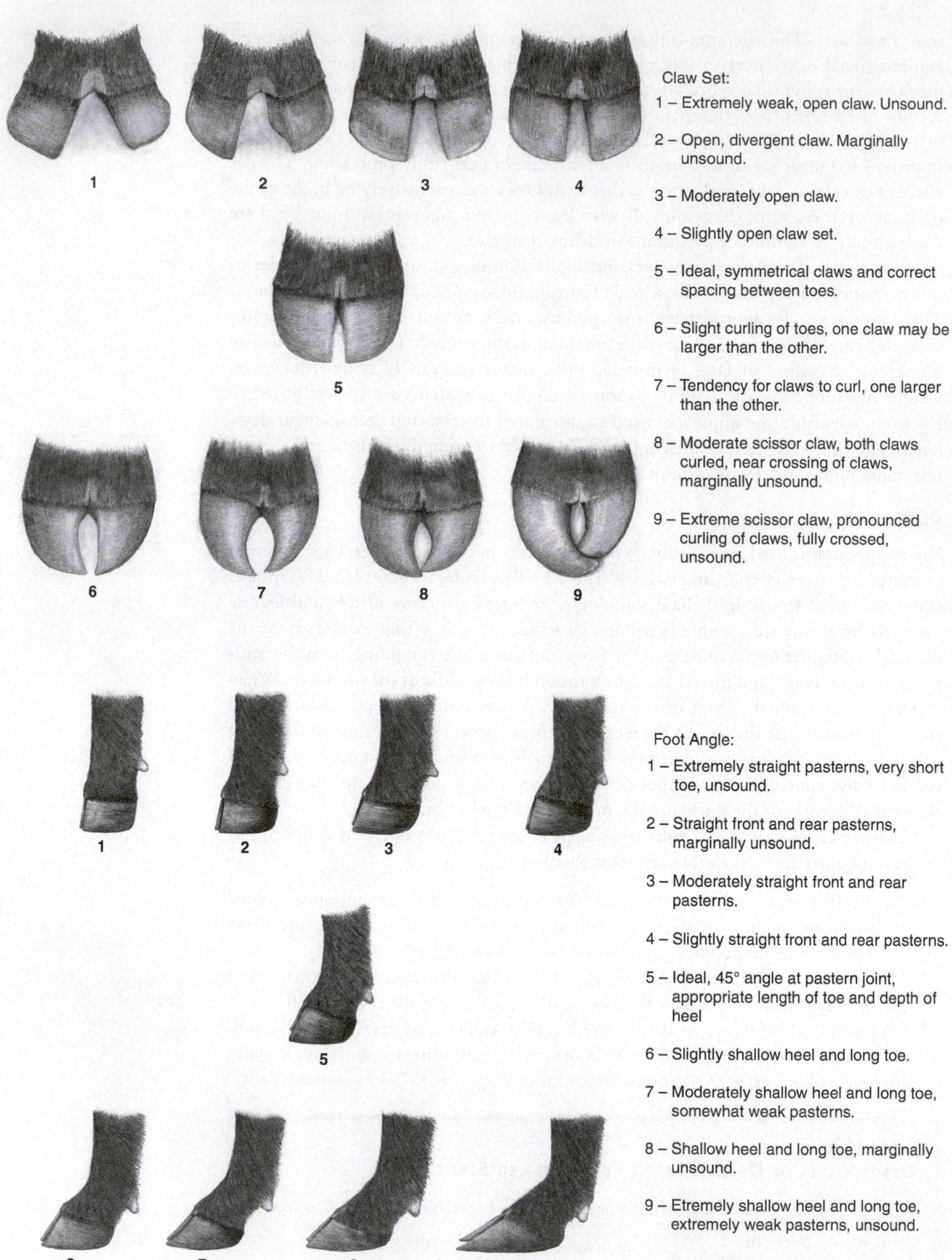

Figure 17.15

Foot and leg structure and related structural soundness scores.

Source: American Angus Association.

Abnormalities of the foot and leg may lead to early culling due to lameness. Some breed associations have begun to utilize objective scoring systems to determine genetic differences in these traits (Figure 17.15).

Reproduction

Jan Bonsma (1983) from South Africa advocated that cattle can be selected visually for reproductive performance, although Wilson et al. (1981) found no conclusive evidence in relating feminine features in cows to fertility. Most of Bonsma's work attempted to predict the number of calves produced by each cow. His appraisal methodology appeared to reflect the expression of hormone levels or hormone imbalances as they affected specific target locations of the neck, shoulders, hair, tailhead, and reproductive organs. This type of appraisal has not proven effective in heifer selection, where hormone expression has been limited. In addition, the use of good performance records in mature cows is far more economically feasible than trying to gain the experience needed to duplicate Bonsma's visual appraisal method.

Visual appraisal of body condition in breeding females is useful in assessing postpartum interval. Visual body scores have been shown to be highly related to carcass fat and carcass energy content ($r = 0.80$) and more useful than weight-to-height ratios.

Udder scoring systems (Figure 17.16) have also been developed to assist breeders in avoiding the propagation of replacement females with dysfunctional udder and teat conformation that leads to higher labor costs and problems associated with nursing.

Merchandising

Visual appraisal can be used to put productive animals into attractive packages. Body form and appearance are generally independent of productivity; however, it may be economically feasible for producers to combine highly productive animals with visual traits that are pleasing to the eye. This is part of the merchandising effort, whether the evaluation is focused on a retail cut, carcass, feeder calf, slaughter steer, or breeding animal. However, caution must be exercised in maintaining productivity as the priority, with only a secondary emphasis on eye appeal.

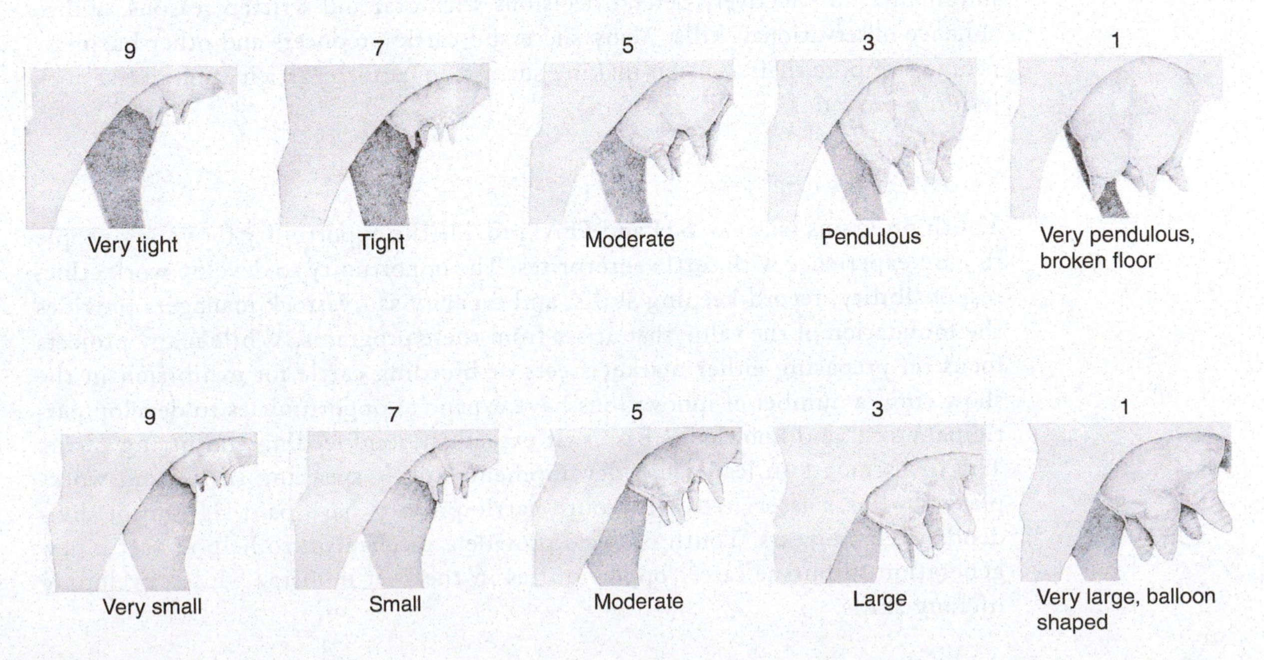

Figure 17.16
Udder scoring system.
Source: Beef Improvement Federation.

THE LIVESTOCK SHOW

The historic role of the livestock show in the development of the purebred cattle industry provided both a marketplace and a forum for discussion and sharing of information. Some cattle breeders were elevated to rather lofty status and were considered to have "the eye of the master"—a nearly mythical capacity to select superior breeding stock on the basis of careful visual appraisal. In the absence of objective selection tools, breeding programs were frequently influenced by the results of the show ring.

As genetic estimates such as breeding values and EPDs came into favor, heated discussions as to the value of the show ring ensued. As traditional notions began to give way to genetic research in the 1940s and 1950s, it became evident that major differences in show ring cattle were created by environmental factors such as how the cattle were fed and fitted. Show ring placings and high sale prices in many cases proved to be negatively related to productivity as measured by the commercial cattle industry.

Nonetheless, the show ring remains popular as a promotion and merchandising tool, a venue to engage urban dwellers who have lost touch with agriculture, and as a place to fuel competitive drive. In today's industry, the show ring has little influence on the commercial cow-calf producer, stocker operator, and cattle feeder. The majority of seedstock producers rely on objective estimates of genetic superiority instead of visual observation of phenotypic differences.

Cattle Judging

Most cattle producers have participated in 4-H, FFA, junior college, and university judging contests that are usually associated with fairs and shows. Most land grant universities have courses in beef cattle and livestock judging. These courses improve students' abilities to make decisions and communicate more effectively rather than make marked improvements in their abilities to identify more productive cattle. It has been demonstrated that livestock judging develops the ability to make complex judgments, to effectively defend decisions with oral and written reasons, and to enhance observational skills. Many successful cattle producers and other businesspeople attribute their decision-making success, in part, to participation in livestock judging programs.

Youth Projects

Youth programs (such as 4-H and FFA) provide the opportunity for young people to gain experience with cattle enterprises. The opportunity to develop work ethic, responsibility, record-keeping skills, and capacity as livestock managers provides the foundation of the value that arises from these programs. While many projects focus on preparing either market steers or breeding cattle for exhibition in the show ring, a number of innovations have expanded opportunities to develop participant skill and knowledge base that extends beyond feeding and fitting cattle. The investments in leadership development, public speaking skills, and workplace readiness associated with youth cattle projects have paid significant dividends over the years. Youth projects provide a mechanism to help attract a new generation to pursue career opportunities in the beef industry while developing lifelong skills.

SELECTED REFERENCES

Publications

Berg, R.T. 1979. *Growth and development of carcass components in cattle*. Fayetteville, AR: Arkansas Agricultural Experiment Station Special Report 12.

Berg, R.T., & Butterfield, R.M. 1976. *New concepts of cattle growth*. New York: Wiley.

Berg, R.T., & Walters, L.E. 1983. The meat animal: Changes and challenges. *Journal of Animal Science*. 57:133.

Boggs, D.L., Merkel, R.A., Doumit, M.E., & Bruns, K. 2006. *Livestock and carcass: An integrated approach to evaluation, grading and selection*. Dubuque, IA: Kendall/Hunt.

Bonsma, J.C. 1983. *Man must measure: Livestock production*. Cody, WY: Agi Books.

Brinks, J.S., Davis, M.E., Mangus, W.L., & Denham, A.H. 1979. *Genetic aspects of hoof growth in cattle*. Fort Collins, CO: CSU Gen. Series 982.

Brungardt, V.H. 1972. *Efficiency and profit differences of Angus, Charolais, and Hereford cattle varying in size and growth*. Research Reports R2397–R2401. Madison, WI: University of Wisconsin.

Currie, W.B. 1992. *Structure and function of domestic animals*. Boca Raton, FL: CRC Press.

Daley, D.A., Tatum, J.D., & Taylor, R.E. 1983a. Accuracy of subjective and objective preslaughter estimates of beef carcass fat thickness. *Journal of Animal Science*. 56:1.

Daley, D.A., Tatum, J.D., & Taylor, R.E. 1983b. Practical methodology of feeder cattle sorting. *Journal of Animal Science*. 57:390, Suppl. 1.

Hedrick, H.B. 1983. Methods of estimating live animal and carcass composition. *Journal of Animal Science*. 57:1316.

Kauffman, R.G., Grummer, R.H., Smith, R.E., Long, R.A., & Shook, G. 1973. Does live-animal and carcass shape influence gross composition? *Journal of Animal Science*. 37:1112.

Prior, R.L., & Lasater, D.B. 1979. Development of the bovine fetus. *Journal of Animal Science*. 48:1546.

Tatum, J.D., & Field, T.G. 1996. *Type changes in the Hereford breed—1950s to 1990s*. Research Report to the American Hereford Association. Fort Collins, CO: Colorado State University.

Trenkle, A.H., & Marple, D.N. 1983. Growth and development of meat animals. *Journal of Animal Science*. 57:273, Suppl. 2.

Wilson, G.R., Reef, J.E., Turner, T.B., & Wilson, G.W. 1981. *Relating visual feminine characteristics to reproductive performance in a herd of Angus cows*. Columbus, OH: Ohio Ag. Res. & Dev. Center Animal Science Series. 81:1:25.

Woodward, R. 1968. *The importance of soundness in the selection of breeding animals*. Charolais Banner (July).

18 Cattle Behavior, Facilities, and Equipment

"We are judged by the animals we keep" was a timeless philosophy that guided William H. Danforth as he built Purina Mills in service of the livestock industry at the turn of the 20th century. Understanding cattle behavior and applying that knowledge in the creation of handling facilities and protocols is central to a productive beef cattle enterprise. The example of the Good Shepherd provides an aspirational model that not only assures quality cattle care but also the fulfillment that accompanies attainment of being recognized as a master stockman.

Effectively designed and well-maintained facilities and equipment are necessary to handle, feed, transport, and market cattle for optimum performance. Poor facilities, rough handling, and wild dispositions of cattle can yield lower weight gains, poorer meat quality, increased injury to humans and cattle, and reduced overall performance. An understanding of cattle behavior is most useful in handling cattle, designing facilities, and choosing appropriate equipment.

BEHAVIOR

Cattle behavior is a complex process involving the interaction of inherited traits and response to the experiences encountered with other cattle, humans, predators, and other environmental influences. Behavioral adaptation enables animals to adjust to changing conditions, improve their chances of survival, and be productive. It is important for producers to observe and understand behavior, for they can influence cattle through improved animal care and management systems, handling, selection decisions, and facilities design.

Disposition

The disposition or temperament of cattle ranges from docile (Figure 18.1) to wild. Assessment of cattle temperament can provide insight as producers make selection decisions and as they assess the effectiveness of their facilities, handling protocols, and employee training programs. The evaluation is often made at calving time when cows are at the peak of their maternal tendencies, during periods when cattle are handled in corrals, pens, chutes, and other working facilities, and when cattle interact with handlers in pasture settings. Typically, cattle with a poor disposition are fearful of people and may exhibit aggressive behavior.

There is evidence that cattle disposition is the result of both inheritance and nongenetic influences such as the handling experience. Heritability estimates for disposition range in the medium-to-high categories, indicating that the trait has a relatively strong genetic basis and is responsive to selection. Some producers cull cattle with poor dispositions because of economic losses (such as broken fences and other facilities damage, risk of injury to both handlers and cattle, and to reduce the excitability of other animals).

Figure 18.1
Docility is a key trait to assure safety of both cattle and people.

The use of a simple temperament rating system based on observing cattle during handling for vaccinations or other procedures can be an effective culling tool. A four-point scale is used to evaluate cattle behavior during chute restraint: 1—stands quietly, 2—is restless, 3—struggles constantly, and 4—demonstrates frenzied activity. Cattle that are scored a 4 should probably be culled.

A nearly 13,000 head study in Iowa demonstrated that docility had a significant impact on feedlot and carcass performance. Cattle were rated as docile, restless, or aggressive and resulting data revealed that "docile" cattle had superior rates of gain of nearly ¼ of a pound compared to their "aggressive" contemporaries and superior USDA Quality Grade performance with the "docile" group having 14% more Choice than the "aggressive" group. Finally, the "docile" cattle had a $60.00 per head higher return than the cattle scored as "aggressive."

Communication

Communication exists when information is exchanged between individuals. This may occur with the transfer of information through any of the senses. A distress call, involving a distinct type of sound, occurs from either the cow or her calf when they become separated (Figure 18.2a). Although the cow and calf may recognize each other's vocal sounds, the most effective way for the cow to recognize her calf is by smell. Cows will more easily adopt a new calf through the transfer of the odor of one calf to another.

Bawling is a typical response of both cows and calves during weaning time. To reduce stress during weaning, calves and cows should be separated by a single fence. It is preferred to accomplish weaning in a pasture in which cattle have familiarity. Calves should not be weaned in a dirt lot to avoid respiratory distress that may result from increased levels of dust. Calves that remain in visual contact with their mothers will bawl less than calves that are moved out of sightline from their dams.

Cattle can learn to respond to vocal calls, whistles, or other sounds made by the producer at feeding time. Cattle soon learn that the stimulus of the sound is related to the feeding time. However, producers should apply judgment in this practice as cattle can become excited by feeding time and rush to the feeding area creating a potential safety issue.

The bull vocally communicates his aggressive challenge to other bulls and intruders to his area through a deep bellow (Figure 18.2b). This behavior is controlled by testosterone. The castrated male seldom exhibits similar behavior. The bull also issues vocal calls to breeding females, especially when he is separated from them but

Figure 18.2
(a) Cows vocalizing as a result of being separated from their calves. (b) A bull will bellow as a way of attracting females or challenging other males.

they are within his sight. A bull with aggressive intent will perform a broadside threat display toward another bull or a person. He will stand sideways and hunch his shoulders to make himself appear large.

Cattle have a range of vision in excess of 300 degrees but have a blind spot extending behind them to which they are particularly sensitive. Cattle have poor depth perception and thus may balk at shadows, changes in the intensity of lighting, or physical barriers at ground level.

Social Behavior

Cattle are herd animals and establish a social dominance structure within the herd. Mature cows often withdraw from the group to find a secluded spot just prior to calving. Cattle sometimes withdraw from the group if they are sick. Early and continuous association of calves within a contemporary group is associated with greater social tolerance, delayed onset of aggressive behavior, and relatively slow formation of social hierarchies. However, the introduction of new animals may disrupt the existing social structure.

Status and social rank typically exist in a herd of cows with certain individuals dominating other, more submissive animals. The presence or absence of horns is important in determining social rank, particularly when strange cows are mixed together. In addition, horned cows usually outrank polled or dehorned cows in situations involving close contact, such as at feeding time.

Large differences in age, size, strength, genetic background, and previous experience have powerful effects in determining social rank. Once rank is established in the cow herd, it tends to be consistent from 1 year to the next. There is evidence that genetic differences exist for social rank, both within and between breeds.

Research also indicates that animals fed together consume more feed than when fed individually. The competitive environment stimulates increased feed consumption. Dairy calves separated from their dams at birth appear to gain equally well whether fed milk in a group or kept separate. There is, however, evidence that they learn to eat grain earlier when group fed compared with being individually fed. Cattle individually fed in metabolism stalls consume only 50–60% of the amount of feed they would eat if group fed.

When fed in a group of older cows, 2-year-old heifers have difficulty getting their share of supplemental feed as they must compete with older cows who have established themselves at the top of the social order. Two-year-old heifers fed separately from mature cows will have approximately twice the weight gain as compared to those commingled with older brood cows. These behavioral differences no doubt explain some of the nutrition, weight gains, and postpartum interrelationships that are age-related when cows of all ages compete for the same supplemental feed.

Dominant cows raised in confinement usually consume more feed and wean heavier calves than submissive cows, whereas submissive cows wean lighter calves (by as much as 25%) and have lower pregnancy rates than aggressive cows. A highly dominant cow may prevent other animals from drinking or eating. In a study by Schake and Riggs (1972), the most dominant animal spent approximately 70% of the hour following feeding at the bunk while the least dominant cow spent only 5.4 minutes eating. Furthermore, the least dominant animal remained in the back half of the pen furthest away from feed for more than half of the hour post feed delivery while the more aggressive contemporaries remained at or within 14 feet of the feed bunk for almost 85% of the time. In breeding situations, a dominant bull may prevent other bulls from mating in a multiple-sire pasture.

Calves rely on their dams and older animals to establish a sense of security. To help alleviate stress at weaning time, place a dry, mature cow with newly weaned calves. Studies demonstrate that calves weaned in the presence of a mature female experience less morbidity than calves weaned without exposure to a mature cow.

Reproductive Behavior

Some profound behavior patterns are associated with the sex or sex condition of cattle. This verifies the importance of the hormonal-directed expression of behavior. Bulls exhibit more aggressive behavior, whereas steers are more docile after losing their source of testosterone following castration.

The bull frequently curls his nose and upper lip (known as an olfactory reflex or **flehmen**) after sniffing the urine or genitals of females. This behavior is associated with inhalation through the upper respiratory tract. The bull, through the use of the vomeronasal organ, identifies chemical substances secreted in the urine of the proestrus or estrus female known as pheromones. In a sexually active group of cows, the bull is attracted to a cow in heat most often by visual means (observing cow-to-cow mounting) rather than by olfactory clues.

When females are sexually receptive, they usually seek out a bull if mating has not previously occurred. Primarily through hormonal influence, the female in estrus will stand for the bull when he mounts. Cows in heat demonstrate male behavior by mounting and being mounted by other cows. A thrusting action is common in the cows as they terminate the mounting action.

The bull may guard a female that is approaching estrus. His success in guarding the female or actually mating with her is dependent on his rank of dominance in a multiple-sire herd. During the mating process, the bull may playfully nudge the female, lick her side and back prior to mounting, and rest his chin on her rump. After mating, most bulls temporarily lose interest in the cow. Some bulls leave the cow and do not make repeated matings, while other bulls may breed the same cow numerous times even in the presence of other cows in estrus. These different behavior patterns may explain the wide variations in the success of individual bulls with varying bull-to-cow ratios.

Recent studies show that many individual bulls have sufficient sex drive and mating ability to fertilize more females than are commonly allotted to them. Some producers use an excessive number of males in multiple-sire herds to offset the few bulls that are poor breeders and to manage the risk resulting from social dominance issues. If low fertility exists in the dominant bull or bulls, then calf crop percentage will be seriously affected even in multiple-sire herds.

Tests have been developed to measure libido and mating ability differences in young bulls. While behavioral differences are evident between bulls, there is no evidence that these differences are manifested in pregnancy rate variation. Some cows eat or chew the afterbirth (placental membranes), and most cows will begin licking the calf after it is born. As the cow licks her calf she obtains the olfactory stimuli

Figure 18.3
A calf is dependent on nursing to attain nourishment early in life. Appropriate maternal behavior is important to assuring calf health and survival.

required to help identify the calf in the future. Exceptions to this bonding behavior occur when the cow has experienced a particularly difficult birth or is sick. However, there are females that do not bond to their newborn and will abandon the calf.

During suckling, the calf usually nurses with its rear end toward the cow's head (Figure 18.3). This allows the cow to smell the calf and decide to accept or reject it. If a calf different than its own attempts to nurse a cow, she will typically bunt it away with her head or kick it away as it approaches her udder.

Cows better accept foster calves when the foster calves are either smeared with amniotic fluid previously collected from the second "water bag" or by putting the skin of the dead calf on the foster calf. Even when fostering is successful, foster calves suckle less frequently, for shorter periods, and have lower weaning weights than non-adopted calves.

Certain cows become very aggressive in protecting their calves shortly after calving. Serious injury can occur to producers who do not use caution with these cows. In rare cases, a cow will become aggressive to her newborn, even to the extent of causing calf death.

Bulls being raised with other bulls commonly mount one another, have a penile erection, and occasionally ejaculate. Individual bulls can be observed arching their back, thrusting their penis toward their front legs, and ejaculating. Bulls can be easily trained to mount objects that provide the stimulus for them to ejaculate. AI studs commonly use restrained steers for collection of semen. Bulls soon respond to the artificial vagina, when mounting steers, which provides them with a sensual reward. Mating behavior has an apparent genetic base, as there is evidence of more frequent mountings in hybrid or crossbred animals.

Bull calves that are raised in a social group with other cattle are less likely to attack people. Bull calves reared in isolation are more likely to become dangerous.

Research indicates that more cows calve during periods of darkness than during daylight hours. Altering when cows are fed, however, can change the calving pattern. Cows fed during late evening will have a higher percentage of their calves during daylight hours.

Suckling Behavior

Cattle have certain instinctive behavior patterns, such as the newborn calf's searching movement of head and neck, sucking of protruding objects, and swallowing of fluids. Obviously, the function of these behaviors is to locate the mammary gland and suckle the teats for milk (Figure 18.3).

The number and duration of suckling incidences have been observed in range beef calves (Odde, 1983). In this study, the average time spent suckling during each of several 24-hour periods was approximately 45 minutes, with an average of five suckles per calf. Peaks in suckling activity occurred from 5:00 A.M. to 7:00 A.M., 10:00 A.M. to 1:00 P.M., and 5:00 P.M. to 9:00 P.M. The most suckles in a single hour occurred between 5:00 A.M. and 6:00 A.M. and the fewest between 10:00 P.M. and 11:00 P.M. Cows that gave more milk nursed less frequently, and heavier calves sucked less frequently. Age, breed (Polled Hereford and Simmental), and sex of calf did not influence the number or duration of suckles.

The calf is usually born away from the herd in a secluded spot, if one is available. The calf remains secluded while the cow grazes or feeds elsewhere, usually with the herd. The cow will return to the calf several times each day to allow the calf to nurse. In a few days, the cow will lead the calf to the herd where both will remain.

Cross nursing is more frequent for cows forced to accept a second calf than for cows raising only a single calf. In the latter situation, some cross nursing of calves may be observed and occasionally a cow will nurse another cow. These behaviors can affect the validity of weaning weight comparisons.

Grazing Behavior

Cattle develop palatability preferences for certain plants and may have difficulty changing from one type of plant to another. Young calves learn to graze the plant types that their dams consume. Cattle kept in barns or sheds have been observed to be awake approximately 20 hours per day and asleep the remaining 4 hours, with 40% of the day spent standing. Cattle tend to sleep less when on pasture than when in familiar buildings.

The behavior of cows grazing native range during winter conditions may be affected by age of cow and changes in the weather. At the Range Research Station at Miles City, Montana, cows grazed less as temperatures dropped below 20°F, and 3-year-olds grazed approximately 2 hours less than 6-year-olds. The colder the temperature, the longer cows waited before starting to graze in the morning. At 30°F, cows started grazing between 6:30 and 7:00 in the morning; at –30°F, they waited until about 10 A.M. to begin grazing.

Refer to Chapter 15 for a more detailed discussion of grazing management and resources.

Feedlot Behavior

The "buller-steer" problem with steers in a feedlot is a unique behavior problem with serious economic significance to feedlot operators. Certain steers are singled out, probably by an olfactory identification through the vomeronasal organ, and ridden continually by other steers. While the exact cause of the buller-steer syndrome is not known, there is evidence that the use of hormonal implants (especially estrogenic ones) have increased the incidence. Other factors affecting the syndrome appear to be keeping large numbers in a group and introducing new cattle into a pen. Providing adequate watering and bunk space may help prevent bullers. The behavior pattern reduces feedlot gains and usually causes the buller-steer to experience serious health problems, including death.

Feedlot steers have different patterns of water consumption throughout the day. Some of these patterns are significantly affected by season of the year. Studies have demonstrated that the percentage of steers drinking peaks at 2 P.M. and again at 8 P.M. during the summer months while winter time water consumption peaks at noon. Summer season water consumption is characterized by progressive increases in percentage of cattle consuming water beginning at 9 A.M. until the 2 P.M. peak. During winter, the percentage of cattle drinking is characterized by subsequent small peaks every 2–3 hours beginning at 8 A.M. and concluding at 5 P.M. When the weather becomes extremely hot, cattle will have dramatically increased water consumption.

The feeding behavior of steers in confined areas with concrete floors appears to be slightly different from steers in drylot, dirt pens. A higher percentage of steers on concrete eat during the night than during the middle of the day (10 A.M. to 3 P.M.).

Bulls can become very aggressive toward one another in riding and fighting. A group of bulls can become extremely aggressive toward another bull that may have been separated from the group for a few days and then reintroduced into the group. These behavioral tendencies are important to managing the bull battery as well as bulls being grown and developed in a feedlot setting.

Behavior During Handling and Restraint

Most cattle are handled and restrained several times during their lifetime for procedures such as vaccinations, health mitigation procedures, pregnancy testing, artificial insemination, or transport. Ease of handling depends largely on cattle's temperament, size, and previous experience as well as the design of the handling facilities. Producers knowledgeable about animal behavior can prevent injury to both people and cattle, minimize stress, and reduce damage to facilities and equipment. There are several principles of cattle behavior that should be acknowledged in both handling and facility design.

- For example, an effective stockman knows how to approach cattle so that they will respond in the desired manner. Cattle are unnerved by complexity; thus, minimization of distractions and sensory input is critical to success.
- Cattle naturally prefer to associate with other cattle and move more easily toward other cattle.
- Cattle tend to circle around pressure and prefer to go back to where they came from.
- Cattle are more calm when they can see the handler.

Because cattle are prey animals, they exhibit behavioral mechanisms to avoid predation. Understanding these behaviors provides the foundation for effective, low-stress cattle handling. Cattle have 300-plus degree panoramic vision with a small blind spot that extends directly behind them. While their vision field is wide ranging, cattle have poor depth perception and limited vertical vision.

Cattle have a "flight zone" (Figure 18.4) equivalent to the comfort zone of humans. When a handler is outside the flight zone, cattle exhibit inquisitive behavior and face the handler. When a person moves inside the flight zone, the animal usually

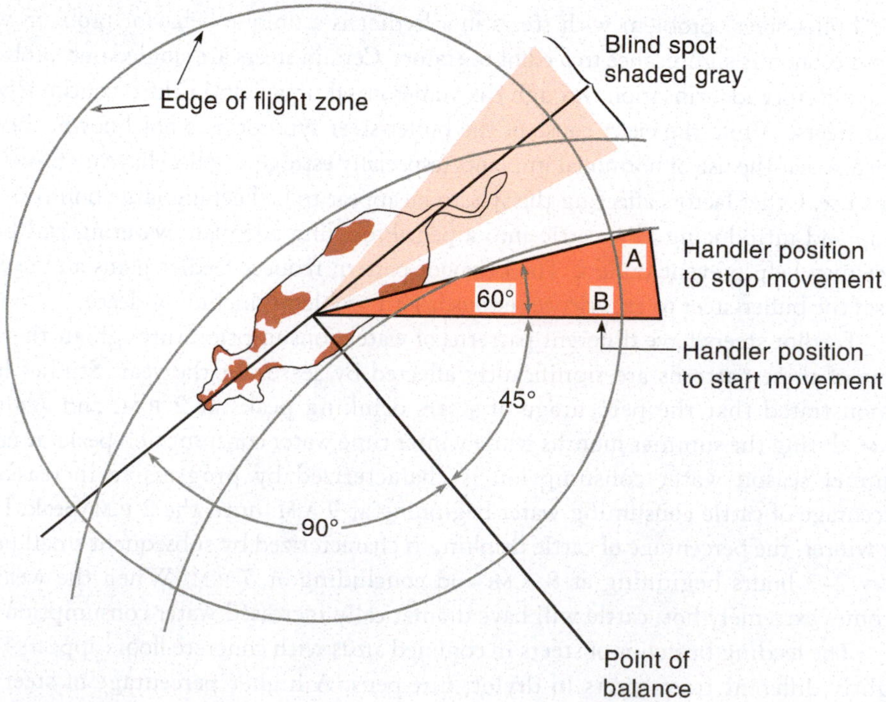

Figure 18.4
Flight zone of cattle with positions of handler to influence movement of the cattle.
Source: Temple Grandin.

Edge of flight zone
Blind spot shaded gray
A
B
60°
45°
90°
Handler position to stop movement
Handler position to start movement
Point of balance

moves away. Flight zone size varies from animal to animal and is influenced by the speed, angle, and noise level of a handler's approach as well as factors such as age and previous experience with humans. Furthermore, stockmen need to understand the principle of point of balance that is an imaginary line extending perpendicularly across the shoulders of the bovine. When the handler is positioned behind the point of balance and calmly enters the flight zone, cattle will move forward with their rump to the handler. When approaching cattle from front of the shoulder, they will back away from the handler. By moving outside the flight zone, cattle will slow and eventually stop. A central lesson is that cattle want to maintain a sight line to the handler.

Cattle can be handled most effectively by the producer who understands their behavior patterns and who provides handling facilities that complement those behaviors in positive ways. Calm cattle are easier to move and sort than excited cattle. Once agitated, it takes cattle about one-half hour to calm down. Following are some important considerations about the behavior of cattle.

1. Because cattle have poor depth perception, they are sensitive to shadows and unusual movements observed at the end of or outside of the chute (Figure 18.5). Thus, cattle usually move with greater ease through chutes that are curved and that have solid sides that minimize shadows and distractions, but still allow cattle to see the animal in front of them as they move through the chute as cattle naturally tend to follow one another (Figure 18.6).
2. Cattle tend to move toward light (except blinding direct sunlight). Loading cattle at night is best accomplished by providing a nonglaring light near the truck or trailer gate. Cattle move more easily into a dark area when in a single-file chute rather than as a group. Loading chutes and squeeze chutes should face north and south because cattle do not move easily into direct sunlight. White translucent skylights will facilitate cattle movement inside buildings. If the interior of a building is too dark, cattle will be hesitant to enter it.
3. Shadows across alleyways, chutes, scales, and load outs should be prevented because cattle are fearful of variations in light and will balk when encountering them rather than maintaining an even flow of cattle movement.
4. Poor dispositions are developed by cows and calves handled in an abusive manner. Cattle remember painful and adverse experiences. Thus, driving and sorting devices should be carefully selected and utilized. A small flag on the end of a stick

Figure 18.5
Shadows that fall across a chute can disrupt the handling of animals. The lead animal often balks and refuses to cross the shadow. Other distractions that will make cattle balk and refuse to move are: moving chain ends, reflections in puddles, seeing people in front of them, fan blades turning in the wind, changes in flooring type, or an object on a fence (such as a coat) that is flapping.

Figure 18.6
Animals move more easily through curved chutes with solid sides.

is useful for moving and sorting cattle. The use of whips should be avoided and electric prod use employed only when absolutely required.

5. Cattle have exceptional hearing with the ability to perceive both lower volume and higher frequency than do humans. However, they have a difficult time determining the source of sound. These characteristics cause cattle to respond negatively to whistling, shouting, and loud noises from equipment. Thus, noisy equipment such as compressors, hydraulic pumps, and motors should be kept as far as possible from cattle or have their sound mitigated. Metal chutes and alleys should be constructed so that loud clanging and banging noises are eliminated.
6. Cattle are creatures of habit; an established, calm daily routine will result in ease of handling.
7. Handle animals in groups. A single animal often resists going into a chute or pen by itself. It may also become excited and injure itself or the handler or damage the facilities.
8. Curved chutes (Figure 18.6) are preferred by some experts over straight chutes because cattle cannot see what is at the end of a curved chute until they are almost to the end. A curved chute also utilizes the natural tendency of cattle to circle around the handler and to go back where they came from. The catwalk should be interior to the curved chute so that the handler is positioned at the best angle for working with the cattle. However, others favor a straight design such as is the case with the "Bud Box."
9. The cattle handler's movements should be slow and deliberate; any sudden movements will frighten cattle and make them difficult to handle.
10. By understanding the flight zone of cattle (Figure 18.4), the handler can effectively work cattle in the corral or pasture (Figures 18.7, 18.8, 18.9, and 18.10). Understanding animal behavior provides a framework to better design facilities such as loadout chutes for cattle transport (Figure 18.11). Cattle can be moved more easily if the handler works on the edge of the flight zone. The handler penetrates the flight zone to start cattle movement and retreats outside the flight zone to stop cattle movement. When the handler is positioned behind the point of balance at the shoulder, the animal will move forward. It will move backward when the handler is in front of the point-of-balance. Figure 18.12 shows how to use the point-of-balance to move an animal into a squeeze chute. The animal moves forward when the handler walks quickly past the point-of-balance at the shoulder.

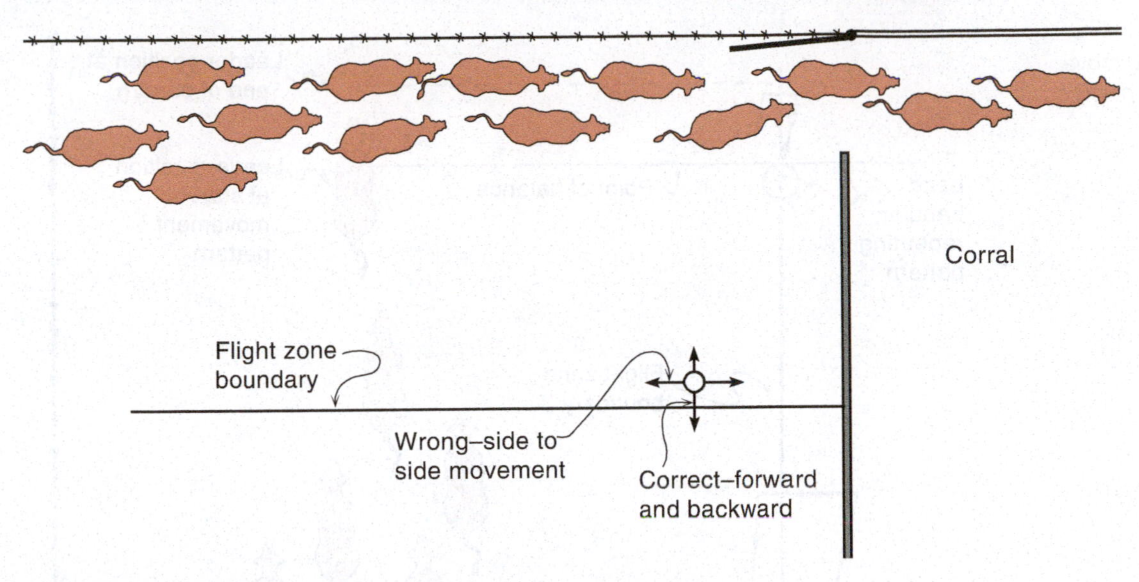

Figure 18.7
Leader handler position for filling corral.
Source: Temple Grandin.

Figure 18.8
Handler positions for emptying a pen and sorting at a gate. The handler should control the animal's movement out of the pen. This is especially important when moving from one pasture to another so as to prevent the cows from leaving their young calves behind.
Source: Temple Grandin.

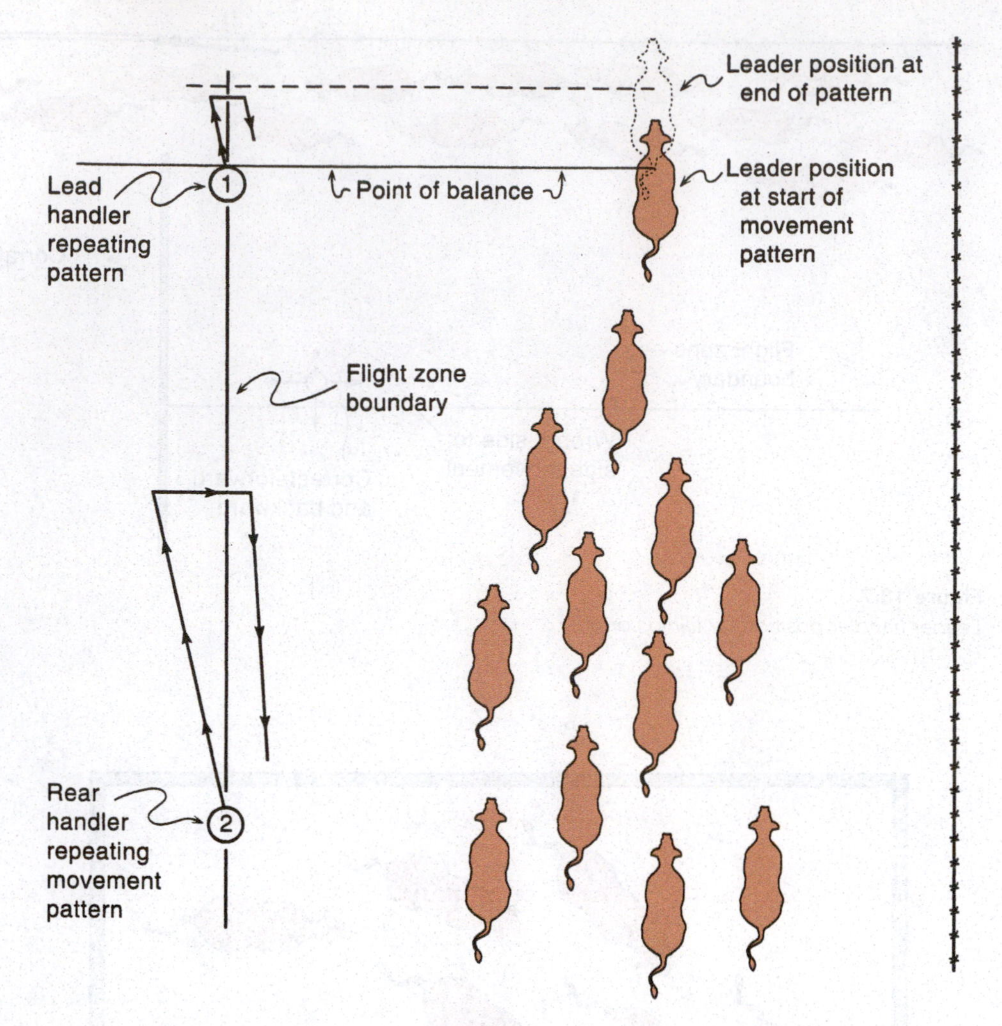

Figure 18.9
Handler positions to move groups of cattle on pasture.
Source: Temple Grandin.

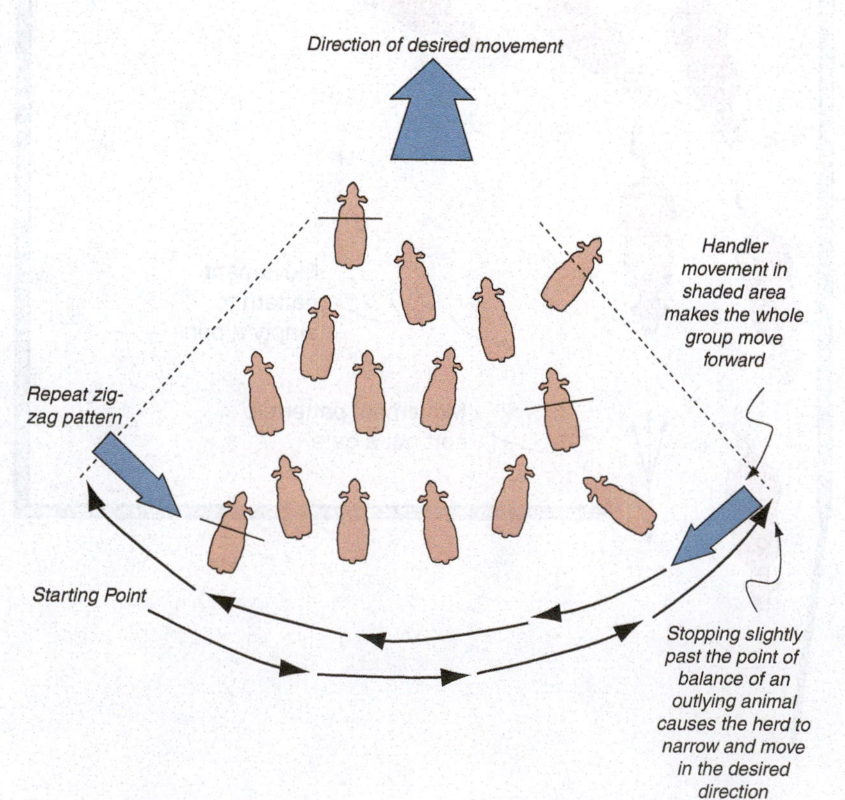

Figure 18.10
Handler zigzag movement pattern for use in open pastures—starting movement in the desired direction. The handler must zigzag back and forth to keep the herd straight. Imagine that the leaders are the pivot point of a windshield wiper and the handler is out on off the blade sweeping back and forth. As the herd narrows and gets good forward movement, the width of the handler's zigzag narrows.
Source: Temple Grandin.

Figure 18.11
Transportation of cattle requires good footing on loading ramps, correct handling technique, trained personnel, and limited use of prodding devices. Furthermore, it is critical that appropriate loading densities are utilized to assure cattle well-being.

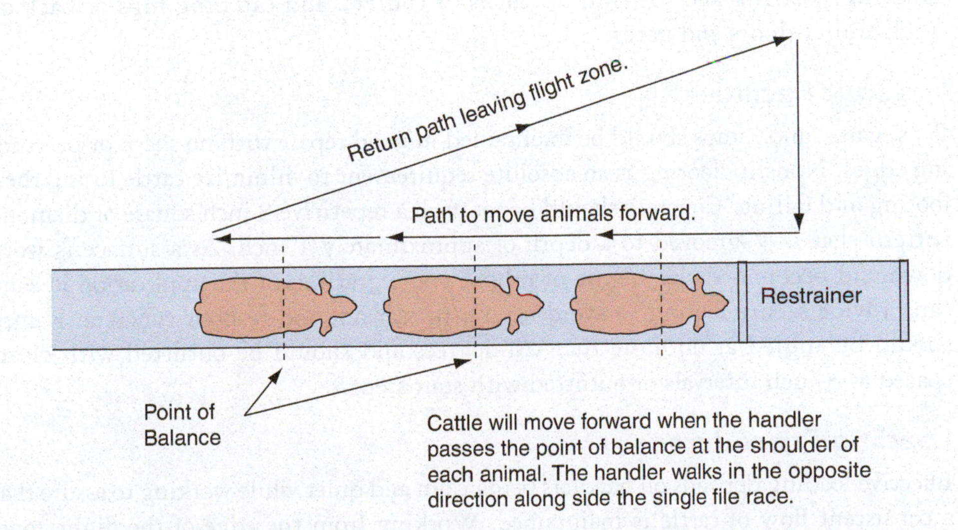

Figure 18.12
Handler movement pattern to keep cattle moving into a squeeze chute or restrainer. Cattle will move forward when the handler passes the point of balance at the shoulder of each animal. The handler walks in the opposite direction alongside the single-file race.
Source: Temple Grandin.

11. If cattle balk and refuse to move through a chute, distractions such as a moving piece of chain or a coat hung on a fence should be removed to facilitate movement.
12. Blood odor appears to be offensive to cattle because it is novel and strange. Therefore, the reduction or elimination of such odors may well encourage cattle to move through handling facilities with greater ease.

Round pens (e.g., those without square corners) enhance cattle movement and prevent injury to more excitable cattle. Head catches are used to immobilize cattle of large sizes. If the head needs to be further restrained once the animal is in the chute, a halter is preferred to nose tongs.

GATHERING CATTLE ON PASTURE AND TRAINING CATTLE

Cattle can be trained to gather in response to an auditory signal such as when a horn on a vehicle is blown. Cattle can also be gathered by using the movement pattern shown in Figure 18.10. This movement pattern will trigger the animals' instinct to bunch together. The handler walks or rides on the edge of the herd's collective flight zone. The handler must *not* circle around the cattle or chase stragglers. The handler must calmly move back and forth until the instinct to bunch is triggered. The principle is to trigger the bunching instinct before any attempt is made to move the cattle forward. Increasing pressure on the collective flight zone moves cattle forward. When the animals start moving, the handler must practice the principle of pressure and release. After the herd starts moving, the handler should back off, and when the herd slows down, the handler can apply more pressure to the flight zone. Practicing pressure and release will help prevent running and breaking fences.

Cattle will be easier to handle when they are moved from a ranch to a feedlot if they have been habituated and trained to different methods of handling. They should become accustomed to vehicles, people on foot, and riders on horseback. Novel experiences may cause stress to cattle and thus new environments or routines should be introduced with an intent to assure positive outcomes.

CATTLE TRANSPORT

Prevention of bruising and injury during transport depends on effective cattle handling, design of facilities and trailers, and training of handlers and truckers (Figure 18.11). The three primary sources of bruising during transport is rough handling, jostling and shifting in trucks or trailers, and catching hips or back on truck/trailer doors and gates.

Loadout Facilities

Pens, gates, and ramps should be maintained in good repair with no sharp or protruding edges. Nonslip flooring is an absolute requirement to minimize cattle losing their footing and falling. Concrete should be set with a repetitive 8 inch square or diamond pattern that is V-grooved to a depth of approximately 1 inch. As a surface is worn down and becomes slick, the use of rubber mats, grating, or the application of sand can provide secure footing to avoid having to tear out and re-pour concrete. Ramps should be angled at no more than 20 degrees and should be outfitted with cleats spaced at 8-inch intervals or outfitted with stair steps.

Loading Procedures

Effective loading depends on handlers being calm and quiet while working to assure that a consistent flow of cattle is maintained. Working from the edge of the flight zone, minimizing the use of physical prods, and avoiding overcrowding of alleys and pens leading up to the ramp are essential. Effective use of the point of balance is also important. For example, handlers moving past the point of balance in the opposite direction of cattle headed up the chute will usually stimulate the cattle to move forward. To avoid injury and other losses resulting from incorrect loading, overloading of trailers, and poor transport management, it is important that producers have a clear line of communication with the cattle transport company and their employees in terms of desired transport handling. The mistakes made in transport can diminish the gains made by producers from correct handling at the ranch or feedyard. Maintaining appropriate loading density during transport is essential. Loading densities for semi-trailers are listed in Table 18.1.

Table 18.1
RECOMMENDED LOADING DENSITIES FOR SEMITRAILERS HAULING POLLED CATTLE

Fed Cattle/Breeding Females (wt)	Per Head Space Requirement (ft^2)
800	10.40
1,000	12.00
1,200	14.50
1,400	18.00

Source: Based on Grandin (2004a).

In addition to loading densities, producers should measure the percentage of cattle that fall or slip during loading/unloading, the speed of cattle entering/leaving the trailer, the percentage of cattle that must be prodded, the time from arrival at the plant or feedlot to the conclusion of unloading, handling of nonambulatory animals, and the general cleanliness of the transport vehicles. These efforts not only improve public perception but also help to avoid costly financial losses.

FACILITIES AND EQUIPMENT

General Facilities Guidelines

Facilities that are properly constructed and maintained in good working order will enhance the producer's time management and safety. Cattle disposition will be calm and docile if facilities are constructed to control or influence their behavior in a positive manner. Tables 18.2 and 18.3 give the space requirements and general facilities guidelines needed by beef cattle. Footing considerations are critical to assure that cattle can move through a working facility without slipping. Good footing helps to calm cattle, protects them from injury, and provides a safer working environment for employees (Figure 18.13).

Methods of Cattle Restraint

Restraining and controlling cattle are necessary for many production and management practices such as health care, identification, marketing, weighing, breeding, pregnancy checking, and implanting. Cattle restraining facilities should be constructed to provide safety and minimize stress for both handlers and cattle. An understanding of cattle behavior is needed to construct or purchase working facilities and equipment that conform to cattle's natural instincts.

Working Chutes

Working chutes fitted with a head catch can be used to restrain cattle to facilitate management practices such as vaccinations, pregnancy checking, and artificial insemination. Many commercial and homemade working chute designs are available. For procedures such as applying pour-on or sprayed parasiticides, cattle may be held as a group head-to-tail in the chute while the procedure is carried out. The best working chutes have the following characteristics:

1. V-shaped sides or an adjustable side (18–30 in.) so that cattle of variable sizes can be handled.
2. The V-shaped sides should measure 14–17 in. wide at the bottom and remain so for the first 2 feet from the bottom, at which point they should flare out to approximately 24–30 in. at hip and head height.

Table 18.2
CATTLE AND FACILITY SPACE REQUIREMENTS

	Corral and Pen Dimensions		
Facility Description	**To 600 lb**	**600–1,200 lb**	**1,200 lb or Cow-Calf**
Holding area (sq ft per head)	14	17	20
Crowding pen (sq ft per head)	6	10	12
Working chute with vertical sides			
Width (in.)	18	22	26
Minimum length (ft)	20	20	20
Working chute with sloping sides			
Width at bottom inside clear (in.)	15	15	16
Width at top inside clear (in.)	20	24	26
Minimum length (ft)	18	18	18
Working chute fence			
Height solid wall (in.)	45	50	50
Depth of posts in ground (in.)	36	36	36
Overall height (7 ft minimum clearance below cross ties to walk under)			
Top rail, medium-sized docile cattle (in.)	55	60	60
Top rail, large-sized or wild cattle (in.)	60	66	72
Corral fence			
Recommended height (in.)	60	60	60
Depth of posts in ground (in.)	30	30	30
Loading chute			
Width (in.)	26	26	26–30
Length (in.)	12	12	12
Rise (in. per ft)	3	3	3
Ramp height for gooseneck trailer (in.)	15	—	—
Pickup truck (in.)	28	—	—
Van type truck (in.)	40	—	—
Tractor trailer (in.)	48	—	—
Double deck (in.)	100	—	—

Source: Based on multiple sources.

3. Straight-sided permanent chutes should measure 26–28 inches in width and the sides should be solid so that cattle cannot see through them and become distracted.
4. The crowding pen and working chute should be curved with solid sides.
5. The working chute exit and squeeze chute entry gate are usually made of bars to give cattle the illusion of being able to escape and thereby encouraging them to enter.

When constructing a working chute, an access gate should be built adjacent to the squeeze chute end so that the working chute can be entered to perform palpations and castrations.

The step-by-step procedure for moving cattle into and through the working chute follows:

1. Move cattle to the working alley and crowding pen that funnels into the working chute. Avoid overfilling the crowding pen; cattle need room to turn. Only fill the crowding pen half full.

Table 18.3
Space Requirements and General Facility Guidelines for Feedlot Cattle

	Feedlot (sq ft per head)
20 sq ft in barn and 30 sq ft in lot	Lot surfaced; cattle have free access to shelter
50 sq ft	Lot surfaced; no shelter
150–800 sq ft	Lot unsurfaced except around waterers, along bunks, and open-front building and a connecting strip between them
20–25 sq ft	Sunshade
	Buildings with Feedlots (sq ft per head)
20–25 sq ft	600 lb to market
15–20 sq ft	Calves to 600 lb
1/2 ton per head	Bedding
	Cold Confinement Buildings (sq ft per head)
30 sq ft	Solid floor, bedded
17–18 sq ft	Solid floor, flushing flume
17–18 sq ft	Totally or partly slotted
100 sq ft	Calving pen
1 pen for 12 cows	Calving space
	Feeders (in. per head along feeder)
All animals eat at once:	
18–22 in.	Calves to 600 lb
22–26 in.	600 lb to market
26–30 in.	Mature cows
14–18 in.	Calves
Feed always available:	
4–6 in.	Hay or silage
3–4 in.	Grain or supplement
6 in.	Grain or silage
1 space per 5 calves	Creep or supplement
	Water Requirements
Gal. per 1,000 lb of cattle	29 gal. per day @ 50°F
	18 gal. per day @ 90°F
Automatic waterers for cattle:	1/2 gpm minimum
20–40 head per waterer	2 gpm preferred
	Waterers
40 head per available water space in drylot	
	Isolation and Sick Pens
40–50 sq ft per head	
Pens for 2–5% of herd	
	Mounds
25 sq ft per head	Minimum
50 sq ft per head	If windbreak on top of mound, 25 sq ft per head each side
	Slopes
Floors, pavements	
1/2–1/4 in. per ft	Bunk aprons with step
1 in. per ft	Bunk aprons without step
1/4–1 in. per ft	Solid floors toward slats, flumes
1/2% or more	Longitudinal bottoms of gutters, flumes
1%	Gravity pipe to lagoon
Earth	
4:1 to 5:1	Mound side slopes
5% maximum	Mound longitudinal
4–6%	Lots
	Daily Manure Production per 100-lb Live weight
60 lb, 1 cu ft	Feces and urine
6.91 lb, 0.12 cu ft	Solids

Source: Based on multiple sources.

A

B

C

Figure 18.13

Even if all other elements of a facility are correctly designed, poor footing can make them unusable. When using a bud box or sorting alley, the use of sand can help provide an environment that slows animals and helps them avoid slipping or falling as they turn (a). Examples of chute floors that minimize slippage and enhance cattle movement are a steel grate system (b) or a concrete floor with a diamond groove pattern (c) that provides stable footing.

2. Start cattle into the working chute and let them follow the leader. It will be more efficient to get one animal started down the chute instead of trying at random to drive all of them at once. If the cattle do not cooperate, look to remove any distractions and ensure that people are in the right place. Shaking plastic streamers next to their heads can easily turn cattle that get turned around in the crowd pen. A plastic garbage bag taped to a broom handle works well. The animal will turn away from the plastic strips because their vision is blocked on one side.

Furthermore, when several animals are in a relatively small area, the producer is at risk of being squeezed, stepped on, or kicked. Avoid using the electric prod or abusive hitting; this will not only make cattle nervous and less cooperative but will also increase the producer's risk of injury. If an animal rears up in the single-file chute, the handler should back up and move away. Cattle typically rear up in an attempt to get away from the handler who is deep in its flight zone.

3. Walking along the chute toward the crowd pen will cause each animal to move forward as the handler moves past the point of balance at each animal's shoulder.
4. Install a one-way entry in the working chute. Cattle can walk through the one-way passage, but the device prevents them from backing up.

Squeeze Chute and Headgate

The combination of squeeze chute and headgate can be used to advantage on any type of cattle farm or ranch to facilitate several management techniques, including dehorning, castrating, branding, implanting, ear tagging, stomach tubing, artificial insemination, and blood testing. A squeeze chute and headgate should be positioned at the terminal end of a working chute.

There are many designs available for the squeeze chute and headgate. A workable combination should consist of the following: a squeeze mechanism, headgate, tailgate, removable solid side panels measuring approximately 24 in. from the ground, and removable side bars for easy access to the animal's side (Figure 18.14).

Three basic headgate designs are used in beef operations: straight-bar headgate, positive-type headgate, and curved-bar headgate. The straight-bar headgate generally is designed to catch an animal automatically as it walks through the chute. The greatest advantage of the straight-bar headgate is its protection against choking the animal. It is also the recommended type if a body squeeze chute is not available. Its main disadvantage is that the animal can move its head up and down easily, which can create problems with techniques that require head immobilization. The positive-type headgate, in contrast, operates somewhat like a guillotine. Its main advantage is almost complete head control, both sideways and up and down. The headgate almost completely immobilizes the head without the necessity of a head and nose bar. Finally, the curved-bar headgate has vertical neck bars that are curved, providing more head control than the straight-bar headgate but with a somewhat greater risk of choking. Curved-bar stanchion headgates are used on most hydraulic squeeze chutes in large feedlots. This headgate is a good compromise for most cow-calf producers and should be used in conjunction with body

Figure 18.14
A squeeze chute and headgate. This piece of equipment is manually operated.

restraint to prevent the animal from lying down and choking. Effective use of a headgate is very much dependent on the skill of the operator coupled with application of stress-free handling techniques.

It is absolutely necessary that the headgate be adjusted to the correct height of the animal being worked in order to prevent choking. In addition, it is mandatory that the animal's head be released from the headgate before the body squeeze is released. This helps prevent the animal from lunging forward and possibly injuring itself. Regardless of the headgate design used, it is important that the headgate be adjusted to the size of the animals being worked to prevent choking or escape. If an animal shows signs of choking, it should be released immediately. Calm handling of cattle prior to their entry into the chute will help prevent most problems at the head catch from occurring. Calves can be worked through a conventional chute, but additional labor is typically required to prevent the animals from turning around.

Vertical Tilt Tables

In many cow-calf operations, vertical tilt calf tables (Figure 18.15) are used for restraining the calf during the branding, vaccination, castration, and dehorning process. The calves are relatively small because they are only a few weeks to a few months old.

Some cow-calf operations, however, continue to restrain calves by roping their hind legs from a horse while another person "tails" or "flanks" the calf to its side and sits on it as others dehorn, castrate, vaccinate, and brand the calf.

Another type of vertical tilt equipment is used for hoof trimming. Hooves can also be trimmed in squeeze chutes and in upright nontilting chutes constructed for hoof trimming.

Cow-Calf Handling Facilities

A corral design for a cow-calf facility is shown in Figure 18.16. As noted earlier in the chapter, cattle move easier through facilities with a crowding area and chutes that are curved with solid sides.

Another design concept that effectively facilitates flow of cattle into an alley leading to chute or into a loading alley is called the "Bud Box" (Figure 18.17). The Bud Box is designed to help assure cattle flow and while construction and dimensions are important, they pale in comparison to the importance of the handler's skill. The goal is to keep a consistent flow of cattle into the alley leading to the chute so that processing can be done

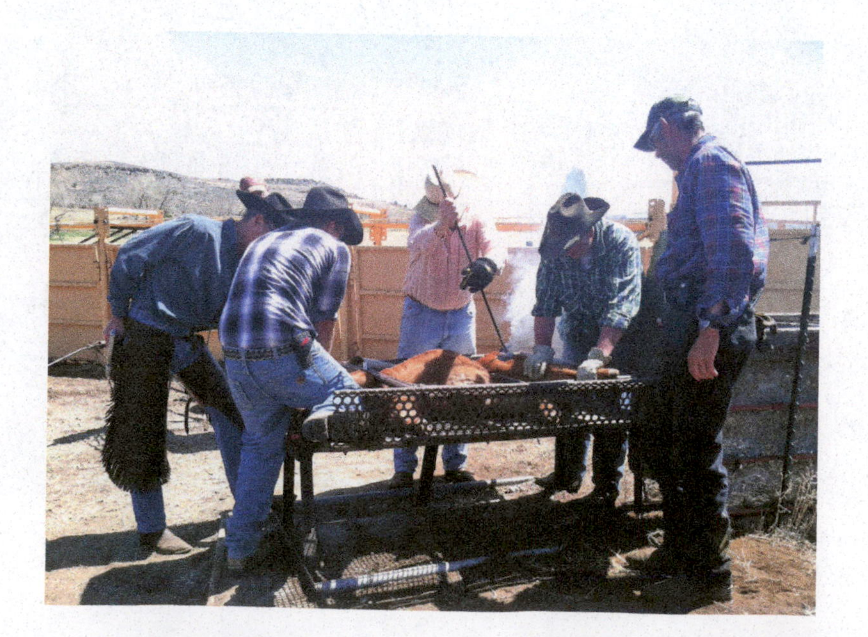

Figure 18.15
Branding a calf using a calf table.

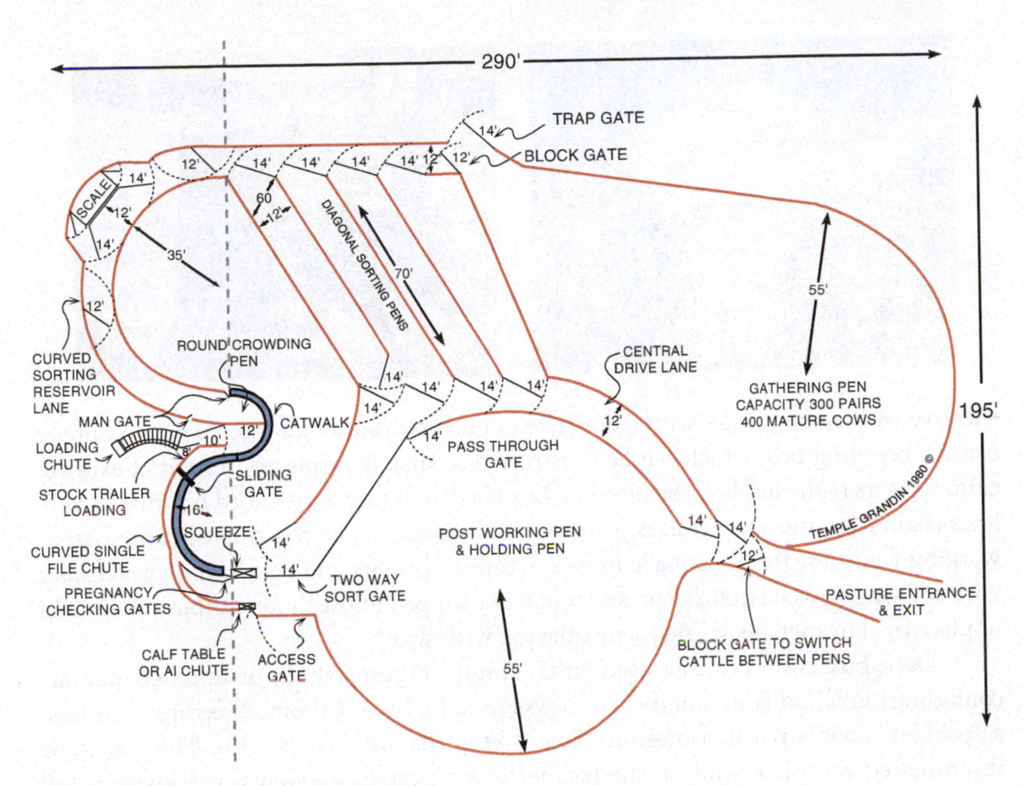

Figure 18.16
Corral system for a cow-calf enterprise.
Source: Temple Grandin.

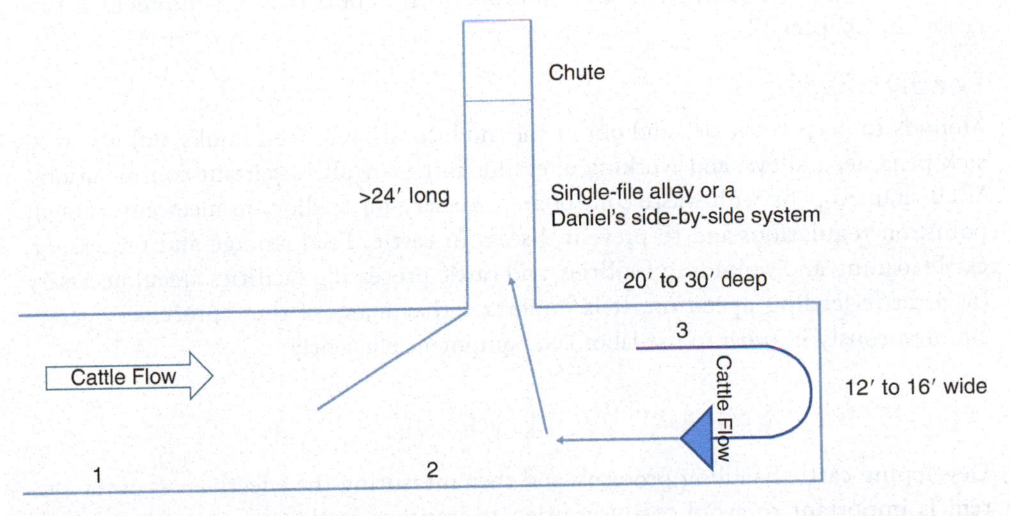

Figure 18.17
Bud box design to enhance loading a chute. The 'box' should be wider and deeper if cattle will be handled from horseback and narrower and less deep if handler is afoot. 1 – handler position as cattle movement toward the Bud Box is initiated. 2- handler position as solid sided gate is closed behind cattle. 3 – handler position as cattle move back they way they came and into the alley leading to the chute."

efficiently. The box should be constructed to hold the number of cattle that are required to fill the alley leading to the chute or in the case of a load out—the number of cattle required to fill a trailer compartment. In this design, the side panels should not be solid. However, a solid-sided gate will enhance the ease of movement.

AI facilities should permit cows to be handled quietly and carefully. Facilities where cows have previously felt pain should not be used for artificial insemination. Pregnancy rates are likely to be higher if handling cattle calmly minimizes stress. AI facilities can be adapted to existing facilities (fences, pens, corrals, and gates) to make them more cost

Figure 18.18
Use of a breeding box to facilitate ease of handling and assuring that cattle are calm during artificial insemination.

effective. A breeding chute with a headgate can be used, although some producers prefer using a breeding box, which can be constructed separately or placed in front of existing chutes. As its name implies, the breeding box is a dark, quiet, solid-sided box that usually has a chain restrainer at its rear. A portable breeding box can be constructed to fit existing working facilities. It can be made from 2 × 6-feet uprights and half-inch plywood sides with a $^{3}/_{4}$-in. plywood front, door, and top. If not slipped into a chute, the box can be held in place by four metal posts tied across the top with wire.

Dark box chutes can be used either singly (Figure 18.18) or in a herringbone configuration when large numbers of cows are to be bred. In some breeding facilities, a "pacifier" cow is put in a breeding box to keep the other cows calm. The cow to be inseminated rests her head on the pacifier cow's rump. Afterward, the inseminated cow is released through a side gate and the pacifier cow remains in the box.

Cow-calf producers need useful equipment and facilities at calving time, particularly to assist first-calf heifers with difficult births. Some of this equipment is discussed in Chapter 11.

Feedlot Facilities

Mounds to keep cattle dry and out of the mud, location of feed bunks and waterers, sick pens, feed alleys, and working alleys for cattle are all important considerations. Well-drained pens with waste control are necessary for feedlots to meet government pollution regulations and to prevent disease in cattle. Feed storage and processing, cattle loading and unloading facilities, and cattle processing facilities are all necessary for a cattle-feeding operation. It is important that many of these processes operate simultaneously in order to use labor and equipment efficiently.

AUDITING ANIMAL WELL-BEING

Developing cattle handling protocols and then measuring the effectiveness of the system is important to avoid costly injuries, to improve cattle performance and well-being, and to assure that facilities and equipment function correctly. As supply chains continue to develop amid increasing pressure from consumers for accountability in regard to animal welfare issues, the incorporation of animal handling guidelines into production systems becomes more important. Large retail and foodservice chains place increasing emphasis on assuring that suppliers meet specified animal handling standards. As such, the development of protocols, training programs, and performance measurement systems will be required at the cow-calf, stocker, feeder, packer, and animal transport levels to assure access to some of these major market outlets.

Critical control points have been identified at the ranch and feedyard levels that should be used to monitor effectiveness of cattle handling. At the ranch, these control points include the following:

- Acceptable body condition
- Humane euthanasia

- Avoidance of advanced stages of cancer eye
- Correct management of nonambulatory animals
- Development of best management practices for handling, dehorning, branding, spaying, and castrating
- Preconditioning prior to sale
- Provision of protection against extreme weather conditions

Feedyard critical control points would include the following components:

- Provision of heat relief and mud control
- Maintenance of clean water troughs
- Development of best management practices for cattle handling and transport
- Humane euthanasia
- Avoidance of knife castration, dehorning, late stage abortion, and branding (except where law dictates otherwise)
- Maintenance of nonslip flooring in alleyways, processing facilities, etc.
- Minimization of founder and hoof problems
- Purchase of preconditioned calves

Once good handling protocols have been developed, a number of measurements can be taken to determine effectiveness of facilities and handling techniques (Table 18.4).

Third party verification systems provide participants of the beef supply chain a process to assure that specific requirements associated with domestic or export brands/labels are met. The use of third party auditing provides transparency and confidence that quality based management systems have been implemented and that marketing claims are valid. Examples of market demand for third party verification systems include programs focused on source and age verification in support of traceability, exclusion of management protocols such as the use of exogenous hormones or antibiotics, breed based protocols, and certification of natural or organic label claims.

The process of developing, executing, auditing, and improving production processes is as follows:

1. Develop the standards
2. Create documentation and measurement process

Table 18.4
Measurement of Critical Control Points for Cattle Handling

Fall/slip (%)	
Excellent	None
Acceptable	Less than 3% slip (interruption of normal action)
Not acceptable	1% fall (body part other than foot hits floor)
Serious problem	2% fall, 15% or more slip
Speed Score (leaving the chute)	
Excellent	>90% move at a walk or trot
Acceptable	75–89% move at a walk or trot
Not acceptable	51–75% move at a walk or trot
Serious problem	<50% move at a walk or trot
Electric Prod Score	
Excellent	0% moved with prod
Acceptable	5% or less moved with prod
Not acceptable	6–20% moved with prod
Serious problem	>20% moved with prod or as result of abusive behavior

Source: Based on Grandin (2004c).

3. Implement training across the organization
4. Conduct periodic internal audits
5. Take corrective action to resolve problems identified
6. Conduct third party audit
7. Review results followed by corrective action as specified in the audit report
8. Determine approval status
9. Repeat the process with a goal of renewing approval

SELECTED REFERENCES

Publications

Battaglia, R.A., & Mayrose, V.B. 2001. *Handbook of livestock management* (3rd ed.). Upper Saddle River, NJ: Prentice Hall.

Beef housing and equipment handbook. 1986. Ames, IA: Midwest Plan Service, Iowa State University.

Blockey, M.A. de B. 1976. Sexual behavior of bulls at pasture: A review. *Theriogenology*. 6:387.

Craig, J.V. 1981. *Domestic animal behavior*. Englewood Cliffs, NJ: Prentice-Hall.

Curtis, S.E., & Houpt, K.A. 1983. Animal ethology: Its emergence in animal science. *Journal of Animal Science*. 57:234: Suppl. 2.

Ewing, S.A., Lay, D.C., & Von Borrell, E. 1999. *Farm animal well-being, stress physiology, animal behavior, and environmental design*. Upper Saddle River, NJ: Prentice Hall.

Grandin, T. 1989. Behavioral principles of livestock handling. *Professional Animal Scientist*. 5:1.

Grandin, T. 1997. Assessment of stress during handling and transport. *Journal of Animal Science*. 75:249.

Grandin, T. 2000. *Livestock handling and transport* (2nd ed.). Wallingford, United Kingdom: CAB International.

Grandin, T. 2004a. *Cattle transport guidelines*. Fort Collins, CO: Grandin Livestock Systems.

Grandin, T. 2004b. *Cattle welfare: Critical control points on feedlots, ranches, and stocker operations*. Fort Collins, CO: Grandin Livestock Systems.

Grandin, T. 2004c. *Welfare of cattle in feedlots: Audit form*. Fort Collins, CO: Grandin Livestock Systems.

Grandin, T. 2004d. *Cattle slaughter audit form*. Fort Collins, CO: Grandin Livestock Systems.

Odde, K.G. 1983. *The postpartum cow: Effect of early weaning and limited suckling on cow and calf performance, and suckling behavior in range calves*. Ph.D. Thesis. Manhattan, KS: Kansas State University.

Rile, R.W., MacNeil, M.D., Jenkins, T.G., & Koong, L.J. 1982. *A simulation model of grazing behavior in cattle*. Proc. Western Section American Society of Animal Science.

Rupp, G.P., Ball, L., Shoop, M.C., Las Cruces, N.M., & Chenoweth, P.J. 1977. Reproductive efficiency of bulls in natural service: Effects of male-to-female ratio and single vs. multiple sire breeding groups. *Journal of the American Veterinary Medical Association*. 171:639.

Schake, L.M., & Riggs, J.K. 1972. *Behavior of beef cattle in confinement*. College Station, TX: Texas Agricultural Experiment Station Technical Report 27.

Sherry, C.J., Klemm, W.R., Sis, R.F., & Schake, L.M. 1982. *Reproductive and feedlot behavior: The role of the vomeronasal organ*. College Station, TX: Beef Cattle Research in Texas, PR 3923.

Voisinet, B.D., Grandin, T., Tatum, J.D., O'Connor, S.F., & Struthers, J.J. 1997. Bos indicus-cross feedlot cattle with excitable temperaments have tougher meat and a higher incidence of borderline dark cutters. *Meat Science*. 46:367–377.

Voisinet, B.D., Grandin, T., Tatum, J.D., O'Connor, S.F., & Struthers, J.J. 1997. Feedlot cattle with calm temperaments have higher average daily gains than cattle with excitable temperaments. *Journal of Animal Science*. 75:892–896.

Appendix

The Metric System 564
Other Weights, Measures, and Sizes 564
Volumes and Weights of Stacked and Baled Hay 564
Round Grain Bin Volumes 568
Measuring Irrigation Water Flow 568
Land Description for Legal Purposes 570
Major Organizations Within or Affecting the Beef Industry 570
Table A.1 Metric System Prefixes, Symbols, and Power 564
Table A.2 Weight-Unit Conversion Factors 565
Table A.3 Average Bushel Weights of Selected Grains and Seeds 565
Table A.4 Determining the Amount of Corn Silage in Trench Silo 566
Table A.5 Silo Volumes per Foot and Pounds per 2-Inch Layer 566
Table A.6 Approximate Silo Capacity in Tons 567
Table A.7 Nail Sizes and Descriptions 567
Table A.8 Discharge in gpm for Various Nozzle Sizes and Pressures 569
Table A.9 Precipitation Rates in Inches per Hour for Various Sprinkler Discharges and Spacing 569
Table A.10 Water Applied per Set in Inches @75% Efficiency 570
Table A.11 Letter Codes for Indicating Year of Birth in Identification Systems 570

Table A.1
METRIC SYSTEM PREFIXES, SYMBOLS, AND POWER

Prefix	Symbol	Power and Meaning
tera	T	10^{12}
giga	G	10^{9}
mega	M	10^{6} 1,000,000 times base
kilo	k	10^{3} 1,000 times base
hecto	h	10^{2} 100 times base
deca	da	10 10 times base
deci	d	10^{-1} tenth
centi	C	10^{-2} hundredth
milli	m	10^{-3} thousandth
micro	µ	10^{-6} millionth
nano	n	10^{-9}
pico	p	10^{-12}
femto	f	10^{-15}
atto	a	10^{-18}

THE METRIC SYSTEM

The metric system has been accepted as the preferred system in nearly all countries of the world. The United States is the only major trading nation worldwide that does not presently use the metric system.

It is useful to understand both measurement systems (U.S. customary and metric) and be able to convert one to the other. Weights and measurements with their conversions are shown in Tables A.1 and A.2.

OTHER WEIGHTS, MEASURES, AND SIZES

Table A.3 lists the bushel weights of grains, while the capacities and volumes of silos are shown in Tables A.4, A.5, and A.6. Nail sizes and descriptions are identified in Table A.7.

VOLUMES AND WEIGHTS OF STACKED AND BALED HAY

Determining the volume and weight of hay is important whenever hay is sold or yields are determined. Volume of hay is expressed in cubic feet. Weight is expressed in pounds for individual bales, or tons for stacks and loads. Converting from volume, which is reasonably easy to measure, to weight requires that density (weight per cubic foot) of the hay be measured or estimated.

Stacks

The formula commonly used for estimating the volume of loose hay in stacks is:

$$V\frac{(O - 5.6W)}{2} \times W \times L$$

V = Stack Volume in cubic feet
O = Average distance Over stack in feet
W = Average stack Width in feet
L = Stack Length in feet

Table A.2
Weight-Unit Conversion Factors

Unit Given	Unit Wanted	For Conversion Multiply by
Lb	g	453.6
Lb	kg	0.4536
Oz	g	28.35
Kg	lb	2.2046
Kg	mg	1,000,000
Kg	g	1,000
G	mg	1,000
G	μg	1,000,000
mg	μg	1,000
mg/g	mg/lb	453.6
mg/kg	mg/lb	0.4536
μg/kg	g/lb	0.4536
Mcal	Kcal	1,000
kcal/kg	kcal/lb	0.4536
kcal/lb	kcal/kg	2.2046
ppm	μg/g	1
ppm	mg/kg	1
ppm	mg/lb	0.4536
mg/kg	%	0.0001
ppm	%	0.0001
mg/g	%	0.1
g/kg	%	0.1

Table A.3
Average Bushel Weights of Selected Grains and Seeds

Grain or Seed	Average Lb/Bushel
Alfalfa	60
Barley	48
Bluegrass	14–30
Clover	60
Corn	56
Oats	32
Orchardgrass	14
Sorghum	56
Soybeans	60
Wheat	60

Table A.4

DETERMINING THE AMOUNT OF CORN SILAGE IN TRENCH SILO (TONS = AVERAGE WIDTH × LENGTH × AVERAGE DEPTH OF SILAGE [IN FEET] × TONS PER CUBIC FOOT OF DEPTH[1])

Depth of Settled Silage (ft.)	Tons/Cu Ft	Depth of Settled Silage (ft.)	Tons/Cu Ft.
1	0.00925	9	0.01320
2	0.00985	10	0.01365
3	0.01040	11	0.01405
4	0.01090	12	0.01445
5	0.01140	13	0.01490
6	0.01190	14	0.01530
7	0.01235	15	0.01565
8	0.01280	16	0.01605

[1]*Example:* A trench silo measures 8 feet wide at the top and 12 feet wide at the bottom, and is 50 feet long. The silage averages 8 feet deep. The calculation for total tons in the trench silo is as follows:

Solution:

$$\frac{8 + 12}{2} \times 50 \times 8 \times 0.0128 = 21.2 \text{ tons}$$

Table A.5

SILO VOLUMES PER FOOT AND POUNDS PER 2-INCH LAYER

Silo Diameter	Volume/Ft of Depth (Cu ft.)	Silage (lb) in 2-in. Layer Based on 40 lb/cubic ft.
12	113.1	755
14	153.9	1,025
16	201.1	1,340
18	254.5	1,696
20	314.2	2,094
22	380.1	2,534
24	452.4	3,015
26	530.9	3,539
28	615.8	4,105
30	706.9	4,712
34	908.0	6,053
38	1,134.1	7,560
42	1,385.4	9,235
50	1,963.5	13,089
60	2,827.4	18,847

Table A.6
Approximate Silo Capacity in Tons

Inside Diameter of Silo (ft.)	Silo Height (ft.)					
	20	30	40	50	60	70
10	33	56	77			
12	48	80	110			
14	66	109	150	193		
16	86	143	196	252		
18		180	248	320	392	
20		223	307	394	483	574
22		270	371	477	585	694
24		321	442	570	697	827
26		377	520	668	818	970
28			600	773	947	1,125
30			690	886	1,087	1,290

Table A.7
Nail Sizes and Descriptions

Size	Length (in.)	Approximate N/lb
Common Nails		
2d	1	847
3d	1¼	543
4d	1½	294
5d	1¾	254
6d	2	167
7d	2¼	150
8d	2½	101
9d	2¾	92
10d	3	69
12d	3¼	63
16d	3½	49
20d	4	31
30d	4½	24
40d	5	18
50d	5½	14
60d	6	11
Spikes		
10d	3	32
12d	3¼	31
16d	3½	24
20d	4	19
30d	4½	14
40d	5	12
50d	5½	10
60d	6	9
5/16	7	6
3/8	8–12	5

The over measurement (O) can be obtained using a tape or string with attached weight that is thrown over the stack. The stack should be checked in about four places, and then those measurements averaged.

Bales

Measuring length, width, and height, then multiplying these together can determine the volume of a stack of baled hay. Another technique used with baled hay is to count the number of bales, then multiply by an estimated or determined weight per bale to determine total weight. The volume of the bales is occasionally needed for storage purposes. With round bales, weighing a few bales and counting the number of bales involved can make an estimate of the total tonnage.

Weight of Hay

The density, or pounds per cubic foot, of both stacked and baled hay varies greatly. The following table gives some guides. It is always better to calculate the density after weighing a few bales and determining the volume. It may also be possible to weigh a stack or portion of a stack. Remember that stacked hay settles over time.

	Weight of Loose Hay (lb/cu ft)	Weight of Baled Hay (lb/cu ft)
Alfalfa	4–5	8–14
Grass hay	3–5	6–10
Straw	2–3	4–6

ROUND GRAIN BIN VOLUMES

The following formula assumes that grain has fallen freely into the bin with the top forming a cone.

$$\text{Volume in bushels} = (0.974 \times d^2) \times \left(h + \frac{d}{20}\right)$$

d = diameter (in feet)
h = height (in feet)

Example: d = 18 h = 12

$$V = (0.974 \times 18^2) \times 12 + \frac{18}{20}$$

V = 4,071 bushels

MEASURING IRRIGATION WATER FLOW

Water measurement is a necessary part of water management. Measurement is also used to verify water rights.

Several devices can be installed for measuring water delivery in open-ditch irrigation. Common devices are weirs and flumes. For accurate water-flow measurements, weirs and flumes must be properly constructed, installed, and maintained.

Inches of Water Applied with Hand Move and Wheel Line Sprinklers

Factors that affect the inches of water applied are pressure, nozzle size, sprinkler spacing, and time. Pressure is measured with a gauge. The size is stamped on the nozzle, but older sprinklers should be verified using a drill bit as a measure. Spacing is the distance between sprinklers and line positions. Use Tables A.8, A.9, and A.10 in sequence, to determine inches of water applied.

Examples:

1. Pressure of 50 psi and 13/64-inch nozzle from Table A.8 read the discharge as 8.71 gpm.
2. Sprinklers spaced 40 feet apart and lines 60 feet apart. From Table A.9 estimate the precipitation rate at 0.35 inches per hour.
3. Sprinklers are run for 12 hours. From Table A.10 estimate the amount of water entering the root zone at 3.15 inches.

Table A.8
Discharge in GPM for Various Nozzle Sizes and Pressures

	Nozzle Diameter (in.)					
PSI	9/64	5/32	11/64	3/16	13/64	7/32
30	3.26	4.01	4.83	5.75	6.80	7.85
40	3.74	4.61	5.54	6.64	7.80	9.02
50	4.18	5.15	6.19	7.41	8.71	10.10
60	4.50	5.65	6.80	8.12	9.56	11.05
70	4.96	6.10	7.34	8.78	10.32	11.95
80	5.29	6.52	7.84	9.39	11.02	12.74

Table A.9
Precipitation Rates in Inches Per Hour for Various Sprinkler Discharges and Spacing

	Sprinkler Discharge (gpm)						
Spacing Feet	3.0	4.0	5.0	6.0	8.0	10.0	12.0
30 × 50	0.19	0.25	0.32	0.38	0.51	0.64	0.76
30 × 60	0.16	0.21	0.27	0.32	0.43	0.53	0.64
40 × 50	0.14	0.19	0.24	0.29	0.38	0.48	0.58
40 × 60	0.12	0.16	0.20	0.24	0.32	0.40	0.48

Table A.10
WATER APPLIED PER SET IN INCHES @75% EFFICIENCY

	Precipitation Rate (in./hr)					
Hours	0.20	0.30	0.40	0.50	0.60	0.70
6.0	0.90	1.13	1.80	2.25	2.70	3.15
8.0	1.20	1.80	2.40	3.00	3.60	4.20
12.0	1.80	2.70	3.60	4.50	5.40	6.30
24.0	3.60	5.40				

Table A.11
LETTER CODES USED TO INDICATE YEAR OF BIRTH IN IDENTIFICATION SYSTEMS

K—2000	Y—2011
L—2001	Z—2012
M—2002	A—2013
N—2003	B—2014
P—2004	C—2015
R—2005	D—2016
S—2006	E—2017
T—2007	F—2018
U—2008	G—2019
W—2009	H—2020
X—2010	J—2021

LAND DESCRIPTION FOR LEGAL PURPOSES

The federal government has established land surveys that are used for the legal description of land. Townships, 6 miles square, are located north or south of standard parallels and east and west of prime meridians.

Each township is divided into 36 sections, each section being a mile square. The sections are numbered 1 through 36, starting with number 1 in the northeast corner of the township and ending with number 36 in the southeast corner.

Sections are further divided into smaller units so that the location of every land parcel can be identified for legal purposes. For example, the legal description of a certain 40 acres may read: The south half of the west half of the southwest quarter of section 6 in township 32, north of range 4 west.

MAJOR ORGANIZATIONS WITHIN OR AFFECTING THE BEEF INDUSTRY

The names, addresses, and functions of the major organizations representing and influencing the beef industry are listed here for reference.

American Association of Bovine Practitioners
P.O. Box 3610
Auburn, AL 36831-3610

Physical Address:
3320 Skyway Drive
Suite 802
Opelika, AL 36801

The American Association of Bovine Practitioners is an international association of veterinarians organized to enhance the professional lives of its members through relevant continuing education that will improve the well-being of cattle and the economic success of their owners, increase awareness and promote leadership for issues critical to cattle industries, and improve opportunities for careers in bovine medicine.

American Association of Meat Processors (AAMP)
1 Meating Place
Elizabethtown, PA 17022
www.aamp.com

Membership includes more than 1,700 medium-sized and smaller meat, poultry, and food businesses: slaughterers, packers, processors, wholesalers, in-home food service business, retailers, deli and catering operators, and industry suppliers.

American Farm Bureau Federation
600 Maryland Ave. SW
Suite 100 W
Washington, DC 20024
www.fb.org

Farm Bureau is an independent, nongovernmental, voluntary organization governed by and representing farm and ranch families united for the purpose of analyzing their problems and formulating action to achieve educational improvement, economic opportunity, and social advancement and, thereby, to promote the national well-being.

American Farmland Trust
1150 Connecticut Ave NW
Suite 600
Washington, DC 20036
www.farmland.org

American Farmland Trust is a private, nonprofit organization founded in 1980 to protect U.S. farmland. AFT works to stop the loss of productive farmland and to promote farming practices that lead to a healthy environment.

American Feed Industry Association (AFIA)
2101 Wilson Blvd., Suite 916
Arlington, VA 22201
www.afia.org

AFIA is the only national organization devoted exclusively to representing the regulatory, commerce, and governmental interests of the animal feed and pet food industries and their suppliers.

American Forage and Grassland Council (AFGC)
P.O. Box 867
Berea, KY 40403
www.afgc.org

A spokesgroup for North America's forage-based agriculture, members are representatives of agencies that provide research, educational, and public services.

North American Meat Institute (AMI)
1150 Connecticut Ave NW
Washington, DC 20036
www.meatami.org
See Chapter 1.

American Meat Science Association
201 W. Springfield Ave
Suite 1202
Champaign, IL 61820
www.meatscience.org

This organization is engaged in meat research, extension, and education in universities, industry, government, and other organizations. It encourages the exchange of ideas and information and seeks to foster education, research, and development in the field of meat science.

American Livestock Breeds Conservancy (ALBC)
P.O. Box 477
Pittsboro, NC 27312
livestockconservancy.org

The ALBC was organized in 1977 to conserve and preserve rare breeds and stocks of livestock and poultry. Some of its members raise the rare breeds. A semen bank has been established, in which Milking Devon, Dutch Belted, Florida Cracker Cattle, Dexter, Red Poll, and Lineback are among the first to be stored.

American National Cattlewomen (ANCW)
200 NW 66th Street
Suite 972 Oklahoma City, OK 73116
www.ancw.org
See Chapter 1.

American Registry of Professional Animal Scientists (ARPAS)
1800 S. Oak St.
Suite 100
Champaign, IL 61820
www.arpas.org

The ARPAS certifies qualified animal scientists as credible experts in their fields based on intense scholarly preparation and practical experience.

American Society of Animal Science (ASAS)
P.O. Box 7410
Champaign, IL 61826
www.asas.org

ASAS works to increase knowledge and understanding of animals, especially farm animals, and to improve the care and productivity of animals both in commercial production and in research. It publishes the *Journal of Animal Science*.

American Veal Association
2900 NE Brooktree Lane

Suite 200
Gladstone, MO 64119
www.americanveal.com

This group represents more than 1,300 family veal farmers, producers, and associated industry members.

The Animal Transportation Association
P.O. Box 3363
Warrenton, VA 20188
www.animaltransportationassociation.org

Organized in 1976 to assist in animal transportation by air, this group promotes safe and humane transportation of animals by land, air, and sea.

Animal Industry Foundation (AIF)
2101 Wilson Blvd
Suite 916-B
Arlington, VA 22201
www.animalagalliance.org

AIF attempts to redefine animal agriculture as high-quality animal care, correcting misinformation and violence aimed at farmers, ranchers, researchers, processors, auction markets, and related businesses. It is working to be the single, clear voice speaking to the American public on behalf of livestock and poultry producers.

Beef AI Organizations

AI organizations are the primary sources of commercial semen used in artificial insemination programs. These organizations publish sire directories, which list bulls and their available semen.

ABS Global, Inc.
1525 River Road
Deforest, WI 53532
www.absglobal.com

Accelerated Genetics
E10890 Penny Lane
Baraboo, WI 53913
www.accelgen.com

Alta Genetics, Inc.
P.O. Box 437
N8350 High Road
Watertown, WI 53094
www.altagenetics.com

Elgin Breeding Service
P.O. Box 68
Elgin, TX 78621
www.elginbreedingservice.com

Genetic Cooperative, Inc.
Cooperative Resources
100 MBC Drive
Shawano, WI 54166

(Contact for affiliates and regional centers.)
www.crinet.com

KABSU
3171 Tuttle Creek
Manhattan, KS 66502
www.asi.k-state.edu/services-and-sales/kabsu/

Select Sires, Inc.
11740 U.S. 42
Plain City, OH 43064
(Select Sires is a group of 10 co-ops located throughout the United States. Contact Select Sires for their addresses.)
www.selectsires.com

There are additional AI organizations that do custom semen collection, while certain organizations sell semen for others. Some of these organizations are:

Androgenics, P.O. Box 183, 11240 26 Mile Road, Oakdale, CA 95361 www.androgenicsinc.com
Cottage Farm Genetics, 971 Old Bells Road, Jackson, TN 3830
www.cottagefarmgenetics.com
Great Lakes Sire Service, Inc., 723 Himebaugh Road, Bronson, MI 49028
www.greatlakessireservice.com
Great Plains Breeders Service/Taurus, Box 468, N. Hwy. 83, Shamrock, TX 79079
www.greatplainsshamrocktx.net
Hawkeye Breeders Service, 3257 Old Portland Road, Adel, IA 50003
www.hawkeyebreeders.com
Hoffman AI Breeders, 1950 S. Hwy. 89–91, Logan, UT 84321
www.hoffmanaibreeders.com
Interglobe Genetics, 14814 N. 1500 E. Road, Pontiac, IL 61764
interglobegenetics.com
JLG Enterprises, P.O. Box 1375 (11116 Sierra Road), Oakdale, CA 95361
www.jlgenterprises.com
Nebraska Bull Service, Inc., P.O. Box 998 (38364 Road 720) McCook, NE 69001
www.nebraskabullservice.com
Nichols Cryo-Genetics, 11745 NE 112th St. Maxwell, IA 50161
www.nicholscryogenetics.com
Nokota Genetics, 6921 Hwy. 83 N., Minot, ND 58701-0241
North American Breeders, 1075 W. Main St, Berryville, VA 22611
ORIgen, 10 West Arrow Creek Road, Huntley, MT 59037
www.origenbeef.org
Reproduction Enterprises., Inc., 908 N. Prairie Road, Stillwater, OK 74074
reproductionenterprises.com
Rocky Mtn. Sire Service, 1616 Manila Road, Bennett, CO 80102
www.rmssbulls.com
Southeastern Semen Services, 16878 45th Street, Wellborn, FL 32094
Taurus Service, Inc., P.O. Box 164, (Grist Flat Road), Mehoopany, PA 18629
www.taurus-st.com
Vogler Semen Centre Lab, Inc., 27104 Church Road, Ashland, NE 68003
www.voglersemen.com

Beef Improvement Federation (BIF)

Jane Parish, Executive Director
NMREC Prairie Research Unit
10223 Hwy 382 P.O. Box 60
Prairie, MS 39756
www.beefimprovement.org

BIF coordinates the performance testing programs of approximately 50 organizations from several states, including the national breed associations. Its primary objective is to establish accurate and uniform procedures for measuring and recording beef cattle performance data.

Breed and Breeder Associations

Members of the following associations are breeders or owners of specific breeds of cattle. Their purposes are to promote the breed, record the performance of the cattle, issue registrations, and keep the herd book. Dairy breeds are included because they are used in some crossbreeding programs, and all cattle eventually produce beef products.

American Breed
American Breed International Assoc., 1530 S. Avenue E, Portales, NM 88130

Amerifax
Amerifax Cattle Association, P.O. Box 149, Hastings, NE 68901

Angus
American Angus Association, 3201 Frederick Boulevard, St. Joseph, MO 64506
www.angus.org

Ankole-Watusi
Ankole Watusi International Registry, 22484 W. 239th Street, Spring Hill, KS 66083
www.awir.org

Aubrac Alliance
611 Sudbury Drive, Columbia, MO 65203

Ayrshire
Ayrshire Breeders Association, 1224 Alton Darby Creek Rd, Suite B, Columbus, OH 43228
www.usayrshire.com

Barzona
Barzona Breeders Association, Alecia Heinz, Executive Secretary, 604 Cedar St. Adair, IA 50002
www.barzona.com

Beefalo
American Beefalo World Registry, 9824 E. YZ Ave. Vicksburg, MT 49097
www.americanbeefalo.org

Beef Friesian
Beef Friesian Society, 25377 Weld County Road 17, Johnstown, CO 80534

Beefmaster
Beefmaster Breeders United, 6800 Park Ten Blvd., Suite 290 West, San Antonio, TX 78213
www.beefmasters.org

Belgian Blue

American Belgian Blue Breeders, Inc., P.O. Box 633404 Nacogdoches, TX 75963
www.belgianblue.org

Belted Galloway

Belted Galloway Society, Inc., N8603 Zentner Rd. New Glarus, WI 53574
www.beltie.org

Bison and Buffalo

National Bison Association
8690 Wolff Ct. #200
Westminster, CO 80031
www.bisoncentral.com

Black Hereford

1704 South Cannon Road
Shelbyville, TN 37160
www.blackhereford.org

Blonde d'Aquitaine

American Blonde d'Aquitaine Assoc., 57 Friar Tuck Way, Fyffe, AL 35971
www.blondecattle.org

Braford

United Braford Breeders, P.O. Box 14100, Kansas City, MO 64101
www.brafords.org

Brah-Maine

International Brah-Maine Society, RR 1, Box 233, Franklin, TX 77856

Brahman

American Brahman Breeders Assoc., 3003 S. Loop West, Suite 140, Houston, TX 77054
www.brahman.org

Brahmousin

American Brahmousin Council, Box 88, Whitesboro, TX 76273
www.americanbrahmousincouncil.org

Bralers

American Bralers Association, Box 75, Burton, TX 77835

Brangus

International Brangus Breeders Association, Inc., 5750 Epsilon, San Antonio, TX 78249
www.int-brangus.org

Braunvieh

Braunvieh Association of America, 5750 Epsilon Suite 200, San Antonio, TX 78249
www.braunvieh.org

British White

British White Cattle Association of America, 6656 45th Ave SW, Pequot Lakes, MN 56472
www.britishwhite.org

Brown Swiss
Brown Swiss Cattle Breeders Assoc., 800 Pleasant St, Beloit, WI 53511
www.brownswissusa.com

BueLingo Cattle Society
National BueLingo Beef Cattle Society 15904 W. Warren Road, Warren, IL 61087
www.buelingo.com

Charbray
American-International Charolais Association, Charbray Division, 11700 NW Plaza Circle, Kansas City, MO 64153
www.charolaisusa.com

Charolais
American-International Charolais Association, 11700 NX Plaza Circle, Kansas City, MO 64153
www.charolaisusa.com

Char-Swiss
Char-Swiss Breeders Association, 407 Chambers St., Marlin, TX 76661

Chiangus
Chiangus, American Chianina Association, 1708 NW Prairie View Rd, Platte City, MO 64079
www.chicattle.org

Chianina
American Chianina Association, 1708 NW Prairie View Rd, Platte City, MO 64079
www.chicattle.org

Chiford
Chiford, American Chianina Association, 1708 NW Prairie View Rd, Platte City, MO 64079
www.chicattle.org

Chimaine
Chimaine, American Chianina Association, 1708 NW Prairie View Rd, Platte City, MO 64079
www.chicattle.org

Corriente
North American Corriente Association, P.O. Box 2698, Monument, CO 80132www.corrientecattle.org

Cracker Cattle[1]
Florida Cracker Cattle Breeders Assoc., 11928 NW 199th Avenue, Alachua, FL 32615
www.floridacrackercattle.org

[1] Also referred to as Florida Scrub or Piney Woods cattle. These cattle are descendants of the Spanish cattle brought to Florida from the islands of Hispaniola, Jamaica, and Cuba. They may also contain some breeding from mixed types of cattle developed in the eastern United States from cattle brought from Europe.

Devon

Devon Cattle Association, Inc., Jeremy Anderson Engh, President, 9272 Big Horn Road, Remington, VA 22734
http://devoncattle.com

Dexter

American Dexter Cattle Association, 1325 W. Sunshine #519, Springfield, MO 65807
www.dextercattle.org

Dutch Belted

Dutch Belted Cattle Association of America, Cornell Upson, President, 3724 Co Hwy 33, Cherry Valley, NY 13320
www.dutchbelted.com

Galloway

American Galloway Breeders Assoc., c/o Canadian Livestock Records Corporation, 2417 Holly Lane, Ottawa, Ontario, Canada.
www.americangalloway.com

Gelbray

Gelbray International, Inc., Rt. 1, Box 273C, Madill, OK 73446

Gelbvieh

American Gelbvieh Assoc., 10900 Dover St., Westminster, CO 80021
www.gelbvieh.org

Guernsey

American Guernsey Organization, 1224 Alton Darby Creek Rd. Suite G, Columbus, OH 43228
www.usguernsey.com

Hereford

American Hereford Association, P.O. Box 014059, Kansas City, MO 64101
www.hereford.org

Herens

American Herens Association, P.O. Box 1250, Lewisburg, WV 24901

Highland

American Highland Breeders Association, Historic City Hall, 22 S. 4th Ave., Ste 201, Brighton, CO 80601
www.highlandcattleusa.org

Holstein-Friesian

Holstein-Friesian Association, 1 Holstein Place, Brattleboro, VT 05302
www.holsteinusa.com

Irish Blacks

(Registrations recorded in Beef Friesian Society Herdbook.)

Jersey

The American Jersey Cattle Assn., 6486 East Main, Reynoldsburg, OH 43068
www.usjersey.com

Limousin

North American Limousin Foundation, 6 Inverness Court East Suite 260, Englewood, CO 80112
www.nalf.org

Lineback

Lineback Cattle Registry Association, Nedra Yoder-Keller, President, 68 Hecktown Road, Shoemakersville, PA 19555
www.americanlinebacks.tripod.com

Longhorn

Texas Longhorn Breeders Association of America, P.O. Box 4430, Fort Worth, TX 76164 76106
www.tlbaa.org

Maine-Anjou

American Maine-Anjou Association, P.O. Box 1100, 204 Marshall Road, Platte City, MO 64079
www.maine-anjou.org

Mandalong Special

Accelerated Genetics, E10890 Penny Lane, Baraboo, WI 53913

Marchigiana

American International Marchigiana Society, Box 198, Walton, KS 67151-0198
www.marchigiana.org

Milking Shorthorn

American Milking Shorthorn Society, 23515 Range Road, Union City, PA 16438
www.milkingshorthorn.com

Murray Grey

American Murray Grey Association, P.O. Box 1222 Shelbyville, KY 40066
www.murraygreybeefcattle.com

Normande

American Normande Association, 748 Enloe Road, Rewey, WI 53580
www.normandeassociation.com

Parthenais

Parthenais Cattle Breeders Assn. of America, Box 778, Arp, TX 75750

Piedmontese

North American Piedmontese Association, 1740 Co. Road 185, Ramah, CO 80832
www.piedmontese.org

Pinzgauer

American Pinzgauer Association, 681 Maple Ridge Road, Harrison, ME 04040
www.pinzgauers.org

Red Angus

Red Angus Association, 4201 I-35 North, Denton, TX 76207
www.redangus.org

Red Brangus
American Red Brangus Association, 1260 Pin Oak Rd. #204 Katy, TX 77494
www.redbrangus.org

Red Poll
American Red Poll Association, P.O. Box 847 Frankton, IN 46044
www.redpollusa.org

Romagnola
American Romagnola Association, Attn: Mary Flanigan, Secretary, 14305 W. 379th St. La Cygne, KS 66040
www.romagnola.org

RX_3
American RX3 Cattle Registry, 4568 T. Avenue, Oelwein, IA 50661

Salers
American Salers Association, 19590 East Mainstreet #202, Parker, CO 80138
www.salersusa.org

Santa Cruz
King Ranch, Box 1090, 201 E. Kleberg, Kingsville, TX 78364

Santa Gertrudis
Santa Gertrudis Breeders International, P.O. Box 1257, Kingsville, TX 78363
www.santagertrudis.com

Senepol
Senepol Cattle Breeders Association, 2321 Chestnut Street, Wilmington, NC 28405
www.senepolecattle.com

Shorthorn/Polled Shorthorn
American Shorthorn Association, 7607 NW Prairie View Road, Kansas City, MO 64151
www.shorthorn.org

Simbrah
American Simmental Association, 1 Simmental Way, Bozeman, MT 59715
www.simmental.org

Simmental
American Simmental Association, 1 Simmental Way, Bozeman, MT 59715
www.simmental.org

South Devon
North American South Devon Association, 19590 E. Main St. Suite 104, Parker, CO 80138
www.southdevon.com

Tarentaise
American Tarentaise Association, 9150 North 216th Street, Elkhorn, NE 68022
www.Americantarentaise.org

Tuli
North American Tuli Association, 10853 Forest Drive, College Station, TX 77845

Wagyu
American Wagyu Association, P.O. Box 3235, Coeur d'Alene, ID 83816
www.wagyu.org

Water Buffalo
American Water Buffalo Association, 2415 N. Mosley Road, Texarkana, AR 71854
www.americanwaterbuffaloassociation.com

Watusi
World Watusi Association, P.O. Box 201 Walnut Springs, TX 76690
www.watusi.org

Welsh Black
Welsh Black Cattle Association

White Park
American British White Park Cattle Association, P.O. Box 409, Myerstown, PA 17067
www.whitecattle.org

Zebu
International Zebu Breeders Association, 1901 Miller Road, Rowlett, TX 75088

Cattle-Fax
Cattle Marketing Information Service, Inc.
9110 East Nichols Avenue
Centennial, CO 80112
www.cattlefax.com

A nonprofit corporation governed by cattle producers, Cattle-Fax has a staff of skilled, knowledgeable market analysts who provide local, state, regional, national, and international market information and analysis as well as analysis of factors affecting the market.

Center for Consumer Freedom
P.O. Box 34557
Washington, DC 20043
www.consumerfreedom.com

The Center for Consumer Freedom is a nonprofit coalition supported by restaurants, food companies, and consumers working together to promote personal responsibility and protect consumer choices.

Centers for Disease Control and Prevention
1600 Clifton Road
Atlanta, GA 30329
www.cdc.gov

The federal agency responsible for protecting the public health of the nation by preventing unnecessary disease, disability, and premature death through the promotion of healthy lifestyles.

Chicago Mercantile Exchange (CME)
20 S. Wacker Dr.
Chicago, IL 60606
www.cme.com

See Chapter 9.

Council for Agricultural Science and Technology (CAST)
4420 West Lincoln Way
Ames, IA 50014-3447
www.cast-science.org

CAST organizes task forces of food and agricultural scientists and technologists from the relevant disciplines to assemble and interpret factual information on food and agricultural issues of public concern. It disseminates this information in a usable and effective form to the public, news media, and government, as appropriate.

Environmental Protection Agency

See U.S. Environmental Protection Agency (EPA).

Federal Trade Commission (FTC)
600 Pennsylvania Avenue NW
Washington, DC 20580
www.ftc.gov

The FTC works to keep the free enterprise system from becoming stifled by monopolies or restraints on trade or corrupted by unfair or deceptive trade practices.

Federation of Scientific Agricultural Societies (FSAS)

The FSAS represents scientific agricultural societies, particularly in identifying national goals and priorities related to education and research in food, agriculture, and natural resources, and in urging executive and legislative support of the priority areas.

Food and Drug Administration (FDA)
Department of Health, Education, and Welfare
10903 New Hampshire Avenue
Silver Spring, MD 20993
www.fda.gov

Activities relate to protecting public health as it may be impaired by foods, drugs, biological products, cosmetics poisons, pesticides, and food additives. FDA ensures that foods are safe, pure, and wholesome.

Food and Drug Law Institute
1155 15th Street, NW, Suite 910
Washington, DC 20005
www.fdli.org

Manufacturers and distributors of food, drugs, and cosmetics are the members of this organization, which promotes knowledge about the laws governing these and other products.

Food Marketing Institute (FMI)
1155 15th Street, NW, Suite 910
Washington, DC 20005 www.fmi.org

See Chapter 1.

Grazing Lands Conservation Institute (GLCI)
501 W. Felix Street, FWFC Bldg. 23
Fort Worth, TX 76115-3494
www.glci.org

GLCI provides high-quality technical assistance to private landowners and works to increase awareness of the importance of grazing resources.

Institute of Food Technologists (IFT)
525 West Van Buren, Suite 1000
Chicago, IL 60607
www.ift.org

IFT is a worldwide society of professional food technologists, scientists, engineers, executives, and educators in the field of food technology. Other individuals interested in food technology because of their work in closely related fields are also members.

International Embryo Transfer Society (IETS)
1800 S. Oak Street, Suite 100
Champaign, IL 61820
www.iets.org

Members of the society are engaged in the practice of embryo transfer in a variety of species, and in research on embryo production, transgenesis and cloning, on mechanisms regulating embryo development, and on development following embryo transfer. Species studied include domesticated and laboratory animals and endangered species.

Leather Industries of America (LIA)
3050 K Street, NW, Suite 400
Washington, DC 20007
www.leatherusa.com

LIA is a trade association that represents tanners and allied industries.

National Institute for Animal Agriculture (NIAA)
13570 Meadowgrass Drive, Suite 201
Colorado Springs, CO 80921
www.animalagriculture.org

NIAA serves as a clearinghouse for all sectors of the livestock and meat industry in sponsoring research and educational programs designed to eradicate diseases among livestock, to promote a safe and wholesome food supply, and to promote best practices in environmental stewardship, animal health, and well-being.

Livestock Marketing Association (LMA)
10510 NW Ambassador Drive
Kansas City, MO 64153
www.lmaweb.com

LMA promotes the maintenance of a free, competitive marketing sector for the livestock industry.

Livestock Publications Council
200 West Exchange Avenue
Fort Worth, TX 76164
www.livestockpublications.com

Among the many objectives of this organization are (1) to promote understanding and cooperation among publications serving the livestock industry; (2) to encourage and support research and activities designed to further the livestock industry; (3) to foster, through cooperative effort, relations among publishers, legislators,

government administrators, and people in all segments of the livestock industry and allied enterprises.

National Academy of Sciences (NAS)
500 5th Street NW
Washington, DC 20001
www.nas.edu

The NAS is a private honorary organization that serves as an independent advisor to the federal government on matters of science and technology. In each case, such counsel is provided as a formal report of the deliberations of a study project and its conclusions and recommendations; all reports are publicly available. The NAS created the National Research Council (NRC) as part of its structure; it is the operating arm of the NAS and currently comprises some 900 committees and panels.

National Agri-Marketing Association
11020 King Street, Suite 205
Overland Park, KS 66210
www.nama.org

This association is made up of persons engaged in agricultural advertising and marketing for manufacturers, advertising agencies, and the media. It promotes high standards of agri-marketing, provides for the exchange of ideas, and encourages study and better understanding of agricultural advertising, selling, and marketing.

National American Agricultural Marketing Officials (NAAMO)
c/o CA Department of Food and Agriculture
778 S. Madera Avenue
Kerman, CA 93630
www.naamo.org

NAAMO membership comprised of state officials responsible for the administration of state agricultural marketing programs work through this organization to improve the marketing, handling, storage, processing, transportation, and distribution of North American agricultural products worldwide.

National Industrial Transportation League
1700 N. Moore Street, Suite 1900
Arlington, VA 22209-1904
www.nitl.org

Truckers, truck brokers, and others concerned with long-haul agricultural trucking are members of this organization.

National Association of Animal Breeders (NAAB)
P.O. Box 1033
Columbia, MO 65205
www.naab-css.org

Farmer cooperatives and private businesses interested in the improvement of farm livestock make up the NAAB, which works to stimulate and encourage research in artificial insemination and reproduction. It regularly establishes guidelines for AI, certified semen services (CSS), disease prevention, CSS health programs, and state and foreign health requirements.

National Auctioneers Association
8880 Ballentine
Overland Park, KS 66214
www.auctioneers.org

This association of professionals works to promote and advance the auction profession and the mutual interests of its members.

National Cattlemen's Beef Association (NCBA)
9110 E. Nichols Ave. #300
Centennial, CO 80112
www.beef.org

See Chapter 1.

National Farm-City Council, Inc.
225 Touhy Avenue
Park Ridge, IL 60068
www.farmcity.org

The purpose of this group is to bring about better understanding between the rural and urban segments of society. Members of the council are organizations and individuals prominent in the agribusiness complex.

National Grain and Feed Association (NGFA)
1250 I Street, NW, Suite 1003
Washington, DC 20005-3922
www.ngfa.org

The NGFA is the national spokesgroup for grain, feed, and related commercial industries. NGFA provides educational, policy and business support to its membership.

National Food Brokers Association (NFBA)
1010 Massachusetts Avenue, NW, 6th Floor
Washington, DC 20001

The NFBA represents qualified firms and persons who sell food and nonfood products (at the wholesale-buying level) for a commission or brokerage fee.

National Grange
1616 H St. NW
Washington, DC 20006
www.nationalgrange.org

Grange is a fraternal organization of rural families. It promotes general welfare and agriculture through legislative, social, education, community service, home economics, youth, aid for handicapped, cooperatives, insurance, and credit-union programs.

National Hot Dog and Sausage Council
1150 Connecticut Ave. NW, 12th Floor
Washington, DC
www.hot-dog.org

Established in 1994, the Council conducts scientific research to benefit hot dog and sausage manufacturers. The Council also serves as an information resource to consumers and media on questions related to quality, safety, nutrition, and preparation of hot dogs and sausages.

National Meat Association (NMA)
1150 Connecticut Ave. NW, 12th Floor
Washington, DC
www.meatinstitute.org

The NMA represents independent meatpackers and meat processors and suppliers to the industry. It has divisions in beef, pork, and processed meats.

National Meat Canners Association
1700 North Moore Street, Suite 1600
Arlington, VA 22209
www.meatami.org

The National Meat Canners Association was founded in 1923 to promote the welfare of the sterile processed meat industry in the United States. Today more than 35 companies, whose primary or secondary businesses lie in sterile processed meat products, comprise the association's membership.

National Restaurant Association
2055 L Street NW, Suite 700
Washington, DC 20036
www.restaurant.org

This association represents restaurants, cafeterias, clubs, contract feeders, drive-ins, caterers, institutional food services, and other members of the food service industry. It supports food service education and research in several educational institutions, conducts traveling management courses and seminars for restaurant personnel, and distributes educational pamphlets, books, and films about the industry.

National Renderers Association, Inc.
500 Montgomery Street, Suite 310
Alexandria, VA 22314
www.renderers.org

This is a trade association of by-product rendering companies that advances its membership via policy, education, and business support efforts.

Nature Conservancy (NC)
4245 North Fairfax Drive, Suite 100
Arlington, VA 22203
www.nature.org

The NC mission is to preserve the plants, animals, and natural communities that represent the diversity of life on Earth by protecting the lands and waters they need to survive. NC works with a variety of partners to acquire key land holdings to advance its mission.

Public Lands Council
1301 Pennsylvania Avenue, Suite 300
Washington, DC 20004
www.publiclandcouncil.org

Established in 1969, the council represents the interests of ranchers who hold leases and permits to graze livestock on public lands in the western states. It is dedicated to the principle of sound management of federal lands for grazing and all other multiple-use purposes. Membership of the council includes both individual livestock producers and sheep and cattle organizations in the 13 western states.

Samuel Roberts Noble Foundation (NF)
2510 Sam Noble Parkway
Ardmore, OK 73401
www.noble.org

NF conducts agricultural, forage, biotechnical, and plant biology research. It also provides farmer-rancher consultation services.

Sierra Club
85 Second Street, 2nd Floor
San Francisco, CA 94105-3441
www.sierraclub.org

This organization is responsible for intensive lobbying efforts that influence government decisions on use of land and natural resources. The Sierra Club is involved with the major conservation issues.

Society for Range Management
6901 S Pierce St Ste 225
Littleton, CO 80128
www.srm.org

This is an educational and research group that promotes the understanding of rangeland and its ecosystems and uses.

State Beef Council

Contact the National Cattlemen's Beef Association (www.beef.org) for addresses for specific states.

State Cattle and Feeders Association

Contact the National Cattlemen's Beef Association for addresses for specific states. Several states have their own state cattle association that includes both cow-calf producers and feeders; other states have separate organizations representing each group. Members discuss and take action on legislation, research, marketing, and other current issues influencing beef producers in their respective states and in the nation. Most state cattle organizations are affiliated with the National Cattlemen's Beef Association (NCBA) (www.beef.org).

The Allan Savory Center for Holistic Management
5941 Jefferson St. NE, Suite B
Albuquerque, NM 87109
www.holisticmanagement.org

The Savory Center is focused on restoration of landscapes and the lives of people dependent on them using practical, low-cost problem solving.

U.S. Beef Breeds Council

No established office—correspondence is addressed to the current president's office. Contact major breed association for current address. The Council functions to create a more unified effort by purebred breeders and to promote beef and the industry.

U.S. Hide, Skin, and Leather Association (USHSLA)
1150 Connecticut Avenue NW, 12th Floor
Washington, DC
www.ushsla.org

USHSLA is the exclusive representative of the hide and skin industry. Its goals are to promote the marketing and processing of these primary by-products.

U.S. Department of Agriculture (USDA)
www.usda.gov

The USDA works to improve and maintain farm income and develop and expand markets abroad for agricultural products. It also helps curb poverty, hunger, and malnutrition, and works to enhance the environment and maintain production capacity by helping landowners protect soil, water, forests, and other natural resources. Rural development, credit, and conservation programs are key resources for carrying out national growth policies. The USDA, through inspection and grading services, safeguards and assures standards of quality in the daily food supply.

USDA—Agricultural Marketing Services (AMS)
www.ams.usda.gov

AMS provides market news reports; develops quality grade standards for agricultural commodities; provides voluntary grading services for livestock, meat, poultry, and more; administers marketing regulatory programs, marketing agreements and orders, and research and promotion programs; administers national organic standards activities; administers federal and state marketing improvement programs, wholesale facilities research programs, and food purchases; and is the coordinator for USDA's pesticide data program activities.

USDA—Agricultural Research Service (ARS)
www.ars.usda.gov

ARS conducts mission-oriented research to ensure adequate protection of food and agricultural products to meet nutritional and other needs of American consumers in animal and plant sciences, including disease and pest controls, soil and water conservation, postharvest processing and storage of commodities, safety of food, human nutrition, and integration of agricultural systems.

USDA—Animal and Plant Health Inspection Service (APHIS)
www.aphis.usda.gov

APHIS administers regulatory programs to control or eradicate animal and plant pests and diseases; enforces domestic and port-of-entry agricultural quarantines; licenses and ensures safety and effectiveness of veterinary biological products; enforces the Animal Welfare and Horse Protection acts; administers programs providing protection to livestock and crops from depredation by rodents, birds, and predatory animals; and conducts cooperative programs to eradicate animal and plant pests and diseases in other countries.

USDA—Economic Research Service (ERS)
355 E Street SW
Washington, DC 20024
www.ers.usda.gov

ERS conducts research in domestic and foreign agricultural economics; analyzes factors affecting agriculture, farm productivity, financing, use of resources, and potentials of rural areas; evaluates marketing potential and development and marketing costs; studies U.S. trade in agricultural products and the role of agriculture in economic development of other countries; and reports agricultural situation and outlook, commodity projections, price spreads, and analysis of U.S. farm commodity programs.

USDA—Extension Service
www.nifa.usda.gov/extension

The Extension Service serves as a partner with state governments through its land-grant universities and the county governments, forming the Cooperative Extension Service. These levels of government finance and conduct educational programs to help the public learn about and apply the latest technologies and management techniques developed through the research of the land-grant universities, the USDA, and other sources. Its major areas of assistance include agricultural production, marketing, natural resources, home economics, food and nutrition, 4-H youth development, and community and rural development.

USDA—Farm Service Agency (FSA)
www.fsa.usda.gov

The Farm Service Agency administers commodity production adjustment and support programs; conservation cost sharing with farmers and ranchers; the conservation reserve program; natural disaster assistance to agricultural producers through payments and cost sharing; and certain national emergency preparedness activities. The agency also provides services for the Commodity Credit Corporation, and for state and county committees.

USDA—Food Safety and Inspection Service
www.fsis.usda.gov

FSIS administers the federal meat and poultry inspection program to assure safety, wholesomeness, and truthful labeling of meat and poultry products. It also conducts food-safety consumer educational programs.

USDA—Foreign Agricultural Service
www.fas.usda.gov

FAS serves as a basic source of information to U.S. agriculture on world crops, policies, and markets; administers agricultural import regulations; assists in the export of U.S. farm products; represents U.S. agriculture in foreign trade matters; and administers USDA's responsibility for P.L. 480, CCC export credit programs, and the reporting of export sales.

USDA—Forest Service
www.fs.fed.us

The Forest Service has the federal responsibility for national leadership in forestry. Some of its goals are generation of forestry opportunity to accelerate rural community growth and encouragement of optimum forest land-ownership patterns.

USDA—Natural Resources Conservation Service (NRCS)
www.nrcs.usda.gov

The NRCS carries out a national soil and water conservation program with cooperation of landowners and operators in local soil and water conservation districts, and with other governmental agencies; administers USDA's Great Plains conservation program, and watershed protection and flood prevention program; leads national cooperative soil survey; and provides USDA leadership in assisting landowners and local groups in resource conservation and development projects.

USDA—Packers and Stockyards Administration (P&SA)
www.gipsa.usda.gov

P&SA enforces the Packers and Stockyards Act, an antitrust, fair-trade practice and payment protection law designed to ensure free and open competition, and to prevent unfair and deceptive practices in the marketing of livestock, meat, and poultry.

USDA—Poison Plant Research Laboratory
1150 E. 14th N., Logan, UT 84321
www.ars.usda.gov/pwa/ppr

This laboratory conducts research on major problems associated with poisonous plants.

USDA—U.S. Meat Animal Research Center (MARC)
P.O. BOX 166
Clay Center, NE 68933

www.ars.usda.gov/main/site_main.htm?modecode=30-40-05-00MARC develops, through research, new technology for meat animal production. Its goals are to improve carcass merit and reduce production costs of cattle, sheep, and hogs. The center is being developed by the federal government and is administered by the Agricultural Research Service of the USDA.

U.S. Environmental Protection Agency (EPA)
1200 Pennsylvania Avenue, NW
Washington, DC 20460
www.epa.gov

The EPA was created to permit coordinated and effective governmental action on behalf of the environment. It endeavors to abate and control pollution systematically by proper integration of a variety of research, monitoring, standard setting, and enforcement activities. The EPA is designed to serve as the public's advocate for a livable environment.

U.S. Meat Export Federation (USMEF)
1855 Blake Street, Suite 200
Denver, CO 80202
www.usmef.org

See Chapter 1.

Glossary

abomasum Fourth stomach compartment of cattle that corresponds to the true stomach of monogastric animals.

abortion Delivery of fetus between time of conception and normal parturition.

abscess Localized collection of pus in a cavity formed by disintegration of tissues.

accrual method of accounting Accounting method whereby revenue and expenses are recorded when they are earned or incurred regardless of when the cash is received or paid.

accuracy or ACC (of selection) Confidence that can be placed in the EPD (expected progeny difference); for example, high (0.70 and above), medium (0.40–0.69), and low (below 0.40).

acetonemia *See* ketosis.

acidosis A high-acid condition in the rumen (pH 5.3–5.7) caused by rapid consumption or overconsumption of readily fermentable feed; may cause digestive disturbance and/or death.

ADG *See* average daily gain.

adjusted weaning weight Weaning weights of calves are adjusted to a standard age (205 days) and age of dam (5–9 years of age).

adjusted yearling weight Yearling weights of calves are adjusted to a standard age (365 days) by adding (160 times average daily postweaning gain) to the adjusted 205-day weight.

ad lib *See ad libitum*.

***ad libitum* (ad lib)** Free choice; allowing cattle to eat all they want.

afterbirth Fetal membranes that are expelled after parturition. *See also* placenta.

AHIR *See* Angus herd improvement records.

AI *See* artificial insemination.

AI certificates Certificates issued by some breed associations that must be submitted before AI calves can be registered.

alliance An organization in the beef industry (horizontal or vertical) designed to improve profitability by improving coordination of beef production, processing, and merchandizing.

American Meat Institute (AMI) Association of meat-packing and processing companies.

American National Cattlewomen (ANCW) Organization of women involved in the promotion of beef through education and consumer relations programs.

AMI *See* American Meat Institute.

amnion Fluid-filled membrane located next to the fetus.

ANCW *See* American National Cattlewomen.

anestrous Period of time when the female is not in estrus; the nonbreeding season.

Angus Herd Improvement Records (AHIR) Performance records program administered by the American Angus Association.

animal unit (AU) A generalized unit for describing stocking density, stocking rate, and carry capacity. Usually accepted to be a 1,000-lb cow with calf or 1.4 yearling cattle.

annual cow cost Cost (dollars) to keep a cow for a year.

animal unit month (AUM) Amount of feed or forage required to maintain one animal unit (e.g., a 1,000-lb cow and calf) for one month.

ante mortem Before death.

anthelmintic Drug or chemical agent used to kill or remove internal parasites.

antibiotic Product produced by living organisms such as yeast that destroys or inhibits the growth of other organisms, especially bacteria.

antibody Specific protein molecule that is produced in response to a foreign protein (antigen) that has been introduced into the body.

antigen Foreign substance that, when introduced into the blood or tissues, causes the formation of antibodies. Antigens may be toxins or native proteins.

appreciation Increase in the value of a capital asset (e.g., land) due to external influences such as inflation.

arteriosclerosis Disease resulting in the thickening and hardening of the artery walls.

artificial insemination (AI) Placing semen into the female reproductive tract (usually the cervix or uterus) by means other than natural service.

artificial vagina Device used to collect semen from a male while he mounts in a normal manner to copulate. The bull ejaculates into this device, which simulates the vagina of the female in pressure, temperature, and sensation to the penis.

as fed Feeding of feeds that contain their normal amount of moisture.

assets Items of value owned by a beef business or producer.

atherosclerosis Form of arteriosclerosis involving fatty deposits in the inner walls of the arteries. *See also arteriosclerosis*.

atrophy Shrinking or wasting away of tissue.

auction Market for cattle through which an auctioneer sells cattle to the highest bidder.

AUM *See* animal unit month.

autopsy Postmortem examination in which the body is dissected to determine cause of death.

average daily gain Pounds of live weight gained per day.

backcross Mating of a two-breed crossbred offspring back to one of its parental breeds.

backfat Amount of fat over the animal's back, usually measured at the 12th to 13th rib.

backgrounding Growing program (grazing or fed harvested feed) for feeder cattle from time calves are weaned until they are on a finishing ration in the feedlot.

balance sheet Financial statement that summarizes assets, liabilities, and net worth at a specific point in time. Also called a *net worth statement.*

balling gun Instrument inserted into the animal's throat to discharge pills.

Bang's disease *See* brucellosis.

barren Incapable of producing offspring.

basis Difference between the cash market price and the futures market price.

BCS *See* body condition score.

BCTRC Boneless, closely trimmed retail cuts from round, loin, rib, and chuck.

beef Meat from cattle (bovine species) other than calves. Meat from calves is called *veal.*

beef belt Area of the United States where commercial beef production, slaughtering, and processing are concentrated.

Beef Breeds Council National organization of beef breed associations.

beef checkoff program Beef Promotion and Research Act established in October 1986. Each time cattle are marketed, $1 per head is paid by the seller to the Beef Industry Council (BIC). Money is used in promotion, research, and education.

Beef Improvement Federation (BIF) A federation of organizations, businesses, and individuals interested or involved in performance evaluation of beef cattle.

Beef Promotion and Research Program *See* beef checkoff program.

Beef Quality Assurance (BQA) Program(s) designed to help beef producers assure that their production methods are not causing defects in beef products.

BIF *See* Beef Improvement Federation.

bioeconomic trait Any biological trait of economic importance.

biological efficiency Ratio of physical input to physical output (e.g., pounds of feed per 100 lb of gain).

biological type Usually refers to size of cattle (large, medium, or small), growth rate, milk production (high, medium, or low), and age at puberty.

biotechnology Use of microorganisms, plant cells, and animal cells or parts of cells (such as enzymes) to produce industrially important products or processes.

birth weight (BW or B. Wt.) Weight of the calf taken within 24 hours after birth.

birth weight EPD The expected average increase or decrease in birth weight (lb) of a bull's calves when compared to other bulls in the same sire summary. A plus figure indicates an increase in birth weight, while a negative value is a decrease. The value is a measure of calving ease. *See also* expected progeny difference (EPD).

birth weight ratio Compares the individual birth weight of a calf to the herd average. Usually calculated within sex.

bloat Abnormal condition characterized by a distention of the rumen, usually seen on the left side, due to an accumulation of gases.

"bloom" Haircoat usually has a luster (shine) that gives the appearance of a healthy animal.

BLUP Best linear unbiased prediction method for estimating the breeding values of breeding animals.

body condition score (BCS) A visual score (usually 1 = thin; 9 = very fat) for body fatness that is related to postpartum interval in beef females.

bolus (1) Regurgitated food. (2) Large pill for treating cattle.

Bos indicus Zebu (humped) cattle, including the Brahman breed in the United States.

Bos taurus Includes most cattle found in the United States and their European ancestors.

bovine Refers to a general family grouping of cattle.

bovine spongiform encephalopathy A degenerative disease that affects the central nervous system of cattle.

bovine viral diarrhea (BVD) Viral disease in cattle that can cause diarrhea, lesions of the digestive tract, repeat breeding, abortion, mummification, and congenital defects.

boxed beef Cuts of beef put in boxes for shipping from packing plant to retailers. These primal (rounds, loins, ribs, and chucks) and subprimal cuts are intermediate cuts between the carcass and retail cuts.

BQA *See* Beef Quality Assurance.

brand (1) Permanent identification of cattle, usually made on the hide with hot-iron or freeze branding. (2) Process of branding.

branded beef product A specifically labeled product that is differentiated from commodity items by its brand name. Certified Angus Beef, Laura's Lean, and Cattlemen's Collection are examples.

breakeven price Volume of output required for revenue to equal the total of fixed and variable expenses.

breaking Cutting carcasses into primal and subprimal cuts.

bred Female has been mated to a bull and is usually assumed to be pregnant.

breech Buttocks; a breech presentation at birth occurs when the rear portion of the fetus is presented first.

breed Cattle of common origin and having characteristics that distinguish them from other groups within the same species.

breed complementarity Combining breeds to take advantage of breed superiority for specific traits.

breeder In most beef breed associations, the owner of the dam of a calf at the time she was mated or bred to produce that calf.

breeding soundness examination (BSE) Evaluation of the reproductive potential of the bull giving the reproductive trait; a physical examination. Measuring involves scrotal circumference and evaluating a semen sample for motility and morphology.

breeding value Value of an animal as a parent. The working definition is twice the difference in performance between a very large number of progeny and the population average when individuals are mated at random within the population and all progeny are managed alike. The difference is doubled because only a sample half (one gene of each pair) is transmitted from a parent to each progeny.

brisket disease Noninfectious disease of cattle characterized by congestive right heart failure. It affects animals residing at high altitudes (usually above 7,000 ft). Sometimes referred to as *high mountain disease* or *high altitude disease*.

British breeds Breeds of cattle, such as Angus, Hereford, and Shorthorn, originating in Great Britain.

brockle-faced White-faced with other colors splotched on the face and head.

broken mouth Some teeth are missing or broken.

broker Individual or firm that buys and sells options, futures contracts, and stocks and bonds for a commission fee.

browse Woody or brushy plants (e.g., sagebrush, shadscale, and other shrubs and bushes). Cattle feed on the tender shoots or twigs.

brucellosis Contagious bacterial disease that results in abortions; also called *Bang's disease*.

BSE *See* breeding soundness examination or bovine spongiform encephalopathy.

budget Financial form used to examine alternative plans for a beef operation and to estimate the profitability of each alternative.

bull Bovine male. The term usually denotes animals of breeding age.

buller-steer syndrome Behavior problem in which a steer is sexually attracted to other steers in the pen. The steer is ridden by the other steers, resulting in poor performance and injury.

bulling Term describing a cow or heifer in estrus.

bullock Young bull, typically less than 20 months of age.

butt-branded Hides from cattle that are hot-iron branded on the hip.

buttons May refer to cartilage on dorsal processes of the thoracic vertebrae. *See also* cotyledon.

BVD *See* bovine viral diarrhea.

bypass protein Feed protein that escapes microbial degradation in the rumen and is digested in the small intestine.

by-product Product of considerably less value than the major product. For example, the hide and offal are by-products, while beef is the major product.

C-section *See* caesarean section.

Caesarean section Delivery of fetus through an incision in the abdominal and uterine walls.

calf Male or female bovine animal under 1 year of age.

calf crop *See* percent calf crop.

calorie Amount of heat required to raise the temperature of 1 g of water from 15°C to 16°C.

calve Giving birth to a calf. Same as parturition.

calving difficulty (dystocia) Abnormal or difficult labor causing difficulty in delivering the calf.

calving interval Time (days or months) between the birth of a calf and the birth of a subsequent calf, both from the same cow.

calving season Season(s) of the year when calves are born.

cancer eye Cancerous growth on the eyeball or eyelid.

carcass evaluation Techniques for measuring components of meat quality and quantity in carcasses.

carcass merit Value of a carcass for consumption.

carotene Orange pigment found in leafy plants (e.g., alfalfa), yellow corn, and other feeds that can be broken down to form two molecules of vitamin A.

carrier Heterozygous individual having one recessive gene and one dominant gene for a given pair of genes (alleles).

carrying capacity The maximum stocking rate that will achieve a target level of animal performance on a particular grazing unit under a specified grazing method. Or, the potential number of animals or live weight that may be supported on a unit area for a grazing season based on forage potential.

case-ready Beef cuts received by the retailer that do not require further processing before they are put in the retail case for selling.

cash flow Cash receipts and cash expenses.

cash-flow budget Detailed estimate of the projected cash receipts and expenses over a future period of time used to evaluate the financial feasibility of a plan.

cash-flow statement Financial statement summarizing all cash receipts and disbursements over a period of time (usually monthly for a year).

cash market price Price that results when cattle go to market.

cash method of accounting An accounting method by which revenue and expenses are recorded when the cash is received or paid.

castrate (1) To remove the testicles. (2) An animal that has had its testicles removed.

cattalo Cross between domestic cattle and bison.

Cattle-Fax Nonprofit marketing organization governed by cattle producers. Market analysis and information is provided to members by a staff of market analysts.

Cattlemen's Beef Board (CBB) Responsible for the management of the beef checkoff program, oversees the collection of $1 per head on domestic cattle as well as the equivalent on imported beef, beef products, and cattle.

CBB *See* Cattlemen's Beef Board.

central test Location where animals are assembled from several herds to evaluate differences in certain performance traits under uniform management conditions. Usually involves breeding bulls, though some slaughter steer and heifer tests exist.

Certified Angus Beef Branded beef product supplied by Angus or Angus crossbred cattle that meets certain carcass specifications.

cervix Portion of the female reproductive tract between the vagina and the uterus. It is usually sealed by thick mucus except when the female is in estrus or delivering young.

checkoff *See* beef checkoff program.

chorion Outermost layer of fetal membranes.

chromosome Rod-like or string-like body found in the nucleus of the cell that is darkly stained by chrome dyes. The chromosome contains the genes.

chronic Regular appearance of a symptom or situation.

chuck Wholesale cut (shoulder) of the beef carcass.

class Group of cattle determined primarily by sex and age (e.g., market class or show-ring class).

clitoris A highly sensitive organ in females located inside the ventral part of the vulva. It is homologous to the penis in the male.

clone Genetically identical organisms produced by nucleus substitution or embryo division.

closed herd Herd in which no outside breeding stock (cattle) are introduced.

cod Scrotal area of steer remaining after castration.

cold shortening Sarcomeres as part of the muscle fiber, shorten too rapidly during the chilling of the carcass, thus decreasing meat tenderness.

collagen Primary protein in connective tissue. Collagen envelops individual muscle fibers and attaches muscles to bones.

collateral relatives Relatives of an individual that are not its ancestors or descendants. Brothers and sisters are examples of collateral relatives.

colon Large intestine from the end of the ileum and beginning with the cecum to the anus.

Colorado branded Hides from cattle hot-iron branded on the ribs.

colostrum First milk given by a female following delivery of her calf. It is high in antibodies that protects the calf from invading microorganisms.

compensatory gain Faster-than-normal rate of gain following a period of restricted gain.

compensatory growth *See* compensatory gain.

complementarity Using breed differences to achieve a more optimum additive and nonadditive breed composition for production and carcass traits of economic value.

composite breed Breed that has been formed by crossing two or more breeds.

composition Usually refers to the carcass composition of fat, lean, and bone.

Compudose® Growth implant containing estradiol and progesterone.

computer Electronic machine that by means of stored instructions and information performs rapid, often complex, calculations or compiles, correlates, and selects data.

concentrate Feed that is high in energy, low in fiber content, and highly digestible.

conception Fertilization of the ovum (egg).

conditioning Treatment of cattle by vaccination and other means prior to putting them in the feedlot.

conformation Physical form of an animal; its shape and arrangement of parts.

congenital Acquired during prenatal life; condition exists at birth. Often used in the context of congenital (birth) defects.

contemporaries Group of animals of the same sex and breed (or similar breeding) that have been raised under similar environmental conditions (same management group).

continental breed *See* European breed.

continuous grazing A method of grazing where animals have unrestricted access to an entire grazing unit throughout a large portion or all of a grazing season.

controlled grazing Grazing management designed to improve utilization of forage either by (1) allocating pasture in subunits with grazing periods typically less than 5 days, or (2) varying stocking rate to match forage growth rate and availability (put-and-take stocking).

cooler A room in packing plant where carcasses are chilled after slaughter and prior to processing.

corpus luteum Yellowish body in the ovary. The cells that were follicular cells develop into the corpus luteum, which secretes progesterone. It becomes yellow in color from the yellow lipids that are in the cells.

correlation coefficient Measure of how two traits vary together. A correlation of +1.00 means that two traits will move in the same direction (either increase or decrease). A correlation of –1.00 means that as one trait increases the other decreases—a perfect negative or inverse relationship. A correlation of 0.00 means that as one trait increases, the other may increase or decrease—no consistent relationship. Correlation coefficients may vary between +1.00 and –1.00.

cost of gain Total of all costs divided by the total pounds gained; usually expressed on a per-pound basis.

cotyledon Area where the placenta and the uterine lining are in close association such that nutrients can pass to and wastes can pass from the circulation of the developing young. Sometimes referred to as *button*.

cow Sexually mature female bovine animal that has usually produced a calf.

cow-calf operation Management unit that maintains a breeding herd and produces weaned calves typically for sale to either a stocker enterprise or a feedyard.

cow hocked Condition in which the hocks are close together but the feet stand apart.

creep Enclosure where calves can enter to obtain feed but cows cannot enter. This process is called creep feeding.

creep feeding *See* creep.

creep grazing The practice of allowing calves to graze areas that cows, with a lower nutritional requirement, cannot access.

crest Bulging, top part of the neck on a bull.

crossbred Animal produced by crossing two or more breeds.

crossbreeding Mating animals from different breeds; utilized to take advantage of hybrid vigor (heterosis) and breed complementarity.

cryptorchidism Retention of one or both testicles in the abdominal cavity.

cud Bolus of feed that cattle regurgitate for further chewing.

cull To eliminate one or more animals from the breeding herd or flock.

currentness Marketing term indicating how feedlots market fed cattle. If current, then feedlots market cattle on schedule. If feedlots are not current, then a backlog of cattle usually results—these cattle typically have higher slaughter weights, poorer yield grades, and usually lower prices.

custom cattle feeding Cattle feeders who provide facilities, labor, feed, and care as a service, but they do not own the cattle.

cutability Fat, lean, and bone composition of the beef carcass. Used interchangeably with yield grade. *See also* yield grades.

cutting chute Narrow chute where cattle go through in single file, with gates such that selected animals can be diverted into pens alongside the chute; also referred to as a *sorting chute*.

cwt Abbreviation for hundredweight (100 lb).

cycling Infers that nonpregnant females are having estrous cycle.

dam Female parent.

dark cutter Color of the lean (muscle) in the carcass has a dark appearance, usually caused by stress (excitement) to the animal prior to slaughter.

deflation General decrease in prices that increases the purchasing power of a dollar.

dehorn To remove the horns of an animal.

deoxyribonucleic acid (DNA) Molecule that comprises the genetic material of animals. Genes are units of DNA. *See also* gene.

depreciation Decrease in the value of an asset due to age, use, and obsolescence; the prorated expense of owning an asset.

dewclaws Hard, horny structures above the hoof on the rear surface of the legs of cattle.

dewlap Loose skin under the chin and neck of cattle.

digestibility Quality of being digestible. If a high percentage of a given food taken into the digestive tract is absorbed into the body, that food is said to have high digestibility.

direct selling Selling cattle from one ranch to another, from ranch to feedlot, or from feedlot to packer.

disease Any deviation from the normal state of health.

DM *See* dry matter.

DNA *See* deoxyribonucleic acid.

DNA markers Areas of the genome at which differences in the DNA sequence can be visually detected. A marker locus by itself may not have a direct effect on a phenotypic trait, but it may be located close to a gene that directly affects a trait. Markers can serve as location reference points for gene mapping and marker-assisted selection.

DNA probe A method to determine an animal's genotype for a particular gene or marker.

dominance One allele masks the effect of another (recessive) allele.

double-entry accounting System of bookkeeping in which every transaction is recorded as a debit in one or more accounts and as a credit in one or more accounts such that the total of the debit entries equals the total of the credit entries.

double muscling A simple recessive trait evidenced by an enlargement of the muscles with large grooves between the muscle systems, especially noticeable in the hind leg.

drench To give fluid by mouth.

dressed beef Carcasses from cattle.

dressing percentage Percentage of the live animal weight that becomes the carcass weight at slaughter. It is determined by dividing the carcass weight by the live weight then multiplying by 100. Also referred to as *yield*.

drop Body parts removed at slaughter, primarily the hide, head, shanks, and offal.

drop credit Value of the drop.

dropped Being born (e.g., "the calf is dropped").

dry (cow) Refers to a nonlactating female.

dry matter Feed after water (moisture) has been removed (100% dry).

dystocia Difficult birth; *see* calving difficulty.

ear mark Method of permanent identification by which slits or notches are placed in the ear.

ear tag Method of identification by which a numbered, lettered, and/or colored tag is placed in the ear.

early maturity Early puberty as the animal begins to fatten early, sometimes before desired slaughter weight is obtained.

EBV *See* breeding value; expected progeny difference (EPD).

economic efficiency Ratio of output value to cost of input.

economic value The net return within a herd for making a pound or percentage change of the trait in question.

edema Abnormal fluid accumulation in the intercellular tissue spaces of the body.

efficiency Ratio of output to input. *See also* biological efficiency; economic efficiency.

80%–20% rule Basic rule of management. Too often managers expend 80% of their effort on "the trivial many" problems that produce only 20% of the results. Effective managers recognize that spending time (20%) on problems or situations that count most will produce 80% of the desired results.

ejaculation Discharge of semen from the male.

emaciation Thinness; loss of flesh such that bony structures (hips, ribs, and vertebrae) become prominent.

embryo Fertilized egg in its early stages of development; after body parts can be distinguished, it is known as a *fetus*.

embryo splitting Dividing an embryo into two or more similar parts to produce several calves from a single embryo.

embryo transfer (ET) Transfer of fertilized egg(s) from a donor female to one or more recipient females.

Endangered Species Act (ESA) A regulatory statute intended to protect threatened and endangered species by preserving the ecosystems on which they depend.

endocrine gland Ductless gland that secretes a hormone into the bloodstream.

energy Force, or power, that is used to drive a wide variety of systems. It can be used as power of mobility in animals, but most of it is used as chemical energy to drive reactions necessary to convert feed into animal products and to keep the animals warm and functioning.

enterprise Segment of the cattle business or an associated business that is isolated by accounting procedures so that its revenue and expenses can be identified.

enterprise budget Detailed list of all estimated revenue and expenses associated with a specific enterprise.

environment Total of all external (nongenetic) conditions that affect the well-being and performance of an animal.

Environmental Protection Agency (EPA) Independent agency of the federal government established to protect the nation's environment from pollution.

enzyme Complex protein produced by living cells that causes changes in other substances in cells without being changed itself and without becoming a part of the product.

EPA *See* Environmental Protection Agency.

EPD *See* expected progeny difference.

epididymis Long, coiled tubule leading from the testis to the vas deferens.

epididymitis Inflammation of the epididymis.

epistasis Situation in which a gene or gene pair masks (or controls) the expression of another nonallelic pair of genes.

equity *See* net worth (equity).

eruction (or eructation) Elimination of gas by belching.

esophageal groove Groove in the reticulum between the esophagus and omasum. Directs milk in the nursing calf directly from the esophagus to the omasum.

estrogen Any hormone (including estradiol, estriol, and estrone) that causes the female to express heat and to be receptive to the male. Estrogens are produced by the follicle of the ovary and by the placenta and have additional body functions.

estrous Adjective meaning *heat* that modifies such words as *cycle*. The estrous cycle is the heat cycle, or the time from one heat to the next.

estrous synchronization Controlling the estrous cycle so that a high percentage of the females in the herd express estrus at approximately the same time.

estrus Period of mating activity in the heifer or cow. Same as *heat*.

ET *See* embryo transfer.

ethology Study of animal behavior.

EU *See* European Union.

European breed Breed originating in European countries other than England (these are called *British breeds*); a larger dual-purpose breed such as Charolais, Simmental, and Limousin; also called *continental* or *exotic breed* in the United States.

European Union Group of 28 countries whose major objective is to coordinate the development of economic activities. Previously called The European Economic community, European community, and Common Market.

eviscerate Remove the internal organs during the slaughtering process.

Excel One of the three largest U.S. beef-packing companies.

exotic breed *See* European breed.

expected progeny difference (EPD) One-half of the breeding value of a sire or dam; the difference in expected performance of future progeny of a sire when compared with that expected from future progeny of bulls in the same sire summary.

F1 Offspring resulting from the mating of a purebred (straightbred) bull to purebred (straightbred) females of another breed.

fabrication Breaking the carcass into primal, subprimal, or retail cuts. These cuts may be boned and trimmed of excess fat.

fat thickness Usually refers to the amount of fat that covers muscles; typically measured at the 12th and 13th rib as inches of fat over the *longissimus dorsi* muscle (rib eye).

FDA *See* Food and Drug Administration (FDA).

feces Bodily wastes; excretion product from the intestinal tract.

fed cattle Steers and heifers that have been fed concentrates, usually for 90–120 days in a feedlot or until they reach a desired slaughter weight.

feed additive Ingredient such as an antibiotic or hormone-like substance that is added to a diet to perform a specific role.

feed bunk Trough or container used to feed cattle.

feed conversion *See* feed efficiency.

feed efficiency (1) Amount of feed required to produce a unit of weight gain or milk. (2) Amount of gain made per unit of feed.

feed markup Per-ton feed cost charged to the customer by the feedyard for the cattle-feeding services it provides.

feeder (1) Cattle that need further feeding prior to slaughter. (2) Producer who feeds cattle.

feeder grades Grouping of feeder cattle to predict the slaughter weight endpoint to a desirable fat-to-lean composition. Frame size and thickness are the two criteria used to determine feeder grade.

feedlot Enterprise in which cattle are fed grain and other concentrates usually for 90–120 days. Feedlots range in size from less than 100-head capacity to many thousands.

feedyard Cattle-feeding facility.

femininity Well-developed secondary female sex characteristics, udder development and refinement in head and neck.

fertility Capacity to initiate, sustain, and support reproduction.

fertilization Process by which a sperm unites with an egg to produce a zygote.

fetus Late stage of individual development within the uterus. Generally, the new individual is regarded as an embryo during the first half of pregnancy and as a fetus during the last half.

fill Contents of the digestive tract.

financing Acquiring control of assets by borrowing money.

finish (1) Degree of fatness of an animal. (2) Completion of the last feeding phase of slaughter cattle.

finished cattle Fed cattle whose time in the feedlot is completed and are now ready for slaughter.

finishing ration Feedlot ration, usually high in energy, that is fed during the latter part of the feeding period.

fitting Proper feeding, grooming, and handling of an animal, usually to prepare it for the show ring.

fixed cost Costs incurred whether or not production occurs (e.g., interest, taxes).

flehmen Pattern of behavior expressed by the bull during sexual activity. The upper lip curls up and the bull initiates the smelling process in the vicinity of the vulva or urine.

flushing Placing females on a high level of nutrition before breeding to decrease postpartum interval and possibly stimulate an increased conception rate.

FMD *See* foot and mouth disease.

FMI *See* Food Marketing Institute.

FOB (or fob) Free on board; buyer pays freight after loading.

follicle Blister-like, fluid-filled structure in the ovary that contains the egg.

follicle-stimulating hormone (FSH) Hormone produced and released by the anterior pituitary that stimulates the development of the follicle in the ovary.

Food and Drug Administration (FDA) U.S. government agency responsible for protecting the public against impure and unsafe foods, drugs, veterinary products, biologics, and other products.

Food Marketing Institute (FMI) National association of food retailers and wholesalers located in Washington, DC, that conducts programs of research, education, and public affairs for its members.

foot and mouth disease (FMD) Highly contagious disease affecting many species of livestock including cattle. This disease is of particular concern in that it can lead to loss of export markets.

foot rot Disease of the foot in cattle.

forage Grazed or harvested herbaceous plants that are utilized by cattle.

forage production The total amount of dry matter produced per unit of area on an annual basis (e.g., lb/acre/year).

forb Weedy or broad-leaf plants (unlike grasses) that serve as pasture for animals (e.g., clover, alfalfa).

forward contracting Future delivery of a specified type and amount of product at a specified price.

founder Nutritional ailment resulting from overeating; lameness in front feet with excessive hoof growth usually occurs.

frame score Score based on visual evaluation of skeletal size or by measuring hip height (from ground to top of hips). This score is related to the slaughter weights at which cattle grade Choice or have comparable amounts of fat cover over the loin eye at the 12th to 13th rib.

frame size Usually measured by frame score or estimated visually.

freemartin Female born twin to a bull (approximately 90% of such heifers will never conceive).

FSH *See* follicle-stimulating hormone.

full sibs Animals having the same sire and dam.

futures market Electronic market through which buyers and sellers trade contracts on commodities or raw materials. Futures contracts are available for a variety of delivery months. However, delivery of actual products seldom occurs. Futures markets are used as a risk management tool or as a speculative venture.

GATT (General Agreement on Tariffs and Trade) An agreement originally negotiated in Geneva, Switzerland, in 1947 among 23 countries, including the United States, to increase international trade by reducing tariffs and other trade barriers. The agreement provides a code of conduct for international commerce and a framework for periodic multilateral negotiations on trade liberalization and expansion.

gene Segment of DNA in the chromosome that codes for a trait and determines how a trait will develop.

gene map A blueprint of the chromosomes of a species, indicating the relative order of location of genes and DNA markers.

generation interval Average age of the parents when the offspring are born.

generation turnover Length of time from one generation of animals to the next generation.

genetic correlation Correlation between two traits that arises because some of the same genes affect both traits. *See* correlation coefficient.

genetic engineering Changing the characteristics of an animal by altering or rearranging its DNA. It is an all-embracing term for several techniques: (1) manipulation at a cellular level (cloning); (2) manipulation of the DNA itself (gene manipulation); and (3) changing the DNA sequence through the selection and mating of cattle.

genome Total number of genes in a species.

genotype Genetic constitution or makeup of an individual. For any pair of alleles, three genotypes (e.g., AA, Aa, and aa) are possible.

genotype–environmental interaction Variation in the relative performance of different genotypes from one environment to another. For example, the superior cattle (genotypes) for one environment may not be superior for another environment.

gestation Time from conception until the female gives birth, an average of 285 days in cattle.

goal Target or desired condition that motivates the decision maker.

gonad Testis of the male; ovary of the female.

gonadotropin Hormone that stimulates the gonads.

grade augmentation Supplementation of traditional USDA visual carcass grading using objective instrumentation.

grade and yield Marketing transaction whereby payment is made on the basis of carcass weight and quality grade.

grading up Continued use of purebred sires of the same breed in a grade herd.

grass tetany Disease of cattle marked by staggering, convulsions, coma, and frequently death that is caused by a mineral imbalance (magnesium) while grazing lush pasture.

grazier A person who manages grazing livestock.

grazing cell A parcel of land subdivided into paddocks and grazed rotationally.

grazing cycle The length or passage of time between two grazing periods in a particular paddock of a grazing unit. One grazing cycle includes one grazing period and one rest period.

gross margin Difference between the revenue and variable production cost for one unit (one acre or one animal) of an enterprise.

growing ration Usually a high-roughage ration whereby gains of 0.25–2 lb per day are anticipated.

growth Increase in protein over its loss in the animal body. Growth occurs by increases in cell numbers, cell size, or both.

grubs Larvae of the heel fly found on the backs of cattle under the hide.

half-sib Animals having one common parent.

hand mating Bringing a female to a male for breeding, after which she is removed from the area where the male is located (same as hand breeding).

hanging tenderloin Part of the diaphragm muscle, not to be confused with the tenderloin of the carcass.

"hard keeper" Term used when an animal does not perform well; it may have hardware, parasites, or shows the effects of disease.

hardware disease Ingested sharp objects perforate the reticulum and cause infection of the heart sac, lungs, or abdominal cavity.

Hazard Analysis Critical Control Point (HACCP) A process used to identify those steps in production where mistakes may critically damage the final performance of the product and to establish a system of monitoring and intervention to avoid these mistakes.

heart girth Circumference of the animal's body, measured just behind the shoulders.

heat *See* estrus.

heat increment Increase in heat production following consumption of feed when an animal is in a thermoneutral environment; includes additional heat generated in fermentation, digestion, and nutrient metabolism.

hedge Risk management strategy that allows a producer to lock-in a price for a given commodity at a specified time.

heifer Young female bovine cow prior to the time that she has produced her first calf.

heiferette Heifer that has calved once and is then fed for slaughter; the calf has usually died or been weaned at an early age.

herd Group of cattle (usually cows) that are in a similar management program.

heredity The transmission of genetic or physical traits of parents to their offspring.

heritability Portion of the phenotypic differences between animals that is due to heredity.

hernia Protrusion of an intestine through an opening in the body wall (also commonly called *rupture*). Two types of hernias—umbilical and scrotal—occur in cattle.

heterosis Performance of offspring that is greater than the average of the parents. Usually referred to as the amount of superiority of the crossbred over the average of the parental breeds. Also called *hybrid vigor*.

heterozygous Designates an individual possessing unlike genes for a particular trait.

hides Skins from cattle.

high mountain disease *See* brisket disease.

hip lock Condition at calving in which the hips of the calf cannot pass through the pelvis of the cow.

homozygous Designates an individual whose genes for a particular trait are alike.

hormone Chemical substance secreted by a ductless gland. Usually carried by the bloodstream to other places in the body, where it has its specific effect on another organ.

hot carcass weight Weight of carcass just prior to chilling.

"hot fat trimming" Removal of excess surface fat while the carcass is still "hot," prior to chilling the carcass.

HRI (hotel, restaurant, and institutional) Used in the context that some beef is supplied to the HRI trade.

hybrid vigor *See* heterosis.

hydrocephalus Condition characterized by an abnormal increase in the amount of cerebral fluid, accompanied by dilation of the cerebral ventricles.

hypothalamus Portion of the brain found in the floor of the third ventricle that regulates reproduction, hunger, and body temperature and performs other functions.

hypothermia Potentially dangerous drop in body temperature; this is of particular concern for newborn and neonatal calves.

immunity Ability of an animal to resist or overcome infection.

implant To graft or insert material to intact tissues.

inbreeding Mating of individuals more closely related than the average individuals in a population. Inbreeding increases homozygosity in the cattle population but does not change gene frequency.

income Difference between revenue and expenses that is referred to as *net income*; *gross income* refers to total income.

income statement Financial statement that summarizes all revenues and expenses and is used to determine the net income or net loss for a given period of time, usually a year.

independent culling level Selection method whereby minimum acceptable phenotypic levels are assigned to several traits.

index Overall merit rating of an animal.

inflation General increase in prices that decreases the purchasing power of a dollar.

insemination Deposition of semen in the female reproductive tract.

intake The amount of feed consumed by an animal per day. Intake is usually expressed as a percent of bodyweight or in pounds per day.

integrated resource management (IRM) Multidisciplinary approach to managing cattle more efficiently and profitably; management decisions are based on how all resources are affected.

integration Bringing together of two or more segments of beef production and processing under one centrally organized unit.

intensive grazing management (IGM) Grazing management where a grazing unit is subdivided into subunits (paddocks) with grazing periods typically less than 5 days.

intensive rotational grazing Synonymous with *intensive grazing management.*

interest rate Charge or fee associated with borrowed money.

intermuscular fat Fat located between muscle systems. Also called *seam fat*.

intramuscular fat Fat within the muscle or marbling.

inter se mating Mating of animals within a defined population; literally "to mate among themselves."

intravenous Within the vein; an intravenous injection is made into a vein.

in vitro Outside the living body; in a test tube or artificial environment.

in vivo Within the living body.

involution Return of an organ to its normal size or condition after being enlarged (e.g., the uterus after parturition). A decline in size or activity of other tissues; the mammary gland tissues normally involute with advancing lactation.

ionophore Antibiotic that enhances feed efficiency by changing microbial fermentation in the rumen.

IRM *See* integrated resource management (IRM).

joint venture Any business arrangement whereby two or more parties contribute resources to and engage in a specific business undertaking.

kidney knob The kidney and the fat that surrounds it.

kidney, pelvic, and heart fat (KPH) The internal carcass fat associated with the kidney, pelvic cavity, and heart expressed as a percentage of chilled carcass weight. The kidney is included in the estimate of kidney fat. Used in the calculation of yield grade.

ketosis Condition characterized by a high concentration of ketone bodies in the body tissues and fluids. Also called *acetonemia*.

kosher meat Meat from ruminant animals (with split hooves) that have been slaughtered according to Jewish law.

labor (1) Parturition or the birth process. (2) Human resource that produces goods or provides services.

lactation Secretion and production of milk.

LEA *See* loin-eye area (LEA); rib-eye area (REA).

lethal gene A gene that causes the death of an individual at some stage of life.

legume Any plant type within the family *Leguminosae*, such as pea, bean, alfalfa, and clover.

leucocytes White blood cells.

LH *See* luteinizing hormone (LH).

liabilities Obligations or debts owed by a business or person to others.

libido Sex drive or the male's desire to mate.

lice Small, flat, wingless insects with sucking mouth parts that are parasitic on the skin of animals.

limited partnership Partnership consisting of at least one general partner, who is responsible for the management and liabilities of the business, and at least one limited partner, whose liability is limited to his or her investment.

linear programming Mathematical technique used to find profit-maximizing combinations of production activities or cost-minimizing combinations of ingredients subject to a number of linear relationships that constrain the activities or ingredients.

linebreeding Form of inbreeding whereby a bull's genes are concentrated in a herd. The average relationship of the individuals in the herd to this ancestor (outstanding individual or individuals) is increased by linebreeding.

linecrossing Crossing of inbred lines.

liquidate To convert to cash; to sell.

liver flukes Parasitic flatworm found in the liver.

load Pounds (number) of cattle that can be hauled on a large cattle truck. For example, pot load is 42,000–52,000 lb (40–42 head of slaughter steers, 72 yearlings, or 100 calves).

locus Place on a chromosome where a gene is located.

loin-eye area (LEA) Area of the *longissimus dorsi* muscle, measured in square inches between the 12th and 13th ribs. Usually referred to as *rib-eye area (REA)*.

long yearling Animal between 18 months and 2 years of age.

longevity Life span of an animal; usually refers to the number of years a cow remains productive.

longissimus dorsi *See* rib-eye area.

lousy Infested with lice.

luteinizing hormone (LH) Protein hormone produced and released by the anterior pituitary that stimulates the formation and retention of the corpus luteum. It also initiates ovulation.

maintenance Condition in which the body is maintained without an increase or decrease in body weight and with no production or work being done.

mammary gland Gland that secretes milk.

management Act, art, or manner of managing, handling, controlling, or directing a resource or integrating several resources.

management systems Methods of systematically organizing information from several resources to make effective management decisions. *See also* integrated resource management (IRM).

marbling Flecks of intramuscular fat distributed in muscle tissue. Marbling is usually evaluated in the rib eye between the 12th and 13th ribs.

MARC *See* Meat Animal Research Center (MARC).

margin (1) "Earnest money" that serves as default protection in a futures transaction. (2) Difference between prices at different levels of the marketing system. (3) Difference between cost and sale price.

marker-assisted selection A method of genetic evaluation that takes into consideration the DNA marker genotype along with conventional selection procedures.

market class Cattle grouped according to their use, such as slaughter, feeder, or stocker.

market grade Cattle grouped within a market class according to their value.

market niche Segment of consumer demand targeted by a specialized production and marketing plant. Examples include the "white tablecloth" restaurant trade, health foods, and convenience foods.

masculinity Well-developed secondary sex characteristics in the neck, chest, and shoulders of the bull.

masticate To chew food.

mastitis Inflammation of the mammary gland.

maternal Pertaining to the female (cow or heifer).

maternal first-calf calving ease Ease with which a sire's daughters calve as first-calf heifers (under 33 months of age). Reported as a ratio or an EPD.

maternal heterosis Heterosis for those traits influenced by the cow genotype. For example, *maternal heterosis of weaning weight* refers to the increase in weaning weight from being raised on a crossbred cow rather than a straightbred cow.

maternal traits All the traits expressed by the cow. A limited definition implies milk and weaning weight production of the cow.

maternal weaning weight Weaning weight of a bull's daughter's calves. The EPD value predicts the difference in average 205-day weight of a bull's daughter's calves compared to daughters of all other bulls evaluated. It can be calculated by adding one-half of the bull's EPD for weaning weight to his milk EPD.

maturity An estimation of the chronological age of the animal or carcass.

maverick Unbranded animal, usually on the range.

M/B or M:B ratio *See* muscle-to-bone ratio.

mean (1) Statistical term for *average*. (2) Term used to describe cattle having bad behavior.

meat Tissues of the animal body that are used for food.

Meat Animal Research Center (MARC) Large U.S. government research center located in Clay Center, NE, that conducts numerous beef cattle research projects.

Meat Export Federation (MEF) *See* U.S. Meat Export Federation (MEF).

MEF *See* U.S. Meat Export Federation (MEF).

melengestrol acetate (MGA) Feed additive that suppresses estrus in heifers; used in estrus synchronization and feedlot heifers.

MERCOSUR (Common Market of the South) A customs union implemented in January 1995, and including Argentina, Brazil, Paraguay, and Uruguay. MERCOSUR represents the culmination of bilateral negotiations started by Argentina and Brazil in 1986.

metabolic body size Weight of the animal raised to the 3/4-power ($W^{0.75}$); a figure indicative of metabolic needs and of the feed required to maintain a certain body weight.

metabolism (1) Sum total of chemical changes in the body, including the "building up" and "breaking down" processes. (2) Transformation by which energy is made available for body uses.

metabolizable energy Gross energy in the feed minus the sum of energy in the feces, gaseous products of digestion, and energy in the urine. Energy that is made available for body uses.

metritis Inflammation (infection) of the uterus.

MGA *See* melengestrol acetate (MGA).

middle meats Rib and loin of a beef carcass. These primals generally yield the highest-priced beef cuts.

milk EPD Estimate of the milking ability of a bull's daughters compared to the average of the daughters of other bulls. Reported in pounds of weaning weight; positive values indicate above-average performance and negative numbers indicate below-average maternal ability. *See also* expected progeny difference (EPD).

mill feed Any feed that is subjected to the milling process.

minimum culling level Selection method in which an animal must meet minimum standards for each trait desired in order to qualify for being retained for breeding purposes.

mites Very small arachnids that can be parasites of cattle.

morbidity Measurement of illness; morbidity rate is the number of individuals in a group that become ill during a specified time period.

mortality rate Number of individuals that die from a disease during a specified time period, usually 1 year.

most probable producing ability (MPPA) Estimate of a cow's future productivity for a trait (such as progeny weaning weight ratio) based on her past productivity. For example, a cow's MPPA for weaning ratio is calculated from the cow's average progeny weight ratio, the number of her progeny weaning records, and the repeatability of weaning weight.

mouthed Examination of an animal's teeth.

MPPA *See* most probable producing ability (MPPA).

muley Term used to describe the polled (hornless) condition.

muscle-to-bone (M/B) ratio Pounds of muscle divided by pounds of bone. For example, 4:1 ratio means that there is 4 lb of muscle to 1 lb of bone (usually on a carcass basis).

muscling Amount of lean meat in a slaughter animal or carcass; estimated on the live animal by thickness of forearm muscle or stifle thickness. Ultimately, it is the ratio of muscle to bone or lean yield of the carcass after fat and bone are removed.

muzzle Nose of cattle.

myofibrils Primary component part of muscle fibers.

NAFTA (North American Free Trade Agreement) A trade agreement involving Canada, Mexico, and the United States, implemented on January 1, 1994, with a 15-year transition period.

National Cattlemen's Beef Association (NCBA) National organization for cattle breeders, producers, feeders, and affiliated organizations with offices in Englewood, CO, and Washington, DC. Previously known as the National Cattlemen's Association or NCA.

national sire evaluation Programs of sire evaluation conducted by breed associations to compare sires on a progeny-test basis. Carefully conducted national reference sire evaluation programs give unbiased estimates of expected progeny differences. Sire evaluations based on field data rely on large number of progeny per sire to compensate for possible favoritism or bias for sires within herds.

native hides Hides from cattle that have not been hot-iron branded.

natural beef Refers to beef from cattle that have not been fed growth stimulants or antibiotics.

natural fleshing Lean meat or muscle.

navel Area where the umbilical cord was formerly attached to the body of the offspring.

NCBA *See* National Cattlemen's Beef Association.

necropsy To perform a postmortem examination.

NEg Net energy for gain.

NEl Net energy for lactation.

NEm Net energy for maintenance.

net energy Metabolizable energy minus heat increment, or the energy available to the animal for maintenance and production.

net income Total revenue earned minus expenses incurred for a given period of time.

net worth (equity) Represents the owner's claim on the assets of a business: net worth = assets – liabilities.

net worth statement *See* balance sheet.

nicking Way in which certain lines, strains, or breeds perform when mated together. When outstanding offspring result, the parents are said to have nicked well.

nipple *See* teat.

NPN (nonprotein nitrogen) Nitrogen in feeds from substances such as urea and amino acids, but not from preformed proteins.

nutrient (1) Substance that nourishes the metabolic processes of the body. (2) End product of digestion.

nutrient density Amount of essential nutrients relative to the number of calories in a given amount of food.

obesity Excessive accumulation of body fat.

offal All organs and tissues removed from inside the animal during the slaughtering process.

off feed Animal refuses to eat or consumes only small amounts of feed.

omasum One of the stomach components of cattle that has many folds.

on full feed Refers to cattle that are receiving all the feed they will consume. *See also ad libitum.*

open Refers to nonpregnant females.

operating expenses Expenses incurred in the usual production cycle, such as seed, fuel, feed, and hired labor costs.

opportunity cost Cost of using a resource based on what it could have earned using it in the next best alternative use.

optimize To make as effective as possible.

optimum Amount or degree of something that is most favorable to some end (e.g., the best combination of resources associated with cattle production yields the highest sustainable net return).

optimum level of performance Performance level of a trait or traits that maximizes net profit. Resources are managed (including a balance of traits) that sustain high levels of profitability.

outbreeding Process of continuously mating females of the herd to unrelated males of the same breed.

outcrossing Mating of an individual to another in the same breed that is not related to it; a type of outbreeding.

ova Plural of *ovum*, meaning "eggs." *See also* ovum.

ovary Female reproductive organ in which the eggs are formed and progesterone and estrogenic hormones are produced.

overhead Expenses incurred in the operation of the business that cannot conveniently be attributed to the production of specific commodities or services.

ovulation Shedding or release of the egg from the follicle of the ovary.

ovum Egg produced by a female.

packing plant Facility in which cattle are slaughtered and processed.

paddock A pasture subdivision within a grazing unit.

palatability Degree to which food (e.g., beef) is acceptable to the taste or sufficiently agreeable in flavor, juiciness, and tenderness to be eaten.

palpation Feeling or examining by hand (e.g., the reproductive tract is palpated for reproductive soundness or pregnancy diagnosis).

parasite Organism that lives a part of its life cycle in or on, and at the expense of, another organism. Parasites of farm animals live at the expense of the animals.

parity Number of different times a female has had offspring.

parrot mouth Upper jaw is longer than the lower jaw.

partial budget Budget that includes only those revenue and expense items that would change as a result of a proposed change in the business.

parturition Process of giving birth.

pasture rotation Rotation of animals from one pasture to another so that some pasture areas have no livestock grazing on them during certain periods of time.

patchy Uneven fat accumulations; usually lumps of exterior fat around the tailhead and pin bones.

paternal Refers to the sire or bull.

pathogen Biologic agent (e.g., bacteria, virus, protozoa, nematode) that may produce disease or illness.

paunch *See* rumen.

paunchy Heavy-middled.

pay weight Actual weight for which payment is made. In many cases, it is the shrunk weight (actual weight – pencil shrink).

pedigree Records of the ancestry of an animal.

pelvic area Size of pelvic opening determined by measuring pelvic width and length and used to predict calving difficulty.

pen rider Person who rides through feedlot pens and checks cattle.

pencil shrink Deduction from an animal's weight, often expressed as a percentage of live weight, to account for fill (usually 3% for off-pasture weights and 4% for fed-cattle weights).

pendulous Hanging loosely.

Percent calf crop The percentage of calves produced within a herd in a given year relative to the number of cows and heifers exposed to breeding.

Per capita Per person.

performance data Records on individual animal's reproduction, production, and possibly carcass merit. Traits included are birth, weaning, and yearling weights; calving ease; calving interval; milk production; and others.

performance pedigree Includes the performance records of ancestors, half and full sibs, and progeny in addition to the usual ancestral pedigree information. The performance information is systematically combined to list estimated breeding values on the pedigrees by some breed associations.

performance test Evaluation of an animal according to its performance.

pharmaceutical Medicinal drug.

phenotype Characteristics of an animal that can be seen and (or) measured (e.g., presence or absence of horns, color, or weight).

phenotypic correlations Correlations between two traits caused by both genetic and environmental influences. *See* correlation coefficient.

pheromones Chemical substances that attract the opposite sex.

photoperiod Time period when light is present.

pituitary Small endocrine gland located at the base of the brain.

placenta Membranes that form around the embryo and attach to the uterus. *See also* afterbirth.

Plains states Includes Texas, Oklahoma, Kansas, Nebraska, South Dakota, and North Dakota and the eastern parts of New Mexico, Colorado, Wyoming, and Montana; often referred to as the "Beef Belt."

pluck Organs of the thoracic cavity (e.g., heart and lungs).

Pneumonia Inflammation or infection of alveoli of the lungs caused by either bacteria or viruses.

polled Naturally or genetically hornless.

pons Accumulation of fat over pin bones.

portion-controlled beef products Retail cuts of beef that meet size and form specifications.

postnatal *See* postpartum.

postpartum After birth.

postpartum interval Days from calving until the cow returns to estrus, or days from calving until cow is pregnant again.

pounds of retail cuts per day of age A measure of cutability and growth combined; it is calculated as follows: cutability times carcass weight divided by age in days.

pounds of calf weaned per cow exposed Calculated by multiplying percent calf crop by the average weaning weight of calves.

preconditioning Preparation of feeder calves for marketing and shipment; may include vaccinations, castration, and training calves to eat and drink in pens.

prenatal Prior to being born; before birth.

prepotent Ability of a parent to transmit its characteristics to its offspring so that they resemble that parent, or each other, more than usual. Homozygous dominant individuals are prepotent. Also, inbred cattle tend to be more prepotent than outbred cattle.

price cycle Traditional or historic changes in prices (usually by months, seasons, or years).

price discovery Process that shows how the specific price for a given quantity and quality of beef is determined.

primal cuts Wholesale cuts—round, loin, flank, rib, chuck, brisket, plate, and shank.

production testing Evaluation of an animal based on its production record.

progeny Offspring of the parents.

progeny testing Evaluation of an animal based on the performance of its offspring.

progesterone Hormone produced by the corpus luteum that stimulates progestational proliferation in the uterus of the female.

prolapse Abnormal protrusion of part of an organ, such as the uterus or rectum.

prostaglandins Chemical mediators that control many physiological and biochemical functions in the body. One prostaglandin, $PGF_{2\alpha}$, can be used to synchronize estrus.

prostate Gland of the male reproductive tract located just behind the bladder that secretes a fluid that becomes a part of semen at ejaculation.

protein Substance made up of amino acids that contains approximately 16% nitrogen (based on molecular weight).

protein supplement Any dietary component containing a high concentration (at least 25%) of protein.

puberty Age at which the reproductive organs become functionally operative.

purebred Animal eligible for registry with a recognized breed association.

purveyor Firm that purchases beef (usually from a packer) and then performs some fabrication before selling the beef to another firm.

qualitative traits Those in which there is a sharp distinction between phenotypes (e.g., red or black color). Usually only one or two gene pairs are involved.

quality Something special about an object that makes it what it is; a characteristic, attribute, excellence. Quality is the composite or attribute of an animal or product that has economic or aesthetic value to the user; meeting or exceeding each customer's expectations at a cost that represents value to the customer every time.

quality grades Grades such as Prime, Choice, and Select that group slaughter cattle and carcasses into value- and palatability-based categories; determined primarily by marbling and age of animal.

quantitative traits Those in which there is no sharp distinction between phenotypes, with a gradual variation from one phenotype to another (such as weaning weight). Usually, many gene pairs are involved, as well as environmental influences.

random mating System of mating whereby every female (cow and/or heifer) has an equal or random chance of being assigned to any bull used for breeding in a particular breeding season. Random mating is required for accurate progeny tests.

ration Feed fed to an animal during a 24-hour period.

REA *See* rib-eye area (REA).

reach *See* selection differential.

realizer Feedlot animal that is removed before the end of the feeding program. Only part of the animal's potential value is realized because of disease, injury, or the like.

recessive gene A gene that has its phenotypic expression masked by its dominant allele when the two genes are present together in an individual.

rectal prolapse Protrusion of part of the large intestine through the anus.

red meat Meat from cattle, sheep, swine, and goats. *See also* white meat.

reference sire Bull designated to be used as a benchmark in progeny testing other bulls (young sires). Progeny by reference sires in several herds enables comparisons to be made between bulls not producing progeny in the same herd(s).

registered Recorded in the herd book of a breed.

regurgitate To cast up undigested food to the mouth as is done by ruminants.

replacement heifers Heifers, usually between 6 months and 16 months of age that have been selected to replace cows in the breeding herd.

replacements Cattle that are going into feedlots or breeding herds to replace those being sold or that have died. *See also* replacement heifers.

reproductive tract score Numerical score based on palpation of the heifer's reproductive tract (1 = not cycling; 5 = heifer cycling).

resource Input or factor used in production, such as cattle, labor, or land.

retail cuts Cuts of beef in sizes that are purchased by the consumer.

retained ownership Usually refers to cow-calf producers maintaining ownership of their cattle through the feedlot.

retained placenta Fetal membranes (afterbirth) are not expelled through the reproductive tract in the normal length of time following calving.

reticulum One of the stomach components of cattle that is lined with small compartments giving a honeycomb appearance.

rib-eye area (REA) Area of the *longissimus dorsi* muscle, measured in square inches, between the 12th and 13th ribs. Also referred to as the *loin-eye area*.

rib-eye area per cwt carcass wt Rib-eye area divided by carcass weight.

risk Possibility of suffering economic loss. Sources of risk include climate, disease, and changes in the marketplace.

risk management Managing risks in ways that allow a desired outcome to be achieved.

rotational crossbreeding Systems of crossing two or more breeds whereby the crossbred females are bred to bulls of the breed contributing the least genes to the females' genotype.

roughage Feed that is high in fiber, low in digestible nutrients, and low in energy (e.g., hay, straw, silage, and pasture).

rugged Big and strong in appearance; usually heavy boned.

rumen A compartment of the ruminant stomach that is similar to a large fermentation pouch where bacteria and protozoa break down fibrous plant material swallowed by the animal. Sometimes referred to as the paunch.

ruminant Mammal whose stomach has four parts: rumen, reticulum, omasum, and abomasum. Cattle, sheep, goats, deer, and elk are ruminants.

rumination Regurgitation of undigested food that is chewed and then swallowed again.

scale (1) Size of cattle. (2) Equipment on which an animal is weighed.

scours Diarrhea; profuse watery discharge from the intestines.

scrotal circumference Measure of testes size obtained by measuring the distance around the testicles in the scrotum with a circular tape; related to the bull's semen-producing capacity and age at puberty of his daughters.

scrotum Pouch that contains the testicles. Also, a thermo-regulatory organ that contracts when cold and relaxes when warm, thus tending to keep the testes at a lower temperature than that of the body.

scurs Small growths of horn-like tissue attached to the skin of polled or dehorned animals.

seam fat *See* intermuscular fat.

seedstock Breeding animals; sometimes used interchangeably with *purebred.*

seedstock breeders Producers of breeding stock for purebred and commercial breeders.

Select USDA carcass quality grade between Choice and Standard; replaced the Good grade in 1988.

selection Differential reproduction (e.g., a bull or cow may leave several, one, or no offspring in a herd).

selection differential (reach) Difference between the average for a trait in selected animals and the average of the group from which they come. Also called *reach*.

selection index Formula that combines performance records from several traits or different measurements of the same trait into a single value for each animal. A selection index combines traits after balancing their relative net economic importance, their heritabilities, and the genetic association among the traits.

self-management Managing oneself as part of human resource management (e.g., time management, information management, self-motivation, honesty).

semen Fluid containing sperm that is ejaculated by the male. Secretions from the seminal vesicles, prostate gland, bulbourethral glands, and urethral glands provide most of the fluid.

seminal vesicles Accessory sex glands of the male that provide a portion of the fluid of semen.

served Female is bred, but not guaranteed pregnant.

service To breed or mate.

settle To become pregnant.

shipping fever Widespread respiratory disease of cattle.

short yearling Animal is over 1 year of age but under 18 months of age.

"show list" or "show pens" Slaughter cattle that are ready for the cattle feeder to "show" to the packer buyers.

shrink Loss of weight; commonly used in the loss of live weight when animals are marketed.

sib Brother or sister.

sick pen Isolated pen in a feedlot where cattle are treated after they have been removed from a feedlot pen; Sometimes referred to as a *hospital pen*.

sickle hocked Hocks that have too much set, causing the hind feet to be too far forward and too far under the animal.

silage Forage, corn fodder, or sorghum preserved by fermentation that produces acids similar to the acids used to make pickled foods for people.

sire Male parent.

sire summary Published results of national sire evaluation programs that give EPDs and accuracies for several economically important traits. Several major breed associations publish their own sire summaries.

size Usually refers to weight, sometimes to height.

skins *See* hides.

skirt Diaphragm muscle in the beef carcass.

software Program instructions to make computer hardware function.

soundness Degree of freedom from injury or defect.

SPA *See* Standard Performance Analysis.

spay To remove the ovaries.

sperm A mature male germ cell.

specifications A detailed description, with numerical designations, of animal performance or product quantity.

spermatogenesis Process of spermatozoa formation.

splay footed *See* toeing out.

stag Castrated male that has reached sexual maturity prior to castration.

standard deviation For traits having a normal distribution characterized by a bell-shaped curve, 68% of the population = mean (average) ± 1 standard deviation, 95% = mean ± 2 standard deviations, and 99% = mean ± 3 standard deviations.

Standard Performance Analysis A systematic methodology to measure beef cattle enterprise performance to allow across herd comparisons and industry benchmarks.

steer Bovine male castrated prior to puberty.

sterility Inability to produce offspring.

stifle Joint of the hind leg between the femur and tibia.

stifled Injury of the stifle joint.

stillborn Offspring born dead without previously breathing.

stocker Weaned cattle that are fed high-roughage diets (including grazing) before going into the feedlot.

stocking rate The number of animals, animal units, or total animal live weight assigned to a grazing unit for an extended period of time. Stocking rates are usually expressed on a per-acre basis.

stocking density The number of animals, animal units, or total animal live weight present at a particular point in time on a defined area (paddock). Stocking density is usually defined on a per-acre basis.

stockpiling The practice of allowing forage to accumulate for grazing at a later date; most commonly done with late summer and fall forage growth for fall and (or) winter grazing.

strip grazing The practice of dividing a larger pasture into strips with movable fences to control grazing access.

straightbred Animal whose parentage has been from one breed.

stress Unusual or abnormal influence causing a change in an animal's function, structure, or behavior.

subcutaneous Situated beneath, or occurring beneath, the skin. A subcutaneous injection is an injection made under the skin.

subprimal cuts Smaller-than-primal cuts, such as when the primal round is split into top round, bottom round, eye round, and sirloin tip. Subprimal cuts are used in boxed beef programs.

success Progressive realization of predetermined, worthwhile goals that are based on true principles.

suckling gain Gain that a young animal makes from birth until it is weaned.

superovulation Hormonally induced ovulation in which a greater-than-normal number of eggs are typically produced.

sweetbread Edible by-product also known as the *pancreas*.

switch Tuft of long hair at the end of the tail.

syndactyly Union of two or more digits; for example, in cattle the two toes would be a solid hoof.

synthetic breeds *See* composite breed.

systems analysis *See* management systems.

tariff A tax imposed on commodity imports by a government. A tariff may be either a fixed charge per unit of product imported (specific tariff) or a fixed percentage of value (ad valorem tariff).

tagging Usually refers to putting ear tags in the ear.

tandem selection Selection for one trait for a given period of time followed by selection for a second trait and continuing in this way until all important traits are selected.

TDN *See* total digestible nutrients (TDN).

teat Protuberance of the udder through which milk flows.

terminal crossbreeding *See* terminal sires.

terminal market Large livestock collection center where an independent organization serves as a selling agent for the livestock owner.

terminal sires Sires used in a crossbreeding system in which all their progeny, both male and female, are marketed. For example, crossbred dams could be bred to sires of a third breed and all calves marketed. Although this system allows maximum heterosis and complementarity of breeds, replacement females must come from other herds.

testicle Male sex gland that produces sperm and testosterone.

testosterone Male sex hormone that stimulates the accessory sex glands, causes the male sex drive, and results in the development of masculine characteristics.

tie Depression or dimple in the back of cattle caused by an adhesion of the hide to the backbone.

time management Manner in which time is utilized to achieve specific goals.

toeing in Toes of front feet turn in; also called *pigeon-toed*.

toeing out Toes of front feet turn out; also called *splay-footed*.

total digestible nutrients (TDN) Sum of digestible protein, nitrogen-free extract, fiber, and fat (multiplied by 2.25).

trait ratio Expression of an animal's performance for a particular trait relative to the herd or contemporary group average. It is usually calculated for most traits as:

$$\frac{\text{Individual record}}{\text{Average of animals in group}} \times 100$$

transgenic An organism or animal whose genome includes "foreign" genetic material. Foreign genetic material would be a DNA sequence or gene that does not normally occur in the species of the host organism or animal.

tray-ready beef Retail cuts that are cut and packaged at the packing plant for retail sales; also referred to as *case-ready*.

tripe Edible product from the walls of the ruminant stomach.

twist Vertical measurement from the top of the rump to the point where the hind legs separate.

type (1) Physical conformation of an animal. (2) All physical attributes that contribute to the value of an animal for a specific purpose.

udder Encased group of mammary glands of the female.

ultrasound Using high-frequency sound waves to show visual outlines of internal body structures (e.g., fat thickness, rib-eye area, and pregnancy can be predicted). The machine sends sound waves into the animal and records these waves as they bounce off the tissues. Different wavelengths are recorded for fat and lean.

umbilical cord Cord through which arteries and veins travel from the fetus to and from the placenta, respectively. This cord is broken when the young are born.

uncoupling Term used to describe separating quality grading and yield grading to allow the use of one without requiring the other.

unsoundness Any defect or injury that interferes with the usefulness of an animal.

urinary calculi Disease that causes mineral deposits to crystallize in the urinary tract.

USDA *See* U.S. Department of Agriculture.

U.S. Department of Agriculture (USDA) An executive department of the U.S. government that helps farmers supply farm products for U.S. consumers and overseas markets. *See* Appendix for organizational structure.

U.S. Meat Export Federation (USMEF) Organization that works to increase consumer demand for red meats and by-products in overseas markets. Members include NCA, state cattle associations, beef councils, farm and commodity groups, packers, and agribusiness companies. Funds come from its members and the USDA.

uterus That portion of the female reproductive tract where the young develop during pregnancy.

vaccination The act of administering a vaccine or antigens.

vaccine Suspension of attenuated or killed microbes or toxins administered to induce active immunity.

vagina Copulatory portion of the female's reproductive tract. The vestibule portion of the vagina also serves for passage of urine during urination. The vagina also serves as a canal through which young pass when born.

value-based marketing Marketing system based on paying for individual animal differences rather than using average prices.

variable costs Costs that change with the amount produced. If the manager decides to cease production, these costs are avoidable.

variance A statistic that describes the variation we see in a trait.

variety meats Edible organ by-products (e.g., liver, heart, tongue, tripe).

vas deferens Ducts that carry sperm from the epididymis to the urethra.

veal Meat from very young cattle (under 3 months of age). Veal typically comes from dairy bull calves.

video image analysis (VIA) A video image is analyzed via sophisticated computer techniques to estimate factors associated with carcass value.

virus Ultramicroscopic bundle of genetic material capable of multiplying only in living cells. Viruses cause a wide range of diseases in plants, animals, and humans, such as rabies and measles.

viscera Internal organs and glands contained in the thoracic and abdominal cavities.

vitamin Organic catalyst, or component thereof, that facilitates specific and necessary functions.

volatile fatty acids (VFA) Group of fatty acids produced from microbial action in the rumen; examples are acetic, propionic, and butyric acids.

vulva External genitalia of a female mammal.

wasty Excessive accumulation of fat.

wattle Method of cattle identification in which 3–6-inch strips of skin are cut on the nose, jaw, throat, or brisket.

weaner Calf that has been weaned or is near weaning age.

weaning (wean) Separating young animals from their dams so that the offspring can no longer suckle.

weaning weight Weight of the calf at approximately 5–10 months of age when the calf is removed from the cow.

weaning weight EPD Estimate of the weaning weight (lb) potential of a sire's progeny. Positive numbers indicate above-average performance while negative values indicate below-average weights when compared to other bulls in the same sire summary. This estimate is for direct growth, as maternal effects are removed in the calculations. *See also* expected progeny difference (EPD).

weaning weight ratio The weaning weight of a calf divided by the herd average; usually done within sex.

weight per day of age (WDA) Weight of an individual animal divided by days of age.

white meat Meat from poultry. *See also* red meat.

white muscle disease Muscular disease caused by a deficiency of selenium or vitamin E.

wholesalers Beef operations that buy and sell beef to other firms; considered the middlemen between the packer and consumer segments.

window of acceptability Identifies the acceptable minimum and maximum amounts of fat in meat on the basis of meat palatability and human health.

with calf Heifer or cow is pregnant.

withdrawal time Amount of time before slaughter during which a drug cannot be given to an animal.

woody Opposite of *bloom*—that is, the animal's hair coat appears dull, not shiny. Associated with unthrifty calves. *See also* bloom.

World Trade Organization (WTO) Established on January 1, 1995 as a result of the Uruguay Round, the WTO replaces GATT as the legal and institutional foundation of the multilateral trading system of member countries.

yardage Per-head daily fee charged by the feedlot to the customer owning the cattle. This fee is usually in addition to the cost of medicine and the feed markup.

yearling Animals that are approximately 1 year old (usually 12–24 months of age).

yearling weight Weight when approximately 365 days old.

yearling weight EPD Estimate of the yearling weight (lb) potential of a bull's progeny compared to progeny from other bulls in the same sire summary. Positive numbers indicate above-average performance while negative values indicate below-average performance. *See also* expected progeny difference (EPD).

yearling weight ratio Yearling weight of a calf divided by the herd average. Usually calculated within sex.

yield *See* dressing percentage.

yield grades USDA grades identifying differences in cutability—the boneless, fat trimmed retail cuts from the round, loin, rib, and chuck.

Index

Note: *Italicized* page numbers indicate illustrations.

A

Abnormal presentations, of the calf, *321*
Abomasitis, *482*
Abortions/stillbirths, 128, 319
Accessory sex glands, 292, 294
Accrual accounting, 72–73
Accuracy (ACC), 95
Accuracy measures, 358
Achondroplasia, 347
Acid detergent fiber (ADF), 404–405
Acid detergent insoluble nitrogen, 465
Acidosis, *482*
Actinobacillosis, *486*
Actinomycosis, *486*
Acute diseases, 482
Adaptability, 349–350
Additives
  in feedlot operations, 199–202
  silage, 406
Aflatoxin, 408
Age
  classifications of, for meat, 222
  teeth relationship and, 523, *524*
Aged beef, 38
Aging, 37
Agriculturalists, production by, 263
Agricultural productivity, 1
Agricultural systems, evaluation of, 1
Air-dry basis, 397
Alfalfa, *440*, *441*, 462, 463
Alfalfa hay, 415, 463
Alimentary canal/tract, 398
Alleles, 333
Alliances
  in beef industry, 218–220, *220*
  breakeven prices and, 113
  marketing, 113, 219
Alopecia, 347
Alsike clover, *440*
Alyceclover, *440*
American Angus Association, 91, 359
American Association of Meat Processors Association (AAMP), 23
American Gelbvieh Association, 107, 391
American Meat Institute (AMI), 22
American National Cattlewomen (ANCW), 22
American Simmental Association, 391
Amerifax, *366*
Amino acids, *396*, *398*
Ammonia buildup, 498, 513
Amniotic membranes, 317
Amphistomes, *507*
Ampulla, 294, 296
Anaplasmosis, *483*
Angus, 362, 363, *364*, *371–372*, 378, 379, 390, 390, 391, *392*, *526*
Angus Herd Improvement Records (AHIR), 91, 371
Angus x Hereford cows, 391, 392–393
Animal feeding operations (AFOs), 186
Animal identification systems, 76, 89–90, 214
Animal products, 285
Animal unit month (AUM), 447, *448*, 471
Animal well-being
  auditing, 560–562, 561
  issues, 24–25
Ankina, *366*
Ankole Watusi, *366*
Ankylosis, 347
Anterior pituitary, 291, 294
Anthelmintic drugs, 508, *508*
Anthrax, *483*
Antibiotics, 55–56, *56*, 175, 517
Antibodies, 491
Antibody fingerprinting, 215
Antigens, 491
Antitoxins, 492
Anxiety, 498
Area of rib-eye muscle (REA), 226
Argentina, 264, 265, 270–271
Arrowgrass, *476*
Arrowleaf clover, *440*
Arthrogryposis, 347
Artificial insemination (AI)
  about, 299–300
  detecting estrus, 303
  of female, 302–303
  proper timing of, 304–306
  semen collection and processing, 300–302
  sexing semen, 302
Artificial vagina, 300, *301*
Aseptic conditions, 496
Ash, 397, 466
Assortative mating, 337–338
Asthma, *483*
Astragulus, *476*
Auction markets (public), 232–233, 234–235
Australia, 265, 266, 267, 270, 271–273
Average daily gain, 521
Avermectins, 509

B

Bacillary hemoglobinuria, *487*
Backfat, 107, 345
Backgrounded cattle, 429
Backgrounding, 165, 166, *167*
Back tags, 215
Backward presentation, 322–323
Bacterins, 492
Bahiagrass, *440*, *441*, 448, 463
Baker's shortening, 57
Bankrupt worm, *507*
Barberpole worm, *507*
Barley, 430, *440*
Barzona, *366*
Beef, as age classification, 222
Beefalo, *366*
Beef Board. *See* Cattlemen's Beef Promotion and Research Board (Beef Board)
Beef buyers, 270, 281–285
Beef cattle
  carcass conformation in, 528–529
  form and function in, 524–525
  growth and development of, 519–523, *524*
  livestock show, 538
  nutrient requirements of, 411, *412*, *413*. *See also* Nutrition
  parts of, 528, *528*
Beef cattle prices, major factors affecting, 243–251
Beef conformation, 519
Beef cows, 7
Beef cuts, size of, 42
Beef demand, 28–29
Beef Demand index, *35*
Beef disappearance, 31
Beef Friesian, *366*
Beef grades, distribution of, 228–229, *229*

Beef industry. *See also* Global beef industry
goals, 16
issues, 23–24
major goal of, 243
organizations, 18–23
overview, 1–4
products generated by, *27*
segments, 4–18
size and scope of, 1
structural change in, 216–218
Beefmaster, 362, *364*, *371*, *373*
Beef-packing companies, *12*
Beef palatability and consumer preferences
color, 42
consumer attitudes and preferences, 43–44
consumer trust, 45
flavor, 38
foodservice, 34–35, *35*
healthy beef products, 45–52
improving value of chuck and round, 42–43
juiciness, 38, 40
lean to fat, 41–42
production technologies, 55–56
retail, 31–34
safe beef products, 52–54
size of beef cuts, 42
tenderness, 37–38, *39*
USDA quality grade, fat, and overall palatability, 40–41
Beef prices
cycles and, 244–246
forecasting, 251–252, *252*
seasonal, 248, *249*
Beef Quality Assurance (BQA), 499–501
Beef Quality Audit (1991), 35–36
Beef Quality Audit (2011), 25
Beef suppliers, 270–276
Beef type, 524–525, *526*, *527*
Beef value fabrication system, 42–43
Beef wholesalers, 13–14
Belgian Blue, *366*, *372*, 391
Belted Galloway, *366*
Bermudagrass, 439, *440*, 448, 464
Berseem clover, *440*
Best Linear Unbiased Prediction (BLUP), 357
Best Manufacturing Practice (BMP), 219
Beta carotene, 425
Big bluestem, *440*, *441*, 462
Bigflower vetch, *440*
Bioavailability, 411
Bioeconomic efficiency, 85, *86*
Biological control, 455
Biological efficiency, 85
Biologicals, 492
Biological time frame, 245
Biological types, 373, *373*, 393
Biometric identification systems, 215
Biosecurity, 268–269, 501
Birdsfoot trefoil, *440*, *441*, 462, 463
Birth, 290, 319, 320–322, 520–521
Birth canal, 290
Birth weight
bull selection and, 104–105
calving difficulty and, 324–325
growth, development, and, 520–521
variation between and within breeds for, 363, *375*
Black grama, 459
Blackleg, *483*
Black medic, *440*
Bladder, 290, 292, 293, 294
Blade tenderization, 38
Bloat, 400, 474, *483*
Blonde d'Aquitaine, *366*
Blood odor, 551
Bloom, 33
Blue grama grass, *440*
Bluestems, 439
Bluetongue, *483*
Body condition score (BCS), 304, 327, *327*, *328*
Bomb calorimeter, 401
Bone
carcass composition and, 521
growth and distribution of, 523
Boneless, closely trimmed, retail cuts (BCTRC), 226
Bone maturity, 223, 224
Bonsmara, *366*
Boran, *372*
Bos indicus cattle, 37, 274, 341, 362
Bos taurus cattle, 37, 341, 349, 350, 362
Bottom round cuts, 42
Bovatec, 175, *200*, 202
Bovine respiratory disease (BRD), 172, 190, 509
Bovine respiratory syncytial virus (BRSV), *484*
Bovine Spongiform Encephalopathy (BSE), 266
Bovine viral diarrhea (BVD), *484*, 513
Boxed beef, 13
Box scrapers, 185
Brachynathia inferior, 347
Bracken fern, *476*
Braford, *365*
Brah-Maine, *366*
Brahman, *364*, *372*, *373*
Brahmental, *366*
Brahmousin, *366*
Brain, 57
Bralers, *366*
Branded beef products, 255–258
Branding, 77, 90
Brands, 215
Brangus, 362, *364*, *371*, *372*, *373*
Branvieh, *373*
Braunvieh, *365*, 381
Brazil, 264, 270, 273–274
BRD. *See* Bovine respiratory disease (BRD)
Breakeven price, 131
Breakeven price analysis, for commercial cow-calf operations, 122–124
Breakeven prices
for feedlot operations, 187–189
in yearling-stocker operations, 168
Breathing, 482
Breech delivery, 323
Breed alliances, 219
Breed-based programs, 257
Breeding
cow-to-bull ratios, 299
improving beef cattle through, 350–355
marker-assisted selection, 358–360
national sire evaluation, 357–358
natural service, 295–296
physical examination, 296
scrotal circumference, 296, *297*
selection programs, 355–357
semen evaluation, 297–299
Breeding box, 305, 560
Breeding cattle, conformation of, 532–537
Breeding soundness evaluation (BSE), 295–296, 298, 409
Brisket Disease, *484*
British breeds, 362, *371*
British White, *366*
Broken-mouthed cows, 523
Brown stomach worm, *507*
Brown Swiss, *372*
Browntop millet, *440*
Browse, 439, *440*, 458, 459
BRSV (bovine respiratory syncytial virus), *484*
Brucellosis, 319, *484*, 514

Bruising, of cattle, during transportation, 552
BSE. *See* Breeding soundness evaluation (BSE)
Budgeting, in yearling-stocker operations, 166–168, *167*
Buffalo, *366*
Buffalo grass, 439, *440*, *441*, 448
Bulbourethral gland, 292, 293, 294
Bull clover, *440*
Bulldog dwarfism, 347
Buller-steer problem, 545
Bullock, as sex classification, 223
Bulls
- communication, 541–542
- costs of, 144–145
- health management of, 513
- locating genetically superior, 107
- marketing, 230–231, *232*
- puberty in, 295
- reproductive organs of, 292–294
- selection of, in cow-calf operations, 140, 142, *142*
- as sex classification, 223
- *vs.* steers, 107–108

Bull testing stations, 107
Bunching instinct, 552
Burclover, *440*
Bureau of Land Management (BLM), 471
Business relationships. *See* Alliances
Buttonclover, *440*
Buy-sell margin, 168
BVD (bovine viral diarrhea), *484*, 509
B vitamins, 46
By-products, 26, 247–248, *248*

C

Calcium, 424
Caleypea, *440*
Calf, as age classification, 223
Calf backward, 323
Calf crop percentages, 126–129, *130*, 543
Calf forward, 323
Calf scours, 514–517
Calf septicemia, *486*
Call option, 254
Calories, 401
Calpains, 37
Calpastatin, 37
Calves
- birth weight, 104–105
- creep feeding, 135, 428, *428*
- dystocia in, 105
- feeding dairy, 180
- growth stimulants for, 134
- management of, at weaning, 512–513
- number of pounds of weaned per cow, 125–126
- pre-conditioned, 190
- preconditioning for, 499
- sick versus healthy, 499, *500*
- weaning and, 541
- weaning weight of, 105–106, 129–135

Calving
- assisting delivery, 322–324
- ease of, 105
- factors affecting cattle losses at/after birth, 319–320
- handling difficulties, 320–322
- losses after birth until weaning, 326
- management programs to reduce losses, 326
- preventing losses, 324–325
- scours and, 515

Calving programs, 10
Camel hair brushes, 56
Campylobacteriosis, *484*
Canada, 266, 267, 268, 276, 277–278
Canary grass, 439
Cancer, 47, 49–50
Cancer eye, *484*
Carbohydrates, 395, 400
Carcass beef production, 3
Carcass composition
- growth, development, and, 521–522

Carcasses
- cuts of beef from, 26, *27*
- specifications for consumer demands, 28, *28*

Carcass grades
- quality grades, 224

Carcass merit, 345–346
Carcass standards, 534–535
Carcass trait, 107, 390–391
Carcass weights, 4, 247, *248*, *377*
Carotene, 410
Carpetgrass, *440*
Carrying capacity, 445, 446–447
Case-ready merchandising, 13, 14
Case-ready packages, 33
Case-ready production, 535
CASH, *366*
Cash accounting, 72, 73
Cash flow (CF), 149, *149*
Cash flow statement, 72, *73*
Castration, 292, 293, 499
Cattalo, *366*
Cattle auctions. *See* Auction markets
Cattle behavior
- communication, 541–542
- disposition, 540–541
- feedlot, 545–546
- gathering cattle on pasture and training, *551*, 552
- grazing, 545
- reproductive, 543–544
- social, 542–543
- suckling, 544–545

Cattle breeders, comparing breeds, 390, 391
Cattle breeds
- defined, 362
- differences, 372–382
- evaluation for commercial producers, 371–391, *386–387*
- evaluation for low cost production, 390–391
- evaluations to improve consumer market share, 391, *392*
- integrated production systems, 392–393
- major U.S., 363–371, *364–370*
- using in crossbreeding systems, 382–390, *385–387*

Cattle-Fax, *20*, *21*, 251
Cattle grubs, 504–505
Cattle identification, 75–77
Cattlemen's Beef Promotion and Research Board (Beef Board), 260
Cattle-On-Feed reports, from USDA, 11
Cattle resources, 74–78
Cattle scab, 504
Cattle transport, 552–553
Caucasian bluestem, *440*
Cells, 331–333
Cells of Leydig, 293
Cell systems, 445
Cellulose, 466
Centers for Disease Control, 52–53
Cereal grains, 265
Certification, 499
Certified Angus Beef™ (CAB®), 44, 257, 371, 534
Certified Hereford Beef, 257
Cervix, 289–290, *289*, 306, 321
Cestodes, 506
Changing day length, 292
Chaparral-mountain shrub, 459
Charbray, *366*
Chargrey, *366*

Charolais, 362, 363, *364*, *366*, *367*, *371*, *372*, *373*, 378, 379, 381, 382, 392, *392*
Charolais Herd Improvement Program (CHIP), 91
Charswiss, *367*
Checkoff program, 260, 261
Cheeseburger fries, 26
Chemical elements, in nutrients, 396
Chemostatic regulation, 193
Chest width, 521
ChiAngus, *367*, *381*
Chianina, *365*, *372*, *373*, *392*
Chicago Mercantile Exchange (CME), 252
Chiford, *367*
Chimaine, *367*
China, 264, 270, 276, 281–282
Choice quality grade, 40–41
Cholesterol, 49
Chorionic membranes, 317
Chromosomes, 381–383
Chronic diseases, 482
Chuck, 26, *28*, 42–43, 226
Circles of influence, 79, *79*
CL. *See* Corpus luteum (CL)
Clean Water Act, 186
Clear scars, 502
Cleft palate, 347, *478*
Clitoris, 290, 291
Cloning, 317
Closed cooperatives, 219
Clovers, 462
CL regression, 291
Coastal bermudagrass, *441*, 463
Coccidiosis, *484*, 514
Coldness, 419
Cold shortening, 37
Colibacillosis, 514
Color, of meat, 42
Color of lean, 224
Color patterns, for identification, 215
Colostrum, 422, 425
Commercial cow-calf producers, 5–10, *20*
Commercial feedlots
 defined, 178
 facilities, 181, 183
 management of, 180
 ownership, 178
Commercial producers, breed evaluation for, 371–390, *386–387*
Commission representatives, 236
Commodity, 253
Commodity Futures Trading Commission, 252
Common curly mesquite, 448
Common liver fluke, *507*
Common scab, 504
Common vetch, *440*
Communication, 66, 541–542
Compass plant, 439, *440*
Complex carbohydrates, 395
Composite crossbreeding, *388*
Composting, 185
Compudose®, 134
Concentrate feeding, 41
Concentrates, 407–408, 428
Concentrate-to-roughage ratios, 430
Confined animal feeding operations (CAFOs), 186
Conformation
 of breeding cattle, 532–537
 in carcass, 525, 528–529
 defined, 524
 of feeder cattle, 530–532, *531*
 of market cattle, 529–530, *530*
 sex and, 532, 533
 as trait, 346–347
Conical flukes, *507*
Connective tissue, 37
Consignment sales, 113, *114*
Consolidation, 216
Consumer markets, 28
Consumers, 7, *8*, 14–15
Consumption, 28–31
Contamination, 53
Contemporary groups, 95
Continental breeds, 362, *371*
Continuous systems, 450
Continuous variation, 336–337, *337*
Controlled internal drug release (CIDR), 308
Control points, 560–561, *561*
Conventional fabrication, 42
Cooperative marketing, 235–236
Cooperids, *507*
Corn, 405, *440*
Corn cobs, 430
Corn silage, 406–407, *441*
Corn Silage Processing Score, 406
Cornstalk aftermath, 10
Corn stalks, 467
Coronary heart disease (CHD), 47, 48
Corpus luteum (CL), 290, 291, 308
Costs
 assessing, in marketing system, 251
 of bulls, 144–145
 of commercial cow-calf operation, 155–156
 in cow-calf operations
  feed, 137, 139–140
  labor, 145
 economic environment and, 145–150
 factors affecting nonfeed, in feedlot operations, 195–197
 feed, in feedlot operations, 191–195
 of feeder cattle, 189–190
 of gain, in feedlot management decisions, 191
 managing annual cow, 135–145
 of respiratory diseases, 509
 of typical feedlot facility, 184
Cottonseed meal, 408
Country of origin labeling, 284
Cow, as sex classification, 223
Cow-calf operations
 benefits of heterosis for, 384
 breeds used in, 371
 brucellosis and, 514
 commercial management decisions
  annual cow costs and returns, 135–145
  breakeven price analysis, 122–123
  creating vision, 119
  establishing, 155–157
  factors affecting pounds of calf weaned, 124–125
  managing percent calf crop, 126–129
  marketing decisions, 151–154
  matching cows to their economic environment, 145–150
  profitability formula, 122–123
  weaning management, 150–151
 economic evaluation of genetic antagonisms, 379
 evaluating, 390–391
 handling facilities, 558–560
 health management programs, 509–517
 injections to calves, 502
 market channels for, 233, *234*
 nutrition in, 420–429
 restraining calves, 558
Cow-calf producers, comparing breeds, 390–391
Cow herd, health management of, 513
Cowpea, *440*
Cowper's gland, 292. *See also* Bulbourethral gland
Cows
 age of, 105
 culling, 108–110
 managing annual costs and returns, 135–145

Cows (*continued*)
  marketing, 230–231, *232*
  matching to their economic environment, 145–150
  mature weight, 106
  reproductive organs of, 289–295
  selecting, 108
Cow-to-bull ratios, 299
Cow weight, variation between and within breeds for, *376*
Crabgrass, *440*
Cracker Cattle, *367*
Creep feeding, 135, 428, *428*
Crested wheatgrass, *440*, 458
Creutzfeld-Jakob disease (vCJD), 58
Crimson clover, *440*
Criollo/Correinte, *367*
Critical control points, 560, *562*
Critical temperature, 418
Crooked calf disease, 348, *478*
Crop residues, 467
Crossbreeding
  to achieve economic efficiency, 143
  EPDs and, 358
  as form of outbreeding, 338
  growth potential through, 131, 131–133
  longevity and, 346
  reasons for, 341
  using breeds in, 382, 382–390, *390–391*
Crownvetch, *441*
Crude fat, 397
Crude fiber, 397, 400, 402
Crude protein (CP), 324, 397, 403, 465
Cryptorchidism, 347
Cryptorchid, 293
Cud, 398
Culling, 108–110, 355, *356*
Curved-bar headgate, 557
Custom cattle feeding, 178
Custom feedlots, 178
Customer service, 111–112
Cutability, 226
Cyanide, 479

D

Dairy breeds, *365*
Dairy calves, feeding, 180
Dairy cows, 7
Dallisgrass, *440*
Dark cutter, 224
Deferent duct, 292, *293*, 294
Deferred systems, 445
Degree of doneness, 40
Dehorning, 499
Delivery date, 253
Delivery point, 253
Deming, W. Edwards, 202, 218
Deoxyribonucleic acid (DNA), 331
Department of Health and Human Services (HHS), dietary guidelines, 50
Dermoid, 347
Devon, *367*, *372*, *373*
Dewormers, 508–509
Deworming, 499
Dexter, *367*
Diarrhea, 516
Dicots, 439
Diet, role of beef in world, 263
Dietary guidelines, 50–51. *See also* Nutrition
Digestibility, of feeds, 400
Digestible dry matter (DDM), 466
Digestible energy (DE), 401
Digestible protein (DP), 404
Digestive system, 398–400
Digestive tract, 398
Diphtheria, *484*
Direct acidification, 406
Direct marketing, 236, 251
Direct selling, 232
Discoloration, 33
Diseases
  administering health products, 496–497
  common, 482, *482–489*
  immunity, 490–492
  quality assurance programs, 499–503
  types of vaccines, 492–493
Disposition, 349, 540
DNA, 519
DNA sequencing, 215
DNA testing, 359, 360
Docility, 107, 349
Dollar-cost averaging (DCA), 149, *149*
Dominant genes, 334
Doramectin, *508*
Double muscling, 348
Droplets, 297
Drought, 86, *86*, 244, 436, 468
Dry-aged beef, 38
Dry matter, 395
Dry matter (DM) basis, 412
Dry matter intake (DMI), 465
Ductus deferens, 292
Dunn, Barry, 119
Dust, 498
Dust management, 185
Dwarfism, 347
Dystocia, 105, 128, 144, 319–324, 390

E

Ear implants, 174
Early weaning, 150
Ear notches, 215
Ear tags, 89–90, 215
Ear tattoos, 89
Eastern gramagrass, *440*
E. coli K99-specific injections, 516
Economically relevant traits, 342
Economic efficiency, 85
Economic records, 72
Economics, associated with grazing, 469, *470*
Economy, beef industry contributions, 4, *5*
Ectoderm, 519
Edible by-products, *27*, 57
Effective ambient temperature, 419
Eggs, 290, 333
80–20 rule, 145
Electroejaculator, 397, 300
Electronic identification systems, 215
Electronic marketing, 238
Embryo, 315
Embryological development, 519–520
Embryonic stage, 318
Embryo splitting, 317
Embryo transfer, 315–317
Embryo transplant, 315
Employees, 67
End meats, 26
Endoderm, 519, *520*
Endophytes, 475
Energy density, of feed, 192
Energy evaluations, of feeds, 400, *401*
Energy feeds, 402
Energy values, 404
Ensiling, 405
Enteric colibaccilosis, *485*
Enterotoxemia, *485*, 514, 515
Environmental issues, 23–24
Environmental management, in feedlot operations, 185
Environmental Protection Agency, 186
Epididymis, 292, 293, 298
Estate planning, 73
Estrogen, 291, 320
Estrous cycle, 291, 292, 294, 303
Estrous synchronization, 306–315
Estrumate®, 308
Ether extract, 397

- European Union (EU), 256, 274–275
- Expected progeny differences (EPDs), 95, 98, 99, 106
- Expenditures, 28–31
- Exports, 267–270
- External genitalia, 296
- External parasite control, 499
- External parasites, 504–505

F

- Fabrication protocols, 42, 43
- Face fly, *487*
- Facilities
  - cow-calf handling, 558–560
  - feedlot, 560
  - general guidelines, 553, *554*, *555*
  - restraining methods, 553–558
- Facilities investment, for feedlot operations, 181–185
- Family relationships, 66–67
- Farm animals, 24
- Farmer-feeder operations
  - advantages and disadvantages of, 180
  - facilities for, 183
  - ownership, 178
- Fatigue, 498
- Fat measurement, in determining yield grades, 226
- Fats, 38, 41, 48–49, 57, 396, 521, 522
- Fat thickness, 529
- Fatty acids, 48–49
- Feather eyes, 347
- Fed-cattle marketings, 11
  - conformation of, 530–532
  - markets for, 222–223
- Feed efficiency, 192–195, 199, 344–345, 390
- Feeder cattle
  - conformation of, 530–532, *531*
  - costs of, 189–190
  - grades of, 229–230
  - premiums and discounts paid for, 168
- Feeders, *20*
- Feedlot behavior, 545–546
- Feedlot operations
  - facilities, 560
  - management decisions
    - breakeven prices, 187–189
    - cost of feeder cattle, 189–190
    - cost of gain, 191
    - environmental, 185–186, *187*
    - facilities investment, 181–185
    - factors affecting feed cost per pound of gain, 191–195
    - factors affecting nonfeed costs, 195–197
    - implants and additives, 199–202
    - total dollars received, 198–199
    - total quality management, 202
    - types of cattle feeding operations, 178–180
  - managing, 181–202
  - turnover rate, 4
- Feedlots, importance of, 6, 7, 10–11
- Feed resources, 74
- Feeds
  - classification of, 402
  - costs of
    - in feedlot operations, 189–190
  - digestibility of, 400
  - energy evaluations of, 400, *401*
  - grains and concentrates, 407–408
  - management for high-nutrient content of, 405
  - mineral supplements, 410–411
  - nutrient composition of, 402–405
  - protein supplements, 408–410
  - proximate analysis of, 396–398
  - silages, 405–407
  - vitamin supplements, 410
- Feedstuffs, 402, 406, 412–420
- Feed supply, weaning weights and, 134–135
- Female infertility, 128
- Females, insemination of, 302
- Female sex cells, 290
- Fence line weaning, 150
- Fermentable carbohydrates, 405
- Fermentation aids, 406
- Fertilization, 331
- Fertilizing, as range improvement, 455
- Fescues, 439
- Fescue toxicity, *537*, 475–478
- Fetal period, 318
- Fetal size, to determine pregnancy, 318
- Fetus, 290, 519
- Financial records, 71, *72*, 74
- Financial resources, 71–74
- Finger foods, 26
- Firmness of lean tissue, 223
- Fixed costs, 77
- Flavor, 38, 391
- Flehmen response, 543
- Flies, *504*
- Flight zone, 546, 548
- Flukes, 506
- Fly control, 185
- Foley catheter, 316
- Follicles, 290
- Follicle-stimulating hormone (FSH), 291, 294, 316
- Food-borne illnesses, 53
- Food Guide Pyramid, 50
- Food Marketing Institute (FMI), 23
- Food Safety and Inspection Service (FSIS), USDA, 56
- Food safety issues, 24–25
- Foodservice, 34–35
- Foot and mouth disease (FMD), 270–271, 274
- Foot rot, 298, *485*
- Forage-fed beef, vs. grain-fed, 38
- Forages
  - crop residues, 467–468
  - drought management, 468
  - grazed resources, 433–439
  - grazing management, 441–455
  - hays, 464–466
  - health problems associated with grazing plants, 474–479
  - market cattle production on grazed, 468–474, *472*
  - measuring quality of, 465–466
  - plant types and distribution, 439–441, *441*
  - roughages and, 402
  - U.S. grazing regions, 455–464
- Forage sorghum, *440*, 479
- Forbs, *440*, 459
- Foster calves, 544
- Founder, *485*, 535
- 4-H youth programs, 538
- Four-wing saltbush, 439, *440*, 459
- Foxtail millet, *440*
- Frame sizes, 229–230, *531*
- Freeze brands, 90
- Freezer burn, 33
- Frontward presentation, 322
- Frostbite, scrotal, 296, 298
- Fully cooked products, 34
- Futures contract, 253
- Futures market, 252, 253–254

G

- Galloway, *367*, *372*, *373*, *392*
- Gelbray, *367*
- Gelbvieh, *364*, *371*, *372*, *373*, *392*
- Generally accepted accounting principles (GAAP), 73
- Generation interval, 354
- Genes, 333, 334. *See also* Genetics
- Genetic correlation, 355, *355*
- Genetic defects, 347–349, *349*
- Genetics. *See also* Breeding
  - in carcass composition, 521

Genetics (*continued*)
factors influencing calf gains, 131–133
mating strategies, 337–341
principles
cells and chromosomes, 331–333
continuous variation and many pairs of genes, 336, 337
genes, 333
inheritance with one pair of genes, 333–334
inheritance with two pairs of genes, 334–336
sex determination, 333
right genetics, 140–145
trait measurement, 342–350
trend for seedstock herd, 110
Genetic variation, 350–352, 374, *373–378*
Genotypes, 334, 336
Germ Plasm Evaluation (GPE) program, 372
Gestation, 317–319
Gestation length, variation between and within breeds for, *375*
Giant liver fluke, *507*
Glandular tubes, 292
Glans penis, 294
Global beef industry. *See also* Beef industry
buyers, 281–285
exports, 267–270
imports, 266–267, *267*
international trade, 265–270
NAFTA partners, 276–281
suppliers, 270–276
world beef consumption, 265
world cattle numbers, 264
world meat production, 265
Goals, 65
Gonadotropin releasing hormone (GnRH), 291
Government, impact on cattle industry, 19
Graafian follicles, 290
Grading up, 338
Grain-fed beef, *vs.* forage-fed, 38
Grain prices, 191
Grains, 407–408, 430
Grain sorghum, *440*
Grama grasses, 439
Grasses, 439, *440*, *441*, 448, 463
Grass staggers, 478
Grass tetany, 425, *476*, 478–479
Grazing
economics associated with, 469, *470*
fees, 471–473
foraging resources, 433–439
health problems associated with, 474–479
major U.S. regions, 455–460, 461–464
management, 441–455
market cattle production on grazed forages, 468–469
wheat pasture, 473–474
Grazing behavior, 545
Grazing fees, 24
Grazing Fee Task Group, 573
Grazing management, 24
Grazing response index, 454
Grid merit EPD, 357
Grid pricing, 238–242, *241*, *242*
Gross energy (GE), 401
Ground beef, 26, 53
Growth, defined, 519
Growth promotants, 199, 275
Growth stimulants, 134, 174
Grubs, *504*, 504, 505

H

Haemophilus, 172
Haemophilus Somnus, *485*
Hairlessness, 348
Hairy cattle louse, *505*
Hairy vetch, *440*
Halfshrub sundrop, *440*, 439
Hamburger, 26
Handling, cattle behavior during, 546–551
Hardinggrass, *440*
Hardware Disease, *485*
Harvested cattle, 12, 13
Haylage, 405
Hays, 464–466
Hays Converter, *367*
Hazard Analysis and Critical Control Points (HACCP), 219, 500
HDLs (high-density lipoproteins), 49
Headgate, 557
Health issues
beef products and, 45–52
in yearling-stocker operations, 168–172
Health products, administering, 496–497
Healthy beef products, as consumer preference, 45–52
Heart, 57
Heat-damaged protein, 465
Heat stress, 193
Hedgers, 253
Hedging, 254
Heel fly, *504*, 504
Heiferette, as sex classification, 223
Heifers
development of replacement, 148–150
puberty in, 294–295
selecting replacements, 108
as sex classification, 223
Heme iron, 46
Hemoglobin, 479
Herbicidal control, 455
Herd health
brucellosis status, 514
calf scours, 514–517
cow herd, replacement heifers, and bulls, 513
diseases and health problems, 482–490
establishing program for, 497–498
parasites, 504–509
stress and, 498–499
Herd sizes, 9–10
Hereford, 362, 363, 370, *364*, *371*, *372*, 378, 390, 390, 392, 525, *527*
Hereford-Angus reciprocal crosses, 370, 372, *373*, *392*
Herens, *367*
Heritabilities, 343
Heritability estimates, 353, *353*
Herringbone configuration, 560
Heterosis, 112, 132, 341, 382, 384, 393
Hide
by-products of, 56
exports and, 267
High palatability beef, 28, *28*, 255
Hip height, 325, 532, 533
Hip width, 521
Holistic Resource Management (HRM), 60. *See also* Management systems
Holstein, *372*, *373*, *392*
Homozygosity, 362
Hoof growth, 535
Hookworm, *507*
Hop clover, *440*
Hormones, 290, 320
Horn fly, *353*
Hot-iron brands, 90
Human resources, 61–71
Hunger, 263, 498
Hydrocephalus, 348
Hydrocyanic acid, 479
Hydrogenation, 49
Hyperplasia, 520

Hypertrophy, 520
Hypomagnesemia, 478
Hypothalmus, 292
Hypothermia, 481
Hypotrichosis, 348

I

IBR (infectious bovine rhinotracheitis), 319, *485*, 513
Identification, 499
Identification systems, 214–215
Immunity, 490–492
Immunization, 498
Immunoglobulin absorption, 515–516
Implants
in feedlot operations, 199–202
of growth stimulants, 134
Imports, 266–267
Inbred lines, development of, 339
Inbreeding, 338–339
Incisors, 523
Independent culling level, 355, *356*
India, 264
Indiangrass, *440*, 462
Indicator traits, 343
Individual cattle management (ICM) systems, 500
Inedible by-products, *27*, 57
Infectious bovine rhinotracheitis (IBR), 319, *485*
Infectious pododermatitis, *485*
Infectious thromboembolic meningoencephalitis, *485*
Information management, 70–71
Infundibulum, 289, *289*, 290
Inheritance, 333–336
Injection site, 500
Injury, of cattle, during transportation, 552–553
Inorganic components of feed, *397*, *398*
Insecticides, 505
Insemination, of the female, 302–303
Insoluble crude protein (ICP), 465
Inspection stamps, 228, *228*
Instrument grading, 229
Insulin, 57
Integrated production systems, 392–393
Integrated Resource Management (IRM), 60, 79. *See also* Management systems
Intensive inbreeding, 338
Intermediate wheatgrass, *440*, 458
Intermuscular fat, 529
Internal inguinal rings, 296
Internal parasites, 506–509
International Beef Quality Audit (1994), 281
International trade, 265–270
Interstitial cells, 293
Intestinal threadworm, *507*
Intramuscular fat, 38
Intramuscular injections, 496
Intranasal vaccines, 496
Intravenous vaccines, 497
Inventory, profitability and, 246
Ionophores, 174, 175, 202
Irish Blacks, *367*
Iris scanning, 215
Iron, 46
Irrigation, 436, *436*
Ivermectin, *508*

J

Japan, 268, 270, 271, 282–284
Jersey, *373*, 378, *392*
Johne's Disease, *486*
Johnsongrass, *440*
J.R. Simplot Company, 185
Juiciness, 38–40, 391
Junk science, 47

K

Kentucky bluegrass, *440*, *441*, 448, 462
Kidney, 57
Kidney, pelvic, and heart (KPH) fat, 226, 522
Kilocalories, 401
Knuckle, 42
Konefal method, 324
Korea, 270
Korean lespedeza, *440*
K. R. bluestem, 448
Kudzu, 439, *440*

L

Labor costs, reducing in cow-calf operations, 145
Lactic acid, 405
Ladino clover, 463
Lameness, 535
Laminitis, *485*
Lancet fluke, *507*
Land and feed resources, 74
Landcorp Farming, Ltd., 356
Lappa clover, *440*
Large stomach worm, *507*
Larkspur, *476*
Lasalocid, 175, 202
LDLs (low-density lipoproteins), 49
Leadership, 62–64
Lean beef, 28, *28*
Lean color, 223, 239
Lean to bone ratio, 41
Lean to fat ratio, 41–42
Least-cost rations, 415
Leather, 56
Legume-grass pasture, 478
Legumes, 405, 439, *440*, 462, 463, 478
Leptospirosis, 319, *486*
Lespedeza, 462
Lesser liver fluke, *507*
Leukosis, *486*
Levamisole, *508*
Libido, 299
Lice, 504, *505*
Lignin, 466
Limousin, 362, *364*, *371*, *372*, *373*, 391
Linear thinking, 79
Linebreeding, 338, 340, *340*
Linecrossing, 338
Linoleic acid, 50
Lipids, 396
Lipoproteins, 49
Listeriosis, *486*
Lite beef, 255
Little bluestem, *440*
Liver, 57
Liver abscess complex, *482*
Livestock show, 538
Liveweight gain, 521
Loading density, during transport, 552, *553*
Locoweed, *476*
Locus, 333
Loin cuts, 26
Loin steaks, 226
Longevity, 346, 391
Longhorn, *367*, *372*, *373*
Longissimus dorsi muscle, 390
Long yearlings, 10
Lump Jaw, *486*
Lungworm, *507*
Lupines, 439, *476*
Lutalyse®, 308, 316
Luteinizing hormone (LH), 291, 294

M

Macro minerals, 397–398
Mad cow disease, 23
Magnesium oxide (MgO), 425, 478
Magnesium tetany, 478
Maine-Anjou, *364*, *372*, *373*, 378, *392*
Male sex cells, 292

Malignant catarrhal fever, *486*
Malignant edema, *486*
Management systems
  approach, 78–80
  biological efficiency *vs.* economic efficiency, 85
  cattle resources, 74–78
  for commercial cow-calf operations, 117–159
  financial resources, 71–74
  human resources, 61–71
  land and feed resources, 74
  market resources, 78
  optimums in, 80–83
  resources and principles, 60–61
  risk management, 86
Mandalong Special, *367*
Mange, *505*
Manure, 185, 199
Mapping analyses, 333
Marble bone disease, 348
Marbling, 38, 223, 224, *225*, 229, *378*, 379, 391
Marbling fats, 38
MARC germplasm study, 379, *380*, *381*
Marchigiana, *368*
MARC I, *368*
MARC II, *368*, 391
MARC III, *368*, 391
Margarine, 57
Marinated products, 33–34
Marker-assisted selection (MAS), 358–359
Market Basket Survey (1988), 41
Market cattle production, on grazed forage, 468–469, *470*
Market channels, 232–238
Marketing
  alternatives in, 113, *114*
  in cow-calf operations, 151–155
  defined, 222
  effectiveness of, 115
  in seedstock program, 112–115
Marketing alliances, 219
Marketing issues, 24–25
Marketing program, 19
Marketing system
  advertising and promoting retail beef, 260–261
  assessing costs, 251
  channels, 232–238
  classes and grades, 222–230
  cows and bulls, 230–231, *232*
  factors affecting cattle prices, 243–251
  futures market, 252–260
  grid pricing, 238–242, *241*, *242*
Market perception, 243–244
Market prices, yearling-stocker management decisions and, 168–169
Market resources, 78
Market segment, 28, *28*
Market share, breed evaluations to improve, 391, *392*
Mastitis, *486*
Mating process, 543
Mating strategies, 337–342
Mature cow weight, 106
Maturity classifications, 223
McGovern, George, 48
Measures, of accuracy, 358
Meat juice, 33
Mechanical control, 455
Mechanical tenderization, 38
Medium stomach worm, *507*
Megacalories, 401
Melengestrol acetate, 308
Memory cells, 491
Merchandising, 537
Mesoderm, 519
Metabolizable energy (ME), 401
Methane gas, 400
Methemoglobin, 479
Methylene blue, 479
Mexico, 267, 270, 278–280
Micro minerals, 396
Middle meats, 26
Milk EPDs, 106
Milk production, variation between and within breeds for, *376*
Milkvetch, *476*
Milkweed, *476*
Milo, 430
Minerals, 396, 424–425
Mineral supplements, 410–411
Mission statements, 64–65, 119
Mites, 504–505
Moisture, 185, 395
Molars, 523
Molasses, 409
Monensin, 175, 202
Monocots, 439
Morantel tartrate, *508*
Morphology, 297
Mountain mahogany, 439, *440*
Mule foot, 348
Multiple sire breed with crossbred females, *388*
Multiple trait selection, 354
Murray Grey, *368*
Muscle
  carcass composition and, 521–522
  distribution of, 523
  growth and development of, 522–523
Muscle fiber color, 391
Muscle structure, 37
Muscle-to-bone ratio, 522–523, 530
Mycotoxin, 408
Myofibrils, 37

N

NAFTA partners, 270, 276–281
National Academy of Sciences, dietary guidelines, 50–52
National Beef Quality Audit (1995), 40
National Beef Tenderness Survey (1990), *39*
National Beef Tenderness Survey (1999), 37
National Cattle Evaluations, 360
National Cattlemen's Beef Association (NCBA)
  alliances and, 218
  Animal Health Committee, 509
  Beef Industry Long Range Plan, 64
  cattlemen principles, 24
  environmental stewardship beliefs, 452
  long-range plans, 20
  profit variability, 17
National Pollutant Discharge Elimination System (NPDES) permit, 186
National Research Council (NRC), 402, 411
National sire evaluation programs (NSEP), 357–358
Natural basis, 397
Natural beef, 255, 258
Natural service, 295, 315
Navel Ill, *486*
Nellore, *370*, *373*, *392*
Nematodes, 506
Neonatal septicemia, *486*
Net energy for gain (NEg), 402
Net energy for maintenance (NEm), 402
Net energy (NE) system, 401, 415–418, 465
Neuraxial edema, 348
Neutral detergent fiber (NDF), 465
New Zealand, 270, 275–276
Nitrate poisoning, 477, 479
Nitrogen, 400

Nitrogen-free extract (NFE), 397
Nodular worm, *507*
Nonfeed costs, 195–198
Non-heme iron, 46
Non-parent EPDs, 357
Nonprotein nitrogen (NPN), 410, 424
Nonrenewable resources, 60
Non-visual identification, 215
Normande, *368*
North American Free Trade Agreement (NAFTA), 276
North American Limousin Foundation (LimFlex), 391
North American Meat Institute (NAMI), 22
Number brands, 90
Nutrient composition, of feeds, 403–405
Nutrients, 395
Nutrition
  beef cattle requirements, 411, *413*
  beef products and, 45–52
  classification and composition of feeds, 402–411
  cow-calf, 420–429
  digestibility of feeds, 400
  energy evaluations of feeds, 400–402
  feedlot cattle, 429–431
  nutrients, 395–396
  preconditioning and, 499
  proximate analysis of feeds, 396–398
  ration formulation, 411–420
  ruminant digestive system, 398–400
  testicular function in young bulls and, 298
  yearling-stocker cattle, 429
  for yearling-stocker operations, 172–173
Nutritional concerns, of consumers, 45
Nutrition and Human Needs committee, 48

O

Oats, *440*
Odor control, 185
Oils, 396
Oleic acid, 49
Oleo stock, 57
Olfactory reflex, 543
Omasum, 400
One-injection system, of prostaglandin, 308
Opiate peptides, 40
Optimization, 218
Optimums, principle of, 80–83
Options, 254
Oral vaccines, 497
Orchardgrass, 439, *440*, *441*, 462, 463
Order buyers, 236
Organic beef, 255
Organic components of feed, *397*, *398*
Organizations, representing or affecting the beef industry, 19, *20–21*
Outbreeding, 338, 341
Outcrossing, 338, 341
Ova, 290, 305
Ovarian hormones, 291
Ovaries, *289*, 290
Ovary, *290*
Ova transplant, 315
Overgrazing, 436, 459
Overhead costs, 77
Oviducts, 289, *289*, 290
Ov-synch system, 312
Ovulation, 290, *290*, 291, 305
Ownership costs, 77
Oxytocin, 320

P

Packaging, 31
Packers, grade stamps and, 228
Packers and producers, *21*
Packing plants, 13
Packing segment, 6, 7, 12–13, 13
Palate-pastern syndrome, 347
Palpation, to determine pregnancy, 318
Parasites
  controlling external, 505–506
  controlling internal, 506–509, *508*
  external, 504–505
  internal, 506, *507*
  as stressor, 545
Parasitic wasps, 185
Paratuberculosis, *485*
Parentage identification, 359
Parent EPDs, 357
Parrot mouth, 347
Parturition, 291, 319, 320–322
Pasteurella, 172, 509
Pasture
  defined areas, 433
  gathering cattle on, 552, *550*
  leases for yearling-stocker operations, 169
Pathogenic bacteria, 509
Pathogens, 490
Pearl millet, *440*
Pearson Square method, 415
Pedigree estimates, 357
Pedigree information, 359
Pelvic measurements, 325
Pelvic plexus, 294
Penis, 291, 292, *293*, 294, 296
Per-capita consumption, 29
Percent calf crop, 126–129, *130*, 543
Percent retail product, variation between and within breeds for, *377*
Perennial peanut, *440*
Performance Registry System, 91
Perpetual resources, 60
Persian clover, *440*
Petite roast, 42
Pharmaceuticals, *27*, 57, *57*
Phenotypes, 334, 335
Phenotypic correlations, 355
Pheromones, 543
Phosphorus, 185, 425
Photo points, 454
Photosensitivity, 348
Physical examination, 295–296
Piedmontese, *368*, *370*, *373*, 391, *392*
Pine needle abortion, *487*
Pinyon-juniper, 359
Pinkeye, *487*
Pinzgauer, *368*, *370*, *373*, *392*
Pitchers sage, *440*, 439
Placenta, 290, 320
Planning process, 67–68
Plant-oil extracts, 185
Plant types, 439
Pneumonia, *487*, 509
Poisonous plants, 474–475
Polioencephalomalacia, *487*
Polled Hereford, *364*, *371*, 545
Polyunsaturated fatty acid, 396
Ponderosa pine needles, *477*
Population, 23
Positive-type headgate, 557
Posterior ventral vagina, 290
Postpartum intervals, 326–328, 452
Potassium, 516
Pot roast, 27
Pre-conditioned calves, 190
Preconditioning, 499
Pregnancy, 317–319
Pregnant mare serum gonadotrophin (PMSG), 316
Prenatal growth, 519–520
Prepuce, 296
Prescribed burning, 455
Preservatives, 406
Price cycles, 244–47
Price determination, 222
Price forecasting, 251
Price–quantity relationship, 29, *30*
Prime quality grade, 40–41

Principles, 60–61, 78
Private treaty sales, 113, *114*
Processed meats, 2
Processing protocols, 196
Production, systems for yearling-stocker operations, 175–176
Production records, 90–93, *93*
Production sales, 113, *114*
Profitability, 16–18
  in cattle feeding operations, 181
  cattle prices and, 243
  inventory and, 247
Profitability formula, for commercial cow-calf operations, 122
Profit-oriented management decisions, 122–124
Progesterone, 291, 320
Progressive bovine myeloencephaly, 348
Promoting, retail beef, 260–261
Prostaglandin, 308
Prostaglandin (PGF2a), 291, 316
Prostamate®, 308
Prostate, 294
Prostate gland, 292, *293*, 294
Protein, 46, 423–424
Protein levels, in pregnant heifers, 324
Proteins, 396
Protein supplements, 402, 408–410
Proteolytic enzymes, 37
Protoporphyria, 348
Prussic acid poisoning, 479
Psoroptic mites, 504
Psychology (market perception), 244
Puberty
  in bulls, 295
  in heifers, 294
  replacement heifers and, 148
  variation between and within breeds for weight at, *374*
Pulse, 481
Purebred breeders, 5
Purebreds, 362
Purge, 33
Purveyors, 6, 13
Put option, 254

Q

Quality
  defined, 25
Quality Assurance Marketing Code of Ethics, 231
Quality assurance programs, 25, 499–503
Quality grades
  marketing and, 223–225, *225*
  USDA, 40–41
Quality of commodity, 253
Quantity of commodity, 253
Quick beef cuts, 26

R

Rabies, *487*
Ralgro®, 134, 174–175
Random mating, 337, 338
Range area, 433
Range improvements, 455
Rangeland, 433
Rangeland desertification, 24
Range monitoring, 453–454
Ranger, *368*
Rank, in cow herd, 542
Rate of gain, yearling-stocker programs, 173–175
Rations, formulation, 411–420
Rat-tail, 348
Rebreeding, 326–327, 343
Recessive genes, 334
Reconstituting milo, 193
Record systems flow chart, 71
Rectal palpation, 296
Red Angus, *365*, 371, *371*, 390, *392*
Red Brangus, *368*
Red clover, *440*, *441*, 463
Red Poll, *368*, *370*, *373*
Red water disease, *487*
Reed canarygrass, *440*, *441*
Registered breeders, 5
Relationship, description of, 338
Relative degree of use, 453
Relative feed value (RFV), 466
Renewable resources, 60
Replacement heifers, 148–150, 385, 428–429, 513–514
Reproduction
  artificial insemination, 299–306
  breeding, 295–299
  calving, 319–326
  cloning, 317
  embryo transfer, 315–317
  organ structure and function of, 289–295
  pregnancy, 317–319
  rebreeding, 326–327
  synchronization of estrous, 306–315
  visual appraisal for, 537
Reproductive behavior, 543–544
Reproductive losses, 127–128
Reproductive organs
  of bulls, 292–294
  of cows, 289–292
Reproductive performance, 127, 343
Republic of Korea, 284
Rescue grass, *440*
Resources, 60. *See also* Management systems
Respiration, 481
Respiratory viruses, 172
Rested-rotational systems, 445
Rested systems, 445
Restraining cattle
  growing chute, 559
  squeeze chute and headgate, 557–558, *559*
  vertical tilt tables, 558, *558*
  working chutes, 553–557
Restraint, cattle behavior during, 546–551, *549*, *550*, *551*
Retail beef products, 26–28, *27*, *28*, 33–34, 260
Retail cuts, 26–28, *28*
Retailers, 6, 7, 13, *21*
Retail prices, 30
Retail product weight, variation between and within breeds for, *377*
Retail store, 28, *28*
Retinal imaging, 215
Rib cuts, 26, 226
Ribeye
  fat measurement, *226*
  size of, 42
Ribeye muscles, *225*
Right genetics, 140–145
Ringworm, *487*
Ripening, 38
Risk management, 251, *525*
Rolling average value (RAV), 49, *149*
Romagnola, *368*
Romosinuano, *368*
Rose clover, *440*
Rotate sire breed, *387*, *4388*
Rotational systems, 445
Rotational terminal sire, *386*, *388*
Roughages, 402, 474
Round cuts, 26, 42, 226
Roundhead, *440*
Roundworm, 506
Rumenitis, *482*
Rumensin, 175, 202
Russian wild rye, *440*, 458
RX3, *368*
Rye, *440*
Ryegrass, *440*

S

Safety, of beef products, 52–44, 269
Sagebrush, 439, *440*

Sahiwal, *372*, *373*
Sale barns, 234
Salers, *365*, *372*, *373*, *392*
Saliva, 398
Salmonellosis, *484*, *486*, 514
Salorn, *368*
Salt, 424
Sanitation, 498
Santa Cruz, *368*
Santa Gertrudis, 362, *365*, *372*, *373*, *392*
Sarcomere degradation, 37
Sarcomere length, 37
Satisfactory breeders classification, 298
Saturated fats, 55–56
Savory grazing method, 451
Scab, 504
Scotch Highland, *368*
Scours, *488*
Scrotal circumference, 107, 295, 296, *297*
Scrotum, 292, 293, 296
Scur, 336
Seasonal prices, 248
Seasonal systems, 445
Seasoned products, 33
Seeding, 455
Seedstock breeders
  affect on beef industry, *20*
  importance of, 5–6, 7, *8*
  marketing decisions, 129–32
  product records, 90–93, *93*
  program goals and objectives, 89–90
  selecting cows, 108–112
  selecting replacement heifers, 108, *109*
  sire selection, 93–108
Selection differential, 354
Selection index, 356
Selection methods, 355–357
Selection programs, 355–357
Select quality grade, 40–41
Select synch system, 312
Selenium-accumulating plants, *477*
Self-management, 61, 70–71
Semen
  in ampulla, 294
  collection and processing of, 300–302
  sexing, 302
Semen evaluation, 297–299
Semen morphology, 297
Seminal vesicles, 292, 294, 296
Seminiferous tubules, 292, 293, 294
Senepol, *369*
Serum cholesterol, 4
Sericea lespedeza, *440*
Sex
  conformation and, *531*, 532, 533
  tissue growth and, 522
Sex chromosomes, 333
Sex classes for cattle, 223
Sex determination, 333
Sexed-semen technology, 302
Sex-linked traits, 333
Shadscale, 439
Shear force values, 40, 502
Shipping, 498
Shipping fever, 509
Short-duration grazing systems, 451
Shorthorn, 362, 363, *365*, *371*, *372*, *373*, *381*, 525
Short yearlings, 10
Shoulder cuts, 42
Shows, livestock, 538
Shrink, 250–251
Sickle hocked condition, 535
Sigmoid flexure, 294
Silages, 405–407, 430
Silica, 466
Simbrah, *369*
Simmental, 362, *365*, *371*, *372*, *373*, 391, 392, *392*, 545
Single-sire rotations, 389
Sire
  importance of, 93–94
  selection of, 93–108
Sire Evaluation Report, *110*
Sire summaries, 95
Skeletal soundness, 535–537
Slender lespedeza, 439, *440*
Small grains, 405
Small stomach worm, *507*
Smith, Gary C., 40
Smooth bromegrass, 439, *440*, *441*, 462, 463
Snorter dwarfism, 347
Social behavior, 542–543
Society for Range Management, 433
Soil classification, 455
Sorghum, *441*
Sorghums, 405
Sorghum silage, 407
Sorghum stubble, 467
Sorghum-Sudan hybrids, *440*, *441*, 463, 479
Source verification systems, 214–215
South Devon, *372*, *473*, *392*
South Korea, 268
Soybean meal, 408, *440*
Soybean residue, 467
Species crossing, 338, 341
Speculators, 253
Sperm, 292, 293, 305, 331
Sperm abnormalities, 300
Sperm cells, 293, 297
Sperm concentration, 300
Sperm morphology, 297
Sperm motility, 296, 297, 298, 300
Squeeze chute, 557–558, *559*
Stag, as sex classification, 223
Standard deviation (SD), 351, 352
Standard quality grade, 40–41
Standing heat, 303
State Beef Councils, 22
Static terminal sire, *386*
Stayability, 107, 346
Steady size (SS), 149, *149*
Steam flaking, 193, 408
Stearic acid, 49
Steers
  bulls vs., 107–108
  as sex classification, 223
Steer shows, 538
Stifled cattle, 535
Stillbirths, 128
Stocking rate, 445, 451, 474
Stomach, 399
Stomach flukes, *507*
Straight-bar headgate, 557
Strategic alliance pilot project, 218–219
Straw, 467
Stress
  BRD and, 172
  health and, 498–499
  reducing, 193, *194*
Striate lespedeza, *440*
Subcutaneous fat, 529
Subcutaneous injections, 497
Subterranean clover, *440*
Suckling behavior, 544
Sudangrass, *440*, 479
Sunburn, scrotal, 298
Superovulation, 316–317
Supply and demand, 243
Supply chain coordination, 218–220, *220*
Surgeon General, dietary guidelines, 50–52
Sweetbread, 57
Sweet clover, 439, *440*
Sweet clover poisoning, *477*
Switchgrass, 439, *440*, *441*, 462
Sympathetic nervous system, 294
Syndactyly, 348

Synovex®, 134, 174
Synthetic insulin, 58
Systems thinking, 79

T

Tall fescue, *440*, *441*
Tandem, 355
Tapeworms, 506, *507*
Tarentaise, *365*, *372*, *373*
Tattoos, 215
T-bone steaks, size of, 42
Teeth, age relationship and, 523, *524*
Temperament, 540–541
Temperature, 381
Tenderness, 37–38, *39*, 346, *378*, 391
Testes, 292
Testicles, 292–293
Testis, *293*
Tetanus, *488*
Texture of lean tissue, 223
Thermoneutral zone (TNZ), 418
Thiabendazole, *508*
Thickness
 of cut, 42
 scores, 230
Thirst, 498
Threadleaf snakeweed, *477*
Thread-necked intestinal worm, *507*
Three-breed rotation, *386*, *387*
Three-breed static terminal sire, *388*
Thriftiness, 229–230, 296
Thymus, 57
Ticks, 504
Time management, 69–70, *69*
Timothy, *440*, *441*, 463
Tissue growth, sex and, 522, *522*
Tongue, 57
Total digestible nutrients (TDN), 400–401, *402*, 414–415, 466
Total dollars received, in feedlot management decisions, 198–199
Total Performance Records (TPR), 91
Total quality management (TQM), 25, 52, 76, 202, 219, 500
Traceback, 214–215
Trace minerals, 396, 410
Training cattle, 552, *550*
Traits, measurement of, 342–350
Trans fatty acids, 49
Trans-Ova, 317
Transportation of cattle, 552–553
Transport costs, 251
Trefoils, 439
Trematodes, 506
Trenbelone acetate (TBA) implants, 201
Trichomoniasis, *488*
Tripe, 57
Triticale, *440*
True clovers, 439
Tuberculosis, *488*
Tuli, *369*, *372*
Tunica dartos muscle, 293
Two-breed rotation, 390, *388*
Two-injection system, of prostaglandin, 308
Two-step weaning, 150
Type, defined, 524–525

U

Ulcers, gastric, *488*
Unavailable nitrogen, 465
Uniformity, 362
United States, 363, 269–270, 280–281
Unsaturated fats, 48
Urea, 409
Urethra, *289*, 290, 292, 294
Urinary bladder, *289*, 290, *292*, *293*, 293–294
Urinary calculi, *488*
Uruguay, 265
U.S. beef breeds, 363–371, *364–369*
USDA
 Cattle-On-Feed reports, 11
 Certified and Verified, 255
 Choice grade, *534*
 dietary guidelines, 50–52
 feeder cattle grades, 229–230
 feeder grades, 531
 Food Safety and Inspection Service (FSIS), 56
 quality grades, 40–41, 222, *223*
 yield grades, 41
U.S. Forest Service (USFS), 471
U.S. Meat Animal Research Center (MARC), 372
U.S. Meat Export Federation (USMEF), 22, 267
U.S. Surgeon General, dietary guidelines, 50–52
Uterine body, 289, *289*
Uterine horns, 289, *289*, 290
Uterine prolapse, *448*
Uterine wall, 290
Uterus, 289, *289*, 290, 316, 321

V

Vaccination, 498
Vaccines, 491–493, 509
Vagina, 289, *289*, 300
Vaginal prolapse, *488*
Value-added products, 26
Variety meats, *27*, 57, 267
Vas deferens, 292–293, 294
Veal, as age classification, 222
Velvetbean, *440*
Vertical tilt calf tables, 558
Vesicular stomatitis, *489*
Veterinarian–client–patient relationship (VCPR), 490
Video auctions, 238
Video-imaging technology, 229
Viral papillomatosis, *489*
Virus vaccines, 509
Vision statements, 64
Visual appraisal, 296, 525, 529, 530–531, 535, 537
Visual identification, 215
Vitamin A, 396, 410, 425–426
Vitamin B, 46
Vitamin B_{12}, 263
Vitamin E, 42
Vitamins, 396, 425–426
Vitamin supplements, 410
Vomeronasal organ, 543
Vulva, *289*, 290, 290

W

Wagyu, *369*, 391
Wal-Mart, 216
Warner–Bratzler shear force values, 346, 391
Warts, *489*
Waste management, 186
Water, 395, 426–428
Water belly, *488*
Waterhemlock, *477*
Water-holding capacity, 40
Water quality, 185
Weaning, 150–151, 499
Weaning weight
 bull selection and, 106
 distribution of, 351
 heterosis for, 341
 managing, commercial cow-calf operations, 129–134
 as trait, 343–344
 variation between and within breeds for, *376*
Weather, 86, 418–420, 498
Weaver calf, 348
Web-based livestock marketing, 238
Weeds, 439, *440*
Weight gain, of pregnant cows, 422
Welsh Black, *369*
Western wheatgrass, *440*

Wet-aged beef, 38
Wheat, *440*
Wheatgrasses, 439
Wheat grazing, 176
Wheat pasture, 473–474
Wheat poisoning, 478
White clover, *440*
White muscle disease, *489*
White Park, *369*
Whole herd reporting, 91
Whole muscle cuts, 27
Wholesale cuts, location of, *529*
Window of acceptability, of fat content in beef, 41–42
Winterfat, 439, *440*
Wintering program, 184, 185
Winter pea, *440*
Winter rye, 464
Winter tetany, 478
Wire worm, *507*
Womb, 289, *289*. *See also* Uterus
Woody calluses, 502
Working chutes, 553–557
World beef consumption, 265
World cattle numbers, 264
World meat production, 265, *264*
World Trade Organization, 275
Written plan, for cattle operation, 67–68

Y

Yearling bull scrotal circumference, 295
Yearlings, 10
Yearling-stocker operations
  areas of, in U.S., 162, *163–164*
  management decisions
    budgeting process, 166–168
    cattle health, 170–172
    computing breakeven prices, 165–166
    market prices, 168–169
    nutrition, 172–173, *174*
    pasture leases, 169, *170*
    production systems, 175–176
    rate of gain, 173–175
  nutrition in, 429
  primary function of, 164
Yearling-stocker operator, importance of, 6, 7, 10
Yearling weight, 106, 344
Yellow body, 290
Yield grades
  marketing and, 226–228, *229*
  USDA, 46
Youth programs, 538

Z

Zebu, *369*
Zinc, 46
Zoological classifications, *341*

Additional Resources

Beef Cattle Management

Ag Manager—Kansas State University
www.agmanager.info

A comprehensive source of information, analysis, and decision-making tools for agricultural producers and agribusinesses, the site addresses topics such as crop and livestock marketing and outlook reports, crop insurance, farm management, agricultural policy, human resources, income tax and law, and agribusiness. Budgeting, long-range planning, and marketing support information and tools are particular strengths of the site.

eExtension
articles.extension.org/beef_cattle

This site is an information clearinghouse of information provided by cooperative extension professionals across the United States. It also serves as a gateway to help users connect to experts in their state or region. Webinars, beef fact sheets, and management tips are available as well as the opportunity to submit beef production-related questions to topic area experts.

Iowa Beef Center—Iowa State University
www.iowabeefcenter.org

It is one of the premier websites that focuses on beef cattle production and management. The site offers useful information and resources to seedstock, cow-calf, stocker, and feedyard managers. Additionally, the site offers a number of decision-making tools, apps, and software solutions including Beef ration and nutrition decisions software (BRaNDS), Estrus Synchronization Planner, and a number of budgeting, selection, marketing, and pricing calculators.

Mississippi State University Beef Extension
extension.msstate.edu/agriculture/livestock/beef

Featuring links to its beef cattle management You-Tube channel, beef management short courses, this site provides excellent resources for producers in southern tier of states along the Gulf Coast region. Cow-calf management, feeder cattle marketing, and forage management are strengths as is the Cattle Calculator app that provides decision support related to reproductive management, animal performance, and management decisions in either Apple or Android devices.

University of Nebraska Beef Site
beef.unl.edu

An excellent site that provides comprehensive management systems information and tools. Information related to grazing, beef forage crop systems, nutritional management, and integrated systems is of particular note. Learning modules using animation and video enhance the user experience. Apps include body condition score, farm records, range and pasture monitoring, feed cost calculators, dry matter conversions, and a corn stalk calculator.

Oklahoma State University Beef Extension
www.beefextension.com

The site features informational videos, production and management information that spans the beef supply chain, and offers particular value on stocker cattle management, receiving cattle into the feedyard, and a host of user-friendly spreadsheet-based calculators addressing topics such as breeding bull prices, cow purchases decisions, culling decisions, across breed EPD evaluation, hay management and marketing, wheat pasture management, and preconditioning.

Oregon State University
beefcattle.ans.oregonstate.edu/

Provides access to a number of valuable resources including the calving school curriculum, the Cattle Producer's Handbook—a comprehensive compilation of fact sheets related to all areas of beef cattle management, and forage management strategies for the northwest region of the United States.

University of Tennessee Beef and Forage Center
utbeef.com

Site provides relevant information on beef cattle and forage management for producers in the mid-South region of the U.S. Features links to the advanced master producer education program, the Tennessee beef heifer development program, and a host of useful fact sheets and research reports unique to the needs of producers in the region.

Texas A&M University
animalscience.tamu.edu/livestock-species/beef

Featuring the video library of Ranch TV, links to the long standing Beef Cattle Short Course (one of the premier annual beef cattle educational events in the United States), and an array of online beef cattle courses, this site provides full supply chain sources of information with an emphasis on ranch management, quality assurance, and improving end product value.

Beef Promotion and Industry Policy

Cattlemen's Beef Board and Federation of State Beef Councils
www.beef.org

A site funded by the Beef Checkoff, it provides access to beef-related research results, the Master of Beef Advocacy program, detailed information for beef retailers, food service operators, and wholesalers, as well as information for consumers about beef preparation, safety, nutrition, and sustainability. Access to the Beef Issues Quarterly report and a number of other publications is also available.

National Cattlemen's Beef Association
www.beefusa.org

The Cattle Learning Center, Cattlemen's College proceedings, and a number of other industry-related resources are available on this site. The policies of the organization, its political action initiatives, grassroots engagement, and annual convention and trade show links can also be found here.

Forage and Pasture Management

American Forage and Grassland Council
www.afcg.org

An international body, the council provides information and training to enhance effective management of forage, pasture, and grassland resources. Best practice guidelines, conference proceedings, and a host of regional and state resources can be accessed from the site.

National Resources Conservation Service
www.usda.nrcs.gov

The National Range and Pasture Handbook is an excellent source of information available from the NRCS as well as other sources in support of prescribed grazing strategies, pasture and rangeland monitoring, and soil conservation practices.

Genetics, Breeding Systems, and Selection

Beef Improvement Federation
www.beefimprovement.org

The Beef Improvement Federation is a collaborative effort of breed associations, cattle breeders, and affiliated enterprise dedicated to utilizing genetic evaluation to the benefit of the industry. The site provides access to the *BIF Guidelines* that provide best practice standards for genetic evaluation and data interpretation, proceedings of the annual research conference on beef cattle genetic evaluation, and a host of relevant fact sheets.

Breeds of Livestock—Oklahoma State University
www.ansi.okstate.edu/breeds/cattle/
Hands down the best single website for learning about beef cattle breeds.

National Beef Cattle Evaluation Consortium
www.nbcec.org

NBCEC was developed through a partnership of the major land grant universities that provided genetic evaluation services to the industry. The site offers valuable information for cattle producers and scientists. The Sire Selection Manual, a number of white papers, and fact sheets focused on emerging technologies can be obtained from the site.

Reproductive Management

Applied Reproductive Strategies in Beef Cattle
beefrepro.unl.edu

A seven-university consortium dedicated to improved reproductive management of beef cattle, this site is the premier source of information related to estrus synchronization, reproductive management protocols, technologies, and related topics. A number of management decision support tools and proceedings of the ARSBC workshops are available. The site serves producers, veterinarians, scientists, and industry professionals.

Estrus Synchronization Planner
http://www.iowabeefcenter.org/estrus_synch.html

This is the premier planning tool to help producers plan and execute an effective estrus synchronization program.

Marketing, Price Trends, and Industry Analysis

CattleFax
www.cattlefax.com

CattleFax is the leading independent source of market information for the beef industry—a membership organization dedicated to providing relevant data, analysis, and forecasting to assist producers in making marketing decisions, developing risk management strategies, and organizing long-term strategic direction. The site provides meaningful trend interpretation and analysis.

Livestock Marketing Information Center
www.lmic.info

The center is a cooperative effort of 28 state extension agencies, 13 industry organizations, and various federal agencies to provide timely market-related resources to livestock producers. The site provides pricing, inventory, production, trade, and associated data as well as effective synthesis and analysis.

Quality Assurance

Beef Quality Assurance
www.bqa.org

Funded by the Beef Checkoff, the BQA site provides access to training and educational resources related to effective beef cattle handling, care, and well-being. Focused on providing enhanced beef quality through deliberate management, the BQA site also offers certification testing and access to the various state BQA coordinators in the United States.

United States Department of Agriculture
www.usda.gov
A sampling of valuable government websites within USDA are listed below:
Agricultural Marketing Service (www.ams.usda.gov)
Agricultural Research Service (www.ars.usda.gov)
Economic Research Service (www.ers.usda.gov)
Foreign Agricultural Service (www.fas.usda.gov)
National Agricultural Statistics Service (www.nass.usda.gov)
National Animal Health Monitoring Service (www.aphis.usda.gov/nahms)

Cow-Calf and Stocker Software and Spreadsheet-Based Decision-Aid Tools Not Previously Listed (Only a Sample—Not a Comprehensive Listing)

Texas A&M University—https://agecoext.tamu.edu/resources/software-tools/
Oklahoma State University—http://beefextension.com/pages/sccalc.html
Montana State University—http://www.montana.edu/softwaredownloads/livestockdownloads.html

Evaluation and Comparison of Cow-Calf Software

Oklahoma State University—http://pods.dasnr.okstate.edu/docushare/dsweb/Get/Document-1926/CR-3279web15.pdf
University of Arkansas—http://www.uaex.edu/publications/pdf/fsa-3108.pdf
University of Nebraska—http://newsroom.unl.edu/announce/beef/3344/18785

Feedyard Software and Data Services Sample

Cattle Expert—http://www.cattlexpert.com/
Hi Plains Systems—http://www.hiplainsystems.com/
Turnkey—https://www.turnkeynet.com/home
Viewtrak—http://www.viewtrak.com/feedlots.htm
Livestock Tracker—http://www.livestocktracker.com/
PCC—http://www.pcc-online.com/

Gestation Table for Cows Based on a 285-Day Gestation Length

Find date of service in upper line. Figure below indicates date due to calve.

Jan	1	2	3	4	5	6	7	8	9	10	11	12	13	14	15	16	17	18	19	20	21	22	23	24	25	26	27	28	29	30	31	
Oct	13	14	15	16	17	18	19	20	21	22	23	24	25	26	27	28	29	30	31	1	2	3	4	5	6	7	8	9	10	11	12	Nov
Feb	1	2	3	4	5	6	7	8	9	10	11	12	13	14	15	16	17	18	19	20	21	22	23	24	25	26	27	28				
Nov	13	14	15	16	17	18	19	20	21	22	23	24	25	26	27	28	29	30	1	2	3	4	5	6	7	8	9	10				Dec
Mar	1	2	3	4	5	6	7	8	9	10	11	12	13	14	15	16	17	18	19	20	21	22	23	24	25	26	27	28	29	30	31	
Dec	11	12	13	14	15	16	17	18	19	20	21	22	23	24	25	26	27	28	29	30	31	1	2	3	4	5	6	7	8	9	10	Jan
Apr	1	2	3	4	5	6	7	8	9	10	11	12	13	14	15	16	17	18	19	20	21	22	23	24	25	26	27	28	29	30		
Jan	11	12	13	14	15	16	17	18	19	20	21	22	23	24	25	26	27	28	29	30	31	1	2	3	4	5	6	7	8	9		Feb
May	1	2	3	4	5	6	7	8	9	10	11	12	13	14	15	16	17	18	19	20	21	22	23	24	25	26	27	28	29	30	31	
Feb	10	11	12	13	14	15	16	17	18	19	20	21	22	23	24	25	26	27	28	1	2	3	4	5	6	7	8	9	10	11	12	Mar
Jun	1	2	3	4	5	6	7	8	9	10	11	12	13	14	15	16	17	18	19	20	21	22	23	24	25	26	27	28	29	30		
Mar	13	14	15	16	17	18	19	20	21	22	23	24	25	26	27	28	29	30	31	1	2	3	4	5	6	7	8	9	10	11		Apr
Jul	1	2	3	4	5	6	7	8	9	10	11	12	13	14	15	16	17	18	19	20	21	22	23	24	25	26	27	28	29	30	31	
Apr	12	13	14	15	16	17	18	19	20	21	22	23	24	25	26	27	28	29	30	1	2	3	4	5	6	7	8	9	10	11	12	May
Aug	1	2	3	4	5	6	7	8	9	10	11	12	13	14	15	16	17	18	19	20	21	22	23	24	25	26	27	28	29	30	31	
May	13	14	15	16	17	18	19	20	21	22	23	24	25	26	27	28	29	30	31	1	2	3	4	5	6	7	8	9	10	11	12	Jun
Sep	1	2	3	4	5	6	7	8	9	10	11	12	13	14	15	16	17	18	19	20	21	22	23	24	25	26	27	28	29	30		
Jun	13	14	15	16	17	18	19	20	21	22	23	24	25	26	27	28	29	30	1	2	3	4	5	6	7	8	9	10	11	12		Jul
Oct	1	2	3	4	5	6	7	8	9	10	11	12	13	14	15	16	17	18	19	20	21	22	23	24	25	26	27	28	29	30	31	
Jul	13	14	15	16	17	18	19	20	21	22	23	24	25	26	27	28	29	30	31	1	2	3	4	5	6	7	8	9	10	11	12	Aug
Nov	1	2	3	4	5	6	7	8	9	10	11	12	13	14	15	16	17	18	19	20	21	22	23	24	25	26	27	28	29	30		
Aug	13	14	15	16	17	18	19	20	21	22	23	24	25	26	27	28	29	30	31	1	2	3	4	5	6	7	8	9	10	11		Sep
Dec	1	2	3	4	5	6	7	8	9	10	11	12	13	14	15	16	17	18	19	20	21	22	23	24	25	26	27	28	29	30	31	
Sep	12	13	14	15	16	17	18	19	20	21	22	23	24	25	26	27	28	29	30	1	2	3	4	5	6	7	8	9	10	11	12	Oct